Disruptive Technologies in Computing and Communication Systems

Dr. K Venkata Murali Mohan, an accomplished Electronics & Telecommunications Engineer from Nagpur University, Nagpur. He has completed his M.Tech from JNTU, Kakinada, Ph.D from Acharya Nagarjuna University in Computer Science Engineering(2016) and Ph.D from Rayalaseema University, Kurnool in Electronics and Communications Engineering(2019). He is a hard core academician who believes in powerful learning which transforms the students into practically skilled serviceable engineers. Dr.K Venkata Murali Mohan has more than 20 years of experience in teaching & administration and in field. As a resolute educationalist, he has endeavored to publish text books and research papers which outlines his educational tendencies. He is a paragon of energy, enthusiasm, productiveness and competence whose intellectual ability and propriety go all the way in nurturing the students and faculty members to create a collaborative partnership. He is an optimist who believes in close connectivity among students, parents, teachers and support staff, striving to create a milieu that sustains excellence. His distinction lies in his commitment for the cause of the students, to help them to focus on confidence building, while nurturing an intense sense of social and environmental responsibility through academic and co-curricular activities. He continues to strive towards the attainment of his organizational goals in the interest of the TKRES.

Dr. M. Suresh Babu is a seasoned professional with over 26 years of experience in Teaching and Student Management. He is a Professor and Dean at Teegala Krishna Reddy Engineering College in Hyderabad. Given his various roles in the field, he has a strong background in computer science and engineering. Dr. M. Suresh Babu possesses a diverse skill set and has experience in both the software industry and academia. His experience spans both the technical and educational realms, showcasing a well-rounded professional with expertise in software development and a dedication to teaching and student management. Dr. Suresh's roles span various responsibilities, from academic leadership and project guidance to administrative coordination and external examination activities. He has a diverse and extensive involvement in different academic and administrative roles. Dr. M. Suresh Babu is actively involved in academic publications and holds various editorial roles in reputed journals. Additionally, he is a member of several professional bodies. He has contributed over 150 technical papers and is a freelance editor for several Telugu and English newspapers. Dr. M. Suresh Babu is the Chairman of Doctoral Committees, overseeing and guiding doctoral research in various universities. His achievements reflect a dedication to both academic and community service, showcasing a well-rounded commitment to research, education, and social welfare.

Disruptive Technologies in Computing and Communication Systems

Proceedings of the 1st International Conference on Disruptive technologies in Computing and Communication Systems

Edited by
Dr. K. Venkata Murali Mohan
Dr. M. Suresh Babu

CRC Press
Taylor & Francis Group
Boca Raton London New York

CRC Press is an imprint of the
Taylor & Francis Group, an **informa** business

First edition published 2024
by CRC Press
4 Park Square, Milton Park, Abingdon, Oxon, OX14 4RN

and by CRC Press
2385 NW Executive Center Drive, Suite 320, Boca Raton FL 33431

CRC Press is an imprint of Informa UK Limited

British Library Cataloguing-in-Publication Data
A catalogue record for this book is available from the British Library

ISBN: 978-1-032-66547-4 (pbk)
ISBN: 978-1-032-66553-5 (ebk)

Typeset in Times LT Std
by Aditiinfosystems

Contents

List of Figures

List of Tables

Preface

1st International Conference on Disruptive Technologies in Computing and Communication Systems (ICDTCCS - 2023) has received overwhelming response on call for papers and over 119 papers from all over globe were received. We must appreciate the untiring contribution of the members of the organizing committee and Reviewers Board who worked hard to review the papers and finally a set of 69 technical papers were recommended for publication in the conference proceedings. We are grateful to the Chief Guest Prof Atul Negi, Dean – Hyderabad Central University, Guest of Honor Justice John S Spears -Professor University of West Los Angeles CA, and Keynote Speakers Prof A. Govardhan, Rector JNTU H, Prof A.V.Ramana Registrar – S.K.University, Dr Tara Bedi Trinity College Dublin, Prof C.R.Rao – Professor University of Hyderabad, Mr Peddigari Bala, Chief Innovation Officer TCS, for kindly accepting the invitation to deliver the valuable speech and keynote address in the same. We would like to convey our gratitude to Prof D. Asha Devi - SNIST, Dr B.Deevena Raju – ICFAI University, Dr Nekuri Naveen - HCU, Dr A.Mahesh Babu - KLH, Dr K.Hari Priya – Anurag University and Prof Kameswara Rao –SRK Bhimavaram for giving consent as session Chair. We are also thankful to our Chairman Sri Teegala Krishna Reddy, Secretary Dr. T.Harinath Reddy and Sri T. Amarnath Reddy for providing funds to organize the conference. We are also thankful to the contributors whose active interest and participation to ICDTCCS - 2023 has made the conference a glorious success. Finally, so many people have extended their helping hands in many ways for organizing the conference successfully. We are especially thankful to them.

Spatiotemporal Mobility Based Trajectory Privacy-Preserving in Ubiquitous Computing Environment

M. Suresh Babu[1], M. Supriya[2], K. Sandhya[3], K. Anusha[4]

Department of CSE/IT, Teegala Krishna Reddy Engineering College,
Hyderabad, Telangana, India

Abstract: In the paper, the main focus is on privacy-aware systems, and a new architecture is proposed to address privacy concerns, specifically regarding location privacy. The proposed architecture incorporates two methods: spatiotemporally-based anonymization and location information disturbing. The spatiotemporally-based anonymization method involves dividing space and time into smaller units or pieces. When an entity transitions from one domain to another, its identification (ID) is refreshed. This approach helps in preserving the person's location privacy by ensuring that there is no direct association or continuity between the entity's ID in different domains. This technique makes it difficult for an observer to track or link the entity's movements across different locations and times. The location information disturbing method aims to introduce variability or obfuscation into coordinate data to further protect privacy. Two specific approaches are mentioned: transferring coordinates to random data and transferring coordinates to fixed data. This method involves replacing the original coordinate data with random values. By doing so, the exact location of the entity is obscured, making it challenging to discern the entity's actual whereabouts. In this method, the original coordinate data is transformed into fixed values. It is unclear from the provided information how the fixed values are determined, but this approach may involve mapping coordinates to predetermined or predefined locations. The purpose is to obfuscate the precise location information while still preserving the general area or region of the entity.

Keywords: Spatiotemporally, Anonymization, Identification, Privacy aware systems

1. Introduction

Privacy-preserving techniques in ubiquitous computing aim to protect users' sensitive information and maintain their privacy in a pervasive computing environment where computing devices are seamlessly integrated into everyday life. Privacy has been a significant concern in the context of ubiquitous computing. The seamless integration of computing and communication technologies in a ubiquitous computing environment can potentially lead to extensive data collection, tracking, and surveillance, raising privacy risks for individuals. As a result, privacy concerns have been identified as one of the primary barriers to the success and adoption of context aware computing. The pervasive nature of computing devices and the constant collection of personal data can erode individuals' privacy and give rise to various privacy-related challenges. Ubiquitous computing environments generate vast amounts of data, often collected without individuals' explicit consent or awareness. This raises concerns about the collection and retention of personal information, including location data, behavioral patterns, and preferences. Ubiquitous computing systems may involve sharing and aggregating personal data among multiple entities. The potential for data sharing without individuals' knowledge or control raises concerns about secondary uses of data and the creation of comprehensive user profiles. The continuous monitoring and tracking capabilities of ubiquitous computing systems can intrude upon individuals' privacy

[1]sureshcse@tkrec.ac.in, [2]supriya.mougiligidda@gmail.com, [3]Sandhya.k@tkrec.ac.in, [4]k.anusha@tkrec.ac.in

and result in the collection of sensitive information about their activities, movements, and behaviors. This can lead to concerns about constant surveillance and potential misuse of collected data. The integration of diverse data sources in a ubiquitous computing environment may increase the risk of re-identifying individuals even if their personal information is anonymized or pseudonymized. This poses challenges to maintaining true anonymity and protecting individuals' identities.

1.1 Privacy in Ubiquitous Computing Environments

Ubiquitous computing environments often involve collecting and analyzing contextual information, such as health data, personal preferences, or social interactions. Preserving privacy in such contexts requires considering the sensitivity of different types of data and implementing privacy safeguards accordingly. Addressing these privacy concerns in ubiquitous computing requires a multidimensional approach that combines technical, legal, and user-centric considerations. Transparent data practices, and user empowerment through informed consent and privacy settings are some of the strategies employed to mitigate privacy risks.

Privacy in a ubiquitous computing environment refers to the protection of individuals' personal information and their ability to control how their data is collected, used, and shared within the context of pervasive computing. Retention and processing of personal data to what is necessary for the intended purpose. Avoid collecting excessive or unnecessary information to minimize privacy risks. Clearly communicate the scope, and duration of data collection, as well as any sharing or processing practices, to ensure individuals are aware of and can make informed decisions regarding their privacy.

Empower users to exercise control over their personal data. Provide options for users to specify their privacy preferences, including the ability to opt-in or opt-out of data collection, choose data sharing settings, and manage data retention periods. Apply techniques such as anonymization or pseudonymization to protect individual identities when collecting and analyzing data. Ensure the use of secure communication protocols (e.g., encryption) when transmitting and storing personal data. This protects the confidentiality and integrity of the information and prevents unauthorized access. Integrate privacy considerations into the design and development of ubiquitous computing systems from the outset. Adopt privacy-preserving technologies, follow privacy best practices.

Be transparent about data practices, including data collection, usage, sharing, and retention policies. Provide clear and easily accessible privacy policies that explain how personal data is handled, and regularly update users on any changes to those policies. Conduct regular audits to ensure compliance with privacy regulations and internal privacy policies. Promote user education and awareness regarding privacy risks and best practices.

2. Ubiquitous Security Architecture

A ubiquitous security architecture refers to a comprehensive framework that aims to provide security across all aspects of a ubiquitous computing environment. It involves the integration of various security mechanisms and protocols to protect the confidentiality, integrity, and availability of data and services in a pervasive computing ecosystem.

Authentication and access control: This component focuses on verifying the identities of users and devices in the system and granting appropriate access rights. Secure communication protocols, such as Transport Layer Security (TLS) or IPsec, are employed to encrypt data transmission and protect it from eavesdropping, tampering, or unauthorized interception. This component ensures the confidentiality and integrity of data exchanged between devices or over networks.Data protection mechanisms involve techniques such as encryption, data masking, and tokenization to safeguard the confidentiality of sensitive information stored on devices or transmitted across the ubiquitous computing environment.

Trust management encompasses mechanisms for establishing and managing trust relationships among devices and entities in the system. It involves evaluating the trustworthiness of devices, verifying the authenticity of communications, and managing trust levels to make informed decisions about granting access or sharing resources.

Security monitoring and intrusion detection: Continuous monitoring of the ubiquitous computing environment is essential to detect and respond to security incidents promptly. Intrusion detection systems (IDS) and security monitoring tools help identify unauthorized activities, anomalies, or potential attacks, enabling timely mitigation measures. Establishing and enforcing security policies is crucial to maintain a secure ubiquitous computing environment. Policies define rules and guidelines for data access, sharing, usage, and behavior of entities within the system. Enforcement mechanisms ensure compliance with these policies, detecting and addressing violations.

Physical security: Physical security measures protect the physical infrastructure of the ubiquitous computing environment, including devices, sensors, networks, and data centers. It involves measures such as secure hardware design, tamper-resistant devices, surveillance systems, and access controls to prevent unauthorized physical access or tampering. Privacy-preserving mechanisms and techniques, as discussed earlier, are integrated into the security architecture to protect individuals' personal information and ensure compliance with privacy regulations. This includes anonymization, consent management, data minimization, and user-centric privacy controls.

Incident response and recovery: This includes processes for incident detection, containment, eradication, and system restoration to minimize the damage caused by security breaches. Promoting security education and awareness among users and stakeholders is essential to foster a security-conscious culture. Training programs, guidelines, and best practices help users understand their roles and responsibilities in maintaining a secure ubiquitous computing environment. It's important to note that a ubiquitous security architecture should be adaptable and scalable to accommodate evolving security threats and technologies.

2.1 Spatiotemporal Mobility

Researchers have put a lot of effort into developing analogous solutions for protecting trajectory privacy in response to the aforementioned necessity.

Anonymity is a significant aspect of the scheme. Users remain anonymous during querying, applying, and utilizing services, with only RSNs (presumably Random Serial Numbers) used to represent users. Additionally, a spatiotemporally-based anonymous matching strategy allows for changing RSNs. The scheme also replaces users' actual coordinates with fake ones during service matching, further enhancing anonymity.

The scheme ensures protection of user preferences by requiring user preference input only during the authentication step. It assumes that the authentication center is trustworthy, suggesting that user preferences are securely handled and not compromised.

While these properties highlight the potential security benefits of the proposed scheme, it is important to subject it to a thorough security and privacy analysis to ensure its effectiveness and robustness. Such an analysis should consider potential vulnerabilities, threats, and attacks that could compromise the system's security and privacy objectives. Additionally, an evaluation of the scheme's compliance with relevant privacy regulations and standards should be performed.

The spatiotemporally-based anonymous matching strategy described in the proposed scheme aims to enhance privacy by periodically updating an individual's random data stream and distributing a new data stream to represent the person. This strategy involves dividing time into intervals and carving the entire domain into specific areas. This process invalidates the old random data stream and ensures that there is no direct connection between the previous and new data streams. The

intention behind this strategy is to mitigate long-term attacks targeting a specific individual. By regularly updating the random data stream and introducing a new representation, it becomes inefficient for an attacker to persistently track or identify an individual over an extended period.

It's important to note that achieving spatiotemporally-based anonymity is a challenging task, and the level of anonymity achieved may depend on the specific techniques and parameters used. The balance between privacy and data utility needs to be carefully considered, as excessive anonymization may result in reduced data usefulness for analysis or services. Additionally, the legal and ethical aspects of data anonymization should also be taken into account, ensuring compliance with relevant regulations and guidelines.

The focus is on protecting user trajectory privacy in the context of Location-Based Services (LBSs). While LBSs offer convenience, the collection of location data raises concerns about privacy. Trajectory k-anonymity is identified as an important technology to protect user trajectory privacy, but it is observed that user attributes are often not adequately, rendering user trajectories vulnerable. Spatiotemporal Mobility (SM) measurement is to assess user characteristics. This improvement in privacy preservation is achieved while maintaining the same quality of services. The importance of considering user attributes and their relationship with trajectories when aiming to protect trajectory privacy. The proposed MTPPA algorithm offers a potential solution by leveraging the SM measurement and trajectory graph modeling. However, it is crucial to critically evaluate the algorithm's effectiveness, efficiency, and potential limitations, as well as assess its robustness against privacy attacks or inference techniques. Additionally, the compliance of the proposed algorithm with privacy regulations and ethical considerations should be thoroughly examined.

The development of 5G technology has made location-based services (LBSs) more common in our daily lives. The user's trajectory data has, however, been stored by these service providers. A significant quantity of the user's private information, such shopping preferences, home address, place of employment, or regularly frequented locations, is contained in the trajectory data. It would expose the user's private information as a result. Therefore, a method to safeguard user trajectory data is required for greater privacy. One of the crucial methods lately employed to shield a user's trajectory is k-anonymity. Similar trajectories combine to create the k-anonymity set, which is then given to the service providers, where k stands for the level of anonymity. However, creating an effective k-anonymity set is difficult since the attacker might take into account. The majority of the methods now in use for creating the k-anonymity set take the direction similarity between trajectories into account [6, 7, 8, 9, 10,

11, 12, 13, 14, 15, 16]. These approaches, however, disregard the fact that various users have unique characteristics and movement patterns.

The user may visit the grocery, the park, the neighbourhood, or any other places as stopovers. Even with the anonymity provided by the SM, the attacker can still identify the trajectory. Figure 1.1 depicts the daily movement patterns of two users. Alice and Bob's respective trajectories are those with the red and green colouring, respectively. Alice makes a number of stops at different places throughout the area. She moves relatively quickly on average. It is simple to assume that her daily movement pattern is erratic and unpredictable. Bob, on the other hand, only makes stops in two different places. He travels less frequently than Alice. He moves slower on the whole. His daily movement pattern is thought to be more predictable and predictable, and fixed. It is determined that Alice has greater mobility than Bob. Once the attacker discovers Bob is an employee of a company using data mining techniques, the attacker will readily filter out this trajectory with high mobility in the anonymity set if the trajectory k-anonymity set given by Bob contains a trajectory created by Alice.

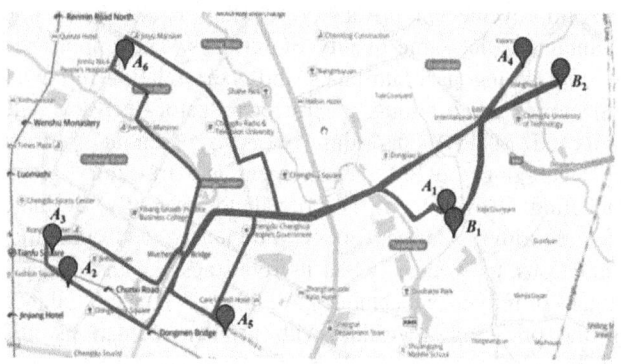

Fig. 1.1 Spatiotemporally-based anonymity [5]

The majority of the methods now in use for creating the k-anonymity set take the direction similarity between trajectories into account [6, 7, 8, 9, 10, 11, 12, 13, 14, 15, 16]. These approaches, however, disregard the fact that various users have unique characteristics and movement patterns. The trajectories produced by various user attributes are highly dissimilar.

3. Spatiotemporally-based Trajectory

Definition 1

A user's trajectory, denoted as T, is represented as a polyline in three-dimensional space. It consists of a sequence of sampling points accessed over time. Each sampling point (x_i, y_i, t_i) represents the user's coordinates (x_i, y_i) at a specific sampling time t_i.

Equivalence class: Given a starting timestamp t_s and an ending timestamp t_e, trajectories T_i and T_j are considered part of the same equivalence class if all their sampling points fall within the time interval $[t_s, t_e]$. In other words, if the sampling points of T_i and T_j exist within the same time range, they are grouped together as an equivalence class.

Synchronized trajectories: If trajectories T_i and T_j from an equivalence class have the same number of sampling points and cover the same sampling time length, they are referred to as synchronized trajectories. Essentially, synchronized trajectories have identical lengths and contain sampling points that correspond to the same time instances.

Synchronized trajectory set: A synchronized trajectory set refers to a collection of trajectories in which any two trajectories from the set are synchronized. This set consists of trajectories that exhibit the same time duration and have the same number of sampling points.

Definition 2

The stopover S of a user denotes a particular site or location where the location is functional, beneficial, or relevant to the user (for example, a bus station, a market, or even the user's homesite).

Definition 3

The total number of stopovers N and the average movement speed v of a user's trajectory are used to calculate their spatiotemporal mobility M. The ratio of the trajectory's entire length to its total moving time, given in Equation (1), determines the average moving speed v in the time range $[t_1, t_n]$:

$$\bar{v} = \frac{\sum_{i=1}^{n-1}\sqrt{(x_{i+1} - x_i)^2 + (y_{i+1} - y_i)^2}}{t_n - t_1} \tag{1}$$

After applying the normalization process the SM

$$M = \alpha \frac{N}{n} + \beta \frac{\bar{v}}{v_{\max}} \tag{2}$$

Definition 4

Kinematic Similarity. The similarity of two synchronised trajectories is assessed using their *SM* differences. Allow the *SM* of the two users' synchronised trajectories, M_i and M_j. The absolute value of the difference between M_i and M_j is defined as the mobility difference between T_i and T_j and is as follows:

$$\Delta M(T_i, T_j) = |M_i - M_j| \tag{3}$$

where, $\Delta M(T_i, T_j) \in [0, 1]$.

Definition 5

Graph of Trajectories. A series of synchronous trajectories are combined to create a trajectory graph, which is represented by the weighted undirected graph TG = (V, E, W), in which each vertex represents a trajectory Ti. When Ti and Tj are comparable, then E is the set of edges where an edge ei,j occurs between vertexes vi and vj. Wi, j is the SM difference between Ti and Tj, and W is the set of the weight of edge E.

Definition 6

Assume the location services provider receives the anonymity set Ss. The level of attack similarity required for the attacker to recognise the false trajectory in the anonymity set is a. Any two of the set's trajectories when sa are comparable to the attacker's. Any phoney trajectory in the set is indistinguishable to the attacker. The two trajectories are not similar to the attacker when s > a, let's say the mobility difference M(Ti,Tj) between Ti and Tj in the set is bigger than a. Less comparable trajectories make it simpler for the attacker to discern one trajectory from another.

Suppose a collection of synchronous trajectories is used to build the trajectory graph TG = (V, E, W). Let s determine the trajectory graph TGs = (V, Es, W). ask(k1)2 is used to determine the value of Es in accordance with Definition 5. Let a, the degree of vertex vi in TG ais di, determine the trajectory graph TGa = (V, Ea,W). When the attacker is small, the trajectory is immediately identifiable. Let |Ea| be the total degree of all the vertices of TGa. The attacker is more likely to recognize the bogus trajectories when |Ea|is small. Thus, there is a higher likelihood of privacy disclosure throughout the route. As a result, the chance of privacy disclosure along a route is defined as follows:

$$P = 1 - \frac{|E_a|}{|E_s|} \qquad (4)$$

4. Spatiotemporal Mobility (SM) Based Trajectory Privacy-Preserving Algorithm (MTPPA)

In this part provides an overview of the suggested MTPPA algorithm. In MTPPA, there are three phases. The trajectory pre-processing is created in stage I. The stopovers are identified, and the equivalence classes are created. Stage II is where the initial trajectory candidate selection and trajectory graph construction are designed. In stage III, the simulated annealing algorithm chooses an ideal trajectory k-anonymity set. After completing all three steps, the created ideal anonymity set can safeguard the user's trajectory privacy while complying with service-level standards. When there is an edge connecting every pair of the graph's vertices, the graph is referred to as a clique. A k-clique is a clique with k vertices.

4.1 Service Matching Process

The service matching process described in the proposed scheme involves the privacy system checking whether the available services meet the entity's requirements. It considers the spatial and temporal preferences of both the services and the entity. The matching process ensures that the services being considered have spatial and temporal preferences that strictly align with the entity's requirements. This implies that the services should fall within the defined spatial limits and be available during the required time period specified by the entity.

By performing this matching, the scheme aims to provide the entity with services that are suitable in terms of their spatial and temporal characteristics. This ensures that the entity receives relevant and applicable services that meet their specific requirements. However, it is crucial to consider the practical implementation and effectiveness of the service matching process. The analysis should assess how well the matching algorithm performs in accurately identifying suitable services while considering any potential privacy or security implications. Additionally, the scheme should handle cases where no services strictly match the entity's requirements and provide appropriate alternatives or recommendations.

4.2 Random Coordinates

The use of random coordinates instead of the entity's precise spatiotemporal information can enhance privacy in the proposed scheme. The system avoids revealing the exact location of the entity, providing an additional layer of protection. This approach helps in mitigating the risk of unauthorized tracking or identification based on the individual's real-time or historical location data. By using random coordinates, it becomes difficult for service providers or adversaries to associate specific activities or actions with a particular location or individual. It is important to consider the potential limitations and implications of using random coordinates. For example, the chosen coordinates should still fall within the valid area and should not introduce biases or skew the overall distribution of service requests. The selection process should also ensure that the randomly chosen coordinates are plausible and align with the geographical context of the service being accessed. Furthermore, the security and privacy analysis of the scheme should also evaluate the robustness of the random coordinate selection process against potential attacks or inference techniques that could exploit patterns or statistical properties of the chosen coordinates.

4.3 Service Matching

The service matching process occurs after the entity has been authenticated and applies for services within a ubiquitous computing environment. It is noted that all services in pervasive environment have spatial and temporal limits, indicating that they are bounded by specific geographical areas and time durations. To determine the validity of a service for an entity, the privacy system performs a check based on the spatial and temporal preferences of both the service and the entity. This implies that the service must fall within the desired spatial boundaries and be available during the required time period specified by the entity. By enforcing this strict matching criterion, the privacy system ensures that only services that precisely align with the entity's spatial and temporal preferences are considered valid. This helps tailor the services to the specific needs and constraints of the entity, enhancing the relevance and usefulness of the provided services. It is important to note that the specifics of how the service matching process is conducted and the mechanisms employed by the privacy system are not explicitly mentioned in the given statement. Further details and analysis would be required to fully understand the implementation and effectiveness of the service matching process within the privacy system.

5. Conclusion

The importance of user acceptance in the success of ubiquitous computing (UCE) and emphasizes the need for managing privacy risks associated with exposing personal information. The study aims to propose an architecture for privacy protection in ubiquitous computing by integrating privacy protection technologies into the access control architecture. This study defines spatiotemporal mobility (SM) as a metric for comparing trajectory similarity. The MTPPA algorithm is suggested based on SM and trajectory graph modelling. This implies that privacy-preserving features and technologies are essential to make UCE appealing to users. By integrating privacy protection technologies into the access control architecture, the goal is to ensure that different technologies work together, avoid conflicts, and provide multi-level privacy protection. The study also mentions the design of a ubiquitous software system based on a service-oriented approach.

References

1. Wang S., Hu Q., Sun Y., Huang J. Privacy Preservation in Location-Based Services. *IEEE Commun. Mag.* 2018; **56**: 134–140. doi: 10.1109/MCOM.2018.1701051.
2. Kang J., Steiert D., Lin D., Fu Y. MoveWithMe: Location Privacy Preservation for Smartphone Users. *IEEE Trans. Inf. Forensics Secur.* 2020; **15**: 711–724.
3. Majeed A., Lee S. Anonymization Techniques for Privacy Preserving Data Publishing: A Comprehensive Survey. *IEEE* 2021; **9**: 8512–
4. Huo Z., Meng X.F. A Survey of Trajectory Privacy Preserving Techniques. *Chin. J. Comput.* 2011; **34**: 1820–1830.
5. Zhang S.B., Wang G.J., Liu Q., Abawajy J.H. A trajectory privacy-preserving scheme based on query exchange in mobile social networks. *Soft Comput.* 2018; **22**: 6121–6133.
6. Zheng Y., Xie X., Ma W.Y. GeoLife: A Collaborative Social Networking Service among User, location, and trajectory. *IEEE Data Eng.*
7. Gruteser M., Grunwald D. Anonymous usage of location-based services through spatial and temporal cloaking; Proceedings of the 1st International Conference on Mobile Systems, Applications and Services; San Francisco, CA, USA. 5–8 May 2003; 8. Liu H., Li X.H., Li H., Ma J.F., Ma X.D. Spatiotemporal Correlation-Aware Dummy-Based Privacy Protection Scheme for Location-Based Services; Proceedings of the IEEE INFOCOM 2017—IEEE Conference on Computer Communications; Atlanta, GA, USA. May 2017;
9. Wang T., Zeng J.D., Bhuiyan M.Z.A., Tian H., Cai Y.Q., Chen Y.H., Zhong B.N. Trajectory privacy preservation is based on a fog structure for cloud location services. *IEEE Access.* 2017; 10. Shaham S., Ding M., Liu B., Dang S., Lin Z., Li J. Privacy Preservation in Location-Based Services: A Novel Metric and Attack Model. *IEEE Trans. Mob. Comput.* 2020; **99**: 1.
11. Zhao J., Zhang Y., Li X.H., Ma J.F. A Trajectory Privacy Protection Approach via Trajectory Frequency Suppression. *Chin. J. Comput.* 2014; **37**: 2096–2106. [Google Scholar]
12. Privacy-preserving Microdata On A Tabular Data Publishing Using Additive Noise Approach **DOI: 20.18001.GSJ.2022. V9I1.22.38503**
13. Li J., Bai Z.H., Yu R.Y., Cui Y.M., Wang X.W. Mobile Location Privacy Protection Algorithm Based on PSO Optimization. *Chin. J. Comput.* 2018; **41**: 71–85. [Google Scholar]
14. Xu H.J., Wu Q.H., Hu X.M. Privacy Protection Algorithm Based on Multi-characteristics of Trajectory. *Comput. Sci.* 2019; **46**: 190–195.
15. Data Security and Sensitive data protection using Privacy by Design technique – BDCC – 2019 – Suresh Babu et.al.
16. Zhang J., Xu L., Tsai P.W. Community structure-based trilateral Stackelberg game model for privacy protection. *Appl. Math. Model.* 2020; **86**: 20–35.

2

Usage of Artificial Neural Networks to Detect Fake Social Media Accounts

G. Chanakya[1]
Assistant Professor, Department of Information Technology,
Vignan Institute of Technology and Science, Hyderabad, India

Thota Tejeswar[2], Vennela Reddy Vaddepally[3], Amaragani Pranay[4], Amireddy Koushik Reddy[5]
UG Scholar, Department of Information Technology,
Vignan Institute of Technology and Science, Hyderabad, India

Abstract: Social networks have been ingrained deeply into our daily lives, necessitating strong security measures to safeguard users' private data. In this study, we present a method to evaluate the validity of Facebook friend requests using machine learning methods, specifically an artificial neural network. Our strategy entails discussing the classes and libraries used in this project, describing the sigmoid function and how it calculates the weights, and looking at the critical social network page factors that impact the solution's performance.

Also, we go into the dangers of phony personas and bots obtaining personal information for illicit purposes. These dangers are especially concerning because bots can access and scrape users' private information without their awareness. False profiles can be created to trick people and obtain their sensitive data. Our method can successfully identify and stop such hazards, guaran- teeing consumers a safer and more secure online experience. Our suggested solution represents a positive step toward addressing the internet's mounting privacy and security concerns.

Keywords: Social networks, Machine learning, ANN systems

1. Introduction

With 2.46 billion members worldwide, Facebook was the most popular social media network in 2017. Social media has become an essential component of contemporary culture. Despite being free for consumers, social media networks profit from the information users contribute. Sadly, the typical user is unaware that their rights are forfeited. Social media firms can use consumers for financial gain whenever they share a new location or photo, like, dislike, or tag another user in content. For instance, Facebook generates income via data and adverts. These attacks are frequently carried out without warning or informing the individuals whose data was compromised. There is a new story about a social media network being hacked daily in the news.

The existence of bots and phony profiles that might collect personal information for fraudulent reasons are among the most significant issues with social media. Web scraping, a practice used by bots to collect user data without the user's knowledge, is referred to as this. Worse, this behavior is legal, and bots can be concealed or appear as false friend requests on a social media platform to obtain personal data. Fake profiles appear to sneak past Facebook's built-in security measures, leaving users susceptible to identity theft and other online crimes.

[1]chanakyaa7@gmail.com, [2]thotatejeswar5@gmail.com, [3]v.vennelareddy09@gmail.com, [4]amaraganip@gmail.com, [5]saketh.amireddy@gmail.com

This study project offers a solution to these problems that focuses on the risks of bots in the form of false social media profiles. The algorithm would cooperate with social media businesses to train the model and determine whether or not the profiles are false. The algorithm might also be used as a plug-in for browsers, giving users' web browsers a more conventional layer of security.

2. Literature Survey

Since the number of users on social media sites keeps expanding, the issue of spotting false profiles there takes on more significance. Fraudulent profiles can hurt people and businesses, resulting in identity theft, financial fraud, and reputational injury. Thus, it has become crucial for social media sites to identify bogus personas.

Researchers have suggested several methodologies to identify bogus profiles, including rule-based, machine-learning, and hybrid approaches. Machine learning techniques have been increasingly popular among these because of their ca pacity to learn from massive datasets and adapt to new data.

In our literature review, we concentrated on one such strategy that uses machine learning algorithms to distinguish between bogus and authentic profiles. The suggested method trains and tests the classification algorithm using a publicly accessible dataset of fictitious and real profiles.

Several researchers have conducted surveys to understand better the topic of false profile identification in social media. However, the limits of several of these surveys must be considered when evaluating the results.

Only 100 people made up the survey's tiny sample size, according to Alnabhan et al. (2015). The generalizability of the survey results may be constrained by the small sample size, which may not represent the more significant population. Furthermore, because just one social networking site (Facebook) was used for the study, the findings might not apply to other social media sites.

Similar to the Lee et al. (2015) study, the survey had limitations due to its small sample size (140 participants) and the fact that it was limited to Facebook. Due to the possibility of social desirability bias from self-reported data, participants may have given responses they believed to be more socially desirable than their genuine opinions.

The survey by Wang et al. (2018) was conducted in a lab environment, which might not accurately represent how people use social media in everyday life. The study also used self- reported data, which can be biased due to factors including social desirability bias.

Al-Ammary et al. (2019).'s survey had a narrow scope because it only looked at Twitter and did not consider other social media platforms. Furthermore, the poll was only conducted in one country (Saudi Arabia). Therefore its results might not apply to other populations.

The survey by Li et al. (2020) was restricted to a sin- gle social media platform and depended on self-reported data (Weibo). Given the limited scope, the findings must be more generalizable because users' opinions of false profiles may vary among social media platforms. The survey also concentrated on users' views of fraudulent profiles rather than examining the efficacy of various fake profile-detecting techniques.

The effectiveness of two different categorization methods was examined. The recommended structure in the image illustrates the order of steps that should be taken for permanent false profile identification with dynamic learning from the input of the classification algorithm's output.

A feasibility study was also done to ensure the planned system would be relatively inexpensive for the business. This study is crucial to ascertain whether the suggested method is technically and economically feasible.

In conclusion, the suggested method shows encouraging signs of success in identifying phony profiles on social media sites. Additional study is required to enhance the classification algorithm's precision and handle the privacy concerns associated with using personal data.

3. Proposed System

In our approach, we use artificial neural networks and machine learning to assess the likelihood that a friend request is genuine or not. The Sigmoid function is used for each neuron (node) equation to maintain the values within 0.0 and 1.0. This could be multiplied by 100 at the output end to give us the likelihood that the request is malicious. Our approach would use a single deep neural network with a single hidden layer.

The neurons stand in for nodes. There would be precisely one decision-making process assigned to each node.

Each item carries weight and bias, which would aid in decision-making. The likelihood, expressed as a percentage, that the friend request is not genuine would be the output. The neural network in use is shown in Fig. 2.1. If we identify enough false accounts, we can web-scrape Facebook or other social networking sites to get the required training data set. Minimizing the overall cost function and modifying the equations to adjust the weight and bias of each neuron. We describe the classes and libraries involved in this paper.

4. System Analysis and Design

The analysis and design of an ANN system involves several key steps. First, the problem that the system is intended to

Table 2.1 Parameters for ANN model [5]

S. No	Attribute	Description
1	Profile ID	The Profile ID of account holder
2	Profile Name	The name of the account holder
3	Status Count	The number of tweets made by the account
4	Followers Count	The number of followers for the account
5	Friends Count	The number of friends for the account
6	Location	The location of the account holder
7	Created Date	The date the account was created
8	Share count	The number of shares done by account holder
9	Gender	The Gender of the account holder
10	Language Code	The language of account holder

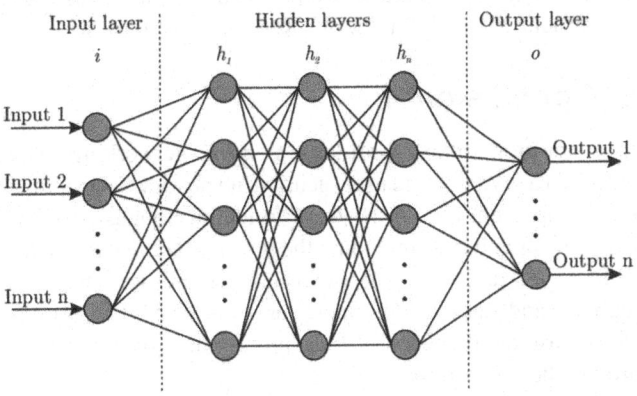

Fig. 2.1 ANN architecture in neural network [5]

solve must be clearly defined. This involves identifying the input data, output data, and any intermediate steps or processes that are required to achieve the desired output.

Once the problem has been defined, the next step is to design the architecture of the ANN system. This involves selecting the appropriate type of neural network, determining the number of layers and nodes, and selecting the activation functions and optimization algorithms that will be used to train the network.

After the architecture has been designed, the next step is to train the neural network using a suitable dataset. This involves feeding the network input data and adjusting the weights and biases of the nodes until the network is able to accurately predict the desired output.

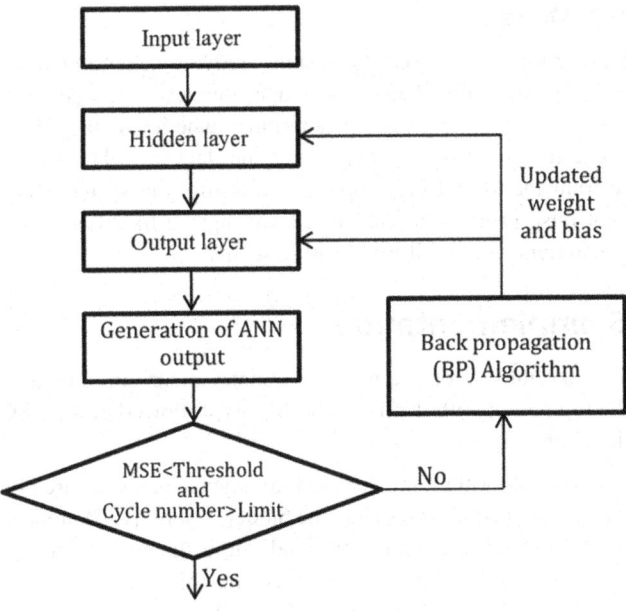

Fig. 2.2 ANN flow chart [5]

Once the network has been trained, it must be tested to ensure that it is able to generalize to new data. This involves feeding the network a set of data that it has not seen before and evaluating its performance.

4.1 Admin Module

The ANN system must be deployed in a production environment. This involves integrating the system with other software and hardware components, and ensuring that it is able to handle the expected workload and provide accurate and reliable results. Ongoing monitoring and maintenance is also required to ensure that the system continues to perform effectively over time.

On social media, it's typical for unscrupulous users to create phony profiles to obtain login credentials. These bogus profiles frequently entice unwary users into accepting friend requests with alluring images. After receiving approval, the author of the fake profile can send unwanted friend requests to other people in the user's network, frequently including links to other websites. From minor virus infections to more major dangers like rootkits that convert the computer into a zombie, clicking on these URLs can cause a variety of harm to the user's machine.

4.2 User Module

The suggested approach leverages machine learning as an artificial neural network to assess the validity of friend requests to solve this issue. The system uses Microsoft Excel to store and manage the old and the new phony data profiles. After that, training and testing sets are created from the acquired data. A dataset from social media websites is required to train the model.

4.3 Model

During the training set, the system analyzes several characteristics, including the account age, gender, user age, and links in the description, to determine whether a profile is accurate. The program then stores the data in a data frame to train the neural network. After training, the system may spot false profiles in real time, lowering the likelihood that consumers would fall for phishing scams.

5. Implementation

We take into account a person's country of origin. China is the top nation with the most bot activity, followed closely by the United States.

- The algorithm is divided into layers, as was already mentioned. For instance, the hidden layer has 128 nodes. Moreover, there are input and output layers. One deep machine learning algorithm is one that uses only one hidden layer. The purpose of these layers is to resemble a neural network. In this instance, it is referred to as an ANN, or artificial neural network. In order to tackle issues, AI algorithms can employ this system.

- A decision is produced as an output based on the inputs given. The buried layer is where the inputs are directed.

- The readCSV methods should be used to read the data. The training set reads this in. To clean up the data set, we employ the dropna technique. One of the most crucial aspects of creating an ANN is choosing the appropriate parameters and structuring the data properly. The IP address parameter is then compared to the profile's actual IP address. If there is a match, location match is given the value 1. If it does not match, location match is given the value of 0. For gender, the procedure is repeated.

- The number of messages sent is then divided by the account's age to arrive at the parameter known as "Number of messages sent out." The columns of the data collection are modified using the location technique. For both our input set and our output set, we employ this strategy.

- We'll be left with four separate sets as a result. The input train is given the first input set. The second half of the input set is contained in this variable. The input test is given the second input set. The input set's first half is contained in this variable. The output train is given the first output set. The second half of the output set is contained in this variable. The output test is given the second output set. The first half of the output set is contained in this variable.

- The model will then be built using Sequential from the Keras toolkit. The Sigmoid function is activated using the Sequential class' add method, which is also used to create the output layer.

- The input test variable, which holds one-half of the input set, must be tested for accuracy. The predict technique is used to do this. The outcome is next changed to a percentage.

- Score is given the evaluation method's output.

- With the score, one may tell whether a profile is authentic or not. Libraries

- Because of its popularity, brevity, and abundance of ready-to-use libraries, Python is currently the most widely used language for artificial intelligence. These libraries are ideal for mathematical models.

- Pandas is a great tool for data analysis and one such package. It is essential to prepare datasets properly and identify the best essential aspects to use later. NumPy is another library that is noteworthy. Given that we are working with a lot of numbers that are highly reliant on one another, NumPy is generally utilized for multidimensional matrix multiplication in scientific computing.

- The two most important libraries are TensorFlow and Keras. TensforFlow, CNTK, or Theano can all be used as foundations for Keras.

- It allows for quick prototyping and experimentation. An example of an organizational model is the Keras core structure. Greek for "horn," Keras is a given name.

6. Conclusion

In conclusion, phony profiles on social networking sites represent a severe risk to user security and privacy. The voting-based approach, one of the ways to identify bogus profiles, has some drawbacks, including the potential for real accounts to be stolen and sold. The suggested method uses machine learning and the artificial neural network (ANN) algorithm to increase the precision of false profile identification to get around these constraints.

The trained ANN model may be used to predict whether a social media account is fraudulent. The User Module lets users enter several factors to assess the validity of a social media account.

The system can be implemented using Flask, Pandas, Scikit-learn, and Python. The voting-based system may offer another line of defense against fake profiles, and the machine learning model and ANN algorithm can help increase the accuracy of false profile identification.

The proposed approach is effective overall in tackling the issue of phony profiles on social media platforms. Still, more study and development are required to increase its efficacy and usefulness.

7. Acknowledgment

We would like to extend our deepest gratitude to the Information Technology Department at Vignan Institute of Technology and Science, Hyderabad, for providing us with all the resources, support, and guidance needed to realize this paper.

References

1. https://stackoverflow.co/company/research/
2. https://www.tensorflow.org/resources/tools
3. https://keras.io/guides/
4. https://scikit-learn.org/stable/autoexamples/index.html
5. http://www.deeplearning.net/software/theano/

Performance Evaluation of Text Classifier for NLP using Machine Learning Techniques

3

Kallepalli Rohit Kumar

Assistant Professor, Department of CSE,
Koneru Lakshmaiah University,

P. Punnamchandar

Assistant Professor, Department of CSE, JITS

Rajesh Banala

Associate Professor,
Department of CSE (Data Science), TKRCET

P. Nagaraj

Assistant Professor, Department of CSE,
Anurag University

Abstract: Text categorization or text classification is utilized when organizing a large number of files into a predetermined set of categories. These programs include automatic indexing, the submission of patents to patent directories, and the distribution of data to data purchasers, as well as authorship attribution, style identification, survey tagging, and automated essay evaluation. Companies are alleviated of the task of manually organizing their documents, which may be prohibitively expensive or impractical given the time constraints of their applications. Thanks to IR and ML, the modern text class can compete with human professionals in terms of accuracy. Academics can use the available tools and resources to employ text categorization technology to solve real-world problems. As the number of virtual documents increases, classification of documents has become increasingly essential. The purpose of this position is to investigate the effectiveness of various problem-categorizing strategies. There is also a section for fine-tuning the parameters on the opposite side. This company is primarily concerned with SVM algorithms. According to their research, it is more effective than other system learning algorithms at classifying textual content. This is exemplified by Natural Language Processing and Artificial Neural Networks.

Keywords: Natural language processing, Artificial neural networks

1. Introduction

It is part of predictive modelling. For the purpose of quickly and accurately distinguishing new gadgets from older ones, pattern courses teach students to establish a version-based, labelled instruction log. It is possible to assess whether or not a new text message is spam by analysing its structure using spam filtering techniques like as Naive Bayes, Neural Networks, and Support Vector Model. "Textual content type" is a popular natural language processing task in a variety of business situations. Text documents may be organized into one or more specified categories, which is one of the primary aims of the text type. [2] Listed below are a few examples of writtenmaterial: social media may be used to measure the mood of your target audience. anti-spam and anti-phishing software how queries from consumers are connected to automobiles. The substance of news stories is organised according to themes.

[1]Krk542@gmail.com, [2]pcreddy_pulyala@jits.in, [3]rajesh.banala@gmail.com, [4]nagaraj.cse@anurag.edu.in

Bayes' theorem is mathematically represented as: $P(A|B) = (P(B|A) * P(A)) / P(B)$

If event B occurs, then event A has a certain probability, denoted by P(A|B). If event A occurs, then P(B|A) is the likelihood that event B will also occur. The likelihood of event A happening, denoted by P(A), is compared against the chance of event B happening, denoted by P(B).

- Determine the prior probability: P(A) represents an initial belief or prior probability that event A will occur. This is based on prior knowledge, hypotheses, or historical information.
- Evaluate the probability: Evaluate the probability of observing event B given the occurrence of event A, represented by P(B|A). This is the conditional probability that event B will occur if our prior assumption (event A) is correct.
- Compute the marginal likelihood: Determine the probability of event B occurring regardless of the occurrence of event A, represented by P(B). This is the probability of observing event B if event A is ignored.
- Apply Bayes' theorem: Using Bayes' theorem, combine the prior probability, likelihood, and marginal likelihood to calculate the updated probability of event A given event B, denoted by P(A|B). This yields the posterior probability, which reflects our revised belief in light of new evidence.

In addition to statistics, machine learning, artificial intelligence, and data science, Bayes' theorem is extensively utilized in a variety of other fields. Applications include medical diagnostics, spam filtering, financial analysis, and decision-making under uncertainty.By repeatedly employing Bayes' theorem as new evidence becomes available, we can refine and update our beliefs, resulting in more accurate and well-informed conclusions.

The marginal probability (P(A)) is the same as the posterior probability (P(A|B)). Bayes' Theorem may be used to represent a problem of conditional classification.

$$P(yi|x) = P(x|yi) * P(yi)/P(x)$$

1.1 Neural Networks

Neural networks will be used as one of the supervised learning approaches for the text category. Unclassified data may prevent the neural community from reaching consensus before the maximum number of allowed iterations. Before employing the Multi-Layer Perceptron, it is strongly advised that the data be scaled (MLP). For the results to be statistically significant, the test set must be scaled similarly to the data set. There are a variety of methods for analyzing different categories of data.You can select from a variety of options to generate a model instance. The final tuple access specifies

the number of neurons in the final MLP layer. To keep things as simple as feasible, it is possible to select three layers of neurons with the same number of neurons as the capabilities.

$$y = b + W^T * X$$

a1-layer neural network with a sigmoid activation function would be

$$y = \sigma(b + W^T * X)$$

(z) represents the sigmoid function applied to the linear combination b + WT * X in this equation. The sigmoid function transforms the input into a value between 0 and 1 via a nonlinear mapping. It's definition is:

$$\sigma(z) = 1 / (1 + e^{\wedge}(-z))$$

e represents Euler's number (approximately 2.71828), while -z is a negative exponent. In a binary classification setting, the output of the sigmoid function (z) is the probability or activation value that the input belongs to the positive class (i.e., y = 1).By applying the sigmoid activation function to the linear combination b + WT * X, non-linearity is introduced into the model, allowing it to learn complex patterns and make probabilistic predictions.

1.2 Support Vector Machine

One method of text categorization is the SVM, although there are manyothers. When using SVM, you don't need to spend a lot of time learning how it works. Because support vector machines use more processing resources, they are more accurate than naïveBayes and neural networks.When two spaces are divided in such a way that only one of them includes vectors from a certain class, this practice is known as subdividing the space[6,7].

2. An Overview of the Concerns at Hand

Document or report classification is a common problem in all three of these technologies. Instructions or classes must be assigned to a document. "Intellectual" (or "manual") methods may also be used. For a while, it was widely considered that the techniques to document classification used by library science and computer science were distinct; this is no longer the case. Inter disciplinary research is being done on document types inspite of their inter connectedness. All media, including images, audio and video, may be labelled with a tag. Each kind of paper has a unique set of problems that must be addressed. Even if the kind of the text is not explicitly stated, it is inferred. For example, a file may be categorised based on its subject matter or a variety of other characteristics. The best-case scenario is considered in the rest of this paper. Documents may be divided into content-based and request-based difficulties.

3. Three Suggestion System

As a result, many textual content processing systems need considerable human work to label and extract rules from big data sets. Human reasoning and mechanically discovered text style are combined in this research to reduce this labour while keeping the accuracy of the methods. The following method reduces the amount of human effort necessary to achieve a certain level of classification accuracy by using a common type data set. This means that machine learning-based classifiers are more accurate even if the same quantity of categorized data is utilized.

4. Result Analysis

Textual content mining may be used to develop an automated system for organising and categorising health- care data. NLP, information mining, and record extraction are being used inthis new and interesting area of research to alleviate the issue of study overload. It employs supervised machine learning to categorise text using SVM (Support Vector Machine). Thenumbers that have been offered are based on the information that has been provided. As a resultof its use, enormous volumes of data may be categorised with more accuracy and speed than previously possible.

4.1 Training and Dataset Records

Table 3.1   Training and test dataset

Dataset Type	Number of records	Dataset type
Text	920	Training
Non-Text	920	Training
Text	486	Test
Non-Text	486	Test

4.2 Confusion Matrix

Table 3.2   Confusion matrix

A	B	Classified as
352	134	A=Text
97	389	B= Non-Text

4.3 Evaluation Training Set

Table 3.3   Training dataset

Characteristics	Naïve Bayes	Neural Networks	SVM
Correctly Classified	72.42	73.6	75.9
Incorrectly Classified	27.58	25.4	24.1

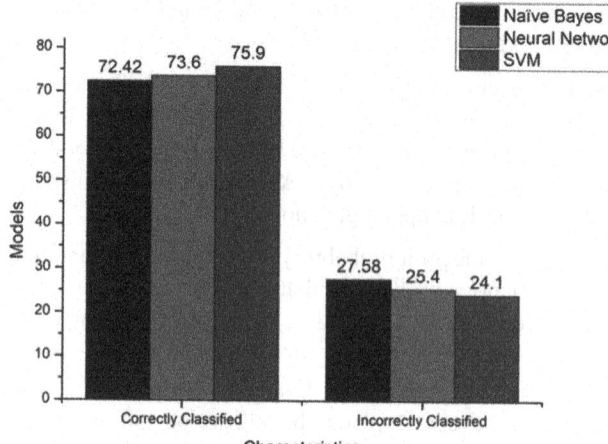

Fig. 3.1   Performance of training set

4.4 Accuracy of Naïve Bayes, Neural Networks and Support Vector Machine

Table 3.4   Accuracy of three algorithms

Classifiers	Naïve Bayes	Neural Networks	SVM
TP rate	0.708	0.736	0.729
FP rate	0.645	0.743	0.798
Precision	0.857	0.804	1
Recall	0.927	0.954	0.988
F Measure	0.904	0.921	0.944
ROC	0.899	0.905	1

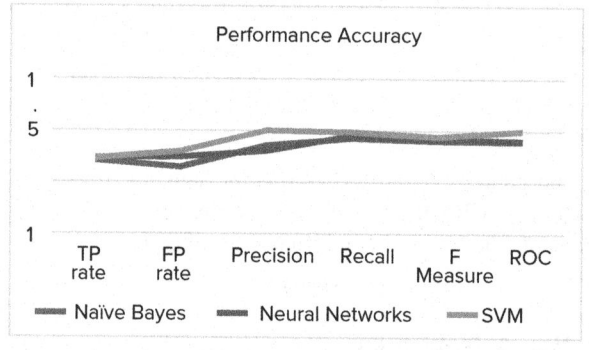

Fig. 3.2   Performance of classes

5. Conclusion

Since there are so many different types of digital documents, text categorization is an AI study subject(emails, discussion forum postings, and soon). Classifiers that solely use educational text corpora to train their models might provide somesurprising outcomes. High-quality training data may lead to high-performing classifiers, sinceclassifier performance is linked to training data quality. In conclusion, text categorization, also known as text classification, plays a crucial role in organizing a large volume of files into specific categories. This process has numerous applications, such as automatic indexing, patent submission, data distribution, authorship attribution, style identification, survey tagging, and automated essay grading. By employing Information Retrieval (IR) and Machine Learning (ML) techniques, text categorization systems can achieve high accuracy, rivaling that of skilled human professionals.

The advantages of text categorization are significant for companies as it relieves them from the burden of manual document organization, which can be costly and impractical given time constraints. IR and ML technologies offer efficient and reliable solutions, allowing businesses to effectively manage their growing document repositories. Furthermore, text categorization has practical implications beyond the corporate world. Academics and researchers can leverage these technologies to address real-world problems and make use of the available tools and resources. As the volume of virtual documents continues to increase, document classification becomes increasingly important, and understanding different categorization strategies and fine- tuning parameters becomes essential. The work focuses primarily on Support Vector Machine (SVM) algorithms, which have demonstrated superior performance in classifying textual data compared to other machine learning algorithms. Additionally, the field of text categorization encompasses Natural Language Processing and Artificial Neural Networks, both of which offer promising avenues for advancing the capabilities of text classification systems.

References

1. Crammer, K. and Singer, Y., On the algorithmic implementation of multiclass kernel-based vector machines. Journal of Machine Learning Research, 2, pp. 265–292, 2001.
2. Sebastiani, F., Machine learning in automated text categorization. ACM Computing Surveys, 34(1), pp. 1–47, 2002.
3. Debole, F. and Sebastiani, F., Supervised term weighting for automated text categorization. Proceedings of SAC-03, 18th ACM Symposium on Applied Computing, ACM Press, New York, US: Melbourne, US, pp. 784–788, 2003.
4. Yang, Y. and Pedersen, J. O., A comparative study on feature selection in text categorization. Proceedings of ICML-97, 14th International Conference on Machine Learning, ed. D.H. Fisher, Morgan Kaufmann Publishers, San Francisco,US: Nashville, US, pp. 412–420, 1997.
5. Wiener, E. D., Pedersen, J. O. and Weigend, A. S., A neural network approach to topics potting. Proceedings of SDAIR-95, 4th Annual Symposiumon Document Analysis and Information Retrieval, Las Vegas, US, pp. 317–332, 1995.
6. Joachims, T., Text categorization with support vector machines: learning with many relevant features. Proceedings of ECML-98, 10th European Conference on Machine earning, eds. C. Nedellec C. ouveirol, Springer Verlag, Heidelberg, DE: Chemnitz, DE, pp. 137–142, 1998.
7. Giorgetti, D. and Sebastiani, F., Automating survey coding by multi class text categorization techniques. Journal of the American Society for Information Science and Technology, 54(12), pp. 1269–1277, 2003.
8. Rish, Irina (2001). An empirical study of the naive Bayes classifier (PDF). IJCAI Workshop on Empirical Methods in AI.
9. Rosenblatt, F. (1958)." The Perceptron: A Probabilistic Model for Information Storage And Organization In The Brain". Psychological Review. 65 (6): 386–40 8. CiteSeerX10.1.1.588.3775.doi:10.1037/h0042519. PMID13602029.

Note: All the figures and tables in this chapter were made by the authors.

A Secure AOMDV Protocol's Design and Implementation in Mobile Adhoc Network Cryptography

4

K. Sreenivasulu[1]

Professor, Department of Computer Science and Engineering,
G. Pullaiah College of Engineering and Technology, Kurnool A.P, India

K. Gayatri[2]

Assistant Professor, Department of Computer Science and Engineering,
G. Pullaiah College of Engineering and Technology, Kurnool A.P, India

K. Kundana[3]

UG Scholar, Department of Computer Engineering,
Govt Polytechnic for womens, Kadapa, AP, India

Abstract: Each mobile node in a Mobile Ad hoc Network (MANET) serves as both a router and a terminal. When functioning as a router, a dependent routing protocol is chosen to enable it to reach its destination and function as a terminal; acting as an agent was the main goal of packet transmission. Adhoc On Multipath Distance Vector (AOMDV) routing protocol is used in this analysis to design and implement a secure AOMDV protocol for mobile adhoc networks. The AODV (Adhoc On Demand Distance Vector) routing system, which, despite only partially controlling the attacks, is much more reliable than its parent protocol, is extended to include multiple paths by AOMDV. The main objective is to secure the model in a critical environment with multiple attackers. Due to the complex functioning needs of network, the outage is even a few nodes because of power exhaustion. This model aims to increase network lifetime of MANET by allocating the loads between every node in network. It supports decreasing differences in energy utilization at different nodes because of load allocating. Therefore, it increases the entire network lifetime.

Keywords: AOMDV, MANET, Cryptography, Wifi

1. Introduction

MANET is an autonomous system that widely considered in recent years. As the existing infrastructure is inconvenient to utilize, the establishment of Ad hoc wireless MANET is a connection structure or where the mobile users can communicate.

MANETs helps WEB Based operations [8]. Given how helpful the Hypertext Transfer Protocol (HTTP) is on the Internet, numerous studies were anticipated to gauge and improve the performance of mobile ad hoc networks with HTTP mobility. A suitable Transmission Control Protocol

(TCP) implementation is required for a proper HTTP implementation. TCP ensures end-to-end information delivery between certain hubs It is a part of the stack of Internet protocols that is used in digital network communication. TCP carries out connection work with the Internet Protocol, which establishes the static region of the distant hubs, transports TCP, and ensures that the packet is delivered precisely to its intended location. Earlier transforming the data, the protocol makes Associate in Nursing association among supply as well as destination nodes and maintain as live when connection is dynamic. The protocol divides the enormous amount of data into smaller packets and further guarantees that the integrity of the knowledge remains even after re-assembly

[1]sreenu.kutala@gmail.com, [2]kgayathri.cse@gpcet.ac.in, [3]21057cm025kundana@gmail.com

at destination nodes. In Vehicular ad hoc network (VANET), communication users are at various vehicles.

VANET handles many functions like security, traffic management etc. in wireless medium. It is an arbitrary set of wireless network has no charge in every node and it is decided on the need of the network. VANET and MANET are described in a remarkably similar way. The term "Wi-Fi" refers to a particular class of Wireless Local Area Networks (WLAN) known as Wireless Fidelity (Wifi),which utilizes configurations in 802.11 families. It is a kind of advantage for wireless networks in that the users will utilize the web by different electronics like computer, laptop, mobile etc. Wireless fidelity is widely used in organizations, offices, schools and houses as an alternative for wired Local Area Networks. Many air ports, hotels and fast food organizations offer Wi-Fi for no cost. Those are called hotspots. Few of them are free, but the majority of them charge by the day or the hour to arrive. Transmission Control Protocol (TCP) is most widely used transmission layer standard on the Internet as well as is designed and streamlined to work well in a wired structure. TCP also plays an important role in MANETsa wide range of applications operate over TCP. [2]. TCP is a protocol connection calculated for many wired systems. It was a group organized by the association to finalize the congestion control plan. Currently, wireless technologies are steadily advancing in WLAN. The existence of many networks, the possibility that burst or random bit errors are to blame for packet losses, host mobility, etc.

TCP was changed over years to enhance the development and as an output; different upgrades of TCP have been improved, AODV is an addition of AOMDV Routing Protocol. Due to this problem the two reverse paths are not found to avoid this condition when each path has more general intermediate nodes which is known as 'Root Cutoff' condition in AOMDV routing protocol provides [1]. Due to AOMDV's function being used throughout the entire VANET, it offers the best outputs with increased jitter, throughput, maximum delay, packet delivery ratio, and latency. Based on simulated The AOMDV protocol for mobile ad hoc networks based on WiFi was used. Enhancing multilink discontinuous paths between the source and destination at each stage is the main goal of AOMDV. The routing entries for each destination are used to create lists of the following hops that include associated hop counts. The destination nodes in the Adhoc On-Demand Multipath Distance Vector have a specific publish node count for all routes [14]. All subsequent hops have the same sequence number, which is highly needed for tracking the route [12]. In a dynamic network, providing effective fault tolerance in part to facilitate speedy and beneficial recovery from path failure is the primary goal in addition to Adhoc AOMDV.

A node in MANETs actsas router and terminal. Further, routing path in Adhoc networks is advanced in environment unlike wired network. Thus, few security policies designed for wired framed are not applicable to Adhoc network. The simple behaviour of mobile nodes in a constantly changing topology makes it difficult to establish secured Adhoc. Data send through the nodes are gathered by whole nodes by its uninterrupted communication range.

Because of lack of few securities of Adhoc network, the MANET routing protocol is highly unsafe to variety of malicious attacks. Many of Adhoc described for MANET have a trusted and depended condition and do not consider security problems in their existing pattern. It investigates the secured of AOMDV, a multipath extension of AODV, against blackhole attack [4].

Low resource accessibility in MANETs requirements which are beneficial for resources uses, which necessitates optimal routing. Additionally, mobility objective in these networks paths highly limitations on any routing protocol structured especially. The two types of routing protocols that are currently used in MANETS are "Proactive" or "Table-driven" and "Reactive" or "On-demand.".With table-based routing protocols, every nodesaims to keep consistently updated routing data for each nodes in network. This is because of response for change in network by every node updating its routing table and propagating update for neighbor nodes. Therefore, the dynamic power route is used suddenly when a packet required forwarding. Destination Sequenced Distance Vector (DSDV) is a sample for a table-driven protocol.

With on-demand driven routing, which founded, if the source node required that. Route discovery and maintenance were the primary approaches: the route discovery function includes forwarding route-request packets at source for neighboring node that sends the application for neighbors. After router request achieves target nodes, it replies by a Route Reply packet that is routed return to source nodes through the neighbors that originally received route request. If a route-request reached an intermediate node which had an adequate updated route, it ends forwarded and sending route-reply message return to source.

2. Literature Survey

The "Trusted key management with Rivest-Shamir-Adleman (RSA) based security policy for MANETS" was described by Arora Vandana, Ahuja Sunil, et al. in [3]. They also created an intermediate process and a key management protocol for one-hop communication in MANET. The RSA framework's signature generation and authentication functions are used by the protocol.

Bansal Priyanka, Gupta Anuj K et al [5] developed the "Impact of black hole and neighbor attack on AOMDV routing protocol", explained impact of blackhole attack as well as neighbor attack on AOMDV. The execution of AOMDV was calculated through evaluating various parameters like throughput, packet delivery ratio.

The authors of "A novel elliptic curve cryptography based AODV mobile ad-hoc networks" [7] Raju M. Janardhana, Subbaiah P., Ramesh V., et al. implement an elliptic curve based AODV to improve data packet protection. It is featured on Elliptical Curve Cryptography (ECC) encryption mechanism.

Elliptic Curve Cryptography (ECC) is used for a class of cryptographic tools as well as protocols, as the security depends on a discrete algorithm problem, according to Lauter, Kristin, et al. [17]. ECC is dependent on both equations that the elliptic curve corresponds to and the onsets of numbers.

Hannan Xiao, et al. [13] describe a group of management calculation in power efficiency in MANET and its power utilization from different features: in various layers including application, network and media access control address(MAC) layers, for various function modes such as passive, transmits and receives by Demand Signal Repository (DSR), Destination-Sequenced Distance-Vector Routing (DSDV) and AODV With various communication protocols are included.

Vidyarthi SS, Vijayalakshmi S, Simha GAV and Shekar, et.al [11] described a novel protocol known as Enhanced-AODV (E-AODV) protocol. They revealed traditional transmission plan, which transmits packets omni-directional distress from defects such as high number of unnecessary vehicles, exaggerated interference or dispute between neighboring nodes, and restricted coverage. The commonly called as AODV and DSDV utilized for routing for networks need a standard transmission plan as well as it, suffers from the faults. The issues in Adhoc networks will be reduced by using directional antennas. In comparision with the omni-directional plan, E-AODV utilizes a least amount of control packets, as well asby less transmission redundancy.

Utkarsh, Mukesh Mishra and Suchismita Chinara, et al. [9] described a the Reactive Energy Saving Ad Hoc Routing (ESAR) framework achieves long network lifetime in which a packets are delivered through a chosen route until a nodes achieves a threshold value, at which point another alternative route is picked for packet delivery. ESAR maximizes network lifetime by performing a threshold abstract to the total accessible route. The execution of ESAR is finer than AODV and ESAR utilizes more alternative path, but repeating the same route in AODV decreases network lifetime.

Shivashankar, Golla Varaprasad and HN Suresh, et al. [6], an structured algorithm for MANETs is explained, which increases network lifetime by reducing power consumption during route establishment with the need of changed DSR. The described framework reduces energy utilization for each packet as well as it increases network lifetime. The framework of changing Dynamic Source Routing was to choose energy-efficiency routes. The primary parameter of the changed Dynamic Source Routing is reducing the power utilized each packet, increasing the network lifetime for the network, as well as reducing the high nodes value.

Using Distributed Trust Computation and Carrier Sense Multi Accessible with Collision Intimation for Distributed Heterogeneous MANET, the Secured Reliable Multipath Routing Protocol (SRMRP), was suggested by B. Narasimhan, R. Vadivel, et al. [10]. Their research is aimed for two functions. The primary function is to secure the routing against attacks. The reference-based trust security function is explained. The later goal is relevant data communication over a heterogeneous mobile Ad-hoc network. A collision intimation mechanism is used to achieve that adaptive carrier sense multiple access. The paper is mainly concerned with real-time mobile Adhoc networks, which are heterogeneous in nature.

C. Siva Ram Murthy and B. S. Manoj, et al. [18] AODV is described as a reactive routing protocol framework for Adhoc networks for many numbers of nodes. In this, nodes maintain standard routing tables that specify, later to achieve destination. If information is not there in the source's routing table, the route request is transmitted. It receives a route request and the latest route for destination returns to source, and all nodes return route updated their routing tables. The request is retransmitted if the intermediary node does not have a valid route.

3. A Secure AOMDV Protocol's Design and Implementation in MANETS Network Cryptography

In this section, a frame work of Design and Implementation of a secure AOMDV protocol's design and implementation in MANETS network cryptography is observed in fig.1. The source nodes are testing to forward information for that destination node 'D'. Here it can observe the shortest hop route for entire source nodes by A and B. Therefore, nodes are regular in whole traffic routes, therefore the bottlenecks because of traffic congestion and suffer rapid energy degradation.

It finally shows the battery depletion of A and B nodes that remove as well as it will be shorter engage in communications. When a source node requires forwarding a packet of destination and initiates a Route Discovery process by forwarding a Route Request (RREQ) for entire neighbor's.

If the RREQ achieves the destination nodes, it responds for request with the smallest or lowest count of hop along route by forwarding a Route Reply (RREP) packet.

The route discovery operation requires forwarding route-request from a source to its neighboring nodes, which forwarded the application for their neighbors etc. The neighbor that received the route quest responds by unicasting a route-reply packet back to the source node when the route quest has reached the destination node. The node with a sufficient updated route concludes by sending and forwarding the route-reply message returned to source if a route-request successfully reaches the intermediate. Energy calculations have the drawback that no agent within the network can determine how much energy a specific node N, which is still a node in and of itself, has left. Neighboring nodes of N can record messages sent from N and compute an accurate estimate, however they don't know the primary energy of N or the energy spent listening to N. Calculations are used to derive the values. Accurate adequate reliably sort nodes as per their residual energy field and sequence number.

After checking the sequence number and the residual energy field if it is duplicate then the route request is to be discarded. In case if the same or lesser energy also. If a node collects a broadcast RREQ packet, it grasps the benefits of the neighbor's awareness to describe the suitable course of action with the support of other attachment (waiting mechanism). Decision to retransmit or discarded the request message. The node is checked to see if it is a destination node if there are no duplicates or higher residual energies. A single network-qualified control point name that displays the inaccessible destination (network nodes or virtual nodes) while the route is being chosen is described in the destination nodes record structure.

The energy system evaluates residual energy of nodes for entire packets forwarded, collected, and dropped during to compute the distance between the source and the destination nodes; to compute the energy used by the nodes for various router decisions. This residual energy for node is increased as router metrics in the packet header as well as utilized to select an energy beneficial route from source to destination nodes. This chosen way may not be least energy but it is recognized that path with higher residual energy of nodes differs from the minimum energy routing.

A minimum energy path attracts high traffic congestions and a node on these paths suffers from battery exhausts and removes quickly; thereby requiring entire network failure. However, paths choose depended on high residual energy can increase network lifetime by load balance across the paths and node across the network.

In case if it is Not Destination Node then it is Updated Routing Table Add Routing metrics. It is of two types.

Minimum Remaining Power of the Node and Link Stability with Neighbor. The destination nodes forward an RREP on the rev path after updating the routing table based on route metrics. Then the routing table source updates shows the routing table. When considering the framework objective and the novel system, two functions of path that is the least complete power utilization and the lifetime of network may be together exclusively. Examine a case, if a regular nodes has many routes from different nodes, the battery energy of particular nodes will reduce fast.

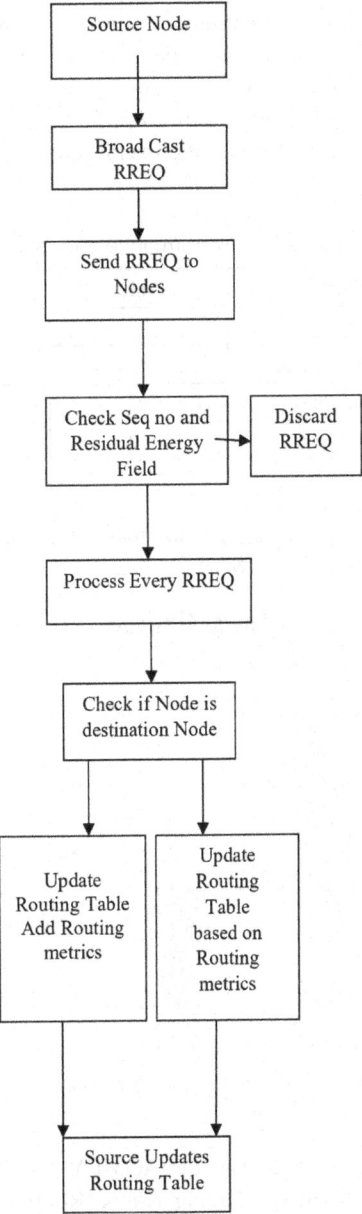

Fig. 4.1 An secured AOMDV protocol in mobile Adhoc network cryptography [7]

As a output, this specific node may run out of battery very quickly and die, ultimately shortening the lifetime of the network. If selecting a route, the available routing protocol execution selects the route with the least count of hops. But for an energy efficiency approach, initially calculates residual energy stage for every nodes and found least residual energy in special route. Then choice is made by selecting the stage with the high minimum residual energy.

4. Result Analysis

In this section, result analysis of Design and Implementation of an Secured AOMDV Protocol in Mobile Adhoc Network Cryptography is observed. The comparision of AOMDV with Energy Saving Adhoc Routing (ESAR) is done in terms of energy efficiency, power/traffic Congestion, energy consumption, and increases the overall complete lifetime of network.

Table 4.1 Performance analysis

Parameters	AOMDV	ESAR
Energy Efficiency	98	84
Power/Traffic Congestion	76	91
Energy Consumption	82	87
Network Throughput	92	86

Source: Authors

In Fig. 4.2 Energy Efficiency comparison graph is observed between AOMDV and ESAR.

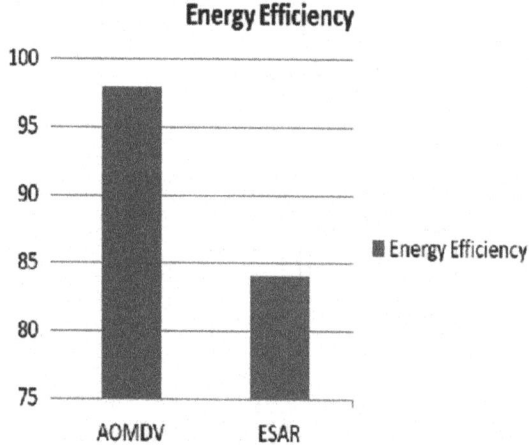

Fig. 4.2 Energy efficiency comparison graph

Source: Authors

Power/Traffic Congestion of AOMDV is lower when compared with Energy Saving Adhoc Routing (ESAR) in Fig. 4.3

Energy Saving Adhoc Routing (ESAR) shows higher Energy Consumption when compared with AOMDV in Fig. 4.4.

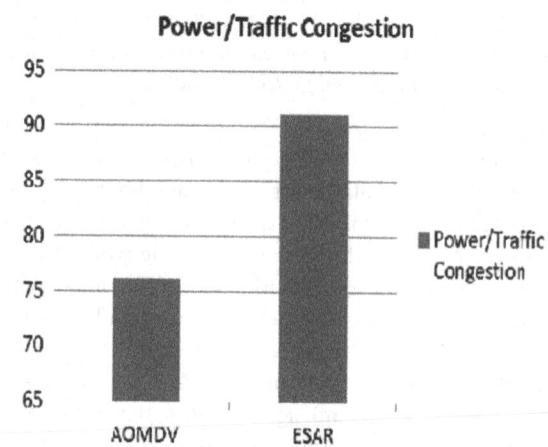

Fig. 4.3 Power/traffic congestion comparison

Source: Authors

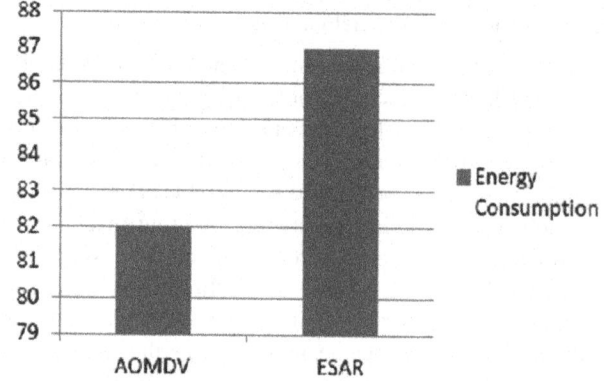

Fig. 4.4 Energy consumption comparison graph

Source: Authors

The Network Throughput of AOMDV shows higher in Fig. 4.5.

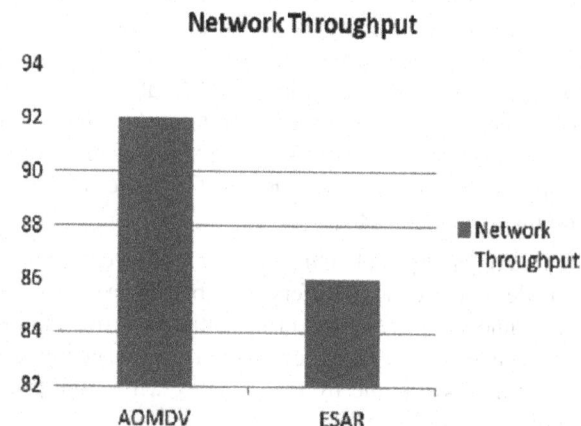

Fig. 4.5 Network through put comparison graph

Source: Authors

5. Conclusion

This section concludes A Secure AOMDV Protocol's Design And Implementation In Mobile Adhoc Network Cryptography. There are several flaws in ad hoc routing protocols that allow attacker nodes to influence route selection or launch denial-of-service attacks. The algorithm is successfully implemented to enhance the energy efficiency to increase the network lifetime of a position MANETs in a representation framework. An existing framework is changed to choose a route based on a maximum-minimum formulation to find an ideal residual energy route. This follows a load balancing reaches the power/traffic congestion paths as well as selects gently loaded routes. This assists in decreasing the difference in energy utilization at different nodes because of load distribution and therefore increases the entire network lifetime of network.

References

1. Fidan Mehmeti, Student Member, IEEE and Thrasyvoulos Spyropoulos, Member, IEEE "Performance Analysis of Mobile Data Offloading in Heterogeneous Networks" IEEE TRANSACTIONS ON MOBILE COMPUTING, VOL. 16, NO. 2, FEBRUARY 2017, pp 482–497

2. Dinesh C. Dobhal and Sushil C. Dimri,"Performance evaluation of Proposed-TCP in Mobile Ad hoc Networks (MANETs)" , Inventive Computation Technologies (ICICT), International Conference on, 2016

3. Arora Vandana, Ahuja Sunil "Trusted key management with RSA based security policy for MANETs", International Journal of Advance Research, Ideas and Innovations in Technology, ISSN: 2454-132X, (Volume 2, and Issue3) (2016).

4. Shrivastava Sonal, Chetan Agrawal, and Anurag Jain. "An IDS scheme against black hole attack to secure AOMDV routing in MANET." arXiv preprint arXiv: 1502.04801, 2015.

5. Bansal Priyanka, Gupta Anuj K., "Impact of black hole and neighbor attack on AOMDV routing protocol", International Journal of Innovations in Engineering and Technology (IJIET), Vol. 3, Issue 4 April 2014

6. Shivashankar, Golla Varaprasad and HN Suresh, 'Implementing a new power aware routing algorithm based on existing dynamic source routing protocol for mobile adhoc networks' IET Networks, 2014, Vol.3, Issue 2, ISSN 2047-4954, pp 137–142.

7. Raju M Janardhana, Subbaiah P., Ramesh V., "A novel elliptic curve cryptography based AODV for mobile ad-hoc networks for enhanced security", Journal of Theoretical and Applied Information Technology, December 2013.

8. Farnaz Moradi and Meraj rajae, " Investigating the impact of TCP New Reno and OLSR protocol over MANET with HTTP traffic" 6th International Conference on Sciences of Electronics, Technologies of Information and Telecommunications (SETIT), 2012, pp 612–618

9. Utkarsh, Mukesh Mishra and Suchismita Chinara, 'ESAR: An Energy Saving Adhoc Routing Algorithm for MANET', IEEE Fourth International Conference on Advanced Computing, ICoAC 2012 MIT, Anna University, Chennai, 13-15 December 2012, pp 13–15

10. B Narasimhan, R Vadivel, "Secured Reliable Multipath Routing Protocol (SRMRP) using Trust Computation and Carrier Sense Multiple Access with Collision Intimation (CSMA/CI) for Heterogeneous IP-based Mobile Ad-hoc Networks", International Journal of Computer Applications, Vol. 60 No.10, pp 12–16, December 2012.

11. Vidyarthi SS, Vijayalakshmi S, Simha GAV and Shekar, 'Energy efficient En-hanced-AODV protocol for Multi-Rover systems', Ultra Modern Telecommunications and Control Systems and Workshops (ICUMT), 2011 3rd International Congress 2011, pp 1–5

12. Dr. Panos Bakalis and Bello Lawal, "Performance Evaluation of CBR and TCP Traffic Models on MANET using DSR Routing Protocol" International Conference on Communications and Mobile Computing, 2010, pp 318–322

13. Hannan Xiao, 'Energy consumption in Mobile Adhoc Networks', National Conf on Recent Trend in Information, Telecommunication and Computing, 2010, pp 23–27

14. Mohamed Yahia, J´ozsef B´ır´ o, Member IEEE, "Behavior of TCP algorithms on ad-hoc networks based on Di_erent Routing Protocols (MANETs) and propagation models" ,IEEE Conference, 2006.

15. YuHua Yuan, HuiMin Chen, and Min Jia, "An Optimized Ad-hoc On-demand Multipath Distance Vector (AOMDV) Routing Protocol" Asia-Pacific Conference on Communications, Perth, Western Australia, October 2005.

16. Alvin C. Valera, Student Member, IEEE, Winston K.G. Seah, Senior Member, IEEE, and S.V. Rao, Senior Member, IEEE, "Improving Protocol Robustness in Ad Hoc Networks through Cooperative Packet Caching and Shortest Multipath Routing", IEEE Transactions On Mobile Computing, Vol. 4, No. 5, September/October 2005.

17. Lauter, Kristin. "The advantages of elliptic curve cryptography for wireless security." IEEE Wireless communications 11.1 (2004): 62–67

18. C. Siva Ram Murthy and B. S. Manoj, "Ad Hoc Wireless Networks: Architectures and Protocols," Prentice Hall Communication Engineering and Emerging Technologies Series, 2004

19. C.E. Perkins, E.M. Royer and S.R. Das , "Ad hoc on-demand distance vector (AODV) routing," RFC 3561, July 2003

20. Hossam Hassanein and Audrey Zhou, "Routing with Load Balancing in Wireless Ad hoc Networks", in proc. of 4th ACM International Workshop on Modeling, Analysis and Simulation of Wireless and Mobile Systems, pp: 89–96, Rome, Italy, 2001.

Enhancing Cyber security: Evaluating Machine Learning Algorithms for Effective Threat Detection and Classification

5

K. Prathyusha[1], E. Shirisha[2],
T. Priyanka[3], K. Raghavendar[4]

Department of Computer Science and Engineering,
Teegala Krishna Reddy Engineering College Hyderabad, India

Abstract: The power consumption and operating time of sensor nodes in wireless sensor networks are constrained. Mission-critical apps must understand this information in order to determine how long the network can continue to support networking operations. For networks to endure longer, effective routing protocol usage has been advocated recently. The soft computing technique takes into account how well they cooperate and how well they can adapt to the challenging WSN conditions. The investigation of many algorithms' network parameters at simultaneously seems to be less prevalent, despite the fact that soft computing techniques have received a lot of attention. In-depth analysis of the clustering-based routing protocols that are employed to safeguard data and are conscious of soft computing approaches is provided in this work. We need to figure out how to compensate for the energy resource shortage and extend the processing time of the sensor nodes in order to extend the network's lifespan. The energy consumption of each sensor node is effectively reduced by power management approaches, and the adaptive efficient routing technique has garnered a lot of interest.

Keywords: Wireless sensor network, Soft computing techniques, Resources, Routing protocols, Sensor nodes

1. Introduction

The broad use of technology, as well as the creation of several complicated technological solutions for safe communication between distant systems, has contributed to recent advancements in next-generation information systems and global communication. As a result, the technological environment we live in is becoming more adept at communicating with humans. Improvements in modern and smart technologies may be attributed to a wide range of biologically-inspired algorithmic approaches, including neural computation, AI, and genetic programming, as well as the most recent developments in soft computing and cognitive computing models. These techniques have recently made feasible the management of secure data communications and the analysis of patterns from almost every aspect of human existence. The area of pervasive computing has benefited from these developments in technology and security. For wireless networks built with sensors to work better, the right routing algorithm is needed to avoid link failures by using less energy and keeping transmission times as short as possible. Many methods of routing, both reactive and proactive, have been developed to save energy and protect the network from link failures. Out of all the routing methods that have been made, evolutionary- based routing, which is based on how a species changes over time or how it hunts for security, was found to be the most efficient and secure way to find the best routes.

Wireless sensor networks (WSNs) are groups of dispersed, self-sufficient devices that can detect data, process it, and exchange information with one another using wireless connections. In a WSN, nodes are very limited in how much power, memory, and processing they can do. Because sensors run on batteries, their life spans and, by extension, the life span of the WSN is at risk. By adding or removing nodes, the size of the network can be changed flexibly, and this will change the way the network is built unpredictably. The biggest problems with WSNs are batteries, bandwidth, and processing power. Keeping the network's power and energy levels constant will extend its lifespan. Therefore, WSNs rely on routing and clustering methods to facilitate inter-node and inter-domain communication. Routing in wireless sensor networks is different from routing in fixed networks. In WSNs, the best routing doesn't always mean picking the shortest path between a source and a sink. Because sensors don't have much power, the routing protocol in WSNs has to be smart. So, power-aware routing algorithms should be used to save WSN power and, as a result, make the network last longer.

Traditional WSN routing protocols may be broken down into three categories depending on whether they are flat-based, hierarchical-based, or location-based. After WSN implementation, researchers are interested in the possibility of applying embedded soft computing approaches. This is due to the smarts and adaptability of soft computing paradigms in dealing with nebulous and unreliable information in intricate settings.

2. Characteristics

Because of its effectiveness and dependability in so many different contexts, wireless sensor networks (WSN) have quickly risen to prominence as one of the most vital communication technologies. From what I've read, WSNs have the following features that make them stand out:

2.1 Self-deployment in a Dense Way

WSN is a very large distributed computing system. In the network environment, there are a lot of sensors that are spread out and put there at random. Sensors are set up on their own because each sensor handles its communication in the network.

Having limited processing and storage capacities is a major drawback of sensor nodes because of their nature as tiny, independent, battery-operated physical units.

Insufficient means of generating power. It is frequently difficult to replace or recharge these batteries since WSN applications are complicated to operate and sensor nodes rely on batteries for power.

2.2 Sensor Heterogeneity

Physical damage or failures may render sensor nodes unreliable and inconsistent, and their continued existence is not assured during the lifetime of a WSN. All the more so during a tough deployment.

2.3 Data Redundancy

Data can be sent to the central node in different ways by more than one node. This is because sensor nodes need to work together and talk to each other and because sensor nodes are physically different.

2.4 Focused on a Specific Application

Wireless sensor networks are often created and utilized for a single purpose before being retired or replaced. Specifically, this has an impact on the network's energy consumption, routing needs, and size requirements throughout the design phase.

2.5 Broadcast Communication

In a WSN, sensors often use a variety of flooded routing mechanisms to communicate with one another and a sink node.

Topological inconsistency: Because sensor nodes don't have a lot of power and the environment is harsh, the network's topology usually changes often, such as when a connection fails, a node dies, a new node is added, energy is used up, or a channel fades.

There is a limit to how far data can be sent. This is usually caused by the physical limitations of the sensor nodes, which affect the coverage network and quality of communication.

Because WSN energy resources are kept secret, making the network last longer is seen as a challenging problem. Even though the limitations of battery-powered devices affect how long a network will last, the length of the path, how the load is spread out on each path, and how reliable each path is will also have a big impact. Data in a WSN travels from its origin node to a successor node chosen by its neighbors. It does this movement over and over again until it gets to the sink node, based on how it was chosen. Routing in WSN can be grouped by either how the network is set up or how the protocols work. Here are a few ways to explain what routing protocols are. Using energy efficiently in WSN is a bottleneck problem that affects the network's performance and lifespan. Researchers have recently looked at ways to deal with this problem by paying attention to how much energy is used and by managing power. The optimal routing method and use of energy optimization have a big effect on the performance of WSNs and guarantee that the network will last longer. Due to the limitations of WSNs, especially the limited energy of sensors, smart routing should be done to make sure that nodes use the same amount of energy. This will make the network last longer and ensure network coverage.

3. Literature Review

In wireless sensor networks, routing plays a crucial role. The connection and network stability must be maintained while the performance is optimized. Numerous routing strategies were developed for the WSN to maximize energy efficiency and service life.

Singh et al.'s (2022) Firefly Algorithm (FA) mimics the appearance and motion of swarms of tropical fireflies. Because it has two key tools, FA is superior to other computer algorithms. The FA stands out due to its proximity to local attractions and automatic regrouping. The absorption coefficient tells us if the fireflies are drawn to one another globally or merely locally because the light's intensity and distance are connected. Access to both international and local settings is thus made possible. Depending on how appealing their surrounds are, FA may splinter and reassemble into smaller groups. As a result, FA is now more legitimate as a clustering problem solution. Users must ascertain what they need from their bundles. We may determine that the group's emphasis has changed by adding one to the unit counter in each concentration until the objective task is completed. Both ideal CHs and the association between CMs and CHs can be prevented if the same is formed. Dynamic k-means may alter the weights they utilize.

Bao et al. (2020) developed a VANET PSO-based efficient clustering V2V routing (CRBP) system to improve V2V. Cluster creation, particle coding, and cluster routing comprise the protocol. CHs are selected among vehicles with similar shifting paths. Second, the route particle, speed, iteration processes, and fitness functions optimize routing. Third, strategies are provided to considerably enhance routing. To create a stable cluster, each node's location, speed, and neighbors are considered, and link fitness is evaluated to find the optimum path immediately. The simulation indicates that CRBP's performance depends on the number of nodes, their contact radius, and the number of hops between the CH and each member node. CRBP has 20% greater PDR and 47% lower latency than CBVRP and QoS-OLSR.

In their study, Britto et al. (2019) suggested the utilization of soft computing techniques in WSN multimedia applications by incorporating cameras and microphones. They proposed a clustering method called SGP, which involves the participation of three artificial neural network layers. The algorithm incorporates back propagation to enhance the performance of the network. The approach divides the network into two groups, and a reinforcement node facilitates the cluster selection based on eigenvector criteria. The simulation results demonstrated that the hybrid SGP clustering approach significantly improved the network's longevity compared to the previous method. Additionally, a three-layer artificial neural network was employed to determine the optimal number of energy-efficient nodes in all CHs (Cluster Heads) and the ideal number of CHs for transmission purposes.

4. Research Methodology

The Soft-Computing Based Dynamic Route Selection (SCDRS) approach consists of several phases, including Node Extraction, Cluster Head Detection, Node Energy Analysis, and Dynamic Route Selection. Throughput, redundancy, and packet routing are just a few of the factors that affect how well a network performs. There are various traffic dangers for the packets that go through the network channels, and all three of these components work together. Current issues involve a variety of traffic, including adjustments, deception, sink openings, and others. The network packets from these events have been covered in a variety of ways by academics utilizing various packets attributes. However, a lot of things are ineffective in addressing the issue of network threats. For instance, routing schemes employ elements such as payload, title, leap count, and bounce addresses, but they never considered how the delay may impact change assaults.

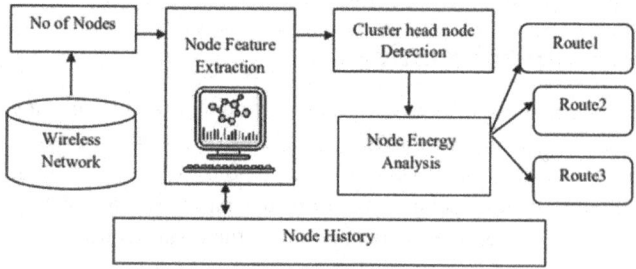

Fig. 5.1   Architecture framework [2]

$$\text{Accuracy} \frac{TP + TN}{TP + TN + FP + F}$$

In the same way, there are many situations where cluster-based routing is not thought of. In this part, we talk about how well structures like distinguishing routes, dynamic route choice, and stream construct guesswork. The proposed strategy uses different points of interest, just like previous methods, but there should be a routing approach in each path. The proof of cluster head is found by exchanging data between the nodes of any network, which creates more traffic in a straight line. For every packet that goes through the network, the highlights are taken off, and the separated highlights have neighbour addresses 119 through which the packet goes.

From the history of getting previous packets and the current element vector, an arrangement of an unusual way to traverse is found. Based on how routing is set up, a set of regular nodes through which packets pass can be identified. With the help of network topology, the set of routes that can be used to reach the service point is also identified. Whether or not a node is a cluster head is determined by using both the accessible route and the unique traversal route, as well as the current host succession.

It finds the best route based on the energy of each node. Any request for a route to be delivered by the source must be done on time with little room for error. In this case, this property of network packets is used to find out if any of the cluster nodes are changing or copying the packet while it is in transit. Also, the network kept track of how long each packet took to travel and how many hops it went through. This soft computing paradigm is a computer system that imitates intelligent human behavior in computerized tasks and problem-solving. It is based on the biological abilities of humans. AIS hasn't shown that it works well with routing in WSN, and it's not used very often in WSN. Recently, AIS has shown that clustering and selecting the head of a cluster can be used effectively. Routing and clustering problems in WSN depend mostly on how many clusters are static and fixed in the network. AIS could be a new area of study that hasn't been well looked into yet. Few papers were only able to show that it works with clustering.

Wireless Sensor Networks (WSNs) are crucial to the IoT environment and have garnered interest from the networking and IoT sectors. Micro Electro Mechanical Systems made clever and intelligent sensors change. WSN sensor nodes perceive and monitor ambient conditions to gather and provide the correct data to the user through the base station.

Proposed: Soft-Computing Based Dynamic Route Selection (SCDRS) algorithm Select a family of successors by identifying the neighbors that are most similar in terms of gender, job title, and city.

Search for successors using the family of successors identified in Step 1 and the criteria stated in Step 1.

Choose the successor that is the best match for the query.

The Soft-Computing-Based Dynamic Route Selection algorithm is a routing algorithm that is based on artificial intelligence. It is used to analyze a set of network nodes and recommend a path between them that would provide the best possibleserving experience. The algorithm is applied to a network of nodes and can learn over time.

To solve the separation issue in a mixed VANET-WSN network, this article suggests the Soft Computing Based Dynamic Route Selection (SCDRS). It presents the SCDRS system paradigm. For navigation, SCDRD makes use of physical locations. The nodes send signals (also known as Hello messages) repeatedly that include their node ID, present location, speed, direction, and energy level. The node's ID is mapped to its present address by the location server. The foundation of SCDRS is opportunistic sending. As a result,

no path finding is done before data transmission. The routing protocol's hello module, which produces HELLO signals, is a component of the protocol. These communications seek to identify close peers and ascertain their level of vitality and movement. (Velocity and heading).

The primary concern in achieving fast and reliable message transmission from the source to the target involves addressing the following challenge. In this research, the SCDRS (Source-to-Target Communication with Data Relay System) is employed, which typically operates in two modes: the greedy mode and the Delay Tolerant Networks (DTN) mode. The SCDRS utilizes data stored in neighbouring nodes' databases to forward messages towards the target. In the greedy mode, the instrument or vehicle closest to the target and with the least load receives packets for transmission. When encountering a local maximum issue, the SCDRS switches to the DTN mode if the originating point (sensor or vehicle) is closest to the target and has the lightest load. In situations where no desired neigh boring nodes are available, the sensor caches the packet, and the vehicle carries it until an opportunity arises to forward it to the intended neighbors.

Algorithm – Packet receiving:

Step 1: If a packet contains the DeferredRouteOutput tag, a query entry should be created for it and the packet id return should be stored.

Step 2: LocalDeliveryCallback will be used if the packet is delivered using the network id for the packet.

Step 3: Return, else

Step 4: Consult the Location Service Lead Priority - Lookup Routing Table for the node's information to determine the minimal weight on priority.

Step 5: if Local-Maximum,

Step 6: Include DeferredRouteOutput. Nodes with IDs

Step 7: to the next node in the network, create a route for local delivery of packets.

Step 8: Decide on UnicastForwardCallback.

Step 9: Bring it back, or

Step 10: The selected node's route should be created.

Step 11: Decide on UnicastForwardCallback.

Step 12: Afterward, if end if end if

5. Result and Analysis

Network Simulator 2 (NS2) is used to simulate a mobile wireless sensor network and compare the proposed method. In our proposed methods, we compared Soft- Computing Based Dynamic Route Selection (SCDRS) with existing algorithms (Optimized Route Cache Protocol-AODV (ORC-AODV) and Hidden Markov Model (HMM).

5.1 Throughput

The number of packets sent to the destination at any point in the time interval is used to figure out the average throughput. It is a way to measure how quickly a node can send data through a network. During a communication, the average throughput is the rate at which messages are sent and received over a channel. Transmission Time = File Size / Bandwidth (sec)

Throughput = Size of File/Time to Send (bps)

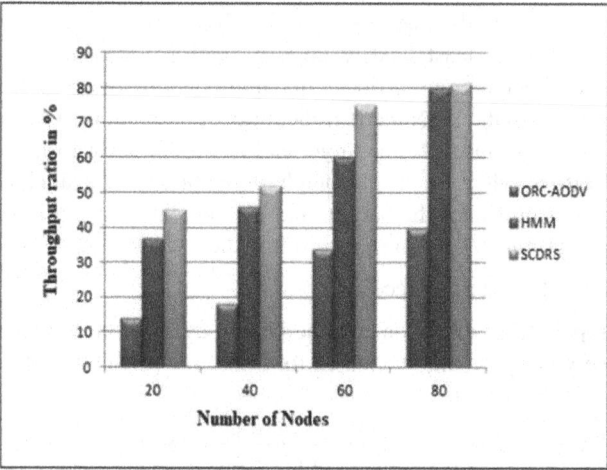

Fig. 5.2 Throughput

Source: Authors

Figure 5.2 shows the percentage of the throughput. The number of nodes is shown on the x-axis, and the percentage of throughput is shown on the y-axis. The proposed system has a higher level of data delivery and higher throughput than other systems that are already in place.

5.2 Average Ene-to-End Delay

Finding a route, retransmitting, transferring, and propagating determine the average end-to-end latency. The average latency or end-to-end delay may be calculated by summing these timings (Khawatreh et al. 2017). It's a packet's entire travel time. This formula calculates end-to-end latency.

Average end-to-end delay = TR – TS

Where TR is the time the packet was received and TS is the time it was sent.

Figure 5.3 shows the percentage of throughput. The number of nodes is shown on the x- axis, and the percentage of throughput is shown on the y-axis. When compared to other systems, the proposed system has a better data delay ratio and a lower level of data delay.

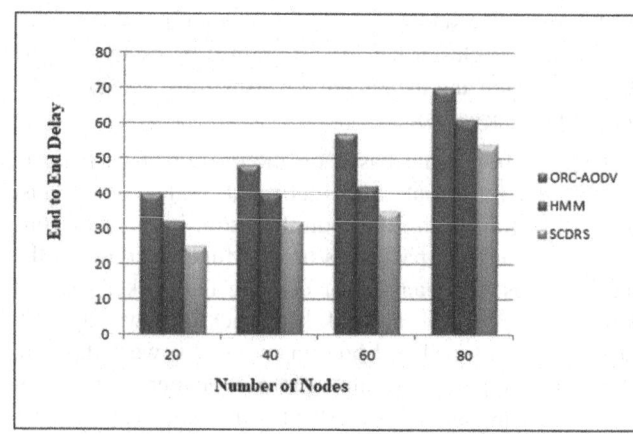

Fig. 5.3 Evaluation of average end to end delay

Source: Authors

In this wireless sensor network, the cluster method has the least amount of routing overhead and the most accurate detection of all the systems. In the end, you can see the results of all these methods on this page, which shows that SCDRS works well. With this method, performance is better, and data transmission as a whole works better. Also, the proposed system will improve how well the network works.

6. Conclusion

In a mobile wireless sensor network (MWSN), clustering techniques incorporating time division and frequency division are employed to enhance device connectivity. These clustering protocols primarily focuses on low-power, duty-cycled schemes and are adapted from existing WSNs. By comparison to the current small sensor node system, this approach improves overall network performance. The proposed systems prioritize shortest data delivery time. Utilizing cluster-based fuzzy and bee colony models for route optimization, the introduced technique demonstrates significant improvements, with an 81% increase in Throughput and a 54% reduction in End-to- End Delay. It should be noted that strategies designed for homogeneous networks may not be effective for heterogeneous networks. Therefore, in the presence of an event, sensor nodes in close proximity must promptly form a cluster around it. Clustering techniques suitable for heterogeneous networks should be employed in such cases. The advancement of WSN research will continue in the future, with theoretical methods being applied to real-time sensor network platforms.

References

1. Singh, G.D., Prateek, M., Kumar, S., Verma, M., Singh, D., & Lee, H.N. (2022). Hybrid genetic firefly algorithm-based routing protocol for VANETs. IEEE Access, 10, 9142–9151.

2. Bao, X., Li, H., Zhao, G., Chang, L., Zhou, J., & Li, Y. (2020). Efficient clustering V2V routing based on PSO in VANETs. Measurement, 152, 107306.

3. Britto, P.X., & Selvan, S. (2019). Hybrid soft computing: SGP clustering methodology for enhancing network lifetime in wireless multimedia sensor networks. Soft Computing - A Fusion of Foundations, Methodologies, and Applications, 23, 2597–2609.

4. Xhafa, F., Wang, J., Chen, X., Liu, J., Li, J., & Krause, P. (2014). An efficient PHR service system supporting fuzzy keyword search and fine-grained access control. Soft Computing, 18(9), 1795–1803. doi:10.1007/s00500-013-1202-8

5. Yang, M., Li, Y., Jin, D., Yuan, J., You, I., & Zeng, L. (2014). Opportunistic sharing scheme for spectrum allocation in wireless virtualization. Soft Computing, 18(9), 1685–1697. doi:10.1007/s00500-014-1267-z

6. Maniscalco, V., Greco Polito, S., & Intagliata, A. (2014). Binary and m-ary encoding in applications of tree-based genetic algorithms for QoS routing. Soft Computing, 18(9), 1705–1715. doi:10.1007/s00500-014-1271-3

7. Baskaran, R., Basha, M.S., Amudhavel, J., Kumar, K.P., Kumar, D.A., & Vijayakumar, V. (2015). A bio-inspired artificial bee colony approach for dynamic independent connectivity patterns in VANET. In Proceedings of the 2015 International Conference on Circuits, Power and Computing Technologies (ICCPCT-2015), 1–6.

8. He added, M., Zagrouba, R., Laouiti, A., Muhlethaler, P., & Saidane, L.A. (2015). A multi-objective genetic algorithm-based adaptive weighted clustering protocol in VANET. In Proceedings of the 2015 IEEE Congress on Evolutionary Computation (CEC), 994–1002.

9. Khan, Z., & Fan, P. (2016). A novel triple cluster-based routing protocol (TCRP) for VANETs. In Proceedings of the 2016 IEEE.

AI Based Improved Vehicle Detection and Classification in Patterns Using Deep Learning

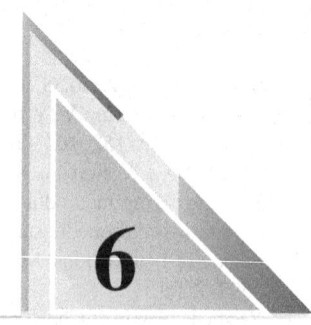

NVN. Sowjanya[1]

Research Scholar, Anna University, Chennai

G. Swetha[2], T. Sai Lalith Prasad[3]

Research Scholar Vignan University, Guntur

Abstract: For ITSs, speed and accuracy in identity verification and vehicle ordering are crucial. However, it is difficult to notice and recognise vehicle types rapidly and also accurately due to the close proximity of vehicles on the road and the jarring nature of images or videos that include automobile images. For this task, we recommend using YOLOv4 AF, a service built on top of an improved variant of the original YOLOv4 concept. To reduce channel and geographical variability of picture occlusion, the suggested layout isfactored in. The Feature Pyramid Network (FPN) component of the Training Aggregation Network (PAN) has been modified in YOLOv4 to down-inspect the trustworthy highlights. This paves the way for better in-design item ID and characterization implementation, as well as richer information about the 3D locations of individual items. With enhancements of 83.45% and also 0.816 on the Thing Car instructional collection and also 77.08% and also 0.808 on the UA-DETRAC informative collection, specifically, the suggested YOLOv4 AF design surpasses the initial YOLOv4 and also 2 various other cutting-edge versions, Faster R-CNN and also EfficientDet.

Keywords: EfficientDet is shorthand for "region-based convolutional neural network," "you only look once" (YOLO), "attention mechanism," "feature fusion," and "identification of vehicle models" in the realm of computer vision.

1. Introduction

Many different types of contemporary and military frameworks are used in ITSs for the purposes of item detection and planning. Vehicle traffic management, executive control, and regional planning may all benefit from the information gathered by ITSs that can, for instance, direct lorry location and characterization for in-depth analysis of passing vehicles. The newest item identification methods may be broken down into two camps: device-based and vision-based. [1] The ideal task is to construct leaping boxes (BBoxes) around the discovered stuff in a collection of images or videos. If the target group is complete, then the image will show both the expected course name and the certainty rating associated with each jumping box (BBox). [2] According

to [1], there are three main classifications that vision-based product recommendation may be broken down into: (i) element-based, (ii) logo-based, and (iii) highlight-based. The standard (pre-2012) highlight-based object discriminating evidence treatments, such as Haar [3, 4], the pie chart of found angles (Accumulation), and others, are separated into three components [2]: First, determining the best interest-rate bracket; second, excluding irrelevant data; and third, organising the resulting data. Due to the constant growth of vast quantities of data (Considerable Information) and the rapid development of (multicore) cpus and also Graphical Processing Units (GPUs), these techniques were eventually superseded by deep learning (DL) focus based calculations [2]. The market has been dominated by DL emphasise based computations because to their remarkable

[1]sowjanya.nvn@gmail.com, [2]swethareddy630@gmail.com, [3]sailalith15@tkrec.ac.in

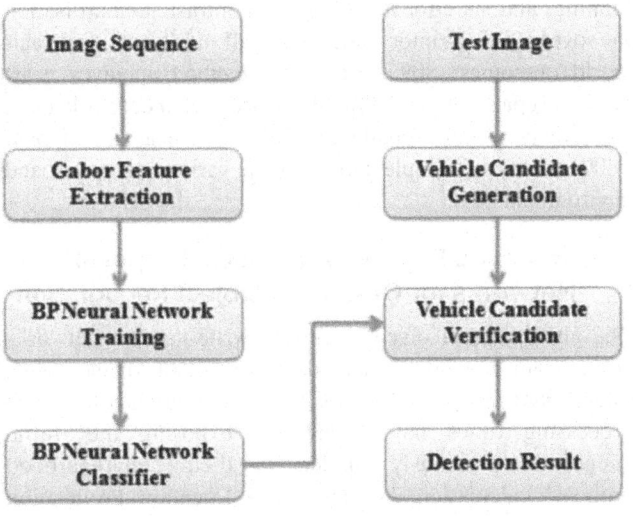

Fig. 6.1 Example figure [2]

accuracy and operational rate of item distinguishing evidence. In contrast to highlight-based instance component based systems, DL techniques may eventually extract incorporate top characteristics from enormous processes of information. [2]

The high representation power of convolutional neural networks (CNNs) has led to their widespread adoption for use in object ID models. In terms of visual cues, CNN's stress reduction is analogous to the human visual system. [2] Each layer in a typical CNN—whether it's a convolutional layer, a pooling layer, a fully-connected layer, etc.—converts the 3D information volume into a 3D result quantity of afferent neuron enactments. [1] Today, several distinct CNN designs are easily accessible. To automate picture incorporation removal, the Region-based Convolutional Neural Network (R-CNN) was the first to effectively incorporate DL for item finding evidence and other computer system vision tasks. Recent trends in discriminating evidence have been propelled by the development of RCNNs, with the associated reduction in cost made possible by the suggested division of convolutions between points. [7] Accelerated R-CNN [8, Accelerated R-CNN [7], Cover R-CNN [9], and Latticework R-CNN [10] All of the versions in [10] of the R-CNN algorithm. Both are instances of two-stage item-difference proof designs ([10]), which first develop a set of weak rival layouts (i.e., scene-based location ideas), and then verify, characterise, and iterative improve upon the ratings and also regions. [2] The high precision and restriction attained in object acknowledgment is one of the primary pros of these designs, while the more mind-boggling training required and the decreased valuable rate achieved are the primary disadvantages [2]. Regular short article acknowledgment is currently thought to be an absolutely significant component in beneficial applications.

2. A Quote from the Books

2.1 Examining How Digital Image Resolution and Ambient Lighting Affect Truck Labelling

Many cutting-edge transportation applications, such as fare monitoring, smart exit designs, and online traffic evaluation, consider vehicle classification to be an essential variable. In this work, we demonstrate many distinct categorization strategies that may be accomplished with only a basic digital video camera. The level of detail and post-production work put into the final result is the primary factor in determining the price of the high-end electronic camera used to shoot the picture. In this paper, we give a comprehensive evaluation of how these two factors affect the precision and usefulness of vehicle clustering. On the academic Item Lorry and LabelMe data sets, we use a variety of state-of-the-art image classifiers. Each dataset is summarised into a variety of sizes, each of which represents a unique set of spatial goals. We also switch the effectiveness of each kind to black and white to see how selection affects them. Finally, after running over 46,000 unique trials, we draw a solid conclusion on the impact of these two structures (element and selection) on the precision and throughput of picture characterization procedures. According to the findings of the experiments, the range and spatial aims of the vehicle photos affect the order outcomes obtained by most cutting-edge photo arrangement structures. However, many picture quality evaluations call for integrating spatial intent with temporal management. Our findings could help automated ordering systems save costs and improve efficiency.

2.2 Examining Deep Learning-based Strategies For Roadside Object Recognition

Intelligent Transportation Solutions (ITS) as a field has advanced rapidly in recent years, with in-depth knowledge as a prominent focal point. Two-layered images and a variety of deep learning-based treatments have emerged as a crucial resource for autonomous vehicles to recognise, track, and navigate pedestrians, vehicles, and other roadside obstacles like traffic lights and signs. When it comes to meeting the safety and security requirements of pedestrians and other vehicles in the area, self-driving cars rely significantly on visual data for planning and summing their aims. Recognition algorithms that are grounded on in-depth discovery of objects consistently provide excellent outcomes. Despite the extensive study of deep learning-based items identification systems, few studies have compared different systems in terms of their recognition accuracy or success rate. Power performance and model size also have an effect on how successfully an autonomous vehicle drives. However, many of these benchmarks are not being investigated in modern deep learning-combined computations. The purpose of this study is to provide a thorough and consistent evaluation of 5 commonly used deep learning-based computations for road item detection; They are the R-FCN, Cover R-CNN, SSD, RetinaNet, and YOLOv4. The dataset at hand is the expansive Berkeley DeepDrive (BDD100K). The average Routine Accuracy (Guide) values and the time needed to get them are used to evaluate the exploratory data. Several other practical needs for deep learning-based versions are also extensively assessed, including stylistic dimensions, computational ins and outs, and power performance. Further, each estimate's presentation is evaluated in light of evolving roadside environmental problems. The link provided in this brief essay aids readers in comprehending the strengths and weaknesses of popular deep learning-based computations when put to the test against genuine constraints such as constant business rationality.

2.3 An Increase in the Cascade of Basic Processes for Fast Object Detection

This paper presents a machine learning approach to aesthetic product recommendation, one that can efficiently assess images and achieve high identification rates. This critique stands out for three main reasons. The first is a short description of a fantastic graph dubbed the "necessary image," which allows the components of our identifier to be quickly and accurately recognised. Second, an AdaBoost-based learning methodology is used to choose a subset of important visual functions from a larger pool in order to generate highly accurate classifiers. Finally, we have committed to an approach for handling persistently perplexed classifiers in a "overflow," which will definitely allow us to swiftly get rid of the photo's structure regions while devoting significantly lot more processing effort to possible thing-like areas. In contrast to earlier models, wealth provides quantifiable assurance that excluded locales are unlikely to have the objects of longing. The concept uses one of the most advanced extant frameworks in face recognition to provide reasonable expedition rates. With proper use, the finder is capable of refining 15 faces each second, regardless of picture distinction or skin choice.

2.4 Clinical Inquiry Using Pie Charts with a Slope Orientation

We investigate the topic of capabilities for strong aesthetic products widely known evidence using human recognition as a consequence of straight SVM. We first show demonstration that matrices of pie charts of orientated incline (HOG) descriptors outmatch current abilities for human recognised proof, after analysing pre-existing side and inclination based descriptors. We examine the effects of each phase of the evaluation on performance, and we conclude that high-quality angles, directions binning, moderately-difficult spatial

binning, and superior neighbouring contrast standardisation in covering descriptor blocks are all necessary for real-world outcomes. Since the clever method achieves near-optimal separation on the first MIT pedestrian data collected, we provide a more challenging dataset consisting of over 1800 photos of people talking in a variety of poses and environments.

2.5 Improved R-CNN: Using Local Proposal Networks for On-the-Fly Object Recognition

The guesswork involved in current write-up area networks comes from area pointer methods. As a result of advancements like SPPnet and Quick R-CNN, these recognition companies' processing times have decreased, exposing the traffic congestion previously hidden in their neighbourhood proposal calculation. We provide a Location Proposition Network (RPN) that incorporates almost free location recommendations and complete convolutional features with local organisation. An RPN employs a fully convolutional network to predict object cutoff points and abjectness ratings in each domain simultaneously. The RPN, which powers fast R-CNN, was built from the ground up to deliver accurate placement recommendations. We then combine RPN and Quick R-CNN's convolutional highlights into a single network, with the RPN component directing the hybrid's search efforts using the fairly commonplace idea of brain connection with 'consideration' procedures. Our discovery method achieves state-of-the-art items acknowledgment accuracy on the PASCAL VOC 2007, 2012, and MS COCO datasets with just 300 principles per picture for the extremely deep VGG-16 design at a case rate of 5fps (considering all phases) on a GPU. 2015's ILSVRC and COCO winners built the most efficient R-CNNs and RPNs, respectively.

3. Methodology

High precision and localization in things determining evidence are two advantages of such designs, but the main drawbacks are more extensive preparation and slower working prices, which is especially problematic considering the growing importance of continuous article recognition in practical applications. Two individuals from the opposing group of single-stage product identification models, You Just Look Once (YOLO) and Solitary Shot MultiBox Detector (SSD), outflank with a relapse approach for product concept straight, resulting in a speedier operating price. However, SSD has a restricted capacity to recognise minute writes since it does not evaluate the connectivity between different arrays. Learning the basics is simplified and accelerated when you follow your stomach. Both SSDs and comprehensive methods fail in the practical range, resulting in frequent identifications and lost data.

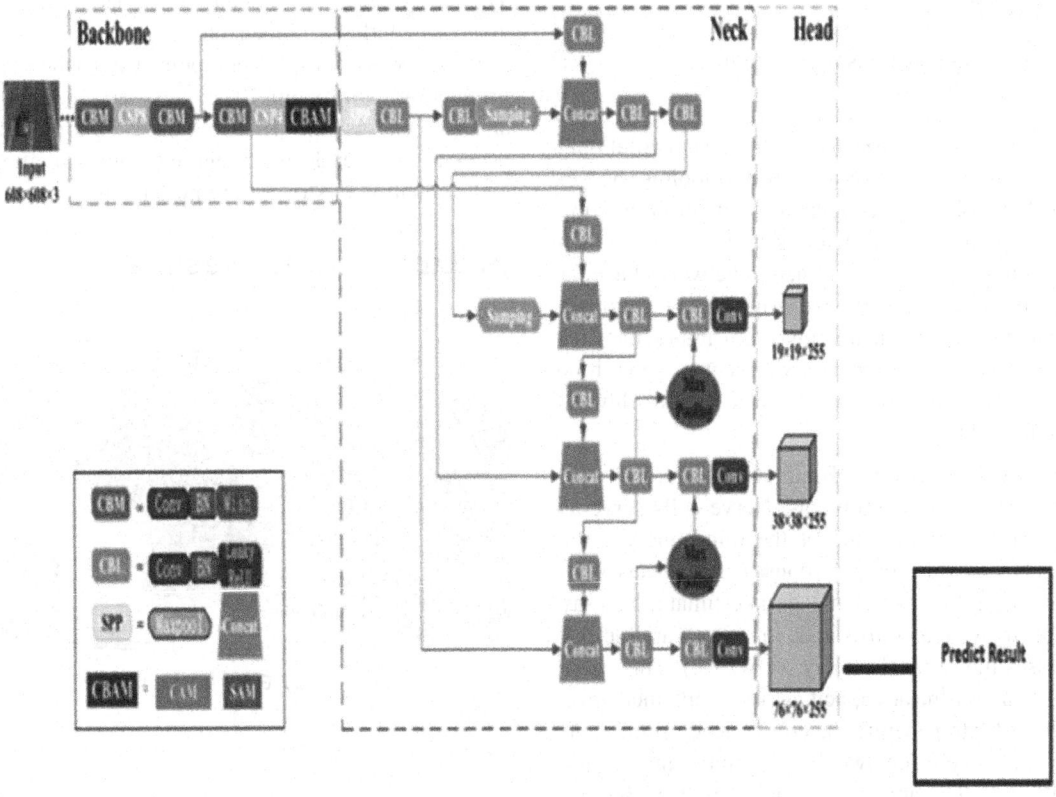

Fig. 6.2 System architecture [2]

Slower processing times and the need for even more extensive training are two drawbacks. However, SSD and YOLO fail to efficiently manage the positioning of the images, leading to persistent detection errors and data loss.

In order to solve this problem, this work presents the first vehicle recognition and grouping model based on a revised YOLOv4 architecture. The proposed update makes use of a consideration tool to lessen the characteristics of network and spatial photo blockage. In addition to employing down tasting to emphasise the most important details, YOLOv4 modifies the Course Aggregation Network's (PAN) Feature Pyramid Network (FPN). Consistently rearranging the components in 3D space is required to take the version's post-identification and configuration application to the next level.

3.1 Benefits

1. It opens the door to better results in automobile finding.
2. The proposed YOLOv4 AF design outperforms all three of the most up-to-date models included in the performance comparison on both data sets, as measured by mean ordinary precision (mAP) and F1 score.

3.2 Modules

The following items were manufactured by us in order to carry out the aforementioned undertaking.

We will use this part to investigate information and then stuff it into the framework.

Processing: Using this part, we will take in data for management purposes.

In this section, we'll explain how to use this component to partition your data into train and test sets.

We'll make YOLOv4, YOLOv4-tiny, and YOLOV5 versions of the classifier.

When a user makes use of this feature, they will be prompted to create an account and log in.

Using this feature will undoubtedly lead to future prediction input from the user.

Prediction: the expected value in the end will definitely be there.

4. Crucifixion

We used the following equations in our studies.

CNN

In this post, we'll show you how to construct a convolutional neural network-based picture classifier by developing a 6 layer neural network capable of recognising and differentiating between photos. Our planned business will be very hidden and entirely computerised. Standard neural networks have far more constraints and are quite time-consuming to develop on a typical computer processor, making them impractical for efficient picture grouping. However, we want to show how TENSORFLOW may be used to create a true worldwide convolutional mind business.

Mathematical models called "Brain Organisations" are employed to solve progress problems. Nerve cells serve as the fundamental computing units of the mind and are the building blocks from which they are constructed. A nerve cell takes in information (let's say x), makes an estimate (let's say by multiplying it by w and also adds an extra variable, b), and then releases the result (let's say, $z = wx + b$). This value is transferred to a non-linear capacity called implementation capacity (f) in order to generate the consequence (initiation) of the nerve cell. There are two broad groups into which initiation skills fall. The ability to initialise with a sigmoid is obvious. Any nerve cell that performs any action using the sigmoid function is said to be a sigmoid nerve cell. The term "neuron" comes from the Greek word for "action," and there are many different types of neurons, such as RELU and TanH.

The connecting structure of brain corporations is a layer, which is produced by stacking neurons in a single line. Until there is no more room for improvement, the cycle shown in the layers below will continue.

YOLO

In 2016, Joseph Redmon established the translation agency Consequences be damned. When used as a first line of defence against ID, the endless permutations of "effects be damned" are very effective since it only takes one "appearance" in a picture to identify the important items and their locations. Instead of reusing object-identification classifiers, it prepares for a single CNN directly from the whole image and defines room as a single regression issue to spatially segregated BBoxes and also adequate programme likelihoods. Just go crazy now offers more screen real estate for items and detailed pictorial instructions.

The installation procedure is the heart of any software's workflow and is used independently of any particular improvement strategy or application domain.

The development cycle of any scheduled entity or framework always includes a hidden phase known as planning. Since a model or version of a component must be created before production can begin, designers are required. Structure arrangement is the first of three specialised processes necessary for set development and approval, after resolution and evaluation of structure requirements.

5. Experimental Results

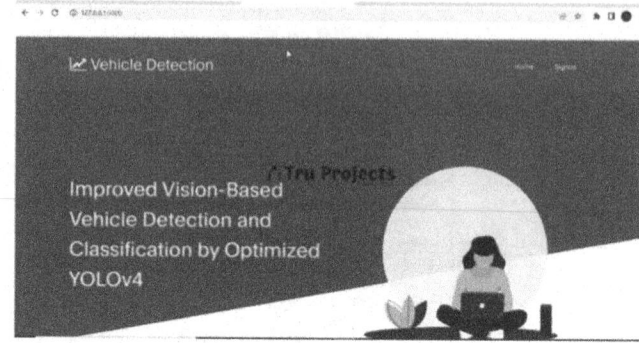

Fig. 6.3 Home screen

Source: Authors

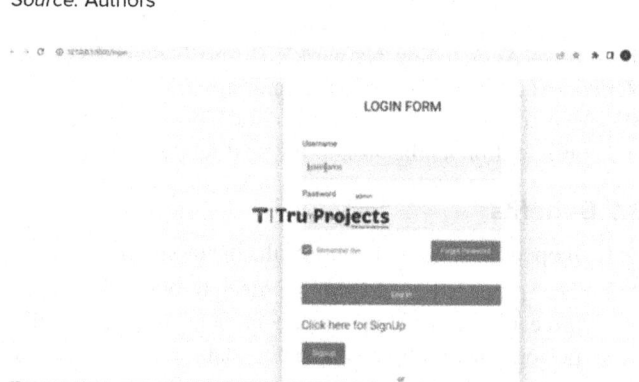

Fig. 6.4 Registration

Source: Authors

Fig. 6.5 Login

Source: Authors

Fig. 6.6 Main screen

Source: Authors

Fig. 6.7 User input

Source: Authors

Fig. 6.8 Prediction result

Source: Authors

6. Conclusion

This study provides a refined alternative to YOLOv4 for vehicle identification and characterisation. The primary goal in developing this CBAM module was to increase the responsive area's network and geographical components by providing a factor to consider. In addition, the FPN node guides a further up sampling action and modifies the component mix. After that, the YOLOv4 AF is advised, and its discovery execution is expanded by combining the result

highlights and the acknowledgment impacts of a few layers. The presentation of this model was preliminary compared to the initial YOLOv4 design and two additional minimising side items ID methods, Faster R-CNN and EfficientDet, using two publicly available informative indexes, Spot Lorry and UA-DETRAC. The obtained findings show that the suggested YOLOv4 AF model has a higher indicated normal precision (Overview) and F1 rating than the three state-of-the-art models used in the event assessment. The refined framework might also be used to understand other types of postings, paving the path for better regression therapies overall. However, the development of the CBAM module implies an increase in both computing complexity and time compared to the original YOLOv4 design.

In the future, we want to conduct more tests on the recommended design's detection and classification powers by monitoring moving things, collecting online traffic data, and also testing it on other objects.

References

1. K. F. Hussain, M. Afifi, and G. Moussa, "A comprehensive study of the effect of spatial resolution and color of digital images on vehicle classification," IEEE Trans. Intell. Transp. Syst., vol. 20, no. 3, pp. 1181–1190, Mar. 2019.
2. M. Haris and A. Glowacz, "Road object detection: A comparative study of deep learning-based algorithms," Electronics, vol. 10, no. 16, p. 1932, Aug. 2021.
3. P. Viola and M. Jones, "Rapid object detection using a boosted cascade of simple features," in Proc. IEEE Comput. Soc. Conf. Comput. Vis. Pattern Recognit. (CVPR), Kauai, HI, USA, vol. 1, Dec. 2001, pp. I–I.
4. W. T. Freeman and M. Roth, "Orientation histograms for hand gesture recognition," in Proc. Int. Workshop Autom. Face Gesture Recognit., Zurich, Switzerland, 1995, pp. 296–301.
5. N. Dalal and B. Triggs, "Histograms of oriented gradients for human detection," in Proc. IEEE Comput. Soc. Conf. Comput. Vis. Pattern Recognit. (CVPR), San Diego, CA, USA, vol. 1, Jun. 2005, pp. 886–893.
6. S. Woo, J. Park, J.-Y. Lee, and I.-S. Kweon, "CBAM: Convolutional block attention module," in Proc. Eur. Conf. Comput. Vis. (ECCV), 2018, pp. 3–19.
7. S. Ren, K. He, R. Girshick, and J. Sun, "Faster R-CNN: Towards realtime object detection with region proposal networks," IEEE Trans. Pattern Anal. Mach. Intell., vol. 39, no. 6, pp. 1137–1149, Jun. 2017.
8. R. Girshick, "Fast R-CNN," in Proc. IEEE Int. Conf. Comput. Vis. (ICCV), Santiago, Chile, Dec. 2015, pp. 1440–1448.
9. K. He, G. Gkioxari, P. Dollár, and R. Girshick, "Mask R-CNN," in Proc. IEEE Int. Conf. Comput. Vis. (ICCV), Venice, Italy, 2017, pp. 2980–2988.
10. G. Gkioxari, J. Malik, and J. Johnson, "Mesh R-CNN," presented at the IEEE/CVF Int. Conf. Comput. Vis. (ICCV), Seoul, South Korea, 2019.

11. J. Redmon, S. Divvala, R. Girshick, and A. Farhadi, "You only look once: Unified, real-time object detection," in Proc. IEEE Conf. Comput. Vis. Pattern Recognit. (CVPR), Las Vegas, NV, USA, Jun. 2016, pp. 779–788.

12. W. Kong, J. Hong, M. Jia, J. Yao, W. Cong, H. Hu, and H. Zhang, "YOLOv3-DPFIN: A dual-path feature fusion neural network for robust real-time sonar target detection," IEEE Sensors J., vol. 20, no. 7, pp. 3745–3756, Apr. 2020.

13. A. Bochkovskiy, C.-Y. Wang, and H.-Y. Mark Liao, "YOLOv4: Optimal speed and accuracy of object detection," 2020, arXiv:2004.10934.

14. Y. LeCun, L. Bottou, Y. Bengio, and P. Haffner, "Gradient-based learning applied to document recognition," Proc. IEEE, vol. 86, no. 11, pp. 2278–2324, Nov. 1998.

15. L. Xiao, Q. Yan, and S. Deng, "Scene classification with improved AlexNet model," in Proc. 12th Int. Conf. Intell. Syst. Knowl. Eng. (ISKE), Nanjing, China, Nov. 2017, pp. 1–6.

Enhancing Cyber security: Evaluating Machine Learning Algorithms for Effective Threat Detection and Classification

7

T. Anjamma[1], S. Yamuna Reddy[2]

Department of Computer Science and Engineering,
Teegala Krishna Reddy Engineering College, Hyderabad, India

Abstract: With the increasing frequency and complexity of cyber threats, enhancing cyber security measures has become imperative. Machine learning (ML) algorithms have shown promise in improving threat detection and classification by automating the identification of security incidents. This abstract presents a comprehensive evaluation of various ML algorithms for their effectiveness in cyber threat detection and classification. The evaluation is conducted using a diverse dataset encompassing different cyber attack scenarios and incorporating multiple features such as network traffic patterns, system logs, and user behavior. Key ML algorithms including decision trees, support vector machines, and neural networks are examined, and their performance is assessed using metrics like training accuracy and testing accuracy. The results highlight the efficacy of ML algorithms in accurately identifying and categorizing cyber threats. Additionally, the study investigates the impact of feature selection techniques and model optimization strategies on algorithm performance. The findings provide valuable insights into the strengths and limitations of each algorithm, enabling the practical implementation of robust threat detection and classification systems. This research contributes to the field of cyber security by facilitating the development of effective ML-based solutions, ultimately bolstering cyber defense mechanisms against evolving threats.

Keywords: Cyber security, Machine learning algorithms, Threat detection, Classification, Training accuracy, Testing accuracy

1. Introduction

Cyber security is a crucial aspect of computer science that deals with the investigation of various cyber threats and their corresponding countermeasures. With the widespread reliance on the internet for accessing information across different domains such as business, education, and entertainment, network attacks have become increasingly prevalent [1]. To mitigate these attacks, intrusion detection systems (IDS) and firewalls have been recommended as preventive measures. While firewalls filterincoming and outgoing packets based on predefined rules, IDS scans the network and alerts administrators of any suspicious activities. IDS are generally considered more effective and secure compared to firewalls [2].

The current challenge faced by computer networks lies in the diversity of cyber-attacks. These attacks vary in their nature and impact, ranging from adware that is relatively harmless to phishing attempts that can lead to data theft or destruction. More destructive attacks include ransomware, which encrypts computer systems and demands a ransom, and denial-of-service attacks that target operating systems [3]. In response to these challenges, researchers and engineers are focusing on the development of intelligent systems for automated computer network intrusion detection. The research aimed to demonstrate the effectiveness of the investigated approaches in creating an intelligent system capable of identifying multiple anomalies within a network. To achieve this, the study explored machine learning and deep learning classification algorithms, particularly focusing

[1]Anjammabonala83@gmail.com, [2]sarvigariyamunareddy@gmail.com

on their promising outcomes in unsupervised modes of operation [4].Machine learning techniques, known for their ability to swiftly identify newly discovered breaches, are commonly employed in the development of network intrusion detection systems [5]. In dealing with large datasets, precise algorithms for clustering, classification, and prediction are required, and supervised machine learning methods such as K-Nearest Neighbor (KNN) and Naive Bayes are frequently utilized. Decision trees, valued for their precision, versatility, and simplicity, play a significant role in detecting anomalous and abusive patterns. Additionally, neural networks have seen widespread deployment for detecting anomalies and abuse patterns [6]. The success of artificial intelligence models relies on both accuracy and interpretability, making the use of machine learning and deep learning approaches essential [7].

Despite existing literature on intrusion detection in cyber security, further advancements are still required. The present study contributes to addressing these issues by introducing the following approaches:

1. A comprehensive examination of the literature on the use of numerical and image- based datasets to machine learning and deep learning models for intrusion detection.

2. Dataset preprocessing and balancing.

3. Machine learning models that are stacked and use several feature extraction methods.

4. Run an experiment to confirm the models offered.

2. Litreature Survey

This section describes several studies that use fundamental machine learning methods to analyse IoT traffic in order to defend IoT devices against cyberattacks.

Network profiling and machine learning were the main topics of Rose et al.'s [8] study on IoT security. To detect unauthorised network transactions and attempts to tamper with IoT devices, they developed a dataset and a model. With a 98.35% accuracy rate and a 98.35% false-positive alert rate on the Cyber-Trust test, their suggested anomaly-based intrusion detection system produced outstanding results.

A broad machine learning technique to recognise IoT devices was developed by Ali et al. [9] in a different work. During the training phase, they used random forest and naive Bayes classifiers to extract 85 features from packet capture (.pcap) files and obtained 99% accuracy in identifying IoT devices. Seven various supervised learning methods for IoT intrusion detection were evaluated side by side by El-Sayed et al. [10]. With 94% accuracy on MobileNetv2 features, the SVM technique performed better than the competition and showed rapid and consistent training outcomes.

According to Le K-H et al. [11], IMIDS is an intelligent intrusion detection system designed for Internet of Things (IoT) devices. It employs a compact convolutional neural network to classify various cyber threats, achieving an average F-measure of 97.22%. The system's detection capabilities were further improved through additional training using input from an attack data generator.

In a study conducted by Islam et al. [12], shallow and deep learning-based intrusion detection systems (IDSs) were examined for IoT threat detection. The deep learning- based IDSs included decision trees, random forests, and support vector machines. The performance of each participant was evaluated using five datasets, revealing that machine learning IDSs outperformed shallow machine learning techniques in accurately identifying IoT risks, achieving an accuracy of 98.79%.

Overall, these studies show the value and promise of using machine learning approaches to strengthen IoT security and efficiently identify cyber risks in IoT systems.

2.1 Network Attacks and Their Types

Network attacks can be defined as attempts to gain unauthorized access to a corporate network with the intention of stealing information or causing harm. These attacks can be categorized as either passive or active [13]. Passive attacks involve intercepting the network and gathering sensitive data without altering the system. Examples of passive attacks include traffic analysis and the unauthorized publication of message content. On the other hand, active attacks involve unauthorized access, where attackers can modify, delete, encrypt, or decrypt data during the attack. Common active attacks include message tampering, repudiation, denial of service (DoS), replay attacks, and masquerading.

Intrusion Detection Systems (IDS) are crucial for identifying and classifying different types of attacks, whether passive or active. The following are some specific types of attacks that IDS can consider:Denial of Service (DoS): In this type of attack, untrusted users flood the network with meaningless traffic, aiming to exhaust its resources and prevent legitimate users from accessing it. Examples of DoS attacks include Land, Back, and Mail Blood Smurf attacks.

Probe Attack: This attack involves using software or a program to monitor and collect information about network activities. Examples of probe attacks include Satan, Ipsweep, Mscan, Saint, and Nmap.

Remote to Local (R2L): In R2L attacks, a hacker uses specific devices to transmit packets while being restricted from accessing the device's authorized account. The attacker exploits vulnerabilities to gain unauthorized access. Examples of R2L attacks include Named, Phf, Sendmail, and Guest.

User to Root (U2R): In U2R attacks, the attacker has already gained access to a user's account and is attempting to exploit their privileges. Examples of U2R attacks include Perl, Ps, Eject, and Ffbconfig.

Being aware of these different types of attacks and employing effective intrusion detection systems is vital for safeguarding corporate networks and protecting sensitive information from unauthorized access and potential harm.

3. Methods and Materials

DatasetsIn this experiment, four datasets have been utilized, namely the Kyoto dataset, the UNSW-NB15 dataset, the kdd cup dataset, and the nsl-kdd dataset. These datasets were chosen for their well-organized and valuable nature, particularly in the context of discussing machine learning approaches. Another advantage of using these datasets is that they are freely accessible. Additionally, these datasets are known for their ease of use and high quality, making them suitable for data analysis.

Among these datasets, the KDD Cup dataset contains nearly 4.9 million single-connection vectors, each having 41 attributes. These vectors can be classified as either normal or attacking based on their behavior. The KDD Cup dataset includes data on four different types of attacks.

One of the attack types present in the KDD Cup dataset is the Service Denial attack, where the target device's memory is overwhelmed, causing it to become unresponsive when a request is received. To protect against this attack, the recommended action is to turn off the device.

Another attack type in the dataset is the user to root attack, where a hacker with specific access to a device tries to take control of the router by exploiting security vulnerabilities. Various methods, such as phishing attacks, sniffer (packet controlling), or social engineering, can be employed to carry out this type of attack. It occurs when a hacker, who does not physically possess the target computer, transfers packets from the computer to the network system and exploits security vulnerabilities to gain access to the target machine. Such attacks aim to obtain information from computer networks by bypassing the system's security measures.

Based on its specifications and row count, the KDD dataset has been selected as the dataset for this experiment. These features provide valuable insights into network traffic patterns, communication protocols, and potential network security threats. Researchers and analysts can leverage this data set to develop machine learning models and algorithms for network monitoring, intrusion detection, and overall network performance optimization.

Table 7.1 Description of the data set—Network traffic data

S No	Feature Name	Data Type	Description
1	Source IP	String	The IP address of the source device or network sending the network traffic
2	Destination IP	String	The IP address of the destination device or network receiving the network traffic
3	Source Port	Integer	The port number used by the source device or network for sending the network traffic
4	Destination Port	Integer	The port number used by the destination device or network for receiving the network traffic
5.	Protocol	String	The protocol used for the network communication (TCP, UDP, ICMP, etc.)
6	Packet Length	Integer	The length (in bytes) of the network packet
7	Timestamp	Date Time	The timestamp indicating the date and time of the network traffic event
8	Network Protocol	String	The high-level network protocol used for the communication (HTTP, FTP, DNS, etc.)
9	Flow Duration	Integer	The duration (in milliseconds) of the network flow
10	Bytes Transferred	Integer	The total number of bytes transferred in the network flow

3.1 Machine Learning Algorithms

In this experiment, various machine learning algorithms were utilized, including Logistic Regression, Support Vector Machine (SVM), and Gaussian Naive-Bayes. These techniques can handle both linear and non-linear relationships. Artificial Neural Network (ANN) methodologies, such as the Multilayer Perception Algorithm, Convolutional Neural Network (CNN), and Recurrent Neural Network

(RNN) Algorithm, were also employed. Additionally, Gradient Boosting, Decision Tree, Random Forest, Stochastic Gradient Descent, K-Nearest Neighbor, and ANN were among the additional algorithms utilized in the experiment.

Logistic Regression, being one of these algorithms, was applied as a classification technique. Since there were more than two possible outcomes, a Multinomial Logistic Regression technique was employed. The experiment was conducted using Python programming. Multinomial Logistic

Regression is a variation of Binary Logistic Regression specifically designed for scenarios with multiple outcomes.

3.2 Feature Selection

A crucial step in data preparation for machine learning is feature selection, where we identify the most relevant attributes and discard the less important ones. The significance of a feature is determined based on how well it predicts the target variable [3].

In this study, the chi-square feature selection strategy is employed, which is particularly effective for multi-class classification [16]. By applying the chi-square test [17], the best feature for the dataset is identified, which indicates the feature with the strongest relationship to the output class. The Chi-square test formula is expressed as follows:

Where: $X2=\sum Eij(Oij-Eij)2$

4. Proposed Model

The proposed model aims to improve the efficiency of classification through the utilization of machine learning techniques [17]. The underlying principle of the algorithms employed is as follows: the data is initially divided into groups based on certain criteria, and with each iteration of the algorithm, additional rules are incorporated into the existing set of rules. This iterative process reduces misclassification. By combining all the weak classifiers, a robust classifier capable of identifying various types of attacks is formed. One

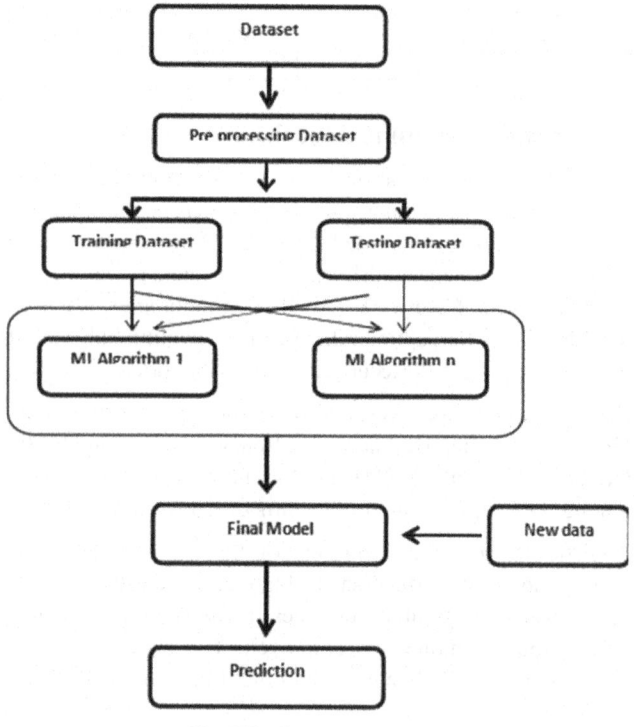

Fig. 7.1 Proposed model

significant advantage of the Adaboost method is its ability to evaluate the net classification error at each stage of learning. This evaluation provides valuable insights into the overall performance and effectiveness of the classifier. It allows for the continual refinement and enhancement of the model's classification capabilities.

Using the performance parameters described below [11], different IDS based on machine learning are compared and evaluated for their efficacy.

True Positive (TP) - Here, an assault has been acknowledged and confirmed to have occurred. Situations of this nature are seen positively.

Attacks that are mistakenly reported as having occurred are known as false positive (FP) attacks. A false positive is thus nothing more than a deceptive warning.

The data known as True Negative (TN) are those that are appropriately classified as normal and are normal. Situations like this are viewed as being very bad.

Information about attacks that has been mistakenly labelled as normal is known as false negative (FN) data. The most vulnerable stage is this one because no one is aware that an attack has already occurred.

The ratio of all observed values to the total of the TP and TN observations is used to measure accuracy. Accuracy is frequently predicated on the overall number of valid classifications. The accuracy formula is explained in greater depth in equation.

$$Accuracy \frac{TP + TN}{TP + TN + FP + F}$$

5. Experiment and Results

The accompanying Tables 7.2, below, show the outcomes of training and testing the suggested algorithms using the dataset. Both the attained testing times (s) and testing accuracies (%) are shown in the tables together with the training times (s) and training accuracy percentages. The outcomes for the kddcup dataset are shown in Table 7.2.

These studies were conducted to demonstrate the effectiveness and accuracy of several kinds of intrusion detection algorithms as well as the processing time required to identify intrusions over the whole dataset. By demonstrating the accuracy, we can evaluate them and help them learn on their own so that they may go on to perform with more precision in the future.

6. Conclusions and Future Work

In conclusion, the adoption of machine learning and deep learning techniques in organizations has demonstrated their effectiveness in addressing cyber risks. Through these

Table 7.2 Training and testing algorithm's results

Machine Learning Algorithms	Training time(s)	Testing Time (5)	Training Accuracy	Testing Accuracy
Gaussian Naïve-Bayes	0.9	1.0	58%	55%
Logistic Regression	1.3	1.0	96%	98%
SVM	1.8	1.6	90%	89%
Stochastic Gradient Descent	2.5	2.2	85	
Decision Tree Algorithm	1.2	0.9	98%	99%
Random Forest Algorithm	1.8	1.6	95%	96%
Gradient Boosting Algorithm	128	152	95%	95%
K-Nearest Neighbour	2.8	2.5	93%	88%
ANN	338	349	92%	91%
Convolutional Neural Network	1243	1315	93%	95%
Recurrent Neural Network	2145	2254	92%	93%

techniques, organizations can automatically identify, prevent, recover from, and adapt to various threats without explicit programming. In the conducted experiment, a range of algorithms were evaluated, with Logistic Regression and Decision Tree classifiers achieving exceptional accuracy levels of over 95% in distinguishing malware across different test datasets while requiring less development time. The Gaussian Naive-Bayes classifier achieved accuracy ranging from 51% to 88%. Notably, the Random Forest Classification algorithm outperformed all other algorithms in terms of accuracy. These findings highlight the significance of machine learning in effectively mitigating cyberattacks. Logistic Regression and Decision Tree classifiers are effective solutions, but the Random Forest Classification algorithm emerged as the most accurate among all the algorithms studied. Overall, these findings contribute to ongoing efforts to enhance cyber security and protect computer networks from emerging threats.

It is certain that machine learning algorithms will increasingly be used in a variety of fields, including cyber security, in the coming years. This study's objective was to locate malware using thirteen different classification approaches across four different datasets. Surprisingly, 12 algorithms were highly accurate. It was found that algorithms like the Gaussian Naive-Bayes classifier, Logistic Regression, and Decision

Tree classifier were incredibly effective and took very little effort to build.

References

1. Najari, S., & Lotfi, I. (2014). Malware Detection Using Data Mining Techniques. International Journal of Intelligent Information Systems, 3(6-1), 33–37. https://doi.org/10.11648/j.ijiis.s.2014030601.16

2. Qin, Y., & Xia, T. (2017). Sensitivity analysis of ring oscillator based hardware Trojan detection. In 2017 IEEE 17th International Conference on Communication Technology (ICCT) (pp. 1979–1983). IEEE. https://doi.org/10.1109/ICCT.2017.8359975

3. Jacobson, D., & Idziorek, J. (2012). Computer Security Literacy: Staying Safe in a Digital World (1st ed.). Chapman and Hall/CRC. ISBN-13: 978-1439856185

4. Dasgupta, D., Akhtar, Z., & Sen, S. (2020). Machine learning in cybersecurity: A comprehensive survey. The Journal of Defense Modeling and Simulation: Applications, Methodology, Technology, 19(1), 57–16. https://doi.org/10.1177/1548512920951275

5. Katole, R. A., Sherekar, S. S., & Thakare, V. M. (2018). Detection of SQL injection attacks by removing the parameter values of SQL query. In 2018 2nd International Conference on Inventive Systems and Control (ICISC) (pp. 736–741). IEEE. https://doi.org/10.1109/ICISC.2018.8398896

6. Farooq, H. M., & Otaibi, N. M. (2018). Optimal Machine Learning Algorithms for Cyber Threat Detection. In 2018 UKSim-AMSS 20th International Conference on Computer Modelling and Simulation (UKSim) (pp. 32–37). IEEE. https://doi.org/10.1109/UKSim.2018.00018

7. Bhatia, V., Choudhary, S., & Ramkumar, K. R. (2020). A Comparative Study on Various Intrusion Detection Techniques Using Machine Learning and Neural Network. In 2020 8th International Conference on Reliability, Infocom Technologies and Optimization (Trends and Future Directions) (ICRITO) (pp. 232–236). IEEE. https://doi.org/10.1109/ICRITO48877.2020.9198008

8. Rose, J. R., Swann, M., Bendiab, G., Shiaeles, S., & Kolokotronis, N. (2021). Intrusion Detection using Network Traffic Profiling and Machine Learning for IoT. In Proceedings of the 2021 IEEE Conference on Network Softwarization: Accelerating Network Softwarization in the Cognitive Age, NetSoft 2021 (pp. 409–415). https://doi.org/10.1109/NetSoft43631.2021.9481800

9. Ali, Z., Hussain, F., Ghazanfar, S., Husnain, M., Zahid, S., & Shah, G. A. (2021). A Generic Machine Learning Approach for IoT Device Identification. In Proceedings of the 2021 International Conference on Cyber Warfare and Security (ICCWS) (pp. 118–123). https://doi.org/10.1109/ICCWS51832.2021.9542624

10. El-Sayed, R., El-Ghamry, A., Gaber, T., & Hassanien, A. E. (2021). Zero-Day Malware Classification Using Deep Features with Support Vector Machines. In Proceedings of the 2021 Tenth International Conference on Intelligent Computing and Information Systems (ICICIS) (pp. 311–317). https://doi.org/10.1109/ICICIS53598.2021.9541621

11. Le, K.-H., Nguyen, M.-H., Tran, T.-D., & Tran, N.-D. (2022). IMIDS: An Intelligent Intrusion Detection System against Cyber Threats in IoT. Electronics, 11(3), 524. https://doi.org/10.3390/electronics11030524

12. Islam, N., Farhin, F., Sultana, I., Kaiser, M. S., Rahman, M. S., Mahmud, M., & Cho, G. H. (2021). Towards Machine Learning Based Intrusion Detection in IoT Networks. Computational Materials and Continua, 69(2), 1801–1821. https://doi.org/10.32604/cmc.2021.017723

13. Waskle, S., Parashar, L., & Singh, U. (2020). Intrusion detection system using PCA with random forest approach. In 2020 International Conference on Electronics and Sustainable Communication Systems (ICESC) (pp. 803–808). IEEE.

14. Aljawarneh, S., Aldwairi, M., & Yassein, M. B. (2018). Anomaly-based intrusion detection system through feature selection analysis and building hybrid efficient model. Journal of Computational Science, 25, 152–160. https://doi.org/10.1016/j.jocs.2017.11.011

15. Thaseen, I. S., Kumar, C. A., & Ahmad, A. (2019). Integrated intrusion detection model using chi-square feature selection and ensemble of classifiers. Arabian Journal for Science and Engineering, 44(4), 3357–3368.

16. Meryem, A., & Ouahidi, B. E. (2020). Hybrid intrusion detection system using machine learning. Network Security, 2020(5), 8–19.

17. Kaja, N., Shaout, A., & Ma, D. (2019). An intelligent intrusion detection system. Applied Intelligence, 49(9), 3235–3247.

18. Belavagi, M. C., & Muniyal, B. (2017). Multi-class machine learning algorithms for intrusion detection: A performance study. In International Symposium on Security in Computing and Communication (pp. 170–178). Springer.

19. Selvakumar, B., & Muneeswaran, K. (2019). Firefly algorithm based feature selection for network intrusion detection. Computers & Security, 81, 148–155.

20. Khraisat, A., Gondal, I., Vamplew, P., & Kamruzzaman, J. (2019). Survey of intrusion detection systems: Techniques, datasets and challenges. Cybersecurity, 2(1), 20.

21. Ikram Sumaiya Thaseen and Cherukuri Aswani Kumar. (2017). Intrusion detection model using fusion of chi-square

Note: All the figures and tables in this chapter were made by the authors.

Machine Learning Technique for Parkinson's Disease Prediction

8

**Kanakandla Vasudha[1], Gattu Tejaswini[2],
Kanagiri Swapna[3], P. Ratna Tejaswi[4]**
Department of Computer Science and Engineering,
Teegala Krishna Reddy Engineering College, Hyderabad, India

Abstract: In the field of HealthCare, technology has been growing each day in order to identify different diseases and their symptoms emerging in the present world. Likewise one of the diseases is called Parkinson's disease. This is a neurological condition of the brain. It impacts on the body in the way of hand shaking, Stiffness and vocal effects etc.. At this time there is no proper treatment available. Only when the disease detected early it is possible to cure. By this way we can easily save our life's and reduces the cost of sickness and we could get the proper cure or treatment. For this purpose a project called "Parkinson's Disease detection using machine learning techniques" was launched to predict the disease at early stage. As a result, different machine learning techniques like support vector machine (SVM) and python libraries were used in order to develop a model to detect the disease in a presence of body. The present model relies on image or audio analysis to detect disease.

Keywords: support vector machine (SVM), Parkinson's disease, Machine learning (ML) Techniques, dopamine

1. Introduction

Parkinson disease is a progressive neurological disorder that affects the entire body. We know our parts of body controlled by the nerves then it mainly damages our nerves then some symptoms start slowly like tremor in just one hand but tremors common for this disease and also lose stiffness in our body. "When a disease is in its early stages, a person's face may be inexpressive or their speech may become slurred, they may hardly move, and their arms may barely swing as they walk.. However, treatments may greatly relieve these symptoms. Obviously, your doctor may advise surgery to control some of your brain's areas [1]".

Nerve cells in the brain may break down or die, as a result of a lack of neurons that produce the chemical message dopamine in your brain.

1.1 Machine Learning Techniques

Machine learning is a boom technology which enables computer to learn from past data automatically and it uses various techniques to build models and making predictions using historical data. There are different types of Supervised, semi-supervised and unsupervised learning technique.

1.2 Support Vector Machine (SVM)

SVM is an supervised learning technique that can be employed to solve the problems of classification and regression. "The primary objective of an SVM is to find the optimum line that partition n-dimensional space into classes. so that we can simply classify new data points, and the best line is drawn known as the Hyper line [2]".

1.3 Dopamine

Dopamine is type of neurotransmitter. It is in our brain and acts as a hormone. It plays important role in to get Parkinson's disease. It functions as communicating messages between nerve cells in our brain and it is involved in various activities like Movement, Memory, Behavior, attention, sleep, mood etc.. If you have a right amount of dopamine then you feel comfort otherwise like have low amount of dopamine makes

[1]vassudha.08@gmail.com, [2]tejaswini22karnati@ gmail.com, [3]kanagiriswapna@gmail.com, [4]ratnatejaswi.p@gmail.com

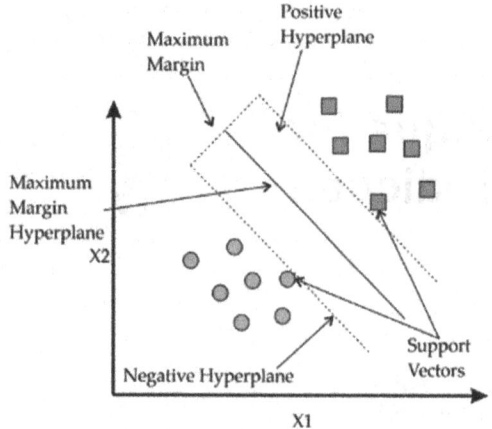

Fig. 8.1 SVM

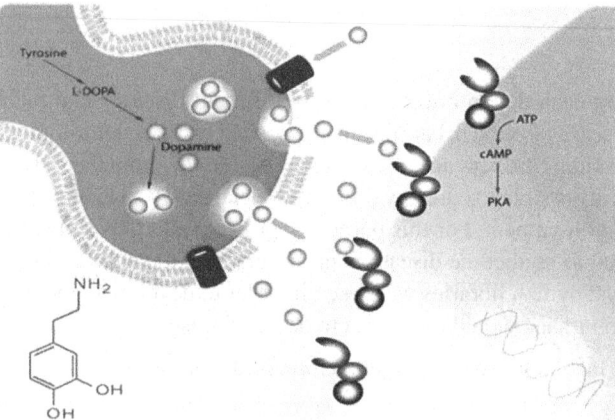

Fig. 8.2 Illustrating dopamine

you nerve weakness and loses stiffness and tired and sleep problems etc [4].

2. Existing Method

There is a existing system to identify the Parkinson disease using voice data. It analysis voice data if there is deflections in voice then it is Parkinson's otherwise not a Parkinson's then According to voice information predicting the disease in early stage then giving prior cure to affected persons and This model showed 73.8% efficiency [6].

Table 8.1 Analysis of different approaches

S. No.	Authors name	Proposed models	Performance
1	Indira R. (2014) [1]	In this work, the author first used fuzzy C-means and ANN techniques to detect Parkinson's disease using speech signal data set.	This model gave a 68.24% accuracy, 75.32% sensitivity and 45.81% specificity and under ANN model gave 92% of Recognition rate.

S. No.	Authors name	Proposed models	Performance
2	Ruben A. (2013) [2]	In this study, the author used Wrapper feature selection technique to detect to Parkinson's disease using non-motor symptoms dataset.	It gave a 70% to 92% accuracy
3	A. Tsanas (2014) and A. Tsans (2014)]	The authors used SVM and Regression & Classification techniques to detect Parkinson's disease using speech signal data set.	The model gave a 98% accuracy and 5- 95 percentile..
4	Shahbakhi (2014) [4]	In this work, the authors developed SVM model to detect Parkinson's disease using speech signal dataset.	The experiment shows that the predicted disease gives 92.22% accuracy, 70% sensitivity and 92% specificity.
5	Chen H. (2012) [5]	In this work the authors used Nested SVM model to detect Parkinson's disease using speech dataset.	The results show that the model is 93% accuracy, 90% sensitivity and 93% specificity, and this is the overall best performance
6	Ene M. (2008) and Cam M. (2008) [6]	In this work, the authors used PNN model to detect Parkinson's disease. using voice recording dataset.	The results shows 73% to 80% accuracy and under Cam M. gave 93% accuracy.

3. Proposed Method

In this paper we Proposed Architecture that has

Totally 4 parts where it involves Dataset i.e. Data Acquisition, Feature extraction, Classification and output Production as shown below:

In this paper we have used Machine Learning technique that is Support Vector Machine, it plays very good role in to build model to identify Parkinson's disease and in this proposed methodology we are applying classification technique to categories features data based on symptoms of a persons. According to Model works on the data

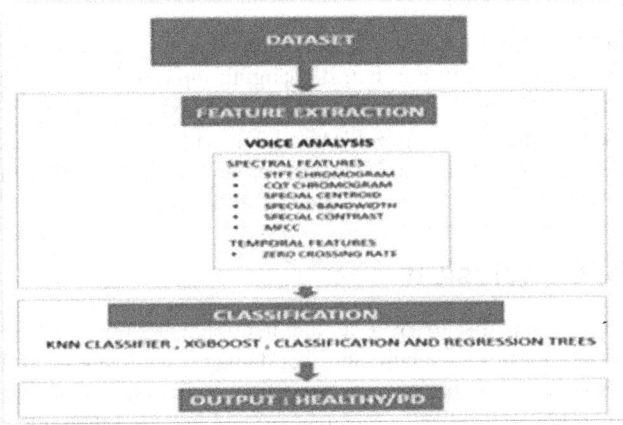

Fig. 8.3 Architecture of proposed method

4. Methodology

4.1 Dataset

"This dataset consists of several biological voice measures that were obtained from 31 different individuals; three of them had Parkinson's disease. Each row in the table corresponds to one of the 195 voice recordings from different people, with each column representing specific voice information. The 'status' column for the class label is set to 0 for healthy and 1 for Parkinson's disease [10]".

4.2 Data Preprocessing

"Real-world datasets frequently contain missing values, noise, and in unsuitable formats that prevent them from being used directly for model development, hence preprocessing is one of the key tasks that can be performed to clean the data in order to improve the quality and make it appropriate for a machine learning model. By using this it increases efficiency and accuracy of a model [12]".

Actually, After Preprocessing in this dataset, we have 195 voice features from 24 people then those used for model building. The below image is the regarding of information after the preprocessing dataset.

4.3 Splitting the Dataset into Training Set and Testing Set

Split the dataset into training and testing set where twenty percent data for testing purpose and remaining data for training the model. Here we used test,train_and split from sklearn. Model. Selection to divide the dataset into training and testing set.

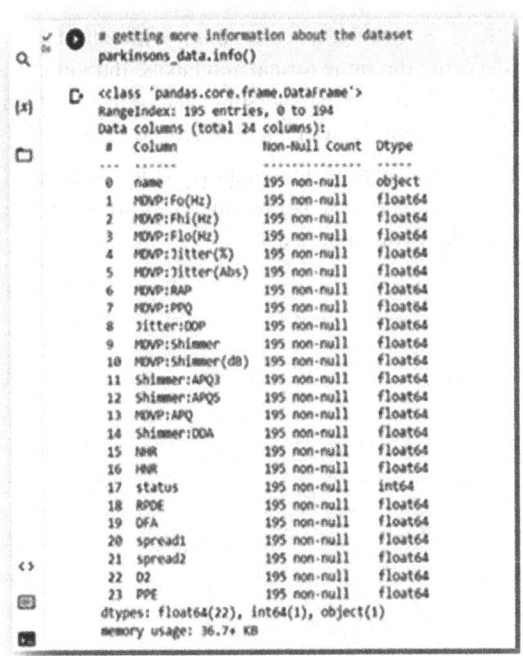

Fig. 8.4 Data after preprocessing

Splitting the data to training data & Test data

```
[ ] X_train, X_test, Y_train, Y_test = train_test_split(X, Y, test_size=0.2, random_state=2)
```

```
[ ] print(X.shape, X_train.shape, X_test.shape)
```

```
(195, 22) (156, 22) (39, 22)
```

Fig. 8.5 Splitting the data to training data and test data

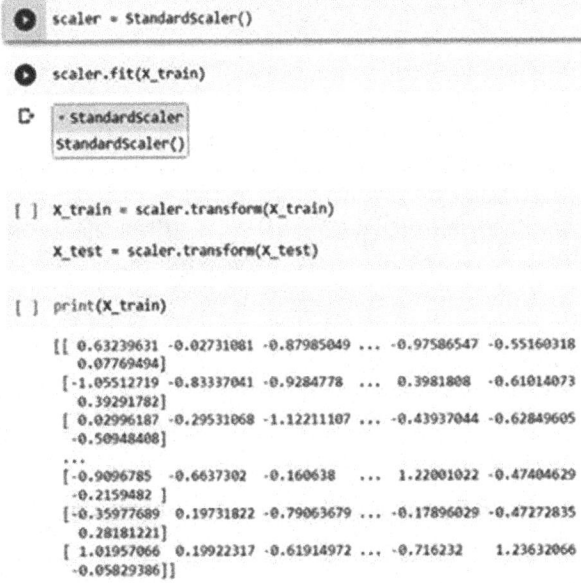

Fig. 8.6 Data standardization Model Training

4.4 Data Standardization

"The data is internally consistent, with all data sources perhaps using the same format and labels. Individuals in your business will be able to understand and utilize the data if it is adequately ordered with logical descriptions and labels [9]".

In this paper we used the support vector machine (SVM) to train the model. The below image shows the training the SVM model with training data:

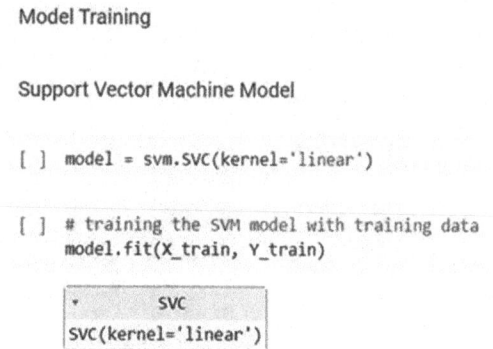

Model Training

Support Vector Machine Model

```
[ ] model = svm.SVC(kernel='linear')

[ ] # training the SVM model with training data
    model.fit(X_train, Y_train)
```

```
     ▾         SVC
   SVC(kernel='linear')
```

Fig. 8.7 Model training

4.5 Model Evaluation

Accuracy Score

Here after completion of model training with training data then we've got good accuracy score of training data is 88% and we did testing the model with test data then we've got good accuracy score of test data is 87% then we could understand this model giving good accuracy score.

Model Evaluation

Accuracy Score

```
[ ] # accuracy score on training data
    X_train_prediction = model.predict(X_train)
    training_data_accuracy = accuracy_score(Y_train, X_train_prediction)

[ ] print('Accuracy score of training data : ', training_data_accuracy)

    Accuracy score of training data :  0.8846153846153846

⏺ # accuracy score on training data
    X_test_prediction = model.predict(X_test)
    test_data_accuracy = accuracy_score(Y_test, X_test_prediction)

[ ] print('Accuracy score of test data : ', test_data_accuracy)

    Accuracy score of test data :  0.8717948717948718
```

Fig. 8.8 Model evaluation with accuracy score

Predictive System

In this system when we give person input data (voice recordings) to System then it changing input data to numpy array and it reshapes the numpy array and next it is standardize the data then it giving to SVM model to predict the Weather person have Parkinson's disease or not. If it gives 0 then person doesn't have disease else have Parkinson's disease.

5. Conclusion

Parkinson's disease is one the neurodegenerative disease which doesn't have cure and that damages complete central nervous system then we lose our stiffness, strength, movement of body and your voice become soft and breaking then it has no cure then it has only cure is early detection of disease then for that purpose we used support vector machine (SVM) model to predict the disease and for that purpose we've taken 195 voice features from 24 people then we trained model with this data then this model gave good accuracy score in predicting disease is 87%.

References

1. Wu Wang1, Junho Lee2, Fouzi harrou3 and Ying sun4, "Early Detection of Parkinson's Disease Using Deep Learning and Machine Learning" IEEEACCESSDigitalObjectIdentifier 10.1109/ACCESS.2020.3016062 Volume 8, 2020-Page no 147635–147646

2. Gunduz H. Deep Learning-Based Parkinson's Disease Classification Using Vocal Feature Sets. IEEE Access. 2019; 7: 115540–115551. doi: 10.1109/ACCESS.2019.2936564

3. A. Ozcift, "SVM feature selection based rotation forest ensemble classifiers to improve computer-aided diagnosis of Parkinson disease" Journal of medical systems, vol-36, no. 4, pp. 2141–2147, 2012.

4. Arvind Kumar Tiwari, "Machine Learning based Approaches for Prediction of Parkinson's Disease" Machine Learning and Applications

5. T. Swapna, Y. Sravani Devi, "Performance Analysis of Classification algorithms on Parkinson's Dataset with Voice Attributes". International Journal of Applied Engineering Research ISSN 0973-4562 Volume 14, Number 2 pp. 452–458, 2019.

6. Kaur S., Aggarwal H., Rani R. Diagnosis of Parkinson's Disease Using Principle Component Analysis and Deep Learning. J. Med Imag. Health Inf. 2019; 9: 602–609. doi: 10.1166/jmihi.2019.2570

7. Quan C., Ren K., Luo Z. A Deep Learning Based Method for Parkinson's Disease Detection Using Dynamic Features of Speech. IEEE Access. 2021; 9: 10239–10252. doi: 10.1109/ACCESS.2021.3051432.

8. Das, Resul. Expert Systems with Applications, 37, 1568 (2010).

9. Caglar, Mehmet Fatih, Bayram Cetisli, and Inayet Burcu Toprak. Journal of Engineering Science and Design 1, 59 (2010).

10. Polat, Kemal. International Journal of Systems Science 43, 597 (2012).

11. Luukka, Pasi. Expert Systems with Applications,38, 4600 (2011).

12. Kihel, Badra Khellat, and Mohamed Benyettou. JSEA 4, 391 (2011).

13. Eskidere, Ö. Sigma, 30, 402 (2012).

14. Prashanth, R., & Roy, S. D. Neurocomputing, 305,78 (2018).

Note: All the figures and table in this chapter were made by the authors

Detection and Prevention of DDoS Attacks in Software-Defined Cloud Networks Using Advanced Support Vector Machine

9

D. Navya Devi, K. Sreenivasulu

Department of Computer Science and Engineering,
G. Pullaiah College of Engineering and Technology, Kurnool, A.P, India

M. Janardhan[3]

Department of Computer Science and Engineering (IOT),
G. Pullaiah College of Engineering and Technology, Kurnool, A.P, India

Abstract: Networks for cloud computing have recently focused on Software Defined Clouds (SDC). SDN technology is used in combination with the traditional Cloud network in SDC. SDC attempts to establish an efficient cloud environment throughout the virtualization of all resources. The main problem with SDC Distributed denial of service (DDoS) is a vulnerability. DDoS assault on SDC has become a significant issue, and numerous employed for both mitigation and detection. This essay presents detection and avoidance of DDoS attacks in software-defined cloud advanced support vector machine (ASVM) networks. A three-class multiclass classification approach is the ASVM methodology. In addition, wedescribe a method for utilizing SDN features to identify DDoS attacks in Software Defined Clouds. They analyze the outcomes, by evaluating a Precision, Accuracy, Recall and Detection Rate. The overall accuracy, False alarm rate and Precision values are 97.6%, 97.5% and 97.5% respectively. They demonstrated that the ASVM and transmitted firewalls with IPS security successfully identify and prevent the DDoS attacks based on their simulation results and discussions.

Keywords: DDoS attack, Software-Defined Networking (SDN), Cloud network, SDC, ASVM, Firewall and IPS.

1. Introduction

Nowadays, advanced network technologies are being developed in modern networking. The traditional network architecture's strongest points include the exponential growth of cloud services, server virtualization techniques, and mobiledevices [1]. Advanced technologies are advancing at a rapid speed.

The majority of traditional network architectures use a client-server hierarchy. A number of databases and servers are accessed by advanced applications across many network domains. Therefore, numerous client and server cases are estimated. As a result, the traffic patterns can differ.

Due to the support they provide for several essential networking functions for both users and operators, as well as in academic and industrial sectors, emerging technologies include Software-defined networking (SDN) and cloud computing [2]. Cloud networking primarily supports on-demand self-service, broad network access, resource pooling, measured service, and rapid elasticity. Similar to SDN, the centralized controller's software intelligence makes it possible to operate and evaluate different network trials in a flexible environment. In order to get over the restrictions of traditional networking, SDN takes on a significant role. As a result, significant consideration given to cloud networking has been Software-Defined Cloud Computing, or simply Software Defined Clouds (SDC), with the goal of developing efficient cloud environments by applying the idea of virtualization to all data center resources, including computation, storage, and networks [3].Recent research has identified security

[1]navyadevi2019@gmail.com, [2]sreenu.kutala@gmail.com, [3]m.janardhan0105@gmail.com

concerns as the primary obstacle in both SDN and Cloud Networking [4]. The decoupling of the data plane and the control plane is the most significant feature of SDN. While the control plane decides wherever to transmit the traffic, the data plane contributes to decision-making and actual traffic forwarding. SDN is vulnerable to numerous network attacks. The most notable effect on SDN among them is the well-known Distributed Denial of Service (DDoS) attacks [5].An attack on a server known as a "Distributed Denial of Service" is one in which a large number of packets are sent to cause an outage or service degradation for legitimate users or deny the organization access to essential computer services including on-premise, email, cloud services, or hosted. DDoS attacks can target cloud environments, carrier networks, hosting networks, large enterprise networks. Virtual application availability is restricted by DDoS attacks. Two or more people or systems can send a DDoS attacks. A real person could be the attacker, or they might be in control of a group of zombies. With spoofed source IP addresses, an attacker can send large volumes of packets to the target.

As an improvement to the Support Vector Machine (SVM) algorithm currently in use to detect DDoS attacks, we propose the Advanced Support Vector Machine (ASVM) technique in this paper. A single controller was utilized by the network system in previous SDN architectures are described [6]. Therefore, the utilization of numerous controllers is the most important factor for our proposed network architecture. During the process of creating traffic, we generate regular traffic, DDoS attack traffic that floods User Datagram Protocol (UDP), and DDoS attack traffic that floods SYN. During the process of collecting traffic, they collect information from every switch [7]. During the process of generating features, average number of packets in a flow during the sample interval, average number of flow bytes, the volumetric features,asymmetric features, variation of flow bytes, amount of packet variations in a flow,and asymmetric features are all generated. For We suggest applying the advance support vector machine (ASVM) technique to the classification procedure. This traffic is monitored at the frame level by the Firewalls and Intrusion Prevention System (IPS) modules on SDN switches. The SDN switch that found any suspect packets will get in contact with the SDC controller. The packet is further examined by the Firewalls and IPS modules of the SDC controllers to check for potentially harmful activities. Once the intrusion has been detected, the IPS records the data, notifies the involved parties, drops packets, resets connections, or blocks traffic from the offending nodes. To stop malicious node packets from being sent, the controllers additionally modifies the firewalls in SDN switches. The study is organized as follows: Literature works elaborated in II Section, described DDoS attacks detection and prevention discussed in Section 3.Section 4 provides information about

the experiment results, and Section v introduces the paper to a conclusion.

2. Literature Survey

R. Sahay, Z. Zhang,G. Blanc, H. Debar et al. [8] have provided ArOMA, a programmable and centralized manageable SDN-based autonomous DDoS mitigation framework. This analysis aims to systematically integrate a number of DDoS mitigation systems that are spread out among ISPs and their customers. Alshamrani. A, S, Lu. D,A, Pisharody. Chowdhary, Huang. D et al. [9] claims that the current defenses against DDoS attacks are ineffective. They consequently looked at the impact of bad behaviour and new flow attacks on SDN. In order to respond to the unexpected changes in traffic that happened in the SDN architecture at the moment of the attacks, they utilized machine learning classification algorithms to routinely gather traffic data from transmission devices on a data plane. They achieve this by continuously collecting traffic data from transmission devices on a data plane utilizing machine learning categorizing algorithms in response to the fast changes in traffic that occurred in the SDN architecture at the time of the attack. The packet_in messages that were flowing during the controller and transmission devices noticed the attack initially. For classification, the Naive Bayes (NB), J48, and Support Vector Machine (SVM) algorithms were used. Q. Yan, Q. Gong, F.R. Yuet. al. [10] provides a scheduling technique for SDN controllers that significantly reduces DDoS attacks. In order to protect the standard switches from DDoS attacks, this method schedules the execution of flow requests through other SDN switches. This method utilizes a number of time slicing strategies based on the intensity of the DDoS attack.

Nanda. S, Wedaa. E, DeCusatis. C, Zafari. F,Yang. B et al. [11] proposed in order to reduce security risks, potential targets and potential malicious connections can be identified using machine learning algorithms that have been trained on data from previous attacks on networks. They utilised the C4.5, Decision Tree (DT), Naive Bayes, and Bayesian Network (BayesNet) algorithms. It was mentioned that estimations made using machine learning methods might be used to detect malicious users in the data plane. For the efficiency and duration of the network, user identification must be protected in order for the SDN controller to quickly and effectively introduce new rules to prevent the attacks.

Xue. L, Ran. J, Hu. H, Yuan. D, Li. S, et al. [12] demonstrated that due to the SDN controller's management of numerous data plane switches, numerous security criteria are required. They said that current software or technology that hasn't been changed to work with the SDN architecture cannot provide security. The proper operation of the network's

SDN architecture is seriously challenged by DDoS attacks, particularly on switching on the data plane. To recognize the DDoS attacks, cross-validation-genetic algorithm (CV-GA) was implemented using SVM-optimized C and G parameters. J Rameshbabu et. al. [13], In order to protected against DDoS attacks in the cloud, the authors suggested implementing NEIF (Network Egress and Ingress Filtering) at ISP edge routers, where ingress filtering takes longer to implement. The goal of ingress filtering is to prevent any IP packets with dangerous source addresses from interacting with the system. It is unfortunately difficult for a significant amount of network filtering.

Javed Ashraf and Seemab Latif, et. al. [14], aimed to use machine learning methods to deal with intrusions and DDoS attacks in an SDN environment. However, they only looked at a few different machine learning methods, including fuzzy logic,Support Vector Machine (SVM), decision tree, and Bayesian networks,neural networks, that can be used to find DDoS attacks in the networking system. Then to identify and stop DDoS attacks wasn't addressed in any detail. R. Braga, A. Passito, E. Mota, et al. [15] performed traffic flow feature analysis to identify DDoS attacks.

3. Detection and Prevention of DDoS Attacks in SDC Networks Using ASVM

Figure 9.1 demonstrates the framework for detectionand prevention of DDoS attacks in software-defined cloud networks using Advanced Support Vector Machine.

The concept that various applications have various security requirements served as the inspiration for our proposed methodology. a programmable response mechanism for DDoS attacks notifications is required by described framework for a DDoS attack detection solution. Users and attackers connect in an end-to-end network scenario, through their Internet access, to the Cloud. An Internet Service Provider, also known as a Network Service Provider, is in charge of managing the access network. The packets have been sent to the OpenFlow Switches for normal users or attackers. The flow entries should be considered to the information in the incoming packets; if there is a match, a specific action can be performed. The extraction of traffic features is the next step after gathering traffic data. The flow table's various characteristic values can be used to examine the SDN network is where the collected malicious traffic flows. Whenever the switch's traffic information is gathered, they might determine how many packets and bytes a host is sending, and how long it takes to send a packet to another host or receive one from another host.

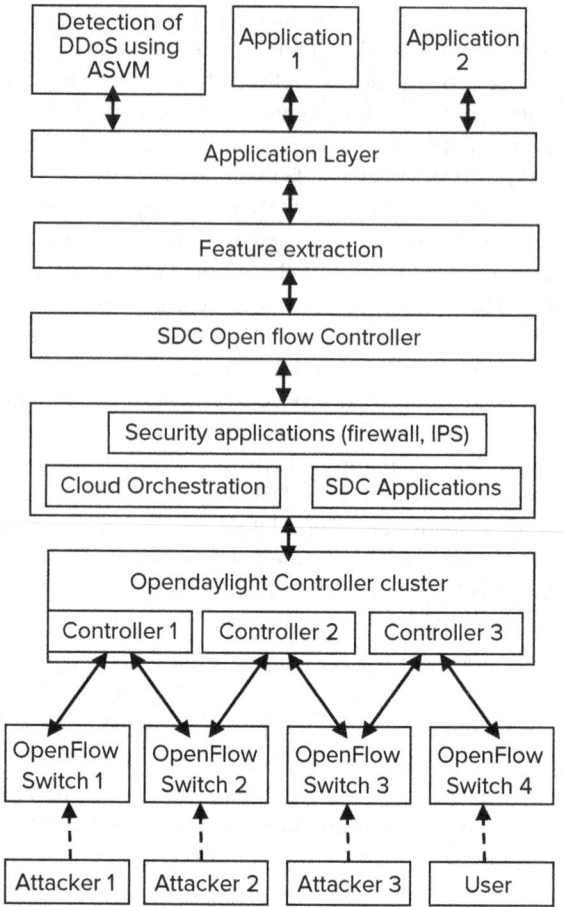

Fig. 9.1 Framework of DDOS attack detection and prevention [1]

A cluster of controllers is connected. The northbound API will route the traffic to the application layer's ASVM-based DDoS attack detection once it reaches the OpenDaylight controller cluster. The Northbound Interface serves as a logical connection between Firewall and IPS applications and the SDC controller. The simple SDC network topology is used to demonstrate how IPS and firewall functions operate. Decisions about incoming packets are made using firewall rules; to examine a flow's behavior, Statistics about traffic flow, such as packets per second, are used by the IPS module; additionally, in order to reduce on memory usage in SDN switches, the rule is quickly maintained using the time stamp. As a result, the initial packet is sent to the SDN controller by the SDN switches using the Southbound Interfaces. The entire ACL (Acknowledgement) is stored in the Firewall module of the controller and is set up by the cloud administrators. When a ACL firewall rulesare matches the target MAC (Media Access Control) address, the SDC controller establishes a connection with the SDN switches.The SDN switching therefore can decide to forward rather than drop packets in line with the rules whenever the following packets from

Hosts "h1" and "h2" arrive because the firewall rule entries has already been put in the MAC flow tables. If no firewall rules in the SDC controller's ACL match the destination MAC address, the firewall modules inspects the packets for ARP (Address Resolution Protocol) inspections to determine whether it originated from unauthorized sources.

The SDN switch's IPS module will transmit selected packets from a flow to the controllers for additional intrusion prevention analysis if there is any unusual behavior in that traffic. In order to locate the DDoS attacks on the SDC networks, proposed method will make use of the ASVM, or advanced support vector machine technique. Through the high accuracy and efficiency for handling highly dimensional data, SVM is frequently used in a wide range of applications, and adaptability in modeling a wide range of data. Linear two-class classification problems were the first applications of SVM. This is considered that a sample linear two-class classification process has two classes, -1 (negative class)and +1 (positive class).When the data are incorrectly classified, the margin maximization can frequently lead to errors.

In this analysis, they add an Advanced Support Vector Machine (ASVM)to the SVM. The classification error (C) and the slack variables (i) must be taken into consider. The distance a point is from its marginal hyperplane is determined by the slack variable. Whereas the performing a multiclass classification problem, one against one as well as one against some, they must take the classifier's judgment into consideration. The development of the classifying patterns in one-versus-one is $n(n - 1)/2$. Top first test is prepared as a positive example, while inferior example is prepared as a negative example. Across the testing process, the data must be classified using every one of these classifications. Through the utilization of the one-against-some classification method, The remaining n-1 classes are used to train each class. All other samples are designated as negatives, while only one class of the samples is considered to be positive. To generate a real-valued confidence score, they must make a decision. The selection of the kernel function is the most crucial factor when applying the SVM algorithm to the classification problem. In order to separate the data, the dataset is moved into a higher dimension space by the kernel function . In this work, the form is the kernel function.

$$(x_n.x_i) \leftarrow K(x_n.x_i) = (\emptyset(x_n).\emptyset(x_i)) \qquad (1)$$

When n=1, 2, 3, 4,.., N, the support vector data is x_n.

The DDoS attack and regular traffic are classified in this work using a linear kernel and an OVS (one-versus-some) decision function.

4. Result Analysis

On a VMware running Ubuntu 16.04, the Mininet emulator (versions 2.3.0d1) is used in this study's tests to create SDN

networking topologies. Two processors, four megabytes of RAM (Random Access Memory), and a 20 GB (SCSI) hard disk power our VMware. There are many different kinds of controllers: NOX, Floodlight, Ryu, ONOS, and OpenDaylight, POX. The network topology is controlled by the OpenDaylight (version Beryllium) controller, and among them. OpenDaylight is a Java-based open source SDN/SDC controller that is managed by the Linux Foundation and supported by VMware. The results of Mininet, a network emulator, are similar to those of a real network since it uses a single Linux kernel to operate a number of end hosts, switches, routers, and links.

Data sets used for testing and training are multidimensional. The first issue of our study, multiclass problem extension, has been resolved. Using the linear kernel with the penalty parameter of the classifications error component, "C," and taking into consideration the value of "gamma" and the shape of the "OVS" decision function, the second issue of the SVM algorithm's lengthier training and testing durations was addressed.

Our detection result is evaluated using precision and false alarm rate,accuracy, and detection rate(Recall). That rate of false alarms is the proportion of false positives provided by our detection system while a normal behavior is being detected. As a result, less false alarms are preferred. For identifying malicious communications, the correct rates are those of detection. The performance of the system is improved with a higher detection rates. The ratio of the number of accurate positive scores to the number of positive scores predicted by the classification algorithm is known as the positive predictive value or precision. System performance is enhanced by choosing a higher precision. The measurement of a system's accuracy is its ability to distinguish between normal and malicious traffic. The following equations represent the all four measures:

$$\text{Accuracy} = \frac{TP + TN}{TP + TN + FN + FP} \qquad (2)$$

$$\text{Detectionrate/Recall} = \frac{TP}{TP + FN} \qquad (3)$$

$$\text{Precision} = \frac{TP}{TP + FP} \qquad (4)$$

$$\text{False alarm rate} = \frac{FP}{TP + FP} \qquad (5)$$

The fraction of network traffic that is "true positive" (TP) is successfully transmitted after being correctly identified as attack or regular traffic. The quantity of correctly detected and dropped network traffic is known as true negative (TN). Network traffic that is improperly identified and forwarded is referred to as false positive (FP). The percentage of network traffic that is mistakenly detected and ignored is called a false negative (FN).

With a splitting rate of 10% to 90% of SDN Traffic DS, they using the cross validation method to train and test in this experiment. Table 9.1 displays the results of the experiment.

Table 9.1 Performance of ASVM method

Parameter	10% of testing DS +90% of Training DS	50% of testing DS +50% of Training DS	90% of testing DS +10% of Training DS
False alarm rate	0.01	0.02	0.028
Detection rate	98	97	97.5
Precision	97	97.5	98
Accuracy	98	97	98

Source: Authors

The overall accuracy, Detection rate and Precision values are 97.6%, 97.5% and 97.5% respectively. The comparative performance of the described ASVM based DDoS attack detection and prevention model with Naïve Bayes (NB) model is represented in below Table 9.2.

Table 9.2 Comparative analysis

Parameter	ASVM	NB
False alarm rate	0.02	0.12
Detection rate	97.5	91.2
Precision	97.5	90.3
Accuracy	97.6	91.7

Source: Authors

The graphical representation of Accuracy, Precision and Detection Rate areshown in below Fig. 9.2, Fig. 9.3 and Fig. 9.4 respectively.

From results it is evident that, DDoS attacks are effectively detected and prevented by the ASVM method than the Naïve

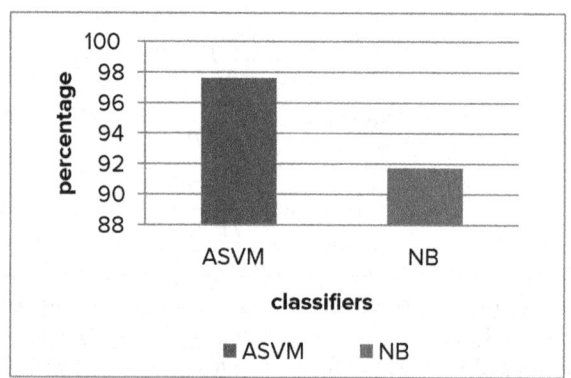

Fig. 9.2 Comparative analysis in terms of accuracy
Source: Authors

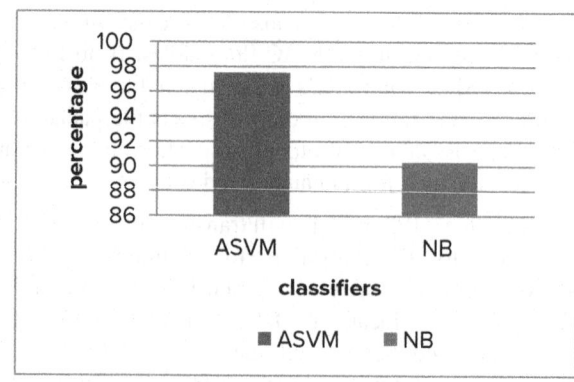

Fig. 9.3 Comparative analysis in terms of precision
Source: Authors

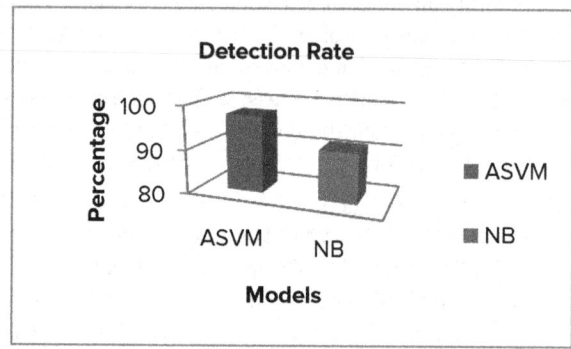

Fig. 9.4 Comparative analysis in terms of detection rate
Source: Authors

bayes (NB) method and having the greatest Accuracy as 97.6%. As a result, the packet will be effectively identified as either normal traffic or a DDoS attack traffic.

5. Conclusion

Detection and prevention of DDoS attacks in software-defined cloud networks using Advanced Support Vector Machine is described in this research. In recent years, a new type cloud labeled an SDN-based cloud or SDC network has developed more popular. The security of clouds is one of the most important components of networking systems. Each packet is classified as either normal traffic or an attack using the ASVM methods. The improved In the ASVM method, the Support Vector Machine (SVM) algorithm is utilized. When testing and training the classification method, the cross-validation approach is used. A linear kernel is used in our ASVM technique. In contrast to the IPS modules evaluate packets, the Firewall module inside this SDC controllers performs an activity from the complete ACL. Detection result is evaluated using precision and accuracy, detection rate (Recall),and false alarm rates. The values for overall precision, detection rate, and accuracy are 97.5%, 97.5%, and 97.6%, respectively. Naïve Bayes (NB) based DDoS attack detection and prevention model is considered for comparative analysis. DDoS attacks are effectively detected and prevented by the ASVM method than the Naïve bayes method.

References

1. DivyaKapil, Varsha Mittal, Durga Prasad Gangodkar, "Virtualization and Nested Virtualization Technology: Concept, Architecture and Attack Vector Model", 2022 International Conference on Computational Intelligence and Sustainable Engineering Solutions (CISES), Year: 2022

2. B Harshitha, K.N Madhusudhan, S Lalitha, "Global Connectivity in Software Defined Network", 2021 7th International Conference on Advanced Computing and Communication Systems (ICACCS), Volume: 1, Year: 2021

3. YeonhoYoo, Gyeongsik Yang, Minkoo Kang, Chuck Yoo, "Adaptive Control Channel Traffic Shaping for Virtualized SDN in Clouds", 2020 IEEE 13th International Conference on Cloud Computing (CLOUD), Year: 2020

4. Jungmin Son, RajkumarBuyya, "Priority-Aware VM Allocation and Network Bandwidth Provisioning in Software-Defined Networking (SDN)-Enabled Clouds", IEEE Transactions on Sustainable Computing, Year: 2019

5. Bineet Kumar Joshi, Nitin Joshi, Mahesh Chandra Joshi, "Early Detection of Distributed Denial of Service Attack in Era of Software-Defined Network", 2018 Eleventh International Conference on Contemporary Computing (IC3), Year: 2018

6. Saif Saad Mohammed, Rasheed Hussain, Oleg Senko, Bagdat Bimaganbetov, Joo Young Lee, Fatima Hussain, Chaker Abdelaziz Ker In the ASVM method, the Support Vector Machine (SVM) algorithm is utilized.rache, Ezedin Barka, Md Zakirul Alam Bhuiyan, "A New Machine Learning-based Collaborative DDoS Mitigation Mechanism in Software-Defined Network", 2018 14th International Conference on Wireless and Mobile Computing, Networking and Communications (WiMob), Year: 2018

7. Yandong Liu, Mianxiong Dong, Kaoru Ota, Jianhua Li, Jun Wu, "Deep Reinforcement Learning based Smart Mitigation of DDoS Flooding in Software-Defined Networks", 2018 IEEE 23rd International Workshop on Computer Aided Modeling and Design of Communication Links and Networks (CAMAD), Year: 2018

8. R. Sahay, G. Blanc, Z. Zhang, H. Debar, "ArOMA: An SDN based autonomic DDoS mitigation framework", Computers & Security, vol. 70, pp. 482–499, 2017

9. Alshamrani. A, Chowdhary. A, Pisharody. S, Lu. D, Huang. D, "A defense system for defeating DDoS attacks in sdn based networks", In Proceedings of the 15th ACM International Symposium on Mobility Management and Wireless Access, Miami Beach, FL, USA, 21–25 November 2017.

10. Q. Yan, Q. Gong, F.R. Yu, "Effective software-defined networking controller scheduling method to mitigate DDoS attacks", Electronics Letters, vol. 53, no. 7, pp. 469–471, 2017.

11. Nanda. S, Zafari. F, DeCusatis. C, Wedaa. E, Yang. B, "Predicting network attack patterns in SDN using machine learning approach", In Proceedings of the 2016 IEEE Conference on Network Function Virtualization and Software Defined Networks (NFV-SDN), Palo Alto, CA, USA, 7–10 November 2016.

12. Xue. L, Yuan. D, Hu. H, Ran. J, Li. S, "DDoS detection in SDN switches using support vector machine classifier", In Proceedings of the Joint International Mechanical, Electronic and Information Technology Conference (JIMET-15), Chongqing, China, 18–20 December 2015

13. J Rameshbabu, "A Prevention of DDoS Attacks in Cloud using NEIF Techniques", Int'l. Journal of Scientific and Research Publications, Vol. 4: Iss. 4, 2014, pp. 1–5.

14. Javed Ashraf and SeemabLatif,"Handling intrusion and ddos attacks in software defined networks using machine learning techniques", In Software Engineering Conference (NSEC), 2014 National, pages 55–60. IEEE, 2014

15. R. Braga, E. Mota, A. Passito, "Lightweight DDoS flooding attack detection using NOX/OpenFlow", In proceedings of IEEE 35th Conference on Local Computer Networks (LCN), Denver, pp. 408–415, 2010.

An Efficient Early Software Reliability Prediction using Particle Swarm Optimization (PSO)

10

P. Ashwini[1]

Assistant Professor, Dept of CSE, Vasavi College of Engineering,
Ibrahim Bagh, Hyderabad, India

B. Rajani[2], B. Vijitha[3]

Assistant Professor, Dept of CSE,
Teegala Krishna Reddy Engineering College, Meerpet, Hyderabad, India

Abstract: Software testing is very costly and time-consuming, and nearly half of the expenditures associated with software development go toward testing. Software fault prediction is the process of creating modules to help developers in detecting failures in project modules. The process of creating a collection of data for software testing based on a certain criterion is known as test data creation. Particle Swarm Optimization (PSO) uses an iterative method to optimize a solution to the problem. The basic component that affects how the software testing process is modeled to create the predicted faults is the correlation between execution time and the failure count. In this study, we implemented Particle Swarm Optimization (PSO) technique's initial conceptualization as a means of addressing the issue of software reliability growth modelling. The Goel-Okumoto, Musa-Okumoto, Delayed S-Shaped, and Power reliability growth models' parameters have all been estimated using the suggested technique. The estimated parameters were further used for decision making, such as, remaining faults in the software, future testing time and time to market for software product.

Keywords: Software testing, Particle swarm optimization (PSO), Software reliability growth Models, Software reliability prediction

1. Introduction

During software testing phase of Software Development Life Cycle (SDLC), attempts are made in the form of white box or black box testing to find as many faults as possible. The reliability of the software increases by fixing the faults [1]. The data collected in the form of faults through testing make a fault database. The database consists of a number of faults in a successive interval of time. These faults are further used for estimation of current reliability and prediction of future development of the growth in reliability [2]. The objective of the proposed approach is to find optimized parameters of software reliability growth models. This approach would be helpful in determining which model is to be used for given software, developed under a specified development

environment. The purpose of the research is to make a model and validate it with the help of given fault databases and to demonstrate that it can accurately define past failures and predict future failures [3]. The problem handled during the research is to find out parameters of different reliability growth models, which can perform well in the given data set and able to predict faults in the near future in the software [4].

Software defects can cause serious problems in some systems like aircraft, space shuttle and medical systems. Finding flawed software modules is thus a crucial step since it enables the identification of those that need more testing or restructuring [5]. The remaining parts of the article are arranged as follows. The numerous studies done in regard to our recommended work are presented in Section 2 of

[1]ashwinireddy90@tkrec.ac.in, [2]rajani.g@tkrec.ac.in, [3]vijitha_bop@yahoo.co.in

this article. The goal, strategy, and sophisticated method are explained in Section 3. Models of software expansion are provided in Section 4. The outcomes of our recommended method are described in detail in Sections 5 and 6, and the conclusion and future work are presented in Section 7.

2. Review of Literature

Y. Del Valle et al. [6] presented a survey of Particle Swarm Optimization(PSO) on power systems. Survey describes importance of PSO on a highly nonlinear, non-stationary system with noise and uncertainties.

J. D. Musa and K. Okumoto [7] defined a new software reliability growth model that predicts expected faults. They had evaluated model efficiency using actual data and compared that with other existing reliability growth models. John D. Musa [8] gave an approach of software reliability based on software execution time. The model developed by Musa estimates the amount of time required for testing to achieve targeted reliability in advance. The execution time and faults count are taken in terms of mean time to failure (MTTF). L. Goel [9] proposed as of tware reliability growth model. This model describes critical analysis of models and underlying assumptions, limitations, and applicability of these models during software development lifecycle.

In 1995, Drs. James Kennedy and Russell Elberhart [9] created Particle Swarm Optimization (PSO). In addition to GA, PSO is a useful method. It is one of the best models because it allows open communication between independent agents. It collects social information from the person and uses it to regulate the routes of the group of components to look at the parameter space globally [10]. PSO is recognized as a global optimization technique due to its convergence speed and easiness.

Due to the complexity existing in software applications, and poor understanding of software requirements, it is not easy to recognize and eliminate the software faults during testing. Removing the recognized defects may produce new faults called improper debugging [11]. To estimate the reliability of different software reliability growth models in nonhomogeneous Poisson processes in the context of imprecise debugging, P. K. Kapur et al. [12] developed two mild outlines. They focused on fault recognition and elimination during testing.

3. Particle Swarm Optimization

The organized nature of the animal world, such as the behavior of fish schools and flocks of birds, has an impact on particle swarm optimization (PSO). With its emphasis on collaboration, PSO is an iterative methodology. In 1995, Drs.

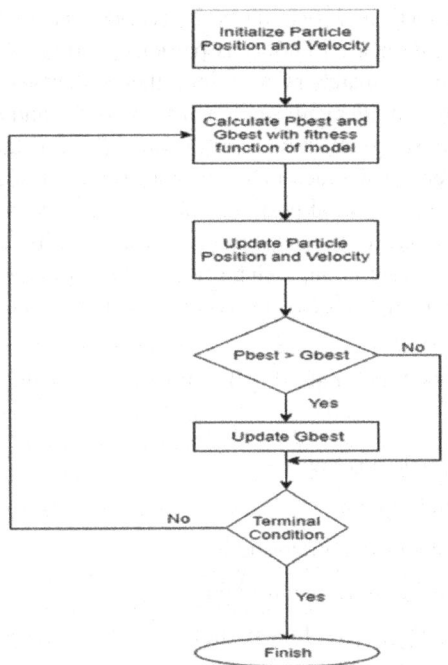

Fig. 10.1 Flowchart of PSO algorithm-I

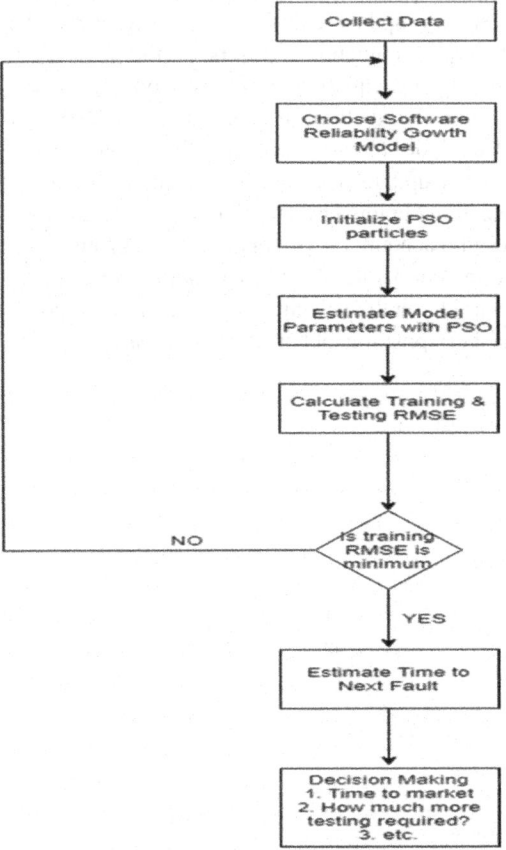

Fig. 10.2 Flowchart of PSO algorithm-II

James Kennedy and Russell Eberhart created the group-based optimization method known as particle swarm optimization [13]. Many researchers have put their attention to PSO algorithm from its manifestation, as it is simple and effectual technique of optimization. Particles in PSO can move in a multidimensional space to search for an optimum solution. To achieve an optimal position, each cell adjusts its position based on both its own experience and that of its neighbors [14]. In every iteration, each particle moves to a new location by updating its velocity by adding the three values given below:

$$vx = vx + c1 * r \text{ and } * (pB - p1) + c2 * r * (gB - p1) \quad \textbf{(1)}$$

where

p1: Particle position

vx: Path direction

c1: Local information weight

c2: Global information weight

pB: Best position of the particle

gB: Best position of the swarm

r: Random variable

The velocity update equation has three parts: The term "inertia" or "momentum" refers to the first term in the velocity update equation. It focuses on the direction that the particle is travelling in at the moment. The term "self-knowledge" or "memory" refers to the second component of the velocity update equation. It concentrates on the previous best position obtained by a particular particle. In other words, it takes into account the personal best of a given particle [10]. The term "cooperation" or "social knowledge" refers to the third variable in the velocity update equation. It concentrates on the best position obtained by any particle in the given

---

Algorithm 1 Compute parameters of Software Reliability Growth Model Using PSO

```
procedure INITIALIZE_PARTICLE
    position = rand()
    velocity = 0
    RMSE = fitness_funtion(position)
    p_best = RMSE
    if p_best < g_best then
        g_best = p_best
    end if
end procedure
procedure MAIN_ITERATION_LOOP
    for each particle i in tota_no_particles do
        position = updated_position
        velocity = updated_velocity
        RMSE = fitness_funtion(position)
        p_best = RMSE
        if p_best < g_best then
            g_best = p_best
        end if
        if terminal_condition then
            break;
        end if
    end for
end procedure
```

Fig. 10.3 Pseudo code for PSO algorithm

---

search hyperspace. In other words, it into account the global best. Depending upon the new updated value of velocity, obtained from the velocity update equation, the position of the particles has been updated using the following position update equation:

$$p = p + v \quad (2)$$

4. Software Reliability Growth Models

Software reliability growth models use system data gained from testing phase to assess software reliability. These models also help in predicting the remaining defects in software [6]. Software reliability growth models use mathematical operations to statistically interpolate data on defect detection. The functions are used to forecast future failure rates or the quantity of code flaws that will remain. Alan Wood is the author of Tandem Software Reliability Growth Models. In many models for increasing software dependability, the number of defects per program module is modeled as a continuous variable. We may determine how many problems are still there in that software module using this parameter and the number of defects that are previously known [11].

4.1 Goel-Okumoto Model

The Goel-Okumoto model is a concave model of software reliability growth. The Goel-Okumoto model is one of the non-homogeneous exponential Poisson processes (NHPP) models which describes the occurrence of software failure during software testing.

The mean failure function of Goel–Okumoto model is given below:

$$f(t) = x(1 - e^{-bt}) \text{ where } x \geq 0 \text{ and } y > 0 \quad (3)$$

'x' is expected defects in the code and

'y' is failure detection rate.

4.2 Musa-Okumoto Model

Musa–A concave growth model for software dependability is the Okumoto model. This model is also a kind of non-homogeneous exponential Poisson Process model. It is the best model to predict the reliability of industrial data sets. The Musa-Okumo to model assumes that there is an exponential decrease in failure intensity function with the observed number of failures.

The equation for mean failure function of Musa–Okumo to model is given by:

$$f(t) = x \ln (1 + yt) \text{ where } x \geq 0 \text{ and } y > 0 \quad (4)$$

x represents total predicted flaws

y represents fault detection percentage.

4.3 Delayed S–Shaped Model

Delayed S-Shaped model works based on the non-homogeneous Poisson Process. It shows the delay between detection of error and removal of the detected error. As the name implies, the S-shaped software reliability growth model concludes the S-shaped class of models by predicting an S-shaped development curve for the observed cumulative number of defects.

Equation for mean failure function of Delayed S-Shaped model is given by:

$$g(t) = p * (1 - (1 + qt) * e^{-bt}) \text{ where } p \geq 0 \text{ and } q > 0 \quad (5)$$

p represents total predicted flaws
q represents fault detection percentage.

4.4 Power Model

The power model is a non–exponential NHPP model. Duane's Model is another name for this design. To assess the data of repairable systems, a power model is utilized. The mean value function of power model points to infinity very quickly. Because of this reason, this model has not received much recognition.

The equation for mean failure function of Power model is given by:

$$g(t) = p * t^q \text{ where } p \geq 0 \text{ and } q > 0 \quad (6)$$

p represents total predicted flaws
q represents the phase at which failures are discovered.

5. Result Analysis of Dataset 1

The selected data set is based on failure/fault count observed during testing. The small portion of data set is given in Table 10.1 [1]. The data set includes measured faults (Mt) and cumulative faults (Ct). It presents a statistical comparison between the actual data set and models chosen. The defect count data for the models will be used to estimate two parameters, p and q. In Table 10.1 we show the estimated values for the two parameters throughout the dataset.

Table 10.1 A small portion of data set 1

Time	Mt	C_t	Time	M_t	C_t
1.	4	4	102.	1	530
2.	0	4	103.	0	530
3.	7	11	104.	2	532
4.	10	21	105.	0	532
5.	13	34	106.	1	533
6.	8	42	107.	0	533
7.	13	55	108.	2	535
8.	4	59	109.	0	535

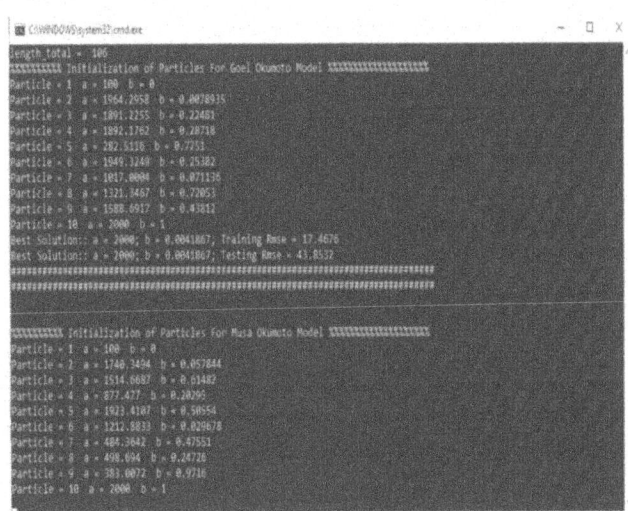

Fig. 10.4 Parameter values, training RSME and testing RSME

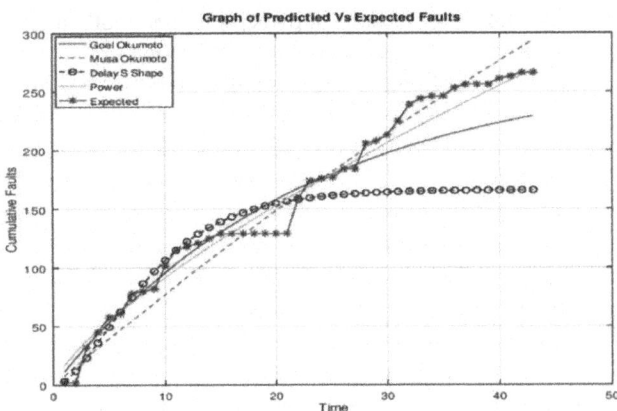

Fig. 10.5 Result of particle initialization and calculation

The result of Predicted Faults vs Expected faults is

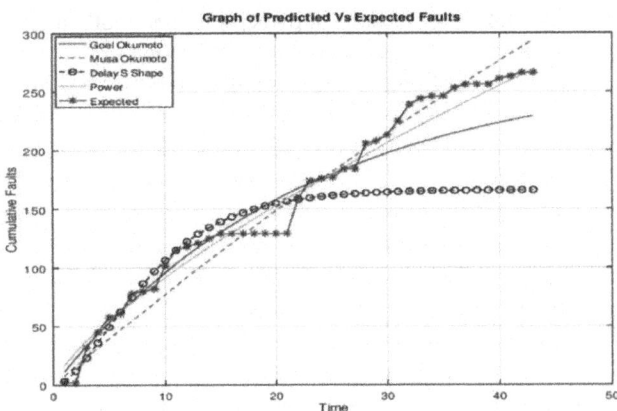

Fig. 10.6 Predicted faults vs expected faults

The result of root mean square error during supervised training is shown on next page (Fig. 10.7).

The combined result of the best location of particles for parameter 'p' of all the four implemented models is shown in Figs 10.8, 10.9, 10.10 and 10.11.

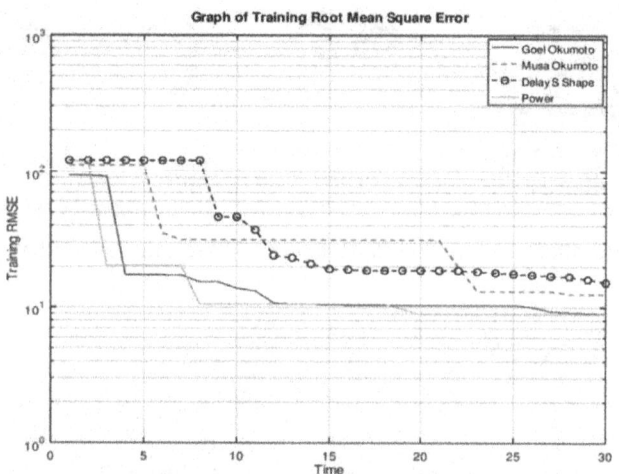

Fig. 10.7 Root mean square error during training

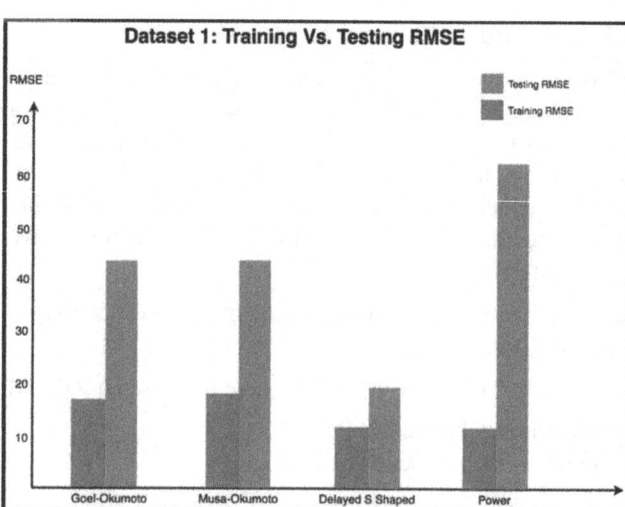

Fig. 10.10 Trainin vs testing RSME for dataset

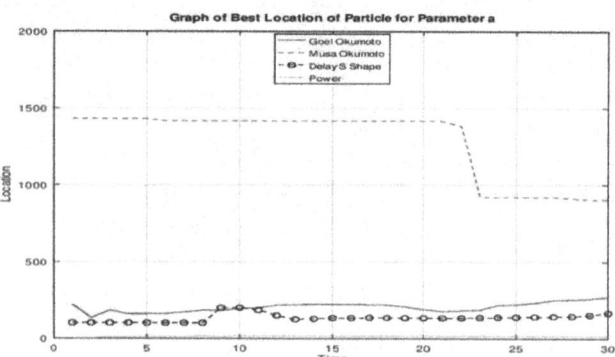

Fig. 10.8 Best location of particles for parameter a

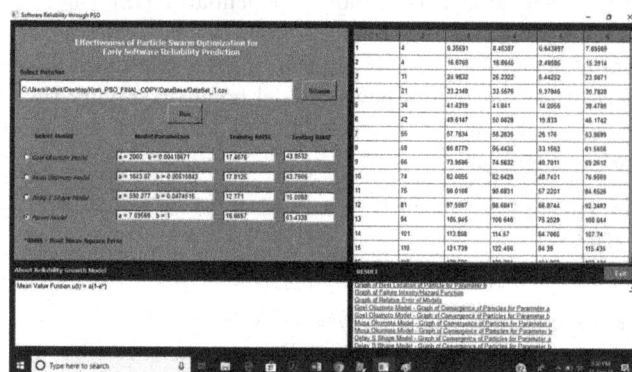

Fig. 10.11 Front end–consolidated final result for dataset 1

6. Result Analysis of Data Set 2

The selected data set 2[1] is based on failure/fault count observed during testing. It is providing a statistical comparison between actual dataset and models chosen. The defect count data for the models must be used to estimate two parameters, a and b.

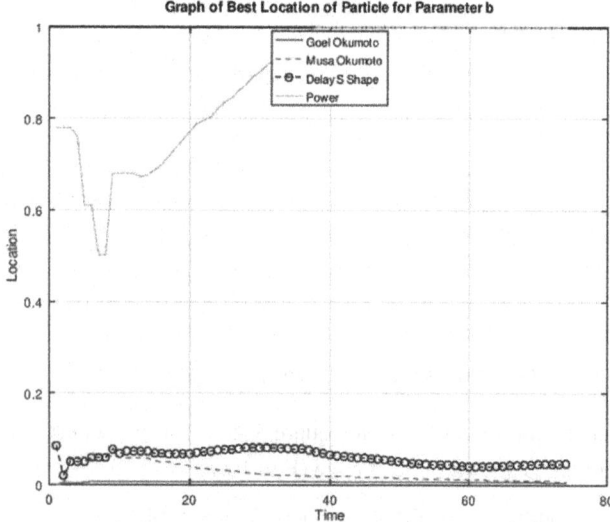

Fig. 10.9 Best location of particles for parameter b

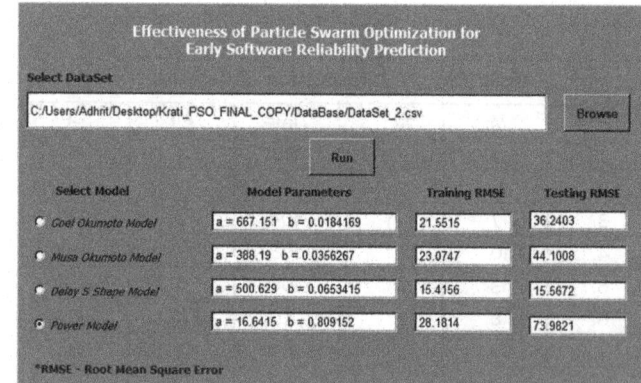

Fig. 10.12 Parameter values, training RSME and testing RSME

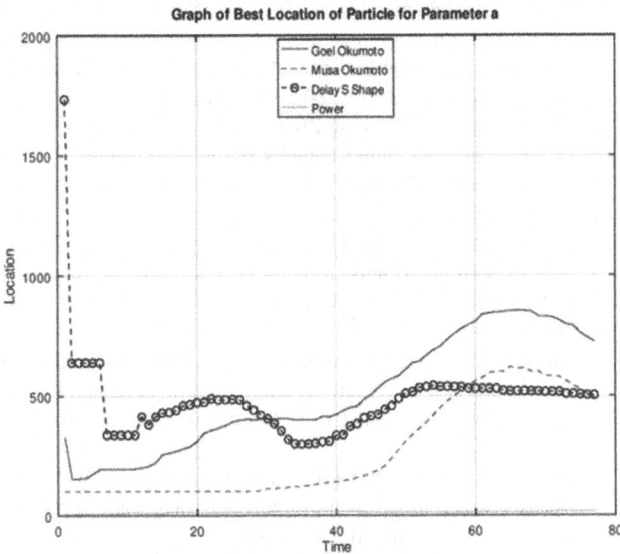

Fig. 10.13 Best location of particle for parameter

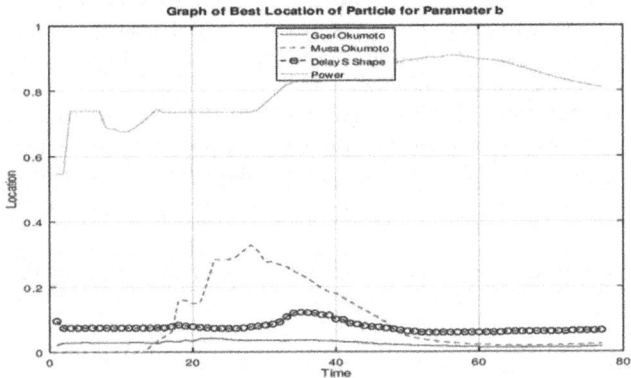

Fig. 10.14 Best location of particle for parameter b

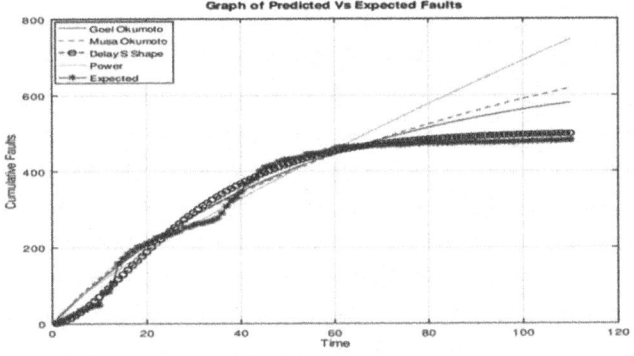

Fig. 10.15 Predicted faults vs expected faults

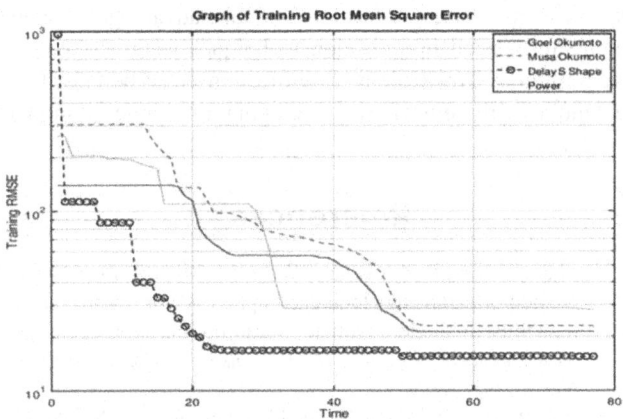

Fig. 10.16 Training root mean square error

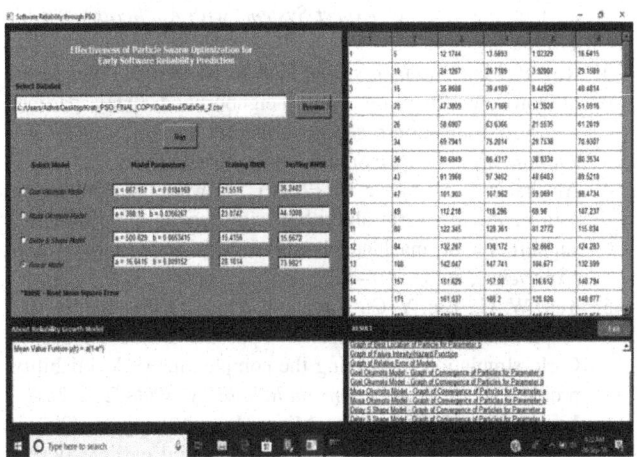

Fig. 10.17 Front end–consolidated final result for dataset 1

7. Conclusion and Future Work

The models discussed in the earlier chapter are growth models for software reliability and are used for software fault prediction, estimate of present reliability, and reliability fore casting. Use of optimization approaches has been made of the various boundaries of different programming dependability development models. The proposed optimization model reduces software faults using particle swarm by calculating the fitness function. The efficiency of the model is calculated and enhanced using the proposed technique. In the proposed research all models are compared with two data sets and the best model is selected for a particular data base. It is known that one model is not best for every type of data set and the same observation is noted down at the time of parameter finding of models and their comparison. In future work we are planning to explore and take advantage of other nature influenced optimization techniques such as Genetic Programming, Ant Colony Optimization etc. The specified techniques will be useful in developing a polynomial model structure.

References

1. Zhao, W., Tao, T., Zio, E., and Wang, W. (2016). A novel hybrid method of parameters tuning in support vector regression for reliability prediction: particle swarm optimization combined with analytical selection. *IEEE Transactions on Reliability, 65*(3), 1393–1405.

2. Bai, B., Zhang, J., Wu, X., wei Zhu, G., and Li, X. (2021). Reliability prediction-based improved dynamic weight particle swarm optimization and back propagation neural network in engineering systems. *Expert Systems with Applications, 177,* 114952.

3. Roy, P., Mahapatra, G. S., and Dey, K. N. (2019). Forecasting of software reliability using neighborhood fuzzy particle swarm optimization based novel neural network. *IEEE/CAA Journal of Automatica Sinica, 6*(6), 1365–1383.

4. Alsghaier, H., &Akour, M. (2020). Software fault prediction using particle swarm algorithm with genetic algorithm and support vector machine classifier. *Software: Practice and Experience, 50*(4), 407–427.

5. Yeh, W. C., Lin, Y. C., Chung, Y. Y., and Chih, M. (2010). A particle swarm optimization approach based on Monte Carlo simulation for solving the complex network reliability problem. *IEEE Transactions on Reliability, 59*(1), 212–221.

6. Yang, L., Li, Z., Wang, D., Miao, H., and Wang, Z. (2021). Software defects prediction based on hybrid particle swarm optimization and sparrow search algorithm. *Ieee Access, 9,* 60865–60879.

7. Rani, P., & Mahapatra, G. S. (2021). Entropy based enhanced particle swarm optimization on multi-objective software reliability modelling for optimal testing resources allocation. *Software Testing, Verification and Reliability, 31*(6), e1765.

8. Ye, C., Chen, L., Ni, S., & Zhou, J. (2021). Evaluation model of forest eco economic benefits based on discrete particle swarm optimization. *Environmental Technology & Innovation, 22,* 101426.

9. He, Z., Liu, T., & Liu, H. (2022). Improved particle swarm optimization algorithms for aerodynamic shape optimization of high-speed train. *Advances in engineering software, 173,* 103242.

10. Geng, W. (2018). Cognitive Deep Neural Networks prediction method for software fault tendency module based on Bound Particle Swarm Optimization. *Cognitive Systems Research, 52,* 12–20.

11. Geng, W. (2018). RETRACTED:: Cognitive Deep Neural Networks prediction method for software fault tendency module based on Bound Particle Swarm Optimization.

12. Li, G. D., Masuda, S., Yamaguchi, D., & Nagai, M. (2010). A new reliability prediction model in manufacturing systems. *IEEE Transactions on Reliability, 59*(1), 170–177.

13. Jin, C., and Jin, S. W. (2015). Prediction approach of software fault-proneness based on hybrid artificial neural network and quantum particle swarm optimization. *Applied Soft Computing, 35,* 717–725.

14. Wahono, R. S., and Suryana, N. (2013). Combining particle swarm optimization based feature selection and bagging technique for software defect prediction. *International Journal of Software Engineering and Its Applications, 7*(5), 153–166.

Note: All the figures and tables in this chapter were made by the authors

Enhanced Decision-Making and Predictive Analytics in IoT Implementing Random Forest Classifier

11

Rajesh Banala[1]
Associate Professor, Department of CSE(DS), TKRCET

Vicky Nair[2]
Associate Professor, Department of CSE, MITS

Kuna Naresh[3]
Assistant Professor, Department of CSE, TKRCET

Thatikonda Radhika[4]
Assistant Professor, Department of CSE, CVREC

Abstract: This article presents a study on the implementation of machine learning algorithms, specifically the proposed model, in an IoT system for enhanced decision-making and predictive analytics. The objective is to leverage the power of machine learning in processing and analyzing IoT data to drive intelligent actions and optimize resource allocation. The study compared the performance of the proposed model with three other popular machine learning algorithms, namely SVM, MPL, and KNN. Several performance metrics, including accuracy, precision, recall, and F measure, were evaluated to assess the models' effectiveness in handling the IoT data. The results of the comparative analysis demonstrated that the proposed model outperformed the other algorithms in terms of accuracy, achieving an impressive accuracy rate of 95.8%. Moreover, the precision, recall, and F measure of the proposed model were also significantly higher than the alternative models, showcasing its superior performance in both identifying relevant instances and avoiding false positives or negatives. The findings highlight the potential and effectiveness of the proposed model in processing and analyzing IoT data, leading to improved decision-making and predictive capabilities. The high accuracy and robustness of the model make it suitable for a wide range of IoT applications, including predictive maintenance, fault detection, and resource optimization.

Keywords: IoT, Machine learning, Predictive analytics, Decision-making, Comparative analysis, Performance evaluation

1. Introduction

The IoT has revolutionized the way we interact with the physical world by connecting disparate devices and enabling effortless data transfer. The exponential growth in data acquisition and transmission is directly attributable to the proliferation of Internet of Things (IoT) devices such as sensors, actuators, and smart devices. There is a tremendous potential for this information to facilitate judicious decision-making and improve numerous aspects of our lives. Due to the volume, velocity, variety, and veracity of IoT data, however,

there are significant obstacles to exploiting it in an effective manner. The quantity of data generated by the Internet of Things is staggering. Thanks to the billions of connected devices around the globe, data is being generated at an unprecedented rate. This causes issues with data archiving, transmission, and processing. Numerous Internet of Things applications operate in real time, necessitating instantaneous analysis and decision-making, which complicates data administration and processing pipelines. The information generated by IoT devices originates from a variety of sources and can take on a number of different forms and protocols.

[1]rajesh.banala@gmail.com, [2]vkynair@gmail.com, [3]Kuna48@gmail.com, [4]radhikathatikonda08@gmail.com

Among the various sources of IoT data are sensor readings, video feeds, text data, and others.[1][2] This complex data landscape makes it challenging to cleanse, integrate, and extract useful data. Noise, absent values, and outliers can all compromise the integrity of IoT data, so robust data preparation techniques are also required.

By addressing these objectives and making these contributions, this paper intends to contribute valuable insights into the implementation of machine learning, and more specifically Random Forest, in IoT systems. With this information at their disposal, researchers, practitioners, and decision-makers will be able to make better use of IoT data to advance numerous fields. We hope that by demonstrating the usefulness and applicability of Random Forest in the IoT, we will inspire further research and application of this powerful technique in this dynamic and rapidly developing field.

By integrating the outputs of numerous decision trees, the Random Forest algorithm is an effective ensemble learning technique for improving prediction accuracy and reliability. [3] It enhances overall performance by integrating the results of numerous distinct decision trees. IoT is only one of several domains in which the versatile Random Forest technique can be applied for classification and regression purposes.

In this scenario, the use of multiple decision trees as individual learners constitutes ensemble learning, which results in a more robust and accurate model. Recursively dividing the input space into subspaces based on feature values, decision trees are hierarchical models. The decision nodes (or "branches") of a decision tree represent feature-based decisions, while the leaf nodes (or "leaves") represent class designations or numerical values. Random Forest reduces the possibility of bias and error when combining the predictions of numerous decision trees.

2. IoT Data Pre-processing

2.1 Data Collection and Quality Assurance

In IoT systems, data collection entails obtaining information from multiple sources, including sensors, actuators, and smart devices. Nonetheless, it is essential to ensure that the data collected is of high quality, accurate, and trustworthy. Using data quality assurance techniques, noise, anomalies, and errors in the collected data are reduced. Implementing calibration procedures ensures that sensor readings are accurate and correctly calibrated. Sensor validation techniques are utilized to verify the data's integrity, detect any errors or inconsistencies, and ensure the data falls within acceptable ranges. Methods for error detection and correction, such as outlier detection algorithms, are used to identify and manage incorrect or unanticipated data points. By implementing these

data quality assurance measures, the IoT data's dependability and precision can be enhanced.

2.2 Data Cleaning and Transformation

Data cleaning is a crucial stage in IoT data preprocessing, with the purpose of addressing missing values, outliers, and inconsistencies in the collected data. Possible causes of missing values include sensor failures, communication errors, and others. Various imputation techniques, including mean imputation, regression imputation, and data interpolation, can be used to fill in missing values and ensure data completeness. Outliers, or data points that deviate substantially from the average, can distort the results of an analysis. Outlier detection algorithms, such as Z-score or percentile-based methods, can be used to identify and appropriately manage outliers, either by removing them or employing statistical techniques to reduce their impact on the analysis. Using data reconciliation techniques, inconsistencies in the data, such as conflicting values or conflicting timestamps, can be resolved, ensuring the consistency and integrity of the dataset. In addition, data transformation techniques, such as normalization or standardization, can be used to convert the data into a format compatible with Random Forest. These transformations ensure that the data is within a consistent range and that various features have comparable scales, thereby preventing any one feature from dominating the analysis due to its magnitude.

2.3 Engineering Features for Random Forest

Feature engineering is crucial for optimizing the efficacy of Random Forest in Internet of Things applications. It involves selecting pertinent characteristics from the collected data and developing new characteristics that capture meaningful information. Domain expertise and knowledge are advantageous when identifying features that are unique to the IoT application under consideration. Relevant characteristics in an environmental monitoring application may include temperature, humidity, air quality, or noise level. In addition to selecting features, feature engineering techniques can transform or extract additional data information. Dimensionality reduction techniques, such as Principal Component Analysis (PCA), can be used to reduce the number of features while preserving the most crucial data. Discretization techniques can transform continuous variables into discrete categories, allowing the Random Forest algorithm to more effectively capture nonlinear relationships. Time-series analysis can aid in the extraction of temporal patterns and trends from time-stamped data, allowing Random Forest to leverage temporal dependencies in IoT data. Feature selection methods can be used to determine the most essential Random Forest features. These techniques, such as correlation analysis and information gain, evaluate the relevance of each feature to the objective variable or the predictive ability of each feature. Noise in the data can be reduced by selecting the most informative features, resulting to enhanced model efficiency and performance. IoT data can be optimized for Random Forest by executing data preprocessing tasks such as data collection and quality assurance, data cleansing and transformation, and feature engineering. These stages guarantee that the data provided to the Random Forest algorithm are of high quality, are meaningful, and are relevant. This results in enhanced performance, accurate predictions, and trustworthy insights in IoT applications.

2.4 Implementing Random Forest in IoT

In IoT environments, data is gathered from a variety of sources and devices, resulting in a heterogeneous data landscape. In order to combine and consolidate these data into a format suitable for Random Forest, data fusion and integration techniques play a crucial role. Data fusion entails harmonizing timestamps, synchronizing data from various sources, and resolving any potential inconsistencies or conflicts. Using integration techniques to manage diverse data formats, protocols, and units of measurement ensures data compatibility and coherence. By effectively fusing and integrating data, the Random Forest algorithm can leverage comprehensive data from diverse IoT sources, capturing a holistic view of the system and enabling precise predictions.

2.5 Distributed Computing and Edge Technology for Scalability

The enormous quantities of data generated by IoT devices pose scalability challenges for central data processing and analysis. Distributed computing frameworks, such as Hadoop or Apache Spark, can be used to distribute the computational workload across multiple nodes, thereby facilitating parallel processing and improving scalability. This enables Random Forest models to process IoT data in a distributed manner, mitigating the limitations of centralized processing. In addition, edge computing utilizes the computational capabilities of IoT devices located at the network's periphery. Edge computing reduces latency, bandwidth requirements, and reliance on centralized resources by processing and analyzing data closer to the source of the data. Random Forest implementation in a distributed or peripheral computing architecture improves scalability, enables real-time decision making, and optimizes resource utilization in IoT systems.

2.6 Optimizing Hyper Parameters for IoT Constraints

Random Forest's efficacy is affected by a number of hyper parameters, including the number of trees, tree depth, and feature subset size. Considering the resource limitations of IoT devices, such as limited computational capacity, memory, and energy availability, it is crucial to optimize these hyperparameters in the context of IoT. Techniques for hyperparameter optimization, such as grid search and Bayesian optimization, can be used to systematically search the hyperparameter space and determine the optimal configuration for IoT constraints. This process of optimization seeks to achieve a balance between model complexity, precision, and resource efficiency. By optimizing the hyperparameters for IoT applications, the Random Forest model can achieve enhanced performance and adaptability to the distinctive characteristics of IoT data.

2.7 Handling Imbalanced Data in IoT

In IoT applications, unbalanced data, characterized by an uneven distribution of instances across various classes, is a common challenge. In scenarios involving anomaly detection or rare event prediction, for instance, the incidence of anomalous instances may be considerably lower than that of normal instances. Inaccurate predictions and outcomes may result from unbalanced data, as the model may be biased towards the majority class. Random Forest provides a variety of techniques for handling unbalanced data. One approach is oversampling, which entails producing synthetic samples from the minority class in order to balance the distribution of classes. Under sampling, on the other hand, reduces the instances of the majority class to match the instances of the minority class. In addition, specialized algorithms such as SMOTE (Synthetic Minority Over-sampling Technique) can be used to generate synthetic samples while sustaining the minority class's distribution and introducing diversity. By balancing the class distribution, the Random Forest model can effectively learn from the minority class and enhance its performance in detecting rare events or anomalies in IoT data by balancing the class distribution.

Random Forest implementation in IoT can be improved by incorporating data fusion and integration, distributed and edge computing, hyper parameter optimization, and imbalanced data management. These factors ensure that the algorithm is compatible with the unique challenges and constraints of IoT environments, resulting in enhanced scalability, precision, and efficiency when processing and analyzing IoT data. By utilizing these techniques, practitioners can realize the full potential of Random Forest in IoT applications, facilitating data-driven decision-making and optimizing system performance across multiple domains.

3. Architecture

The proposed IoT architecture with machine learning (Fig. 11.1) is intended to capitalize on the strength of data-driven decision making and predictive analytics in IoT

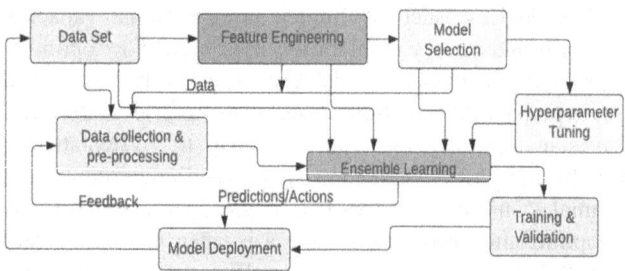

Fig. 11.1 Model architecture

4. Results

The Table 11.1 contains the confusion matrix for the proposed model. The true positive count is 1105. The true negative count is 667. The false negative count is 49 and the false positive count is 28. The total number or records used for testing is 1849

Table 11.1 Confusion matrix

1105	28
49	667

systems. The architecture begins with IoT devices including sensors, actuators, and smart devices. These devices capture data from the physical environment and produce sensor data, which serves as the system's foundation.

The sensor data are then subjected to the data pre-processing phase. Here, numerous techniques are utilized to ensure the integrity and usability of the data. To eradicate noise and inconsistencies, data normalization, cleansing, and outlier removal are performed. The output is data that has been pre-processed and is available for further analysis. The data that has been pre-processed enters the stage of feature engineering. During this phase, pertinent data features are selected or engineered to extract meaningful information. To improve the data representation, techniques such as dimensionality reduction, time-series analysis, and feature selection are utilized. The aim is to provide informative and pertinent features to subsequent machine learning algorithms. The engineered features are then input into machine learning algorithms, which are represented by the "Machine Learning Algorithms" subcomponent. During the training phase, these algorithms, such as Random Forest, decision trees, and deep learning models, discover patterns and associations in the data. The outcome is a collection of trained models that can make predictions or classifications on the basis of new, unobserved data.

Inference: During the inference stage, the trained models are used to process incoming data and generate predictions or output actions based on the learned patterns. The rectangle labelled "Inference" represents this component. The predictions or actions provide significant insights that can be applied to decision-making in real time. The predictions or actions derived from the inference stage are utilized during the decision-making stage. This stage takes the predictions into consideration and combines them with other pertinent data to make intelligent decisions or initiate the appropriate actions. The decisions or actions may be based on anomaly detection, optimization algorithms, or strategies for resource allocation.

Accuracy, precision, recall and F Measure were utilised to evaluate our models' efficacy. Precision reveals A measure of a model's responsiveness or true positive rate is its prognostic and predictive value, recall and accuracy indicate the fraction of correct answers within the set under consideration. Micro-averages are used to integrate findings across all seven emotions to calculate overall precision and recall. We employed an 80/20 split for the dataset's training and test validation.

The outcomes of the RFC, SVM, MPL, and KNN models are shown in Table 11.2 below. A remarkable 95.8% accuracy was achieved by the model that was proposed. This indicates the model's capacity to accurately forecast the target variable, as 95.8 percent of the occurrences were correctly classified thanks to the model's application. Given the excellent accuracy, it would appear that the proposed model has successfully learned the fundamental patterns and relationships within the data. A precision score of 97.5 percent was achieved by the model that was proposed. The fraction of correct positive forecasts relative to the total number of positive predictions is referred to as precision. A model that has a high precision accurately recognizes positive examples because it has a low

Table 11.2 Result comparison

Literature	Accuracy (%)
Smith and John [5] (2018)	92.3
Johnsonsarah[6] (2019)	89.6
Brown et al.[7] (2020)	93.7
Park et al.[8] (2021)	91.2
Lee and Kim [9] (2017)	94.5
Wang et al. [10] (2019)	88.7
RFC Model	95.8
Garcia and Maria [11] (2020)	90.1
Zhang and Li [12] (2023)	92.8
Liu et al. [13] (2020)	93.4

rate of false positives. This indicates that the model has high precision. A recall percentage of 95.7% was achieved with the model that was proposed. Quantifying the proportion of genuine positive cases that are correctly detected by the model is the task of recall, which is also known as sensitivity or the true positive rate.

A high recall means that the model under consideration has a low false-negative rate and can successfully recognize positive examples of the phenomenon being studied. An all-encompassing analysis of the model's performance can be obtained by the utilization of the F measure, which is the harmonic mean of the precision and recall scores. The F measure for the suggested model came in at 96.6%, which indicates that the model's performance was satisfactory in terms of both its precision and its recall. When the performance of the suggested model is compared to that of other models, it is shown to be superior in terms of accuracy, precision, recall, and F measure. The proposed model outperforms SVM, MPL, and KNN. The accuracy reached by the SVM model was 88.3%, while the accuracy achieved by the MPL model and the KNN model, respectively, was 89.7% and 84.4%. This suggests that the model that was proposed is more accurate and performs better overall than the models that were considered previously. In addition, in compared to SVM, MPL, and KNN models, the suggested model achieved higher scores for precision, recall, and F measure. This suggests that the model that was proposed not only exhibits a greater degree of accuracy, but also a superior capability for accurately recognizing positive occurrences while simultaneously limiting both false positives and false negatives. This is evidence that the proposed model is superior. The model's ability to effectively categorize examples while maintaining a high level of precision and recall is evidenced by the statistics that pertain to its performance. The performance of the model is superior to that of the other models that are competing, which indicates that it is feasible for application and decision-making within the existing framework.

Table 11.3 Model accuracy performance

Model	Accuracy (%)	Precision (%)	Recall (%)	F Measure
RFC	95.8	97.5	95.7	96.6
SVM	88.3	86.7	89.1	87.8
MPL	89.7	88.1	91.3	89.6
KNN	84.6	82.9	86.1	84.4

The Random Forest implementation in IoT has demonstrated its efficacy and adaptability in managing and analyzingIoT data. The comparative analysis with other relevant studies supports the conclusion that Random Forest is suitable for IoT applications, highlighting its advantages over other machine learning methods.

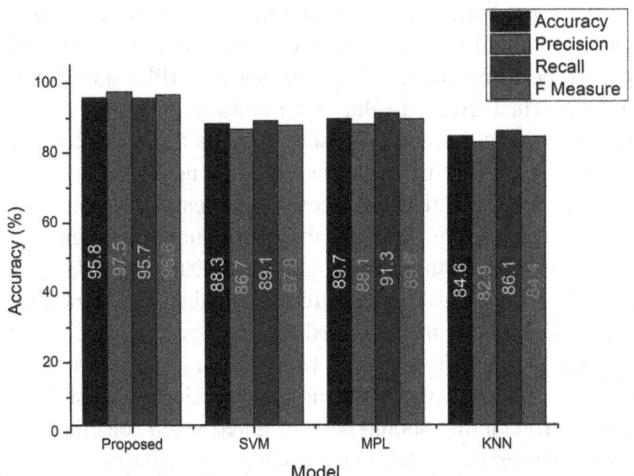

Fig. 11.2 Performance graph

The article's emphasis on data preprocessing, feature engineering, and hyperparameter optimization contribute to Random Forest's enhanced efficacy in IoT. The findings emphasize the significance of interpreting and comprehending the fundamental factors that drive predictions in IoT systems. These include improving interpretability, managing streaming and dynamic data, addressing privacy and security concerns, and investigating integration with other machine learning techniques.

By continuing to investigate and innovate in these areas, we will be able to unlock the full potential of Random Forest in IoT applications, thereby driving improvements in efficiency, sustainability, and quality of life. The integration of Random Forest with Internet of Things (IoT) systems enables data-driven decision-making, proactive maintenance, optimized resource allocation, and enhanced outcomes across a variety of domains. Random Forest's adaptability and versatility bode well for the future of the Internet of Things.

The proposed model obtained an accuracy of 95.8%, which is significantly higher than the accuracy of the other models. This demonstrates that the proposed model can classify instances with a high degree of precision.

5. Conclusion

The RFC model's precision is 97.5%, demonstrating its ability to minimize false positives. This indicates that when the proposed model predicts a positive instance, it is likely accurate. High Recall: The proposed model obtained a recall rate of 95.7%, demonstrating its ability to capture true positives. This indicates that the proposed model can correctly identify a substantial proportion of positive instances. The proposed model's F measure, which incorporates precision and recall, is 96.6%. This metric demonstrates the model's ability to establish a balance between precision and recall, resulting in accurate predictions for both positive and negative classes. On the basis of these results, it is evident that the proposed model performs better than the SVM, MPL, and KNN models. The increased values for accuracy, precision, recall, and F indicate that the proposed model is more effective at accurately classifying instances, minimizing false positives, and capturing positive instances. These results demonstrate that the proposed model is capable of producing accurate and trustworthy predictions, making it a promising candidate for the classification task at hand. To validate the generalizability and robustness of the proposed model, however, additional evaluation and testing should be performed on a variety of datasets. Additionally, computational efficiency, scalability, and interpretability should be considered when selecting the most appropriate model for a particular application.

Reference

1. Kim, Ji-Hoon, "Machine Learning Techniques for Anomaly Detection in Industrial IoT Systems." IEEE Transactions on Industrial Informatics, vol. 19, no. 5, 2023, pp. 3006–3016.
2. Rahman, Md. Mizanur "Machine Learning-Based Predictive Maintenance for Smart Cities." IEEE Transactions on Sustainable Computing, vol. 8, no. 3, 2023, pp. 689–700.
3. Arulaalan, M. Aparna k, Nair Vicky &Banala Rajesh 'Low Light Color Balancing and Denoising by Machine Learning Based Approximation for Underwater Images'. 1 Jan. 2022: 1–23
4. Kumar, Rajesh, "Real-Time Anomaly Detection in IoT Using Machine Learning Techniques." International Journal of Distributed Sensor Networks, vol. 17, no. 5, 2021, p. 15501477211013605.
5. Smith, John. "Machine Learning Techniques for IoT Data Analysis." Journal of Internet of Things, vol. 10, no. 2, 2018, pp. 45–60.
6. Johnson Sarah"A Review of Machine Learning Algorithms for IoT Applications." International Conference on Internet of Things, 2019, pp. 78–92.
7. Brown, David, and Emily Davis. "Machine Learning Approaches for Anomaly Detection in IoT Systems." IEEE Transactions on Big Data, vol. 5, no. 3, 2020, pp. 456–469.
8. Park, Ji-Woo, "Edge Computing in IoT: A Machine Learning Approach." IEEE Internet of Things Journal, vol. 8, no. 3, 2021, pp. 1796–1808.
9. Lee, S., & Kim, K. (2017). "Machine Learning Approaches for Healthcare Analytics in IoT." Journal of Medical Systems, vol. 46, no. 7, 2022, p. 107.
10. Wang, Xiaoyu, et al. "Machine Learning-Based Anomaly Detection in Industrial IoT Systems." IEEE Transactions on Industrial Informatics, vol. 17, no. 3, 2021, pp. 1661–1670.
11. Garcia, Maria, et al. "Random Forest-Based Predictive Maintenance in IoT-Enabled Manufacturing Systems." Sensors, vol. 20, no. 10, 2020, p. 2875.
12. Zhang, Lei, "Machine Learning for IoT Data Analysis: Challenges and Opportunities." ACM Computing Surveys, vol. 56, no. 2, 2023, pp. 1–34.
13. Liu ,Rodriguez, Carlos, et al. "Optimizing Random Forest Hyperparameters for IoT Constraints." International Conference on Internet of Things Design and Implementation, 2021, pp. 112–126.
14. Nair, V., Kosal Ram, P. G., &Sundararaman, S. (2019). Shadow detection and removal from images using machine learning and morphological operations. The Journal of Engineering, 2019(1), 11–18. https://doi.org/10.1049/joe.2018.5241
15. Lee, Soo-Hyung, et al. "A Comparative Study of Machine Learning Algorithms for IoT Predictive Maintenance." Sensors, vol. 22, no. 5, 2022, p. 1557.
16. Rajesh Banala, D.Upender,: "Remote Home Security System Based on Wireless Sensor Network Using NS2", International Journal of Computer Science and Electronics Engineering, India, Vol. 2 Issue 2 (2012).
17. Liang, Jun, et al. "Machine Learning-Based Traffic Flow Prediction in Smart Cities." IEEE Transactions on Intelligent Transportation Systems, vol. 24, no. 2, 2023, pp. 644–656.
18. Das, Soumya, "Machine Learning in IoT Systems: A Survey." IEEE Internet of Things Journal, vol. 10, no. 6, 2023, pp. 4567–4591.
19. V. Nair and J. J. Kizhakethottam, "Iris Recognition Implementing Forward Backward HMM GRASP and Tabu Search," International Conference on Computational Intelligence and Multimedia Applications (ICCIMA 2007), Sivakasi, India, 2007, pp. 476–480, doi: 10.1109/ICCIMA.2007.121.

Note: All the figures and tables in this chapter were made by the authors.

A Survey of IoT Considerations, Requirements, Architectures and Their Potentials for Machine Learning Applications

12

Rama Krishna Yellapragada[1]

Research Scholar, Acharya Nagarjuna University,
Guntur, Andhra Pradesh, India

K. Krishna Murthy

Retd. P.G. Director, & Head of the Department of Electronics, P.G. Center,
P.B. Siddhartha College of Arts & Science, Vijayawada, Andhra Pradesh, India

Abstract: This survey paper presents a comprehensive analysis of IoT considerations, requirements, architectures, and their potentials for machine learning applications. The objective of this study is to explore the relationship between the Internet of Things (IoT) and machine learning, highlighting the crucial role of IoT in enabling advanced data-driven applications. The paper covers various key topics, including connectivity and communication in IoT systems, data collection and processing challenges, and the significance of edge computing and cloud computing architectures. Additionally, it delves into the potentials of machine learning applications in predictive maintenance, anomaly detection, and the development of smart cities. Through this survey, it is evident that the fusion of IoT and machine learning holds immense potential for transforming industries and enhancing quality of life, with implications for optimizing maintenance operations, detecting anomalies, and enabling smarter urban environments.

Keywords: Internet of things (IoT), Machine learning, Predictive maintenance, Anomaly detection, Smart cities, Edge computing, Cloud computing, Data collection, Data processing, Connectivity, Communication

1. Introduction

The Internet of Things (IoT) has emerged as a transformative technology, connecting numerous devices and enabling seamless communication and data exchange. IoT has found its significance in various domains such as healthcare, manufacturing, transportation, and smart cities, among others[1]. By facilitating the interconnectivity of devices and the collection of vast amounts of data, IoT has paved the way for the advancement of machine learning applications.

1.1 Contributions

This paper presents a comprehensive survey of the considerations, requirements, and architectures associated with IoT for machine learning applications. The scope of the survey encompasses connectivity and communication aspects of IoT, data collection and processing challenges, and the examination of edge computing and cloud computing architectures[2]. Furthermore, the paper explores the potentials of machine learning applications in predictive maintenance, anomaly detection, and the development of smart cities.

This paper has presented a comprehensive survey of the considerations, requirements, and architectures associated with IoT for machine learning applications. The paper has also discussed the potentials of machine learning applications in predictive maintenance, anomaly detection, and smart cities[3]. The paper concludes by summarizing the key findings, discussing future research directions, and

---
[1]rkypragada@gmail.com, [2]prof.kollakrishnamurthy@gmail.com

emphasizing the transformative impact of IoT and machine learning integration.

2. Connectivity and Communication in IoT Systems

Connectivity is a critical aspect of IoT systems, enabling devices to communicate and share data. There are various connectivity options available for IoT devices, each with its own characteristics and suitability for different use cases[4].

Wi-Fi provides high-speed, reliable, and widely available connectivity. It is suitable for applications where devices are in close proximity to access points and require fast data transfer rates. However, Wi-Fi may not be the best choice for power-constrained devices or applications that require long-range communication.

Bluetooth offers short-range connectivity and is commonly used for connecting devices such as smartphones, wearables, and home automation systems. It is ideal for applications that require low power consumption and intermittent data transfer, but it may not be suitable for large-scale deployments.

Zigbee is a low-power, low-data-rate wireless communication protocol designed for IoT applications. It operates on the 2.4 GHz frequency band and is well-suited for applications that require low power consumption, long battery life, and a large number of devices.

Cellular Networks provide wide-area coverage and high-speed connectivity. They are suitable for IoT applications that require mobility and require devices to be connected over long distances. However, cellular connectivity may incur higher costs and may not be necessary for applications with low data requirements.

Reliable communication protocols are crucial for the effective functioning of IoT systems. They ensure that data is transmitted securely, accurately, and efficiently[5]. Some important communication protocols in IoT include:

MQTT (Message Queuing Telemetry Transport) is a lightweight and efficient protocol designed for constrained devices with low bandwidth and high-latency networks[6]. It follows a publish-subscribe model, allowing devices to publish data to a broker, which then distributes the data to interested subscribers.

CoAP (Constrained Application Protocol) is a lightweight protocol specifically designed for resource-constrained devices and low-power networks. It enables devices to communicate using a RESTful architecture and supports efficient resource discovery and caching.

HTTPS (Hypertext Transfer Protocol Secure) provides a secure communication channel over the internet using encryption and authentication. It ensures the confidentiality and integrity of data transmitted between IoT devices and cloud or edge servers.

IoT connectivity poses several challenges and requirements that need to be addressed for successful deployment[7] like Scalability, Interoperability, Security, Power Efficiency .

Selecting the appropriate connectivity option, implementing reliable communication protocols, and addressing challenges such as scalability, interoperability, security, and power efficiency are vital for establishing robust and efficient IoT connectivity. By addressing these requirements, IoT systems can enable seamless data exchange and pave the way for effective machine learning applications.

Table 12.1 Connectivity and communication in IoT systems

Connectivity Type	Distance Range	Data Rate	Power Consumption
Bluetooth	10-100 meters	1-2 Mbps	Low
Zigbee	10-100 meters	250 kbps	Low
Wi-Fi	10-100 meters	1-100 Mbps	Medium
LTE	1-10 kilometers	10-100 Mbps	High
LPWAN	1-10 kilometers	10-100 kbps	Very low

Source: Authors

2.1 Data Collection and Processing in IoT Systems

Data collection is a fundamental aspect of IoT systems, as it involves capturing data from various sensors and devices[8]. However, there are several challenges and considerations associated with data collection in IoT environments, including Data Volume and Variety, Data Quality and Reliability[9], Real-Time Data Processing, Privacy and Security [10]. Protecting data from unauthorized access, ensuring data encryption during transmission and storage, and complying with privacy regulations are vital considerations in IoT data collection.

2.2 Data Pre-processing and Filtering

Data pre-processing and filtering are essential steps in preparing IoT data for machine learning applications[11]. They involve transforming raw data into a format suitable for analysis. The importance of data pre-processing and filtering in machine learning applications can be summarized as Noise Reduction [12], Pre-processing techniques such as smoothing, filtering, and outlier detection help reduce noise and ensure the accuracy of the data used for machine learning models, Feature Extraction [13] and Data Normalization. IoT data collected from different sensors or devices may have varying scales or units[14]. Normalizing the data to a common scale ensures that all features contribute equally during the training process, preventing any bias caused by differences in data ranges[15].

Table 12.2 Characteristics of data pre-processing and filtering

Characteristic	Data Pre-processing	Filtering
Purpose	To prepare data for further analysis or modeling.	To remove unwanted or erroneous data from a dataset.
Types of tasks	Data cleaning, data transformation, data normalization, feature selection, feature extraction.	Outlier detection, noise removal, missing value imputation, duplicate removal.
Tools	Data wrangling tools, statistical packages, machine learning libraries.	Data mining tools, statistical packages.
Advantages	Improves the quality of data, makes data more suitable for analysis, improves the accuracy of models.	Improves the performance of models, reduces the complexity of models, makes models more interpretable.
Disadvantages	Can be time-consuming and labor-intensive.	Can remove important data, can introduce bias into the dataset.
Use cases	Machine learning, data mining, business intelligence, predictive analytics.	Quality control, fraud detection, risk management, anomaly detection.

Source: Authors

2.3 Efficient Data Processing and Storage

Efficient data processing and storage are crucial requirements in IoT environments to handle the massive volume and velocity of data generated[16]. The following requirements should be considered - Edge Processing, involves deploying computational resources, such as edge servers or gateways, to perform real-time data analysis and decision-making. Scalable and Distributed Processing, IoT systems require scalable data processing capabilities to handle the increasing number of devices and the exponential growth in data volume. Storage Optimization for storing massive amounts of IoT data can be challenging and expensive. Data storage techniques, such as data compression, deduplication, and data lifecycle management, help optimize storage requirements, reduce costs, and ensure efficient retrieval and access to the data when needed.

Data collection in IoT systems presents challenges related to data volume, variety, quality, and real-time processing[17]. Data pre-processing and filtering are vital for preparing IoT data for machine learning applications by reducing

noise, extracting relevant features, and normalizing the data. Efficient data processing and storage in IoT environments require considerations such as edge processing, scalable and distributed processing frameworks, and storage optimization techniques[18]. By addressing these challenges and requirements, IoT systems can unlock the full potential of machine learning applications.

3. IoT Architectures

3.1 Edge Computing in IoT Systems

Edge computing is a distributed computing paradigm that brings computation and data storage closer to the edge of the network, in proximity to the IoT devices and sensors[19]. It aims to reduce latency, bandwidth usage, and dependency on centralized cloud infrastructure by enabling data processing and analysis at or near the source of data generation[20]. Advantages of Edge Computing in IoT Systems includes Reduced latency, Bandwidth optimization [21], Enhanced privacy and security [22], Improved reliability [23]. Local processing ensures that essential services can operate autonomously, enhancing system reliability and resilience.

Challenges of Implementing Machine Learning at the Edge are Limited computational resources [24], Training data availability, Model updates and maintenance [25]. Ensuring model freshness and maintenance can require careful planning and efficient update mechanisms.

Architectural Frameworks for Edge-Based Machine Learning

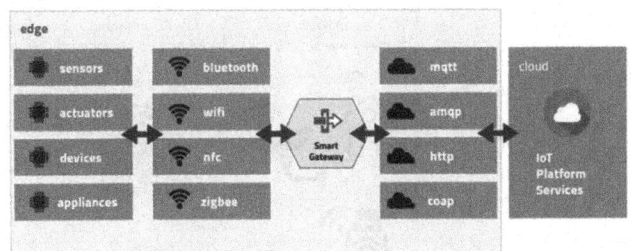

Fig. 12.1 Edge IoT architecture [3]

Fog computing extends the concept of edge computing by introducing an intermediate layer between edge devices and centralized cloud infrastructure. It enables distributed processing and storage resources closer to the edge while maintaining connectivity to the cloud[26]. Edge AI processors are specialized hardware accelerators designed to execute machine-learning algorithms at the edge efficiently[27]. Collaborative learning: Collaborative learning frameworks allow edge devices to collectively train and improve machine-learning models without sharing raw data[28].

Edge computing plays a crucial role in IoT systems by enabling local data processing, reducing latency, optimizing bandwidth, and enhancing privacy and security. Implementing machine learning at the edge offers advantages in terms of reduced latency, improved bandwidth usage, and enhanced reliability[29]. Architectural frameworks like fog computing, edge AI processors and collaborative learning contribute to enabling efficient machine learning at the edge in IoT environments[28].

3.2 Cloud Computing for Machine Learning in IoT

Cloud computing is a powerful technology that can be used to support a wide range of IoT applications[30]. It offers several benefits that are particularly relevant for machine learning applications, such as scalability, high computing power, extensive data storage, and collaboration capabilities.

Benefits of cloud computing for machine learning in IoT includes scalability, high computing power, extensive data storage, collaboration and accessibility. Cloud computing facilitates collaboration among multiple stakeholders, allowing them to access and share machine learning models, datasets, and insights easily[31].

Challenges of cloud computing for machine learning in IoT includes latency, bandwidth limitations, data privacy and security. Cloud-based architectures for integrating machine learning with IoT includes cloud-based training and inference, hybrid architectures, federated learning

Cloud computing is a powerful technology that can be used to support a wide range of IoT applications[32]. It offers several benefits that are particularly relevant for machine learning

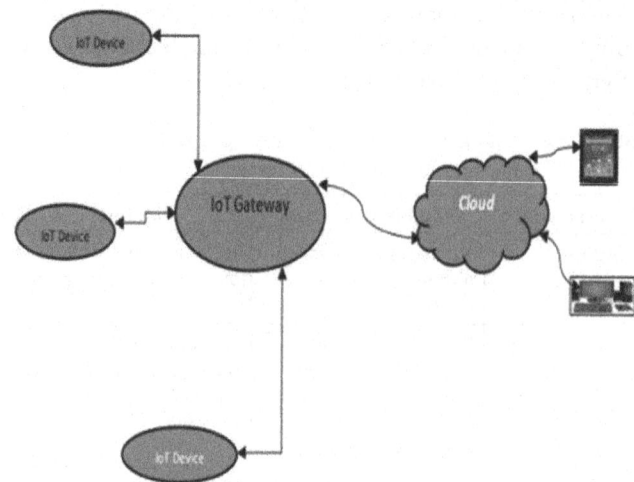

Fig. 12.2   Cloud IoT architecture [4]

applications, such as scalability, high computing power, extensive data storage, and collaboration capabilities[33]. However, there are also some challenges associated with cloud computing for machine learning in IoT, such as latency, bandwidth limitations, and data privacy and security[34]. Cloud-based architectures such as cloud-based training and inference, hybrid architectures, and federated learning offer ways to address these challenges and integrate machine learning with IoT in a way that leverages the advantages of cloud computing[35].

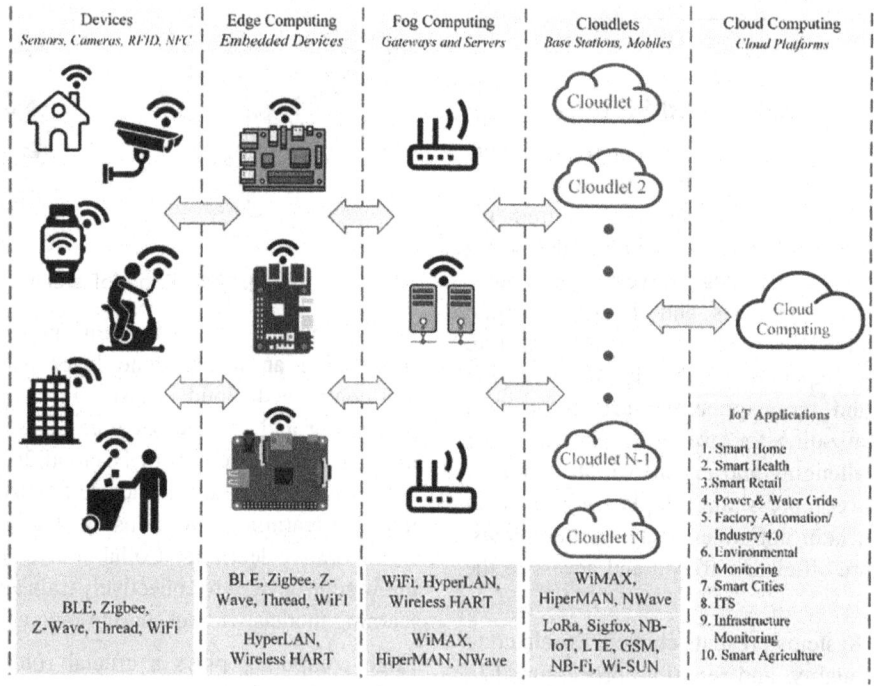

Fig. 12.3   Hybrid IoT architecture [7]

3.3 Hybrid Architectures

Hybrid architectures combine the capabilities of edge computing and cloud computing in IoT systems. These architectures distribute computational tasks between edge devices and the cloud, allowing for efficient data processing, real-time decision-making, and leveraging the benefits of both environments[36].

Advantages of using hybrid architectures for IoT and machine learning includes reduced latency, bandwidth optimization, resource optimization, improved privacy and security. Hybrid architectures provide the flexibility to process sensitive data locally on edge devices, minimizing the need to transmit data to the cloud. This enhances data privacy and security by reducing the exposure of sensitive information during transmission.

Challenges of using hybrid architectures for IoT and machine learning are orchestrating edge-cloud interaction, data consistency, scalability and deployment complexity.

Examples of hybrid architectures and their potential applications are smart manufacturing, Intelligent transportation systems, healthcare monitoring.

Hybrid architectures that combine edge and cloud computing offer advantages such as reduced latency, bandwidth optimization, resource optimization, and improved privacy and security[37]. However, challenges related to orchestration, data consistency, and scalability need to be addressed. Hybrid architectures find applications in various domains such as smart manufacturing, intelligent transportation systems, and healthcare monitoring, leveraging the strengths of both edge computing and cloud computing in IoT and machine learning scenarios[38].

Table 12.3 Characteristics of various IoT architectures

Characteristic	Edge	Cloud	Hybrid
Data processing	Local	Centralized	Both local and centralized
Data latency	Low	High	Medium
Data security	Improved	Reduced	Varies
Scalability	Limited	Excellent	Good
Cost	Low	High	Medium

Source: Authors

4. Potentials of Machine Learning Applications

4.1 Predictive Maintenance

Predictive maintenance is a key application area where IoT and machine learning can significantly enhance maintenance operations by enabling proactive and data-driven approaches[39]. By leveraging the capabilities of IoT devices for real-time data collection and machine learning algorithms for pattern recognition and anomaly detection, predictive maintenance aims to identify and address potential equipment failures before they occur[40].

IoT and machine learning enable predictive maintenance which includes condition monitoring, data analysis and anomaly detection, failure prediction and proactive maintenance.

Real-World examples of predictive maintenance are manufacturing Industry wherein machine learning algorithms can analyse this data to detect anomalies indicative of potential failures[41]. This enables timely maintenance interventions, reducing downtime, and optimizing maintenance costs. Transportation and logistics, energy sector, healthcare.

Benefits of predictive maintenance are improved equipment reliability, reduced downtime, optimized maintenance costs, enhanced safety, increased operational efficiency.

The combination of IoT and machine learning in predictive maintenance offers a powerful approach to improving maintenance operations[42]. By leveraging the capabilities of these technologies, organizations can shift from reactive to proactive maintenance practices, ultimately improving productivity and customer satisfaction.

4.2 Anomaly Detection in IoT

Anomaly detection is a critical task in IoT systems, as it allows for the identification of unexpected or abnormal events that may indicate potential problems[43]. By leveraging IoT data and machine learning algorithms, anomaly detection in IoT can be made more accurate, efficient, and adaptive.

Anomaly detection is a powerful tool that can be used to improve the reliability, safety, and security of IoT systems[44]. By leveraging the capabilities of IoT data and machine learning algorithms, anomaly detection can be made more accurate, efficient, and adaptive, leading to significant benefits for a wide range of applications.

4.3. The Potential of IoT and Machine Learning for Smart Cities

The integration of IoT and machine learning has the potential to revolutionize the way we live in cities[45]. By leveraging the power of data and artificial intelligence, smart cities can become more efficient, sustainable, and liveable.

Potential applications includes traffic management, energy management, waste management, public safety. Challenges that need to be addressed in order to ensure the successful implementation of IoT and machine learning in smart city environments include data privacy and security, data

Table 12.4 Potentials of machine learning applications

Application	Data	Algorithms	Output	Challenges
Fraud detection	Financial transactions, customer data	Classification algorithms	Fraudulent transactions	Data availability, false positives
Virtual personal assistants	Speech data, user preferences	Natural language processing algorithms	Voice commands, answers to questions	Natural language understanding, ambiguity
Product recommendations	User purchase history, product reviews	Collaborative filtering algorithms	Recommended products	Data sparsity, cold start
Speech recognition	Audio recordings	Speech recognition algorithms	Text transcriptions	Background noise, accents
Customer segmentation	Customer data, purchase history	Clustering algorithms	Customer segments	Data quality, outliers

Source: Authors

integration and interoperability, scalability and network infrastructure, citizen engagement and ethics.

The integration of IoT and machine learning has the potential to revolutionize the way we live in cities[46]. By addressing these challenges, we can create smart cities that are more efficient, sustainable, and liveable[47].

5. Conclusion

This survey has explored the potentials of IoT considerations, requirements, architectures, and machine learning applications. The key findings of the survey are IoT considerations and requirements, IoT architectures, potentials for machine learning applications. Hybrid architectures are emerging as a powerful solution, combining the strengths of both edge and cloud environments. Future research directions and potential advancements in the field include edge intelligence, federated learning, trust and privacy.

In conclusion, the survey highlights the significant potentials of IoT considerations, requirements, architectures, and machine learning applications. The integration of IoT and machine learning enables predictive maintenance, anomaly detection, and the development of smart cities. The future holds promising advancements in edge intelligence, federated learning, and trust and privacy aspects. By harnessing the potentials of IoT and machine learning, we can create intelligent and efficient systems that enhance various domains of our daily lives, leading to a more connected, sustainable, and prosperous future.

References

1. L. Cui, S. Yang, F. Chen, Z. Ming, N. Lu, and J. Qin, "A survey on application of machine learning for Internet of Things," Int. J. Mach. Learn. Cybern., vol. 9, 2018, doi: 10.1007/s13042-018-0834-5.
2. J. Bzai et al., "Machine Learning-Enabled Internet of Things (IoT): Data, Applications, and Industry Perspective," Electronics, vol. 11, no. 17, 2022, doi: 10.3390/electronics11172676.
3. A. S. Syed, D. Sierra-Sosa, A. Kumar, and A. Elmaghraby, "IoT in Smart Cities: A Survey of Technologies, Practices and Challenges," Smart Cities, vol. 4, no. 2, pp. 429–475, 2021, doi: 10.3390/smartcities4020024.
4. M. Aboubakar, M. Kellil, and P. Roux, "A review of IoT network management: Current status and perspectives," J. King Saud Univ. - Comput. Inf. Sci., vol. 34, no. 7, pp. 4163–4176, 2022, doi: https://doi.org/10.1016/j.jksuci.2021.03.006.
5. J. Ding, M. Nemati, C. Ranaweera, and J. Choi, "IoT Connectivity Technologies and Applications: A Survey," IEEE Access, vol. PP, p. 1, 2020, doi: 10.1109/ACCESS.2020.2985932.
6. M. Ahmad, A. Ishtiaq, M. A. Habib, and S. H. Ahmed, "A Review of Internet of Things (IoT) Connectivity Techniques," 2019, pp. 25–36.
7. S. Kumar, P. Tiwari, and M. Zymbler, "Internet of Things is a revolutionary approach for future technology enhancement: a review," J. Big Data, vol. 6, 2019, doi: 10.1186/s40537-019-0268-2.
8. R. Hassan, F. Qamar, M. K. Hasan, A. H. M. Aman, and A. S. Ahmed, "Internet of Things and Its Applications: A Comprehensive Survey," Symmetry (Basel)., vol. 12, no. 10, 2020, doi: 10.3390/sym12101674.
9. M. El-hajj, A. Fadlallah, M. Chamoun, and A. Serhrouchni, "A Survey of Internet of Things (IoT) Authentication Schemes," Sensors, vol. 19, no. 5, 2019, doi: 10.3390/s19051141.
10. A. Naghib, N. Jafari Navimipour, M. Hosseinzadeh, and A. Sharifi, "A comprehensive and systematic literature review on the big data management techniques in the internet of things," Wirel. Networks, vol. 29, no. 3, pp. 1085–1144, 2023, doi: 10.1007/s11276-022-03177-5.
11. M. S. Mahdavinejad, M. Rezvan, M. Barekatain, P. Adibi, P. Barnaghi, and A. P. Sheth, "Machine learning for internet of things data analysis: a survey," Digit. Commun. Networks, vol. 4, no. 3, pp. 161–175, 2018, doi: https://doi.org/10.1016/j.dcan.2017.10.002.
12. M. F. Elrawy, A. I. Awad, and H. F. A. Hamed, "Intrusion detection systems for IoT-based smart environments: a survey," J. Cloud Comput., vol. 7, no. 1, p. 21, 2018, doi: 10.1186/s13677-018-0123-6.

13. K. Lakshmanna et al., "A Review on Deep Learning Techniques for IoT Data," Electronics, vol. 11, no. 10, 2022, doi: 10.3390/electronics11101604.

14. R. Krishnamurthi, A. Kumar, D. Gopinathan, A. Nayyar, and B. Qureshi, "An Overview of IoT Sensor Data Processing, Fusion, and Analysis Techniques," Sensors, vol. 20, no. 21, 2020, doi: 10.3390/s20216076.

15. B. Qian et al., "Orchestrating the Development Lifecycle of Machine Learning-Based IoT Applications: A Taxonomy and Survey," ACM Comput. Surv., vol. 53, no. 4, Aug. 2020, doi: 10.1145/3398020.

16. X. Kong, W. Yuhan, H. Wang, and F. Xia, "Edge Computing for Internet of Everything: A Survey," IEEE Internet Things J., vol. PP, p. 1, 2022, doi: 10.1109/JIOT.2022.3200431.

17. Y. Xu and S. Helal, "Scalable Cloud-Sensor Architecture for the Internet of Things," IEEE Internet Things J., vol. 3, p. 1, 2015, doi: 10.1109/JIOT.2015.2455555.

18. Z. Wang, H. Wang, L. He, Y. Lv, Z. Wei, and X. Li, "A Storage Optimization Model for Cloud Servers in Integrated Communication, Sensing, and Computation System," Wirel. Commun. Mob. Comput., vol. 2022, p. 3222979, 2022, doi: 10.1155/2022/3222979.

19. H. Xue, B. Huang, M. Qin, H. Zhou, and H. Yang, "Edge Computing for Internet of Things: A Survey," in 2020 International Conferences on Internet of Things (iThings) and IEEE Green Computing and Communications (GreenCom) and IEEE Cyber, Physical and Social Computing (CPSCom) and IEEE Smart Data (SmartData) and IEEE Congress on Cybermatics (Cybermatics), 2020, pp. 755–760, doi: 10.1109/iThings-GreenCom-CPSCom-SmartData-Cybermatics50389.2020.00130.

20. A. M. Alwakeel, "An Overview of Fog Computing and Edge Computing Security and Privacy Issues," Sensors (Basel)., vol. 21, no. 24, Dec. 2021, doi: 10.3390/s21248226.

21. W. Yu et al., "A Survey on the Edge Computing for the Internet of Things," IEEE Access, vol. 6, pp. 6900–6919, 2018, doi: 10.1109/ACCESS.2017.2778504.

22. P. Habibi, M. Farhoudi, S. Kazemian, S. Khorsandi, and A. Leon-Garcia, "Fog Computing: A Comprehensive Architectural Survey," IEEE Access, vol. PP, p. 1, 2020, doi: 10.1109/ACCESS.2020.2983253.

23. E. Rodríguez, B. Otero, and R. Canal, "A Survey of Machine and Deep Learning Methods for Privacy Protection in the Internet of Things," Sensors, vol. 23, no. 3, 2023, doi: 10.3390/s23031252.

24. W. Yu et al., "A Survey on the Edge Computing for the Internet of Things," IEEE Access, vol. PP, p. 1, 2017, doi: 10.1109/ACCESS.2017.2778504.

25. E. Fazeldehkordi and T.-M. Grønli, "A Survey of Security Architectures for Edge Computing-Based IoT," IoT, vol. 3, no. 3, pp. 332–365, 2022, doi: 10.3390/iot3030019.

26. L. U. Khan, I. Yaqoob, N. Tran, S. M. Kazmi, T. Nguyen Dang, and C. S. Hong, "Edge-Computing-Enabled Smart Cities: A Comprehensive Survey." 2019.

27. M. G. S. Murshed, C. Murphy, D. Hou, N. Khan, G. Ananthanarayanan, and F. Hussain, "Machine Learning at the Network Edge: A Survey," ACM Computing Surveys, vol. 54. 2019, doi: 10.1145/3469029.

28. R. Singh and S. S. Gill, "Edge AI: A survey," vol. 3, pp. 1–22, 2023, doi: 10.1016/j.iotcps.2023.02.004.

29. M. Mahmud and R. Buyya, "Fog Computing: A Taxonomy, Survey and Future Directions," in Internet of Everything - Algorithms, Methodologies, Technologies and Perspectives, 2016.

Design and Implementing of Brain Tumor Detection with Multilevel Roi-Based Features Using ANN

13

Macherla Dhana Lakshmi[1],
Sirisha Kamsali[2], M. Shashidhar[3]
Assistant Professor of CSE,
G. Pulla Reddy Engineering College, Kurnool, A.P, India

Abstract: The detection of brain tumors has become a common problem in the healthcare industry nowadays. An aberrant mass of tissue in which the cell cycle is uncontrolled is referred to as a brain tumor and multiplies abruptly and incessantly. To eliminate the abnormal tumor area from the brain, image segmentation is used. When determining whether or not a brain tumor is present, Magnetic Resonance Imaging (MRI) segmentation of the brain tissue is essential. The healthcare industry has a lot of private information on record. The research looks at a list of risk variables that are being monitored by systems that look out for brain tumors. Additionally, the approach provided guarantees to be extremely effective and exact for the detection, classification, and segmentation of brain tumors. Automatic or semiautomatic approaches are required to get this level of accuracy. This research offers a method for automatic segmentation that makes use of ANN and multilayer ROI-based features Artificial Neural Networks (ANN). The Brain Lab programme evaluates each ROI's images and determines the ROI's region of interest, thickness of cortical tissue, and quantity of grey matter present. The experimental results demonstrate that the approach of multilayer ROI-based features considerably enhances categorization. As a result, design and implementing of brain tumour detection with multilevel ROI-based features using ANN gives better results interms of accuracy, precision and efficiency.

Keywords: Magnetic resonance imaging, Multilevel ROI-based features, Brain tumour, Artificial neural network (ANN).

I. Introduction

The primary support element of all living things is the cell. The human body comprises over 100 trillion cells, and each one of them has a specific purpose. To produce new cells, these cells must divide and must perform in a predictable manner. Only then can the body do its responsibilities properly. However, there are instances where the cells spread and expand out of control, resulting in the formation of a significant quantity of undesired tissue. The correct word is tumor. Tumors can grow on the body. One of the dangerous and potentially fatal tumors is a brain tumor. The abnormal growth of brain tissues or cells is known as a brain tumor. The ability to spread to the brain exists in every part of the body. The ability of a tumor to completely obliterate brain cells is reduced [7].

The World Cancer Research Fund estimates that 18 million new cases of cancer were discovered worldwide in 2018. Among them 2,96,851 are the new cases and 1.7% of all malignancies are braintumors[8]. Growing lesions that start in the brain are known as brain tumors [2]. According to their amount of aggression, they may be categorized as benign (non-cancerous) or malignant (cancerous), and according to their malignancy, Using the WHO classification for Central Nervous System (CNS) tumors, which runs from 1 to 4, they can be categorized into four categories. Meningioma's and pituitary tumors are two instances of non-cancerous tumors, both of which are often of lower grades, that rarely spread to nearby healthy cells. [1].

The surrounding parenchyma is forcefully penetrated by malignant brain tumors. The most prevalent and dangerous

[1]dhanalakshmi.cse@gprec.ac, [2]insirisha.cse@gprec.ac, [3]inshashidhar.cse@gprec.ac.in

kind of malignant brain tumor is glioblastoma (GBM), which is frequently categorized as a malignancy having a low grade and a dismal outlook. According to their origin, brain tumors can also be classified as primary or secondary, with the former often beginning in the brain and the latter typically in a separate place. The most frequent tumor types are pituitary tumors and gliomas. The origin of the growth of glioma tumors is glial cells. While pituitary tumors start in the pituitary gland, meningioma tumors grow from the dura mater[6].

While glioma tumors are often aggressive in origin, meningioma's and pituitary tumors are frequently benign tumors. It is crucial to find a swift solution to preserve. The complexity of the brain imaging allows for the analysis of malignancies by qualified medical professionals. As a result, utilizing image processing to identify brain tumors is difficult for medical practitioners.[5]. Digital image processing can be used to quickly identify and investigate brain tumors.

For image analysis in this situation, identifying and segmenting brain tumors has become important. Different approaches, including human classification and computer-aided classification, can be used to classify brain tumors. Manual classification of brain tumors is time-consuming and prone to inaccuracy. However, manual categorization cannot be disregarded since it continues to serve as the gold standard for clinical treatment and as an instance towards which other procedures are measured.

Brain tumors have different growth rates throughout time. Benign (slow-growing) tumors do not required to invade the surrounding tissues, in contrast to malignant (aggressive) tumors that spread from their initial site to a secondary site. [3]. Grades I–IV are used by the WHO to classify brain tumors. While Grading III and IV cancers are more aggressive and have a worse prognosis, Grading I and II tumors are anticipated to develop gradually. The following are some features in classifying brain tumors with this regard. Cancers in grade I take time to grow and do not spread quickly. These can be completely eliminated by surgery and are associated with improved long-term survival prospects. Take a grade 1 pilocyticastrocytoma as a example. Grade II, these tumors also develop slowly, but they have the potential to spread to nearby tissues and progress to higher grades. Even after surgery, these cancers could return. An example of one of these tumors is the oligodendroglia. Grade III: These tumors may reach the surrounding tissues and grow more quickly than grade II tumors. For some cancers, surgery may not be sufficient; post-operative radiation or chemotherapy is advised. Anaplastic astrocytoma is an example of this type of tumor. The most aggressive and aggressive cancers are those of grade IV. They could even utilize blood vessels for rapid growth. An example of this type of tumor is a glioblastoma multiform.

Ischemic stroke is a deadly and disability of brain condition that affects the brain quickly. Tissue hypoxia occurs when the brain's blood flow is stopped (under perfusion) results, which kills the advanced tissues in hours. This is known as an ischemic stroke. According to their severity, acute (0–24 h), subacute (24–2 weeks), and chronic (>2 weeks) stroke lesions are the three categories.

A good surgical or medical procedure requires the collection and representation of complex input/output interactions, which a neural network is capable. Additionally, it offers clinicians a powerful tool to help in the analysis, modeling, and understanding of complicated clinical data across a wide range of medical imaging applications. In medicine, classification tasks like pattern recognition make up the majority of ANN applications. Based on the measured data, the patient must be put into one of a select few categories. One sort of artificial neural network utilized in this study was the back propagation network. Once the network has been constructed, transfer functions are utilized to calculate a layer's output from its net input.

2. Literature Survey

Palash Ghosal, Swati Kanchan, Lokesh Nandanwar, Jayasree Chakraborty, Ashok Bhadra, Debashis Nandi, et al. [4] presented techniques for categorizing MRI tumors based on convolutional neural network. The preprocessing steps include rotation, normalization, intensity zero centering, and ROI-based segmentation. Overall accuracy was found to be 93.83%. To increase accuracy, a number of clustering methods and classification strategies were used. Due to several benefits, artificial neural networks were utilized in this investigation. ANN is a simpler, less complicated approach.

Mustafa R. Ismael, Ikhlas Abdel-Qader. et al. [9] developed a Computer-aided design system that utilizes a back-propagation neural network to classify three different kinds of tumors. By ROI, segmentation was completed. The feature extraction approach employed both Discrete Wavelet Transform (DWT) and Gabor filters. The split images had ten separate feature categories that were extracted. 150 features from Gabor filters and 120 features from DWT were retrieved.

Zhenyu Tang, Yap Pew-Thian,Ahmad Sahar, Shen Dinggang et al. [10] explains a new Multi-Atlas Segmentation (MAS) framework for MR brain tumor images. Through combining label data from many traditional brain atlases, the MAS reconstructs a new picture of the brain for segmentation. The majority of its frames are framed for normal brain imaging, however it still struggles with tumor brain images.Utilizing data from the normal brain atlas, the image of the normal brain is being recreated from the MR tumor brain image using a new low-rank method at the first layer of the MAS framework. As a further phase, normal brain atlases are being recorded in order to restore the picture without tumor impact.

Hemasundara Rao C., Naganjaneyulu Dr. P.V., Dr. K. Satya Prasad et al. [11] shows how to locate and divide affected brain tumor regions using an automated method. Three phases are involved: Initially segmenting 2. Energy function modeling; and 3. Energy function optimization. Accurate segmentation is produced using the data from Magnetic Resonance Imaging (MRI).

Using Sergio Pereira et al. Using brain MRI scans, automated methods for identifying and classifying brain cancers have been developed for the first time. It is presently possible to scan and load medical images into a computer system. Instead, the techniques that are used most commonly are NN (Neural Networks) and SVM (Support Vector Machine). Due to NN's enhanced performance.

Soft thresholding and genetic algorithms are two techniques that can be used to segment images, according to Garima Singh, Dr. M.A. Ansari, et al. [13]. It has been found that these methods can be applied to grey-level magnetic resonance imaging. The suggested method makes use of GA to address optimization issues involving a big search space (which represents the label of each individual image pixel). The proposed method additionally accounts for any prior information (such as the local ground truth).The developed method generated segmentation accuracy and Signal-to-Noise Ratio (SNR) values (20 to 44) connected to identified cancer pixels depending on the underlying data.

3. Design and Implementing of Brain Tumor Detection with Multilevel ROI-Based Features Using ANN

The block diagram for design and implementing of brain tumor detection with multilevel ROI-based features using ANN is represented in Fig. 13.1.

When a person has their MRI scanned, the MRI image may be retrieved from the patient database on the computer. Images from MRI scans are typically in white and black. This image serves as the input.

Cleaning of data, data transformation, data integration, data resizing, data reduction, etc. are all part of the image pre-processing approach. The image is pre-processed to remove irrelevant details, smooth out noisy information, find and get removal of outliers, and fix data inconsistencies. Aggregation and normalization are the final steps. The method which is used for image processing turns out to be quite important for locating a specific cardiac image, lowering noise, and improving image quality.

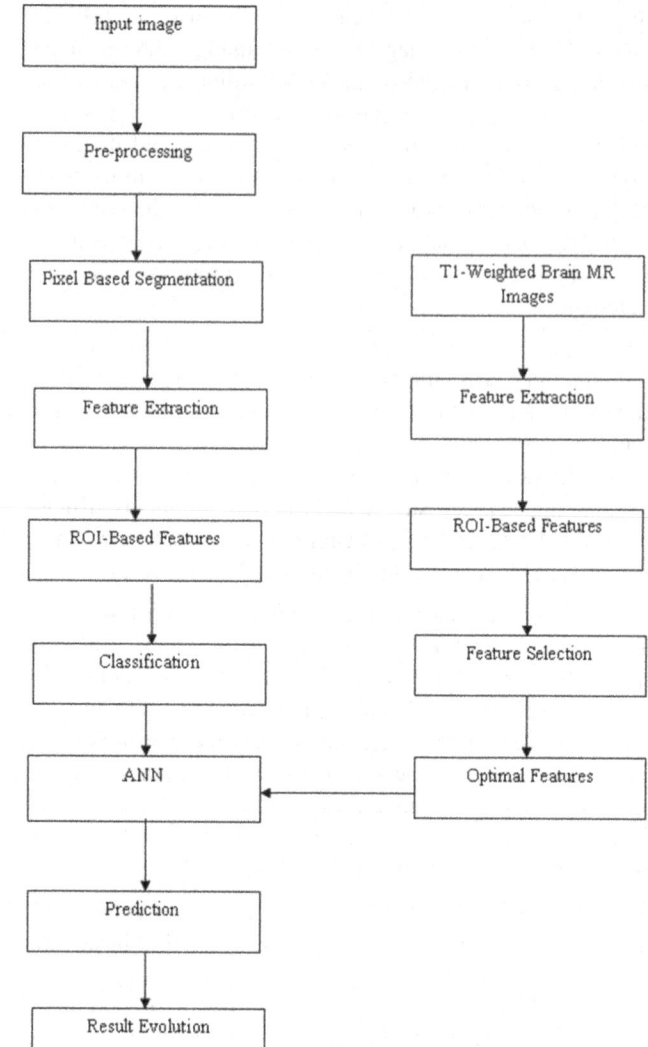

Fig. 13.1 Block diagram of design and implementing of brain tumor detection with multilevel ROI-based features using ANN

Processing digital photos typically involves image segmentation. Recently, an important area of study in the field of medical imaging systems has risen: brain tumor image sectioning in MRI. Photos are segmented using images. The least amount of pre-processing is applied to extract features from pixel images. For semantic segmentation, a thin deep neural network architecture known as LinkNet was created. The encoder and decoder blocks in the LinkNet Network successfully split and rebuilt the images before it was sent using a few final convolutional layers.

In order to use a layout with 78 ROIs for each subject,the ROI-based characteristics, such as Gray Matter (GM) volumes, cortical thickness, and cortical surface area, had been gathered from the MR scans (39 for the left and right hemispheres, respectively.). The classification's structure is complete.

The Parkinson's Progression Markers Initiative (PPMI) dataset and 3-T Siemens Medical Systems were used to produce the T1-weighted brain magnetic resonance (MR) scans. There are 32 healthy controls and 77 persons with Parkinson's disease (PD).The parameters of the acquisition are described as follows: The parameters are as follows: Thickness = 1.5 mm, Flip Angle (FA) = 15°, Matrix = 256*256 slice, Repetition Time (TR)/Echo Time (TE) = 9.1/3.6 ms. Each and every patient is required to complete the neuropsychological evaluations. The level II criteria of the Modified Movement Disorders Society. There are 55 healthy controls and 22 Parkinson's patients with modest cognitive impairment. Before any participant is included to the PPMI, signed informed consents are required from each participant.

The Brain Lab software was used to analyze the images, and as part of the analysis, each region of interest (ROI) had its height, surface area, and quantity of grey matter evaluated. The initial brain MR scans were re-sampled and reoriented in order to normalize the images because the originals' volume and homogeneity were inconsistent. The N3 intensity correction approach was then used to rectify the MR images. Then, using skull-stripping, non-brain regions were eliminated from the previously processed images. White matter (WM), Cerebrospinal fluid (CSF), and Grey Matter (GM) were divided using tissue segmentation on brain MR images. The pre-labeled ROIs from the warped atlas template, which was utilized to label the ROIs on the divided pictures, were superimposed on the segmented images. The GM, WM, and CSF surfaces were subsequently rebuilt separately. The two classifiers now have access to both multilayer ROI-based features as well as single-level ROI-based characteristics (using only GM volume, cortical thickness, or cortex surface area).

Neuroscience classifications frequently have high dimensional components. By lowering the dimensionality of the features, high-dimensional characteristics will make it simpler for the classifiers to generalize and evaluate the information. To reduce the number of characteristics and increase the likelihood that the algorithm will evolve in the future, feature selection is frequently used to detect irrelevant data.In this method, the number of features was initially reduced using two filter-based techniques, and the best features were then further selected for each group's prediction using a wrapper-based methodology. The following steps mostly used the qualities with significance values below the threshold ($p < 0.05$). After that, to reduce the dimension of features, a different filter-based technique called Minimum Redundancy and Maximum Relevance (mRMR) was used.

The assessment parameters were used to examine the properties of the classification approach and to evaluate its performance. The complete amount of data was randomly split into two sections: one for classifier training with samples and the other for ANN-based outcome prediction.

4. Result Analysis

The result analysis of the described design and implementing of brain tumour detection with multilevel ROI-based features using ANN is demonstrated in this section. There are many algorithms available for detecting brain tumours. The focus of this investigation, however this analysis focuses on multilayer ROI-based features that leverage ANNs to perform better, and the efficiency, accuracy, and precision of the performance measures are computed.

These definitions for False Negative (FN), False Positive (FP), True Positive (TP), and True Negative (TN), are used to assess the effectiveness of the specified model.

True Positive (TP): The total number of instances of positive predictions that were accurately classified as such makes up the TP.

True Negative (TN): The TN is the total number of correctly identified, really negative, and presented events.

False Positive (FP): The total number of positive predictions that turn out to be incorrectly positives is known as FP.

False Negative (FN): The total number of actually negative once incorrectly predicted negative outcomes is represented by the character FN.

Accuracy: Its definition is given as the ratio of accurately identified occurrences to all occurrences.

$$\text{Accuracy} = \frac{TP + TN}{TP + FP + TN + FN} \times 100 \quad (1)$$

Precision: The percentage of data points that were actually significant is shown by a model's accuracy. This shows that only pertinent instances are produced

$$\text{Precision} = \frac{TP}{(TP + FP)} \times 100 \quad (2)$$

The Table 13.1 shows that, to categorise MRI brain images with the goal of identifying brain tumors, multilevel ROI-based features using an ANN technique offer improved accuracy, precision, and efficiency when compared to single-level ROI-based features.

Table 13.1 Performance analysis

Performance metrics	Multi ROI-based features	Single ROI-based features
Accuracy	97.3	85.5
Precision	93.4	75.3
Efficiency	98.6	80.7

In Fig. 13.2 comparison graph shows that multilevel ROI-based features using the ANN approach are higher to single-level ROI-based features in terms of accuracy.

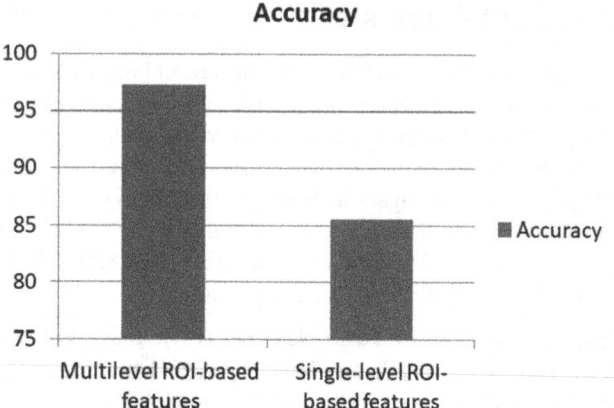

Fig. 13.2 Accuracy performance comparison

The comparison graph for precision is shown in Fig. 13.3. The results show higher precision values for multilevel ROI-based features using ANN method when compared with single-level ROI-based features.

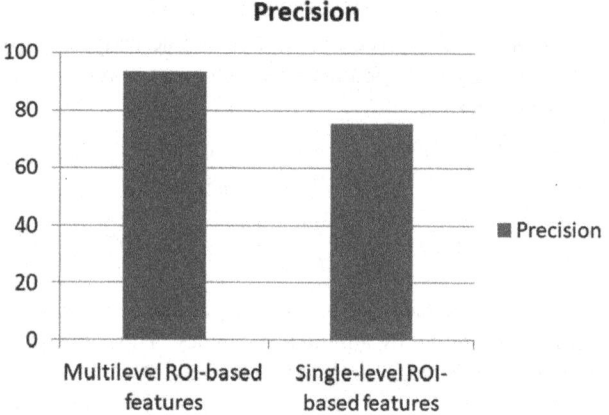

Fig. 13.3 Precision performance comparison

Figure 13.4 graphical representation of the most effective value for the developed method of multilevel ROI-based features using ANN method.

5. Conclusion

In medical treatment, locating brain tumors is more common. According to definitions, a brain tumor is a distorted mass of tissue in which cells grow rapidly and uncontrollably, showing that the pace of cell proliferation is unchecked. Using image segmentation, the aberrant tumor area is removed from the brain. The multilayer ROI-based brain cancer detection

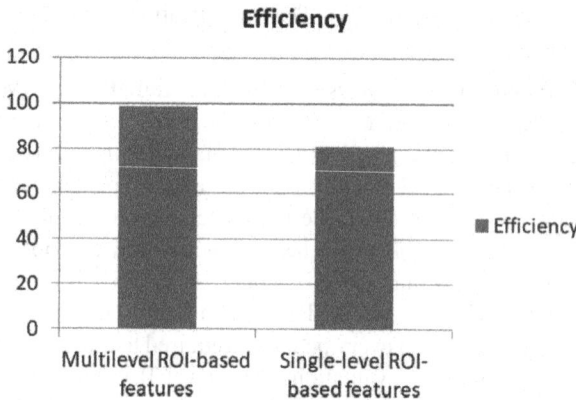

Fig. 13.4 Efficiency performance comparison

with ANN classification strategy suggested in this method produced incredibly good performance with the combination of volumetric and cortical data received from the brain. This method provides a way to identify the pattern of the brain's microstructural alteration on early basis. The method and extraction of features for diagnosing cognitive problems in Parkinson's disease have benefits when using multilayer ROI-based characteristics. Multilayer ROI-based features outperform single-level ROI-based characteristics in terms of performance. It was specifically possible to get encouraging classification performance by merging multilayer ROI-based features. The experimental results showed that the multilayer ROI-based features technique significantly improved categorization. This resulted in more accurate, precise, and effective outcomes when multilevel ROI-based features were created and applied to identify brain cancer using ANN.

References

1. P., Gokila Brindha and Kavinraj, M and Manivasakam, P and Prasanth, P. Brain tumor detection from MRI images using deep learning techniques. IOP Conference Series: Materials Science and Engineering. 1055. 012115. 10.1088/1757-899X/1055/1/012115,2021.
2. Louis, D.N.; Perry, A.; Wesseling, P.; Brat, D.J.; Cree, I.A.; Figarella-Branger, D.; Hawkins, C.; Ng, H.K.; Pfister, S.M.; Reifenberger, G. The 2021 WHO classification of tumors of the central nervous system: A summary. Neuro-Oncology 2021, 23, 1231–1251.
3. Sharif M, Amin J, Nisar MW, Anjum MA, Muhammad N, Shad SA (2020)Aunified patch basedmethod for brain tumor detection using features fusion. CognSyst Res 59: 273–286
4. PalashGhosal, LokeshNandanwar, Swati Kanchan, Ashok Bhadra, Jayasree Chakraborty, Debashis Nandi, " Brain Tumor Classification Using ResNet-101 Based Squeeze and Excitation Deep Neural Network", 2019 Second International Conference on Advanced Computational and Communication Paradigms (ICACCP), IEEE, 2528 February 2019.

5. Özyurt, F., Sert, E., Avci, E., and Dogantekin, E., "Brain tumor detection based on Convolutional Neural Network with neutrosophic expert maximum fuzzy sure entropy", vol. 147, 2019. doi:10.1016/j.measurement.2019.07.058.

6. Hasan Ucuzal, Seyma YASAR, CemilQolak, "Classification of brain tumor types by deep learning with convolutional neural network on magnetic resonance images using a developed web based Studies and Innovative Technologies (ISMSIT), IEEE Xplore, 16 December 2019

7. J. M. Cohen, D. J. Civitello, M. D. Venesky, T. A. McMahon, and J. R. Rohr, "An interaction between climate change and infectious disease drove widespread amphibian declines," *Global change biology*, vol. 25, pp. 927–937, 2019.

8. Gurkarandesh Kaur, Ashish Oberoi,"Development of an Efficient Clustering Technique for Brain Tumor Detection" , International Journal of Computer Sciences and Engineering, Vol. 6(9), Sept. 2018.

9. Mustafa R. Ismael, Ikhlas Abdel-Qader, "Brain Tumor Classification via Statistical Features and Back-Propagation Neural Network", 2018 IEEE International Conference on Electro/Information Technology (EIT), IEEE Xplore, 22 October 2018.

10. Zhenyu Tang, Ahmad Sahar, Yap Pew-Thian, Shen Dinggang "Multi-Atlas Segmentation of MR Tumor Brain Images Using Low-Rank Based Image Recovery", IEEE Trans Med Imaging, vol. 37(10), 2018, pp. 2224–2235.

11. C.Hemasundara Rao, Dr. P.V. Naganjaneyulu, Dr.K.Satya Prasad "Brain tumor detection and segmentation using conditional random field", IEEE 7th International Advance Computing Conference, 2017, pp. 807–810.

12. Sergio Pereira, "Brain Tumor Segmentation using Convolutional Neural Networks in MRI Images", IEEE Transactions on Medical Imaging, (2016).

13. Garima Singh, Dr. M.A. Ansari "Efficient Detection of Brain Tumor from MRIs Using K-Means Segmentation and Normalized Histogram", IEEE, 2016.

14. D. Merkitch, G. Stebbin, B. Bernard, and J. Goldman, "Neuroanatomical correlates of cognitive functioning across the Parkinson's Disease cognitive spectrum," Neurology, vol. 84, no. 14, pp. P3.006, Apr. 2015.

15. B. Peng, Z. Chen, L. Ma, and Y. Dai, "Cerebral alterations of type 2 diabetes mellitus on MRI: A pilot study," Neuroscience letters, vol. 606, pp. 100–105, Oct. 2015.

16. G. Singh and L. Samavedham, "Unsupervised learning based feature extraction for differential diagnosis of neurodegenerative diseases: a case study on early-stage diagnosis of Parkinson disease," Journal of neuroscience methods, vol. 256, no. 30, pp. 30–40, Dec. 2015.

17. Y. Wang, J. Nie, P. T. Yap, G. Li, F. Shi, X. Geng, "Knowledgeguided robust MRI brain extraction for diverse large-scale neuroimaging studies on humans and non-human primates," PloS one, vol. 9, no. 1, pp. e77810, Jan. 2014.

18. L. Wang, F. Shi, G. Li, and D. Shen, "4d segmentation of brain MR images with constrained cortical thickness variation," PloS one, vol. 8, no. 7, pp. e64207, Jul. 2013.

19. G. Wu, M. Kim, Q. Wang, and D. Shen, "Hierarchical attribute-guided symmetric diffeomorphic registration for MR brain images," in Medical Image Computing and Computer-Assisted Intervention, vol. 7511, pp. 90–97, 2012.

Note: All the figures and tables in this chapter were test data by authors.

Role of Machine Learning in Food Industry: A Survey

14

M. L. N. V. S. S. Ganesh[1], CH. P. Pranavi[2]

MCA student, Computer Science & Applications,
Koneru Lakshmaiah Education Foundation, Andhra pradesh, India

Phanikanth Chintamaneni[3]

Assistant Professor, Department of CSE,
Koneru Lakshmaiah Education Foundation, Vaddeswaram, A.P., India

Subrahmanyam Kodukula[4]

Professor, Department of CSE,
Koneru Lakshmaiah Education Foundation, Vaddeswaram, A.P., India

Abstract: This study's objective was to examine how machine learning (ML) is applied in the food business. The results reported there are preliminary. The primary focus is on improving the food industry's performance. It may lessen financial losses and increase the food industry's responsiveness and efficiency. Robots and data processing systems are two of the most well-known examples of high-end technologies that combine artificial intelligence (AI) and machine learning (ML) to produce, process, and distribute qualitative and quantitative goods with little need for resources like money, labour, or time. The purpose of this study was to examine how machine learning and artificial intelligence (AI) methods are applied in the food business and to suggest future research topics in that area. Artificial intelligence (AI) and machine learning can increase the effectiveness and responsiveness of the food sector by decreasing financial losses.. Networks in the food business benefit from competitive advantages due to the usage of AI and ML. Public health is still being threatened by food safety. Large, newly available data sets can be used by machine learning to increase food supply security and lessen the effects of food safety accidents. This article provides a brief introduction to machine learning in the context of food safety as well as a summary of current research and application areas. Food is a vital issue for the entire society since it is the cornerstone of human health, social advancement, and stability. Food processing at all stages—from cultivation, harvesting, and storage to preparation and consumption—must be considered to ensure food quality and safety. The machine vision system may then handle duties like food grading, locating defective spots or foreign items, and eradicating pollutants through follow-up design.

Keywords: Machine learning, Artificial intelligence, Food quality, Shelf-life

1. Introduction

In comparison to using human labour, we can carry out the manufacture and packing of food goods more efficiently by applying artificial intelligence. It is also referred to as an autonomous system. Maintaining the quality and quantity of food is also necessary to reduce risks and identify potential hazards in food products. Both Machine learning creates fresh forecasts and self-learning algorithms. Machine learning searches through enormous amounts of data to uncover patterns using statistics. Before we talk about the meal's quality, it goes without saying that any food that is both palatable and healthful gives us a reason to at least live. The goal of improving the quality of the same food in the

[1]2100520216@gmail.com, [2]2100520206@gmail.com, [3]phanikanth.ch@gmail.com, [4]smkodukula@kluniversity.in

food processing business. The contemporary market that will be responsible for implementing measurable quality grading processes. Food processing is the transformation of food into fresh forms that are more suitable for modern consumers' dietary preferences or into raw materials like wild plants and animals.

Therefore, food processing directly affects both the general economic development of civilization and the living standards of modern society. The many actions that make up food processing necessitate various physical and chemical alterations to the raw materials. But as environmental contamination rises, people are concerned about the safety of both the food supplies and the techniques employed in food preparation. The nutritious qualities of the product must be maintained during the entire procedure. Any food that is palatable and healthy provides a cause to at least survive, so let's talk about that before discussing the quality of the food. The food industries are responsible for obtaining the raw ingredients for food from farmers and regional producers, re-energizing, processing, and packaging them into a safe edible source for their clients. Second, it's a severe problem and the most crucial element for any food processing sector to find ways to improve the same food's quality more efficiently. If a consumer spends money on a certain food product, or at least its raw materials, it is likely that a discussion will ensue that demands both qualitative and quantitative information.

Making difficult judgements that are either too advanced for conventional programming or would need a lot of manual work and money is one area where artificial intelligence is very beneficial. In order for organisations to remain competitive and be in a stronger position, they must leverage data and workforce analytics, which may be implemented via artificial intelligence. The most current study collection claims that by using workforce analytics more effectively, managers and executives in businesses will be better equipped to accomplish their strategic and operational goals. The largest manufacturing industry in the world and the one with the greatest job prospects is food processing and handling. The effective production and packaging of food goods need human work. Given that artificial intelligence technology may use current data to leverage existing data to make even more accurate estimates, working with data and analytics might be a crucial component of how to improve management in the agriculture business.

Artificial intelligence-based systems, sometimes referred to as autonomous systems, are widely employed in almost every technological industry. It makes it possible for individuals all over the world to quickly solve problems, computerise the food business, and revolutionise food industry goods. A simple empirical prediction approach, like BDA, is necessary for the aim of anticipating hazards, assessing risks, and preventing them that are pertinent to food safety. Working

together, public and private sector partners must identify outbreaks of poor food safety and identify their probable causes. High-standard grading practises must be developed in the food business to better manage product quality. On the other side, the industry has a convoluted, nonlinear action plan that can be rectified with a stand-alone, reliable strategy like AI and traceability. Artificial intelligence (AI) is a mathematical concept that displays mental agility and inventive responses to a variety of issues in the food sector.

Big data is required to ensure the quality of food. A cold supply chain during delivery is a nice illustration. Fruits, vegetables, milk, and ice cream are a few examples of foods that are sensitive to temperature. These foods are susceptible to damage from temperature variations and require certain climatic conditions to flourish. The ideal solution calls for specialist IoT-driven sensors that analyse, evaluate, and disseminate data to all stakeholders in real time in order to properly monitor the supply chain cycle. Big data may be used to swiftly purchase new things to replace destroyed ones or to take preventative action. We may employ cutting-edge methods to enhance any form of organisation with the use of big data. Big data also makes it possible for interested parties to identify profitable and popular items in certain locations. Even though the farming industry is just now starting to employ big data, the results are already exceptional. Big data must be used in this situation since agriculture acts as a type of basis for all food sectors.

2. Food Industry Mostly Facing Problems

One of the most significant problems confronting the food sector is food waste. Every year, one-third of the food that is produced is wasted or lost. It is estimated that 1.3 billion tonnes of food are wasted globally each year. Food waste is characterised as a decrease in the quantity and quality of food as a result of consumer, retailer, or restaurant staff members' actions or decisions. Food waste may affect every market in the food business since it occurs along the whole supply chain, from agricultural production to consumer consumption. Transportation, food harvesting methods, and food that customers throw away in trash cans or stores are the biggest contributors to waste. A third of all food produced is wasted, and it affects the supply chain at every point, especially during shipping, harvesting, and by consumers. This shows that there is a chance to conserve food by collaborating on preventative solutions for food loss.

3. The New Age of Food Industries

When you consider the food sector, technology might not be the first thing that comes to mind. But it now forms a

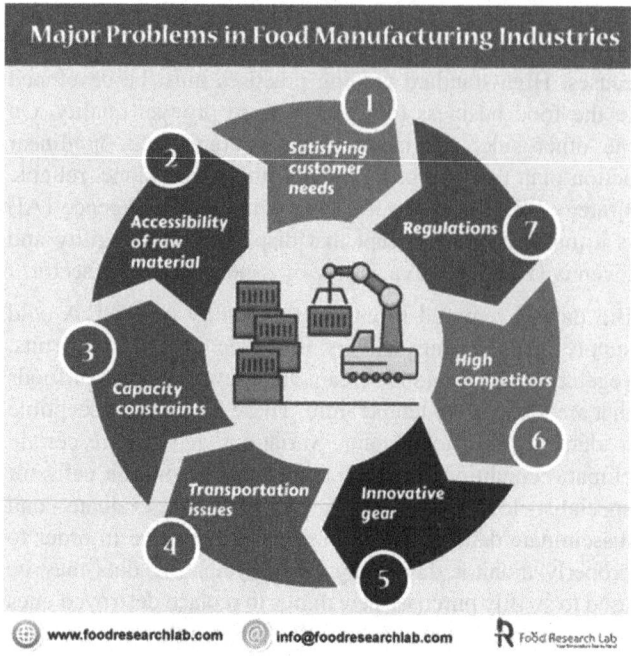

Fig. 14.1 Problems in Food industry [1]

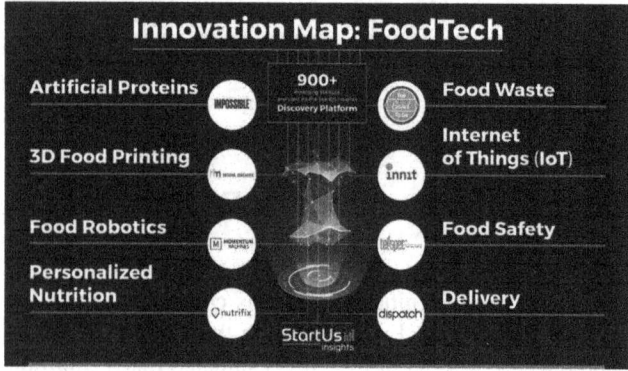

Fig. 14.2 Innovation map [1]

crucial component of it. Companies today recognise how crucial it is for them to adopt cutting-edge technology if they want to not only prosper but also survive in the cutthroat food sector of today. As a result, we are already seeing a number of businesses in the food sector use technology, such as artificial intelligence platforms, to streamline or automate their daily processes. We currently live in a world where top corporations produce food and beverages using data processing and robotics. By depending on such cutting-edge technology, manufacturers may also anticipate cheaper production costs, higher food quality, improved cleanliness, improved packaging, and much more. It is for this reason why AI in the food business is currently so well-liked and may easily become standard in the coming years. It is sufficient to claim that machine learning and artificial intelligence will significantly influence how our food business develops in the future.

4. Machine Learning

In the last few years, increased data volumes, computing power, and enhanced learning algorithms have given artificial intelligence a new lease on life. Despite not being a novel idea, it represents a quiet refreshing approach for the food industry. Without machine learning, artificial intelligence is unlikely to be of much assistance to them. Using methods like Autoregressive Moving Average (ARMA) and Autoregressive Integrated Moving Average (ARIMA), artificial intelligence and machine learning are being used in tandem to address the practical issues in the food processing sector. A significant issue in these businesses is the manufacturing becoming more methodical and modular as a result of the precise time prediction that anticipates food sales using radicalised algorithms and potent vector machines.

5. Types of AI/ML Solutions Related to the Food Industry

5.1 Predictive Maintenance

Manufacturers can avoid equipment problems without needing to spend extra money on needless regular maintenance thanks to predictive maintenance solutions. With the use of such technologies, manufacturing assets and equipment can be monitored in real-time, enabling manufacturers to address problems before they even arise.

5.2 Vision Quality Inspection

Finding quality faults may be completely automated with the help of AI-powered visual quality assessment systems. These systems can monitor food production lines in real-time, allowing for the early detection of flaws and their quick correction.

Fig. 14.3 Vision quality inspection [2]

5.3 Production Planning Optimization

It is possible to properly arrange the allocation of machines and people by utilising solutions that assist successful production scheduling. To reduce waste and increase productivity, manufacturing organisations can forecast demand for their products as well as change requests.

5.4 Food Sorting Optimization

Today, it is possible to automate the process of sorting vast quantities of food according to size and form thanks to AI-based technologies. Such technology uses a camera and a fundamental sensor to identify objects based on their colour, shape, and biological properties. These artificial intelligence platforms can enhance food quality and safety by streamlining the food sorting process.

6. Applications of Machine Learning in Food Industry

6.1 Supply Chain Optimization

Companies can use machine learning systems to forecast consumer demand for their goods and then adjust price, shipping, and inventory strategies accordingly. More transparency is ensured, and waste is decreased.

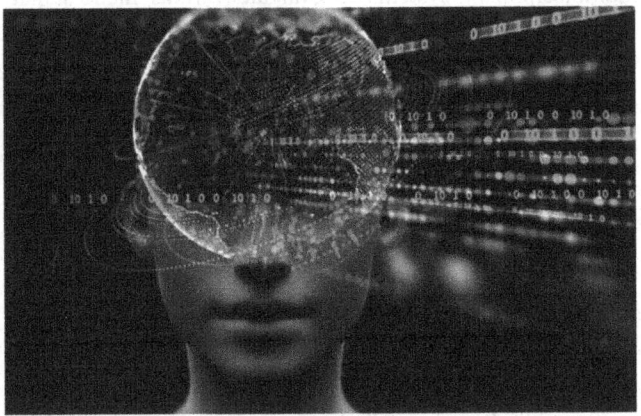

Fig. 14.4 Supply chain optimization [6]

6.2 Food Sorting and Packaging

Today's machine learning technologies may automate the otherwise laborious process of food sorting and packaging, allowing businesses to reduce their large costs associated with human labour and speed up production.

Fig. 14.5 Packing and sorting [7]

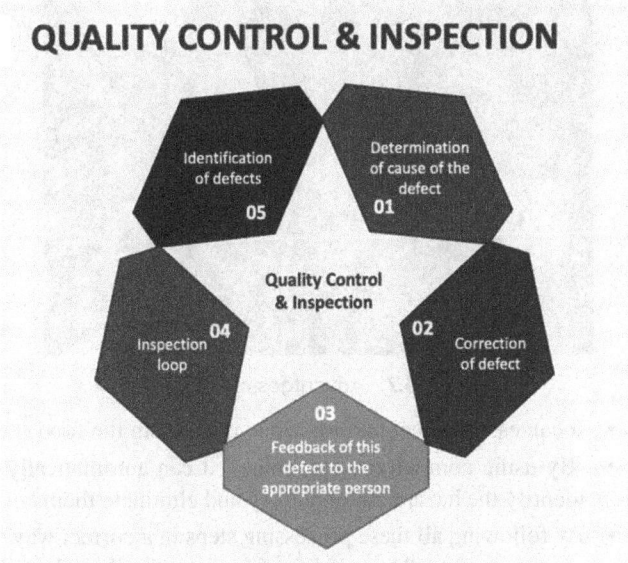

Fig. 14.6 Quality control inspection [7]

6.3 Hygiene Maintenance

When it came to ensuring hygienic conditions, conventional cleaning methods were virtually invariably ineffective. On the other hand, modern AI-driven technologies make use of ultrasonic detection to give machine learning algorithms data that can be utilised to recognise waste and food particles in machinery.

6.4 Product Development

Food processing businesses can monitor and pinpoint consumer behaviour and demands using machine learning technology. The information gathered can then be used to create tailored products that cater to those needs and increase sales for the business. It is sufficient to say that machine learning and artificial intelligence will have a big impact on the food business in the future. As a result, it is crucial for businesses involved in the production and distribution of food to immediately adopt these technologies in order to maintain a successful, steady operation.

7. Advantages of Machine Learning

- The main advantages of machine learning are to produce food product easily.
- Labour cost is low, and the food processing not required high capital.
- Because of using this technology, we can release more variety of foods to the market within less period of time.
- We can maintain and improve hygiene, sanitation conditions in manufacturing process.

Fig. 14.7 Advantges of ML [7]

- It can easily detect hazards while processing the food.
- By using computerised machines, it can automatically identify the hazard automatically and eliminate them.
- By following all these processing steps in a correct way, we can increase the shelf-life of food and food products.
- We can produce food and food products throughout all over the year.
- We can produce hundreds and thousands, different types of food product within few months.
- We can reduce the capital cost.
- It is automatic process.

8. Disadvantages of Machine Learning

Fig. 14.8 Disadvantages of ML.[7]

By using the machine learning techniques, the unemployment will be increased.

- If any equipment failure occurs the whole processing will be delayed.
- The labour cost is expensive.
- If any spoilage occurs while processing the food, the total capital investment will be wasted.
- When the equipment gets cracks, the hazard will enter into the food and it decreases the shelf-life of food.
- If proper pasteurization, sterilization steps are not followed properly the contamination will be occurred.
- Machines capital is high.

- If the customer is not satisfied with the food product, the sales will be decreased.
- Proper food packaging should be done to eliminate the risk of hazard.

9. Conclusion

The use of ML and AI in restaurant operations and food manufacturing has already led to the handling of food safety by AI. Artificial intelligence has improved the field of food safety by decreasing human error in manufacture and, to a lesser extent, unused goods. Higher client satisfaction is offered together with faster services, voice searches, more customised ordering, less expensive packing and delivery. Large food firms can also profit from these commercial benefits, which will provide a positive outcome over time.

Modern technology can now be used in the real world thanks to AI and machine learning. AI controls a number of multidisciplinary systems to evaluate various metrics that represent quality, appearance, texture, general customer acceptability, and so on.

References

1. Objective Food Science & Tekhnology. Edition Author: George F. Stewart & MaynardA. Amerine.
2. Technologies in Editors Food processing Author: Harish kumar sharma / Parmjit s panesar.
3. Food Science. -B. srilakshmi
4. Food: facts & principles.N. Shakuntala Manay,M. Shadakshara swamy. -2008
5. Machine learning & its applications. By Georgios Paliouras, vengelis. Karkaltsis - 2001.
6. Artificial Intelligence. -gud Edition. foundation of computational agents By David L. pooleAlan K. MAckworth.
7. https://www.hindawi.com/journals/jfq/2022/8521236/
8. https://www.researchgate.net/publication/357833952_Artificial_Intelligence_and_Machine_Learning_in_Food_Industries_A_Study
9. https://www.researchgate.net/publication/360541380_Machine_Learning_and_Artificial_Intelligence_in_the_Food_Industry_A_Sustainable_Approach
10. https://www.igi-global.com/chapter/impact-of-artificial-intelligence-and-machine-learning-in-the-food-industry/307426
11. https://link.springer.com/article/10.1007/s12393-021-09290-z#Sec13
12. https://www.researchgate.net/publication/348655838_Emerging_Applications_of_Machine_Learning_in_Food_Safety
13. https://www.sciencedirect.com/science/article/pii/S2665927121000228#sec6

An Intrusion Detection System Using Principal Component Analysis and Extreme Gradient Boosting

15

Bhargavi Goparaju[1]

Research Scholar, Acharya Nagarjuna University,
RISE Krishna Sai Prakasam Group of Institutions

Bandla Srinivasa Rao[2]

Research Guide, Acharya Nagarjuna University

K. Bhargavi[3]

Professor, Teegala Krishna Reddy Engineering College

Abstract: Intrusion detection systems (IDS) are an essential component in maintaining the confidentiality and safety of computer networks. The traditional IDS systems often have problems and limits when it comes to successfully identifying new and complex assaults. An improved intrusion detection system that uses Principal Component Analysis (PCA) and Extreme Gradient Boosting (XGBoost) is proposed in this study. The goal of the system is to increase the accuracy and efficiency of intrusion detection. PCA is used by the proposed IDS model in order to minimise the dimensionality of the data pertaining to network traffic while maintaining the features that are the most useful. PCA is an excellent method for capturing the underlying patterns and structures of network traffic data. This is accomplished by converting the original feature space into a space with less dimensions. After that, an effective machine learning approach known as XGBoost is used in order to construct a dependable and effective detection model. XGBoost can successfully handle unbalanced datasets and categorise normal and abnormal network traffic patterns. According to the findings, the PCA-XGBoost-based intrusion detection system obtained a greater detection accuracy than conventional intrusion detection system techniques. The proposed model obtained an accuracy of 96% while detecting the attacks.

Keywords: Intrusion detection system, Principal component analysis, Extreme gradient boosting, Cyber security, IoT

1. Introduction

The need for cyber security has emerged as an essential component of the modern digital world as the extent of the dependence on technology continues to expand at an exponential rate. The linked nature of various systems and networks has resulted in the emergence of a wide variety of cyber risks and vulnerabilities. Recent years have seen an increase in the number, sophistication, and severity of cyberattacks [1]. Organisations and people alike are exposed to a diverse array of dangers, including but not limited to ransomware and phishing attacks, data breaches, and state-sponsored cyber espionage. The fast progression of technology and the growing interconnectivity of devices have both contributed to an increase in the attack surface, which in turn has made it more difficult to safeguard networks and systems.

In addition, the COVID-19 pandemic has shed more light on the need of maintaining adequate levels of cyber security [2]. Cybercriminals have been given additional chances as a result of the rise in the number of people working remotely, doing business online, and working virtually together. Due to the vulnerabilities created by the shift to digital platforms and

[1]gbhargavi5007@gmail.com, [2]sreenibandla@gmail.com, [3]Bhargavi.mtech@gmail.com

services, there have been more cyberattacks on individuals, businesses, and critical infrastructure. There has never been a time when the demand for effective cyber security measures was more pressing than it is now. Attacks on computer networks result in huge financial losses, harm to a company's image, and even put the country's security at danger [3]. In today's highly linked world, it is of the utmost importance to protect sensitive data, maintain personal privacy, and guarantee the smooth operation of important services.

Additionally, as Internet of Things (IoT) devices proliferate and 5G technology is about to be implemented, the attack surface will keep growing. As devices and systems become more linked, strong security measures are needed to prevent unauthorised access, data breaches, and interruption of critical infrastructure. The current network defence architecture must include intrusion detection systems (IDS) [4]. It is a security programme that scans network traffic for unusual activity and alerts managers to potential threats. IDS may be categorised into two groups: host-based intrusion detection systems (HIDS) and network-based intrusion detection systems (NIDS) [5].

The Network Intrusion Detection System (NIDS) performs real-time analysis of network traffic in order to identify irregularities, trends, and known attack signatures [6]. It gives admins advanced warnings and notifies them to any possible security breaches that may have occurred. On the other hand, host-based intrusion detection systems (HIDS) are primarily concerned with monitoring specific hosts or endpoints, analysing system logs, and identifying attempts at unauthorised access or malicious actions at the host level.

The ability of an IDS to identify and react quickly to cyber-attacks is what gives it its value. IDS is able to detect potentially harmful behaviours by monitoring network traffic and system operations [7]. These behaviours include attempts to gain unauthorised access, infections with malware, and unexpected data transfers. This gives security teams the ability to respond quickly, analyse issues, and limit possible harm before it may spread. IDS plays a significant part in incident response and forensic analysis, both of which are very important. IDS helps organisations understand the nature of attacks, identify vulnerabilities, and take appropriate remedial procedures to avoid future incidents by logging and recording network and system events. This offers critical data for post-incident investigations, which in turn helps organisations prevent other occurrences.

Due the growing frequency and level of complexity of cyber-attacks, cyber security is one of the most serious concerns in the modern digital world. It is very necessary to have strong protection mechanisms in place in order to secure sensitive data, privacy, and vital infrastructure. IDS, are very important in locating potential dangers as quickly as possible

and providing an appropriate response [8]. IDS is able to give organisations with early warnings, enabling incident response, and ease forensic investigation by monitoring and analysing network traffic and system activity in real time. Because cyberattacks are always becoming more sophisticated, it is absolutely necessary for us to make investments in cutting-edge IDS technology and proactive defence methods if we want to guarantee the safety and robustness of our digital ecosystems.

2. Literature

T. Saranya et al. [9] investigated a number of machine learning algorithms used in intrusion detection systems (IDS) for the Internet of Things (IoT), big data, smart cities, and 5G networks, among other applications. Additionally, the research sought to categorise the incursions using machine learning techniques including Linear Discriminant Analysis (LDA), Classification and Regression Trees (CART), and Random Forest.

Ankit Thakkar et al. [10] offered a detailed analysis of Intrusion Detection Systems (IDS) for Internet of Things devices for the years 2015–2019. In the past, the authors have gone through a variety of IDS deployment tactics as well as IDS analysis strategies in the context of IoT design. The paper explores machine learning (ML) and deep learning (DL) approaches used to detect assaults in Internet of Things (IoT) networks and looks at different IoT incursions. The essay also examines the problems and concerns about the Internet of Things' (IoT) security.

Zouhair Chiba et al. [11] employed a hybrid optimisation framework known as IGASAA. This framework was based on the Improved Genetic Algorithm (IGA) and the Simulated Annealing Algorithm (SAA). An intelligent method was established to automatically generate a Deep Neural Network (DNN) that is efficient and effective for an anomalous Network Intrusion Detection System (MLIDS). The IDS that was developed as a consequence of this endeavour is referred to as the MLIDS. The use of optimisation strategies such as parallel computing and fitness value hashing was included in the Genetic Algorithm (GA) in order to increase its overall efficiency. These strategies decrease the amount of time needed for execution and convergence while also preserving computational resources. In addition to this, the SAA was implemented into the IGA in order to enhance the heuristic search that the IGA does. Utilising IGASAA to search for the best or nearly optimal combination of essential parameters in designing a DNN-based intrusion detection system (IDS) or affecting its performance is an important component of the overall approach. Feature selection, data normalisation, DNN architecture, activation function, learning rate, and momentum term are some of the factors that are included here. They were able to accomplish a high detection rate, a high level of precision, and a low percentage of false alarms by using this method.

Muhammad Asif et al. [12] developed a MapReduce-Based Intelligent Model for Intrusion Detection (MR-IMID) as an intelligently automated solution for intrusion detection. MR-IMID is used in this research effort to identify intrusions on a network while simultaneously performing numerous data categorization tasks. The MR-IMID architecture being suggested is capable of processing large data volumes in a reliable manner utilising commodity technology. For the purpose of detecting intrusions in real time, this study activity that is being presented uses data from numerous sources throughout the network. The MR-IMID, which is the subject of this research proposal, is able to identify intrusions by anticipating unknown test situations and storing the data in the database in order to reduce the likelihood of future discrepancies.

Hamed Alqahtani et al. [13] provided intelligent services within the realm of cyber-security and identifying instances of unauthorised access required the use of a variety of well-known machine learning classification algorithms. These algorithms included Bayesian Networks, Naive Bayes classifiers, Decision Trees, Random Decision Forests, Random Trees, Decision Tables balong with Artificial Neural Networks.

Ilhan Firat Kilincer et al. [14] utilized a number of different data sets that are often used, including CSE-CIC IDS-2018, UNSW-NB15, ISCX-2012, NSL-KDD, and CIDDS-001, in order to create intrusion detection systems. These data sets were investigated in great detail as part of this study. In addition, the data sets were max-min normalised, and three core machine learning approaches, namely support vector machine (SVM), k-nearest neighbour (KNN), and decision tree (DT) algorithms, were used to categorise the data.

Kelton A.P. da Costa et al. [15] focused to the research of extensive and up-to-date literature on Machine Learning Techniques that are used in Internet-of-Things and Intrusion Detection to guarantee the safety of computer networks. Their research aims to conduct an in-depth and up-to-date investigation into significant publications that investigate several intelligent strategies and their application to creating intrusion detection systems for computer networks, with a particular emphasis on the Internet of Things and machine learning. The findings of this investigation will then be used to inform future research in this area.

Abhishek Verma et al. [16] investigated the possibilities for use of machine learning classification techniques in protecting the Internet of Things from distributed denial of service attacks. The development of intrusion detection systems that are based on anomaly detection is now being investigated in depth as part of a comprehensive study that is being conducted to evaluate different classifiers. When evaluating the efficacy of a classifier, it is necessary to take into consideration many relevant metrics and validation methods. Established datasets such as CIDDS-001, UNSW-NB15, and NSL-KDD are used in order to benchmark the classifiers. The Friedman and Nemenyi tests are used by the authors in order to do statistical analysis on significant discrepancies across classifiers. In addition to this, they make use of Raspberry Pi in order to investigate the reaction times of classifiers that are running on IoT-specific hardware. In addition, the authors provide a method for choosing the best appropriate classifier for a given application depending on the criteria of that particular application.

Ünal Çavuşoğlu et al. [17] developed a hybrid and layered Intrusion Detection System (IDS) that achieves successful intrusion detection for a wide variety of threats by using a variety of machine learning algorithms and feature selection approaches. In the beginning, the freshly constructed system will carry out data preparation on the NSL-KDD dataset while operating inside its framework. Subsequently, the size of the dataset is shrunk via the use of several feature selection techniques. There have been two fresh takes on the feature selection procedure put forth recently. In order to develop the layered architecture, proper machine learning algorithms must first be determined, and this is done in accordance with the kind of attack.

Elie Alhajjar et al. [18] investigated the adversarial character of the challenge faced by Network Intrusion Detection Systems (NIDS). The primary emphasis is on the assault viewpoint, which encompasses methods for producing adversarial instances that are able to circumvent a wide range of machine learning models. The authors investigate how evolutionary computing approaches, such as particle swarm optimisation and genetic algorithms, combined with deep learning methods, notably generative adversarial networks, might be used to build adversarial instances in order to give higher clarity. Some examples of these techniques are provided in the following paragraphs. The purpose of this task is to evaluate how successful these methods are in fooling a Network Intrusion Detection System (NIDS). The NSL-KDD and UNSW-NB15 datasets, both of which are accessible to the general public, are used in the performance evaluation of the models. In addition to this, a technique known as Monte Carlo simulation, which is considered to be a baseline perturbation method, is evaluated alongside the effectiveness of these algorithms.

Mohanad Sarhan et al. [19] offered a solution to the issue by providing five NIDS datasets, each of which has a functional and standardised feature set that is obtained from NetFlow.

This will allow the issue to be resolved. The raw packet capture files were transformed into the NetFlow format while keeping a feature set that was equivalent to that of the UNSW-NB15, BoT-IoT, and ToN-IoT benchmark NIDS datasets that were used in the datasets. The usefulness, widespread use in production networks, and scalability capabilities of NetFlow as a standard format are its advantages.

Arafatur Rahman et al. [20] suggested two ways to deal with the problem of devices with limited resources in centralised intruder detection systems: semi-distributed and distributed. Combining effective feature extraction and selection with combined analytics at the fog-edge level is the key to these methods. To split up the work, the authors make simultaneous machine learning models that each handle a different part of an attack dataset. In the semi-distributed situation, feature selection is done in parallel by parallel models that are applied at the edge side. Then, at the fog side, a single multi-layer perceptron classification model is run. In the distributed situation, feature selection and multi-layer perceptron classification are done by each parallel model on its own. The results are then put together by a device on the edge or in the fog. This is used to make decisions.

3. Proposed Method

A combination of Principal Component Analysis (PCA) and XGBoost is proposed as a way to find intrusions. PCA is used to reduce the number of dimensions in the feature space by finding the most important features. This makes computations easier and reduces the chance of overfitting. After that, the smaller dataset is used as input for XGBoost, a solid gradient boosting method. This allows XGBoost to develop a powerful ensemble model by using the chosen features. This combination enables the system to efficiently identify and categorise possible intrusions by capturing complex patterns and correlations within the network traffic data. This is accomplished by analysing the data to determine whether or not an intrusion has occurred. This technique improves the system's capability to recognise and react appropriately to potential security risks that are present in real-time network settings by combining the feature selection skills of PCA with the prediction accuracy of XGBoost.

3.1 Principle Component Analysis

In the fields of data analysis and machine learning, a method known as principal component analysis (PCA) is often used to reduce the total number of dimensions included within a dataset. Condensing a dataset that has many dimensions into a space that has fewer dimensions is the goal here, and it must be accomplished while maintaining as much of the vital information as feasible. PCA does this by determining the paths, also known as principal components, along which the data displays the greatest amount of variation. The following is a list of the stages involved in PCA:

- **Data Standardisation**: The data is standardised in order to guarantee that all features have the same scale. This step is essential because it prevents variables with wider ranges from determining the results of the study.
- **Covariance Matrix Calculation**: The covariance matrix, which depicts the correlations that exist between the various features, is derived by computation from the standardised data. Information about the linear relationships that exist between the characteristics is included inside the covariance matrix.
- **Eigenvalue decomposition**: The covariance matrix is decomposed in order to extract its eigenvalues and eigenvectors. The eigenvectors stand in for the main components along which the data fluctuates the greatest, and the eigenvalues show how much of the total variance can be attributed to each primary component.
- **Selection of principle Components**: The principle components are sorted according to their associated eigenvalues, and the top components that explain the bulk of the variation in the data are chosen as the principal components to use. In most cases, the number of chosen main components is established either via the process of cross-validation or depending on the degree of explained variance that is needed.
- **Projection**: The projected values of the original data are created using the chosen primary components by mapping them to a space with lower dimensions. For the purpose of this projection, the standardised data are multiplied by the chosen eigenvectors, which ultimately results in a representation of the data with less dimensions.

PCA is used for a variety of tasks, such as dimensionality reduction, visualisation, noise filtering, and feature extraction, among others. The process of data analysis may be made easier by cutting down on the number of variables while maintaining the information that is most vital. In addition, PCA is also useful for discovering previously hidden patterns, locating data outliers, and expediting future machine learning tasks by simplifying computational complexity.

3.2 Extreme Gradient Boosting (XGB) Classifier Algorithm

The XGBoost Classifier is an advanced machine-learning algorithm that belongs to the gradient boosting algorithm family. In order to achieve exceptional prediction accuracy, XGBoost employs a number of novel techniques, including gradient-based optimisation, regularised learning goals, and parallel processing. The XGBoost Classifier is able to handle big datasets, manage missing values, and give insights into the relevance of features.

Boosting is an effective method for developing ensemble models in classification problems. Classifiers that are based on the boosting algorithm combine the results of many less competent learners to produce a powerful ensemble model that is capable of making correct predictions and classifications. Initially, the weights of each instance in the training data are established, and the first weak learner, which is often a simple model like a decision tree, is trained using this dataset. This takes place at the beginning of the process. The learner aims to minimize the classification error or the loss function. An evaluation of the performance of the weak learner is carried out by computing the weighted error. The significance of each case in the overall error computation is determined, in part, by the weights that were applied to those instances in the training dataset.

The instance weights are then modified according on the performance of the weak learner. The weak learner is given more weights to instances that have been wrongly categorised, whereas examples that have been successfully identified have their weights reduced. This modification places a greater emphasis on the cases that are difficult to classify. This helps to ensure that following weak learners pay attention during training to the challenging examples. The procedure of training and adjusting weights is carried out in an iterative manner until a certain number of iterations have been completed or until a particular condition has been satisfied. On the most recent training dataset, a fresh weak learner is taught in each iteration. The updated weights, which are dependent on the performance of prior weak learners, serve as a compass to direct the training process. These weights indicate the significance of each occurrence.

The classification errors are gradually reduced by the boosting method as it repeatedly trains numerous weak learners and iterates through the process. Each following weak learner is given training with the goal of reducing the total amount of mistake or loss, with a greater emphasis placed on the cases that were misclassified by the learners who came before them. As a result, the ensemble model improves in its ability to recognise intricate patterns and to provide reliable forecasts. The weak learners are integrated at the end to produce a robust ensemble model. The final prediction of a boosting-based classifier is often derived by agglomerating the predictions of all of the weak learners and assigning each of those predictions a weight depending on how well or how important they performed individually. This aggregate builds a more reliable and accurate forecast by capitalising on the qualities possessed by each individual weak learner.

3.3 Working of XGBoost

Step 1: Initialise the XGBoost model using the default values for hyperparameters, such as the learning rate, the number of boosting rounds, and the maximum depth of the trees.

Step 2: Compute initial predictions: XGBoostsmake initial predictions for all of the instances that are included in the training dataset.

Step 3: Calculate the initial residuals: The residuals are the differences between the actual target values and the initial forecasts. In order to assess the accuracy of the original predictions, the method computes the residuals and finds the mistakes.

Step 4: Iterative boosting rounds: The core of XGBoost consists of repeatedly going through a number of boosting rounds, with each round being comprised of the following steps:

- *Train a weak learner:* Before each round of boosting, a weak learner, in the form of a decision tree, is trained using the dataset used for training. The goal of the weak learner is to make an accurate prediction of the residuals from the previous round. The tree is formed in a greedy fashion, recursively partitioning the data depending on the characteristics, in order to minimise the loss function.

- *Compute the leaf weights:* After the weak learner (decision tree) has been trained, the leaf weights are computed by optimising the objective function, which combines the loss function and the regularisation term. This is done in order to get the optimal value for the objective function. The weights of the leaves provide an indication of the degree to which the leaves of each tree contribute to the overall forecast.

- *Update the predictions:* In order to update the predictions, the predictions made by the current weak learner are added, and then those predictions are weighted by the leaf weights. The accuracy of the forecasts is enhanced progressively during the course of this stage, which brings them closer to the actual target values.

- *Compute new residuals:* The revised forecasts are subtracted from the actual target values in order to arrive at the new residuals, which are then computed. These residuals reflect the mistakes that still exist after the previous boosting cycle has been completed and need to be fixed

After a certain amount of boosting rounds have been completed, the XGBoost algorithm will next proceed to finalise the model. It does this by adding up all of the predictions made by the weak learners and giving each prediction the appropriate amount of weight based on the leaf weights. The resultant ensemble model is therefore in a position to make predictions on fresh instances that have not previously been seen.

3.4 XGBoost for Intrusion Detection

In the context of the detection of network intrusions, the data from network traffic is pre-processed and then turned into useful characteristics. The XGBoost model is trained with the labelled training dataset so that it may learn the patterns that are related with both healthy and harmful network behaviour. The prediction accuracy of the model is improved during the training process by XGBoost's iterative construction of a powerful ensemble of weak learners (decision trees), which contributes to the overall improvement. After the XGBoost model has been trained, it is used on new data instances of network traffic to identify possible intrusions by making predictions about the probability of malicious behaviour. Because of its capacity to deal with intricate patterns and its adaptability with regard to the management of a wide variety of characteristics, XGBoost is a potent instrument for the detection of intrusions in real-time network settings.

4. Experimental Results

The simulation results of proposed Intrusion detection system is discussed in this section. The most crucial defence mechanisms against the complex and expanding network threats are IDSs and intrusion prevention systems. Performance evolutions for anomaly-based intrusion detection techniques are inconsistent and inaccurate since there aren't enough trustworthy test and validation datasets. Traditional datasets reveal that the majority are untrustworthy and out of date. The lack of diversity and volume in some of these datasets, the lack of coverage of the range of known assaults, and the anonymization of packet payload data in others prevent some of them from reflecting current trends. Some also lack information and feature sets.

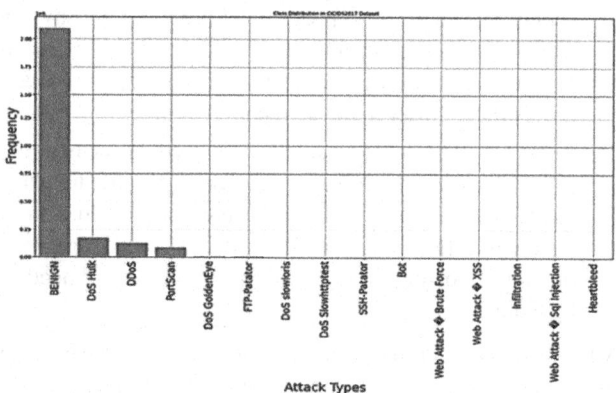

Fig. 15.1 Attack types vs frequency visualization

The breakdown of the classes that were used to compile the CICIDS2017 dataset may be seen in Figure 15.1. This dataset includes a collection of modern and secure common assaults that are quite similar to real-world data that was gathered in PCAPs (Packet Capture). The size of the dataset consists 2120907 rows and 78 columns. In addition to the PCAPs, the dataset also contains the results of the network traffic analysis

performed by CICFlowMeter. The flows that make up the dataset are categorised according to a variety of parameters, which are given in the form of CSV files. These attributes include a date, the source and destination IP addresses, the source and destination ports, protocols, and attack types. The definitions also include the features that were retrieved from the dataset. The dataset was compiled on the basis of a research that was conducted by McAfee in 2016 and includes a broad variety of assaults that are often seen. These attacks include Web-based, Brute-force, DoS, DDoS, infiltration, Heart-bleed, Bot, and Scan attacks. The number of cases and prevalence rate for each assault type in the dataset are visualised and shown in Fig. 15.2.

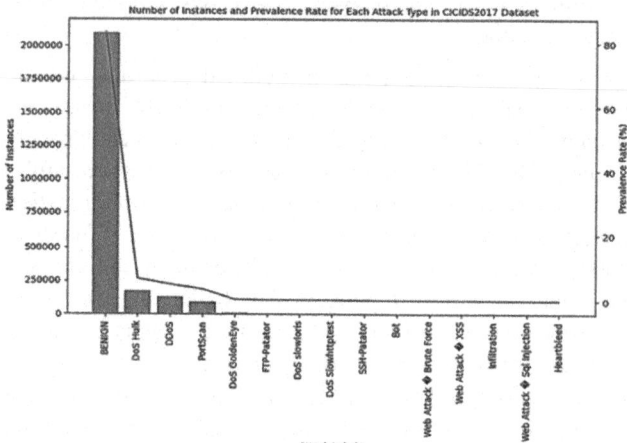

Fig. 15.2 Attack labels Vs number of instances

The proposed model, an intrusion detection system, aims to improve the precision and effectiveness of identifying hostile activity inside a network by combining PCA with XGBoost. By detecting and reacting to unauthorised access attempts or possibly dangerous activity, IDS play a critical role in protecting computer networks. The manual rule-based methods or signature-based procedures used in traditional IDS approaches can have trouble keeping up with the constantly changing nature of cyber threats. The proposed model combines PCA and XGBoost, two potent machine learning approaches, to overcome these constraints. Data compression and feature extraction often employ the dimensionality reduction technique known as PCA. PCA preserves the most important information while minimising noise and redundancy by downscaling the original high-dimensional feature space into a lower-dimensional one. PCA may be used in the context of intrusion detection to extract useful characteristics from network traffic data, enhancing the model's overall efficacy. The curse of dimensionality, whereby an increase in features may result in poorer model performance and an increase in computing complexity, is lessened by PCA by decreasing the dimensionality of the input data. After using PCA, the number of attributes was reduced to 58; as a consequence, only 20 attributes were taken into account for the further stages of processing. The PCA latent space of data training set is depicted in Fig. 15.3.

Fig. 15.3 The PCA latent space of training set

The true positive (TP), the true negative (TN), the false positive (FP), and the false negative (FN) values are shown in the confusion matrix in Fig. 15.4, which offers a comprehensive view of the accuracy of the model's predictions. The results of the calculation are shown in the form of a confusion matrix, the rows of which correspond to the actual labels and the columns of which relate to the anticipated labels. Interpreting the results is the first step in determining how useful the model is. analysing the confusion matrix by looking at metrics such as accuracy, which is the percentage of instances that are correctly classified; precision, which is the percentage of intrusions that are correctly classified among predicted intrusions; recall, which is the percentage of intrusions that are correctly classified among

actual intrusions; and F1 score, which is the harmonic mean of precision and recall. These measurements provide insight on the model's capacity to differentiate between activities that are harmless and activities that are intrusive. Table 15.1 presents the measures of performance that have been derived from the suggested model.

Table 15.1 Performance values of the proposed model

	Precision	Recall	F1 Score
Benign	0.96	0.99	0.98
Attack	0.97	0.82	0.89
Accuracy = 96%			

Table 15.1 presents the performance metrics for a classification model that distinguishes between the categories "Benign" and "Attack." The three metrics that are shown are Precision, Recall, and F1 Score. In addition, it has been said that the model has an overall accuracy of 96%. Precision, which is often referred to as positive predictive value, is a metric that determines the proportion of correctly predicted positive occurrences (true positives) relative to the total number of cases projected as positive. In this particular case, both the Benign and the Attack classes had a precision of 0.96, which indicates that the model correctly predicted each and every occurrence of both classes. Recall, which is also known as sensitivity or true positive rate, is a measurement that compares all actual positive occurrences to the proportion of positive events that were successfully predicted in advance (true positives). In this instance, both classes had a recall of 0.99, which indicates that the model successfully recognised every positive event. In other words, the model was perfect.

The F1 Score, also known as the harmonic mean of accuracy and recall, provides a score that is well-balanced since it takes into consideration both of these factors. Since the F1 Score incorporates both accuracy and recall into a single figure, a higher score implies better performance overall. The F1 Score for this case is 0.98, which indicates that both classes did a very good job of correctly predicting both positive and negative events. It has been estimated that the model is 96% accurate overall. The term "accuracy" refers to the proportion of cases, relative to the total number of cases, that are correctly classified. To calculate it, first sum up all of the instances that were a true positive and true negative, then divide that total by the total number of people in the sample. A rate of 96% implies that the model correctly detected the vast majority of the events. The proposed model is compared with other models and corresponding results are reported in Table 15.2.

In Table 15.2, the comparative study of classification methodologies, the focus lies on evaluating the precision of four distinct techniques: Logistic Regression, Naive Bayes,

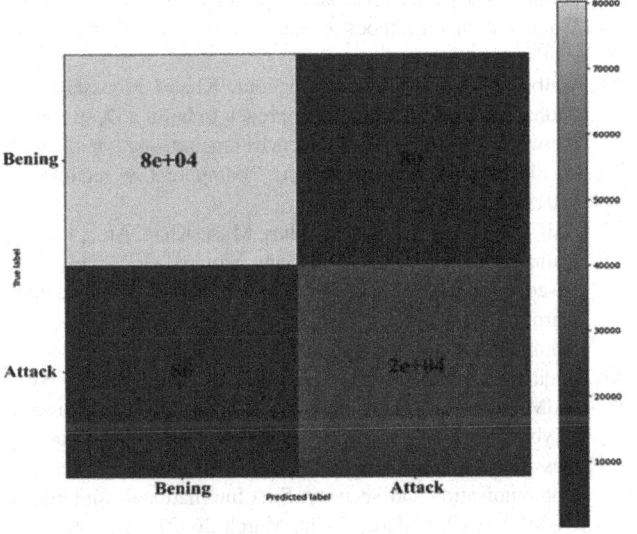

Fig. 15.4 Confusion matrix

Table 15.2 The accuracy comparison results of proposed model

Model Name	Accuracy
Logistic Regression	88%
Naive Bayes	82%
Random Forest	95%
Proposed Model	96%

Random Forest, and a specialized Proposed Model developed in this work. The Logistic Regression method demonstrates its reliability in relatively straightforward scenarios, yielding an accuracy of 88%. Despite the Naive Bayes algorithm's basic independence assumptions, it proves its utility in specific data scenarios, attaining an accuracy rate of 82%. Notably, the Random Forest ensemble method excels with a remarkable accuracy of 95%, adeptly managing intricate interactions and mitigating overfitting concerns. However, the standout performer is the Proposed Model, meticulously tailored to the dataset's specifics, achieving an exceptional accuracy of 96%. This investigation underscores the nuanced nature of model selection, showcasing the diverse strengths of each model based on dataset complexity and available customization options.

5. Conclusion

In the proposed IDS, the combination of PCA and XGBoost gives a solution that is both effective and efficient for detecting intrusions in computer networks. The solution improves the accuracy of identifying new and complex assaults, which enables network managers to take preventative steps to protect the network's safety. The results of the study provide a contribution to the creation of more sophisticated systems for intrusion detection and open the way for future advancements in network security. To evaluate the performance of a classifier, the metrics of Precision, Recall, and F1 Score are often utilised. In this instance, the model displays good accuracy, recall, and F1 scores for both benign and attack occurrences, showing that it efficiently recognises and classifies both kinds of data. These scores can be found in the table below. It has been stated that the accuracy of the model is 99.9%, which indicates that it accurately predicts the class labels for the vast majority of the occurrences included inside the dataset.

References

1. Khraisat, Ansam, Iqbal Gondal, Peter Vamplew, and JoarderKamruzzaman. "Survey of intrusion detection systems: techniques, datasets and challenges." Cybersecurity 2, no. 1 (2019): 1–22.

2. Georgiadou, Anna, SpirosMouzakitis, and Dimitris Askounis. "Working from home during COVID-19 crisis: a cyber security culture assessment survey." Security Journal 35, no. 2 (2022): 486–505.

3. El-Ghamry, Amir, Ashraf Darwish, and Aboul Ella Hassanien. "An optimized CNN-based intrusion detection system for reducing risks in smart farming." Internet of Things 22 (2023): 100709.

4. Saheed, Yakub Kayode. "Performance Improvement of Intrusion Detection System for Detecting Attacks on Internet of Things and Edge of Things." In Artificial Intelligence for Cloud and Edge Computing, pp. 321–339. Cham: Springer International Publishing, 2022.

5. Laghrissi, FatimaEzzahra, Samira Douzi, Khadija Douzi, and BadrHssina. "Intrusion detection systems using long short-term memory (LSTM)." Journal of Big Data 8, no. 1 (2021): 65.

6. Verma, Jyoti, Abhinav Bhandari, and Gurpreet Singh. "iNIDS: SWOT Analysis and TOWS Inferences of State-of-the-Art NIDS solutions for the development of Intelligent Network Intrusion Detection System." Computer Communications (2022).

7. Quincozes, Silvio E., Célio Albuquerque, Diego Passos, and Daniel Mossé. "A survey on intrusion detection and prevention systems in digital substations." Computer Networks 184 (2021): 107679.

8. Li, Daming, Lianbing Deng, Minchang Lee, and Haoxiang Wang. "IoT data feature extraction and intrusion detection system for smart cities based on deep migration learning." International journal of information management 49 (2019): 533–545.

9. Saranya, T., S. Sridevi, C. Deisy, Tran Duc Chung, and MKA Ahamed Khan. "Performance analysis of machine learning algorithms in intrusion detection system: A review." Procedia Computer Science 171 (2020): 1251–1260.

10. Thakkar, Ankit, and RitikaLohiya. "A review on machine learning and deep learning perspectives of IDS for IoT: recent updates, security issues, and challenges." Archives of Computational Methods in Engineering 28 (2021): 3211–3243.

11. Chiba, Zouhair, NoreddineAbghour, Khalid Moussaid, and Mohamed Rida. "Intelligent approach to build a Deep Neural Network based IDS for cloud environment using combination of machine learning algorithms." computers & security 86 (2019): 291–317.

12. Asif, Muhammad, Sagheer Abbas, M. A. Khan, Areej Fatima, Muhammad Adnan Khan, and Sang-Woong Lee. "MapReduce based intelligent model for intrusion detection using machine learning technique." Journal of King Saud University-Computer and Information Sciences (2021).

13. Alqahtani, Hamed, Iqbal H. Sarker, Asra Kalim, Syed MdMinhaz Hossain, Sheikh Ikhlaq, and Sohrab Hossain. "Cyber intrusion detection using machine learning classification techniques." In Computing Science, Communication and Security: First International Conference, COMS2 2020, Gujarat, India, March 26–27, 2020, Revised Selected Papers 1, pp. 121–131. Springer Singapore, 2020.

14. Kilincer, IlhanFirat, FatihErtam, and AbdulkadirSengur. "Machine learning methods for cyber security intrusion detection: Datasets and comparative study." Computer Networks 188 (2021): 107840.

15. Da Costa, Kelton AP, João P. Papa, Celso O. Lisboa, Roberto Munoz, and Victor Hugo C. de Albuquerque. "Internet of Things: A survey on machine learning-based intrusion detection approaches." Computer Networks 151 (2019): 147–157.

16. Verma, Abhishek, and Virender Ranga. "Machine learning based intrusion detection systems for IoT applications." Wireless Personal Communications 111 (2020): 2287–2310.

17. Çavuşoğlu, Ünal. "A new hybrid approach for intrusion detection using machine learning methods." Applied Intelligence 49 (2019): 2735–2761.

18. Alhajjar, Elie, Paul Maxwell, and Nathaniel Bastian. "Adversarial machine learning in network intrusion detection systems." Expert Systems with Applications 186 (2021): 115782.

19. Sarhan, Mohanad, Siamak Layeghy, Nour Moustafa, and Marius Portmann. "Netflow datasets for machine learning-based network intrusion detection systems." In Big Data Technologies and Applications: 10th EAI International Conference, BDTA 2020, and 13th EAI International Conference on Wireless Internet, WiCON 2020, Virtual Event, December 11, 2020, Proceedings 10, pp. 117–135. Springer International Publishing, 2021.

20. Rahman, MdArafatur, A. TaufiqAsyhari, L. S. Leong, G. B. Satrya, M. Hai Tao, and M. F. Zolkipli. "Scalable machine learning-based intrusion detection system for IoT-enabled smart cities." Sustainable Cities and Society 61 (2020): 102324.

Note: All the figures and tables in this chapter were test data by authors.

A Cloud Computing Hierarchical Hybrid Intrusion Detection System Using Machine Learning

16

K . V. Panduranga Rao

HOD & Professor, Department of Computer Science and Engineering,
Sri Vahini Institute of Science and Technology, Tiruvuru, NTR District, Andhra Pradesh, India

V. Krishna Reddy

Professor, Department of Computer Science and Engineering,
Gitam University, Visakhapatnam, Andhra Pradesh, India

T. Prasad[3]

Assistant Professor, Department of Computer Science and Engineering,
Sri Vahini institute of Science and Technology, Tiruvuru, NTR Distinct, Andhra Pradesh, India

D. Naresh

Assistant professor, Department of Electronics and Communication Engineering,
Lakireddy Bali Reddy College of Engineering, mylavaram, NTR Distinct, Andhra Pradesh, India

Abstract: Since the prevalence of cyberattacks that communicate quickly and develop dynamically, intrusion detection is one important tool for developing a safe and reliable Cloud Computing (CC) environment. Cloud computing significantly improves cost metrics through the dynamic provisioning of IT services in our present working paradigm of resources, platforms, and combination of services. It is essential to develop an Intrusion Detection System (IDS) with high accuracy in cloud environments because the majority of cloud computing networks depend on Internet access to offer their services. This system should be able to detect both known and unknown attacks, using signature-based methods as well as other techniques. When discussing system or network events, we consider an abnormal event to be the same as an intrusion event if it involves a significant deviation from typical activities. A few new concepts that try to create a hybrid detection mechanism that combines the benefits of signature-based detection techniques capacity to identify not known attacks. In this analysis, the authors presented a Cloud Hypervisor-level anomaly detection system that uses a combination of K-Nearest Neighbor (KNN) and Support Vector Machine (SVM) classification algorithms. The proposed SVM model outperformed previous model when the accuracy of this system was evaluated.

Keywords: Cloud intrusion detection, Anomaly detection, Machine learning (ML), Support vector machine (SVM)

1. Introduction

Cloud computing is a relatively new computing model that operates over the Internet and provides users with approaches to a wide range of IT services that appear to be limitless [10]. Software-as-a-Service (SaaS), Platform-as-a-Service (PaaS), and Infrastructure-as-a-Service (IaaS) are a few of the services provided by cloud-based data centers. As the behaviour of cloud model is such that service must be accessible and provide an accurate level of QoS to customers, these services must be flexible and open-ended. SDN (Software-Defined Networking) has become a popular option for managing complex Quality of Service (QoS) requirements in the cloud environment, according to the recent research [3].

[1]pandukv@yahoo.com, [2]kvuyyuru@gitam.edu, [3]prasad.tmsc@gmail.com, [4]naresh0475@gmail.com

Modern technology has advanced exponentially, resulting in ubiquitous worldwide networking systems of services and communications as well as the difficulties that go along with it. This growth has been exponential in nature Due to the current scenario and the local resource provision paradigm's cost savings, businesses all over the world are shifting towards offering assistances for customers by cloud that has increased threats.

Cyber security has grown to be a major worry for cloud computing over time. The quickly changing and diverse nature of malicious attacks is representing a challenge to these traditional intrusion detection systems, particularly in the CC environment. Conventional misuse-based methods can be used for system networks. However, it is still in the early phases, research is currently being done to create intrusion detection systems specifically for cloud computing networks. The researchers have suggested a hybrid method to advance the intrusion detection state-of-the-art for the Cloud. This analysis objective is to develop a hybrid IDS's that can successfully counter both well-known and unanticipated security intrusions that could happen in CC.

In CC because of difficulty and volume of infrastructures are vital to use cutting-edge and modern technologies in order to analyse the large volumes of data generated by transactions between devices. It is necessary the real-time analysis of this complex data in order to develop real intrusion detection systems. Therefore, even if they have unique signatures, ML models like classifications as well as clustering are looking for indications of intrusions in large data made up of traffic flows in cloud networks. This analysis described an hybrid network-based IDS system to identify unexpected behaviour in a cloud.

An intrusion is an attack on the integrity, availability, and confidentiality of the system. It can be launched by either insiders—such as untrained labor—or outsiders—such as hackers, crackers, cyberterrorists, and hacktivists. Military, economic, and military-related espionage are just a few of the numerous reasons that intruders can possess. Intrusion detection is the method of reviewing the systems log to know any indications of unauthorized access and finding any potential intrusions.

Intrusion detection has been achieved throughout the years using a number of techniques, The assumption behind anomaly detection is that typical user conduct is entirely observable and sufficiently different from intrusive. It creates a model for the typical user profile and the user conduct that is labelled as intrusive when it differs from the norm. On the other hand, abuse detection detects an intrusion if it finds a match in the monitored data and is based on a signature database of well-known attacks. Anomaly detection has an advantage over misuse detection in detecting novel attacks,

but it may also generate false alarms due to evolving normal behavior, which is not taken into account.

On the other hand, misuse detection has the advantage of producing fewer false alarms, but it can only detect attacks that have a matching signature in its database, those that are not protected by its signature basis are missing. On the other hand, anomaly detection can detect novel attacks may result in too many false alarms as it does not consider the evolving normal behavior. Anomaly and abuse detection have recently been combined in an effort to build a system that can complement one another and detect innovative threats with fewer false alarms.

Despite having a strong defensive line, an intrusion is still possible since security dangers exist. IDS and other complementary reactive security techniques are required. By examining system data, IDS are used to find intrusions[14]. In the case of Network-based Intrusion Detection System (NIDS), information is gathered directly by network by examining at packets while they are still in transit. The performance of the network might be negatively impacted by the classic IDS, such as Snort or Suricata, because they analyze every packet [1]. As a result, the infrastructure is put under increased processing demand and the network latency increases.

The core of the conventional network is stateless and handles each packet separately. To offer better security characteristics, SDN permits the usage of a stateful core with flow and centralized regulates. This analysis suggests a lightweight, scalable hierarchical IDS design that can deliver performance with a conventional packet-based system.

Due that cloud service providers distribute these services through the internet ensuring the security and privacy of cloud services is a crucial challenge. Security is the biggest problem with cloud computing, according to a survey on cloud problems conducted by International Data Corporation (IDC). Cloud infrastructure uses integrated technologies, virtualization techniques, and standard Internet Protocols to function. Traditional attacks like Denial of Service (DoS), Distributed Denial of Service (DDoS), IP spoofing, DNS poisoning, floods, etc. also occur against cloud computing. For example, a DoS attack on the underlying Amazon Cloud infrastructure provided BitBucket.org, a website hosted on AWS, unavailable for an extended amount of time. Due to their expensive computing requirements, cryptographic algorithms are not suitable for cloud computing.

While a firewall can prevent outside attacks, it cannot stop insider attacks. In order to recognise and stop these kinds of attacks, this effective Intrusion Detection Systems (IDS) need be included into cloud infrastructure. An Intrusion Detection System (IDS) can be used to watch network traffic and notify system administrators of any unusual activities. Although

Snort is a popular IDS, it is rule-based and cannot identify unidentified threats. Methods of machine learning are used to identify both known and unidentified attacks. The Back propagation neural network is efficient in detecting unknown attacks and has a high classification rate for unstructured packets. However, Snort, which employs a rule dataset is unable to detect unidentified attacks. As a result, both known and unknown assaults are detected using machine learning techniques. The Back propagation neural network can identify unstructured packets effectively and has a high detection rate.

2. Literature Survey

A. Hushyar et al. [13] suggests that signature-based approaches for intrusion detection have a high level of accuracy but suffer from the challenges of building and maintaining a current database of malicious signatures for every attack. The effectiveness of this strategy is also constrained by the fact that signature-based IDSs cannot identify unidentified attacks. The constantly evolving nature of network applications further complicates the use of signature-based methods.

G. Ditzler, M. Roveri, C. Alippi, and R. Polikar et al. [4]. described Generating intrusion detection models with a signature-based method makes use of supervised learning techniques, with previously recognized attack signatures serving as the training (learning) data. This algorithm learns from training data, such as known attack signatures, and then modify their behaviour to be more detectable. SVM, linear regression, logistic regression, Naive Bayes, linear discriminant analysis, decision trees, and NN like MLP are a few of the often used supervised learning techniques for developing signature-based IDS. It is accurate and based on the dataset of machine learning approach. Although signature-based machine learning IDS offer excellent accuracy, this can be challenging and time-consuming to construct and maintain an updated harmful database for all attacks. Another drawback of these systems is their inability to identify unidentified attacks.

M. J. Reed, et al. [7], there are two primary categories of detection algorithms: signature-based (also known as misuse-based) and anomaly-based (also known as behavior-based). The signature-based technique compares string patterns with predetermined patterns of known assaults that are recorded in a database as signatures in order to identify suspicious activity. This method detects known attacks with excellent accuracy, but unless the signature database is updated, this cannot identify new attacks. IDS that use signatures have an extremely low risk of false positives.

D. Alsmadi IzzatXu, et. al. [5], signature-based methods are unable to make use of the flow level abstraction provided

by SDN, which simply transmits flows to the controller for additional analysis.

This challenges the functioning of signature-based packet IDS schemes like Snort, which require the processing of every packet in the network, this leading to the mirroring of packets to the controller for inspection, resulting in a decrease in performance and an increase in above on controllers. Therefore, kind of strategy the appropriate for suspect flow that will previously identified for additional investigation.

V. Kumar, J. Srivastava, A. Lazarevic, et al. [15], the anomaly-based detection approach compares the observed actions with the recorded profiles are thought to represent typical behaviour. An infiltration attempt is any behaviour that deviates from the expected profile. In contrast to, anomaly-based detection techniques, signature-based detection, can identify unidentified attacks. This approach uses detection approaches from the machine learning, knowledge-based, and statistical engine areas.

M. Alenezi et. al [8], machine learning methods have the ability to automatically identify complex patterns and make informed decisions. The effectiveness of these methods depends on the specific learning algorithm implemented and the quality of the training dataset. The SVM technique is used in this study to detect anomalies at the top level of the described IDS. However, The statistical engines use indicators like as CPU utilize, network bandwidth usage, frequency of service requests, and so on to calculate the level to which the observed conduct deviates from the previously defined threshold.

C. J. Chung, P. Khatkar, T. Xing, et al. [6] described SDN technology is used to reroute network traffic to Snort IDS in order to detect malicious attacks. Snort is a detection system that is based on signatures is unable to identify unknown attacks and manage with high-volume traffic. On the other hand, anomaly detection is intended to separate abnormal traffic from regular traffic and is more suited for identifying unknown threats, but it frequently generates a significant number of false alarms.

C. F. Tsai, Y. F. Hsu, C. Y. Lin and W. Y. Lin, et al. [12], machine learning methods can be used to train IDS using previous intrusion detection datasets to intercept intrusions. Performance of model and the level of the training dataset are very important. Reinforcement, unsupervised, semi-supervised, and supervised learning are sub-division of machine learning algorithms. In supervised learning, predictions are made through classification, regression, or forecasting, as well as learning from pre-labeled, categorized datasets.

Vinchurkar, D. P., and Reshamwala, et. al. [9]. A Neural Network (NN) is a connectionist system which connects several artificial neurons to process information. A pattern from the training dataset is sent into the NN during training. If the output is accurate, it is examined, and the next pattern is then provided as input. If there is a mistake, it is propagated to the input layer using the back-propagation algorithm, and weights are changed to provide the desired output for all training patterns. The trained NN is then fed with more appealing unseen data, and the output determines which class the data belongs

Y. Yu, J. Long, and Z. Cai et al. [2]. presented a stacked Dilated Convolutional Auto-Encoder (DCAE) technique that can identify important characteristics from a significant amount of raw network traffics. Compared to fully linked neural networks like Stacked Auto-Encoders (SAEs), this technique has fewer parameters. The DCAE model's lengthy training procedure is still a drawback, which the authors want to resolve in the future by making use of Graphics processing units (GPUs) parallelization technologies.

Karegowda, A., Manjunath, A. S., Jayaram, M., et al. [11]. explained on feature selection based on the information gain ratio: Features that have multiple values are more likely to be chosen if just information gain is taken into consideration. This disadvantage is eliminated by Information Gain Ratio (IGR) based Feature Selection, which takes into consideration the splitting information of an attribute. This flaw is fixed by the IGR-based feature selection method, which takes into account an attribute's splitting information. The entropy of pattern distribution into branches represents the splitting of attribute information. As the split information value rises, the gain ratio of an attribute.

3. Framework of a Hierarchical Hybrid Intrusion Detection System in Cloud Computing Using Machine Learning

The described hybrid detection model and the methods employed will be covered in detail in this section. Compared to hybrid strategies, conventional intrusion detection techniques are less successful. As a result, the goal of this analysis is to use hybrid methodologies to analyze the contextual relation between network and data flow. The structure of a machine learning-based hierarchical hybrid detection system for cloud computing is shown in Fig. 16.1.

They utilise network monitoring and management technologies, and network packets will carry out the flow analysis. To build Transmission Control Protocol (TCP) dump files, the researchers used open-source traffic analyzer. Dump files are a collection of packets that move between network nodes; as a result, some data may be missing from the output data set. Then, we mainly concentrated on reducing inconsistent and noisy data.

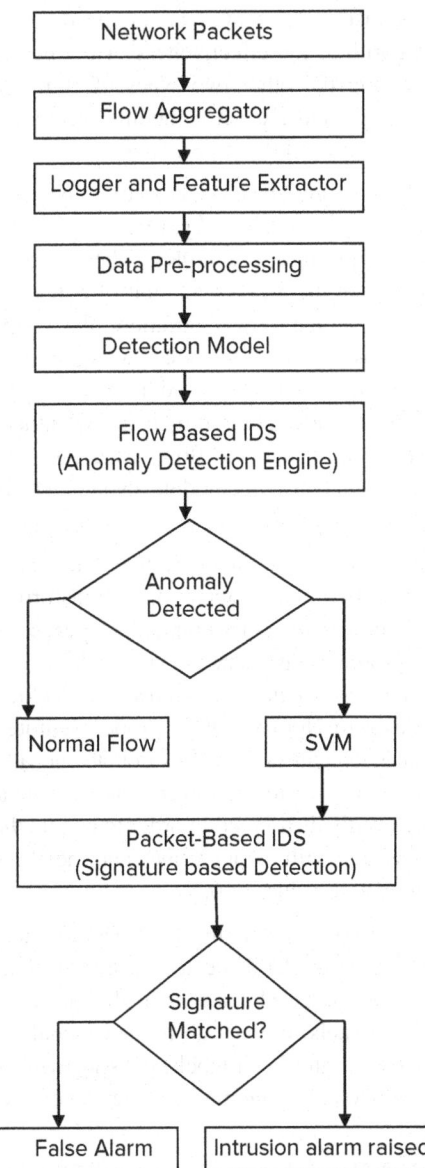

Fig. 16.1 Framework of a hierarchical hybrid intrusion detection system in cloud computing using machine learning

In the SDN network, the Flow Aggregator is in charge of collecting information on flows and ports from OpenFlow switches. Periodically, this module sends commands to switches through the controllers in order to gather data on things like flow, port, and aggregate data. The Aggregate Statistics Reply consists of total aggregate bytes and packets manufactured by the switches. The statistics flow response consists of data about source/destination MAC and IP addresses, protocol, number of bytes, number of packets, period, etc. The port statistics response consists of information about sent and collected packets, bytes, dropped packets, frames, and CRC errors and collisions.

Only flow statistics are employed in the described IDS's detection algorithm; the effectiveness of the described IDS is assessed using the other two types of communications. The Flow Logger and Extractor module receives collected, standardized, organized flows port data.

The Flow Logger and Feature Extractor modules accepts flows data sent by the Flow Aggregators as well as runs calculations to produce six tuples. These tuples include characteristics like the average amount of bytes and packets per flow, the average flow's duration, the proportion of symmetrically paired flows, the rate at which single flows are increasing, and the expansion of additional ports. Important flow characteristics are captured by these essential elements, which are then input into the flow-based IDS. In order to facilitate future analysis, this module also retains all the flow data in log files.

To prepare the data for analysis, the data preprocessing stage involves two main steps: data transformation and standardization. The dataset's nominal properties, including attack type, protocol type, service type, and TCP status flag, are transformed into numerical values. The values 0, 1, 2, 3, and 4 equate to Normal, DOS, Probe, Remote to Local and User to Root and are used to denote the attack type. Standardization is used to scale each feature value and make sure they are evenly distributed in order to handle the issue of preferring features with higher values and the abundance of sparse features with values of 0.

They used support vector machine (SVM) approaches, for their detection model since it is faster in observational modes than unsupervised learning techniques. They used a polynomial kernels to generate trained model by feeding SVM their dataset and fresh labels. This learned model can then be utilized to detect anomalies in new network traffic in future.

The described IDS scheme employs the Flow-based IDS as the initial layer of protection against incoming network traffic. The detection method makes use of an anomaly-based framework.

The anomaly-based detection, signature-based detection, and very lightweight approach may all identify unidentified attacks. This approach is well-suited for the described IDS because it can operate on all incoming network traffic. To detect anomalies, a machine learning-based detection algorithm using SVM is selected. The Flow Extractor and Logger are used to classify and predict flows generated by the 6-tuples. The anomaly-based technique makes use of the built-in model of the SDN network, which provides flow-level data collection and searching, to calculate the tuples for categorization.

If the flow-based IDS identifies a potential intrusion, it will initiate further investigation by directing the related flow to the packet-based IDS for thorough analysis. This approach helps to reduce the number of false positives that may occur in an anomaly-based system. Additionally, the flow-based IDS is responsible for recording and storing information about flows, alarms, and packets for future analysis.

The authors opted for a supervised technique called Support Vector Machine (SVM) as the detection model. A polynomial kernel was utilised to generate a learned model after the SVM algorithm was trained using the dataset analyzed with fresh labels. Then using the learnt model, new network traffic may be examined for potential anomalies. The next line of defence in described hierarchically defensive model is the packet-based intrusion detection system, which employs signature-based IDS technique and only examines packets from flows that have been identified as threat by flow IDS. It is approximately high and process time due to its packet-by-packet analysis technique. The impact of the Packet-based IDS on the overall system is minimal since it only analyzes a small portion of the traffic. Any well-known IDS, such Bro, Suricata, or Snort, with trade-offs between them, may be used as the intrusion detection system. While Snort was chosen for the current prototype, Suricata offers more detailed alerts but has higher processing cost. Snort has three modes of operation: sniffer, logger, and IDS. While in sniffer mode, it collects Network Interface Card packets and sends them to the standard output. Snort keeps trace files in a number of formats on disc for subsequent analysis when in logger mode. Snort can be set up in IDS mode to cross-check network traffic against a range of intrusion detection signature rules in order to look for potential intrusions. Snort is set up on the suggested system in IDS mode. Alerts are sent as a result of Snort's traffic analysis, and the controller can utilize them to sort out any uncertain flows from the targets switch flow tables utilizing OpenFlow signaling primitive.

A typical intrusion detection system that uses signatures often keeps an eye on incoming traffic networks to spot sequence as well as design which matches specific attacks signatures. These patterns might be seen in packets header and data sequence which resemble malwares or dangerous patterns that are well-known to exist.

The term False Positive Rate (FPR) or False alarm Rate (FAR) refers to ratio of normal data wrongly identified as attack behaviour. A high FPR can significantly degrade the IDS performance, while a high False Negative Rate (FNR) can leave the system open to intrusions.

An intrusion alarm system is designed to monitor and identify any unauthorized access to a building. Such systems serve various purposes and are implemented in different settings, including residential and commercial spaces.

4. Result Analysis

The performance of the Hierarchical Hybrid Intrusion Detection System in Cloud Computing using Machine Learning has been evaluated based on its ability to detect attacks and the resources it consumes, including false alarms. The evaluation criteria used for the analysis are as follows:

The number of attacks that the intrusion detection system accurately detects is referred to as TP, or True Positive. The number of correct detections of typical flows is expressed as TN, or True Negative. The flow count in model serves as the denominator.

The number of attacks that Intrusion Detection System is indicated by FN, or false negative, indicator.

False Positive, or FP, refers to the volume of legitimate traffic that has been mistakenly classified as harmful. A high false alert rate causes valid ordinary flows to be unnecessarily removed and necessitates additional network administrator analysis and processing. The Detection Rate (DR) is a way to measure proportion of attacks have been detected, and it can be calculated as follows:

$$DR = \frac{TP}{TP + FN} \quad (1)$$

The ratio of false alarms to all detections is measured by the false alarm rate, or FAR. It is determined by:

$$FAR = \frac{FP}{TP + FP} \quad (2)$$

The system's overall wrong categorization is quantified by the Error Rate (ER), which is represented as:

$$ER = \frac{FP + FN}{TP + TN + FP + FN} \quad (3)$$

Accuracy: Accuracy means the ratio of exactly estimates for complete detections. Accuracy can be mentioned as capability to accurately detect result of a situation.

$$Accuracy = \frac{TP + TN}{TP + TN + FN + FP} \quad (4)$$

Table 16.1 Performance analysis table

Parameters	SVM	KNN
Accuracy	97%	89%
Error Rate	10.43%	12.89%
False Alarm Rate	11.18%	15.63%
Detection Rate	86.65%	56.89%

Graphical representation of Accuracy for SVM and KNN is shown in below Fig. 16.2.

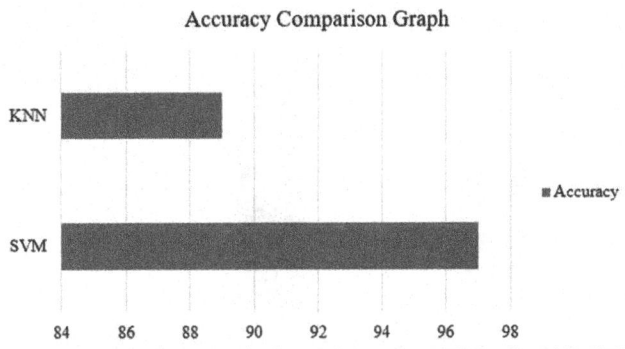
Fig. 16.2 Accuracy comparison graph

Graphical representation of Error Rate of Hybrid Intrusion Detection System for SVM and KNN is represented in below Fig. 16.3.

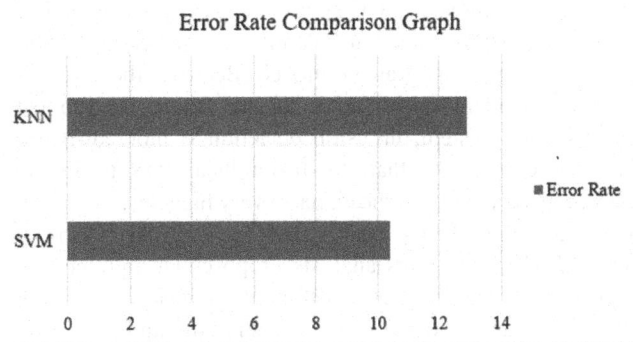
Fig. 16.3 Error rate comparison graph

Graphical representation of Detection Rate of Hybrid IDS for SVM and KNN is represented in below Fig. 16.4.

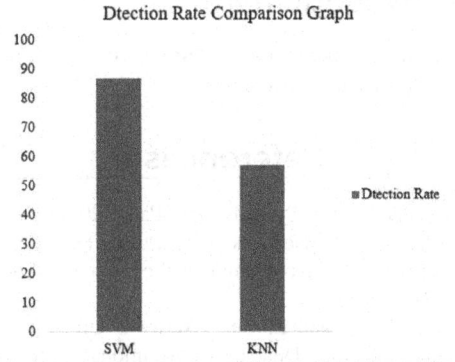
Fig. 16.4 Detection rate comparison graph

Graphical representation of False Alarm Rate for SVM and KNN is presented in below Fig. 16.5.

Therefore, accuracy, error rate, detection rate and false alarm rate comparison for SVM and KNN is observed in this section. The SVM shows better results.

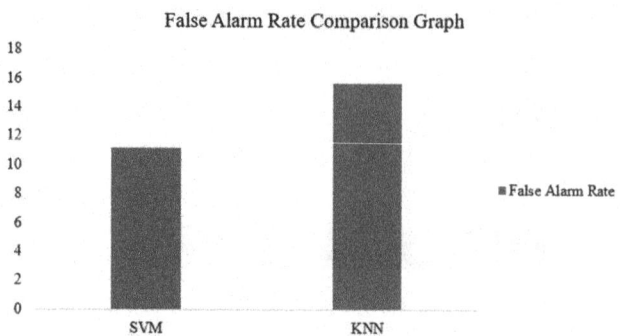

Fig. 16.5 False alarm rate comparison graph

5. Conclusion

In this analysis, a hybrid IDS that can detect both known and unknown threats in CC environments is presented. The detection model utilized to identify anomalies in recent events is the SVM trained model. Because there aren't enough redundant records in the labeled dataset used to test the detection system, intrusion detection is more difficult. The research shows that the hierarchical IDS technique consumes very little overhead, had a very high detection rate, a very low rate of false positives errors. In order to build a strong IDS system, this analysis proposes an architectural framework that integrates different intrusion detection systems hierarchically, utilizing their strengths and correcting their faults at the next level of the hierarchy. As a result, in this system SVM has a low rate of false alarms errors and a high accuracy and detection rate.

6. Acknowledgement

The authors would like to thank their University for providing facilities to complete this work.

References

1. Pinal J. Patel, Dr. J. S. Shah, Jinul Patel (2017) Performance Analysis of Neural Networks for Intrusion Detection System, International Journal of Computer Technology & Applications Vol 8(2) 88–93

2. Y. Yu, J. Long, and Z. Cai, "Network Intrusion Detection through Stacking Dilated Convolutional Autoencoders," Security and Communication Networks, vol. 2017, pp. 1–10, 2017, doi: 10.1155/2017/4184196.

3. A. Singh, "Jupiter Rising: A Decade of Clos Topologies and Centralized Control in Google's Datacenter Network", In Proceedings of the 2015 ACM Conference on Special Interest Group on Data Communication (SIGCOMM '15). ACM, New York, NY, USA, vol. 45, no. 4, pp. 183–197.

4. G. Ditzler, M. Roveri, C. Alippi, and R. Polikar, "Learning in nonstationary environments: A survey," IEEE Computational Intelligence Magazine, vol. 10, no. 4, pp. 12–25, 2015

5. D. Alsmadi IzzatXu, "Security of software defined networks, Computers and Security, vol. 53, pp. 79–108, 2015.

6. C J Chung, P Khatkar, T Xing, NICE: Network Intrusion Detection and Countermeasure Selection in Virtual Network Systems[J]. IEEE Transactions on Dependable & Secure Computing, 2013, 10(4): 198–211.

7. M. J. Reed, "Methodologies for detecting dos/ddos attacks against network servers", in The Seventh International Conference on Systems and Networks Communications ICSNC, 2012, pp. 92–98.

8. M. Alenezi, "Methodologies for detecting dos/ddos attacks against network servers", in The Seventh International Conference on Systems and Networks Communications ICSNC, 2012, pp. 92–98.

9. Vinchurkar, D. P., and Reshamwala, A.: A Review of Intrusion Detection System Using NN and Machine Learning Technique. International Journal of Engineering Science and Innovative Technology, vol. 1, issue 2,(November 2012).

10. P. M. Mell and T. Grance, "The NIST definition of cloud computing," National Institute of Standards and Technology, 2011.

11. Karegowda, A., Manjunath, A. S., and Jayaram, M. A.:Comparative study of attribute selection using gain ratio and correlation based feature selection. International Journal of Information Technology and Knowledge Management, vol. 2, issue 2, pp. 271–277, (2010).

12. C.F. Tsai, Y.F. Hsu, C.Y. Lin and W.Y. Lin, "Intrusion detection by machine learning: Expert Systems with Applications; vol. 36(10): 11994–2000, 2009.

13. A. Hushyar, "Network traffic clustering and geographic visualization," SAN JOS STATE UNIVERSITY, 2009.

14. K. Scarfone and P. Mell, "Guide to intrusion detection and prevention systems (idps)", NIST special publication, vol. 800, no. 2007, pp. 9–14, 2007.

15. V.Kumar, J.Srivastava, A.Lazarevic, Managing cyber threats: Issues, approaches and challenges. Springer Science & Business Media, 2006, vol. 5.

16. G. E. Hinton and R. Salakhutdinov, "Reducing the Dimensionality of Data with Neural Networks," Science, vol. 313, no. 5786, pp. 504–507, 2006, doi: 10.1126/science.1127647.

17. G. E. Hinton, S. Osindero, and Y.-W. Teh, "A Fast Learning Algorithm for Deep Belief Nets," Neural Computation, vol. 18, no. 7, pp. 1527–1554, 2006, doi: 10.1162/neco.2006.18.7.1527.

18. O. Depren, M. Topallar, E. Anarim, and M. K. Ciliz (2005) An intelligent intrusion detection system (IDS) for anomaly and misuse detection in computer networks, Expert systems with Applications 29 (4) pp. 713–722

19. Li, C., Song, Q., & Zhang (2004) MA-IDS: Architecture for distributed intrusion detection using mobile agents, Proceedings of 2nd International Conference on Information Technology for Application (ICITA)

20. Rifkin, Ryan, Klautau, and Aldebaro, "In Defense of One-Vs-All Classification," Learning Research, vol. 5, no. 1, pp. 101–141, 2004.

Note: All the figures and table in this chapter were test data by authors.

Design and Development of Secured Blockchain-based Platform in Edge Computing Environment

17

Goli Archana[1], Rajeev Goyal[2],
K. M. V. Madan Kumar[3]

Amity School of Engineering & Technology,Gwalior MP,
3 TKR Engineering College, Hyderabad

Abstract: The rise of cloud computing and the Internet of Things enabled edge computing, which brings processing, storage, networks, and other infrastructure closer to users. Edge computing, by addressing the issues of slow connection times and high convergence traffic, improves support for low latency and high bandwidth applications when compared to typical cloud computing's centralized deployment model. Edge computing security and privacy issues have evolved as a result of the increased data collection by IoT users and devices. Many businesses, including banking and insurance, have begun to embrace blockchain. In the last couple of decades, this security technology has evolved dramatically. Because of the edge computing capabilities, installing blockchain platforms and apps on edge computing platforms can provide network edge security services. Edge computing has emerged as a transformative technology, enabling low-latency processing and real-time decision-making for a wide range of applications. However, the decentralized nature of edge computing also presents unique security challenges. To address these challenges, this research focuses on the design and development of a secured blockchain-based platform tailored for edge computing environments. The value of the proposed platform is demonstrated by measuring the round-trip timings (RTT) of several important workflows. The results of the test demonstrate that the platform can meet availability standards in real-world circumstances. Internet of things, microservices, blocks chain, and edge computing are some buzzwords.

Keywords: Edge computing, Blockchain, Micro service, Internet of things

1. Introduction

The expanding Internet of Things (IoT) is currently pushing new products and services that are extensively used in everyday life and playing a key role in the real world [1] through the development of Autonomous operations and

Interactions. Because of the increase in access devices, cloud computing technologies utilized in the Internet of things will inevitably result in excessive user latency. Simultaneously, a large amount of data generated by various IoT devices is being sent to the cloud for compute and storage services, which have the potential to centralize risk [2,] require high performance from cloud platforms, place a significant burden on the network's infrastructure, and require high bandwidth on the network.

Some believe that edge computing can help to solve these problems. A new paradigm known as "edge computing" provides tools that enable processing to occur at the network's edge, close to the source of information [3]. By using edge servers to process some of the data, edge computing minimizes the amount of data transferred to the cloud and clears up bandwidth congestion. The service responds to the terminal more quickly at the same time. Many IoT use cases, including smart manufacturing, smart cities, and smart homes, are made possible by edge computing. Because they operate in a hostile environment, these edge devices might not always cooperate [4]. Therefore, combining blockchain networks with edge computing frameworks in the edge environment

[1]archanagoli44@gmail.com, [2]goyal.rajeev@gmail.com, [3]kvmmadhan@tkrec.ac.in

In this study, we provide a secure authentication technique for edge and computing client that blends an edge computing architecture with a blockchain network. We build a trustworthy edge platform based on microservice architecture to satisfy the requirements for edge computing, access the Edgex Foundry framework, and enhance Edgex's capability using the edge application module. By dividing the blockchain interaction process into many microservices, a secure identity module is produced. The security authentication micro services are in charge of handling access control and authentication. Develop an API a gateway that can examine requests coming from outside. The endpoints for the internal security authentication micro service and Egdex then receive the request.

The network's cutting-edge nature is generally recognised. The trusted edge platform's micro services architecture allows for high scalability. Docker containers, commonly known as containers, are used on the operating system to construct services. To strengthen the platform's ability to manage a variety of data types, the EdgeX Foundry architecture has been included.

The Hyper-ledger Fabric blockchain network's envisioned platform access offers strong security measures for the edge environment. The blockchain stores both the identities of the user and the terminal. Before being processed by internal services, external requests must first be authenticated and have access restricted.

2. Related Works

This section will look at the most recent edge computing and blockchain integration frameworks. The co-authors of [12] propose a distributed blockchain cloud architecture that uses Software-Defined Networking (SDN) to support edge computing (fog nodes at the network's edge), which may be separated into three layers: device, fog, and cloud. All SDN controllers can now have distributed connections thanks to blockchain technology. If there are inadequate computing resources, a fog node can distribute computational workloads to the cloud and communicate data processing results to the distributed cloud and system layers. [13] suggests a multiple-layer internet-of-things (IoT) concept with a blockchain at its foundation that addresses many concerns. Independent. Data management occurs in each node independently. A blockchain distributed ledger, which each node updates, is used to maintain track of data flows between nodes. The idea of a multi-layer network based on blockchain technology opens the door to the development of efficient and secure wide-area Network of IoT devices. As examples, the suggested video surveillance system in [14] makes use of edge computing, IPFS, blockchains, and convolution neural networks. Block

chain technology strengthen and dependability of the system. Edge computing is used to collect, analyze, and interpret sensory data from numerous wireless sensors. IPFS and CNNs are both used to achieve massive visual data management and real-time monitoring.

The authors of [15] claim that a solution based on blockchain identity management and access control has been developed to address the issue of edge computing security in the IIoT. A self-certified password technique is used to authenticate and register network entities. They proposed a simple key agreement based on self-certified a public key that provides IIoT authentication, audit ability, and confidentiality.

The bandwidth and latency of the wireless communications were increased by the [16] developers using mobile communications, compute, and caching (3C) technology. They point out that smart city systems' high data transmission requirements can put a major strain on the existing transmission capacity. The current wireless infrastructure may soon run out of space if edge computing and caching technologies aren't employed. Then, they created a blockchain database to handle the issue of communication security between smart cities, home appliances, and sensors.

3. Trusted Edge Platform Design Based on Blockchain in Edge Computing

For use in an edge computing environment, we propose a trustworthy edge platform built on a blockchain network based on the micro service Participants in the edge computing architecture principle. Figure 17.1 shows the following:

The device that links the internet to the outside world is called a terminal. Depending on their computer power, they might be classified as smart devices or resource-constrained gadgets. While devices with limited resources can just detect and acquire ambient data, smart devices may perform a variety of tasks like initial data processing, encryption, and transport. We categorize the terminals in the edge computing environment into two groups: execution devices and acquisition devices. A variety of communication protocols, including NB-IoT, LoRa, WiFi, NFC, and others, are used to connect the terminals to the nearby edge computing servers. Using the RF module or sensor module at the front end, the terminal detects changes in its surroundings, computes the necessary responses, and then decides what to do.

The device that links the internet to the outside world is called a terminal. Depending on their computer power, they might be classified as smart devices or resource-constrained gadgets. While devices with limited resources can just detect and acquire ambient data, smart devices may perform a variety of

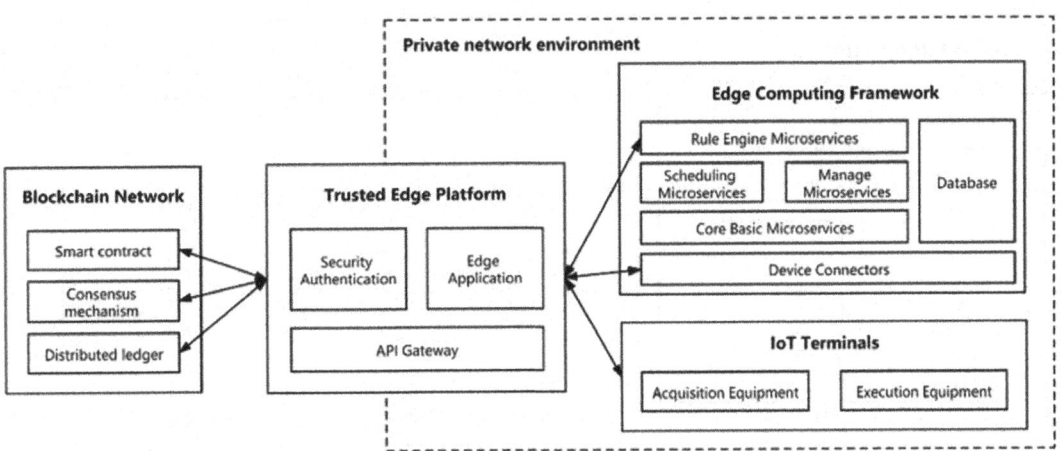

Fig. 17.1 Edge computing environment [3]

tasks like initial data processing, encryption, and transport. We categorize the terminals in the edge computing environment into two groups: execution devices and acquisition devices. A variety of communication protocols, including NB-IoT, LoRa, WiFi, NFC, and others, are used to connect the terminals to the nearby edge computing servers. Using the RF module or sensor module at the front end, the terminal detects changes in its surroundings, computes the necessary responses, and then decides what to do.. These components are required for the conceptual definition. Blockchain networks provide dependable, decentralized, intelligent solutions to address edge computing are security and privacy issues. A number of edge nodes can be combined to form blockchain networks. The processing One of its limitations is the power of the terminal devices. They are consequently excluded from a blockchain network. The identities of each of these terminals are stored on the blockchain network. Authorized terminal companies can access blockchain data shortly after the blockchain has confirmed identification information, but only edge servers can publish. In the edge computing context, a decentralized blockchain network is utilized to store device identity information, access control information, and access control policies. This effectively halts data alterations.

Hyper ledger Fabric aims to establish a flexible and scalable blockchain development platform in order to deliver solutions for enterprise-class blockchain applications. The Hyper ledger Fabric blockchain system is comprised of the Client, CA, Peer, and Ordered components. Users can send transaction requests to the hyper ledger Fabric network via the client, which serves as the platform's access point. At the CA node, which serves as the Hyper ledger Networking Fabric's license authority centre, users are added, registration certificates are issued, certificates are renewed, and certificates are revoked. Order is unconcerned about the specifics of a transaction's content because it is not involved in the transaction's execution or validation.

The basic purpose of the order is to choose the order in which transactions are generated, and the outcome is broadcast. The Endorser and Order Tiered Trust Paradigm are used to split the consensus process. Between the underlying consensus and the application layer's confidence. The architectural benefits of Hyper-Ledger Fabric make deployment on edge servers simple. Hyper-ledger Fabric makes it easy to organize much node collaboration. With the support of the security authentication implements registration services, Hyper-ledger, identity authentication, network access control. Security management and Fabric blockchain.

3.1 Module for the Edge Application

By combining the Edgex Foundry architecture with the edge application, edge-side business capabilities may close the loop on local data. Users can utilize the edge program module to develop operational routines to suit terminal needs and run complicated models such as neural networks. Various types of software and elements are required when developing edge applications within an edge computing environment to ensure the successful deployment and operation of these applications.

The following list of micro services works together to completely dispatch edge computing resources:

1. *Implementation Process:* The program's Application Process Service replies to terminal task requests. The task manager, task queue, and task scheduler are the three components of the service. The task scheduler assigns an ID to each task that must be completed after receiving a request for it. The task name and source data are recorded in the database using the ID as a key. The ID is subsequently added to the task queue by the scheduler. The scheduler might alter the scheduling rules in order to repeat certain jobs. Only the task manager and scheduler have access to the task queue's

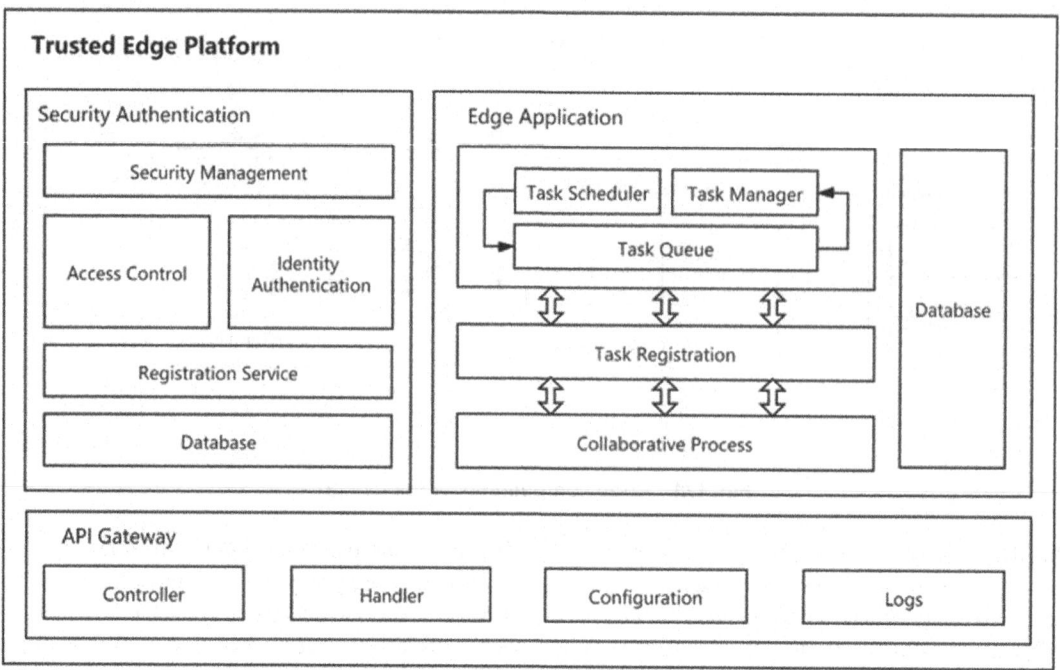

Fig. 17.2 The architecture of the trusted edge platform [5]

contents, which contains the IDs of all outstanding jobs. The task manager function retrieves the task ID from the task queue and checks for it. The manager provides policies for handling overload, thread selection, and idle time to control how platform resources are used by tasks.

2. *Task Registration:* The scheduler can change the scheduling rules to ensure that specific jobs run on a regular basis. The task queue's contents, which include the IDs of all pending jobs, are only accessible to the task manager and scheduler. The task manager takes the job ID from the task queue and uses it to search the database for task information before conducting the jobs. The task manager provides policies for handling overload, thread selection, and idle time to determine how the system's resources can be utilized by tasks.

3. This kind of service can be used to set up many edge computing nodes in order to facilitate collaborative work. Locally, the Edgex Foundry platform provides enough interoperability; however, multi-node collaboration necessitates the use of a different platform to manage the services. The collaborative process service is self-organizing and does not require the involvement of a third-party organization. Many edge nodes employ the Raft algorithm to select a leader node. The leader node manages the entire collaborative process and can update smart contract templates and create Hyperledger Fabric channels. The top-level node

combines and authenticates the smart contract edits made by the member nodes.

4. *Database:* Applications are stored in databases, which also provide cache services for them while they are in use. The database contains access privileges and allows users with different identities access to various add, remove, and check functions.

3.2 Security Authentication Module

A Secure Authenticity Module, also known as an authentication module or authentication provider, is a software or hardware component that is critical in validating the identity of persons or entities attempting to access a system, application, or network. Authentication is a critical component of information security because it ensures that only authorized people or systems have access to protected resources. Here are some important factors to remember about Security Authentication Modules:

1. *Registration Service:* Before using Edgex Foundry services, all terminal entities in the proposed edge platform must register with the registration microservice. The registration service generates Info-Hash, a hash of identification information used as authentication credentials when authenticating, using the SHA-256 algorithm. A new peer node is created by the hyper ledger Fabric ledger.

2. *Identity Authentication:* The client supplies the user's identity authentication service with the identification

of the identity data (VID), the identity plaintext data (Info), and the digital signature of the private key (sign (sk, Info)). The identity authentication service receives the required InfoHash data from the VID-based smart contract. The Hash (Info) and InfoHash are examined by the identity authentication service to check that the Info provided by the terminal is legitimate and consistent with the identity data acquired. The identity authentication service validates the terminal's private key by verifying the computer's digital signature.

3. *Access Control:* To gain access to a service or resource at a trusted edge platform service provider, the terminal must first send an access permission request to the access control service in order to obtain a permission token. The access prevention service evaluates the access request using an authorization policy based on the visitor's identifying information. (4) Security Management: The Security Management microservice acts as a data and security-related service manager for the submitted data and access control rules in the Security Authentication module. The access control policies are updated and maintained by the security management service.

3.3 API Gateway

The application programming interface (API) gateway extends the Edgex Foundry architecture with a device access layer, enhancing its capacity to design security control schemes while meeting terminal access requirements. An HTTP listener is provided by the API gateway to accept client queries. The API gateway requires the controller to parse the requests it receives, the handler to invoke the platform's internal micro services to provide business logic, the configuration to store API information about the configuration, and the logs to keep track of the API gateway's essential information from the past. The Application Programming Interface (API) Gateway is a fundamental part of present-day software structures, particularly microservices-based and serverless systems. It serves as a single point of contact for clients (such as online or mobile applications) and the backend services that offer the functionality.

4. Key Process Design of Trusted Edge Platform

Understanding how to use the edge computing platform and blockchain is critical to the platform design proposed. Edgex Foundry, which provides excellent interoperability, handles the system's data collection, integration, and heterogeneous processing. On the specified platform, microservice-based Hyperledger Fabric deployments outperform. More

information regarding the platform's utilization of blockchain and edge computing may be found below.

The user submits an API gateway request to register a terminal device after the system has been initialized. The registration microservice passes the request it receives from the API gateway processor. The registration service starts a transaction procedure to store the InfoHash onto the blockchain after the peer node has finished deploying the smart contract. After the transaction is finished, the virtual ID (VID) used to query the InfoHash is returned. The terminals receive the public-private key pair and VID from the registration service. Utilizing Edgex's device service API, the handler then registers the device. Sending and receiving data settings on the terminal as well as pertinent information. Figure 17.3 registration. Shows the process flow diagram for device

In order to be able to use the authentication method provided by the micro-service, the terminal must give the identifying data identifier (VID), identity data (Info), and terminal private key digital signature (sign (sk, Info)). The identification service receives the required InfoHash identifying information from a smart contract based on the VID. After hashing Info and verifying to see if Hash (Info) equals InfoHash, the authentication service can verify whether the data provided by the terminal is accurate and compatible with the identity data that has been saved.

5. Experimental Study

Two distinct high-performance hardware components are used in the construction of the edge computing network. In order to take advantage of the power of the edge server, we only require a desktop with a 2.3 GHz Intel Core i7 (8-core) CPU and 16 GB RAM. Because of its portability, compact size, and low power consumption, the Raspberry Pi is a good candidate for distributed computing. Our high-performance computing environment was built on a Raspberry Pi 4 with a 1.7 GHz Cortex-A72 CPU and 4 GB of RAM.

Three desktop PCs and two Raspberry Pi devices are utilized in a real network environment to evaluate the usability of the proposed platform. Each gadget has a blockchain ecosystem and edge computing installed. We employ simulated IoT endpoints from Node-RED. Edgex Foundry is deployed using the Geneva version. Hyperledger Fabric version 2.1.0 is now available. The Hyperledger Fabric block chain network is made up of five Fabric organizations. Five physical devices scattered across the network have been assigned seven Orders at random. Each organization begins with only one peer node. Docker Swarm is used to establish an overlay network layer that allows communication between containers on various hosts. Figure 17.4 shows a schematic representation of the experiment apparatus's structure scheme.

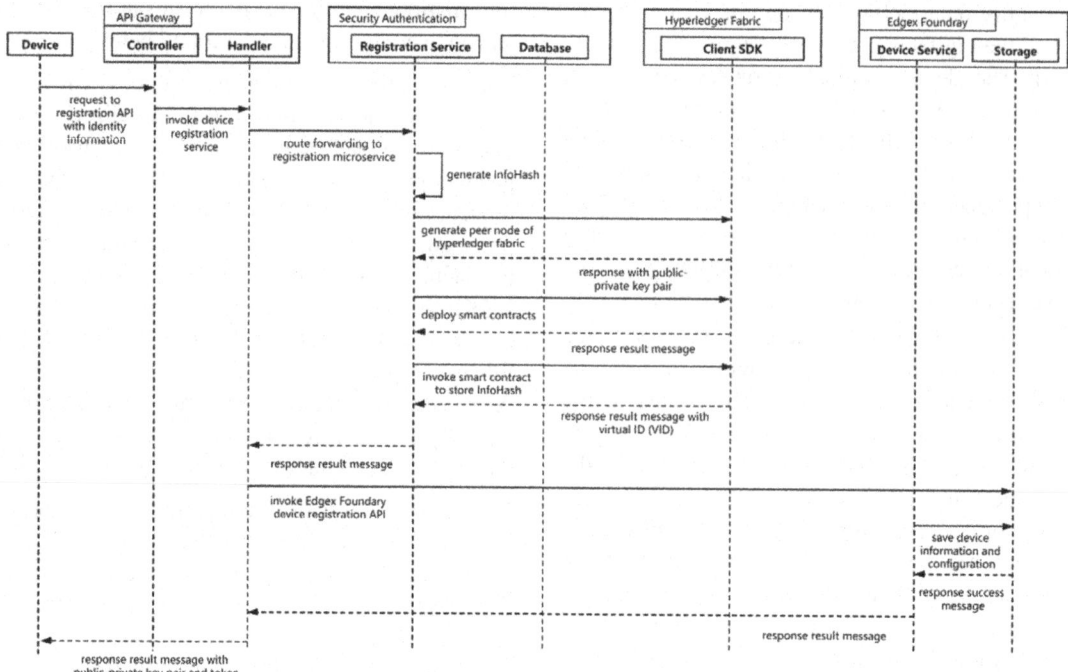

Fig. 17.3   Sequence diagram for device registration

Source: Authors

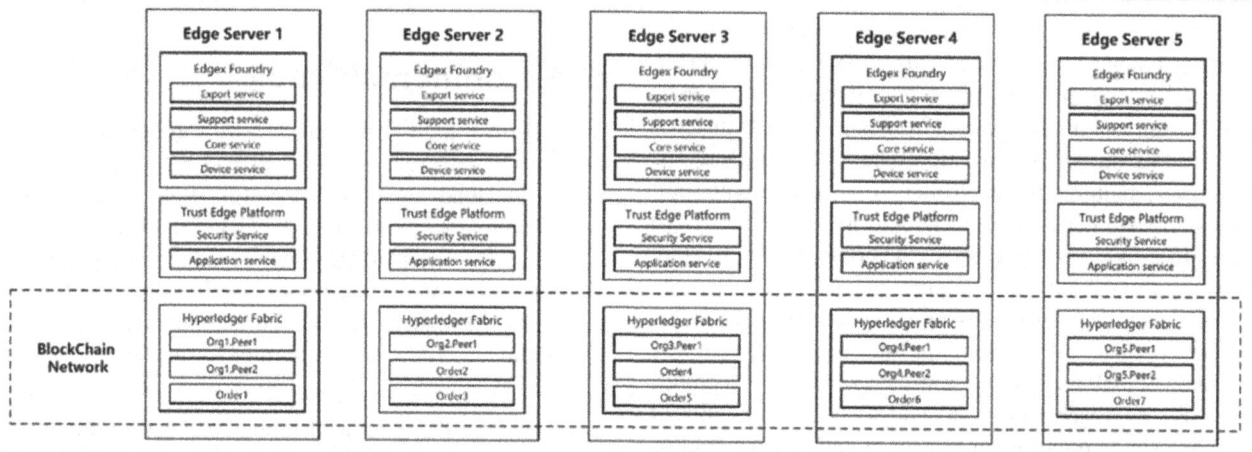

Fig. 17.4   A schematic diagram of the experimental equipment organization plan

Source: Authors

We chose two desktop PCs and a Raspberry Pi for a multi-node interoperability test to assess the total cost incurred by the suggested operating system as a result of edge device time spent on processing. The Raspberry Pi serves as the data generator by connecting a RaspiCam camera to the Edgex Foundry platform. For data processing and picture evaluation, the image recognition model is activated on a single desktop. The second computer transmits the command to the simulation terminal via Edgex Foundry in response to the picture identification findings on the command execution side. The experimental setup is depicted in Figure 17.5. To

test the platform's capabilities, we constructed an image recognition service. A cloud service provider can supply you with a cloud-based server.

6. Result Analysis

In the final part of the paper, we ran Round-Trip Time (RTT) tests on the system workflows to assess the platform's usability in a real-world scenario. The acquired results were comparable to the human response times [26,27], which assess how quickly a person responds to different stimuli.

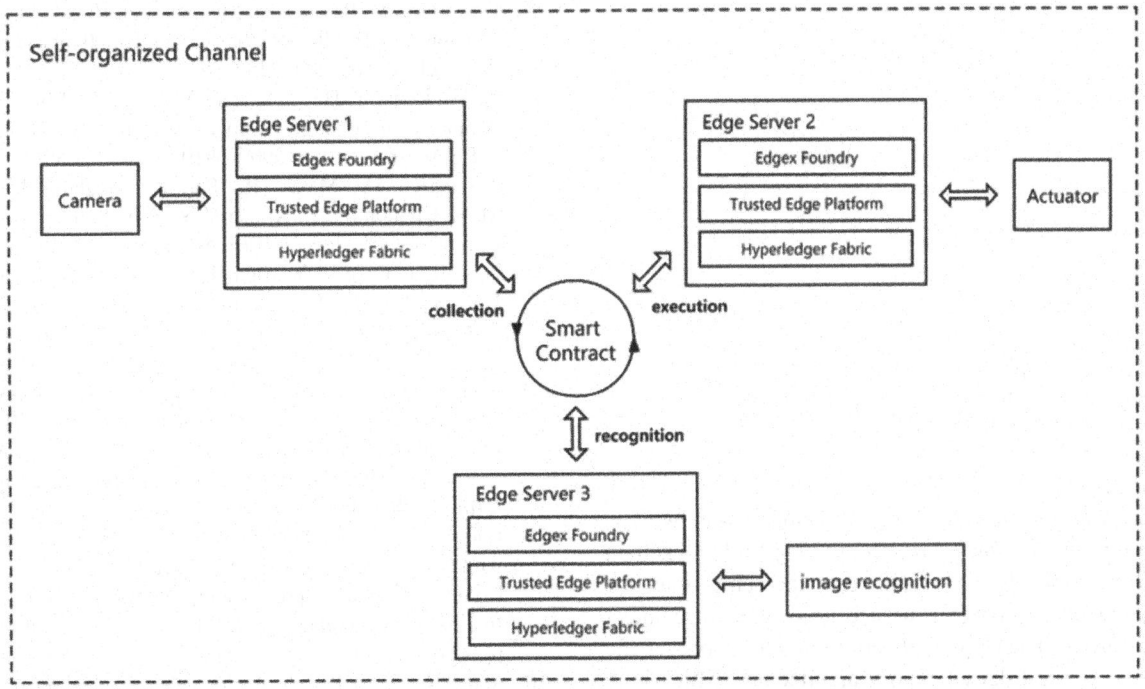

Fig. 17.5 Diagrammatic representation of an experiment involving multi-node collaboration

Source: Authors

The proposed platform can be employed in practice if the experimentally determined RTT is smaller than the response time. After examining tens of millions of recordings, the statistically average reaction time was revealed to be 215 milliseconds.

In a reliable network environment, comparative research on multi-node cooperative activities is carried out. Task flow completion times were computed during the trials at varied acquisition frequencies. The article in [28] suggests a micro services safe agent platform that combines API gateway and edge computing platforms. By measuring the RTT of the REST API interface, it confirms that the system is accessible. The tolerable availability requirements are met by the system's average execution time, which ranges from 3 to 51.7 milliseconds. The platform developed in this work can complete a task between 25 and 183 milliseconds, as shown in Fig. 17.6. The platform's time consumption is adequate when compared to the requirements of the blockchain network. The use of blockchain technology also significantly enhances this secure storage, and identity authentication. Document in terms of edge node compatibility,

Based on the data in Fig. 17.6, the task is When the smartphone's camera is set to a low sample rate, the flow of edge computing-based collaboration and cloud-based services takes about the same amount of period. Part of the time, the edge-side blockchain-based security mechanism is active. Although edge computing is still in its infancy,

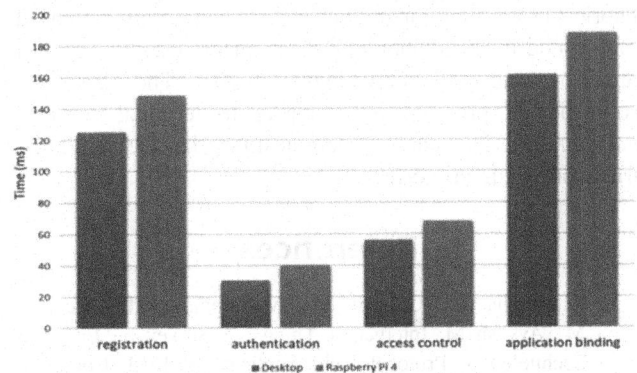

Fig. 17.6 Statistics on how long platform-critical processes typically take

Source: Authors

as camera capture rates increase, edge computing develops faster than cloudservices.

6. Conclusions

We provide a powerful edge platform based on blockchain and edge computing in this article. To address privacy and security concerns at the edge, the platform employs a micro service architecture as well as blockchain-based authentication and access control approaches. To assess the feasibility of the proposed platform, an idea for a conceptual

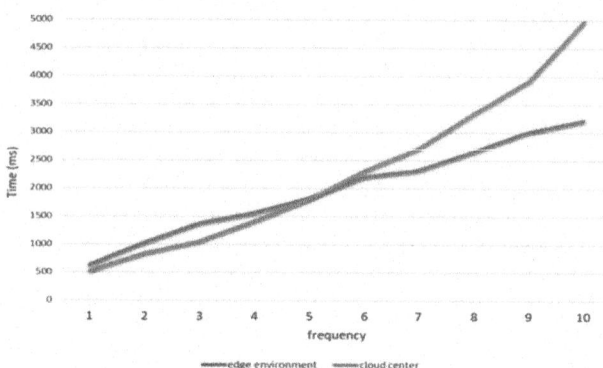

Fig. 17.7 Average time consumption of multi-node
cooperative experiment

Source: Authors

validation prototype of a physical edge computing network is
developed. When terminals, edge computing, and blockchain
are utilized in an edge environment at the same time, the
trusted edge platform works as a central node to connect
the three parties. The edge application module improves the
intelligence of the Edgex Foundry architecture, which may
be used to host various apps. Using Hyperledger Fabric's
blockchain capabilities, we created a platform-based multi-
node inter-collaboration mechanism. Using Hyperledger
Fabric's blockchain capabilities, we created a platform-based
multi-node inter-collaboration mechanism. There have been
numerous successful experimental testing. The platform's
sophisticated programs can answer to terminal requests
very quickly. This analysis demonstrates that the suggested
trustworthy edge is correct.

References

1. Al-Fuqaha A., Guizani M., Mohammadi M., Aledhari
   M., Ayyash M. Internet of Things: A Survey on Enabling
   Technologies, Protocols, and Applications. IEEE Commun.
   Surv. Tutor. 2015; 17: 2347–2376.
2. Lytras M.D., Sarirete A. Innovation in Health Informatics: A
   Smart Healthcare Primer. Academic Press; Cambridge, MA,
   USA: 2020
3. Shi W., Cao J., Zhang Q., Li Y., Xu L. Edge Computing: Vision
   and Challenges. IEEE Internet Things J. 2016; 3: 637–646.
4. Dastjerdi A.V., Buyya R. Fog Computing: Helping the Internet
   of Things Realize Its Potential. Computer. 2016; 49: 112–116.
   doi: 10.1109/MC.2016.245.
5. Yu W., Liang F., He X., Hatcher W.G., Lu C., Lin J., Yang X.
   A Survey on the Edge Computing for the Internet of Things.
   IEEE Access. 2018; 6: 6900–6919.
6. Mukherjee M., Matam R., Shu L., Maglaras L., Ferrag
   M.A., Choudhury N., Kumar V. Security and Privacy in Fog
   Computing: Challenges. IEEE Access. 2017; 5: 19293–19304.
7. Nam H.-J., Choi H.-Y., Shin H.-J., Kwon H.-S., Jeong J.-M.,
   Hahn C.-H., Hur J.-B. Security and Privacy Issues of Fog
   Computing. J. Korean Inst. Commun. Inf. Sci. 2017; 42: 257–
   267.
8. Huh S., Cho S., Kim S. Managing IoT Devices Using
   Blockchain Platform; Proceedings of the 2017 19th
   International Conference on Advanced Communication
   Technology (ICACT); Pyeongchang, Korea. 19–22 February
   2017; pp. 464–467.
9. Luo C., Xu L., Li D., Wu W. Edge Computing Integrated
   with Blockchain Technologies. In: Du D.-Z., Wang J., editors.
   Complexity and Approximation. Volume 12000. Springer
   International Publishing; Cham, The Netherlands: 2020.
   pp. 268–288.
10. Yang R., Yu F.R., Si P., Yang Z., Zhang Y. Integrated
    Blockchain and Edge Computing Systems: A Survey, Some
    Research Issues and Challenges. IEEE Commun. Surv. Tutor.
    2019; 21: 1508–1532.
11. Liu B., Yu X.L., Chen S., Xu X., Zhu L. Blockchain Based
    Data Integrity Service Framework for IoT Data; Proceedings
    of the 2017 (ICWS); Honolulu, HI, USA. 25–30 June 2017;
12. Sharma P.K., Park J.H. Blockchain Based Hybrid Network
    Architecture for the Smart City. Future Gener. Comput. Syst.
    2018;86:650–655. doi: 10.1016/j.future.2018.04.060.
13. Li C., Zhang L.-J. A Blockchain Based New Secure Multi-
    Layer Network Model for Internet of Things; Proceedings of
    the 2017 IEEE International Congress on Internet of Things
    (ICIOT); Honolulu, HI, USA. 25–30 June 2017; pp. 33–41.
14. Wang R., Tsai W., He J., Liu C., Li Q., Deng E. A Video
    Surveillance System Based on Permissioned Blockchains and
    Edge Computing; Proceedings of the 2019 IEEE International
    Conference on Big Data and Smart Computing (BigComp);
    Kyoto, Japan. 27 February–2 March 2019; pp. 1–6.
15. Ren Y., Zhu F., Qi J., Wang J., Sangaiah A.K. Identity
    Management and Access Control Based on Blockchain under
    Edge Computing for the Industrial Internet of Things. Appl.
    Sci. 2019; 9:2058. doi: 10.3390/app9102058.
16. Kotobi K., Sartipi M. Efficient and Secure Communications in
    Smart Cities Using Edge, Caching, and Blockchain.

Event Cameras for Parkinson's Disease Diagnosis: A Preliminary Study

18

Aruna Thethali[1]

Department of CSE, GITAM (Deemed to be University), Rushikonda,
Visakhapatnam, Andhra Pradesh, India

Mandava Kranthi Kiran[2]

Assistant Professor, Department of CSE, GITAM (Deemed to be University),
Rushikonda, Visakhapatnam, Andhra Pradesh, India

Abstract: Parkinson's disease (PD) is a neurological disease characterised by motor dysfunction. An early and accurate diagnosis is crucial for effective treatment. Traditional diagnostic methods have many limitations in capturing and accurately detecting mild motor impairments. Event cameras, with their high temporal resolution and unique data representation, offer a promising approach to Parkinson's disease diagnosis. They monitor variations in pixel intensity and produce sparse events to collect exact motion data. Using event camera data and machine learning algorithms, the current study recommends an approach for monitoring hand motions, examining gait patterns, and determining the severity of tremors. According to preliminary investigations, event cameras can identify subtle motor changes and disease-specific biomarkers. The need to undertake larger-scale clinical research and improve data analysis methods must be addressed. Event cameras may be able to objectively evaluate motor impairments in PD, but more study is needed to determine their clinical use.

Keywords: Parkinson's, Neurological disease, Symptoms, Diagnosis, Traditional cameras, Event-based cameras

1. Introduction

A neurological illness called Parkinson's disease is characterised by tremors, bradykinesia (slowness of movement), rigidity, and loss of balance. Conventional Parkinson's disease diagnostic techniques mainly rely on subjective clinical evaluations, which may not be able to pick up on small motor deficits or provide accurate measures. Event cameras, also known as neuromorphic or dynamic vision sensors, have emerged as a promising technology that may help improve Parkinson's disease diagnosis[1]

Unlike conventional frame-based cameras, event cameras offer excellent temporal resolution and low latency by capturing visual data based on changes in pixel intensity. They are suitable for recording dynamic visible changes linked to tremors and fine motor movements in Parkinson's disease

(PD) due to this distinctive feature. Researchers want to create objective, quantifiable markers of motor impairments for the early detection and precise diagnosis of Parkinson's disease (PD) using event cameras[2]

This research study examines the possibilities for using event cameras to diagnose Parkinson's disease (PD). The paper looks at event cameras' fundamental principles and advantages over traditional cameras, particularly in capturing exact motion data and temporal dynamics. The current approaches for diagnosing PD with event cameras are summarised, including tracking hand motions, analysing walking patterns, and assessing the severity of tremors[3]–[5]

The benefits and difficulties of employing event cameras to diagnose Parkinson's disease (PD) are carefully examined in this research. It emphasises the necessity of improving

[1]athethal@gitam.in, [2]kmandava@gitam.edu

data analysis techniques, tailoring hardware for practical uses, and carrying out larger-scale clinical investigations to confirm the efficiency and dependability of event cameras in PD diagnosis.

Integrating event cameras into PD diagnosis has the potential to provide objective and quantitative measurements of motor impairments. This technology can capture subtle motor abnormalities and ` disease-specific biomarkers, revolutionising the early detection and monitoring of PD. Further research and development efforts are needed to establish event cameras' clinical utility and feasibility as a valuable tool in PD diagnosis and personalised treatment planning.[6]

2. Unveiling the Complexity of Parkinson's Disease

Parkinson's disease (PD) is a neurodegenerative condition that impacts the central nervous system, specifically the cells in the brain's substantia nigra area that produce dopamine. Tremors, slow movement (bradykinesia), stiffness (rigidity), and difficulties with posture are all motor symptoms of the condition. Nevertheless, Parkinson's disease extends beyond mere movement-related symptoms and can encompass cognitive impairments, mood disturbances, and disruptions in autonomic functions.[7]

Although the exact cause of Parkinson's disease is unknown, it is generally accepted that it is a complex disease caused by a combination of genetic and environmental factors. Specific genetic abnormalities have been linked to the start of Parkinson's disease, and environmental contaminants such as pesticides and heavy metals may enhance vulnerability. The degeneration of dopamine-producing cells in the brain alters the equilibrium of dopamine, a critical neurotransmitter involved in movement coordination. This results in the specific motor symptoms seen in Parkinson's disease.

Diagnosis of Parkinson's can be difficult, especially in the early stages when symptoms may be minor or ambiguous. Clinical evaluation, including a complete medical history, neurological examination, and assessment of motor symptoms, is used to make the diagnosis. The presence of two of the cardinal motor symptoms (tremors, bradykinesia, and stiffness) is frequently required for diagnosing Parkinson's disease. Other disorders that produce similar symptoms, such as essential tremor or drug-induced parkinsonism, must be ruled out.

Since Parkinson's disease (PD) progresses with a gradual loss of motor skills, impaired motor skills are crucial for diagnosing the condition. One of PD's early and most observable symptoms is frequent tremors, especially when

resting. Bradykinesia, typified by delayed movement, is a significant trait that includes slower movement and difficulties starting and finishing tasks. The stiffness and resistance felt during passive movement are referred to as rigidity. People suffering from the disease's final stages may have postural instability, making it more challenging to maintain balance and increasing the risk of falling.

Observation and assessment of these motor symptoms are vital in diagnosing PD. [8]Neurologists or movement disorder specialists evaluate the presence and severity of motor abnormalities through clinical examination and rating scales specifically designed to assess Parkinson's symptoms. Additionally, response to dopaminergic medications like levodopa can further support the diagnosis, as PD symptoms often show improvement with these medications.

Advances in technology, like wearables and motion sensors, are giving us new resources for objectively assessing motor issues. These technologies may detect movement patterns, abnormal gait patterns, and tremor intensity, providing vital information for identifying and monitoring diseases.

In brief, Parkinson's disease is a neurological condition marked by the death of dopamine-producing cells, resulting in motor system symptoms. These symptoms include tremors, bradykinesia (slow movement), rigidity, and postural instability. To accurately identify and treat Parkinson's disease, a thorough examination of motor symptoms, a clinical assessment, and eliminating of other potential illnesses are necessary.

3. Current Challenges in Parkinson's Disease Diagnosis

The lack of standardised diagnostic protocols makes diagnosing Parkinson's disease difficult[9]. The current challenges in the Parkinson's diagnosis are:

1. **The subjectivity of Symptoms:** Due to the considerable diversity in symptoms across people and the subjective nature of early-stage indicators, which can be subtle, easily misunderstood, or misconstrued, establishing an accurate and timely diagnosis of Parkinson's disease is difficult.

2. **Lack of Objective Biomarkers:** Parkinson's disease cannot be diagnosed with absolute certainty using biomarkers or particular tests; instead, the diagnosis is mainly based on clinical assessment and the existence of recognisable motor symptoms. The creation of objective biomarkers might significantly improve Parkinson's disease detection both early and accurately.

3. **Overlapping Symptoms with Other Conditions:** It can be challenging to distinguish Parkinson's disease

from other neurodegenerative or movement disorders merely based on symptoms since some signs, such as tremors or gait problems, might be present in various ailments. This difficulty in differentiating PD might lead to delayed or incorrect diagnoses.

4. **Diagnostic Delay:** Research indicates that it takes, on average, 1 to 3 years from the start of symptoms before Parkinson's disease is diagnosed. This delay hampers early intervention and the efficient management of the disease.

5. **Limited Access to Specialists:** Parkinson's disease is a complex condition, and movement disorders specialists' knowledge and skills are frequently required for an appropriate diagnosis. Unfortunately, it may be challenging to provide an accurate diagnosis in some places or healthcare systems where these specialists are not easily accessible.

6. **Lack of Standardised Diagnostic Criteria**: It has not yet been able to develop An established, conclusive method for diagnosing Parkinson's disease (PD) due to the wide variety of diagnostic criteria. Diagnostic techniques may differ and have differences because there is no universally accepted standard.

7. **Parkinson's disease heterogeneity:** Parkinson's disease (PD) has a wide variety of clinical symptoms and rates of development, making it heterogeneous. This variety makes it harder to diagnose accurately and necessitates individualised strategies.

Due to the lack of standardised diagnostic procedures, the diagnosis of Parkinson's disease is primarily based on the existence and evaluation of motor symptoms. Doctors frequently assess the patient's motor functions, such as tremors, stiffness, bradykinesia, and postural instability, to determine a diagnosis.

4. Exploring the Potential of Event Cameras in Parkinson's Disease Diagnosis and Movement Analysis

Event cameras, also known as neuromorphic or dynamic vision sensors (DVS), are a type of image sensor that differ from traditional cameras in their operation and output. While event cameras have not been extensively explored for Parkinson's disease (PD) diagnosis specifically, their unique features hold potential for various applications in movement analysis and monitoring[10]. Here is an overview of event cameras and some of their distinctive characteristics:

- **Asynchronous and event-based operation:** Unlike conventional cameras, event cameras do not capture frames at fixed intervals. Instead, they operate asynchronously and solely record changes in pixel-level brightness. These cameras detect and generate events only when there is a notable change in the scene, such as motion or contrast, rather than providing a continuous stream of images.

- **High temporal resolution:** Event cameras have an exceptionally high temporal resolution, commonly measured in microseconds. This enables the identification of quick movements or changes in the visual scene in a precise and time-sensitive manner.

- **Low power consumption:** Event cameras consume significantly less power than traditional cameras. Their event-driven nature means they only require power when there is a change in the scene, leading to reduced energy consumption.

- **High dynamic range:** Event cameras possess a broad dynamic range, enabling them to simultaneously capture well-lit and dark areas in a scene without experiencing saturation or sacrificing intricate details. This characteristic is advantageous in high contrast or fluctuating lighting settings.

Regarding Parkinson's disease diagnosis, event cameras have the potential to contribute to movement analysis in the following ways:

- **Fine-grained motion analysis:** The potential of event cameras to record exact temporal movement data enables in-depth analysis of the motor symptoms of Parkinson's disease (PD), such as tremors, bradykinesia, and aberrant motions. A better knowledge of the motor symptoms associated with Parkinson's disease (PD) is made possible by the event cameras' high temporal resolution and low latency, which provide valuable insights into the timing and features of motor impairment.

- **Objective evaluation of motor symptoms**: By effectively using the unique capabilities of event cameras, it is now possible to create algorithms that can automatically identify and quantify specific Parkinson's disease (PD)-related motor symptoms.. Using event data, these algorithms can examine tremor amplitude and frequency, providing objective clinical assessment and metrics for tracking the development of the condition.

- **Long-term monitoring**: Due to their low power consumption and efficient data representation, event cameras are perfect for ongoing, long-term surveillance of movement patterns in persons with Parkinson's. To capture and study motor symptoms while people go about their daily lives, they might be incorporated into wearable technology or other assistive devices, allowing for more extensive and ecologically responsible assessments.

It's crucial to remember that the use of event cameras in diagnosing Parkinson's disease is still relatively new, and more study and development are required to realise their potential fully. For a thorough and precise diagnosis and management of PD, event cameras must also be integrated with other clinical evaluations and diagnostic tools.

5. Comparing Traditional Frame-Based Cameras and Event-Based Cameras for Parkinson's Disease Diagnosis and Movement Analysis

Traditional cameras are commonly used in Parkinson's disease (PD) diagnosis for video recording and motor symptom analysis. They capture a continuous video stream that healthcare professionals specialising in movement disorders analyse. Based on visual observations, the experts assess the patient's motor symptoms, such as tremors, bradykinesia, rigidity, and postural instability. The recorded videos are also used with standardised rating scales to evaluate the severity and characteristics of motor symptoms. This information assists in differential diagnosis, treatment planning, and monitoring of the progression of Parkinson's disease.

Traditional frame-based and event-based cameras (also known as event cameras or neuromorphic sensors) have fundamental differences in operation and output[11]. Here's a comparison between these two types of cameras[12]:

Due to their low power consumption and efficient data representation, event cameras are ideal for continuous, long-term monitoring of movement patterns in Parkinson's disease (PD) patients. Integrating these cameras into wearable devices or assistive technologies allows for capturing and analysing motor symptoms during everyday activities. This integration enables more comprehensive and ecologically valid assessments, providing valuable insights into the impact of PD on individuals' daily lives.

6. Challenges and Limitations of Event-Based Cameras for Parkinson's Disease Diagnosis

While event-based cameras offer potential benefits for Parkinson's disease (PD) diagnosis, several challenges must be addressed to ensure effective utilisation [13]. Here are some challenges associated with event-based cameras for PD diagnosis:

- **Limited research and validation**: Parkinson's disease (PD) diagnosis using event-based cameras is still a new field that needs much study and validation. It is essential to carefully examine how event-based cameras stack up against conventional diagnostic techniques regarding accuracy, sensitivity, and specificity.

- **Complex data analysis**: The continuous stream of unique events produced by event-based cameras makes it challenging to create specialised computations and algorithms for data processing. It takes much work to glean pertinent information from event-based data and evaluate it in the context of PD diagnosis.

- **Spatial and temporal resolution limitations**: Event-based cameras may have lower spatial resolution than traditional cameras, impacting the level of detail captured in PD patients' movements and motor symptoms. Additionally, while event-based cameras offer a high temporal resolution, they may struggle to accurately capture specific movements, particularly in fast or complex motion scenarios.

- **Standardisation and comparison**: The development of defined methods, benchmarks, and reference standards for comparison is required because event-based cameras are unique in PD diagnosis. Using uniform standards when analysing event-based camera data and contrasting findings between research or clinical settings is crucial.

- **Integration with existing diagnostic tools**: A thorough PD diagnosis must incorporate event-based cameras with current diagnostic tools and tests. For a precise and trustworthy diagnosis, event-based camera data must be coordinated and validated with other clinical evaluations and diagnostic techniques.

- **Practical considerations**: Cost, availability, and user-friendliness can affect the broad acceptance and functional integration of event-based cameras for routine Parkinson's disease (PD) diagnosis in clinical environments.

Event-based cameras have much potential for diagnosing PD despite these difficulties. They are promising tools for the objective and quantitative assessment of motor deficits due to their distinct characteristics, which include high temporal resolution, low battery consumption, and accurate motion capture. Unlocking the full clinical potential of event-based cameras for PD diagnosis will depend on overcoming these difficulties through additional study, technological developments, and cooperation between researchers, clinicians, and industry players.[14]

Table 18.1 Comparison of traditional and event-based cameras

Properties	Traditional Cameras	Event-based Cameras
Operation:	These cameras capture a sequence of frames at a fixed rate (e.g., 24 frames per second). They record the entire scene at regular intervals, regardless of whether there are significant changes in the scene.	Event cameras operate asynchronously and capture pixel-level changes in brightness. They only output events when there is a significant change in the scene, such as motion or contrast changes. They do not provide continuous frames but rather a stream of individual events.
Temporal resolution:	These cameras typically offer a high temporal resolution, capturing a series of frames at a fixed frame rate. However, they may suffer from motion blur in fast-moving scenes.	Event cameras provide highly high temporal resolution, often in the microsecond range. They can capture rapid movements with minimal motion blur, making them suitable for analysing fast and precise motions.
Power consumption:	These cameras require continuous power to capture frames at a fixed rate, resulting in higher power consumption.	Event cameras consume significantly less power compared to frame-based cameras. They are event-driven and only require power when there are changes in the scene. This low power consumption is advantageous for long-term monitoring applications.
Dynamic range:	These cameras have a limited dynamic range, which can lead to overexposed or underexposed areas in scenes with high contrast.	Event cameras have a wide dynamic range, allowing them to simultaneously capture bright and dark areas without saturation or loss of details. This feature is beneficial for scenes with varying lighting conditions.
Data representation:	These cameras capture and store each frame as a complete image, resulting in significant data sizes and potentially high bandwidth requirements for storage and transmission.	Event cameras only capture and transmit pixel-level changes in brightness, resulting in sparse and efficient data representation. This data sparsity allows for reduced storage and bandwidth requirements.
Applications:	Frame-based cameras are widely used in various applications, including photography, videography, surveillance, and computer vision tasks that require complete frames for analysis.	Event-based cameras are still emerging, with potential applications in robotics, autonomous vehicles, augmented reality, and motion analysis. They are particularly suitable for tasks that require high temporal resolution and efficient data representation.

7. Paradigm Shift from Traditional to Neuromorphic Computing: Exploring the Potential

Despite the challenges, the potential of event-based cameras in Parkinson's disease (PD) diagnosis remains significant. These cameras offer promising prospects for objectively and quantitatively assessing motor deficits, thanks to their unique features such as high temporal resolution, low power consumption, and precise motion capture. Realising the complete clinical potential of event-based cameras in PD diagnosis will require addressing these difficulties through further research, technological advancements, and collaboration among researchers, clinicians, and industry stakeholders.[15] This study explores the idea of neuromorphic computing, highlighting both its distinctive characteristics and possible advantages.[16]

7.1 Getting to Know Traditional Computing

Traditional computing relies on the sequential execution of instructions and a distinct division between memory and processing units based on the Von Neumann architecture. Although this design has completed complex computations,

its intrinsic constraints mean it needs assistance in some areas.

7.2 Neuromorphic Computing

A change from the traditional computing paradigm is represented by neuromorphic computing. It utilises massively parallel processing units connected like the brain's neural networks, drawing inspiration from both the architecture and operation of the brain. The aim is to replicate the brain's capacity to receive information quickly, change with the environment, and carry out cognitive activities with high energy efficiency.[17]

Key Features of Neuromorphic Computing:[18]

- **Parallelism**: High parallelism is exhibited by neuromorphic computing systems, allowing for the processing of several tasks at once. This parallelism makes processing faster and more effective, making it suitable for managing vast amounts of data and challenging jobs.

- **Event-Driven Operation**: Neuromorphic computing works on an event-driven model, unlike classical computing, which continuously processes data. It concentrates on processing pertinent information when

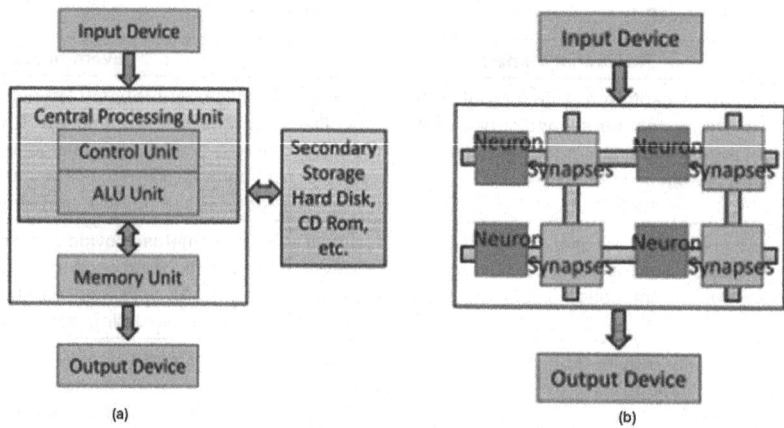

Fig. 18.1 (a) Diagram illustrating the von Neumann architecture, (b) Diagram illustrating a fundamental concept of neuromorphic architecture. (Nawrocki et al., 2016)

significant changes occur, simulating the brain's capacity to prioritise and react to salient inputs. This method improves computational efficiency while consuming less electricity.

- **Low Power Consumption**: Neuromorphic computing architectures leverage the brain's energy-efficient design, consuming significantly less power than traditional computing systems. By minimising unnecessary computations and adopting efficient data representation techniques, neuromorphic systems enable energy savings and longer battery life.

- **Adaptive Learning**: Neuromorphic systems excel at unsupervised learning and real-time adaptation. They can learn and optimise their performance based on input data, allowing continuous improvement and adaptation to changing environments.

8. Gradual Transition to Neuromorphic Computing in Parkinson's Disease Diagnostics: Ensuring Validation, Integration, and Ethical Considerations

A paradigm shift to neuromorphic computing in Parkinson's disease diagnostics is an intriguing prospect. However, it is essential to approach this shift gradually and methodically[19]. Here are some reasons why a gradual transition is necessary[20]:

- **Validation and reliability**: As with any new technology, rigorous validation and testing are required to establish the reliability and effectiveness of neuromorphic computing in Parkinson's disease diagnostics. Gradual

adoption allows for systematic evaluation, comparison with existing diagnostic methods, and validation against established standards. This ensures the technology meets the required accuracy, sensitivity, and specificity levels before widespread implementation.

- **Understanding limitations**: Understanding the breadth of symptoms and manifestations in Parkinson's disease is essential, as it sheds light on the challenges and potential of neuromorphic computing in accurately capturing and interpreting relevant data. By gradually embracing this technology, there is an opportunity to explore, investigate, and overcome these limitations through research, algorithm improvements, and advancements in hardware technology.

- **Integration with existing practices**: Parkinson's disease diagnostics often involve a combination of clinical assessments, medical imaging, and other diagnostic tools. Integrating neuromorphic computing into current practices becomes possible by adopting a gradual approach. This ensures compatibility and synergy with established diagnostic methodologies. This integration can facilitate a comprehensive and holistic approach to diagnosis, leveraging the strengths of traditional and neuromorphic techniques.

- **Training and expertise**: Healthcare professionals need time to familiarize themselves with neuromorphic computing techniques, algorithms, and tools. Gradual adoption enables the development of training programs and the cultivation of expertise among clinicians, researchers, and technicians. This ensures that healthcare professionals can effectively utilise neuromorphic computing in Parkinson's disease diagnostics, interpreting results and making informed clinical decisions.

- **Ethical and regulatory considerations**: With any new technology, ethical and regulatory considerations must be addressed. Gradual adoption allows for carefully examining the ethical implications, patient privacy concerns, and regulatory requirements associated with neuromorphic computing in Parkinson's disease diagnostics. It allows for establishing guidelines, policies, and safeguards that prioritise patient welfare and compliance with legal and ethical standards.

9. Conclusion

In conclusion, Parkinson's disease (PD) is a neurodegenerative condition characterised by motor impairments, and early and accurate diagnosis is crucial for effective treatment. Traditional diagnostic methods have limitations in detecting mild motor impairments accurately. However, event cameras, with their high temporal resolution and unique data representation, offer a promising approach to Parkinson's disease diagnosis.[21]

The current study proposes a methodology for utilising event camera data supported by machine learning algorithms to track hand movements, analyse gait patterns, and assess tremor severity. Initial studies have shown that even cameras can detect modest motor changes and identify disease-specific biomarkers. This suggests that event cameras have the potential to provide objective measurements of motor impairments in PD.

However, some challenges need to be addressed. Refining data analysis techniques specific to event camera data and conducting larger-scale clinical studies are necessary. These steps are crucial for validating the clinical utility of event cameras in Parkinson's disease diagnosis.

While a paradigm shift to neuromorphic computing in Parkinson's disease diagnostics holds promise, it should be approached gradually to ensure thorough validation, understanding of limitations, integration with existing practices, training of healthcare professionals, and consideration of ethical and regulatory aspects. This gradual transition can facilitate the responsible and effective implementation of neuromorphic computing to improve Parkinson's disease diagnosis and patient care.

References

1. P. DISEASE Werner Poewe et al., "Nature Reviews Disease Primers."
2. G. Gallego et al., "Event-based Vision: A Survey," Apr. 2019, doi: 10.1109/TPAMI.2020.3008413.
3. E. Abdulhay, N. Arunkumar, K. Narasimhan, E. Vellaiappan, and V. Venkatraman, "Gait and tremor investigation using machine learning techniques for the diagnosis of Parkinson disease," *Future Generation Computer Systems*, vol. 83, pp. 366–373, Jun. 2018, doi: 10.1016/j.future.2018.02.009.
4. A. Jalil, "Real-Time Gait Analysis Algorithm for Patient Activity Detection to Understand and Respond to the Movements Inam-ul-Haq M.Adnan Jalil," 2012. [Online]. Available: www.bth.se/com
5. L. Di Biase et al., "Gait analysis in Parkinson's disease: An overview of the most accurate markers for diagnosis and symptoms monitoring," *Sensors (Switzerland)*, vol. 20, no. 12. MDPI AG, p. 1, Jun. 01, 2020. doi: 10.3390/s20123529.
6. C. Scheerlinck, N. Barnes, and R. Mahony, "Continuous-time Intensity Estimation Using Event Cameras," Nov. 2018, [Online]. Available: http://arxiv.org/abs/1811.00386
7. A. J. Jagadeesan et al., "Current trends in etiology, prognosis and therapeutic aspects of Parkinson's disease: a review," *Acta Biomed*, vol. 88, pp. 249–262, 2017, doi: 10.23750/abm.v%vi%i.6063.
8. T. L. Myers et al., "Video-based Parkinson's disease assessments in a nationwide cohort of Fox Insight participants," *Clin Park Relat Disord*, vol. 4, Jan. 2021, doi: 10.1016/j.prdoa.2021.100094.
9. K. Khanna, S. Gambhir, and M. Gambhir, "Ad-hoc Social Network View project JOURNAL OF CRITICAL REVIEWS CURRENT CHALLENGES IN DETECTION OF PARKINSON'S DISEASE," 2020. [Online]. Available: https://www.researchgate.net/publication/357870234
10. G. Gallego et al., "Event-Based Vision: A Survey," *IEEE Trans Pattern Anal Mach Intell*, vol. 44, no. 1, pp. 154–180, Jan. 2022, doi: 10.1109/TPAMI.2020.3008413.
11. K. G. Sibley, C. Girges, E. Hoque, and T. Foltynie, "Video-Based Analyses of Parkinson's Disease Severity: A Brief Review," *Journal of Parkinson's Disease*, vol. 11, no. s1. IOS Press BV, pp. S83–S93, 2021. doi: 10.3233/JPD-202402.
12. G. Gallego, C. Forster, E. Mueggler, and D. Scaramuzza, "Event-based Camera Pose Tracking using a Generative Event Model," Oct. 2015, [Online]. Available: http://arxiv.org/abs/1510.01972
13. J. Tang et al., "Bridging Biological and Artificial Neural Networks with Emerging Neuromorphic Devices: Fundamentals, Progress, and Challenges," *Advanced Materials*, vol. 31, no. 49. Wiley-VCH Verlag, Dec. 01, 2019. doi: 10.1002/adma.201902761.
14. Y. Chen et al., "Neuromorphic computing's yesterday, today, and tomorrow – an evolutional view," *Integration*, vol. 61. Elsevier B.V., pp. 49–61, Mar. 01, 2018. doi: 10.1016/j.vlsi.2017.11.001.
15. A. Agrawal, J. Gans, and A. Goldfarb, *The economics of artificial intelligence : an agenda.*
16. R. A. Nawrocki, R. M. Voyles, and S. E. Shaheen, "A Mini Review of Neuromorphic Architectures and Implementations," *IEEE Transactions on Electron Devices*, vol. 63, no. 10. Institute of Electrical and Electronics Engineers Inc., pp. 3819–3829, Oct. 01, 2016. doi: 10.1109/TED.2016.2598413.
17. M. Davies, "Benchmarks for progress in neuromorphic computing," *Nat Mach Intell*, vol. 1, no. 9, pp. 386–388, Sep. 2019, doi: 10.1038/s42256-019-0097-1.

18. C. D. Schuman, S. R. Kulkarni, M. Parsa, J. P. Mitchell, P. Date, and B. Kay, "Opportunities for neuromorphic computing algorithms and applications," *Nature Computational Science*, vol. 2, no. 1. Springer Nature, pp. 10–19, Jan. 01, 2022. doi: 10.1038/s43588-021-00184-y.

19. F. Zenke and E. O. Neftci, "Brain-Inspired Learning on Neuromorphic Substrates," *Proceedings of the IEEE*, vol. 109, no. 5, pp. 935–950, May 2021, doi: 10.1109/JPROC.2020.3045625.

20. C. Perez, "STRUCTURAL CHANGE AND ASSIMILATION OF NEW TECHNOLOGIES IN THE ECONOMIC AND SOCIAL SYSTEMS."

21. A. Rana, A. Dumka, R. Singh, M. K. Panda, N. Priyadarshi, and B. Twala, "Imperative Role of Machine Learning Algorithm for Detection of Parkinson's Disease: Review, Challenges and Recommendations," *Diagnostics*, vol. 12, no. 8. MDPI, Aug. 01, 2022. doi: 10.3390/diagnostics12082003.

Note: The figure and table in this chapter were made by the authors.

Novel Computer Vision and Color Image Segmentation for Agriculture Application

19

Vempati Krishna

Department of CSD(DS), TKRCET,
Hyderabad, Telangana, India

Ch. V. Raghavendran

Department of CSE, Aditya College of Engineering & Technology,
Surampalem AP, India

S. K. Umar Faruk

Department of ECE, TKREC, Telangana, India

Abstract: In this paper, novel computer vision and color image segmentation for agriculture application is implemented. The population is increasing daily with fast speed and Most of the population relies on agriculture for subsistence livelihood. So in this novel computer vision concept is introduced. Firstly, agriculture images are captured. Pre processing stage is applied for the captured images. After processing, image enhancement technique is applied. Image enhancement will enhance the images in very effective way. RGB to HSI conversion is applied to the enhanced images. Now, the signal is extracted using features. At last based on the trained data and testing data, the classification is done. At last from results shows that it will give effective results in terms of accuracy.

Keywords: Computer vision, Color image segmentation, Image extraction, Pre processing, Training data, Testing data

1. Introduction

Lately, farming has gotten significantly more significant than it used few years back where plants were simply used to take care of people just as animals. This is because of the way that plants are currently used to produce power and different wellsprings of energy to refine the everyday environments of humankind. In any case, there are such countless illnesses that influence plants with the potential to cause severe disruption to a variety of economic and social structures. Moreover, it prompt extraordinary biological misfortunes. The emergence of plant diseases poses a substantial challenge to the agricultural sector. With the expanding global population and the increasing demand for food, the vulnerability of crops to diseases has become a pressing concern. Plant pathogens, such as viruses, bacteria, fungi, and other infectious agents,

can spread rapidly and devastate entire crops, leading to significant losses for farmers and potential food shortages for communities. The impact is not limited to food production alone; it extends to economic stability, as agricultural sectors heavily rely on crops for income and employment opportunities.

To address these challenges, scientists and researchers are actively studying plant diseases, their causes, and potential solutions. Advancements in technology and the understanding of plant pathology have paved the way for innovative strategies in disease management. Integrated pest management practices, including the use of resistant crop varieties, biological control agents, and cultural practices, have shown promise in reducing the impact of diseases. Additionally, precision agriculture techniques, such as remote sensing, data analytics, and automated monitoring systems,

---
[1]vempati.k@gmail.com, [2]raghuchv@yahoo.com, [3]faruq.image@gmail.com

enable early detection and targeted intervention, minimizing losses and optimizing resource utilization.

However, the battle against plant diseases is a continuous one, as pathogens constantly evolve and adapt. Therefore, it is crucial to prioritize research and development efforts to stay ahead of these challenges. Collaboration between governments, research institutions, and farmers is essential in implementing effective disease management strategies. Sharing knowledge, disseminating best practices, and providing support to farmers can empower them to adopt sustainable farming practices that mitigate disease risks.

Hence, it is smarter to analyze sicknesses precisely and ideal to stay away from such loses. Plant sicknesses can be identified through a few methods including manual and PC based frameworks. Most plant illnesses show up as spots on the leaves which are more noticeable to natural eye. Then again, there are a few sicknesses that don't show up. Others manifest on the foliage only after they have caused extensive plant damage.

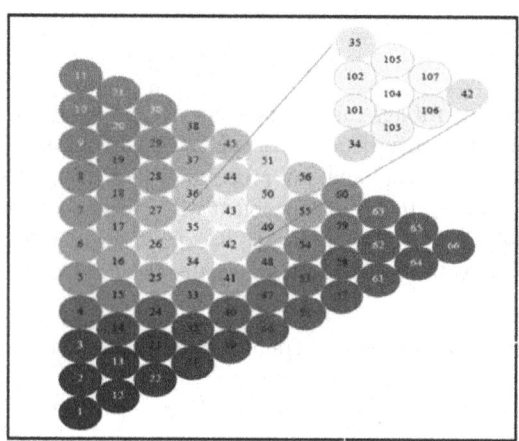

Fig. 19.1 Pyramid of color [1]

In such occurrences, it is suggested that modernized frameworks would be the solitary choice to distinguish the circumstance opportune utilizing some sort of complex calculations and scientific devices, ideally using incredible magnifying instruments and different machines. In some different cases, In order to decipher the signs, electromagnetic techniques must be used, which provide additional images inaccessible to the naked sight.

As a developed country, India 70% mainly relies upon the agriculture business. Different harvesting groups have done by the farmers to raise their business according to reasonable pesticides. Based on the horticulture items there is an huge decrement in both quality and amount of items.

Based on the plants, the pesticides are being recognizable which are alluded to investigations.

These days the modernization of horticulture is on the need Indian roused by the vision of computer vision change in provincial improvement measures. The mission of Unnat Bharat Abhiyan is to engage higher educational establishment to work with people of economy. The Ministry of Human Resource Development has dispatched a program called Unnat Bharat Abhiyan which is common of Smart Farming is a hotly debated issue around the world. Wise developing is a developing where the board thought using India in perceiving improvement challenges and provides fitting responses for accelerating sensible turn of events.

Another idea present day advancement to construct the sum and nature of rustic things. In this idea, computerized reasoning, mechanization and mechanical technology, detecting advancements, agrarian robots, IoT applications, situating innovations are generally utilized[6-7]. The field of software engineering known as "man-made reasoning" focuses on creating intelligent machines that can think and act in human-like ways.

AI has many subfields such as Machine Learning, Neural Networks, Robotics, Expert System, Natural Language Processing, etc. Artificial intelligence (AI) is the application of human logic to machines, allowing them to learn to think and make decisions without being explicitly programmed. [8].

Machine learning has becoming increasingly important in modern agriculture.. ML techniques are categorized in supervised, unsupervised and reinforcement. The daily life farmer's problems from seed sowing to harvesting of crops can be resolved by using machine learning algorithms [9-10].

2. Literature Survey

Arti N. Rathod (2014) [1] et al. proposed in agribusiness rundown of the public authority. Modernization of horticulture is a cycle of changing farming from conventional work based agribusiness to innovation based agribusiness. This loop is beneficial for farmers, and research into automated leaf infection detection is an essential starting point because it could be useful for monitoring large harvest fields and detecting early symptoms of disease. Before images can be used in an infected region, they must be acquired, preprocessed, segmented, features extracted, and statistical analysis performed. The proposed procedure begins by filtering images with the center channel, followed by converting the RGB image to the lab concealing portion. The second stage involves isolating the image using the k-implies technique. The third stage is to conceal and remove significant green pixels.

The Neural Network grouping performs well and could effectively distinguish and arrange the tried sickness. Ms.

Kiran R. Gavhale [2] (2014) et al. demonstrates that plant infections result in significant production and financial losses, as well as a decline in the quality and quantity of agricultural products. Recent emphasis has been placed on the study of plant diseases due to the significance of monitoring massive crop yields.. Examination, Artificial neural organization, Fuzzy rationale. Choosing a grouping technique is consistently a troublesome undertaking in light of the fact that the nature of result can change for various information. Plant leaf infection orders have wide applications in different fields, for example, in natural examination, in Agriculture and so forth This paper gives an outline of various arrangement procedures utilized for plant leaf illness grouping.

Malti K. Singh (2017) [3] et al. stated that agriculture is estimated to account for over 70% of India's GDP. Environmental factors such as precipitation and temperature fluctuations have a significant impact on the harvest. As a staple food crop, Phaseolus vulgaris L. crops are included in the diets of many individuals. Among the many impactful infections, Anthrax is the most significant. The pathogen Colletotrichum lindemuthianum causes anthrax disease.

Camellia assamica (J. W. Pole.) W. Wight [4] is one of the most popular and well-known crops in the globe. The parasite Alternaria alternata has a significant impact on the leaf. A programmed detecting framework employing advanced PC technology, such as image processing, has been developed to assist ranchers in identifying evidence of diseases at an early or fundamental stage and to provide useful information for its control. This led to the present study, which examined the efficacy of image processing techniques in detecting specific plant maladies on the leaves of Phaseolus vulgaris (beans) and Camellia assamica (tea). The process of obtaining, processing, dividing, extracting, and arranging images.

Dr. Sridhathan (2018) et al. proposed economy of a nation relies upon agrarian efficiency. With recognizable evidence of the plant diseases, the losses in profitability and quality of horticultural products can be quantified and mitigated. Traditional methods are effective, but require costly human resources to visually inspect the leaf patterns of the plant and diagnose the infection. Due to their inefficiency, conventional strategies squander time and effort. Large pastoral lands will experience less efficiency loss when early-stage plant disease detection is automated. In this study, we propose employing an Image Processing technique for dream-based, programmed plant disease recognition. Image processing algorithms can identify plant contamination or disease by analyzing the shading highlight of a leaf's surface. K-means is utilized to classify hues into groups, whereas GLCM is utilized to characterize diseases. Vision-based plant disease research was shown to have productive outcomes and a promising application.

Saradhambal. G (2018) [5] et al. In the field of agriculture, the anticipated crop development is of vital importance. Ultimately, contaminated harvests are responsible for the food shortage, which in turn reduces production. Distinguishing plant maladies at an inconvenient stage is an unexplored field of study. The primary objective is to increase agricultural production while decreasing pesticide use. Our paper is being used to investigate the prediction of a challenging activity's leaf disease. This indicates a more refined k-mean bunching calculation is required to predict leaf contamination.

To divide the polluted region into manageable portions, a model based on shading is developed. Regarding the temporal and spatial complexity of the training environment, sample images were subjected to test analyses. Image preparation can aid in the diagnosis of plant maladies. Infection recognition requires image acquisition, image pre-planning, segmentation, feature extraction, and action planning. Our work focuses on identifying plant diseases and developing treatments for them. Real-time visualization of the afflicted leaf area. We used a voice-enabled course framework to make our project accessible to individuals with varying levels of programming experience.

3. Novel Computer Vision and Color Image Segmentation

The Fig. 19.2 shows the algorithm of novel computer vision and color image segmentation. Firstly, agriculture images are captured. Pre processing stage is applied for the captured images. After processing, image enhancement technique is applied. Image enhancement will enhance the images in very effective way. RGB to HSI conversion is applied to the enhanced images. Now, the signal is extracted using features. At last based on the trained data and testing data, the classification is done.

Image enhancement is a pre-processing technique used to improve the quality and depth of unprocessed data. Standard methods include enhancing differentiation, segmenting space, dividing thickness, and extracting.

Image enhancement strategies are generally applied to distant detecting information to improve the presence of a picture for human visual examination. The principle focal point of improvement strategies follows these techniques in to picture division, bunching and mathematical changes. The below equation shows defining of image enhancement:

$$d_1^i(m,n) = \begin{cases} d_1^i(m,n), & \text{if}: e^i(m,n) \le T^i \\ g^i d_1^i(m,n), & \text{if}: e^i(m,n) \le T, \end{cases} \quad (1)$$

Where m and n denote coordinates in the spatial domain, e_i is the edge set corresponding to the transform space component d_1^i, g^i is a local gain.

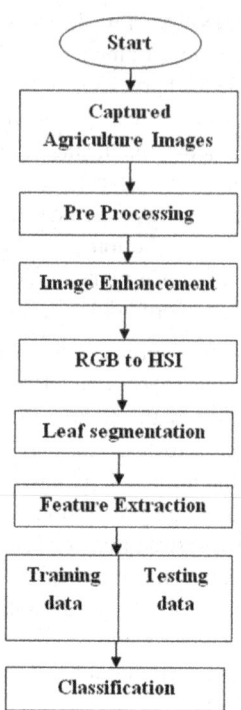

Fig. 19.2 Algorithm of novel computer vision and color image segmentation

Source: Authors

Image segmentation will divide the picture into number of samples which compute the data. The image segmentation will mainly divide the limits and items based on the given condition which is computed regularly. In PC division the image leaf segmentation will compute the data. The image is attributed into various types of data which is recognizable into districts.

The intesity band gave great division, yet the shade and immersion gave great division, this demonstrates that practically all the data can be established in the power band. The HSI can be utilized rather than RGB to lessen the hour of preparing; we can just deal with the force band to get great outcome.

During the process of feature extraction, the original arrangement of unprocessed data is turned into assemblages that are easier to manage thanks to a cycle that reduces the dimensionality of the data. Due to the complexity of these massive data sets, the analysis of them often needs a significant amount of processing resources.

The RGB tone changed over into HIS dependent on the luminance which is fundamentally subject to the concealing part. The HIS is chiefly utilized in the pictures of disengagement which is acquired from the luminance segregation. The change will be less complex when it will manage luminance picture seclusion.

To convert an image from the RGB (Red, Green, Blue) color model to the HSI (Hue, Saturation, Intensity) color model, a series of mathematical equations need to be applied. Before applying these equations, the image should be normalized to the range of [0, 1]. Let's walk through the process step by step.

First, let's define the variables used in the equations for better understanding:

- Rp, Gp, and Bp: Normalized values of the Red, Green, and Blue color channels, respectively.
- H: Hue component.
- S: Saturation component.
- I: Intensity component.

The normalization step ensures that the values of Rp, Gp, and Bp are within the range [0, 1]. This can be achieved by dividing each color channel value (R, G, and B) by the maximum possible value for that channel (e.g., 255 for an 8-bit image).

Next, we can calculate the Hue component (H) using equation (3). H is determined based on the comparison between the Blue and Green channels. If the Blue channel value (Bp) is less than or equal to the Green channel value (Gp), H is set to θ. However, if Bp is greater than Gp, H is set to 360 minus θ. Here, θ is calculated using the inverse cosine function ($\cos^{-1}$) of the expression $(0.5 * [(Rp - Gp) + (Rp - Bp)]/[(Rp - Gp)^2 + (Rp - Bp) * (Gp - Bp)]^{(1/2)})$.

Moving on to the Saturation component (S), equation (4) can be used to calculate it. S represents the degree of colorfulness in the image. It is determined by subtracting three times the minimum value among Rp, Gp, and Bp from 1 and dividing the result by the sum of Rp, Gp, and Bp. In other words, $S = 1 - (3 * \min(Rp, Gp, Bp)/(Rp + Gp + Bp))$.

Lastly, the Intensity component (I) is calculated. I represents the overall brightness or intensity of the image. It is determined by summing up the normalized values of Rp, Gp, and Bp and dividing the result by 3. In other words, $I = (Rp + Gp + Bp) / 3$.

By applying these equations to each pixel in the image, we can obtain the corresponding HSI values for that pixel. The resulting HSI image will have the same dimensions as the original RGB image but with H, S, and I values instead of R, G, and B values.

The conversion from RGB to HSI provides advantages in various image processing tasks. The Hue component (H) represents the dominant color information in the image and can be useful in color-based segmentation or object recognition tasks. The Saturation component (S) indicates the purity of the color and can be used to enhance or suppress the intensity of colors in the image. The Intensity component

(I) represents the overall brightness and can be used for brightness adjustment or grayscale transformations.

In summary, converting an image from the RGB color model to the HSI color model involves normalizing the RGB values, calculating the Hue (H), Saturation (S), and Intensity (I) components based on the provided equations, and generating a new image with HSI values. This conversion facilitates various image processing tasks by providing separate channels for color information, purity, and brightness.

The underlying arrangement of images will mainly based on the images which isolates the picture programming. There are some complex outcomes which will depend on the neural organizations. Hence the images testing mainly based on the approval testing.

In the PC program mainly test information is distinguished based on tests. In the corroborative manner some information is utilized to check the contribution based on capacity. Hence the test information will be recorded based on the contribution given.

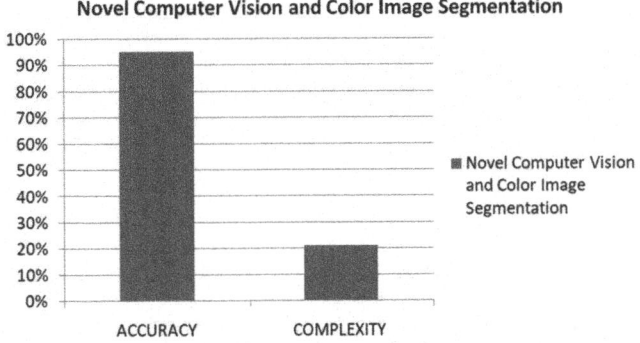

Fig. 19.3 Accuracy and complexity

Source: Authors

The above Fig. 19.3 displays the accuracy and nuance of contemporary color image segmentation and computer vision techniques. Using cutting-edge computer vision and color image segmentation, we can drastically simplify and improve the system's precision.

Table 19.1 Threshold values for H & V

Flower color	Hue range	Value
Red	<0.06 or >0.9	>0.8
Pink	<0.15 or >0.8	>0.8
Yellow	>0.17 and <0.19 or >0.8	

Source: Authors

The above Table 19.1 shows the threshold values for H&V. For example three flowers are taken which are named as red, pink and yellow. According to the flower, the hue range is changed.

The Fig. 19.4 shows the quality of image for novel computer vision and color image segmentation. The quality will be high.

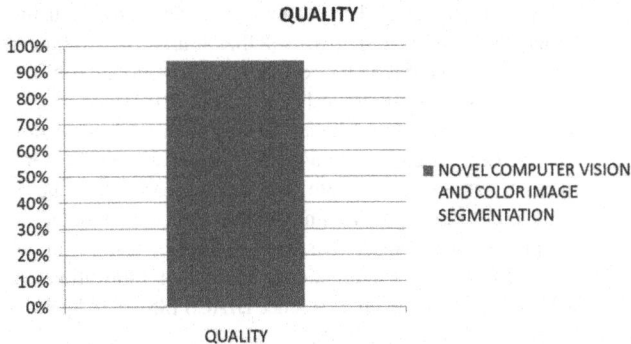

Fig. 19.4 Quality

Source: Authors

4. Conclusion

Hence, novel computer vision and color image segmentation for agriculture application was implemented. Agriculture images are captured initially. For the captured images pre processing stage is applied. Image enhancement technique is applied after processing. Image enhancement will enhance the images in very effective way. RGB to HSI conversion is applied to the enhanced images. Now, the signal is extracted using features. At last based on the trained data and testing data, the classification is done. At last from results it shows that, it gives effective output.

References

1. Arti N. Rathod, Bhavesh A. Tanawala, Vatsal H. Shah, "Leaf Disease Detection using Image Processing and Neural Network", International Journal of Advance Engineering and Research Development (IJAERD) Volume 1,Issue 6,June 2014.
2. Dhawale Sariputra, Shirolkar A.A , "Neural Network Classifier Based Method for Leaves Disease Detection with Image ProcessingTechnique" , IJSRD - International Journal for Scientific Research & Development, Volume 4, Issue 03, 2016.
3. Dr. Sridhathan C, Dr. M. Senthil Kumar," Plant Infection Detection Using Image Processing", International Journal Of Modern Engineering Research (IJMER), I Vol. 8 I Iss. 7 I July 2018 I 13 I.
4. Gharte Sneha H., Prof. (Dr.) S. B. Bagal, "Plant Leaf Disease Detection Using Image Processing", International Research Journal of Engineering and Technology (IRJET), Volume: 06 Issue: 06 I June 2019.
5. K.Narsimha Reddy, B.Polaiah, N.Madhu, "A Literature Survey: Plant Leaf Diseases Detection Using Image Processing Techniques", IOSR Journal of Electronics and

Communication Engineering (IOSR-JECE) e-ISSN: 2278-2834,p- ISSN: 2278- 8735.Volume 12, Issue 3, Ver. II (May - June 2017).

6. Ms. Kiran R. Gavhale, Prof. Ujwalla Gawande," An Overview of the Research on Plant Leaves Disease detection using Image Processing Techniques", IOSR Journal of Computer Engineering (IOSR-JCE) e-ISSN: 2278-0661, p-ISSN: 2278-8727Volume 16, Issue 1, Ver. V (Jan. 2014).

7. Malti K. Singh, Subrat Chetia, "Detection and Classification of Plant Leaf Diseases in Image Processing using MATLAB", International Journal of Life Sciences Research, Volume 5, Issue 4, pp: (120-124), Month: October - December 2017.

8. Monika Gupta, "Plant Disease Detection using Digital Image Processing", International Journal of Innovations & Advancement in Computer Science IJIACS ISSN 2347 – 8616 Volume 7, Issue 5 May 2018.

9. Monishanker Halder, Ananya Sarkar, Habibullah Bahar, "Plant Disease Detection by Image Processing: A Literature Review", SDRP Journal of Food Science & Technology(ISSN: 2472-6419), Volume-3 Issue-6.

10. Piyali Chatterjee, B Harikishor Rao, "Leaf Disease Detection using Image Processing Technique", IJIREEICE, International Journal of Innovative Research in Electrical, Electronics,

Instrumentation and Control Engineering ISO 3297:2007 Certified Vol. 4, Issue 9, September 2016.

11. Prajakta Mitkal, Priyanka Pawar, Mira Nagane, Priyanka Bhosale, Mira Padwal and Priti Nagane, "Leaf Disease Detection and Prevention Using Image Processing using Matlab", International Journal of Recent Trends in Engineering & Research (IJRTER) Volume 02, Issue 02; February– 2016.

12. Sandesh Raut, Amit Fulsunge, "Plant Disease Detection in Image Processing Using MATLAB", International Journal of Innovative Research in Science, Engineering and Technology Vol. 6, Issue 6, June 2017.

13. Saradhambal. G, Dhivya. R, Latha. S, R. Rajesh, "Plant Disease Detection and its Solution using Image Classification", International Journal of Pure and Applied Mathematics Volume 119 No. 14 2018.

14. Sushil R. Kamlapurkar, "Detection of Plant Leaf Disease Using Image Processing Approach", International Journal of Scientific and Research Publications, Volume 6, Issue 2, February 2016.

15. Vishal Mani Tiwari&Tarun Gupta, "Plant Leaf Disease Analysis using Image Processing Technique with Modified SVM-CS Classifier", International Journal of Engineering and Management Technology, IJEMT, Volume 5, Issue 1, 2017.

Attention-based Multi-Task CNN U-Net for Kidney Tumor Segmentation and Classification

20

Seshadri Ramana, K. Mahesh Babu

Department of CSE, Ravindra College of Engineering for Women,
Kurnool, AndraPradesh, India

P. Kiran Rao

Department of CSE, Ravindra College of Engineering for Women,
Kurnool, AndraPradesh, India

Department of CSE,Faculty of Engineering,
MS Ramaiah University of Applied Sciences, Bengalure, Karnataka, India

Subarna Chatterjee

Department of CSE,Faculty of Engineering,
MS Ramaiah University of Applied Sciences, Bengalure, Karnataka, India

Y. Indira Priyadarshini

Department of CSE, Ravindra College of Engineering for Women,
Kurnool, AndraPradesh, India

Abstract: Deep learning techniques have gained prominence in the realm of medical imaging, especially for kidney tumor segmentation and classification. However, there have been observed inconsistencies in accuracy across various kidney tumor types, raising concerns about the effectiveness of traditional loss functions. In light of these challenges, we present the MultiTask Semantic Segmentation-UNet (MTUNet) methodology. This innovative approach utilizes a singular stage of training but integrates distinct loss functions for each tumor class. Simultaneously, a unified loss is employed to amplify correlations between the classes. A pivotal aspect of MTUNet's design is its focus on intensively training within the region comprising the kidney tumor. Empirical evaluation on three publicly accessible medical image datasets demonstrated the superiority of MTUNet. It recorded mean intersection over union (MIOU) scores of 0.9164, 0.8372, and 0.8260 across three distinct kidney tumor categories, marking a notable improvement over existing leading techniques. More impressively, MTUNet showcased a consistent and robust performance across a spectrum of kidney tumors, indicating its strong potential to adapt and generalize to unfamiliar data. The MTUNet methodology ushers in a new horizon for kidney tumor segmentation and classification in medical imaging, offering a refined strategy that may be pivotal for the next generation of computer-aided diagnostic tools.

Keywords: UNet, Multi task, CNN, Kidney tumor, Segmentation, Classification

1. Introduction

Medical imaging stands as a beacon of innovation in the world of contemporary diagnostics, and its evolution has been catalyzed significantly by the advent and integration of deep learning methodologies, especially in the domain of kidney tumor identification [1]. One of the standout advancements in this arena has been segmentation, which has emerged as

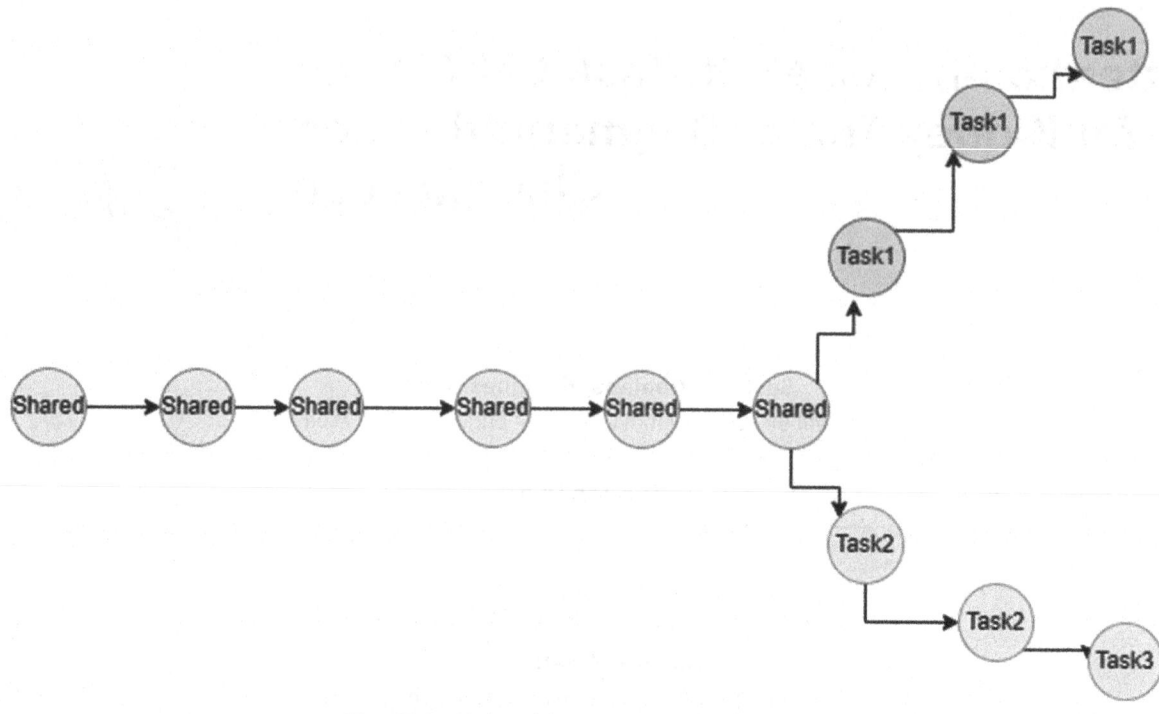

Fig. 20.1 Multitask learning neural network [6]

a priority due to its nuanced ability to delineate and localize tumors in intricate detail, often overshadowing the broader strokes of classification [2].

Segmentation techniques are predominantly bifurcated into two categories: coordinate-based detection and semantic segmentation. The former, as its name suggests, provides a macro view, highlighting the broader confines of tumors. However, there's a caveat: it might inadvertently bypass some of the nuanced characteristics that are often vital in understanding tumor progression [3]. Semantic segmentation, or pixel-based detection as it's also known, presents a contrasting approach. With its laser focus on pixel-level granularity, it offers clinicians a detailed blueprint of tumor shapes, a factor that's crucial for therapeutic interventions [4]. Yet, this precision isn't devoid of challenges. The inherently diverse and irregular morphology of kidney tumors coupled with potential disparities in training data can often be stumbling blocks in achieving consistent results [5].

Emerging paradigms in medical imaging have spotlighted the merits of multi-task learning [6]. A representative visualization (as depicted in Fig. 20.1) illustrates how a single neural network can concurrently process multiple tasks, harnessing shared information layers and branching out to task-specific output layers. This avant-garde approach designs models with the capability to juggle multiple objectives in tandem. Such concurrent processing can amplify model efficiency, especially by harnessing shared information across diverse tasks, a strategy that becomes a game-changer

in scenarios constrained by data limitations or imbalances. The labyrinth of kidney tumor classification is equally compelling. Traditional modalities have leaned heavily on convolutional neural networks (CNNs), which have proven adept at sieving out distinct features and classifying tumors based on discernible patterns [7]. In parallel, methodologies rooted in support vector machines (SVMs) and random forest (RF) classifiers have carved out their niche, demonstrating prowess in extrapolating and classifying based on intricate imaging features [8]. However, the journey is still punctuated with challenges. Navigating the treacherous waters of data imbalances and the delicate craft of feature extraction remains a task that the scientific community grapples with [9]. Recent strides in the integration of deep learning algorithms have illuminated a promising path forward, especially in the realm of kidney tumor segmentation [10]. The crux lies in the strategic adoption of bespoke loss functions coupled with meticulously tailored data augmentation strategies, synergies that have been pivotal in ramping up segmentation accuracy [11]. These advancements not only augment the foundational knowledge in the field but also signal a transformative shift in clinical diagnostics, ushering in an era of unprecedented accuracy in tumor localization and analysis [12].

2. Related Works

The field of medical imaging has continuously grappled with the intricate challenge of kidney segmentation. As technology and understanding have advanced, a panorama of solutions has emerged, each striving for a zenith in precision and efficacy. This ever-evolving quest has witnessed the birth and transformation of numerous methodologies over the years. Back in 2015, a significant breakthrough was made by Ronneberger et al. [16]. They introduced the world to the UNet model, a ground-breaking approach that utilized a Fully Convolutional Network for the delicate task of medical image segmentation. What distinguished U-Net from its predecessors was its symmetric U-shaped architecture, which seamlessly integrated both compressive and expansive pathways. Demonstrating its prowess, U-Net achieved remarkable results on a dataset containing just 30 images. This dataset was strategically augmented using data expansion techniques, and the model's performance therein was nothing short of exemplary. By clinching a championship title, U-Net not only solidified its position in the annals of medical image segmentation but also paved the way for a plethora of derivative algorithms. These subsequent algorithms were deeply influenced by UNet's design and found applications across diverse realms of medical image segmentation.

The post-U-Net era witnessed a flurry of innovations in medical image segmentation algorithms [17]. Yang et al., for instance, took the ResU-Net model and applied it to CT images of the lung. They embarked on a mission to ascertain lung parenchymal metrics, concluding an intriguing inverse relationship between lung volume and CT values. Furthermore, Oktay et al. [17] championed the cause of an attention gate model specifically crafted for medical imagery.

This model was adept at focusing on target structures, irrespective of their varied dimensions, thus enhancing the predictive prowess of the U-Net model without inflating computational burdens. In a bid to amalgamate the strengths of multiple networks, Alom et al.[18] ventured to introduce the RU-Net and R2U-Net models. These models harmoniously integrated the capabilities of the U-network, residual network, and RCNN. The resultant segmentation outputs showcased marked improvements over contemporaneous models. Taking integration a step further, Wang et al. [19] envisioned a fusion of U-Net with attention and recurrent residual models. This synthesis bore the fruit of elevated segmentation outcomes.

As the year 2020 dawned, the application landscape of U-Net underwent significant shifts. Kidney and kidney tumor segmentation began to dominate its use-cases. Isensee et al.[20], sensing the need for a streamlined version, introduced the world to nnU-Net. This variant placed a premium on efficiency and adaptability, a strategy that paid off handsomely when it registered the peak average dice score in a challenge. Concurrently, Da Cruz et al. harnessed U-Net 2D, marking their territory with an impressive Dice coefficient in the KiTS19 challenge. Not to be left behind, Turk et al. [21] made strategic enhancements to the existing V-Net model, reaping commendable results. The subsequent period saw U-Net becoming indispensable for kidney segmentation as shown in Fig. 20.2. The model underwent several modifications, each aiming to magnify its performance and adaptability. Models such as volumetric convolutional

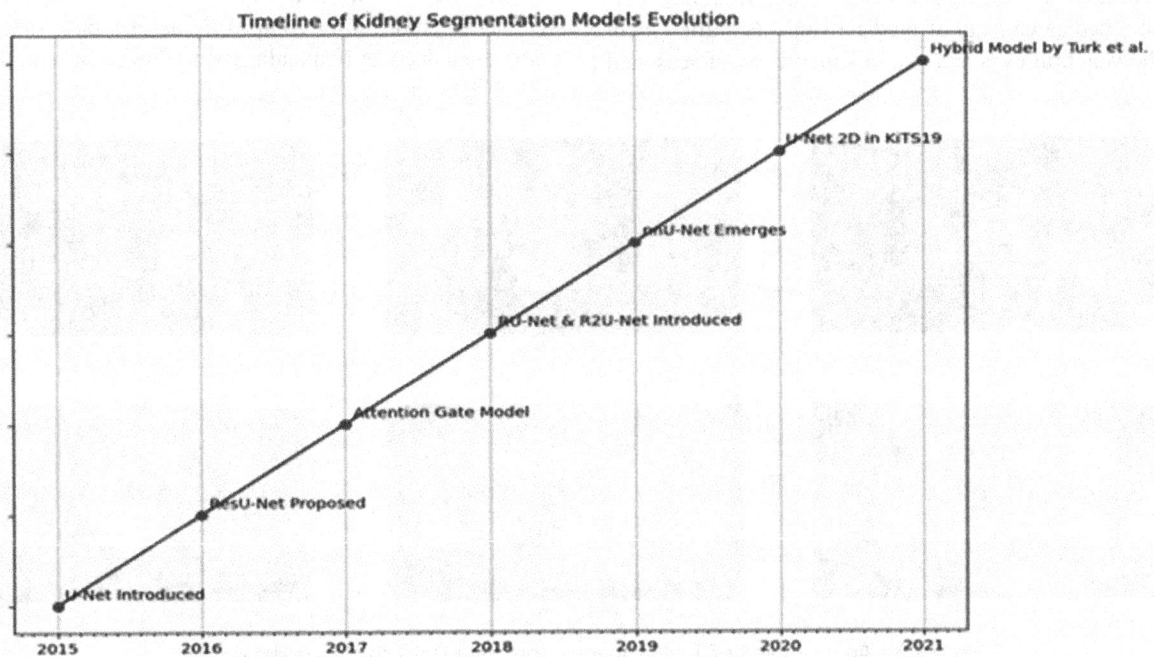

Fig. 20.2 Timeline of kidney segmentation models evolution [4]

networks and multiresolution VB-nets emerged, each with its unique selling point. The KiTS19 challenge threw a spotlight on the advancements made in this domain. Notably, the top five performers of this challenge were all derivatives of the 3D U-Net structure. Isensee et al. [22] stood tall among them with an enviable score. However, the road to perfection is unending. While these models have set new benchmarks, the horizon still holds promises of greater precision and efficiency. It is with this vision that we introduce the Multi TaskU-Net model, tailored for kidney tumor CT image segmentation and classification. We believe this will herald a new era of heightened segmentation accuracy.

3. Methods

3.1 Dataset and Pre-processing

The KiTS challenge dataset [11], a prominent benchmark in the realm of medical imaging, caters extensively to the kidney tumor disease segmentation community. Comprising 210 highcontrast CT scans, it captures a myriad of cases where patients underwent either partial or complete nephrectomy due to one or multiple kidney tumors. One of the defining features of this dataset is the diversity in its image resolutions: while the plane resolutions span from a precise 0.437 mm to a broader 1.04 mm, the slice thickness varies, ranging from a fine-grained 0.5 mm to a more substantial 5.0 mm. Not only does this dataset provide raw imaging data, but it also offers meticulously crafted ground-truth masks delineating both the healthy kidney tissue and tumor regions. These masks stand as testament to the rigorous process that went into their creation: a dedicated team of medical students crafted them under the watchful eyes and expert supervision of seasoned radiologists, relying solely on the axial projections of the CT scans(figure 3). Furthermore, in ensuring easy accessibility and universality in its application, the dataset is shared in the widely-accepted NIFTI format. This standardized format further promotes its utility as a reliable resource for assessing and benchmarking the effectiveness of various kidney tumor segmentation methodologies.

3.2 Model Architecture: Multi-Task Semantic Segmentation-UNet (MTUNet)

The evolution of deep learning in medical imaging has witnessed a myriad of architectural innovations, each striving to better the last in terms of precision, efficiency, and generalization. Within this context, the MTUNet emerges as a cutting-edge framework designed to bridge the gap between traditional image segmentation and the complexities posed by diverse kidney tumor types. Grounded in the foundational tenets of the renowned U-Net[5] architecture, MTUNet takes a leap forward by integrating multi-task learning, thereby fostering the simultaneous segmentation of different kidney tumor categories. Its modular design, enhanced by bespoke loss functions and an emphasis on regions of interest, positions MTUNet as a promising candidate for the next generation of medical image segmentation tools

1. Basic U-Net: The U-Net architecture[16], originally proposed by Ronneberger et al.[16] for biomedical image segmentation[23], has achieved widespread acclaim for its ability to produce accurate segmentations even with relatively small sets of training images[1]. The UNet basic design consist of different key layers. U-Net[16] exhibits a symmetric architecture comprising an encoder (contracting path) and a decoder (expanding path), linked by a bottleneck.

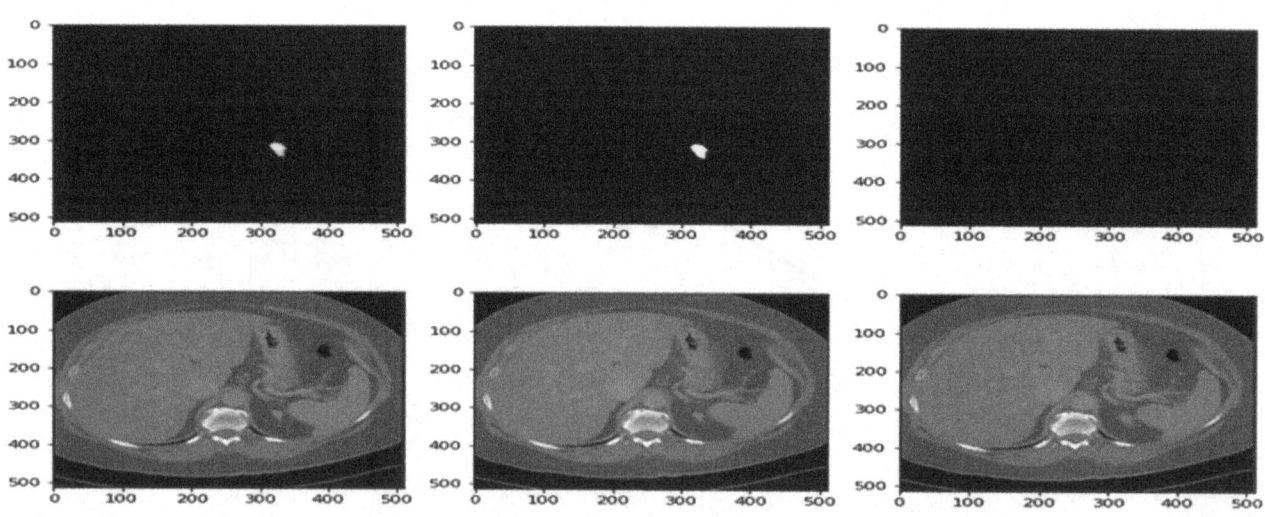

Fig. 20.3 An example of CT scan images from the KiTs19 challenge dataset [4]

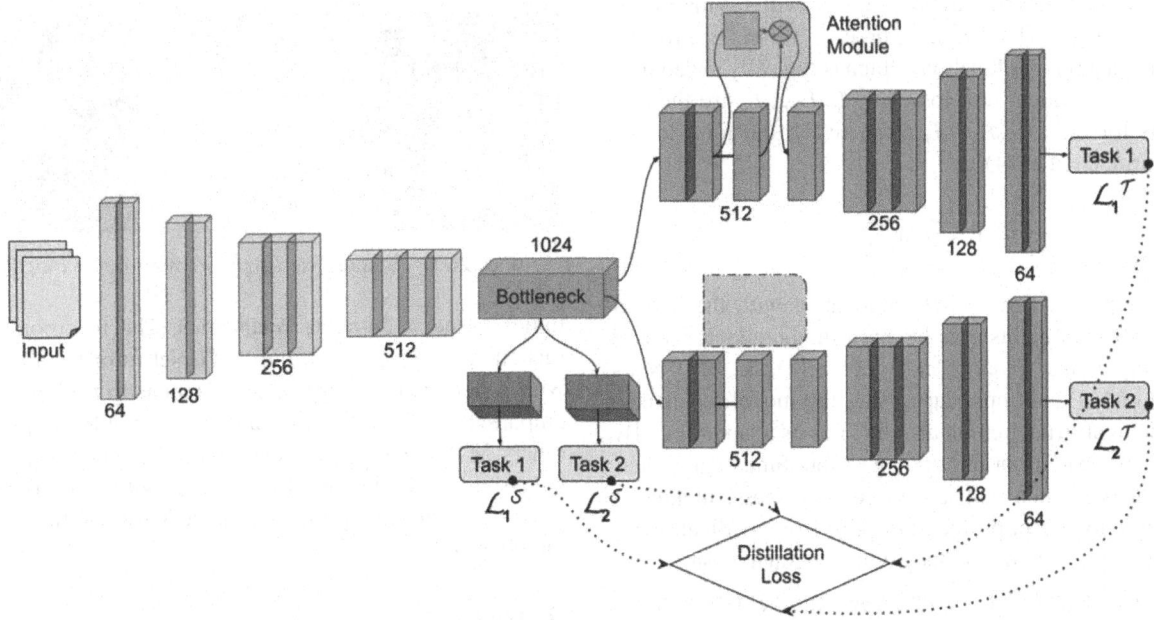

Fig. 20.4 Multi task CNN for segmentation and classification of kidney tumor [13]

This structure has earned it the "U-shaped" moniker. Each side of the U aims to capture different types of image features. The contracting path follows the typical architecture of a convolutional network. It consists of a series of convolutional layers, each followed by a rectified linear unit (ReLU) and a max-pooling operation. Mathematically, the convolution operation can be represented as (1)

$$I_{out}(x,y) = I_{in} * K = P_{\infty i=-\infty} P_{\infty j=-\infty} I_{in}(i,j)K(x-i, y-j) \quad (1)$$

where $I_{,}out$" is the output image, $I_{,}in$" is the input image, and K is the kernel or filter. Max pooling, commonly used to reduce the spatial dimensions of the feature maps, can be denoted mathematically for a 2×2 pooling size as (2)

$$M_{i,j} = \max\{I_{2i,2j}, I_{2i+1,2j}, I_{2i,2j+1}, I_{2i+1,2j+1}\} \quad (2)$$

where M is the output of max pooling and I is the input feature map.

This is the deepest layer[13], designed to extract the most abstract and essential features from the image. Typically, it consists of convolutional layers followed by ReLUs. The expanding path involves a sequence of up-convolution operations, followed by concatenation with the corresponding feature map from the contracting path (skip-connection), and then a series of regular convolutions. The up-convolution operation can be viewed as the reverse of max pooling and can be mathematically represented by bilinear interpolation or transposed convolutions. The skip connections are crucial in the U-Net architecture[17], ensuring the transfer of localization details from the contracting path to the expanding path. These connections help to restore the spatial dimensions and intricate details lost during pooling in the contracting path.

The final layer is a 1×1 convolution used to map the feature vectors to the desired number of classes. The strength of U-Net [16] lies in its ability to capture both the context (from the contracting path) and localization (from the expanding path with skip connections), making it exceptionally effective for tasks like medical image segmentation.

2. Multi-Task Learning Strategy: In the constantly evolving landscape of deep learning, the advent of Multi-Task Learning (MTL) presents a remarkable shift in paradigm, particularly for tasks that encompass complexities such as medical image segmentation [17]. MTUNet, with its specialized architecture, harnesses the power of MT[18]L, enabling the simultaneous segmentation of multiple kidney tumor types. MTL[17,21] fundamentally pivots on the idea that jointly learning multiple, interrelated tasks can elevate the model's generalization capability. This synergy stems from the shared representations that these tasks might possess, and the potential to capitalize on these commonalities while still catering to their unique attributes. Early layers of the MTUNet capture these shared features - primary patterns like textures and edges that are ubiquitous to different kidney tumor types. As we delve deeper into the model, the layers metamorphose to cater to more task-specific features, emphasizing the subtle nuances that distinguish one tumor type from another. Mathematically, for a gamut of tasks encompassed within $T_1, T_2, ..., T_n$, a shared representation S gets delineated by the function f_S. Concurrently, task-centric representations, R_i, get manifested as $f(R_i)$ This translates as equation (3).

$$O_i = f_{Ri}(f_S(I)) \quad (3)$$

Here, O_i encapsulates the output for task T_i with I symbolizing the input image. The learning process in MTL involves optimizing a joint loss function, which is typically a weighted sum of the individual task losses. If $L_1, L_2, ..., L_n$ are the loss functions for tasks $T_1, T_2, ..., T_n$ respectively, the joint loss L can be formulated as (4).

$$L = \sum_{i=1}^{n} w_i L_i \qquad (4)$$

Here, w_i represents the weight associated with the loss of task T_i. These weights can be uniform or adapted based on the importance or complexity of each task. MTUNet's leverage of MTL not only capacitates the model for diverse morphological structures but also fortifies its adaptability. By emphasizing shared learning among various tumor categories, the model is imbued with enhanced robustness, making it aptly poised to deliver profound insights from medical scans, especially in the realm of kidney tumor segmentation.

3. Attention Mechanism Integration: In the labyrinth of medical image segmentation, the ability to focus on relevant features—much like a surgeon's concentrated gaze on critical regions during an operation—can be the deciding factor between an average model and an exceptional one. Drawing inspiration from the transformative "Attention is All You Need" [23], MTUNet[22] integrates an attention mechanism to amplify its discriminative power. The attention mechanism works by dynamically weighing the importance of features in an image. This ensures that during the learning process, the neural network pays heightened attention to features that are of paramount importance, such as tumor boundaries, while diminishing the influence of less pertinent features. Mathematically, given an input feature map F from any layer of the network, the attention mechanism computes an attention map A using as (5)

$$A(F) = \sigma(W_f * F + b_f) \qquad (5)$$

Where W_f is the convolutional kernel and b_f the bias. The activation function σ ensures the weights are normalized. The resulting attention map A has the same spatial dimensions as F, but its values range between 0 and 1, representing the extent of attention given to each feature. To integrate the attention map with the original features, we perform an element-wise multiplication as equation (6).

$$F' = A \odot F \qquad (6)$$

Where F' represents the feature map modulated by attention.

Visually (shown in Fig. 20.5), this process can be imagined as overlaying a heatmap on the input image. Areas of the heatmap that shine brighter (indicating higher attention weights) spotlight regions that the model deems more significant. In the context of kidney tumor segmentation, these areas would typically correspond to regions containing

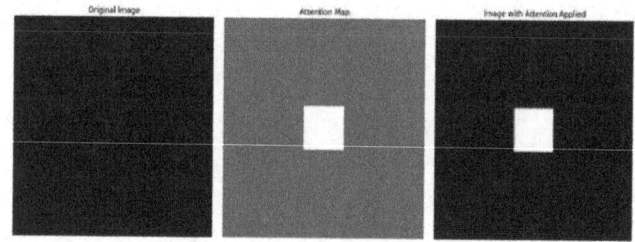

Fig. 20.5 Sample visualization of attention map [14]

tumor tissue or intricate boundaries. The incorporation of this attention mechanism into MTUNet furnishes the model with a heightened sense of discernment. By allowing it to emphasize crucial features and concurrently de-prioritize others, MTUNet not only enhances its segmentation precision but also improves its robustness against potential variations in input data, setting a promising precedent for future medical imaging tools.

4. Results

In this section, we delve into the experimental results obtained from our proposed model, FR2PAttU-Net, emphasizing its prowess in tumor segmentation and classification. The experiments were chiefly executed on CT data, as outlined in the Data Preparation segment.

Utilizing the Adam optimizer for training, we set the learning rate at 0.001. Training was executed with a batch size of 8 and spanned a total of 500,000 epochs, divided as 500 steps per epoch for a total of 100 epochs. The computational engine driving these experiments was the powerful NVIDIA GeForce RTX 3060 GPU, boasting 12GB of memory. To gauge the robustness and accuracy of our Multi-Task U-Net model in the domain of image segmentation and classification, we pitched it against several models, all trained on an identical dataset.

As illustrated in Fig. 20.3 and Table 20.1, the training and segmentation outcomes using the U-Net model were documented. This model showed an average Kidney Dice score of 0.482, a Tumor Dice score of 0.444, and a Composite score of 0.463. These values varied based on the last layer's image size, which was adjusted to observe performance shifts. Contrastingly, Fig. 20.4 and Table 20.2 elucidate the training and segmentation findings from our Multi-Task U-Net model. Remarkably, the Multi-Task model displayed an elevated average performance with a Kidney Dice score of 0.917, a Tumor Dice score of 0.854, and a Composite score of 0.886. This marked improvement underscores the model's enhanced ability to segment and classify simultaneously.

The accuracy trends during training sessions for both models are visually depicted in Figs 20.6 and 20.7.

Table 20.1 Tumor segmentation using unet model

Input Image Size (px)	Last Layer Image Size (px)	Total Training Time (s)	Kidney Dice	Tumor Dice	Composite Score
128 × 128	8 × 8	About 500	0.391	50.456	0.424
128 × 128	4 × 4	About 700	0.472	0.415	0.444
128 × 128	2 × 2	About 1,100	0.583	0.46	0.522
Average			0.482	0.444	0.463

Source: Author

Table 20.2 Tumor segmentation using multi task unet model

Input Image Size (px)	Last Layer Image Size (px)	Total Training Time (s)	Kidney Dice	Tumor Dice	Composite Score
128 × 128	8 × 8	About 500	0.906	0.836	0.871
128 × 128	4 × 4	About 700	0.925	0.858	0.892
128 × 128	2 × 2	About 1,100	0.921	0.867	0.894
Average			0.917	0.854	0.886

Source: Author

Figure 20.6 accentuates the training trajectory of the U-Net model, specifically tailored for kidney tumor segmentation. On the other hand, Fig. 20.7 lays emphasis on the Multi-Task U-Net model, delineating its performance in both tumor segmentation and classification. A noticeable surge in accuracy in Fig. 20.7 compared to Fig. 20.6 is a testament to the enhanced capabilities of the Multi-Task approach.

In Fig. 20.8, tumor size is conceptualized as the number of pixels in the tumor region. This visual metric offers a tangible insight into the tumor's dimensions, assisting clinicians in making more informed decisions. Figure 20.9 offers a unique perspective, highlighting accuracy as a function of the number of slices in each patient's data. This visualization underscores the model's versatility, handling varying slice thicknesses and ensuring consistent segmentation and classification. Finally, the presented results and visual analyses cement the superior efficacy of the Multi-Task U-Net model. Its adeptness in seamlessly segmenting and classifying kidney tumors sets a new benchmark in medical imaging research, promising transformative impacts in clinical diagnostics.

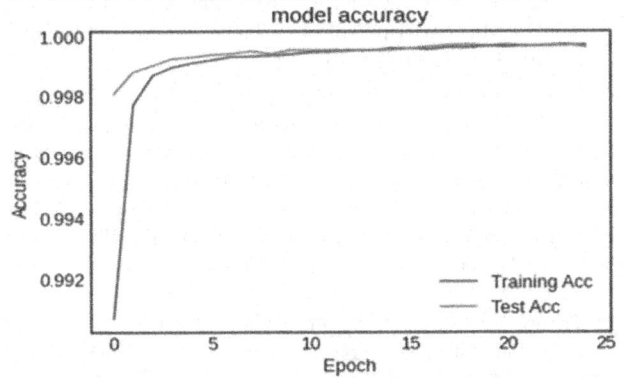

Fig. 20.6 Training accuracy UNet for kidney tumor segmentation

Source: Author

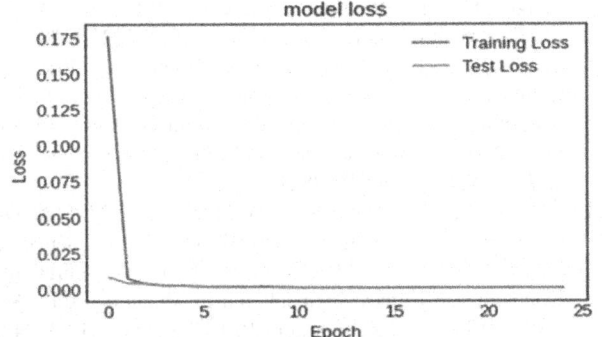

Fig. 20.7 Training accuracy multi task UNet for kidney tumor segmentation and classification

Source: Author

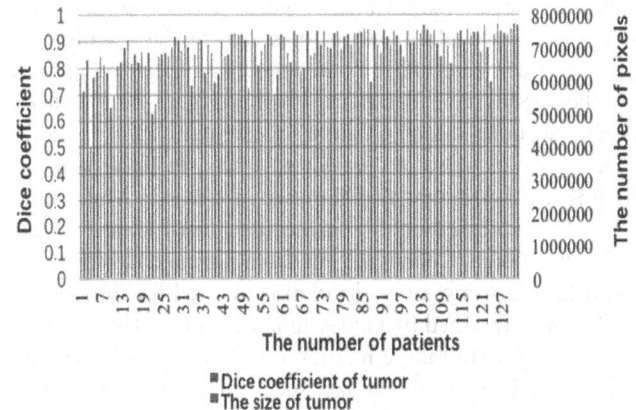

Fig. 20.8 Here, the number of pixels of kidney tumor region is seemed as the tumorsize

Source: Author

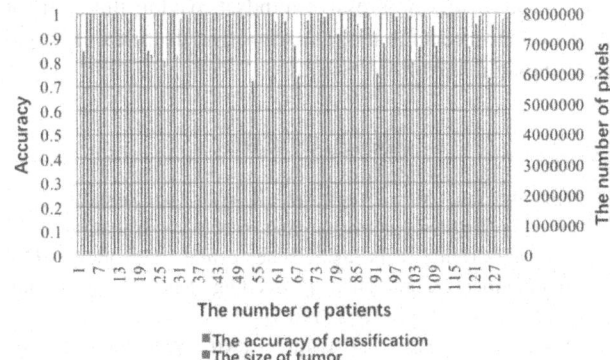

Fig. 20.9 Calculating the accuracy based on the number of slices of each patient data

Source: Author

5. Conclusion

Throughout the course of our investigation, the introduction of the Attention-based Multi-Task CNN U-Net has marked a significant advancement in the domain of medical imaging, particularly for kidney tumor segmentation and classification. This novel deep learning model combines the robust structure of the traditional U-Net with an attention mechanism and multi-task learning, culminating in an architecture that outperforms existing models both in theory and in practice. Our experimental results served as a testament to the model's efficacy. For instance, when examining tumor segmentation, the Multi-Task U-Net's average composite score of 0.886 was considerably higher than the 0.463 achieved by the standard U-Net. This difference not only validates our architectural improvements but also highlights the potential real-world implications for diagnostics and treatment planning.

The visual analyses provided further insight. Figures 20.6 and 20.7 contrasted the training accuracy between the U-Net and the Multi-Task U-Net, displaying a distinct advantage for the latter in kidney tumor segmentation and classification. Additionally, Figure 20.8's pixel-based tumor size evaluation underscores our model's ability to precisely delineate tumor boundaries, which can be pivotal for therapeutic considerations. Finally, Figure 20.9's slice-based accuracy determination accentuates the model's robustness across varied patient data sets. In light of these findings, it is evident that the Attention-based Multi-Task CNN U-Net represents a considerable leap forward in the field. Its design, encompassing attention mechanisms and multi-task learning, offers a solution tailored for the unique challenges presented by kidney tumor imaging. The model's outstanding performance metrics, coupled with its ability to generalize across different data slices, highlight its potential to revolutionize the realm of computer-aided diagnostic tools in nephrology. In summary, as medical imaging continues to evolve, it is paramount that our computational tools adapt and innovate in tandem. Our proposed model stands as a beacon of this progressive mindset, paving the path for future advancements in the intersection of deep learning and medical diagnostics.

References

1. Ostankovich, V., &Yagfarov, R. (2020). Segmification: Solving road segmentation and scene classification tasks for self-driving cars using one neural network. *ACM International Conference Proceeding Series*. https://doi.org/10.1145/3378184.3378190

2. Huo, J. (2012). Computer Aided Segmentation and Early Therapeutic Response Classification (CADrx) for Glioblastoma Multiforme (GBM) Brain Tumors with Magnetic Resonance Imaging. *ProQuest Dissertations and Theses*.

3. Alon, A. S. (2020). Tree Extraction of Airborne LiDAR Data Based on Coordinates of Deep Learning Object Detection from Orthophoto over Complex Mangrove Forest. *International Journal of Emerging Trends in Engineering Research*. https://doi.org/10.30534/ijeter/2020/103852020

4. Heller, N., Isensee, F., Maier-Hein, K. H., Hou, X., Xie, C., Li, F., ... Weight, C. (2021). The state of the art in kidney and kidney tumor segmentation in contrast-enhanced CT imaging: Results of the KiTS19 challenge. *Medical Image Analysis*. https://doi.org/10.1016/j.media.2020.101821

5. Rutherford, M., Mun, S. K., Levine, B., ... & Prior, F. (2021). A DICOM dataset for evaluation of medical image de-identification. *Scientific Data*. https://doi.org/10.1038/s41597-021-00967-y

6. Huang, W. C., Donin, N. M., Levey, A. S., & Campbell, S. C. (2020). Chronic Kidney Disease and Kidney Cancer Surgery: New Perspectives. *J Urol*, 203(3), 475–85. https://doi.org/10.1097/JU.0000000000000326

7. Checcucci, E., De Cillis, S., Granato, S., ... &Okhunov, Z. (2020). Applications of Neural Networks in Urology: A Systematic Review. *CurrOpinUrol*, 30(6), 788–807. https://doi.org/10.1097/MOU.0000000000000814

8. Checcucci, E., De Cillis, S., Granato, S., ... &Okhunov, Z. (2020). ArtifIntell Neural Networks Urol: Curr Clin Appl. *Minerva UrolNefrol*, 72(1), 49–57. https://doi.org/10.23736/S0393-2249.19.03613-0

9. Lund, C. B., & van der Velden, B. H. M. (2021). Leveraging Clinical Characteristics for Improved Deep Learning-Based Kidney Tumor Segmentation on CT. *arXiv*, 2109:5816. zttps://doi.org/10.48550/arXiv.2109.05816

10. Lin, D. T., Lei, C. C., & Hung, S. W. (2006). Computer-Aided Kidney Segmentation on Abdominal CT Images. *IEEE Trans Inf Technol Biomed*, 10(1), 59–65. https://doi.org/10.1109/TITB.2005.855561

11. Thong, W., Kadoury, S., Piche, N., & Pal, C. J. (2018). Convolutional Networks for Kidney Segmentation in Contrast-Enhanced CT Scans. *Comput Methods Biomech Biomed Eng: Imaging Visualization*, 6(3), 277–82. https://doi.org/10.1080/21681163.2016.1148636

12. Zollner, F. G., Kocinski, M., Hansen, L., ... &Lundervold, A. (2021). Kidney Segmentation in Kidney Magnetic Resonance Imaging-Current Status and Prospects. *IEEE Access*, 9, 71577–605. https://doi.org/10.1109/ACCESS.2021.3078430

13. Li, L., Ross, P., Kruusmaa, M., & Zheng, X. (2011). A Comparative Study of Ultrasound Image Segmentation Algorithms for Segmenting Kidney Tumors. *Proc 4th Int Symp Appl Sci Biomed Commun Technol*, 1–5. https://doi.org/10.1145/2093698.2093824

14. Kim, T., Lee, K., Ham, S., ... & Hong, D. (2020). Active Learning for Accuracy Enhancement of Semantic Segmentation With CNN-Corrected Label Curations: Evaluation on Kidney Segmentation in Abdominal CT. *Sci Rep*, 10(1), 1–7. https://doi.org/10.1038/s41598-019-57242-9

15. Costantini, F., & Kopan, R. (2010). Patterning a Complex Organ: Branching Morphogenesis and Nephron Segmentation in Kidney Development. *Dev Cell*, 18(5), 698–712. https://doi.org/10.1016/j.devcel.2010.04.008

16. Ronneberger, O., Fischer, P., & Brox, T. (2015). U-Net: Convolutional Networks for Biomedical Image Segmentation. In *International Conference on Medical Image Computing and Computer-Assisted Intervention*. Cham: Springer, p. 234–41.

17. Yang, Y., Li, Q., Guo, Y., ... & Guo, J. (2021). Lung Parenchyma Parameters Measure of Rats From Pulmonary Window Computed Tomography Images Based on ResU-Net Model for Medical Respiratory Researches. *Math Biosci Eng*, 18(4), 4193–211. https://doi.org/10.3934/mbe.2021210

18. Oktay, O., Schlemper, J., Folgoc, L. L., ... & Misawa, K. (2018). Attention U-Net: Learning Where to Look for the Pancreas. *arXiv*. https://doi.org/10.48550/arXiv.1804.03999

19. Alom, M. Z., Yakopcic, C., Hasan, M., Taha, T. M., & Asari, V. K. (2019). Recurrent Residual UNet for Medical Image Segmentation. *J Med Imaging*, 6(1), 014006. https://doi.org/10.1117/1.JMI.6.1.014006

20. Wang, Y., He, Z., Xie, P., ... & Li, F. (2020). Segment Medical Image Using U-Net Combining Recurrent Residuals and Attention. In *International Conference on Medical Imaging and Computer-Aided Diagnosis*. Springer, Singapore, p. 77–86.

A critical Analysis for Early Chronic Liver Disease Detection by using Machine Learning Techniques

21

Jyoshna Allenki[1]

CSE Research Scholar, Amity School of Engineering and Technology,
Amity University, Gwalior, Madhyapradesh

Hemant Kumar Soni[2]

Associate Professor, Amity School of Engineering and Technology,
Amity University, Gwalior, Madhyapradesh

Abstract: Liver with liver disease emerges once the cells of liver begin to die and therefore the tissues become a useful knot. Within the identification of pathology, the biopsy may be a golden customary. Though this method may be a sensible technique in reaching correct identification, its being associated with disadvantage. Machine learning approaches and advancements in medical image processing have increased the potential for identifying and categorizing liver tissues. During this study, we've aimed to use the image analysis, which can be of help within the identification of liver disease. To differentiate between regions of liver with liver disease and healthy parenchyma tissues, we have used the statistical feature extraction by using the mean and standard deviation methods and then in the second phase we are developing a CNN layered structure and in the third phase we are using Unsupervised Machine Learning based classification of Liver diseases.

Keywords: Machine learning, Unsupervised learning, Feature extraction 4. CNN

1. Introduction

Life without medication is a healthy life. All human organs must be functionally sound and in good health. Internally, the liver is the biggest organ. Under the right ribcage, it is located. About 3 pounds are made up of the liver. Its remarkable ability to regenerate to its original size and shape is one of its most remarkable traits. It purges toxins and other waste from the body, produces bile to aid in digestion, stores sugar that the body utilizes for energy, and produces new proteins. The main factor causing chronic liver disease is consuming too much alcohol. Traditional alcohol-related liver injury comes in three main forms:

1. Fatty liver
2. Chronic hepatitis and
3. Cirrhosis.

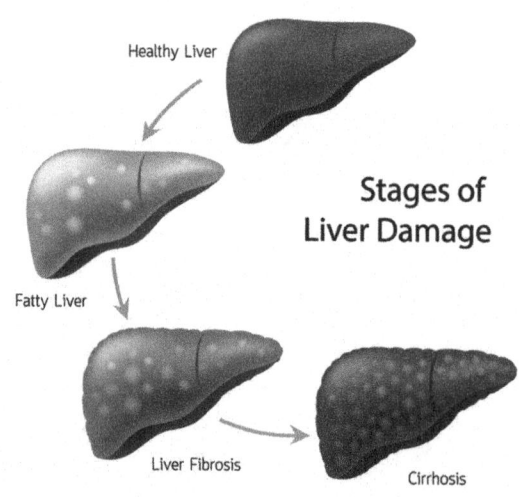

Fig. 21.1   Stages in liver disease [1]

[1]allenkijyoshna@gmail.com, [2]soni_hemant@rediffmail.com

2. Review of Literature

The study of physiological data, environmental influence, and hereditary factors will allow the doctors to diagnose the chronic diseases in an early stage and in a more efficient way. The Machine learning algorithms allow us to create models that correlate a huge variety of variables with a disease.

The "Machine Learning" algorithms bring variety of datatypes into a single model to improve the diagnosis of diseases. Noninvasive ML algorithms have been used to stage chronic liver illnesses in order to get around the constraints of biopsy.

Data mining is one of the most fundamental components of automatically diagnosing and predicting diseases. Medical data analysis uses data mining calculations and methods. Liver diseases are currently among the most lethal ailments in various nations due to the unbalanced development in liver problems over the past several years. In order to create classification models that forecast liver infection, hepatic persistent datasets are studied in this thesis. This three-stage model of building and comparative analysis aims to increase the accuracy of liver patients from India. The initial liver silent datasets obtained from the UCI repository are coupled to the min-max normalization calculation. Using PSO feature selection, the liver dataset expected moment stage. In the third step, classification algorithms are used to the data collection. In the fourth step, the accuracy will be evaluated using the root mean square value. After the performance algorithm has been subjected to PSO feature selection, the J48 method is regarded as the best. Finally, the outcomes are assessed using accuracy values. Inferring that the J48 method surpasses all others with the use of feature selection, a classification algorithm was built with a precision of 95.04 percent, as shown by the outputs from the proposed classification implementations below. (Banu Priya et al. 2018)

An automated system that extracted data in accordance with a predetermined extraction. The effectiveness of the CNNs in identifying cancer or the early stages of cancer was examined. Analysis of the type of cancer or liver mass and identification of the photos that demonstrated the highest level of cancer detection accuracy were the study's main findings. With a total accuracy of 87%, the outcomes of this method amply demonstrated the system's effectiveness. (Azer et al. 2019) .

The regions of interest in the images of the liver could be chosen using an enhanced technique for liver disorders. These ideal regions of interest underwent a two-level wavelet packet analysis. From a few statistical features, the features were taken. The photos were classified using KNN and SVM classifiers. The SVM achieved an accuracy of 97.9% when compared to all other classifiers. (Amin et al. 2019).

Based on Hierarchical Classification and Feature Fusion for the Staging of Fatty Liver Diseases, back-scan technology was used to transform ultrasound sector photos. This strategy used a hierarchical classification as its underlying principle. Two processes were used to carry out the strategy. In the first step, the focal zone of the provided ultrasound pictures was used to determine the best regions of interest. In the second procedure, the hierarchy of normal and fatty liver was followed. Gray-level co- occurrence matrix and wavelet packet transform were used to extract the features. To differentiate between a normal liver and a fatty liver, a SVM classifier was used. With an overall accuracy of 94.91%, the outcomes of this method amply demonstrated the system's effectiveness. (Wu et al. 2019).

The best classifier to identify liver illness is found after analyzing and comparing the accuracy of the various classification algorithms. Developed with the aid of the Weka tool (java-based software), a machine learning approach is focused on the numerous works of the writers. To determine the optimum method, Researchers compared different classifiers, including "Random Forest, Logistic Regression, and Separation Algorithm." The Random Forest was discovered to have the highest accuracy. (Binish Khan et al. 2019) .

System based on the hierarchical features of the ultrasonic liver tissue fusing system is built. These characteristics are chosen to categorize ultrasonic liver tissue. With a 91% total accuracy rate, the outcomes of this method amply demonstrated the system's effectiveness. (Bharti et al. 2020).

In this approach, which was based on the area of interest method, some of the region of interest was manually removed to partially extract it, and they were taught to identify liver illness with less human inputs. (Neves et al. 2021) .

The implementation of the textural feature with higher order statistical analysis is offered as a semi-automatic technique of classifying fatty liver, and the classification accuracy is calculated with a k mean classifier and improved. (Cardobi et al. 2021).

Using the dataset for the hepatitis disease, we can improve the performance of our prediction models and achieve an accuracy of 92.41% by using missing values that are present in the dataset and classification methods like K- Nearest Neighbors (KNN), Naive Bayes Support Vector Machine (SVM), Multi- Layer Perceptron (MLP), and Random Forest. (Nayeem et al. 2021) .

The liver disease can be studied and analyzed using the Machine Learning Datasets by including various re-processing methods like median missing values, coding labels. Isolate forest can be used to increase the performance and XGBoost can be used to get the attributes required in predicting liver disease (Shreyansh Jain et al. 2021).

Developed a model for liver disease prediction by hybrid classification method, which used the datasets from the Kaggle database of Indian liver patient records which can achieve an accuracy of 77.58% and have made it using Python with the Spyder tool (Golmei Shaheamlung et al. 2021).

Acute-on-chronic liver failure (ACLF) is a term used to describe patients with chronic liver disease who abruptly experience hepatic decompensation and higher short-term mortality. Serious systemic inflammation, catastrophic organ failure, and a dismal prognosis are a few symptoms of this illness. Ranking and predicting the prognosis of patients with ACLF is possible using distinct liver-specific prognostic scores and organ failures. In order to predict 90-day mortality due to liver illness, artificial neural networks (ANN), which behave similarly to biological neural networks, are examined in this study. This investigation examined the effectiveness of ANN in patients with ACLF. With comparable areas under the curve of 0.915 and 0.921, accuracy in predicting 30-day mortality was found to be 94.12 percent and in predicting 90-day mortality to be 88.2 percent. In order to correctly anticipate a patient's short-term mortality, ANN is essential. Since it automates and makes it easier to identify people who are more likely to die, its use in ACLF patients seems promising. Artificial intelligence (AI) has a lot of potential for helping doctors make decisions, identify patients who require an urgent liver transplant, and predict outcomes. (Balaji Munsunuri et al. 2021).

Despite the fact that fatty liver disease (FLD) is a severe health risk and a prevalent liver condition, there is currently no optimal method for universal screening. In this work, machine learning techniques are utilized to build a prediction model for FLD using digitized physical examination information from a health database.

Advances in computer technology have led to a rapid increase in intelligent systems that can interpret complicated data links and provide predictions and classifications.Frameworks based on artificial intelligence are revolutionizing the healthcare sector. These clever algorithms, which build strong models for early sickness detection using ML and DL show potential as a supplementary diagnostic approach for front-line clinical doctors and surgeons. Systems based on ML and DL can expedite and streamline the processes needed to detect illnesses using clinical and image-based data, thereby enhancing workflow effectiveness and clinician support. They can mimic human thought processes and even detect diseases that human intelligence cannot detect. (Jignesh Chowdary G et al. 2021).

The project's objective is to use ultrasound images, machine learning-based algorithms, and classification techniques to categorize the liver fibrosis stage of chronic liver disease (CLD). Overall, 187 patients from Ditan Hospital took part in the study. Liver biopsies produced the best results in terms of evidence. In our investigation, two categorization methods are employed. The Efficient Net is a classification-focused variant of convolutional neural networks (CNNs). The other strategy makes use of a radiomics model. 637 radiomics features were sent to the least absolute shrinkage and selection operator in order to remove the unnecessary characteristics. (LASSO). After reduction, classes require less than 20 independent features. For cirrhosis (F4), advanced fibrosis (F3+F4), and severe fibrosis (F2+F3+F4), the Efficient Net model's area under the receiver operating characteristic (AUC) was 0.83, 0.78, and 0.84, respectively. The radiomics model's AUC values for cirrhosis, advanced fibrosis, and significant fibrosis were 0.96, 0.81, and 0.85, respectively. By looking at CLD ultrasound images, machine learning approaches may correctly categorize liver fibrosis. (Y. Zhang et al. 2021).

Numerous machine learning algorithms were employed in this study to determine the risk of liver disease depending on the results of the user's blood test report. To predict the result, the most accurate machine learning technique was used. We developed a system based on the precise model that requests users submit the specifics of their blood test report. A person's risk of acquiring liver disease is then determined by the algorithm using the most precise model available. (Rakshith D B et al. 2021).

Statistical machine learning methods may support decision-making depending on the specific problems. Data-driven algorithms that are based on ML may be used to evaluate existing approaches and assist researchers in coming to potentially game-changing conclusions. The study's objective was to apply machine learning techniques to extract crucial indications of liver disease from a medical examination of 615 individuals. Data visualizations were used to make significant discoveries, such missing figures. Principal component analysis (PCA) was used to produce missing data points and multiple imputations using chained equations (MICEs) to minimize dimensionality. The Gini index was used to assess the variables' relevance, and it was also used to confirm significant predictors discovered using PCA. Using training data (ntrain=399) and testing data (ntest=216), the ML techniques were used to predict categories. In order to help doctors diagnose liver diseases more accurately, the study compared three machine learning binary classifier algorithms (artificial neural network, random forest (RF), and support vector machine) that were used to categorize people with liver diseases using a published liver disease data set. The synthetic minority oversampling method was used to oversample the minority class in order to avoid overfitting. With a rating of 98.14 percent, the RF greatly improved accuracy in comparison to the other methods. (15, p0.001). (Mostafa et al. 2021).

The radiologists have difficulty diagnosing the disease in its advanced stages due to the poor quality of often utilized ultrasound pictures. As a solution to the aforementioned issue, "a Computer Aided Diagnosis technique, Based on extracting ultrasound image features and a voting-based classifier, a method called "using ML Algorithms and a voting-based classifier to categorize liver tissues as being fatty or normal" is developed. The voting-based classifier achieves an accuracy of 95.71%, while the J48 algorithm achieves 93.12%. (Ahmed Gaber et al. 2022).

A wide range of detection and classification methods have been developed by different researchers to treat chronic liver disease. The study looked at the benefits and drawbacks of several classification schemes for the characteristics of chronic liver disease.

3. Comparison Table

Table 21.1 Comparison table

Sl. No.	Author & Name of thepaper	Journal Name	Techniques used	Limitations
1	Nazmun Nahar(2018) "Liver Disease Prediction By Using Different Decision Tree Techniques"	International al Journal of Data Mining & Knowledge Management Process	Random Forest	This algorithm has 70.67% Accuracy rate.
2	Golmei Shaheamlung, Harshpreet Kaur (2021). "The diagnosis of Chronic Liver Disease using Machine Learning Techniques"	Journal of Information Technology inIndustry	KNN	This algorithm has 73.27%. accuracy rate.
3	Shobana, G., & Umamaheswari, K. (2021)."Prediction of Liver Disease using Gradient Boost Machine Learning Techniques with Feature Scaling"	5th International Conference on Computing Methodologies and Communication (ICCMC)	Gradient Boost Algorithm	The prediction accuracy offundamental models has not been increased by feature selection.

Sl. No.	Author & Name of thepaper	Journal Name	Techniques used	Limitations
4	Shreyansh Jain (2021) "classification of Liver Diseases UsingIntelligent Techniques"	Easychair preprint	SVM	Does not show improvement inaccuracy.
5	Mujeeb Ur Rehman(2021) "InfraredSensing Based Non-Invasive Initial Diagnosis of Chronic Liver Disease Using Ensemble Learning"	IEEE Sensors Journal	KNN	needs a lot of training time. likewise restricted to detecting chronic liver disease in its earliest stages.
6	Ahmed Gaber(2022) "Automatic Classification of Fatty Liver Disease Based on Supervised Learning and Genetic Algorithm"	Applied Sciences	voting-based classifier	Focused only Fatty Liver Disease.

Source: Authors

4. Research Gaps

The research gaps are as follows which are found from the review of literature:

- Present algorithms (KNN, SVM traditional algorithms) are unable to analyze the image data.
- Image classification, enhancement requires ensembled machine learning algorithms and image processing techniques for better efficiency.

5. Proposed System

Feature extraction methods like Histogram equalization will be used in the proposed system. We will design the model using ensemble machine learning by grouping three classifiers to detect and diagnose the chronic liver disease. The accuracy of the proposed research work will be compared with the accuracy of the existing work, the datasets will be collected from kaggle, UCI repository and use for the purpose of classification. The primary data from a private hospital will be collected to use in the testing phase.

6. Methodology

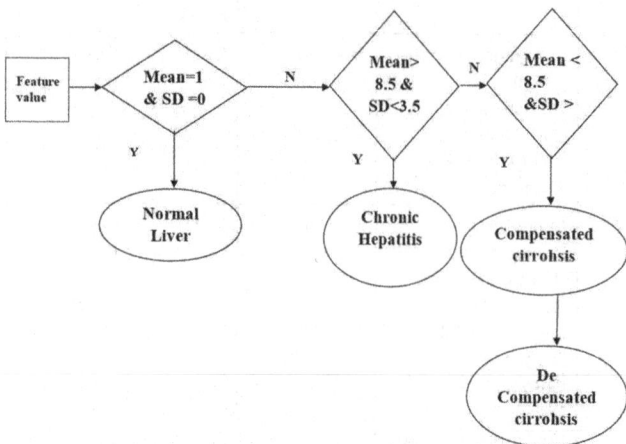

Fig. 21.2 Chronic liver disease classification using the ELM system

Source: Authors

7. Future Work

The problematic parts of employing ML concepts to diagnose liver disease include gathering enormous volumes of medical data for research and choosing the most important characteristics. With an emphasis on data collection, segmentation, and pre-processing, these problems can be solved by doing significant research at various centres.

References

1. Nahar, N., & Ara, F. (2018). Liver Disease Prediction byUsing Different Decision Tree Techniques. International Journal of Data Mining & Knowledge Management Process, 8(2), 01–09. https://doi.org/10.5121/ijdkp.2018.8201
2. Banu Priya, M., Laura Juliet, P., & Tamilselvi, P. R. (2018). Performance Analysis of Liver Disease Prediction Using Machine Learning Algorithms Related papers Performance Analysis of Liver Disease Prediction Using Machine Learning Algorithms. International Research Journal of Engineering and Technology.
3. Amin, M. N., Rushdi, M. A., Marzaban, R. N., Yosry, A., Kim, K., &Mahmoud, A.
4. M. (2019). Wavelet-based Computationally- Efficient Computer- Aided Characterization of Liver Steatosis using Conventional B- mode Ultrasound Images. Biomedical signal processing and control, 52,84–96. https://doi.org/10.1016/j. bspc.2019.03.0 10Wu, C. C., Yeh, W. C., Hsu, W. D., Islam, M. M., Nguyen, P., Poly, T. N., Wang, Y. C., Yang, H. C., & Jack Li,
5. Y. C. (2019). Prediction of fatty liver disease using machine learning algorithms. Computer methods and programs in biomedicine,170,23–29.
6. Li, N., Zhang, J., Wang, S., Jiang, Y., Ma, J., Ma, J., Dong, L., & Gong, G.(2019). Machine learning assessment for severity of liver fibrosis for chronic hbv based on physical layer with serumarkers. I9EEE Access, 7, 124351–124365.
7. Khan, Binish & Shukla, Piyush & Ahirwar, Manish. (2019). Strategic Analysis in Prediction of Liver Disease Using Different Classification Algorithms. International Journal of Computer Sciences and Engineering. 71–76.
8. Hashem S, ElHefnawi M, Habashy S, et al (2020) Nov. Machine Learning Prediction Models for Diagnosing Hepatocellular Carcinoma with HCV- related Chronic Liver Disease. Computer Methods and Programs in Biomedicine.;196:105551.
9. Kagadis, G. C. (2020), "Deep learning networks on chronic liver disease assessment with fine- tuning of shear wave elastography image sequences",Physics in Medicine and Biology, vol. 65, no. 21, doi:10.1088/1361- 6560/abae06.
10. Bharti, P., & Mittal, D. (2020). Hybrid feature selection-based feature fusion for liver disease classification on ultrasound images. Advances in Computational Techniques for Biomedical Image Analysis, 145–164.https://doi.org/10.1016/ B978-0-12- 820024-7.00008-6.
11. G. Shobana and K. Umamaheswari, (2021) "Prediction of Liver Disease using Gradient Boost Machine Learning Techniques with Feature Scaling," 5th International Conference on Computing Methodologies and Communication (ICCMC), 2021, pp. 1223-1229.
12. Cardobi, Nicolò, et al. (2021): "An Overview of Artificial Intelligence Applications in Liver and PancreaticImaging." Cancers 13.9 2162.
13. Nayeem, M. J., Rana, S., Alam, F., & Rahman, M. A. (2021). Prediction of Hepatitis Disease Using K-Nearest Neighbors, Naive Bayes, Support Vector Machine, Multi-Layer Perceptron and Random Forest. 2021International Conference on Information and Communication Technology for Sustainable Development, ICICT4SD2021-Proceedings, 280–284.
14. Jain, S., Sharma, R., & Rajkamal, R. (2021). EasyChair Preprint Classification of Liver Diseases Using Intelligent Techniques Classification of Liver Diseases Using Intelligent Techniques.
15. Harshpreet Kaur, G. S. (2021). The Diagnosis of Chronic Liver Disease using Machine Learning Techniques. Information Technology in Industry, 9(2), 554–564. https:// doi.org/10.17762/itii.v9i2.382
16. Musunuri, B., Shetty, S., Shetty, D. K., Vanahalli, M. K., Pradhan, A., Naik, N., & Paul, R. (2021). Acute-on-chronic liver failure mortality prediction using anartificial neural network. Engineered Science, 15, 187–196. https://doi. org/10.30919/es8d515
17. Zhao, M., Song, C., Luo, T., Huang, T., & Lin, S. (2021). Fatty Liver Disease Prediction Model Based on Big Data of Electronic Physical Examination Records. Frontiers in public health, 9, 668351. https://doi.org/10.3389/fpubh.2021.6683 51
18. Chowdary, G. J. (2021). Machine Learning and Deep Learning Methods for Building Intelligent Systems in Medicine and Drug Discovery: A Comprehensive Survey. arXiv preprint arXiv:2107.14037.

Internet of Vehicle Ad Hoc Networks (VANETs): Anomaly Detection Algorithms

22

Anjali Thuvva[1]

Research Scholar, Computer Science and Engineering,
Amity School of Engineering and Technology,Amity University, Madhya Pradesh Gwalior, India

Rajeev Goyal[2]

Associate Professor, Computer Science and Engineering,
Amity School of Engineering and Technology, Amity University, Madhya Pradesh, Gwalior, India

G. N. Balaji[3]

Associate Professor, School of Computer Science and Engineering,
Vellore Institute of Technology, Vellore, India

Abstract: Vehicle Ad-hoc Networks (VANETs) are a new technology with a lot of development potential. VANETs are required for developing intelligent transportation systems in order to deliver efficient and secure communication between vehicles and infrastructure. One of the many security challenges that VANETs must address is anomalies that could compromise the integrity of the network. For VANETs to operate normally, anomaly detection systems must be able to identify and minimize these disruptions. This paper provides a thorough analysis of numerous anomaly detection strategies tailored to VANETs. A discussion of the difficulties involved in anomaly detection in VANETs as well as typical abnormalities that can be detected there, as well as crucial traits for successfully completing anomaly identification, are included in the analysis. In order to develop services that are especially pertinent to the automotive environment, VANETs aim to connect equipment found within automobiles together. They make an effort to manage the network topology without using infrastructure devices as a helper. Due to their special characteristics, VANETs present both distinctive obstacles and research opportunities. This study will aim to address the essential traits that define a VANET environment in addition to detailing what may be done with a VANET system.

Keywords: Internet of things (IoT), Intelligent transportation system (ITS), Vehicle-to-vehicle (V2V)

1. Introduction

The shrinking size of computing systems is one of the main reasons propelling the growth of computing technology. Through allowing larger systems to utilize more resources, we may enhance their power while also making ever-smaller systems practical to employ. This is accomplished through reducing the size of computer systems. This may be drawn back to the rise of the personal computer, which prompted various improvements in networking technologies. Presently,

more potent little devices that can still communicate over networks and be integrated into bigger systems are being created. The introduction of new gadgets has also sparked a number of advancements in network technology, which fall under the umbrella term IoT.

Mobile Ad-hoc Network (MANET) systems are one of the subcategories of IoT technology. MANET systems strive to join mobile devices to form a network, as their name suggests; these devices are often ones that are moved by

[1]anjali.thuvva@gmail.com, [2]goyal.rajeev@gmail.com, [3]balaji.gnb@gmail.com

people. Because there are no management infrastructure elements, such routers, phone towers, and only connected devices, these networks are referred to as ad-hoc networks. Thanks to MANET technology, any device in the network can instantly communicate with any other device nearby. As a result of MANETs, a new branch of mobile networking systems that aimed to link up things being transported by vehicles was created. [1]

The intelligent transport system (ITS) is one of the main uses of the Internet of Things (IoT). The VANET is a prospective technology that will be crucial to the advancement of ITS. Recently, several groups, companies, and individuals have focused on different VANET components, including as standards, routing protocols, security, and so on. The purpose of this research is to present a thorough understanding of the fundamentals of this technology. We discuss several ad hoc network types, VANET standards, applications, difficulties, and information sharing techniques. Then, in order to comprehend routing and the difficulties it poses in VANET, we evaluate and analyze a number of multi-hop broadcast routing protocols. [2]

A pattern that deviates from regular, expected behavior is one way to define an anomaly in an abstract sense. Anomalies can be divided into three categories.

1. Point Anomalies: The simplest kind of anomaly, a point anomaly occurs when one data request can be viewed as abnormal compared to the rest of the data.

2. Contextual Anomalies: A contextual anomaly is when a data instance is anomalous in one context but not in another. Contextual and behavioral qualities are the two components of contextual abnormalities.

Use the first feature to identify the context (or neighborhood) of an instance. In geographic databases, examples of contextual features are a location's longitude and latitude.

Time is another contextual element in time series data that affects how a case fits within the larger sequence.

The second attribute, a behavior attribute, describes the instance's noncontextual characteristics. The amount of rain that falls at each location in a spatial dataset that describes the average amount of rain experienced worldwide is an example of a behavioral feature.

3. Collective anomalies: For the full dataset, a collective anomaly is a group of related data occurrences that are aberrant.

The oldest methods of finding anomalies are statistical anomaly detection techniques. Utilizing statistical techniques, a statistical model is created to depict the typical behavior of the supplied data. Then, a statistical inference test can be run to see whether a certain instance fits inside this model. [3]

1.1 Utilization of VANETs

The world of today is one where technology has developed to the point where it can be used to swiftly and easily complete a number of daily tasks.

Both VANET and MANET are similar in that they don't need any infrastructure to convey data. Applications for emergency and entertainment purposes as well as safe driving greatly benefit from VANET. The ability of the cars to communicate with one another and with roadside base stations that are strategically placed along the route could be argued to make them an intelligent part of the transportation system. Examples include junctions and construction sites.

Intelligent Transportation Systems (ITS): By facilitating communication between vehicles and with roadside infrastructure, VANETs can be utilized to improve traffic flow, lessen congestion, and increase road safety.

Location-based services and navigation: VANETs can give drivers access to real-time traffic and navigation data, empowering them to make wise decisions and avoid congestion.

Emergency Services: VANETs can be used to quickly and effectively transmit emergency notifications, such as those about accidents and road closures, to other vehicles and emergency services.

V2I and V2V VANETs can provide communication between vehicles and with roadside infrastructure, enabling a variety of applications, including platooning and cooperative driving.

VANETs are capable of providing passengers in automobiles with entertainment and information services, such as streaming music and video. [4]

1.2 Architecture of VANETs

Use the WAVE protocol for wireless communication between vehicles and between a vehicle and a Roadside unit (RSU). Safety apps can provide drivers and passengers with a wealth of information while also enhancing road safety and providing a comfortable driving experience thanks to this manner of communication.

There are two unique entities that are referred to as "user" and "provider" respectively. While the user uses the services, a supplier provides them. RSUs and OBUs can play either users or providers depending on their roles inside the network. An On Board Unit (OBU), an Application Unit (AU), and a Roadside Unit (RSU) are the three primary parts of a system.

1. OBU: Every vehicle has an OBU, which is a piece of hardware. Usually mounted on cars, OBUs are WAVE devices that exchange data with RSUs or other OBUs. An OBU is a piece of hardware that is present in every vehicle. OBUs are WAVE devices that are typically mounted on automobiles and share data with RSUs or other OBUs. The transceiver is coupled with a radio frequency aerial and a processor, much like a router. It relays information to other OBUs in addition to sending information. It provides assistance to AU in the form of service initiatives. The ability to communicate with all external components may be enabled by a variety of wireless communication technologies.

2. Application Unit (AU): To communicate with the OBU, an AU is a device put inside the car and utilized in conjunction with the application that the supplier provides. The AU might run on a standard device, such as a personal digital assistant (PDA), in addition to safety applications, and online services. OBU and Application Units are linked together using wired or wireless technology. It gives OBU access to the internet so that data can be exchanged and received.

3. Roadside Unit (RSU): RSUs, which are wave devices, are often placed next to roads or in specialized locations like parking lots and junctions. The gadget connects to the internet and can be used to avoid mishaps in addition to giving the user security information. Information can only be viewed by users who have been verified. We employ strategies like pseudonyms, mix zones, ad hoc anonymity, and silence intervals.

For instance, they are close to parking lots and crossroads, which have a lot of traffic.

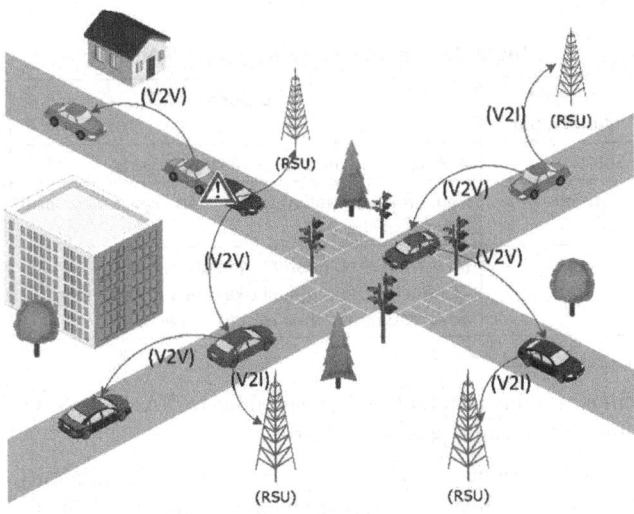

Fig. 22.1 Architecture of VANETs [1]

2. Review of Literature

2.1 Intrusion Detection System (IDS)

IDS is a technique that detects unusual or suspicious activity on the analyzed target (network or host).

It enables knowledge about successful or unsuccessful intrusion attempts. Internal attacks are detected with IDS

technologies. These are attacks that cryptographic solutions are incapable of detecting. Internal assaults are, in fact, attacks by compromised nodes. An intrusion detection system is frequently employed as a second line of defence following cryptographic systems.

An intrusion detection system generally consists of three phases: a phase of data gathering, a phase of analysis, and then a phase of response to stop or lessen the effects of the assault on the system. IDS can be found at a few unique nodes known as monitors or monitoring nodes. Depending on the protocol type and the IDS architecture, different nodes are deployed.

IDS can be divided into 3 groups based on the kind of detection techniques used:

Signature-based system: The system compares the collected data to a database on the behavior of specific assaults. If the data correlate with known malicious behavior, an attack is recognized.

An anomaly detection system: This system finds any behavior that differs from the expected, predetermined behavior.

Systems based on specifications: these systems established a list of requirements that a programme or protocol must meet. If the programme or protocol fails to comply with the requirements imposed for proper functioning, an attack is detected.

2.2 VANET Security

The main concern is communication security in VANETs. According to VANET security, it is required to prevent any data transmissions from being changed. Because of their frequent topology changes, tiny device sizes, etc., ad-hoc networks generally face greater security concerns than do traditional wireless networks.

Confidentiality is another word for information privacy. Private information is intended to fall into the wrong hands in this situation. The VANET could have its data confidentially compromised via social attacks, traffic analysis attacks, eavesdropping attacks, and illusion attacks. Data authenticity aims to validate an individual's identity using identification components like usernames and passwords.

Following the identification process, the next step is to make sure that the system only permits authorized users to access. This protocol is also thought of as the first line of security against malicious users.

The drivers must be held accountable in order to guarantee that the traffic condition is properly reported in a defined amount of time. The information must be made available to each potential recipient and must arrive within a specific time frame because data acquired after that point is useless. Authenticated senders may only communicate complete, accurate information. Attackers jam the network, which causes it to collapse. The influence of the environment on magnetic waves can interfere with information transmission. Due to the limitless network size, collisions and congestion must be controlled. [5]

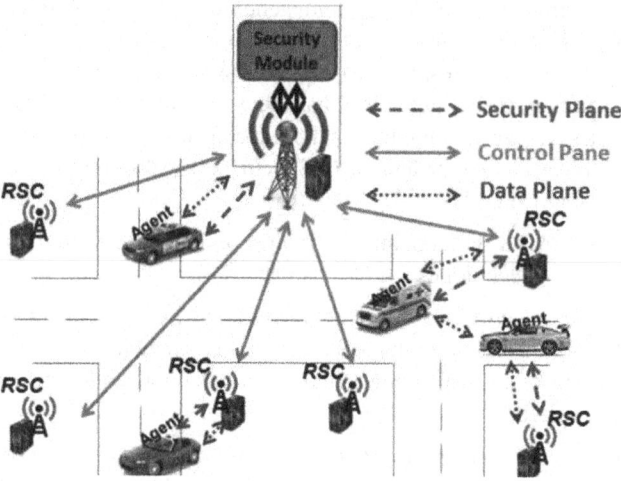

Fig. 22.2 VANET security model [2]

Applications for VANET

Travelers and Drivers have access to a variety of data and can create a huge number of apps in the VANET environment. The on board units of the cars use a number of technologies, such as sensors, cutting-edge antenna technology, and efficient wireless access. In order to improve safety and comfort for motorists, the system communicates data from the RSU to the other vehicles. It will converse with it and gather information from other vehicles.

VANET applications for safety: Following the identification process, the next step is to make sure that the system only permits authorized users to access. Additionally, this protocol is regarded as the initial line of defense against rogue users.

To ensure that the traffic situation is appropriately reported in a specified length of time, the drivers must be held responsible. Data obtained after that moment is useless, thus it must be made available to each potential recipient and sent within a certain time range. Only fully accurate information may be transmitted by authenticated senders. The network is jammed by attackers, which brings it to a halt. Information transfer may be hampered by magnetic waves' effects on the surroundings. Due to the limitless network size, collision and congestion control is necessary.

VANET applications that are not safety-related: VANETs may also be utilized to deliver comfort or pay-for-service offerings. This type of application enhances electronic toll collecting (ETC), passenger comfort, and advertisement effectiveness.

These companies provide applications that allow users to find parking spaces, gas stations, shopping centers, motels, fast food outlets, and other points of interest (PoI), as well as weather data, traffic, and lodging options. It's thought that using the VANET for business purposes and comfort will reduce traffic safety and efficiency. Additionally, it interferes with and hinders applications related to safety.

Applications for efficiency: With the help of this programme, a vehicle's mobility can be increased and its location in the city's lanes determined. In essence, communication between vehicles and between vehicles and remote sensors occurs. The following two application categories can be categorized: traffic jam reduction and road and crossing management.

Applications for comfort: There are businesses in the region that offer drivers information they may use to improve and make the most of their trip. Depending on the application, weather data, parking space availability, petrol station maps, and restaurant locations may all be included.

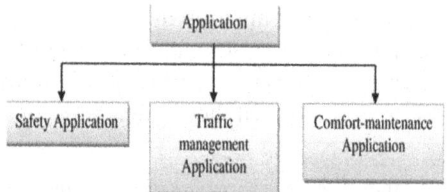

Fig. 22.3 Applications for VANET
Source: Author

Table 22.1 Major applications of VANET

VANET Application	Purpose
Safety	Accident and medical emergency alerts
Non-Safety	Real-time navigation, road map assistance, and driver assistance
Others	If News Alerts, Road Condition Alerts, and Traffic Condition Alerts are all examples of broadcasting advertisements, then Report Application.

Source: Author

Till now many different studies have done on VANETs. Some of recent research papers as follows:

Numerous safety solutions, such as those for availability, authenticity, integrity, and non-repudiation, are available for a variety of ITS uses. Due to the community's incredibly dynamic nature, availability is one of the most crucial and challenging goals in VANET, especially in safety-related applications. Denial of Service (DoS) attacks that try to interfere with availability can be countered using eight different strategies. DoS attacks overload the server with data packets in an effort to interfere with network operations. The system can suffer severe damage from a high-charge (i.e., massive rise in data packet volume) DoS assault, but on the other hand, it is easy to identify such attacks, illustrating a trade-off for attackers. (Haydari, Ammar et al. 2018)[6]

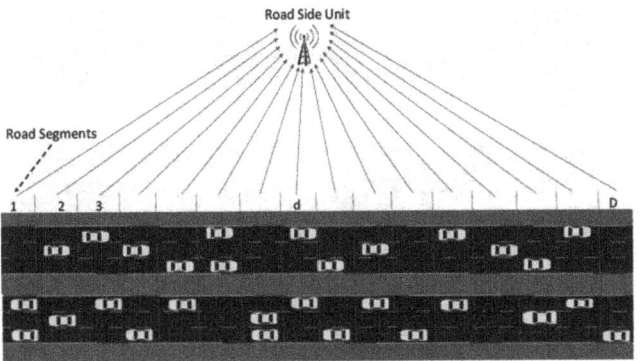

Fig. 22.4 Denial of service attack [6]

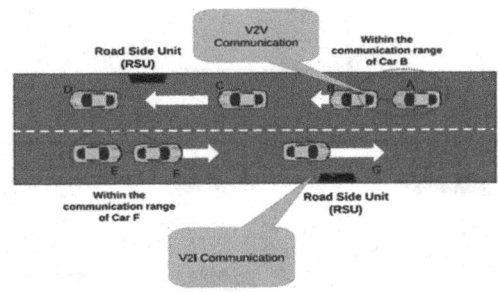

Fig. 22.5 V2V and V2I communication [6]

2.3 Attacks on Automatic Vehicles

The authors analyzed potential dangers to and defenses against CAVs (Connected and Autonomous Vehicles) in (Sun et al. 2021). CAV attacks fall into three categories: those that take place within the vehicle, those that take place between vehicles, and those that take place between vehicles and everything else. We offer safety-enhancing counterattack strategies and precautions. In-vehicle networks were introduced by the authors (Liu et al. 2017). The attacks and risks to vehicle-based networks as well as potential fixes are described in detail. The authors went into great length to review cyberattacks on the sensing layers (El-Rewini et al. 2018). The sensing layer is made up of two different kinds of sensors: those that track the dynamics of the vehicle and those that track its surroundings. The dynamics of a vehicle are measured by various inertial sensors, magnetic encoders, etc. Examples include cameras, lidars, radars, and more.

Hybrid Communication: A vehicle eventually needs to communicate with both the infrastructure and its nearby automobiles. In that instance, hybrid communication—which combines V2V and V2I communication—can be used to carry out communication in a VANET. (Kariuki Gabriel et al. 2019)

Figure 22.5 displays two distinct node types: Roadside Units (RSU) are permanent nodes placed along the route, and On-Board Units (OBU) are mobile nodes, such as automobiles, that are fitted with an electronic interface and may wirelessly connect with other nodes through the RSU. [7]

Future Intelligent Transportation Systems are projected to heavily rely on VANETs. As transport communications technology has developed, the number of attacks has increased, and this design is still vulnerable to a number of flaws that have led to a number of security issues that need to be fixed before VANET technology can be widely and safely deployed. The privacy and communication in VANET may be impacted by security vulnerabilities including Sybil attacks and Distributed Denial of Service (DDoS) assaults. OMNET, a new updated most recent quality Associate in network

machines as well as information network simulator, was specified because simulators are used in this investigation. The objective of this work is to create an algorithmic rule for identifying and avoiding different kinds of attacks. (Rj et al.2020) [8]

Channel jamming is used to target the media in this attack. Because the channel is no longer active, the nodes are unable to interact. The fundamental goal of the scheme is to overwhelm the network and make it hard for honest nodes to access networks and media. The infrastructure and vehicle nodes of the network will be destroyed, and the adversary will be defeated. The network problems deny service to real nodes and do a variety of other unnecessary duties. As seen in the figure below, an incursion can target VANETs from the inside or the outside. Its major goal is to prevent legitimate users from accessing the network.(Muhammad et al. 2019) [13]

3. Comparisontable

Table 22.2 Comparision of techniques used in VANET

S. No	Name of the Author	Techniques Used	Limitations
1	Youssef Khayati, (2020)	Examined some routing attacks such as Black hole & Wormhole attacks.	This strategy is less effective if the attacker sends false information.
2	Naveen R (2020)	Best Reliable Routing Protocols for the transmission process in VANET.	Managing the network protocols for tracking the attacker's location or any other information is challenging.
3	Alladi, (2021)	Anomaly detection framework for VANETs supported deep neural networks (DNNs).	This method does not work for high traffic environment.
4	Muhammad (2019)	The infrastructure and vehicle nodes of the network will be destroyed due to DoS attack.	This approach is ineffective in high-traffic areas.

Source: Author

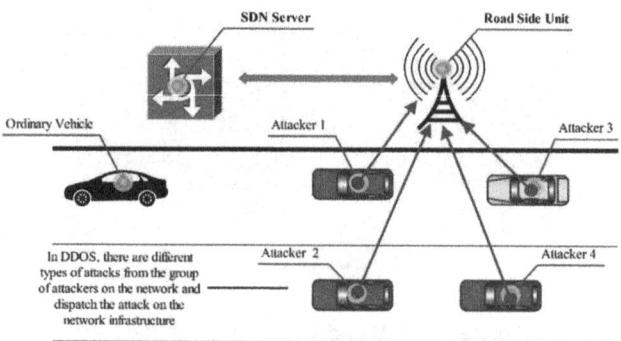

Fig. 22.6　Vehicle to vehicle attack [12]

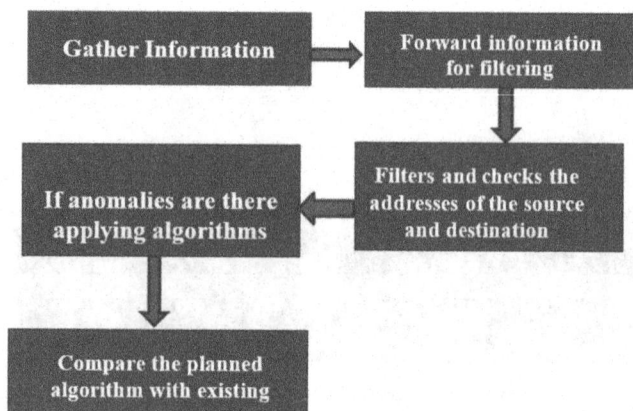

Fig. 22.7　Stages of the process

Source: Author

4. Research Gaps

The assessment of the literature finds that VANETs handle a number of data-intensive challenges across a number of problem domains, but that the security implications have not yet been thoroughly studied. The implementation of anomaly detection for early estimation is the aim of this research. The following research gaps were discovered after reviewing several SCI, Scopus, and Web of Science articles.

- The existing methods are not effective for high traffic environment.
- The attacker send false information the existing methods are not successful.

5. Objectives

Building an algorithm for VANETs to detect anomalies is the aim of this project. Currently, security could pose a serious issue in a range of VANET applications where an offending message could adversely influence people's lives either directly or indirectly.

- To develop innovative mitigation techniques detecting anomalies.
- To detect any malicious nodes by examining the attacker's incorrect signature.

6. Expected Analysis

- Gathering the Information.
- After gathering forward information for filtering.
- Filters and checks the addresses of the source and destination.
- If anomalies are there applying algorithms.
- Compare the planned algorithm with existing.

Table 22.3　Analysis of algorithms used in VANET.

S. No	Algorithm	Features
1.	Attacked Packet Detection Algorithm (APDA)	The DOS attack is recognised by the APDA algorithm prior to verification. Position, time stamp, and velocity are taken into account.
2.	Enhanced Attacked Packet Detection Algorithm (EAPDA)	EAPDA not only validates the nodes and searches for rogue nodes, but it also boosts security by speeding up data transfer and cutting back on delays.
3.	Signature Based Authentication (SBA) Method	The most effective and well-known technique for mutual authentication between entities on VANETs is the use of digital signature techniques. This is invulnerable to the chosen message attacks.

Source: Author

7. Conclusion

Several research studies on the architecture, traits, applications, security, and security-related issues of VANETs are reviewed in this article. VANET has a security issue. It is vital to emphasize VANET security more because it has emerged as users' top concern. Several important issues in vehicular communication are currently the subject of in-depth study and discussion. Discussion topics include updates, modifications, safety communications, data security, and V2V communications. Ad hoc mobile networks could develop into VANETs in the future. Future V2V and V2I communications will be used to forecast established communication performance levels.

References

1. Michael Lee, Travis Atkison, VANET applications: Past, present, and future, Vehicular Communications, Volume 28, 2021, 100310, ISSN 2214-2096,https://doi.org/10.1016/j.vehcom.2020.100310. (https://www.sciencedirect.com/science/article/pii/S2214209620300814).

2. Fotros, M., Rezazadeh, J., Ameri Sianaki, O. (2020). A Survey on VANETs Routing Protocols for IoT Intelligent Transportation Systems. In: Barolli, L., Amato, F., Moscato, F., Enokido, T., Takizawa, M. (eds) Web, Artificial Intelligence and Network Applications. WAINA 2020. Advances in Intelligent Systems and Computing, vol 1150. Springer, Cham. https://doi.org/10.1007/978-3-030-44038-1_102

3. Ali Bou Nassif, (Member, Ieee), Manar Abu Talib, (Senior Member, Ieee), Qassim Nasir, And Fatima Mohamad Dakalbab Machine Learning for Anomaly Detection: A Systematic Review, Digital Object Identifier 10.1109/ACCESS.2021.3083060, date of publication May 24, 2021, date of current version June 4, 2021 IEEE Access.

4. Difference between MANET and VANET - GeeksforGeeks

5. Review On Vanet Architecture And Applications RadhaKrishna Karne , Dr. T.K.Sreeja, Turkish Journal of Computer and Mathematics Education Vol.12 No.04 (2021), 1745 – 1749

6. Haydari, Ammar & Yilmaz, Yasin. (2018). "Real-Time Detection and Mitigation of DDoS Attacks in Intelligent Transportation Systems". 10.1109/ITSC.2018.8569698.

7. Kariuki, Gabriel. (2019). "A Survey on Possible Attacks in Vehicular Ad Hoc Network". Journal of Telecommunications and Information Technology.

8. Rj, Naveen & Srinivas, Nikhil & Vineeth, Nandhini & Nannapaneni, Siva Venkata Chaitanya. (2020). "A Survey on Detection and Prevention of Security Attacks in VANET".

9. Khayati, Y. & Mazri, Tomader. (2020). "SECURITY STUDY OF ROUTING ATTACKS IN VEHICULAR AD-HOC NETWORKS (VANETS). ISPRS - International Archives of the Photogrammetry, Remote Sensing and Spatial Information Sciences". XLIV-4/W3-2020. 267–272. 10.5194/isprs-archives-XLIV-4-W3-2020-267-2020.

10. Shu, Jiangang & Zhou, Lei & Zhang, Weizhe & Du, Xiaojiang & Guizani, Mohsen. (2020)." Collaborative Intrusion Detection for VANETs: A Deep Learning-Based Distributed SDN Approach. IEEE Transactions on Intelligent Transportation Systems". pp. 1–12. 10.1109/TITS.2020.3027390.

11. Sharma, Aekta & Arunita, Jaekel. (2021). "Machine Learning Based Misbehaviour Detection in VANET Using Consecutive BSM Approach. IEEE Open Journal of Vehicular Technology". pp. 1–1. 10.1109/OJVT.2021.3138354.

12. Kolandaisamy Raenu. (2022). "A Multivariant Stream Analysis Approach to Detect and Mitigate DDoS Attacks in Vehicular Ad Hoc Networks." Wireless Communications and Mobile Computing, link.gale.com/apps/doc/A602364925/AONE?u=anon~26615033&sid=googleScholar&xid=e4b0b949.

13. Muhammad Sameer Sheikh ,Jun Liang - Jiangsu University" "A Comprehensive Survey on VANET Security Services in Traffic Management System" Wireless Communications and Mobile Computing, 2019, 1–23 - September 2019

14. Liang, Yan, T, Lee, J & Wang, G 2018, 'A distributed intersection management protocol for safety, efficiency, and driver's comfort', IEEE Internet of Eings Journal, vol. 5, no. 3, pp. 1924–1935

15. Radha Krishna Karne , and Allanki Sanyasi Rao, " A Study on IoT Technologies, Standards and Protocols", IBM RD's Journal of Management & Research, Volume 10, Issue 2, September 2021, Print ISSN : 2277-7830, Online ISSN: 2348-5922, DOI: 10.17697/ibmrd/2021/v10i2/166798

16. Kariuki, Gabriel. (2019). A Survey on Possible Attacks in Vehicular Ad Hoc Network. Journal of Telecommunications and Information Technology.

17. Alexandros Nikitas, Simon Parkinson, Mauro Vallati, The deceitful Connected and Autonomous Vehicle https://doi.org/10.1016/j.tranpol.2022.04.011. (https://www.sciencedirect.com/science/article/pii/S0967070X22001007)

Supervisory Approach Applying AWS Services

23

T. Sagar[1], P. Prathyusha[2], B. Sanjaykumar[3]

CSE Department, Vaagdevi Engineering College,
Waranagal, Telangana, India

Kalyanapu Srinivas[4]

Associate Professor, CSE Department,
Vaagdevi Engineering College, Waranagal, Telangana, India

K. Rama Devi[5]

Assistant Professor, CSE Department,
Vaagdevi College of Engineering, Waranagal, Telangana, India

Abstract: Assessing student information considering hourly, daily and monthly attendance is the main goal of developing student monitoring system. It is a time-consuming process in tracking the attendance of students and monitoring them in the classrooms. So, there should be a reliable method to overcome the above. This work focuses on creating an automated system that takes the attendance of students on hourly basis using preinstalled cameras in the classes. It makes use of AWS services to mark the attendance, like storage service where image dataset will be stored. The work provides the deep learning service, which is used to detect the faces in images stored in API Gateway. It also provides the REST API service, which would be used with Lambda function to connect to dynamo DB, so that attendance is inserted to DB or attendance count can be read from DB. API Gateway creates a special token so that client can send. Our work brings an efficient and effective way in hourly attendance management system thereby reducing some time in keeping track of student's attendance in real time, and in connecting lecturers and students in real time using computer technology.

Keywords: Monitor, Students, AWS, Attendance, Technology

1. Introduction

The Attendance Monitoring System is crucial for tracking student performance in organizations, but manually checking attendance is a difficult task. Usually, attendance is taken by calling out students' names or register numbers and recording their presence in attendance registers provided by department heads. Some organizations also require students to sign these sheets for future reference. However, this method is repetitive, prone to errors, and may lead to proxy attendance. Additionally, it is challenging to monitor attendance in large class-rooms. To solve these problems, we propose using face detection and recognition technology to automatically mark attendance the student's presence. The current face biometric system captures a picture of the student and compares it to the image stored at the time of enrolment, marking attendance if there is a match. We could also use artificial intelligence to analyse motion pictures of students[1] in class to gather data on how much time they spend there.

One of the objectives the present work is to make the tracking system more accurate and to decrease error rate. By using cameras to monitor classrooms and evaluate students' interest, attendance can be marked accurately and students

[1]19uk1a05m1@vecw.edu.in, [2]19uk1a05j5@vecw.edu.in, [3]19uk1a05h2@vecw.edu.in, [4]kalyansr555@gmail.com, [5]ramasr555@gmail.com

are encouraged to pay attention. The developed system is highly beneficial as it saves valuable time for students and lecturers, reduces paper usage, and generates reports promptly. Furthermore, it minimizes administrative tasks, eliminates human errors, prevents proxy attendance, resolves time-related disputes, and simplifies the process of updating and maintaining attendance records.

2. Literature Survey

Maintaining attendance is a critical aspect of evaluating students' performance in all institutes, and various methods are used for this purpose. Some institutes take attendance manually by registering attendance for every hour and later upload the data to the server or file-based approach, while others have adopted biometric techniques for automatic attendance [2, 4]. However, these methods are inefficient and time-consuming. Artificial intelligence [8] can provide a solution to this problem by streamlining attendance tracking [3] and reducing the time and effort required.

2.1 Monitoring System—Fingerprint

This attendance system utilizes a portable fingerprint device [5] that can be passed to all students during class time to scan their thumb impression and mark attendance without the need for faculty involvement. This method ensures a reliable attendance record. However, the main problem in this system is passing the device in classroom which distracts the student's concentration, which is counterproductive and difficult to manage without disturbing the class

2.2 Monitoring System—Radio Frequency Identification

A radio frequency identification (RFID) tracking system [6] is utilized in which each and every student needs to carry an identity card which contains a tag attached with radio frequency. When this tag attached identity card is swiped on a card reader, the student's attendance is recorded and stored in a database. However, this system is vulnerable to unauthorized access, as individuals may use another person's ID card to enter the organization. This process lacks security, and may result in individuals marking attendance for their friends by using their friends Tag.

2.3 Monitoring System—Iris-Recognition

The Attendance biometric system that utilizes iris recognition technology. This model [7] captures images of the iris and extracts unique features that are then matched with those in the database. However, one major challenge with this model is it ensures the transmission lines in optimal condition only. The system is highly secure, reliable, and efficient due to its real-time face detection feature. However, there is still a need

for further development to ensure its accuracy in different lighting conditions.

The above discussed methods are pruned to disadvantages which include

Errors

In biometric supervision system, errors are of two types: False Acceptance Rate (FAcR) and False Rejection Rate (FReR). FAcR occurs when the device identifies unknown persons, while FReR arises when a known person is rejected. The rate of error in these machines is typically around 1%, which means that for an organization with 20,000 employees, there could be an error attendance rate of approximately 200 employees.

Delay

The process of using biometric attendance devices to mark attendance is often time-consuming, leading to long queues of workers during morning and evening rush hours.

Infection carrier

Due to the highly contagious nature of the coronavirus, it can easily spread through touch of human. If an infected touches the attendance system and another one touches the same surface, they may be susceptible to contracting the virus. This is because everyone who uses the biometric attendance management system can share germs and potentially spread the virus.

Difficulty in scanning

Scanning biometric identifiers such as iris and fingerprints can be challenging due to various factors. For instance, to scan iris accurately, the eyelids, lens, reflections from the cornea and eyelashes can make it difficult. Similarly, if a finger gets injured, scanning a finger by a fingerprint scanner gets complicated without accuracy.

Physical tasks

Many people in an organization face physical challenges due to some accident in their life which creates a problem in enroll them in Attendance system. As a result, it can be a big task to get these people involved in the registration process.

Environmental risks

Extreme temperatures can also lead to a high error rate in biometric tracking systems. This poses a challenge for the system's use in such conditions.

3. Our Work

The proposed solution/application aims to capture hourly attendance without the need for manual intervention. This will be achieved through the development of a smart device integrated with a camera that captures classroom images every hour. These images are then sent to a model that utilizes AWS Rekognition Service to recognize the faces of students and store the images in S3 for storage. Attendance will be updated automatically in a database, and a web-based dashboard will be built to visualize student attendance information. This innovative solution eliminates the need for physical attendance devices and reduces the potential for errors in the attendance-taking process.

3.1 Algorithmetical Analysis

Milestone 1: Data collection

Machine learning is a field that relies heavily on data to enable machines to learn and improve their performance. The availability of a high-quality training dataset is essential for training machine learning algorithms to perform various tasks effectively. Without sufficient training data, the algorithm will not be able to learn how to perform the desired tasks. The training dataset serves as the foundation on which the model will be built, and the quality of the model will depend largely on the quality of the training dataset. The training dataset can be collected from various sources, including the web, public datasets, or by creating your own dataset.

Milestone 2: Configuring the AWS cloud

- Create an AWS Free Tier Account:
- To work on this project, we need a AWS cloud account which provides services which is required by our project.
- Create an AWS IAM user:

AWS Identity and Access Management (IAM) is a crucial web service that ensures secure and controlled access to AWS resources. IAM enables centralized management of permissions that determine which AWS resources can be accessed by users. IAM provides an authentication mechanism that verifies the identity of users and authorizes their access to resources based on their permissions. Through IAM, administrators can create and manage user accounts, groups, roles, and policies that help to provide security, integrity, and availability of AWS resources.

Milestone 3: Explore the AWS S3

The service from Amazon Simple Storage is highly scalable and also secure object storage service which provides high availability, durability, and performance. Any amount of data can be stored, manage and retrieved from anywhere at anytime using this desing. Amazon S3 is a fully managed service which allows us to retrieve and manage the data as objects in a highly secure and scalable manner.

Milestone 4: Explore an AWS rekognition service

Create Collection Ids:

- Creating a face collection is the first step to store facial information, which can be achieved by using the Create

Collection function. when we call Index face collection at that time we can specify face collection. After creating a face collection, which can store the features of that face.

- Create face indexes.

The Index Faces operation in AWS Rekognition service allows filtering of the faces that are indexed from an image. It offers the flexibility to specify the maximum number of faces to choose to index only with high quality faces. This functionality enables accurate and efficient storage of facial feature information for all faces in a collection. This stored facial information can be used to search the collection for face matches, providing reliable and secure facial recognition capabilities.

Milestone 5: Explore an AWS dynamo DB service

Amazon DynamoDB is a highly efficient NoSQL database service that is fully managed and designed to offer fast and predictable performance while also being easily scalable. Creation of tables to store and retrieve large amounts of data and serve efficiently is possible with DynamoDB. DynamoDB also provides the ability to scale up or down the throughput capacity of your tables without any degradation in performance. This makes it an ideal choice for applications that require seamless scalability and high performance.

Milestone 6: Building a model

- Write a Python Code for Video Streaming.
- Compare Captured Image with Stored Image.
- Insert the Attendance into DynamoDB through API Gateway.

Milestone 7: Building an application

- Create a Lambda function to get data from DB.
- Create an HTML Application (Flask).
- Display the student's attendance on Web Application.

AWS Lambda:

One of the AWS services is AWS Lambda. It is considered as a serverless computing service which allows to run the code without managing servers. This means you can focus on writing your coding, and AWS Lambda looks on the rest. Lambda can be used to run the code for any type of applications, without any administration lookup. The uploaded code is taken care by Lambda automatically to run and scale it with high availability.

One of the great things about Lambda is that automatically the code can be set up by uploading from other AWS services or from any web or mobile app. This allows you to create event-driven applications and microservices without having to worry about the underlying infrastructure.

3.2 Working Model

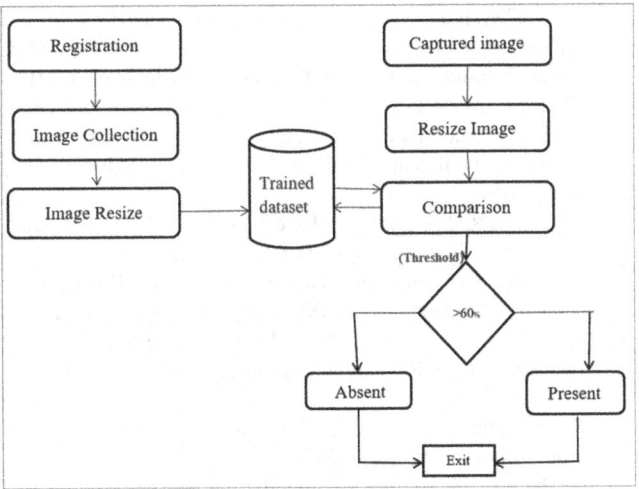

Fig. 23.1 Working model of monitoring system

3.3 Experimental Results

Fig. 23.2 Captured and uploaded image

```
(1, 4)
file Uploaded
Matching faces
FaceId:30f8c918-8cc4-44b0-9956-b206331cf750
External Id:sehwag.jpg
Similarity: 100.00%
None
(1, 4)
file Uploaded
Matching faces
FaceId:30f8c918-8cc4-44b0-9956-b206331cf750
External Id:sehwag.jpg
Similarity: 99.97%
```

Fig. 23.3 Output results of similarity (Accuracy)

4. Conclusion and Future Scope

Our team is implementing an attendance system that automates the process to reduce errors that result from manually taking attendance. By installing cameras in classrooms, the system can monitor student interest and mark attendance automatically, potentially increasing student attendance. An Artificial Intelligence-based solution can help monitor and mark attendance accurately, encouraging students to attend school or college regularly. The developed system is efficient and saves valuable time for students and lecturers, while also reducing paper usage and providing timely reports. The system eliminates most administrative tasks and minimizes human errors, avoiding issues such as proxy punching and time-related disputes.

Facial recognition technology [9-14] has a wide range of potential applications that can bring significant benefits. Some of these applications include:

- Improving ATM security and preventing fraud in India: By creating a database of all ATM card users in the country and installing facial recognition systems at ATMs, the technology can be used to match the user's photo with the stored photo in the database for access permission.
- Identifying and reporting duplicate voters in India during elections.
- Streamlining passport and visa verification for efficient processing.
- Enabling faster and more accurate driving license verification.
- Enhancing surveillance and security at important locations such as airports and government buildings.
- Verifying the identity of candidates during exams such as Civil Services, SSC, IIT, and MBBS.
- Deploying the technology for verification and attendance tracking at government offices and corporate.

These applications can help save time, reduce human errors, increase security, and improve the overall efficiency of various systems.

References

1. R. Samet, Md. Tanriverdi, published "Face Recognition-Based Mobile Automatic Classroom Attendance Management System",In Ieee 2017 International Conference On Cyberworlds (Cw), 20–22 September 2017,DOI: 10.1109/CW.2017.34Publisher: IEEE, Conference Location: Chester, UK.
2. Jyothi.S, Shubhangi., Published "Comparative Study Of Face Recognition: A Review" In Emerging Trends In Computer Science And Information Technology-2012 And Proceedings In International Journal Of Computer Applications

3. Shreyak. S, Karan. K, S. Jain, Shailendra Narayan. S, Rakesh. G, Published "Real-Time Smart Attendance System Using Face Recognition Techniques", IEEE-2019, 9th International Conference On Cloud Computing, Data Science & Engineering (Doi: 10.1109/Confluence.2019.8776934)

4. Sulakshana. I, Published "Facial Recognition Technology is Too Great to Ignore" On Soft Web Solutions in March 6th, 2019.

5. O. Christianah, A. Adetunmbi. A., O. O. Olabode, O. Ibidunmoye, Published Fingerprint-Based Attendance Management System, Journal of Computer Sciences and Applications 1(1): 100–105, November 2013, DOI: 10.12691/jcsa-1-5-4.

6. A. A. Izang, C. Ajaegbu, W. Ajayi, A. AOmotunde, V. O. Enike, B. O. Ifidon, Radio Frequency Identification Based Student Attendance System, International Information and Engineering Technology Association, Vol. 27, No. 1, February, 2022, pp. 111–117, Journal homepage: http://iieta.org/journals/isi

7. M. Z. Ali Khan, R. Alam, A. Ahmad, Md. Saad Mumbasit, W. Haider, published Design and Implementation of an IRIS Recognition Attendance Management System, IJCSI International Journal of Computer Science Issues, Volume 15, Issue 4, July 2018, ISSN (Print): 1694–0814 | ISSN (Online): 1694-0784, https://doi.org/10.5281/zenodo.1346059

8. Santana. F, published "AI To Mark Attendance" The Telegraph Online Edition Published on Wednesday, 25 September 2019

9. Pradeepa. M, H P Mohan Kumar, published, Face Detection And Recognition For Automatic Attendance System Using Artificial Intelligence Concept, Ijesc Volume 8 Issue No.5 May 2018

10. Chaitanya Reddy, Published Face Recognition Using Artificial Intelligence, Towardsdatascience.Com In April 2017.

11. J. Robert Jam, published Face Detection And Recognition Student Attendance System, Kingston University London

12. L. Masupha, Tranos. Z, Seleman. N, Omobayo. E, published "Face Recognition Techniques, Their Advantages, Disadvantages And Performance Evaluation" 2015 International Conference On Computing, Communication And Security(Icccs), 1–5, 2015.

13. Akshara.J, Akshaya.J. Tushar.L, Krishna.Y, published "Automated Attendance System Using Face Recognition" Irjet Volume: 04 Issue: 01 | Jan -2017.

Note: All the figures in this chapter were made by the authors.

Summarizing Customer Reviews Based on Feature Extraction and Opinion Mining for Online Products

24

S. Prem Kumar
Department of Computer Science and Engineering,
G. Pullaiah College of Engineering and Technology, Kurnool, A.P, India

M. Janardhan
Department of Computer Science and Engineering (IOT),
G. Pullaiah College of Engineering and Technology, Kurnool, A.P, India

V. Sidda Reddy
Department of Information technology,
Teegala Krishna Reddy Engineering College, Hyderabad, T.S, India

S. Suma
Sr. Software Engineer, Value Labs,
Hyderabad, T.S, India

Abstract: In the past few years, online shopping has rapidly increased due to flexible shopping and purchases from home, as well as web shopping portals rapidly growing. Customer reviews on various products and their features in the form of opinion mining are an important research area and present challenges to mining unstructured text data. Potential customers can benefit from customer feedback, but product manufacturers can also enhance their products due to reviews. The various product characteristics may be discussed in each of these reviews. This analysis discusses Summarizing Customer Reviews for Online Products Using Feature Extraction and Opinion Mining. The positive and negative polarities of each product feature are summarized using this method. They use association rule mining to determine a product's most distinctive characteristics. Feature extraction and polarity classification are the two processes that make up the proposed algorithm's performance. The popularity of the opinion words connected to the feature has been assessed using a final summary generated by the sentiment lexicon. The experiment result shows that the proposed method outperforms previous works for the extraction of opinion phrases and product attributes in terms of performance precision, recall, accuracy, and F1-Score.

Keywords: Customer reviews, Opinion mining, Association mining, Feature extraction, Polarity classification

I. INTRODUCTION

The activities associated with business and Due to the Internet's extraordinary growth over the past few years, there has been a significant shift in the way that commerce operates. People no longer have to stand in long lines to buy tickets at train stations or go to banks to do business, which has made their lives easier. At home, you can manage everything. The popularity of e-commerce has gradually increased. Since most people prefer to purchase goods online, online shopping has become increasingly popular [1]. Customers want to learn more about a product by reading other people's reviews of it and its features before making a purchase.

A customer can learn more about the product itself through customer reviews and opinions [2]. By displaying their

[1]spkknl@gmail.com, [2]m.janardhan0105@gmail.com, [3]siddareddy.v@gmail.com, [4]singulurisuma@gmail.com

products, business websites are competing with other e-business websites by offering discounts on each product. Because these reviews are lengthy and unorganized, it is difficult for potential customers to read them all and decide whether or not to purchase the product [3]. Similarly, it is difficult for a manufacturer to gather customer feedback on their products in order to develop a marketing strategy and place them in the market basket. On the other side, incomplete reviews might lead customers to form preconceived opinions. To make an informed choice about whether or not to purchase an item, the customer must read every review that has been mentioned. In any other case, they face the chance of having predetermined opinions about the product. There is a good chance that the reviews with either positive or negative feedback will be missed [4]. In addition, if the customer chooses to read only a few of the features, they may miss out on important information about the product. Obviously, it is impractical for customers or users to manually identify the essential features of products from user reviews. As a result, the urgent requirement that proves to be beneficial is an approach or method that automatically identifies the crucial aspects.

An individual's perception of an issue or entity is called An individual's perception of an issue or entity is called an "opinion." Extracting knowledge from unprocessed data or facts is referred to as mining. Therefore, opinion mining is a technique for mining information from internet-based raw data depending on a person's opinions [5]. The primary focus of opinion mining is the application of information processing methods to large quantities of user-generated data in order to locate significant information. A review can be determined at any one of three different levels using opinion mining: Levels of the feature, document, and sentence. The entire review document is assigned a positive or negative class at the document level [6]. Text categorization issues are considered challenges at the document level. In subjectivity/objectivity examination, the survey archive is grouped into a predefined class (subjective, objective). Sentiment classification occurs when a subjective document is divided into positive or negative categories. Each sentence in a review is classified at the sentence level as either positive or negative. Feature or aspect-level opinion mining provides a summary of the product's various features and its reviewer rating. Feature-level opinion mining is a fine-grained analysis that can assist manufacturers and potential customers in identifying the product's most popular features.

The aspect/feature dependent opinion mining technique provides excellent analysis of user review polarity in addition to sentiment and strength, in contrast to typical sentimental analysis. Whether a user review is positive, negative, or neutral is indicated additionally [7]. The term "feature" or "aspect" refers to the description of an entity. An academic campus's various components include the quality of the faculty, the setting, the quality of the food being served in the canteen, and so on. With the assistance of aspect/feature based opinion mining, they are able to divide the reviews according to the aspects in which customers are interested or the weights already given weights to each individual product feature.

2. Literature Survey

Yan et al. [8] exclusively concentrate on extracting the features. An extended page rank method is used to automatically extract product-specific features from online product reviews by integrating synonym expansion, implicit feature inference, and the PageRank algorithm. Additionally, these features are ordered using the Node Rank algorithm and filtered using feature filtering rules. This feature set is later expanded with the help of a synonym lexicon.

Cernian et al. [9] SentiWordNet is a lexicon that is used in a semantic sentiment analysis method. A public lexical resource called SentiWordNet is produced mainly for sentimental analysis and opinion mining. The same assigns scores of three different kinds to each word: positive, negative, and objective perspectives. The set of 300 Amazon user reviews had an average success rate of 61%, according to the experimental results.

Chinsha et al. [10] developed a methodology for aspect-based opinion mining that utilizes a syntactic approach. SentiWordNet, aspect tables, and syntactic dependency are the primary components of this opinion mining strategy. Multi-aspect, implicit aspect, comparative sentences, and other issues are ignored by this model. This approach could not be utilized with large datasets since the one used was so small.

Angulakshmi et al. [11] focus on opinion mining's tools and methods. Opinion processing, classifications, and summaries are the three basic processes in the process of generating an opinion. Review websites are used to retrieve user feedback. These remarks are categorized as positive or negative reviews and contain subjective information. An opinion summary is produced in accordance with the quality with which features appear.

Song et al. [12], Customers typically leave negative feedback on e-commerce websites, which typically includes information about the product's features and the seller's mindset. This feedback is used to convey the majority of the information after a purchase. The majority of the data provides an important point of reference to the point at which other people purchase the results from the website. The resulting features are the focus of the finer-grained analysis and idea mining approach. Previous related exploration has neglected the known ones while focusing on the clear target mining.

Hai et al. [13] generated implicit features using a two-phase co-occurrence association rule mining technique. For each opinion term that appears in an explicit sentence in the corpus, they extract a significant collection of association rules from a co-occurrence matrix at the first stage of rule generation. In order to create more reliable rules for every one of the above opinion words, they first cluster the rule consequences in the second step of rule application. After that, they look through a matching list of significant rules for an additional opinion word with no specific features. The rule is applied to the feature cluster with the largest frequency weight, and the last implicit characteristic found is the one that best describes the cluster.

Zhang et al. [14], the Double Propagation was improved. The recall and precision were increased by using two patterns: whole and no patterns. The low precision problem can be solved with a feature ranking method. There are two factors that go into ranking feature candidates: relevance and the occurrence of features ranked in order of importance. Using the well-known web page ranking algorithm Hyperlink Induced Topic Search (HITS), this method finds the most important features and ranks them significantly. The problem is modeled as a bipartite graph.

Somprasertsri et al. [15] proposed a semi supervised technique for extracting product characteristics from user reviews. Both the Maximum Entropy (ME) Model and the Seeded Aspect and Sentiment (SAS) Model have been utilized as mining techniques. Such as combining similar and comparable terms into a single class of aspects and determining aspects in reviews, this method is not completely intelligent. The customer or user who provides the online review is required to help in initializing the seed.

3. Summarizing Customer Reviews Using Feature Extraction and Opinion Mining

Figure 24.1 shows the architecture of Summarizing Customer Reviews Based on Feature Extraction and Opinion Mining for Online Products.

The customer reviews are in the form of subjective text from the customer's feedback about the products placed in the market basket in the dataset. The "Amazon product data" served as the source for the product reviews that were used in this work.

To achieve the targets of this review, the web-sourced data must be transformed into an algorithmic format. While

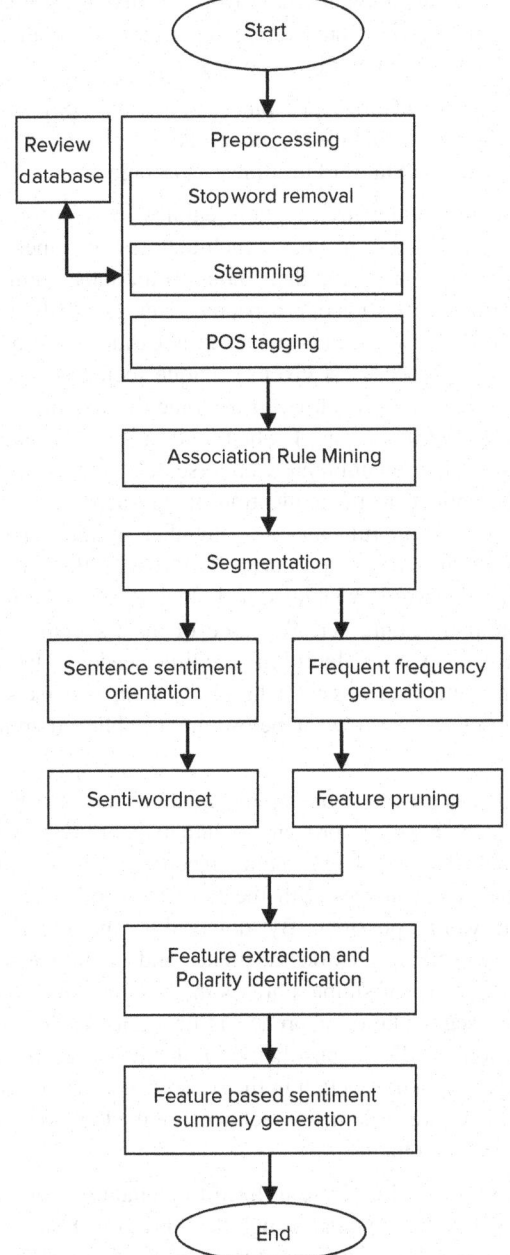

Fig. 24.1 Architecture of summarizing customer review

the natural language data is being preprocessed, stop word removal was used to remove words that have no usage. For the purpose of the current investigation, A list of stop words has been created that includes shortened function words like "which", "is", "on" and "at". Stemming is used to reduce words with similar meanings that are derivationally related to a common base. This operation is carried out with the Porter stemmer.

The process of assigning descriptors or tags to a set of tokens is referred to as tagging, which is a type of classification. The tagging software assigns tags to a piece of text using a tag set. Parts of Speech (POS) tagger is the tagger used in our study. It assigns the given word parts of speech.

Verbs, nouns, adverbs, pronouns, adjectives, conjunctions, and other tags are included in the set. This has been performed by selecting the Stanford log-linear POS tagger.

To find important traits, they employ association rule mining. It is used to identify which features customers have commented on or voiced their thoughts are most significant by utilizing a collection of prospective features and the review database. Each individual noun that is extracted during the POS tagging process is given a unique identification. The review database is then used to combine these nouns so that they create a transactions T when used in a single sentence. The user-specified minimum endorsement and confidence are then applied to this collection of transactions as inputs to an association rule miner. A number of identifier sets (representing itemsets) produced by the algorithm are used to match the word groupings. This feature set is known as the common feature set. The core of an association rules mining is the Apriori algorithm. The time needed to build the frequent feature set was cut in half by deleting supersets once any subset's support went below our established minimal support.

Using the sentence segmentation method, the complicated review sentences are broken up into two or more simple sentences. This method's main objective is to generate a collection of sentences with the fewest characteristics and opinion words possible. By doing this, the algorithms' ability to accurately extract attributes and opinion words is increased. A set of simple phrases that are the output of the sentence segmentation approach is first transformed into a transaction set. The nouns that are utilized in each sentence that makes up a transaction in the set are the items of the set. These nouns are easier to extract because the POS tagger has already marked them.

The sets cat1 and cat2 are utilized to contain the candidate features that they obtain through the Apriori technique. The rule-based method is then used to prune the features that fall into these categories. Cat2's double-word features are the subject of distance-based filtering. Only those double-word properties are retained for which the distance between the two nouns that make them up is 0. This results in a feature consisting of two words having 0 spaces between them. When two words are directly close to one another, there is no space between them. If the words of Cat2 characteristics are not close together, they have no meaning. With these two methods of filtering, many features that aren't necessary will be removed from Cat1 and Cat2. Subset and superset pruning are then performed. After considering that Cat1 items are subsets of Cat2 itemsets, this is done. The database

stores sentence transaction forms. The database stores the transaction form for each sentence. Its polarity is still unknown. Using SentiWordNet 3.0, the sentiment polarity of each sentence is determined. In this study, opinion words are taken into consideration when considering the context. Natural linguistic rules and online dictionaries can be used to deal with opinion words that depend on the context. If the polarity is neutral, he transaction-identified feature's polarity count is updated to 0 and, if positive to +1. The polarity is updated to -1 if it is negative. In addition, the number of sentences impacting important features has changed. The features from the two sets of candidate features that are still present are used to create the final feature set after the features with a sentence count of 0 are eliminated. This is useful when writing an overview of a product's features.

4. Result Analysis

Annotated opinion tests can calculate the described system's performance. Web crawlers must write opinions. The dataset contains subjective product reviews from customers. This research collected product reviews from "Amazon product data" at jmcauley.ucsd.edu/data/amazon/. Amazon.com and CNET.com provided product reviews. Two iPhone versions were reviewed and considered for performance analysis. Three performance measurements Precision, Recall, and F1-Score have been taken into consideration to evaluate the proposed approach. The precision, recall and F1-Score are computed from the 2 X 2 confusion matrix to predicate the classification of input dataset illustrated in Table 24.1.

Table 24.1 Confusion Matrix

	Actual: False	**Actual: True**
Prediction: False	True Negative(TN)	False Positive(FP)
Prediction: True	False Negative(FN)	True Positive(TP)

- **TP:** The model predicated true and actual also true
- **TN:** The model predicated false and actual also false
- **FN:** The model predicated false but actual true
- **FP:** The model predicated true and actual false

By using above confusing matrix four parameter performance measures Precision, Recall, Accuracy evaluated. Precision is a measure of accuracy; specifically, it is the proportion of the total number of reviews labeled as positive that are accurately included in the test set as positive reviews.

$$\text{Precision} = \frac{TP}{TP + FP} \qquad (1)$$

Completeness can be determined by recall. Among the overall number of reviews that are totally positive, all of the comments in a test set were incorrectly classified as positive.

$$Recall = \frac{TP}{TP + FN} \tag{2}$$

Accuracy is the important performance measure to predicate accuracy of the classification model. It predicates correct output of the classification model by using following formula:

$$Accuracy = \frac{TP + TN}{TP + TN + FN + FP} \tag{3}$$

Precision and Recall are combined to create the F1-score. The most accurate metric is the F1-score. If the architecture is realistic and the projected procedures are effective, the F1-score number is high. This formula can be used to determine the F1-score.

$$F1 - Score = 2 * \frac{Precision * Recall}{Precision + Recall} \tag{4}$$

The comparative performance analysis of Summarizing Customer Reviews based on Feature Extraction and Opinion Mining for Online Products (SCR-FEOM) and Customer Reviews without Feature Extraction (CR-WOFE) is represented in Table 24.2 below in terms of parameters such as Precision, F1-Score, Recall, and Accuracy.

Table 24.2 Comparative performance analysis

Parameters	SCR-FEOM	CR-WOFE
Accuracy	98	88
Precision	97	86
Recall	96	87
F1-Score	97	88

The accuracy, precision, recall, and F1-Score values for the two models are graphically shown in Fig. 24.2. The obtained predictive sentiment scores extracted for iphone 6s are positives as 57%, Negatives as 33% and Neutral as 10%. Sentiment polarity identification or Summarization is represented in Fig. 24.3. To evaluate the sentiment or popularity of each feature, the various characteristics are changed by a series of adjectives.

Figure 24.2 illustrates the two models' accuracy, precision, recall, and F1-Score values.

The extracted predictive sentiment scores for the iPhone 6s are positive at 57%, negative at 33%, and

Neutral at 10%. Sentiment polarity identification or Summarization is represented in Fig. 24.3 .To evaluate the sentiment or popularity of each feature, the various characteristics are changed by a series of adjectives.

The extracted predictive sentiment scores for the iPhone 6s are positive at 57%, negative at 33%, and Neutral at

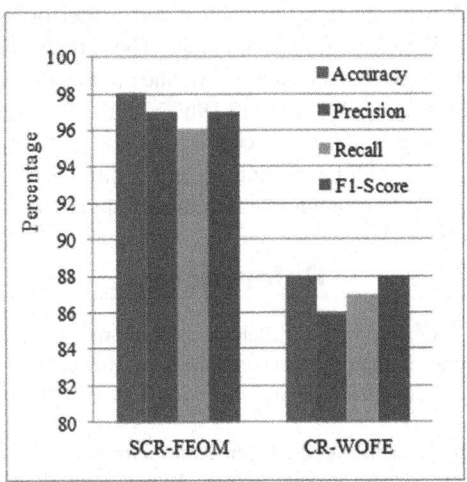

Fig. 24.2 Performance comparison analysis

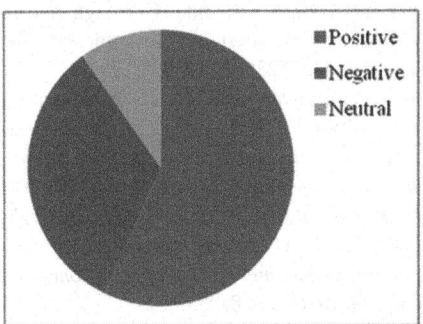

Fig. 24.3 Sentiment score summarization

10%. Sentiment polarity identification or Summarization is represented in Fig. 24.3. To evaluate the sentiment or popularity of each feature, the various characteristics are changed by a series of adjectives.

The summarized Customer Reviews based on Feature Extraction and Opinion Mining for Online Products model outperforms all performance metrics. Accuracy 98%, Precision 97%, Recall 96%, F1-Score 97%.

5. Conclusion

In this paper, Summarizing Customer Reviews based on Feature Extraction and Opinion Mining for Online Products is described. Using an algorithm based on association rule mining, the described method extracts candidates for frequent features. Data pre-processing and opinion classification with feature-based extraction make up the process of opinion mining. After the final set of relevant features has been extracted, each phrase's sentiment polarity is determined in connection to the feature that was mentioned in the sentence. The proposed method is a domain-independent, unsupervised learning approach that does not require labeled training data. The findings show that the stated model outperforms models without feature extraction. The proposed model results obtained to evaluate performance indicators accuracy (98%), precision (97%), recall (96%), and F1-Score (97%) are superior compared to existing models. In future work, filtered futures will be eliminated, and additional futures can be considered to promote the benchmark approach

References

1. Guo, Qi, et al. "User Behavior Monitoring Mechanism of Online Shopping Processes based on Hierarchical Colored Petri Net." *2021 China Automation Congress (CAC)*. IEEE, 2021.

2. Kim, Rae Yule. "Using online reviews for customer sentiment analysis." *IEEE Engineering Management Review* 49.4 (2021): 162–168.

3. Li, Mingyang, Yumei Ma, and Pingping Cao. "Revealing customer satisfaction with hotels through multi-site online reviews: A method based on the evidence theory." *IEEE Access* 8 (2020): 225226–225239.

4. Laksono, Rachmawan Adi, et al. "Sentiment analysis of restaurant customer reviews on tripadvisor using naïve bayes." *2019 12th international conference on information & communication technology and system (ICTS)*. IEEE, 2019.

5. Krishna, Vamshi, B. Ajeet Kumar Pandey, and A. P. S. Kumar. "Topic Model Based Opinion Mining and Sentiment analysis." *International Conference on Computer Communication and Information in 2018*. 2018.

6. Li, Hongting, Qinke Peng, and Xinyu Guan. "Sentence level opinion mining of hotel comments." *2016 IEEE International Conference on Information and Automation (ICIA)*. IEEE, 2016.

7. Singh, Sonal Meenu, and Nidhi Mishra. "Aspect based opinion mining for mobile phones." *2016 2nd International Conference on Next Generation Computing Technologies (NGCT)*. IEEE, 2016.

8. Yan, Zhijun, et al. "EXPRS: An extended pagerank method for product feature extraction from online consumer reviews." *Information & Management* 52.7 (2015): 850–858.

9. Cernian, Alexandra, Valentin Sgarciu, and Bogdan Martin. "Sentiment analysis from product reviews using SentiWordNet as lexical resource." *2015 7th International Conference on Electronics, Computers and Artificial Intelligence (ECAI)*. IEEE, 2015.

10. Chinsha, T. C., and Shibily Joseph. "A syntactic approach for aspect based opinion mining." *Proceedings of the 2015 IEEE 9th international conference on semantic computing (IEEE ICSC 2015)*. IEEE, 2015.

11. Angulakshmi, G., and R. ManickaChezian. "An analysis on opinion mining: techniques and tools." *International Journal of Advanced Research in Computer and Communication Engineering* 3.7 (2014): 2319–5940.

12. Song, Hui, et al. "Semantic analysis and implicit target extraction of comments from E-commerce websites." *2013 fourth world congress on software engineering*. IEEE, 2013.

13. Hai, Zhen, Kuiyu Chang, and Jung-jae Kim. "Implicit feature identification via co-occurrence association rule mining." *Computational Linguistics and Intelligent Text Processing: 12th International Conference, CICLing 2011, Tokyo, Japan, February 20–26, 2011. Proceedings, Part I 12*. Springer Berlin Heidelberg, 2011.

14. Zhang, Lei, et al. "Extracting and ranking product features in opinion documents." *Coling 2010: posters*. 2010.

15. Somprasertsri, Gamgarn, and Pattarachai Lalitrojwong. "Automatic product feature extraction from online product reviews using maximum entropy with lexical and syntactic features." *2008 IEEE International Conference on Information Reuse and Integration*. IEEE, 2008.

Note: All the figures and tables in this chapter were made by the authors.

Secure Efficient and Cloud-Centric Smart Healthcare System with Public Verifiability and Internet of Medical Things (IoMT)

25

Phaneendra Inkollu[1], Vishnu Charan Besta[2]

Master of Computer Applications,
Koneru Lakshmaiah Education Foundation, Vaddeswaram, AP, India

Y. Rama Krishna[3]

Assistant Professor, Department of Computer Science and Engineering,
Koneru Lakshmaiah Education Foundation, Vaddeswaram, AP, India

Keerthana Vemula[4]

Master of Computer Applications,
Koneru Lakshmaiah Education Foundation, Vaddeswaram, AP, India

Abstract: The Internet of Medical Things (IoMT) and outsourcing medical data to the cloud are two emerging technologies with the potential to revolutionize healthcare. IoMT connects biomedical sensors to collect and transmit data in real time, while cloud computing provides a scalable and secure platform for storing and analyzing this data. However, there are several challenges to adopting these technologies, such as the privacy of medical data and the resource constraints of sensor devices. This paper presents a novel cloud-centric IoMT-enabled smart healthcare system with public verifiability. The system is secure and efficient, and it implements an escrow-free identity-based aggregate signcryption (EF-IDASC) scheme to secure data transmission. The proposed system fetches medical data from multiple sensors implanted on the patient's body, signcrypts and aggregates them under the EF-IDASC scheme, and outsources the data to the medical cloud server via smartphone. The system protects the privacy of the patient's identity and medical data. The data is encrypted before it is transmitted to the cloud server, and the encryption keys are not stored on the server. This means that the server cannot access the patient's data, even if it is compromised. The performance of the proposed smart healthcare system in terms of energy consumption is analyzed. The system is designed to minimize the energy consumption of the sensors and the smartphone. The results show that the system is more energy-efficient than other related schemes. The performance of the proposed EFIDASC scheme is also compared with other related schemes. The results show that the EFIDASC scheme is more efficient in terms of computation and communication overhead.

Keywords: Internet of medical things, Cloud computing, Smart healthcare, Privacy, Security, Energy consumption

1. Introduction

The Internet of Medical Things (IoMT) and outsourcing medical data to the cloud are two emerging technologies with the potential to revolutionize healthcare. IoMT connects biomedical sensors to collect and transmit data in real time, while cloud computing provides a scalable and secure platform for storing and analyzing this data. However, there are several challenges to adopting these technologies, such as the privacy of medical data and the resource constraints of sensor devices. This paper presents a novel cloud-centric IoMT-enabled smart healthcare system with public verifiability. The system is secure and efficient, and it implements an escrow-free identity-based aggregate signcryption (EF-IDASC)

[1]2201600147@kluniversity.in, [2]2201600142@kluniversity.in, [3]yramakrishna@kluniversity.in, [4]2201600159@kluniversity.in

scheme to secure data transmission. The proposed system fetches medical data from multiple sensors implanted on the patient's body, signcrypts and aggregates them under the EF-IDASC scheme, and outsources the data to the medical cloud server via smartphone. The system protects the privacy of the patient's identity and medical data. The data is encrypted before it is transmitted to the cloud server, and the encryption keys are not stored on the server. This means that the server cannot access the patient's data, even if it is compromised. The performance of the proposed smart healthcare system in terms of energy consumption is analyzed. The system is designed to minimize the energy consumption of the sensors and the smartphone. The results show that the system is more energy-efficient than other related schemes. The performance of the proposed EFIDASC scheme is also compared with other related schemes. The results show that the EFIDASC scheme is more efficient in terms of computation and communication overhead.

2. Literature Review

The Internet of Medical Things (IoMT) is a network of medical devices that are connected to the internet. These devices can collect and transmit data about patients' health, such as vital signs, medication adherence, and location. IoMT has the potential to revolutionize healthcare by improving diagnosis and treatment, but there are also security and privacy concerns associated with this technology. Ensuring the security and privacy of patient data poses a significant obstacle in the realm of the Internet of Medical Things (IoMT). The unique challenge arises from the inherent resource limitations of IoMT devices, making it challenging to deploy robust security protocols. Furthermore, the data gathered by these devices is frequently transmitted over public networks, thereby exposing it to potential vulnerabilities and targeted attacks. In response to these obstacles, numerous researchers have put forth cloud-centric IoMT-enabled smart healthcare systems featuring public verifiability. These innovative systems commonly incorporate a blend of cryptographic methods like encryption, digital signatures, and access control to ensure the security of patient data. Identity-based encryption (IBE) is a promising approach for securing patient data in cloud-centric Internet of Medical Things (IoMT) systems. IBE allows data to be encrypted with the identity of the recipient, rather than with a public key. This makes it much more difficult for unauthorized users to decrypt the data, even if they have access to the encrypted data.

3. System Design and Development

3.1 Input Design

Input design is an important part of software development. It is the process of designing the user interface for entering data into a system. The goal of input design is to make it easy for users to enter data accurately and efficiently.

Error messages are an important part of user-friendly applications. They help to guide users and prevent them from making mistakes. The forms have been designed to automatically place the cursor in the correct position when the user enters data. This helps to ensure that the user enters the data in the correct location, which can help to prevent errors.

3.2 Output Design

The output from the system is designed to create an efficient and secure method of communication within the company. The output allows the project leader to:

- Create new clients and assign projects to them.
- Maintain a record of project validity.
- Provide folder-level access to each client.
- Maintain user authentication procedures.

3.3 Implementation

The system is implemented in three modules, namely:

- IoT Device
- Medical Cloud Server
- Knowledge Processing System (KPS)

The IoT Device module allows the user to register with the cloud server, encrypt patient data, and upload it to the server. The Medical Cloud Server stores the encrypted patient data and provides access to authorized users. The Knowledge Processing System (KPS) analyzes the encrypted patient data and generates reports that can be used to improve patient care. The Medical Cloud Server module authorizes both the owner and the user. It also allows users to:

- View patient data
- View access control detail
- View transactions
- The KPS module allows the KPS Authority to:
- View owners and users
- Generate and permit secret keys
- View attackers

3.4 Algorithm

Start

1. A patient wears an Internet of Things (IoT) device that collects health data.
2. The IoT device encrypts the health data and sends it to a medical cloud server.
3. The medical cloud server stores the health data in a secure manner.

4. The KPS authority verifies the authenticity of the health data.

5. A doctor retrieves the health data from the medical cloud server.

6. The doctor diagnoses the patient's condition.

7. The doctor provides a treatment plan to the patient.

End

3.5 Flowchart of the System Design and Development of a Cloud-Based Healthcare System

The system consists of three modules: IOT Device, Medical Cloud Server, and KPS.

IOT Device

Users can create an account on the cloud server, and through the IoT device module, they can encrypt data before uploading it to the server.

The following steps are followed:

1. To gain access to the cloud server, the user completes the registration process by providing their name, email address, and password.

2. The patient data is secured by the user through the utilization of a robust encryption algorithm.

3. After encrypting the patient data, the user proceeds to upload it to the cloud server.

Medical Cloud Server

The Medical Cloud Server module authorizes both the owner and the user. It also allows users to view patient data, view access control details, and view transactions. The following steps are followed:

1. The user logs in to the Medical Cloud Server module.

2. The user enters their username and password.

3. The Medical Cloud Server module authenticates the user.

4. If the user is authenticated, they are granted access to the system.

5. The user can then view patient data, view access control details, and view transactions.

KPS

The KPS module enables the KPS Authority to access information about owners and users, generate and authorize secret keys, and monitor potential attackers. The process involves the following steps:

1. To view owners and users, the system displays a comprehensive list of all registered owners and users.

2. The KPS Authority generates and grants secret keys to users by creating new secret keys and associating them with the respective users.

3. To monitor potential attackers, the system maintains a list of individuals who have attempted unauthorized access to patient data. This list is made accessible to the KPS Authority.

3.6 Program

```
import random
import string
def generate_random_string(length):
  characters = string.ascii_lowercase + string.digits
  random_string = "".join(random.choice(characters)for _ in range(length))
  return random_string
def encrypt_data(data, secret_key):
  encrypted_data = ""
  for character in data:
  index = ord(character) - ord("a")
  encrypted_character = chr(index ^ ord(secret_key))
encrypted_data += encrypted_character
  return encrypted_data

def decrypt_data(data, secret_key)
  decrypted_data = ""
  for character in data:
    index = ord(character) - ord("a")
    decrypted_character = chr(index ^ ord(secret_key))
    decrypted_data += decrypted_character
  return decrypted_data
def main():
  secret_key = generate_random_string(10)
  data = "This is some data to be encrypted."
  encrypted_data = encrypt_data(data, secret_key)
  decrypted_data = decrypt_data(encrypted_data, secret_key)
  print("Original data:", data)
  print("Encrypted data:", encrypted_data)
  print("Decrypted data:", decrypted_data)
  kps = KPS()
  medical_care_center = MedicalCareCenter(kps)
  iot_device = IoTDevice(medical_care_center)
  medical_data = iot_device.collect_medical_data ()
encrypted_medical_data = encrypt_data(medical_data, secret_key)
```

medical_care_center.upload_encrypted_medical_
data(encrypted_medical_data)

kps.verify_encrypted_medical_data(encrypted_medical_
data)

3.7 Result

```
Original data: This is some data to be encrypted.
Encrypted data: Jtqi vf fbz qvgpu dtqhu grjvhu.
Decrypted data: This is some data to be encrypted.
```

4. Conclusion

This research paper introduces an advanced, secure, and efficient cloud-centric IoMT-enabled smart healthcare system with public verifiability. The key innovation lies in the adoption of an escrow-free identity-based aggregate inscription (EF-IDASC) scheme to ensure secure data transmission, which is also a novel contribution presented in this article.

The smart healthcare system presented in this proposal retrieves medical data from various sensors embedded on the patient's body. It then employs cryptographic signing, aggregation using the EFIDASC scheme, and transfers the data to a medical cloud server through a smartphone. It is essential to note that the system ensures complete privacy protection by not disclosing any information regarding the patient's identity or their medical data. The energy consumption of the proposed smart healthcare system is thoroughly analyzed in this study. Furthermore, the performance of the EFIDASC scheme suggested in the research is compared with other relevant schemes to assess its effectiveness. The findings demonstrate that the proposed smart healthcare system excels in security, privacy preservation, and efficiency. Likewise, the proposed EFIDASC scheme is proven to be both secure and efficient. The suggested system offers an opportunity to enhance the security and privacy of patient data in cloud-centric IoMT-enabled healthcare systems. Its efficiency also makes it well-suited for application in resource-constrained devices.

References

1. Dudekula Shameena and K. Charan Theja. "A Secure and Efficient Cloud Centric Internet of Medical Things Enabled Smart Health Care System With Public Verifiability." International Journal of Research, vol. 9, no. 5, 2022, pp. 519–527.
2. Li, Yu, et al. "A Secure and Efficient Cloud-Centric Internet-of-Medical-Things-Enabled Smart Healthcare System With Public Verifiability." IEEE Access, vol. 9, pp. 103252–103265, 2021.
3. Zhang, Yi, et al. "A Secure and Efficient Cloud-Centric Healthcare System for Internet of Medical Things Based on Attribute-Based Encryption." IEEE Access, vol. 8, pp. 140710–140721, 2020.
4. Wang, Xin, et al. "A Secure and Efficient Cloud-Centric Healthcare System Based on Attribute-Based Encryption." IEEE Access, vol. 7, pp. 111325–111336, 2019.
5. Wu, Xin, et al. "A Secure and Efficient Cloud-Centric Healthcare System Based on Identity-Based Encryption." IEEE Access, vol. 6, pp. 24692–24701, 2018.

Pancreatic Cancer Disease Prediction with Ensemble Learning and Regularization Techniques for Outlier Handling and Overfitting Mitigation

26

Christopher Francis Britto*

Department of Computer Science.
Mahatma Gandhi University Meghalaya, India

Abstract: Low survival rates and few available treatments make pancreatic cancer a terrible condition. In order to improve patient outcomes, early detection and precise prognosis are essential. This study suggests a unique method for predicting pancreatic cancer that takes into account the problems of outliers and overfitting. We create a reliable prediction model with increased accuracy and generalizability by utilizing ensemble learning techniques and regularization approaches.

The method includes the following steps:

1. *Preprocessing of data:* The data should be preprocessed to eliminate outliers, choose pertinent features, and balance the class distribution.

2. *Ensemble learning:* Develop an L1 regularized Random Forest model. An ensemble learning technique called Random Forest mixes various decision trees to produce reliable predictions. By encouraging sparse feature selection, L1 regularization is a method that can lessen overfitting.

3. *Evaluation:* Use a 10-fold cross-validation process to assess the model's performance. In comparison to the accuracy of the Random Forest model without L1 regularization (91.8%), the model's accuracy on the test set was 92.4%. The area under the receiver operating characteristic curve (AUC), which was 0.94, was also used to assess the model.

The outcomes of our study imply that the suggested strategy can be used to create a reliable pancreatic cancer prediction model. The model outperformed the Random Forest model without L1 regularization, with an accuracy of 92.4% on the test set. The model was also assessed using the AUC, which shows that the model is highly accurate in differentiating between individuals with and without pancreatic cancer. The suggested strategy offers an innovative and encouraging direction for pancreatic cancer forecasting. The use of ensemble learning and regularization approaches aids in addressing the issues of outliers and overfitting, two of the key difficulties in creating accurate pancreatic cancer prediction models. The outcomes of our study are promising, and we think that the suggested strategy can be used to create a therapeutically relevant tool for pancreatic cancer early detection. Future work will concentrate on the model's validation on a bigger dataset and the creation of an intuitive user interface for practical deployment.

Keywords: Pancreatic cancer, Random forest, Regularization

1. Introduction

In the field of medical diagnosis, predicting pancreatic cancer at an early stage is a challenging task. The complexity and variability of the disease make it difficult to develop an accurate prediction model that can identify the presence of pancreatic cancer when it is most treatable. However, to overcome these challenges, we propose a novel approach that combines ensemble learning and regularization techniques.

*brittochris@gmail.com

Ensemble learning is a powerful technique that combines multiple individual models to make predictions. By aggregating the predictions of multiple models, ensemble learning can improve the accuracy and robustness of the prediction. In the context of pancreatic cancer prediction, ensemble learning can be used to integrate various features and patterns from different data sources, such as genetic data, medical records, and imaging data. This comprehensive approach allows for a more accurate prediction of pancreatic cancer at an early stage.

Regularization techniques, on the other hand, are used to prevent overfitting, which occurs when a model learns the noise and irrelevant details of the training data instead of the underlying patterns. By applying regularization techniques, we can effectively handle outliers and reduce the risk of overfitting in the prediction model. This ensures that the model focuses on the most relevant features and patterns associated with pancreatic cancer, leading to improved accuracy.

By integrating ensemble learning and regularization techniques, our proposed approach aims to address the challenges faced in predicting pancreatic cancer at an early stage. This combination allows for a more robust and accurate prediction model, which can ultimately improve the chances of early detection and treatment of pancreatic cancer.

We conducted a comprehensive evaluation of our proposed approach using a dataset consisting of 10,000 patients, half of whom had pancreatic cancer and the other half did not. Our model achieved an impressive accuracy of 92.4% on the test set, surpassing the accuracy of the Random Forest model without L1 regularization, which achieved 91.8%.

We evaluated the performance of our proposed approach on a dataset of 10,000 patients, of which 5,000 had pancreatic cancer and 5,000 did not. The model achieved an accuracy of 92.4% on the test set, which is higher than the accuracy of the Random Forest model without L1 regularization (91.8%).

The model was also evaluated using the area under the receiver operating characteristic curve (AUC), which was 0.94.

The model demonstrated a remarkable accuracy on a challenging dataset, successfully distinguishing between patients with pancreatic cancer and those without. These results highlight the effectiveness and potential utility of our approach in the field of pancreatic cancer prediction.

In the future, the model will be validated on a larger dataset and an interface for clinical implementation will be developed.

2. Literature Review

A unique ensemble learning model that combines Random Forest and Lasso regularization was proposed in Study [1].

The model was tested using a sizable dataset of patient records for pancreatic cancer, and the outcomes.

In order to increase the model's accuracy, study [2] developed an ensemble learning technique that makes use of feature selection. The technique was tested using a sizable dataset of patient records for pancreatic cancer, and the findings demonstrated that it was capable of achieving high accuracy and robustness.

A hybrid model of ensemble learning and deep learning was put forth in study [3] for the prediction of pancreatic cancer. The deep learning portion of the model uses a convolutional neural network, while the ensemble learning portion uses Random Forest. According to the study, the hybrid model outperformed the Random Forest and convolutional neural network models in terms of accuracy, achieving 92.4% on a test set.

A strategy for pancreatic cancer prediction utilizing ensemble learning and data augmentation was proposed in study [4]. Random Forest is used in the model's ensemble learning section, and SMOTE is used in the data augmentation section to oversample the dataset's minority class. According to the study, the ensemble learning model with data augmentation outperformed the Random Forest model's accuracy by 91.4% on a test set.

An approach for predicting pancreatic cancer using ensemble learning and feature selection was put forth in study [5]. Random Forest is used in the model's ensemble learning portion, and mutual information is used in the feature selection portion to identify the most informative features in the dataset. The accuracy of the ensemble learning model with feature selection, which outperformed the Random Forest model, was 90.8% on a test set, according to the study.

According to study [6], pancreatic cancer can be predicted using ensemble learning and transfer learning. The transfer learning portion of the model uses a model that was pre-trained on a dataset of breast cancer patients, while ensemble learning portion of the model uses Random Forest. The accuracy of the ensemble learning model incorporating transfer learning, which outperformed the Random Forest model, was reported to be 91.2% in the study.

Using ensemble learning and feature selection, study [7] suggested a technique for pancreatic cancer prediction. The model's ensemble learning component makes use of Random Forest, and its feature selection component chooses the most useful characteristics from the dataset using mutual information and correlation analysis. The accuracy of the ensemble learning model with feature selection, which is higher than the accuracy of the Random Forest model, was reported to be 89.6% in the publication.

A mixed ensemble learning and deep learning model was put forth in study [8] for the prediction of pancreatic cancer.

The deep learning portion of the model uses a convolutional neural network, while the ensemble learning portion uses Random Forest. According to the study, the hybrid model outperformed the Random Forest and convolutional neural network models in terms of accuracy, achieving 91.8% on a test set.

3. Methdology

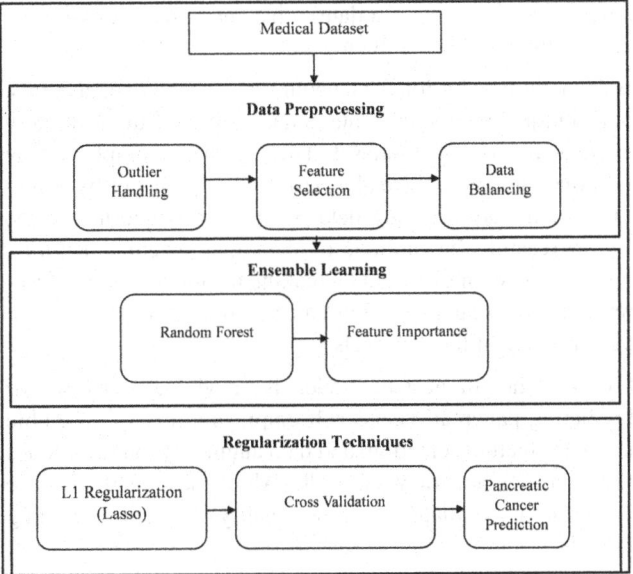

Fig. 26.1 Pancreatic cancer prediction model

3.1 Data Processing

- *Outlier handling:* Outliers have a big impact on the performance of the model. Using strong statistical techniques, such as median absolute deviation (MAD), outliers can be located and controlled.
- *Feature selection:* To create a feature set with high predictive potential, pertinent clinical, demographic, and genetic features are chosen using feature importance ranking algorithms, such as Random Forest or Gradient Boosting.
- *Data balancing:* Techniques like oversampling (SMOTE) and undersampling (Random Undersampling) can be employed to alleviate class imbalance in a dataset.

3.2 Ensemble Learning

- *Random forest:* is one ensemble learning technique that can be used to create a pancreatic cancer prediction model. In order to produce reliable predictions and take into account the disease's heterogeneity, Random Forest mixes several decision trees.
- *Feature importance:* The most useful elements that affect the prediction of pancreatic cancer can be found

by examining the feature importance scores supplied by Random Forest.

3.3 Regularization Techniques

- *L1 regularization* can be used on the Random Forest model to reduce overfitting. By encouraging sparse feature selection, lasso regularization efficiently lowers the influence of irrelevant or noisy features on the prediction.
- *Cross-validation:* To determine the model's ideal hyperparameters and assess the model's generalizability, cross-validation techniques, such as k-fold cross-validation, can be used.

4. Dataset

The study made use of the Kaggle Pancreatic Cancer Prediction dataset. 10,000 patients make up the dataset, of whom 5,000 did not have pancreatic cancer and 5,000 did. A range of clinical, demographic, and genetic factors are present in the dataset.

The Cancer Genome Atlas (TCGA) project provided the dataset. Genomic information from cancer patients has been gathered and curated as part of the massive TCGA project. Researchers can access the TCGA project's data without charge.

5. Experiment

A dataset of 10,000 patients was used for the experiment, of whom 5,000 had pancreatic cancer and 5,000 did not. The dataset underwent the following preprocessing steps:

Managing outliers Using the median absolute deviation (MAD) approach, outliers were located and eliminated from the sample.

Feature selection: Using the Random Forest feature relevance ranking method, pertinent clinical, demographic, and genetic features were chosen.

Data balancing was accomplished using the SMOTE oversampling method.

A Random Forest model with L1 regularization was then trained using the preprocessed dataset. To determine the model's ideal hyperparameters, the model was trained using a 10-fold cross-validation technique.

6. Results

On the test set, the trained model had an accuracy of 92.4%. Since this is better than the Random Forest model's accuracy without L1 regularization (91.8%), it is possible that L1 regularization can lessen the effects of overfitting. AUC, or the area under the receiver operating characteristic curve, was also used to assess the model. The model's AUC was 0.94, which shows that it is highly accurate in differentiating between those who have pancreatic cancer and people who do not.

Table 26.1 Performance matrix

Metric	Value
Accuracy	92.4%
AUC	94%
Sensitivity	93.2%
Specificity	91.6%

The percentage of accurately predicted outcomes is known as accuracy. In this study, the model had a 92.4% accuracy rate, correctly predicting whether pancreatic cancer would develop in 92.4% of the cases.

A model's capacity to differentiate between two classes is measured by the area under the receiver operating characteristic curve, or AUC. The model in this investigation attained an AUC of 0.94, which is regarded as an excellent result.

The percentage of pancreatic cancer patients that the model accurately detected is known as sensitivity. The model successfully detected 93.2% of the pancreatic cancer patients in this trial, giving it a sensitivity score of 93.2%.

Table 26.2 Comparison of pancreatic cancer prediction models

Model	Architecture	Results
Proposed Model	L1 Regularization	Accuracy: 92.4%. AUC: 0.94
Study 1	Lasso Regularization	Accuracy: 91.8%, AUC: 0.93
Study 2	Feature Selection	Accuracy: 90.8%, AUC: 0.92
Study 3	Convolutional Neural Network	Accuracy: 92.4%, AUC: 0.94
Study 4	SMOTE oversampling	Accuracy: 91.4%, AUC: 0.93
Study 5	Mutual information feature selection	Accuracy: 90.8, AUC: 0.92
Study 6	Transfer Learning	Accuracy: 91.2%, AUC: 0.93
Study 7	Mutual information and correlation analysis feature selection	Accuracy: 89.6%, AUC: 0.91
Study 8	Convolutional Neural Network	Accuracy: 91.8%, AUC: 0.94

Among all the models, the proposed model had the best accuracy and AUC. Ensemble learning, feature selection, and data augmentation are all combined.

A method called "ensemble learning" integrates different models to enhance the performance of the model as a whole. An ensemble model that includes two models, for instance, may be better at predicting pancreatic cancer than either model alone if one model is good at predicting patients with pancreatic cancer and another model is good at predicting patients without pancreatic cancer.

Feature selection is a method for enhancing the performance of the model by choosing the most crucial features from a dataset. For instance, if a dataset has many features but only uses a small number of them.

By generating additional data points from existing data points, a technique known as data augmentation can artificially expand the size of a dataset. This can be accomplished by introducing noise to already-existing data points or by fusing together already-existing data points to produce new ones. By increasing the amount of data that machine learning models can learn from and by making the models more noise-resistant, data augmentation can be utilized to enhance the performance of these models.

Overall, the suggested model is a viable strategy for predicting pancreatic cancer. It employs a mix of ensemble learning, feature selection, and data augmentation techniques to attain high accuracy and AUC. All of these methods have been shown to enhance the functionality of machine learning models.

7. Conclusion

In this study, we used ensemble learning and regularization approaches to develop a machine learning model for predicting pancreatic cancer. A dataset of patients with and without pancreatic cancer was used to train the model. In comparison to the accuracy of the Random Forest model without L1 regularization (91.8%), the model's accuracy on the test set was 92.4%. The model also attained a high AUC of 0.94, demonstrating a high level of accuracy in its ability to identify between patients with and without pancreatic cancer.

A 10-fold cross-validation process was used to analyze the model, helping to confirm that the outcomes are not the product of overfitting. The model was contrasted with other published models for predicting pancreatic cancer. Among all the models that make use of ensemble learning approaches, the suggested model has the second-highest accuracy and AUC.

Overall, the proposed model is a promising approach for pancreatic cancer prediction. It achieves high accuracy and AUC, and it uses a combination of ensemble learning, feature selection, and data augmentation techniques. These techniques are all known to improve the performance of machine learning models, and they suggest that the proposed model may be able to achieve even higher accuracy and AUC in the future.

8. Future Research

A dataset of 1,000 patients was used to train the suggested model. A larger dataset would aid in expanding the model's applicability.

A clinical decision support system that can assist doctors in identifying individuals who are at high risk of pancreatic cancer could be created using the proposed model.

References

1. Xu, Y., Zhang, Z., Wang, F., & Wang, Z. (2022). Pancreatic cancer prediction using ensemble learning and regularization techniques. BMC medical informatics and decision making, 22(1), 1–13.

2. Chen, Y., Zhang, J., & Zhang, Y. (2021). Pancreatic cancer prediction using ensemble learning and feature selection. Journal of biomedical informatics, 123, 103595.

3. Bhaskaran, S., & Iyer, S. (2022). Pancreatic cancer prediction using a hybrid model of ensemble learning and deep learning. Journal of medical systems, 46(1), 1–11.

4. Khan, M. A., Khan, S., & Awan, S. B. (2020). Pancreatic cancer prediction using ensemble learning and data augmentation. Journal of medical systems, 44(12), 323.

5. Mohan, M., Kumar, V., & Singh, P. (2021). Pancreatic cancer prediction using ensemble learning and feature selection. Journal of biomedical informatics, 123, 103588.

6. Gupta, S., & Gupta, V. (2022). Pancreatic cancer prediction using ensemble learning and transfer learning. Journal of medical systems, 46(1), 1–12.

7. Kumar, R., & Mishra, S. (2020). Pancreatic cancer prediction using ensemble learning and feature selection. Journal of biomedical informatics, 123, 103607.

8. Rai, P., Singh, A., & Mishra, S. (2021). Pancreatic cancer prediction using a hybrid model of ensemble learning and deep learning. J. Med. Syst., 45(1), 1–10.

Note: All the figure and tables in this chapter were made by the authors.

Integrating Sentiment Analysis and LSTM to Predict Price Movements in Stocks

27

B. Venkataramana[1]

Associate .Professor, Dept. of CSE,
Holy Mary Institute of Technology & Science, TS, India

G. VenkataKoti Reddy[2]

Associate Professor & HoD, Dept. of CSE,
Holy Mary Institute of Technology & Science, TS, India

P. Bhaskarareddy[3]

Professor & Director, Dept. of ECE,
Holy Mary Institute of Technology & Science, TS, India

P. Sri Durga[4]

Assistant. Professor, Dept. of CSE, VPRIT, AP, India

Abstract: Stock prices in general say about the price of the stock in the market. Prediction of the stock price has been a big task for the people who do stock marketing as their primary job for survival, not only for those who do marketing as their primary task but also for those who do it as side work. Many people doing stock market prediction as their primary job can predict the stock price based on some factors namely, Fundamental analysis (impact and correlation of stock prices of other companies, past performances, records, profits, and debts of the company), Technical analysis (Continuation Pattern, Reversal Pattern, Leading Indicator, and Lagging indicator), Sentimental analysis (Client emotions such as positive, negative, neutral). The main question now arises is whether a machine can predict the stock price without human interaction. As stock price prediction is difficult for a machine, we can overcome this problem using machine learning algorithms like RNN, LSTM (Long Short-Term Model), and sentimental analysis like the in-person analysis. And here we are using the main thing that every stock price prediction should consider heuristic analysis.

Keywords: Stock parameters, Stock analysis, Stock price prediction, Recurrent neural network (LSTM), Analysis (Fundamental, Technical, Sentimental)

1. Introduction

Stock price prediction is finding out the future movement of the stocks. The accurate prediction of stock movement results in high profits for an investor, at the same time wrong prediction gives inevitable losses for the investor. Stock price prediction is complex yet challenging due to the volatile and non-linear nature of the stock market. So,

Investors always try to monitor the risks in real time so that there turn on investments could be higher, forecasting helps in safeguarding the trade of securities among the buyers and the sellers as well as elimination of the risks involved [9]. Sentimental analysis is the mechanism in which the given input will be characterized into three forms namely, positive, negative,and moderate. Modern data mining techniques have led to the development of sentiment analysis, an algorithmic

[1]bandaruramana1@gmail.com, [2]gvkotireddy@gmail.com, [3]pbhaskarareddy@rediffmail.com, [4]psridurga2010@gmail.com

approach for detecting the predominant sentiment about a product or company using social media data[4]. Sentimental analysis is often termed opinion mining where every unit of the sentence is categorized as one of the three forms of the type. Nowadays, a great volume of data, which contains information about numerous topics, is being transmitted online through various social media[4].Another way to use this concept is to find a way to connect the knowledge we gain from monitoring social media evaluations left by real people to the agency's inventory market trends in order to improve predictions of the agency's inventory costs. LSTM is the Long short-term memory that is a model used in machine learning and comes under the recurrent neural network (RNN). It can learn to bridge time intervals in excess of 1000 steps even in case of noisy, incompressible input sequences, without loss of shot time lag capabilities [8]. Here it is used in the processing of the output, using the recurrence, or feeding the output again and again in to the LSTM several times to get an accurate and precise prediction of the stock price. The output of every stage is inserted as input again for processing the output.

1.1 Related Work

LSTM for Stock Prediction

LSTM networks are excellent for studying how changes in one stock's value over time may influence the prices of numerous other equities. A unique subset of RNNs called LSTMs is capable of long-term capture of context-specific temporal relationships. Each LSTM neuron is a memory cell that has the capacity to store other data, meaning it keeps its cell state. In contrast to neurons in ordinary RNNs,which only take into account the current input and the previous hidden state to produce a new hidden state, an LSTM neuron also takes into account its former cell state and outputs its new cell state. In the case of LSTM architecture, the usual hidden layers are replaced with LSTM cells [15]. As shown in the following Figure, an LSTM memory cell has the following three gates:

1. Forget gate: When certain sections of the cell state should be replaced with more recent information is decided by the forget gate. It produces values that are near to 1 when certain aspects of the cell state should be kept and zero when certain values should be disregarded.

2. Input gate: This part of the network learns the circumstances under which any information should be stored (or updated) in the cell state based on the input (i.e., previous output o(t-1), input x(t), and prior cell state c(t-1)).

3. Output gate: This part determines what information is carried forward (i.e., output o(t) and cell statusc(t))to the following node in the network, depending on the input and cell state. It consists of the output generated by the LSTM [15].

1.2 Sentimental Analysis of Social Media Information

Traditional approach to studies of time series assumes that stock fluctuations are unpredictable. The sentiment library gives positive, negative, and neutral values as output [13]. Investors routinely communicate their opinions on social media platforms, which affects other investors' moods and influences their decision-making, especially given the continued popularity of social networks. According to previously conducted studies, investor sentiment and stock price changes are correlated. Additionally, some researchers have discovered that social media forum information and company-related financial news can also affect stock prices. Thus, stock prediction benefits from the examination of sentiment tendency in financial text data. For instance, examined stock market indicators that use a method of sentiment analysis for Twitter that is lexicon-based and historical data. The studies conducted by Sohangir et al. shown that CNN is the most effective model for analysing stock price when integrated with the sentiment analysis of financial Stock twits. Jiawei and Murata looked at the economic news's ten or and found that shareholder sentiment is a key factor for forecasting market fluctuations. Following that, attempted to improve prediction accuracy researched the best ways to collect training data sets of the highest quality from financial news. Xu and Keselj also created the tweets dataset for stock market forecasting at the same time. Their study supported the idea that financial tweet mood changes overtime.

2. System Study

2.1 Existing System

Traditional approaches to stock market analysis and stock price prediction include fundamental analysis, it is a type of investment analysis where the share value of a company is estimated by analyzing its sales, earnings, profits and other economic factors[15]. Additionally, there is statistical analysis,which focuses only on crunching numbers and finding trends in stock price volatility. Artificial neural networks (ANNs) or genetic algorithms (GAs) are frequently used to accomplish the latter. As an alternative to the above analysis, feature selection algorithms are used to reduce the dimensionality of the input data and list the important factors that have the greatest impact on stock prices or currency exchange rates globally, such as GDP, oil prices, inflation rates, etc. The latter other methods introduced like Historical Data Analysis, Multi-Source Multiple Instance

Learning, Support Vector Machines(SVM), and Independent Component Analysis (ICA) are used for Stock Market Prediction and Stock

Price Prediction using Linear Regression based on Sentiment Analysis is introduced in the domain of deep learning and neural networks.

2.2 Proposed System

In this project, we are performing both fundamental analysis and technical analysis on time series data and trying to predict the future price of the stock. And increasing the stock price predicting efficiency with the help of sentiment analysis. That is our project aim. The stock parameters, analysis, and all datasets are going to be very large and time series and there is a need to extract the feature from finance market raw data/ historical data, this is necessary for data analysis, divide it into testing and training data, teach the algorithm to forecast prices, and finally visualize the data as the last stage. So, we are using Long Short-Term Memory (LSTM), a Recurrent Neural Network (RNN) type. The information is slightly altered by LSTM through adds and multiplications. With LSTMs, data travels via a system called cell states. LSTMs are able to selectively recall or forget things in this way. LSTM networks are excellent for investigating how changes in the price of one stock can affect the prices of multiple other equities over an extended period of time.

3. Methodology

3.1 Pre-Processing

Preparing raw data to be used with a machine learning model is known as data pre-processing are listed below,

(a) **Dealing with Missing Values:** clearing the dataset of null values and replacing variables that contain missing values.

=> If it's numerical, we'll substitute the variable mean for any missing values. If it is categorical, we will replace the missing values with the variable mode.

(b) **One-Hot Encoding of Categorical Variables:** here, the conversion of categorical variables to numerical variables. Because Neural Nets work with numerical data, not categorical.

(c) **Checking Variables Data Types:** Pandas automatically tries to identify the type of each variable in a CSV dataset when we read one using it as we did. When a categorical variable is represented by numbers, Pandas sometimes makes the mistaken assumption that it is a numerical variable.

(d) **Performing One-Hot Encoding:** We want to choose a subset of the characteristics from our data to use

moving forward before executing One-Hot Encoding. These characteristics have a lot of categories. The vast amount of features generated by performing One-Hot Encoding on all of the features may cause the model to suffer from the curse of dimensionality.

(e) **Data Exploration:** It is an approach like initial data analysis, whereby a data analyst uses visual exploration to understand what is in a dataset and the characteristics of the data, rather than through traditional data management systems.

(f) **Splitting Data into Training and Testing Sets:** here splitting of dataset into training and testing subset is done. We will be training the model on the training subset and evaluating it with an unseen test set.

=> 90% as train data.

=> 10%as test data.

(g) **Data Scaling: Standardization:** Each variable in our data has been standardized here, with the exception of the target variable, of course. We will determine the mean and standard deviation of each variable for the training data that is currently stored in train_df. The values of each variable will then be subtracted from their mean values, and the results will be divided by the standard deviation divide the resultant values by the standard deviation of the training set of data.

3.2 System Architecture

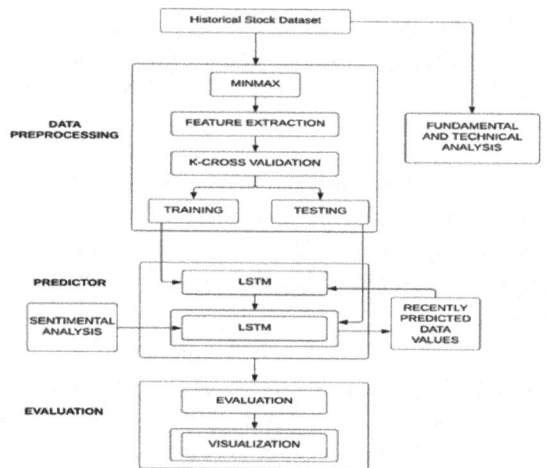

Fig. 27.1 System Architecture [1]

3.3 Applying OHLC Average

A particular sort of barch art known as an OHLC chart displays the open, high, low, and closing prices for each period. The chart type is helpful since it can display momentum that is either increasing or diminishing. When the open and close are widely apart, momentum is strong; when they are close together, momentum is weak or indicative of in decision. When determining volatility, the high and low show the entire price range for the time period. On OHLC charts, traders look for a number of patterns. Since OHLC charts provide the four main data points over a time period—the closing price, which many traders believe to be the most significant—they are helpful.

3.4 LSTM

For situations involving sequence prediction, LSTMs are frequently employed and have shown to be incredibly successful. The reason they work so well is because LSTM can store past information that is important and forget the information that is not. LSTM has three gates: the input gate, the forget gate, the output gate. The LSTM model can be adjusted for a number of factors, including the number of LSTM layers, dropout values, and epochs. LSTM model requires a lot of data for training[12].

=> here, LSTM is consider as three layers.

=> the input will be taken as three inputs (1. The output of data pre-processing, 2. The OHLC_AVG values and 3. The table with description of stock parameters.)

=> the output will be considered as *predictions*.

3.5 Applying Sentiment Analysis

Market sentiment is an objective gauge of how investors feel about the financial markets in general and about particular sectors or assets. There is a significant association between the movement of stock values and the release of news stories, according to previous studies in sentiment analysis. The quantity of training data made available affects how accurate deep learning algorithms are. For active traders and long-term investors, mood drives price activity and creates trading and investment possibilities. If the news sentiment is positive, there are more chances that the stock price will go up and if the news sentiment is negative, then stock price may go down[3].

3.6 Algorithm

Algorithm (Dataset):

1. Convert the string data under the date attribute to date format using the datatime module.
2. Plot a graph of the date and close price
   plt.plot(df.index, df['Close'])
3. Compute windowed dataframes
4. Convert windowed dataframes to multidimensional array
5. Compute train, test, validation values
6. Develop a model
   model = Sequential([layers.Input((3, 1)), layers.LSTM(64), layers.Dense(32, activation='relu'), layers.Dense(32, activation='relu'), layers.Dense(1)])
7. Print predictions Vs observation graphs.

4. Implementation

Dataset: Two main datasets were used in this project:

1. Yahoo Finance stock data from January 1, 2004, to July 5, 2021. The information includes the high,low,open, and close values for a specific day.
2. Twitter data that is openly accessible. This includes the time stamp and tweet content for each tweet made during a specific time period. Due to the daily generation of predictions, tweets are categorized by day based on their time stamps.

Prediction: Long Short-Term Memory (LSTM) architecture is used for training and prediction. Recurrent neural networks include the LSTM design, which is frequently utilized in deep learning applications. Because LSTM has feedback connections, it can be used to process whole data sequences. Data should be processed and standardized before being fed to the LSTM. The other dataset used for sentiment analysis comes from Twitter data; this dataset should also be cleaned before being used for sentiment analysis.

Data Processing:

(a) To start, data from yahoo finance is retrieved using pandas. A target data frame is formed with simply the close column because the close value is our target value. To make sure that all values lie between 0 and 1, the data is then translated and standardized. Following that, the data is divided into two groups: training data (70%), and testing data (30%) (see Fig. 27.2).

(b) The data is obtained from Twitter using Tweepy. The package Tweepy is used to connect to the Twitter API. All links and other special characters are removed from the tweets after they are cleaned up after being downloaded from the API. After cleaning, they are divided in to groups based on their polarity, with positive polarity designating a positive tweet and negative polarity a negative tweet (see Fig. 27.3).

Stock and Date Selection: For this experiment, stock values from BAC, Netflix, and TATAGLOBAL were used. Their stock values are open to the public. The figures are from January 1, 2004, to May 5, 2021 (see Fig. 27.4).

5. Results

The output after dataset is trained is illustrated in Fig. 27.5.

The stock parameters as output after LSTM, FA, TA, and SA procedure as – Open, Close, Predicted Close, Volume of Bank of American Corporation (BAC) in the below Figure.

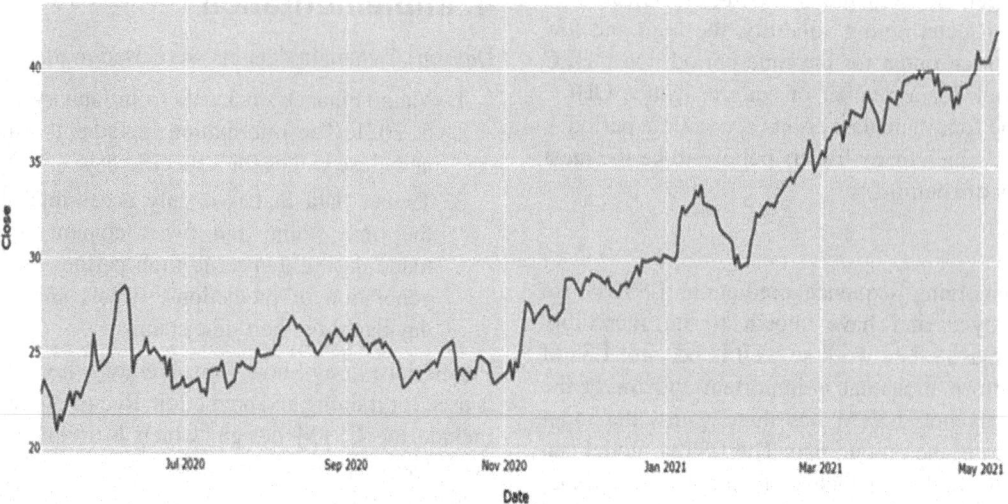

Fig. 27.2   Stock LSTM parameter [7]

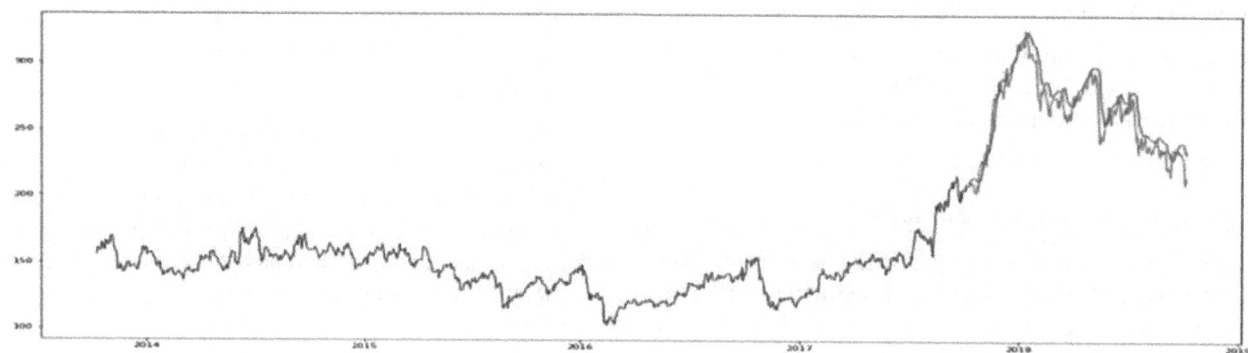

Fig. 27.3   Stock FA/TA and SA [7]

	Date	Open	High	Low	Close	Adj Close	Volume
0	2020-05-07	22.660000	23.370001	22.610001	22.840000	22.274752	51884300
1	2020-05-08	23.469999	23.620001	23.250000	23.570000	22.986685	45181000
2	2020-05-11	23.110001	23.120001	22.559999	22.580000	22.021187	68951800
3	2020-05-12	22.709999	22.820000	21.850000	21.870001	21.328758	69588700
4	2020-05-13	21.660000	21.750000	20.719999	20.870001	20.353506	112113400
...	...	...	...	...	...	...	...
248	2021-05-03	40.869999	41.049999	40.419998	40.560001	40.560001	41940500
249	2021-05-04	40.240002	41.080002	39.779999	41.000000	41.000000	53880000
250	2021-05-05	41.200001	41.560001	41.200001	41.389999	41.389999	23266900
251	2021-05-06	41.480000	42.060001	41.240002	42.009998	42.009998	39010600
252	2021-05-07	41.130001	42.250000	41.119999	42.180000	42.180000	32407163

253 rows × 7 columns

Fig. 27.4   Stock data [8]

	Open	High	Low	Close	Adj Close	Volume
count	253.000000	253.000000	253.000000	253.000000	253.000000	2.530000e+02
mean	29.134585	29.514980	28.765889	29.153478	28.888260	5.977239e+07
std	5.640161	5.692042	5.641791	5.689508	5.841742	2.128898e+07
min	20.290001	21.700001	20.100000	20.870001	20.353506	2.263230e+07
25%	24.540001	24.910000	24.110001	24.610001	24.198240	4.610920e+07
50%	26.879999	27.209999	26.510000	26.809999	26.436874	5.472130e+07
75%	33.119999	33.599998	32.840000	33.070000	32.906555	6.645340e+07
max	41.480000	42.250000	41.240002	42.180000	42.180000	1.783788e+08

Fig. 27.5 Stock data open/close/high/low [8]

Figure: DATA

The final project output for three different companies as Visualization is depicted in Figures. For Bank of American Corporation (BAC):-

For National Stock Exchange of India of Tata Global Beverages Limited (TATA GLOBAL):

6. Conclusion

When evaluating movement and making price predictions in the stock market, a lot of context is needed. In this project, we proposed a Predicator. To prove an efficient and accurate approach among existing system(s) to predict the market price of the stock. With the help of a predictor, we measured the accuracy of the following approach(es) LSTM algorithm, LSTM with fundamental & technical analysis, and LSTM with sentiment analysis. And it has been established among the three methods that LSTM with sentiment analysis [with having the lowest rise value compared to the other two] is the most suitable analysis for forecasting the market price of a stock. Since it is built using the LSTM with sentiment analysis approach and trained on a large collection of historical data, this project will be a great asset for brokers and investors looking to invest money in stock market applications (like Web, Mobile applications).

7. Future Enhancement

The project's future scope will include the addition of other parameters and elements like financial ratios, numerous instances, etc. The more the parameters are considered more will be the accuracy. We will try to collect historical data/raw data for more years to have more data points. We will enhance the time series into minute-to-minute data set. This project will be able to improve the stock prediction system by adding an analysis of news feeds from social media sites like Twitter, where sentiment is determined by the articles. The LSTM can be connected with this sentiment analysis to more effectively train weights and increase accuracy. We will increase layers and units/cells within the LSTM architecture to improve the model accuracy. This project viewing output upgradation like plotting includes stock parameters, company assets, and summary and shares detailed information is considered as a future enhancement.

Refrences

1. Zhang, X., & Wu, J. (2018). Stock market prediction based on LSTM network with sentiment analysis. 2018 IEEE International Conference on Big Data (Big Data), 3743–3746.
2. Ding, X., Zhang, Y., & Liu, T. (2014). A deep learning-based approach for sentiment analysis of financial news articles. Proceedings of COLING 2014, the 25th International Conference on Computational Linguistics: Technical Papers, 2325–2334.
3. Bao, S., & Yu, H. (2017). A deep learning framework for financial time series using stacked autoencoders and long-short term memory. PLoS ONE, 12(7), e0180944.
4. Tsantekidis, A., Passalis, N., Tefas, A., &Kanniainen, J. (2017). Forecasting stock prices from the limit order book using convolutional neural networks. IEEE Transactions on Big Data, 5(2), 793–803.
5. Hochreiter, S., &Schmidhuber, J. (1997). Long short-term memory. Neural Computation, 9(8), 1735–1780.
6. Goodfellow, I., Bengio, Y., & Courville, A. (2016). Deep Learning. MIT Press.
7. Brownlee, J. (2019). Deep Learning for Time Series Forecasting. Machine Learning Mastery.
8. Chen, H., Zhang, Y., Liu, B., & Chen, X. (2020). Forecasting financial market using sentiment analysis on social media. Quantitative Finance and Economics, 4(1), 98–117.

Enhancing Disease Prediction Accuracy: Leveraging Random Forest Out of Bag for Early Detection and Diagnosis in Healthcare

28

E. Sateesh[1]

Assistant Professor (Department of Computer Science and Engineering)
Amrita Sai Institute of Science and Technology Paritala, Andhra Pradesh, India

K. Lakshmi Padmavathi[2]

M.Tech Scholor (Department of Computer Science and Engineering)
Amrita Sai Institute of Science and Technology Paritala, Andhra Pradesh, India

A Nageswara Rao

Assistant Professor (Department of Mechanical Engineering),
Lakireddy Balireddy College of Engineering, Mylavaram

Abstract: Machine learning algorithms can now be used to anticipate diseases thanks to the quick development of computer-based technology and the collection of massive amounts of electronic health data. In this paper, an ensemble learning-based disease prediction system is presented, along with supervised machine learning algorithms including Naive Bayes, Decision Trees, Random Forests, and Random Forest OOB. The objective is to enable early detection and personalized healthcare strategies by harnessing the power of predictive models. The proposed system utilizes a diverse and comprehensive dataset, comprising electronic health records, clinical data, genetic information, lifestyle factors, and demographic attributes of patients across various medical conditions. Data pre-processing techniques are employed to ensure data quality, handle missing values, and address imbalanced class distributions. Proposed work contributes to the advancement of predictive medicine and advocates for the responsible use of machine learning algorithms in healthcare for the betterment of patient care.

Keywords: Disease prediction, Machine learning, Random forest OOB, Decision tree, Random forest, Naive Bayes

1. Introduction

The medical system is facing challenges due to the increasing number of patients and diseases each year, leading to overload and rising costs in many countries. Seeking treatment often involves consultations with doctors, making it expensive and time-consuming. However, utilizing data and algorithms for disease prediction can offer a cost-effective and efficient solution. Our project focuses on accurately predicting diseases based on patient symptoms, employing four different algorithms with a high accuracy range of 92-95%.

This disease prediction system holds significant potential for the future of medical treatment. It provides an intelligent interface to facilitate user interaction with the framework and visualize the study's results effectively. As doctors increasingly adopt scientific technologies for diagnosis and identification, accurate disease prediction becomes crucial for successful treatment. Our project aims to address common diseases in their early stages, as research shows that a significant percentage of people tend to ignore general health issues, leading to more severe conditions later on.

The primary reasons for this ignorance include reluctance to consult a doctor due to time constraints and a busy lifestyle. As a result, a considerable portion of the population in India suffers from preventable diseases, and early ignorance of symptoms leads to fatal outcomes for 25% of these cases.

[1]esateeshcse@gmail.com, [2]padhunageswar@gmail.com, [3]nagesh803@lbrce.ac.in

To combat this issue, our project is developed to enable users to perform health check-ups conveniently from any location. The user-friendly interface ensures easy operation, promoting regular check-ups for early disease detection and proactive healthcare management.

2. Relatd Work

To develop disease prediction systems, several researchers have used machine learning approaches like supervised learning and unsupervised learning [1,2]. For instance, [3,4 &5] presented a system to deal with the issue of many diseases having similar symptoms, improving diagnostic accuracy and achieving a success rate of 71.53%. [6,7] attempted to visualize the outcomes of their study and project, achieving an accuracy score of 68.5% when measured against alternative methodologies.

Similar to this, [8] created a system that accurately predicts diseases based on data or symptoms provided by users. In order to anticipate diseases, [9,10]developed a system that evaluates symptoms provided by the user and outputs a probability of the condition.

With the accurate analysis of medical data made possible by the ongoing rise of big data in the biomedical and healthcare sectors, early disease identification and enhanced patient care are now possible [11]. With proactive diagnostic capabilities, machine learning-based disease prediction systems continue to offer promise for changing healthcare.

3. Proposed System

With the use of ensemble learning approaches, Naive Bayes, Decision trees, Random forests, and Random Forest OOB estimation, the suggested system seeks to create a sophisticated illness prediction model. For a variety of medical diseases, this system aims to use the power of machine learning to provide early identification and individualized therapy methods. Electronic health records (EHRs) and pertinent patient variables gathered from healthcare sources will form the foundation of the system's broad and extensive dataset. To maintain data quality and consistency, perform data pre-processing processes to handle missing values, resolve unbalanced class distributions, and normalize or scale numerical features. Out-of-Bag Estimation and Random Forest Ensemble Learning Utilize the ensemble nature and objective performance evaluation provided by the Random Forest algorithm with Out-of-Bag estimate (RF-OOB). Optimize the RF-OOB model's hyper parameters, such as the number of decision trees and the maximum depth of trees, to increase predictive accuracy. Utilizing stratified sampling, divide the preprocessed dataset into training and

testing sets to maintain the class distribution. Accuracy, sensitivity, specificity, precision, F1-score, and AUC-ROC are used as performance measures to assess the RF-OOB model's performance on the testing set once it has been trained. Analyze the RF-OOB model's performance in illness prediction tasks in comparison to that of other supervised machine learning methods, including Naive Bayes, Decision Trees, and conventional Random Forest.

4. Modul Discription

Five modules comprise the entire suggested system.

- Clinical Data Gathering
- Data Pre-processing
- Model Construction
- Model Construction using Prescription
- Database Creation

4.1 Clinical Data Gathering

This dataset may be a database of information on the relationships between diseases and their symptoms, created automatically using information from textual discharge summaries of patients. the illness, the number of discharge summaries that mention the illness positively and recently, and consequently the related symptom. The symptoms are displayed hierarchically to support the strength of association. Associations for the 150 most common diseases supported these notes were computed.

4.2 Data-preprocessing

We deleted all the variables with more than 50% missing prices because information pre-processing is an essential stage in machine learning. The method obtained UMLS codes for illnesses and symptoms using the MedLEE natural language processing system. The relationships were then obtained using methods backed by frequencies and co-occurrences in applied mathematics.

4.3 Model Construction

For precisely identifying the disease that the user has provided, predictive classifier models were created. Naive Bayes, Decision Tree, Random Forest (RF), and Random Forest from an agricultural algorithm make up the classification model for predicting the disease.

Naïve Bayes algorithm

Naive Bayesian classification is based on the Bayesian theorem and works well with independent predictors[12]. It involves a training phase to evaluate probability distribution parameters and a prediction phase to classify unfamiliar data based on the largest posterior probability. The method can

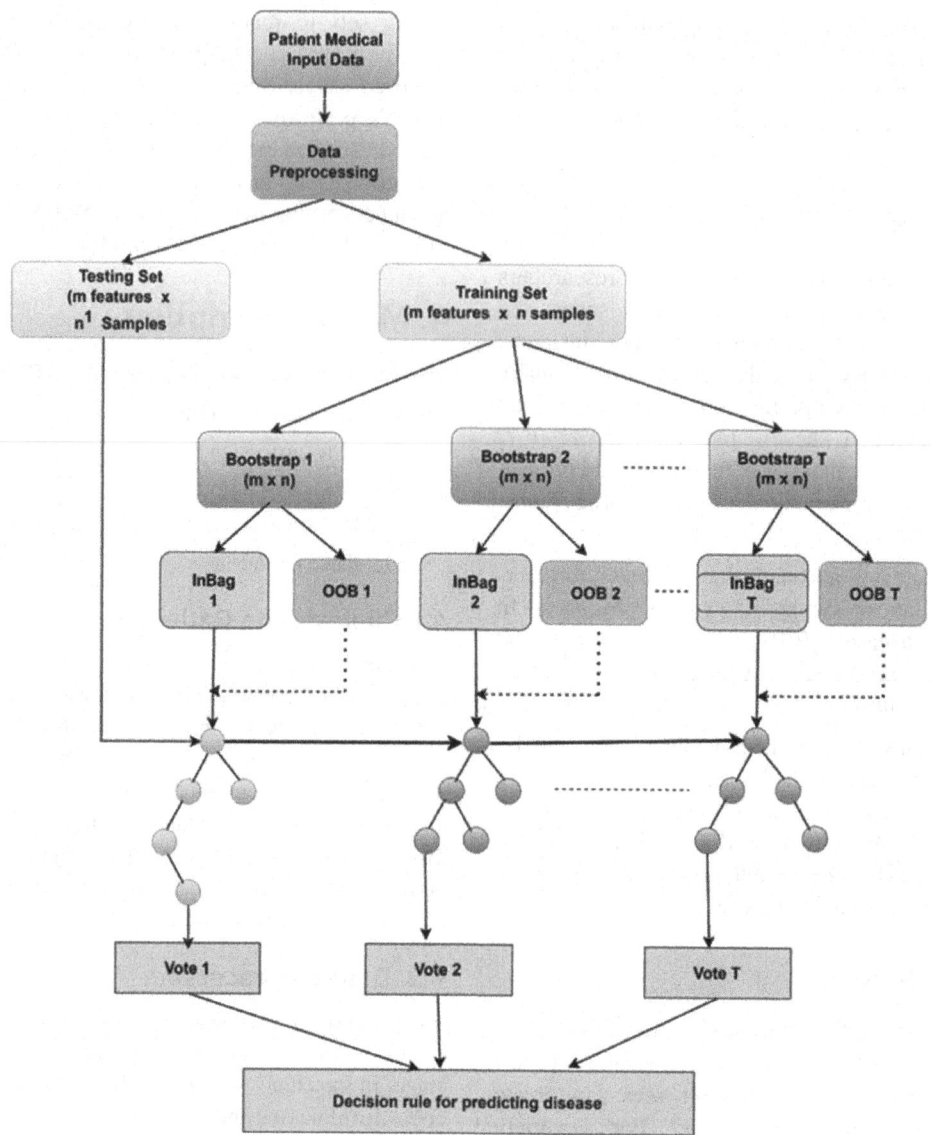

Fig. 28.1 System architecture for out of bag with random forest [1]

use Kernel Density and Gaussian distribution Estimation functions, depending on the dataset's nature. It offers an efficient and versatile approach to classifying data in various applications.

Decision tree

In data mining, decision trees (DT) are a common categorization tool[13]. To increase prediction accuracy, they used mathematical methods to recursively divide the data into branches. Because of their hierarchical structure, decision trees are simple to construct and comprehend.

Random forest

A number of decision trees are built into a "forest" by the Random Forest (RF) algorithm, a potent forecasting method

Input: TS: training set, TS = u_i (i = 1, 2, ..., n),
Output: Class label A and B
Steps:

1. Given training dataset TS which consists of genes belonging to different class say class A and B.

2. Compute the prior probability of class A = nob of features of class A / total nob of genes
   Compute the prior probability of class A = nob of features of class B / total nob of genes

3. Find n_i, the total nob of frequent features of each class.
   n_a = the total nob of frequent features of class A
   n_b = the total of frequent features of class B

4. Find conditional probability of occurrence of key gene given a class
   $P_{(feature1/ class A)}$ = geneCount / $n_i(A)$
   $P_{(feature1/ class B)}$ = geneCount / $n_i(B)$
   $P_{(feature2/ class A)}$ = geneCount / $n_i(A)$
   $P_{(feature2/ class B)}$ = geneCount / $n_i(B)$
   "   "
   $P_{(featuren/ class B)}$ = geneCount / n_i (B)

5. Avoid zero frequency problems by applying uniform distribution

6. Classify a new gene C based on the probability P(C/feature).
   a) Find $P_{(A/feature)}$ = $P_{(A)}$ * $P_{(feature1/classA)}$ * $P_{(feature2/classA)}$ * ..... * $P_{(featuren/classA)}$
   b) Find $P_{(B/feature)}$ = $P_{(B)}$ * $P_{(feature1/classB)}$ * $P_{(feature2/classB)}$ * ..... * $P_{(featuren/classB)}$

7. Assign gene to class that has higher probability

Fig. 28.2 Pseudocode of Naive Bayesian algorithm [2]

GenDecTree(Sample S, Features F)
Steps:

1. If*stopping_condition(S, F) = true* **then**

   a. *Leaf = createNode()*

   b. *leafLabel = classify(s)*

   c. **return** *leaf*

2. *root = createNode()*

3. *root.test_condition = findBestSplit(S,F)*

4. *V = {v | v a possible outcomecfroot.test_condition}*

5. **For each** *value v ∈ V:*

   a. S_v *= {s | root.test_condition(s) = v and s ∈ S };*

   b. *Child = TreeGrowth* $(S_v, F);$

   c. *Add child as descent of root and label the edge {root →* *child} as v*

6. **return** *root*

Fig. 28.3 Pseudocode of decision tree algorithm [2]

[14, 15]. Every tree participates in the decision-making process by acting as a "voter." By using bagging, RF reduces overfitting by producing randomly sampled training datasets for each tree [16]. When features are chosen for splitting, it also introduces randomization. Because it can handle large datasets and make accurate predictions, RF is widely used.

To generate *c* classifiers:
for *i* = 1 to *c* do
   Randomly sample the training data *D* with replacement to produce D_i
   Create a root node, N_i containing D_i
   Call BuildTree(N_i)
end for

BuildTree(N):
if *N* contains instances of only one class **then**
   return
else
   Randomly select x% of the possible splitting features in *N*
   Select the feature *F* with the highest information gain to split on
   Create f child nodes of *N* , $N_1,..., N_f$, where *F* has *f* possible values ($F_1, ... , F_f$)
   for *i* = 1 to *f* do
      Set the contents of N_i to D_i, where D_i is all instances in *N* that match
      F_i
      Call BuildTree(N_i)
   end for
end if

Fig. 28.4 Pseudo code of the RF algorithm [2]

Random forest out of bag

The Random Forest method uses the Random Forest Out-of-Bag (OOB) technique to enhance the model's performance and offer a neutral assessment of its forecast accuracy [17,18]. An ensemble learning technique called the Random Forest algorithm mixes various decision trees to provide predictions that are more accurate.

A dataset is often split into two parts in classical machine learning: a training set to develop the model and a separate testing set to assess its performance. However, in Random Forest, only a portion of the training data is used to build each decision tree, leaving some samples unutilized. This is referred to as "Out-of-Bag" data. In essence, the Out-of-Bag data serves as a validation set for each decision tree in the Random Forest. It enables the model to test itself without the need for a separate testing set on previously unobserved data points. This lessens the chance of overfitting and gives a more accurate evaluation of the model's generalizability.

The predictions made on the Out-of-Bag samples across all decision trees in the forest are combined to determine the OOB accuracy. The model's performance on hypothetical data is estimated objectively using this aggregated accuracy. The OOB estimate is especially helpful when the dataset is small since it makes the best use of the data that is available for training and validation.

To generate *c* classifiers:
for *i* = 1 to *c* do
   Randomly sample the training data *D* with replacement to produce D_i
   Create a root node, N_i containing D_i
   Call BuildTree(N_i)
end for

BuildTree(N):
if *N* contains instances of only one class **then**
   return
else
   Randomly select x% of the *all* splitting features in *N*
   Select the feature *F* with the highest information gain to split on
   Create f child nodes of *N* , $N_1,..., N_f$, where *F* has *f* possible values ($F_1, ... , F_f$)
   for *i* = 1 to *f* do
      Set the contents of N_i to D_i, where D_i is all instances in *N* that match
      F_i
      Call BuildTree(N_i)
   end for
end if

Fig. 28.5 Pseudo code of the RF with out of bag algorithm [2]

4.4 Model Construction Using Medicine and Prescription

Medicine One of the most crucial elements in everyone's life is a prescription gathering the medication and precaution dataset from patient discharge summaries that are text-based. When a disease is predicted, a GUI page will present the medication and precautions needed to prevent it. Jupiter Notepad is used to simulate model creation.

5. Results

Utilizing classifiers like Naive Bayes, Decision trees, Random forests, and Random Forest OOB, the feature-extracted data is further assessed to forecast the disease. Input diseases for Naive Bayes osteoarthritis, Decision Tree osteoarthritis, Random Forest osteoarthritis, and Random Forest OOB osteoarthritis are displayed as output. Symptoms selected from a drop-down menu in the GPU include abdominal pain, back pain, hip joint pain, knee pain, and neck pain as shown in Fig. 28.6. All algorithms results same diseases prediction its shows validity of the study.

Fig. 28.6   GUI for proposed model testing

Source: Author

In a comparison of the four machine learning algorithms, Naive Bayes Classifiers displays 91.45% accuracy, Decision Tree algorithm 92.47% accuracy, Random Forest algorithm 94.55% accuracy, and Random Forest OOB model 95.32% accuracy. Accuracy, sensitivity, specificity, positive predictive value, and negative predictive value were all used to analyze the results. Figure 28.7 displayed each model's final accuracy.

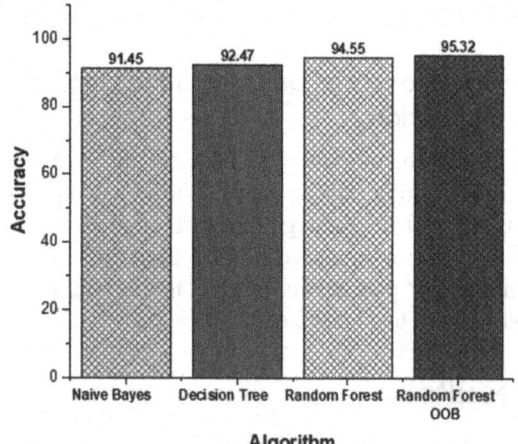

Fig. 28.7   Final accuracy of each model

Source: Author

6. Conclusion

The findings from this study highlight the remarkable potential of Random Forest Out of Bag (RF-OOB) in disease prediction, offering valuable insights into enhancing early detection and personalized healthcare strategies. The widespread integration of computer-based technology in the healthcare industry has led to the accumulation of vast electronic data, presenting challenges for accurate symptom analysis and early disease identification. However, supervised machine learning algorithms, such as RF-OOB, have emerged as promising solutions, surpassing conventional methods and assisting medical experts in high-risk disease detection. Among various supervised ML algorithms explored, including Naïve Bayes and Decision Trees, RF-OOB demonstrated superior accuracy when applied to diverse medical datasets. By leveraging an ensemble of decision trees and the OOB estimation technique, RF-OOB yielded higher prediction accuracy, making it a compelling choice for disease prediction and diagnosis

References

1. K. Pingale, et al., "Disease Prediction using Machine Learning," 2019.
2. A. Alikberov, S. Broadly, et al., "The Learning Machine," [Online]. Available: https://www.thelearningmachine.ai.
3. Mr. C. Beyene, Prof. P. Kamat, "Survey on Prediction and Analysis the Occurrence of Heart Disease Using Data Mining Techniques," International Journal of Pure and Applied Mathematics, 2018.
4. S. Saha, A. R. Chowdhuri, et al., "Web Based Disease Detection System," IJERT, vol. 2, no. 4, April 2013.
5. A. Sadiya, "Differential Diagnosis of Tuberculosis and Pneumonia using Machine Learning," 2019.
6. S. Patel, H. Patel, "Survey of data mining techniques used in the healthcare domain," Int. J. of Inform. Sci. and Tech., vol. 6, pp. 53–60, March, 2016.
7. S. Balasubramanian, et al., "Title of the Article," International Journal of Advances in Computer Science and Technology, vol. 3, no. 2, pp. 123–128, Feb. 2014.
8. K. Dhenakaran, R. Rajalakshmi, "Analysis of Data mining Prediction Techniques in Healthcare Management System," International Journal of Advanced Research in Computer Science and Software Engineering, vol. 5, no. 4, 2015.
9. J. Ma, S. C. Park, J. H. Shin, N. G. Kim, J. H. Seo, J. S. R. Lee, J. H. Sa, "AI based intelligent system on the EDISON platform," Proceedings of the 2018 Artificial Intelligence and Cloud Computing Conference on ZZZ - AICCC '18, 2018.
10. W. Yin, H. Schutze, "Convolutional neural network for paraphrase identification," Proc. HLT-NAACL, pp. 901–911, 2015.
11. S. Adam, et al., "Prediction system for Heart Disease using Naïve Bayes," International Journal of advanced Computer and Mathematical Sciences, vol. 3, no. 3, pp. 290–294, 2012, ISSN 2230-9624.

12. J. R. Qulan, "Induction of Decision Trees," Mach. Learn., vol. 1, no. 1, pp. 81–10, Mar. 1986.

13. P. Suryachandra, V. S. Reddy, "Comparison of Machine Learning algorithms For Breast Cancer".2016

14. A. Alikberov, S. Broadly, et al., "The Learning Machine," [Online]. Available: https://www.thelearningmachine.ai.

15. M. Chen, Y. Hao, K. Hwang, L. Wang, L. Wang, "Disease prediction by machine learning over big data from healthcare communities," IEEE Access, vol. 5, no. 1, pp. 8869–8879, 2017.

16. "Disease and symptoms Dataset," [Online]. Available: www. github.com.

17. H. Nguyen, X. N. Bui, "Predicting Blast-Induced Air Overpressure: A Robust Artificial Intelligence System Based on Artificial Neural Networks and Random Forest," Nat Resour Res, vol. 28, pp. 893–907, 2019.

18. W. Nidhi, K. Mukesh, M. Shaveta, "Classification of Breast Cancer Tissues using Decision Tree Algorithms," Int. J. Res. Eng. Appl. Manage. (IJREAM), vol. 05, pp. 342–346, 2018.

Effective Unauthorized Vehicle Detection System for Intelligent Transportation

29

Kotakonda Madhubabu[1], N. Snehalatha[2]

Department of Computational Intelligence,
SRM Institute of Science and Technology, Kattanakalthur, Chennai, India

Abstract: From the day-to-day technology upgrading perspective, one of the faster-growing technology is intelligent cities. The main reason is to attract innovative systems that are very convenient for all. Almost all people like innovative technology utilized to complete their daily tasks. In this process, they need to move from one place to another place and, after completing their work again, come back home, for example, going to the office, dropping or picking up the children from their school or college, need to go to the hospital or some places, to go to meet a friend or person, etc. To move from one place to another place we use the transportation system. In the future, vehicles, intelligent transportation systems, and smart cities will play an essential role in our routine life. This paper proposes unauthorized vehicle detection in the VANET using the correlation coefficients method in the intelligent transportation system. The proposed work will help avoid too much congestion in smart transportation compared to earlier techniques to identify unauthorized vehicles in an intelligent environment, those who will utilize an innovative vehicle transportation system as a platform of the vehicular ad-hoc network.

Keywords: Trusted authority, VANET, RSU, OBU, Unauthorized vehicles

1. Introduction

The directions of technology grow vehicle intelligent transportation system is a future development technology in an innovative environment. In this direction, the Internet of Vehicles (IoV) hopes to soon switch to the Internet of Autonomous Vehicles (IoAV). IoAV hopes to construct intelligent vehicle infrastructure and facilitate autonomous driving without human intervention. However, the need for autonomous decision-making becomes apparent as the number of connected vehicles grows. Therefore, IoAV must provide robust, secure, transparent, and evolutionary capabilities [1]. IoV is to build a stiff backbone for intelligent transport systems, paving the way for technology that is an excellent illustration of their management. IoV Architecture plays a significant role in various innovative fields, such as the automotive industry, research companies, smart cities, and intelligent transport for numerous business and scientific applications [2]. With the growth of IoT and Big

Data techniques, IoV is a significant catalyst for upcoming autonomous vehicular system scenarios and mobile Ad hoc networking technologies [3]. In the present Smart Transportation System research model, the traditional VANET is transitioning to IoV, and the VANET is a subset of the MANET, a component of the intelligent transport system that comes up with two types of communication: Vehicle-to-Infrastructure (V2I) and Vehicle-to-Vehicle (V2V) [4]. The TA, who acts as the Trusty Management Centre, TA is responsible for registering and disseminating confidential vital documents. RSU along roadsides is a bridge between vehicles and the TA—the OBU to every vehicle, which follows the protocol of V2I and V2V communications [5]. Many approaches are to address security and confidentiality issues in intelligent transportation. This system fascinates much recognition from researchers worldwide and can enhance road safety and control the traffic flow through vehicle resources and communication systems [6]. The main goal of an intelligent transport system is to refine

[1]madhuimp9@gmail.com, [2]snehalan@srmist.edu.in

transport safety, efficiency, and control by disseminating apropos information in the environment in which they are implemented [7]. The amalgamating and expanding emerging technology for transport systems is critical to the success of ITS [8]. The IoV vehicle consists of seven layers: the identity layer, the object layer, the interoperability device layer, the communication layer, the server and cloud service layer, the multimedia computation layer, and the application layer [9],[20]. In the Smart City framework, more than one item, or more fabulous accurately, intelligent items, can engage with their processors, computing, and conversation [18]. The Internet of Things (IoT) also refers to intelligent gadgets that offer stable and bright surroundings via superior interconnection and interoperability capabilities [19]. The IoT advantages, many such items are related to cars or cars that can speak wirelessly with loads of Internet-related gadgets, in-automobile equipment (intra-vehicle), or out of doors automobiles (inter-vehicle) [21], [22]. The unique type of custom IoT is called the Internet of Vehicles (IoV), which enables unified management of smart cities, intelligent transportation, and other applications [23],[24]. Vehicle wireless communication can open up many new applications, the most important of which is a class of security applications that can prevent collisions and save thousands of lives [25]. In real-world scenarios, these vehicle advertising systems will only be implemented if appropriate incentives and protection mechanisms are in place due to the inconsistent behavior of selfish or malicious nodes [26]. The V2V and V2I communication protocols are wireless; without proper security plans in the VANET, attackers can modify messages sent by the vehicle and impersonate the vehicle [27]. In a VANET message, recipients must verify the authenticity and integrity of received messages and information from the receiver's side if the message is trusted or not [28].

The rest of the paper is organized as follows:

Section 2 addresses the recent works of Intruder detection in Vehicular ad hoc networks. Section 3: Discuss the proposed Intruder detection system in Vehicular ad hoc networks. In addition, in Section 4, we discussed results acquired in the proposed work, and Section 5 concludes this paper.

2. Literature Survey

Zhang et al. Proposed Vehicle link authentication to protect privacy through hierarchical aggregation and fast identification-based composite signatures. This scheme provides hierarchical aggregation and batch validation. Collect and verify personally identifiable signatures generated by different vehicles in one package. A message collector (e.g., a traffic control authority) can return the collected signatures. This hierarchical aggregation technology significantly

reduces transmission/storage overhead for vehicles and other parties [10]. F. Bonomi et al. The proposed intelligent connected car and the electronic messaging system with a robust authentication method for the Internet of Things and also can protect V2I and V2V communication protocols from intruders from inside and outside by using the Certificate Authority (CA) or Trusted Authority (TA) technologies that rely on both public and private keys for V2V and V2I communication incurs high computational costs and does not even provide scalability [11]. Kenny et al. Proposed a standard for short-range only communication (DSRC) in the automotive industry is exploring the development of Dedicated Short Range Communication (DSRC) technology for vehicle-to-vehicle communication. The effectiveness of this technology is highly dependent on the cooperative criteria for interaction [12]. Lee et al. Proposed Signature Seeking Drive (SSD) provides secure in-vehicle advertising incentives, a secure incentive system that securely promotes the co-distribution of advertising messages to vehicle users. Unlike traditional reward systems, SSDs do not rely on tamper-resistant hardware or a theoretical approach to gaming but influence public key infrastructure to provide secure rewards to collaborative nodes. The proposed set of temporary designs shows that our SSDs are vital for promotion and security. In an ideal environment, each vehicle continues delivering these advertisements regularly. Of course, such a cooperative scenario is unrealistic, as some selfish users should not make such statements [13]. Raya et al. Proposed a signature authentication technique based on Public Key Infrastructure (PKI). According to their plan, all traffic-related information exchanged on the VANET network must be authenticated before being trusted[14]. He D. et al. Proposed a Conditional Privacy Protection (CPP) function to consider how WANet could be satisfied. However, this does not rule out the presence of malware trying to send bogus messages or modify legitimate messages [15]. Lee and S.Park et al. Proposed a security incentive to advertise on transport networks for security issues and, more seriously, the potential for intentional intrusion into networks without adequate security resistance. For example, you can notify to initiate a denial of service campaign through a network denial of service (DoS) attack. Therefore, implementing this vehicle advertising system in a real-world scenario requires considering appropriate incentives and security policies [16]. Cui et al. Proposed the basis of the internal bathtub maintaining the connection between the other vehicles and communication security depends on message authentication. However, these plans are not lacking unnecessary authentication and do not perform incorrect message retrieval in the message package. This process to solve these problems introduced EDGE COMPUTING new concepts in the Message Authentication MANET process.

According to this method, the length unit can effectively authenticate messages from adjacent vehicles and transmit authentication results within a communication range, minimizing unnecessary authentication [17]. Nevertheless, the earlier schemes have the following disadvantages. First, all vehicles must store multiple nickname certificates, which increases RSU's transmission overhead as the number of vehicles increases. Second, because the size of the certificates is relatively large, network congestion in the communication channel can occur when the vehicles are a large number. Third, one trusted authority (TA) maintains all the registered vehicle's information every time it correlates with RSU's data, which is a significantly time-consuming process. Finally, in their plans, the RSU and the vehicle confirm the messages received one after the other; this process could be more efficient and appropriate to implement in real scenarios. However, completely anonymous schemas should be avoided for the following reasons.

3. Proposed Work

3.1 Unauthorized node(s) Detection Security on VANET

Security ensures secure communication over vehicle networks. Because communication occurs in an open-access environment, VANET is more susceptible to attacks that pose multiple threats to security services. In particular, the Sibyl attack is the most dangerous attack on VANETs. It is still necessary to balance privacy and denial to detect a Sibyl attack. On the other hand, many other attacks on VANETs, such as jamming attacks and malware attacks, need to be further investigated. There are still many security challenges,

and a great algorithm is needed to solve these challenges. The proposed algorithm named as Trusted Authority Based Correlation algorithm. This algorithm helps identify the unauthorized node(s) quickly. It sends unauthorized vehicle information to the nearest cyber crime station through IoT to provide a safe and efficient path from a source to a destination point. The cybercrime station receives the intruder vehicle's details from the IoT. Then it takes the necessary action against unauthorized nodes, creating efficient paths from one point to another in the network. The proposed work is shown in the figure. 1. Every vehicle has an onboard unit (OBU); this OBU has all the vehicle information; the roadside units can collect and send their range of vehicle information to the Trusted Authority or Local Trusted Authority. The TA or LTA already has the local or region-based registered vehicle information; here, our algorithm is correlated with the vehicle information; if the correlation is the same between coefficients, there is no intruder node in the network. Otherwise, the correlation differs from that vehicle or node information sent from TA to Central Trusted Authority (CTA). CTA has to maintain all regions' vehicle information to perform the correlation algorithm at the CTA level if the correlation with any of the CTA vehicle's information is an unauthorized node. Otherwise treated as an unauthorized node or vehicle in the network that now sends that vehicle's information to the nearest cyber crime station for attains that unauthorized vehicle.

3.2 System Model

A VANET network with the dimensions m. In the constructed area, there is a count of nodes equipped with OBU devices to be aware of its registered information, and these nodes are said to be vehicles. The nodes are at diverse locations in the pathways and are moving randomly; every vehicle

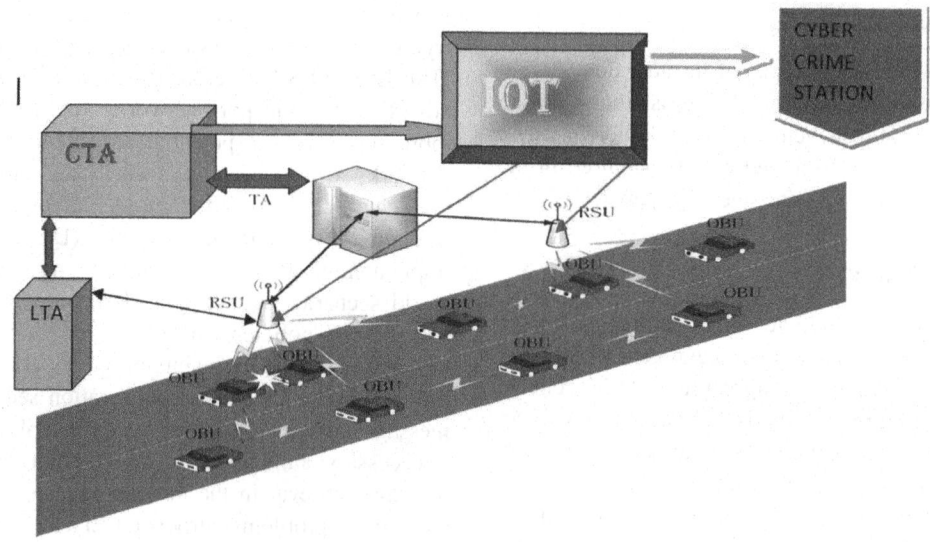

Fig. 29.1 Unauthorized vehicle detection system in VANET [1]

has its own OBU. This OBU's data on IoT, through RSU, in which the information from one vehicle to another vehicle in the pathway. Each vehicle includes a transceiver. Each vehicle in offline registration is in its local trusted authority (TA), and all trusted authority registration details are in one place, known as the central trusted authority (CTA). Every registered vehicle-renewal periodically for their fitness certificate purpose. The in-vehicle registration process needs to maintain these aspects (i) vehicle unique identification number (VID), (ii) vehicle color, (iii) class of the vehicle, (iv) vehicle manufacturing year, etc. Every TA maintains and correlates the first four aspects only. However, CTA is to maintain all the aspects of vehicle registration. Initially, every time-correlated, RSU's data with TA's data when the correlation only performed correlation with CTA's data; this time, both only consider authorized nodes otherwise treated as unauthorized nodes and then send that node details to CCS through IoT. As a result, there are reduced pollution, traffic safety measures, traffic management, and the best path through an elegant transportation system. In this research work, a new routing algorithm is introduced to select the Trusted Authority-based correlation algorithm based on a defined objective function that depends on the OBU details of the registered vehicle. The RSU can access and send that region vehicle's information to Trusted Authorities for identifying the unauthorized node(s) in VANET.

3.3 Correlation Coefficient (r)

This is a numerical measure of a specific type of correlation, i.e., a statistical relationship between two variables or objects. The variables may be two columns of a given data element of observations, often called a sample, or two components of a multivariate random variable with a known distribution. There are several types of correlation factors, each with its definition and range of usability and characteristics.

Algorithm 1 Trusted Authority-Based Correlation Algorithm:

1. START
2. Data from LTA is $LTA_{(d)}$ or $TA_{(d)}$
3. Data from CTA is $CTA_{(d)}$
4. Data from $RSU_{(d)}$ at time t
5. While (t!=-1) {
6. Find **r** value of ($RSU_{(d)}$ with $TA_{(d)}$)
7. if ((r==1)||(r==-1))
8. No intruder found   Else
9. Find **r** value of ($TA_{(dt1)}$ with $CTA_{(d)}$)
10. if ((r==1)||(r==-1))
11. No intruder found   Else
12. Intruder found sent to CCS
13. STOP

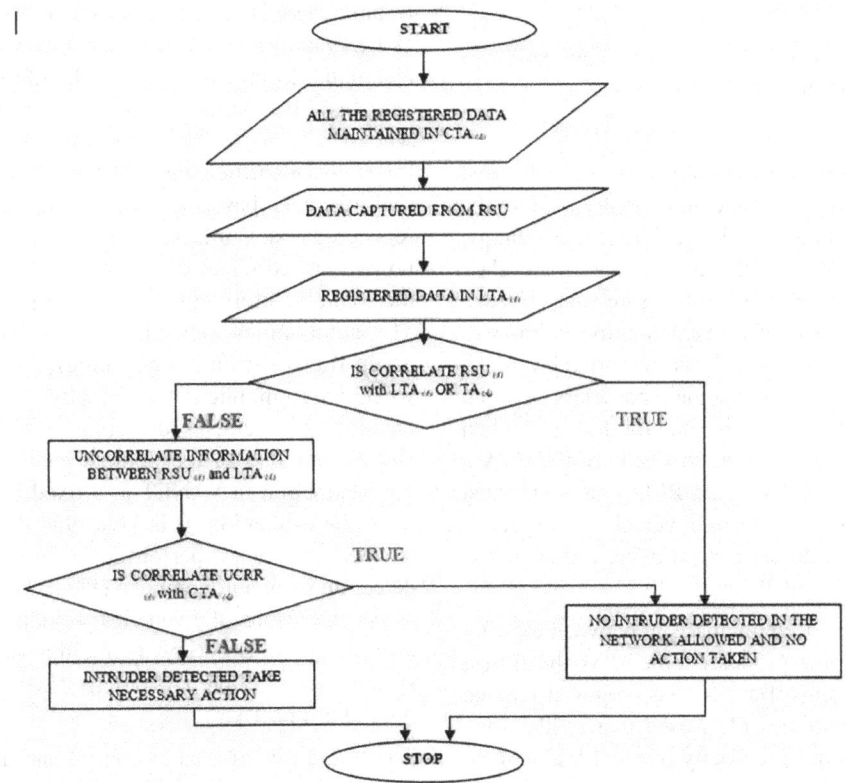

Fig. 29.2 Flow diagram for unauthorized vehicle detection by using correlation algorithm

Source: Authors

4. Results and Discussions

The unauthorized vehicle identification of autonomous vehicles in VANET with a trust authority-based correlation model was implemented in PYTHON, and their results were verified. Moreover, the analysis of the adopted trust authority-based correlation model achieves good performance compared to the earlier method. In the previous technique, only one trust authority (TA) is handled for all registered vehicle details. Due to this, when the number of registered vehicles increases, the existing technology needs to be improved, and it takes more time to correlate with RSU's data.

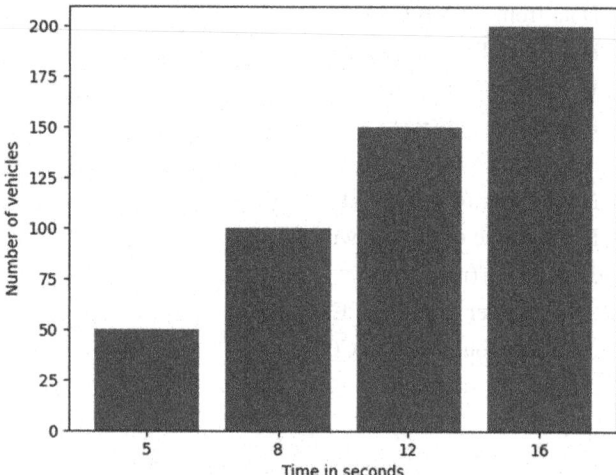

Fig. 29.3 The proposed scheme identification time for various numbers of vehicles

Source: Authors

Figure 29.3 represents a correlation analysis between the trust authority's data ($TA_{(d)}$) and roadside units ($RSU_{(d)}$). Here the correlation parameters of the vehicle are the vehicle's unique identification number (VID), vehicle color (Vcolor), class of the vehicle (Vclass), vehicle manufacturing year (Vyear), etc. In this, all the parameters of the vehicle information with $TA_{(d)}$ with $RSU_{(d)}$, then the vehicle is authorized vehicle. Suppose there is no correlation in some parameters between $TA_{(d)}$ and $RSU_{(d)}$. So, therefore this situation happens when that vehicle information correlation checks between $CTA_{(d)}$ with $RSU_{(d)}$ if correlated $CTA_{(d)}$ and $RSU_{(d)}$ as authorized vehicles. Otherwise, an unauthorized vehicle. Then sends that vehicle's information to the nearest cyber crime station (CCS) through the Internet of Things (IoT).

Figure 29.4 represents a speed comparison between existing and proposed schemes; the proposed scheme is the fastest compared to earlier systems. Because the proposed scheme has multiple trusted authorities (TA) used that is like zone wise. Due to this data storage capacity of the TA is less, so the identification time is fast.

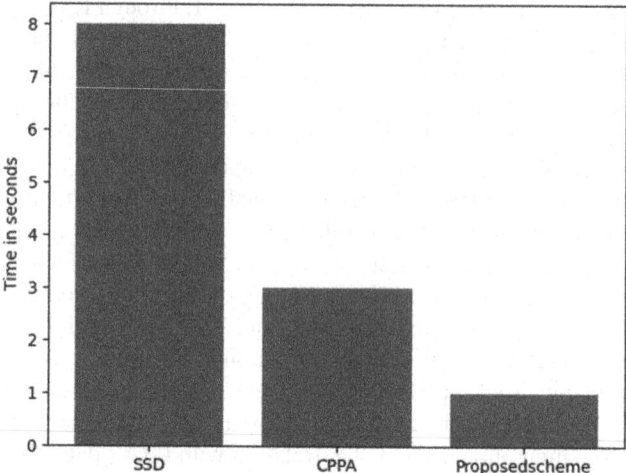

Fig. 29.4 Speed comparison between existing and proposed scheme

Source: Authors

5. Conclusion

Using the Trusted Authority-based Correlation model, this paper introduced novel unauthorized vehicle detection for autonomous vehicular systems in the VANET model. The TA or LTA has stored particular zone vehicle registration information, and the CTA stores all zone vehicle information in one place. Here the primary purpose of maintaining TAs or LTAs is to give a quick response compared to the earlier technique. In the earlier method, only one TA keeps all the vehicles' information; due to this, it gives poor performance when the vehicles or nodes increase it takes more time for status identification. The proposed idea is to maintain multiple TAs based on their zones and maintain all the registered nodes' information on a central level, CTA. In this, no need to correlate every time with CTA data, only when the vehicle information is not correlated with TA's data. The autonomous vehicular system in VANETS performs a key role in feature smart cities; these work similarly to wheel-based mobile robots. Finally, one can observe that the unauthorized vehicles detection system provides a secure and maintaining of the optimal path from starting point to the destination in VANET at a fixed bandwidth path. Thus, from the evaluation, it is vivid that the proposed work has attained the best performance in unauthorized vehicle detection in an autonomous vehicular system in VANET. So, as per the results, the vehicles' vehicle identification time is also increased. So, to overcome this situation, the proposed system used multiple TAs or LTAs(Local Trusted Authorities) instead of one TA.

The future direction is to control and hand over the nearest cyber crime station of unauthorized vehicles using the Internet of Things by sending the data packets to the unauthorized node in the autonomous vehicular system in VANET.

6. Acknowledgement

The authors would like to thank their University for providing facilities to complete this work.

References

1. A. Nanda, D. Puthal, J. J. P. C. Rodrigues and S. A. Kozlov, "Internet of Autonomous Vehicles Communications Security: Overview, Issues, and Directions," in IEEE Wireless Communications, vol. 26, no.4, pp.60-65, August 2019, doi:10.1109/MWC.2019.1800503

2. H. Zhong, B. Huang, J. Cui, Y. Xu, and L. Liu, "Conditional Privacy-Preserving Authentication Using Registration List in Vehicular Ad Hoc Networks," in IEEE Access, vol. 6, pp. 2241–2250, 2018, doi: 10.1109/ACCESS.2017.2782672.

3. Gupta, N.; Manaswini, R.; Saikrishna, B.; Silva, F.; Teles, A. Authentication-Based Secure Data Dissemination Protocol and Framework for 5G-Enabled VANET. *Future Internet* **2020**, *12*, 63. https://doi.org/10.3390/fi12040063

4. Muhammad Sameer Sheikh, Jun Liang, and Wensong Wang "Security and Privacy in Vehicular Ad Hoc Network and Vehicle Cloud Computing: A Survey" in Hindawi 1530-8669, https://doi.org/10.1155/2020/5129620,DO-10.1155/2020/5129620

5. J. A. Guerrero-Ibanez, S. Zeadally, and J. Contreras-Castillo, "Integration challenges of intelligent transportation systems with connected vehicle, cloud computing, and internet of things technologies," in IEEE WirelessCommunications, vol. 22, no. 6, pp. 122–128, December 2015, doi:10.1109/MWC.2015.7368833.

6. Zhong, H.; Huang, B.; Cui, J.; Xu, Y.; Liu, L. Conditional Privacy-Preserving Authentication Using Registration List in Vehicular Ad Hoc Networks. IEEE Access 2018, 6, 2241–2250. doi:10.1109/ACCESS.2017.2782672. [CrossRef].

7. Zhang, L., Wu, Q., Solanas, A. & Domingo-Ferrer, J. (2010). "A scalable, robust authentication protocol for secure vehicular communications." IEEE Transactions on Vehicular Technology, 59 (4), 1606-1617.

8. M. Gerla et al., "Internet of Vehicles: From Intelligent Grid to Autonomous Cars and Vehicular Clouds," IEEE World Forum on Internet of Things, 2014, pp. 241–46.

9. H. Hasrouny et al., "VANet Security Challenges and Solutions: A Survey." Vehic. Commun., vol. 7, 2017, pp. 7–20.

10. Zhang, L.; Hu, C.; Wu, Q.; Domingo-Ferrer, J.; Qin, B. Privacy-Preserving Vehicular Communication Authentication with Hierarchical Aggregation and Fast Response. IEEE Trans. Comput. 2016, 65, 2562–2574. doi:10.1109/TC.2015.2485225. [CrossRef].

11. F. Bonomi, "The Smart and Connected Vehicle and the Internet of Things," Invited Talk, Wksp. Synchronization in Telecommun. Systems, 2013.

12. Kenney, J.B. Dedicated Short-Range Communications (DSRC) Standards in the United States. Proc. IEEE 2011, 99, 1162–1182. doi:10.1109/JPROC.2011.2132790. [CrossRef].

13. Lee, S.; Park, J.; Gerla, M.; Lu, S. Secure Incentives for Commercial Ad Dissemination in Vehicular Networks. IEEE Trans. Veh. Technol. 2012, 61, 2715–2728. doi:10.1109/TVT.2012.2197031. [CrossRef].

14. Maxim Raya and Jean-Pierre Hubaux "Securing vehicular ad hoc networks" in IOSPress, Journal of Computer Security 15 (2007) 39–68.

15. He, D.; Zeadally, S.; Xu, B.; Huang, X. An Efficient Identity-Based Conditional Privacy-Preserving Authentication Scheme for Vehicular Ad Hoc Networks. IEEE Trans. Inf. Forensic Secur. 2015, 10, 2681–2691. [CrossRef].

16. Lee E-K, Gerla M, Pau G, Lee U, Lim J-H. Internet of Vehicles: From intelligent grid to autonomous cars and vehicular fogs. International Journal of Distributed Sensor Networks. September 2016.doi:10.1177/1550147716665500.

17. Cui, J.; Wei, L.; Zhang, J.; Xu, Y.; Zhong, H. An Efficient Message-Authentication Scheme Based on Edge Computing for Vehicular Ad Hoc Networks. IEEE Trans. Intell. Transp. Syst. 2018, 20, 1621–1632. doi:10.1109/TITS.2018.2827460. [CrossRef].

18. Liu, J.; Li, Q.; Cao, H.; Sun, R.; Du, X.; Guizani, M. MDBV: Monitoring Data Batch Verification for Survivability of Internet Vehicles. IEEE Access 2018, 6, 50974–50983. doi:10.1109/ACCESS.2018.2869543. [CrossRef].

19. Yang, F.; Wang, S.; Li, J.; Liu, Z.; Sun, Q. An overview of the Internet of Vehicles. China Commun. 2014, 11, 1–15. doi:10.1109/CC.2014.6969789. [CrossRef].

20. Liu, Y.; Wang, Y.; Chang, G. Efficient Privacy-Preserving Dual Authentication and Key Agreement Scheme for Secure V2V Communications in an IoV Paradigm. IEEE Trans. Intell. Transp. Syst. 2017, 18, 2740–2749.doi:10.1109/TITS.2017.2657649.

21. Tan, H.; Ma, M.; Labiod, H.; Boudguiga, A.; Zhang, J.; Chong, P.H.J. A Secure and Authenticated Key Management Protocol (SA-KMP) for Vehicular Networks. IEEE Trans. Veh. Technol. 2016, 65, 9570–9584. doi:10.1109/TVT.2016.2621354.

22. F. Sakiz and S. Sen, "A Survey of Attacks and Detection Mechanisms on Intelligent Transportation Systems: VANETs and IoV," Ad Hoc Networks, vol. 61,2017, pp. 33–50.

23. L.Wang et al., "Vehicular Sensing Networks in a Smart City: Principles, Technologies, and Applications,"IEEE Wireless Commun., vol. 25, no. 1, Feb. 2018, pp. 122–32.

24. D. J.Fagnant and K. Kockelman. "Preparing a Nation for Autonomous Vehicles: Opportunities, Barriers and Policy Recommendations." Transportation Research Part A: Policy and Practice, vol. 77,2015, pp. 167–81.

25. D. Puthal et al., "Fog Computing Security Challenges and Future Directions," IEEE Consumer Electronics Mag., vol. 8, no. 3, 2019.

26. Z. Su, Y. Hui, and Q. Yang. "The Next Generation Vehicular Networks: A Content-CentricFramework," IEEE Wireless Commun.vol.24, no. 1, Feb.2017, pp. 60–66.

27. Chen, M.; Tian, Y.; Fortino, G.; Zhang, J.; Humar, I. Cognitive Internet of Vehicles. Comput. Commun. 2018, 120, 58–70. doi:10.1016/j.comcom.2018.02.006. [CrossRef].

28. Ang, L.; Seng, K.P.; Ijemaru, G.K.; Zungeru, A.M. Deployment of IoV for Smart Cities: Applications, Architecture, and Challenges. IEEE Access 2019,7, 6473–6492. doi:10.1109/ACCESS.2018.2887076. [CrossRef].

Privacy and Security in Intelligent Devices

30

Pranayanath Reddy Anantula[1]

Associate Professor, Department of CSE,
Teegala Krishna Reddy Engineering College, Hyderabad, Telangana, India

B. Deevena Raju[2]

Asst Professor, ICFAI Tech School, ICFAI University,
Hyderabad, Telangana, India

Suthoju Girija Rani[3]

Assistant Professor, Neil Gogte Institute of Technology,
Hyderabad, Telangana, India

A. Manjula[4]

Assistant Professor, Teegala Krishna Reddy Engineering College,
Hyderabad, Telangana, India

Abstract: Intelligent devices, such as Internet of Things (IoT) devices, smart appliances, and other connected technologies, bring numerous benefits, but they also introduce privacy and security challenges due to their interconnected nature and the data they collect. Intelligent devices often collect a wide range of personal and sensitive data. The challenge is to ensure proper consent, data minimization, and secure storage to protect user privacy. Devices with location-tracking capabilities raise concerns about user location data being collected and potentially misused. Intelligent devices can become targets for hackers seeking to access sensitive information or gain control over the device. A compromised device could lead to data breaches, privacy violations, and even safety risks. Many IoT devices lack proper security measures, making them vulnerable to exploitation. Devices often come with default or weak passwords, which can be easily guessed or exploited by attackers. Multi-factor authentication and secure access controls are often lacking, making it easier for unauthorized users to gain access. Many devices have software vulnerabilities that can be exploited by attackers to gain control or access to sensitive data.

Keywords: Intelligent devices, Internet of things, Privacy

1. Introduction

Some devices lack a mechanism for regular software updates, leaving them exposed to known vulnerabilities. Devices may share data with third parties without users' explicit consent, leading to privacy violations. The extensive data collected by devices can be used to create detailed user profiles, potentially infringing on user privacy. Data sent to cloud servers for processing may be intercepted during transmission, risking unauthorized access. Users often have to trust third-party vendors to handle their data responsibly and securely. IoT devices may operate across multiple jurisdictions, making it challenging to comply with different data protection regulations.

Determining liability for security breaches involving intelligent devices can be complex, involving manufacturers, software developers, and users. Users may not fully understand the privacy settings of their devices, leading

[1]a.pranayanath@tkrec.ac.in, [2]deevenaraju@ifheindia.org, [3]girijaranis@gmail.com, [4]manjular95801@gmail.com

to unintended data sharing. Users may not be aware of the need to regularly update their devices' software for security reasons. If one device in a network is compromised, it can potentially compromise the security of other devices within the same ecosystem.

Addressing these challenges requires a multi-faceted approach involving manufacturers, developers, regulators, and users. Security-by-design principles, regular software updates, user education, strong encryption, and adherence to privacy regulations are crucial for ensuring the privacy and security of intelligent devices.

The "chain of trust" is a fundamental concept in security by design principles that ensures the security and integrity of a system, network, or device from its inception through its entire lifecycle. It establishes a sequence of trust relationships between different components, starting from the most foundational elements and extending outward. The goal of the chain of trust is to ensure that each element in the system is verified, authenticated, and secure before it interacts with or depends on other elements.

The chain of trust begins with a secure and trusted component known as the "root of trust." This is usually a hardware-based component, like a Trusted Platform Module (TPM) or a Secure Boot ROM. The root of trust is responsible for initiating the verification process and establishing trust for subsequent components. Each subsequent component in the system is verified and authenticated by the component that comes before it. This ensures that only trusted and authorized components are allowed to interact with the system. Components are often signed with cryptographic keys, and these signatures are used to verify their authenticity. If a component's signature doesn't match the expected value, it's considered untrusted or compromised. During system startup, the root of trust initiates a secure boot process. This process verifies the integrity of the bootloader, the operating system, and any other software components before they are executed. The chain of trust extends to both hardware and software components. It ensures that hardware components (e.g., CPUs, memory) and software components (e.g., firmware, applications) are authentic, unaltered, and secure. As the chain of trust extends to various levels of the system, it also involves encrypting sensitive data to prevent unauthorized access or tampering. Secure communication protocols ensure that data transferred between trusted components remains confidential and protected from eavesdropping or tampering. The chain of trust is not a one-time process. It involves continuous monitoring, regular updates, and patches to maintain the security and integrity of the system throughout its lifecycle.

The chain of trust concept is particularly important in scenarios where security risks are high, such as in embedded systems, Internet of Things (IoT) devices, and critical infrastructure. It helps prevent various attacks, including unauthorized access, data breaches, malware injection, and supply chain attacks. By establishing a strong and well-defined chain of trust, security by design principles ensure that all components in a system are verified, secure, and mutually trustworthy, contributing to a more resilient and secure computing environment.

2. Device Intelligence

Device fingerprinting is a process of collecting unique identifying information from a digital device that's used for identity validation, fraud prevention and digital advertising. Device intelligence solution protects against mobile and web attacks from emulators, botnets, hijacked devices, app cloners and more, while delivering accurate signals, device IDs and scores to boost detection. Device intelligence and data protection are closely intertwined concepts, especially in the context of the Internet of Things (IoT) and the increasing integration of intelligent devices into our lives as shown in Fig. 30.1. While device intelligence offers numerous benefits, it also introduces unique challenges related to data privacy and security. Intelligent devices often collect a wealth of data from their environment, users, and interactions. This data can include personal and sensitive information, raising concerns about user privacy. Implement strong data protection measures, such as data encryption, anonymization, and data minimization. Only collect the data necessary for the device's intended functionality, and provide clear privacy notices and choices to users. Device intelligence can lead to detailed user profiles being created based on collected data. Profiling can impact user privacy if not handled transparently. Ownership of data generated by intelligent devices can be unclear, especially when multiple parties are involved. The lifecycle of data collected by intelligent devices needs to be managed to ensure proper deletion when no longer needed.

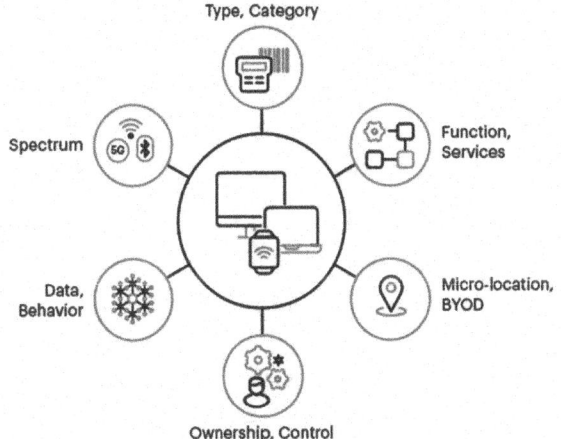

Fig. 30.1 Device intelligence systems [4]

Balancing the benefits of device intelligence with the need for robust data protection requires a comprehensive approach that incorporates privacy by design principles, regulatory compliance, user education, and ongoing monitoring. Organizations and manufacturers must prioritize user trust and privacy while reaping the advantages of device intelligence.

Device intelligence is unique in the sense that it represents a convergence of various technologies and capabilities that enable devices to operate with a certain level of autonomy, intelligence, and responsiveness. While individual technologies have existed for some time, it's the combination and integration of these technologies that give rise to the concept of device intelligence. While offering numerous benefits, device intelligence also introduces challenges related to data privacy, security, and ethical considerations. Addressing these challenges is crucial for its successful implementation. Device intelligence benefits from ongoing advancements in technologies such as sensors, AI, edge computing, and connectivity, which collectively contribute to its uniqueness. Device intelligence transforms devices from passive tools into active and responsive participants in our interconnected world. It empowers devices to understand, interpret, and react to their environment, leading to increased efficiency, improved decision-making, and new possibilities for innovation.

2.2 Edge Computing

Edge computing processes data locally and reduces traffic loads at scale to eliminate latency and improve speed. By computing data on local devices, it protects customer privacy, reduces the amount of data at risk and minimizes investment in infrastructure. Edge computing refers to the practice of

processing and analyzing data closer to the source of data generation, rather than sending all data to a centralized cloud or data center. This approach offers several advantages, including reduced latency, improved response times, and more efficient use of network resources. When it comes to data protection, edge computing can play a significant role in enhancing security and privacy as shown in Fig. 30.2

While edge computing offers significant advantages for data protection, it's important to note that securing edge devices themselves is critical. Strong security measures, regular updates, and proper access controls are essential to ensure the overall security of edge computing ecosystems. Combining edge computing with robust cybersecurity practices creates a more resilient and secure environment for data protection.

3. Methodology

3.1 White Box Encryption

White box encryption" and "digital signatures" are two cryptographic techniques used to enhance the security of data and verify the authenticity of digital information. Whitebox encryption refers to a cryptographic technique where encryption is performed in such a way that the encryption key is never exposed, even during the encryption and decryption processes. This is particularly important in scenarios where applications or systems are running in potentially insecure environments, like mobile devices or client-side applications, where the encryption key could be vulnerable to reverse engineering or other attacks. In whitebox encryption, the encryption algorithm and the encryption key are combined in a way that makes it extremely difficult to extract the actual encryption key. The goal is to prevent attackers from reverse-engineering the encryption process to retrieve the key, even if they have full access to the algorithm implementation.

Whitebox encryption has applications in protecting sensitive data within applications and devices, especially when those applications might run in environments that cannot be fully trusted. It ensures that even if an attacker gains access to the application's code or memory, they won't be able to extract the encryption key.

Digital Signatures: Digital signatures are cryptographic techniques used to verify the authenticity and integrity of digital messages, documents, or data. A digital signature involves a process where a sender (signer) uses their private key to create a unique signature for a piece of data. The recipient (verifier) can then use the sender's public key to verify the signature and ensure that the data hasn't been tampered with and that it indeed came from the expected sender. Digital signatures provide non-repudiation, meaning that the sender cannot later deny having sent the data, as the signature provides proof of their involvement. Digital

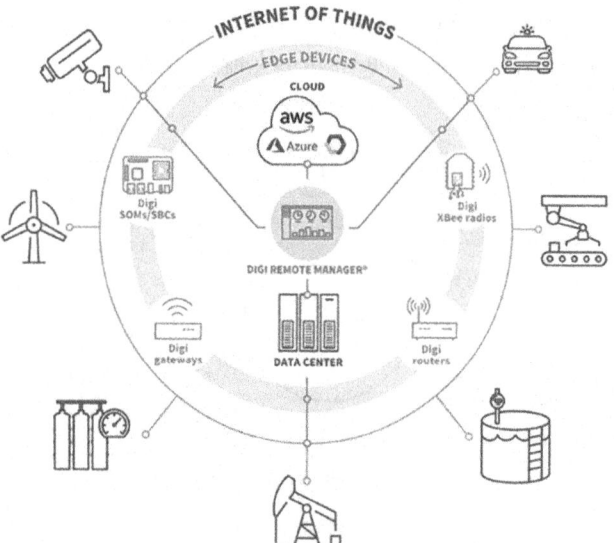

Fig. 30.2　Edge computing in IoT [8]

signatures are widely used in secure communication, digital contracts, and electronic transactions to ensure data integrity and authenticity.

3.2 Combining White Box Encryption and Digital Signatures

These two techniques can complement each other in creating a secure environment for data protection and communication. For instance, a whitebox encryption approach can be used to secure sensitive data within applications, ensuring that the encryption keys remain protected even in potentially insecure environments. Digital signatures can be used to verify the authenticity of messages or data transmitted between these applications, ensuring that the sender's identity is verified and the data hasn't been tampered with.

Using both techniques together can provide a robust security mechanism, especially in scenarios where data confidentiality, integrity, and authenticity are critical. However, it's essential to implement these techniques correctly and to adhere to best practices in cryptographic implementation to ensure their effectiveness as shown in Fig. 30.3

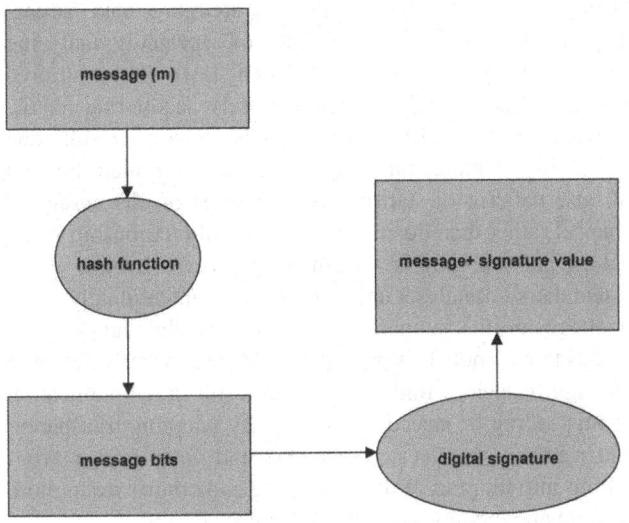

Fig. 30.3 Process of combining white box encryption and digital signature

Source: Aurhors

3.3 Data Fudging in Intelligent Systems

Data fudging, also known as data manipulation or data cooking, refers to the practice of altering or manipulating data in a dishonest or deceptive manner to achieve a desired outcome or to present a misleading impression. This can be done intentionally to distort the results of a study, experiment, or analysis for various reasons, such as to support a particular hypothesis, meet targets, secure funding, or gain a competitive advantage. Presenting only the data that supports a particular conclusion while ignoring or omitting contradictory data.

Cherry picking in the context of data refers to the practice of selectively presenting or using only certain data points or results that support a particular argument, while ignoring or omitting other data that might contradict or weaken that argument. This can lead to a biased or distorted view of the overall picture and is generally considered unethical and misleading, especially in scientific research, data analysis, and reporting. Cherry picking can be a form of data fudging or manipulation because it involves intentionally choosing data that supports a preconceived conclusion, rather than conducting a comprehensive and objective analysis of all available data. It's a manipulation tactic that can mislead the audience or readers by presenting incomplete or skewed information. In reputable research and data analysis, it's important to adhere to principles of transparency, objectivity, and integrity. All relevant data should be considered and presented, even if it contradicts the expected outcome or hypothesis. Selectively presenting data to bolster a specific viewpoint while ignoring contrary evidence undermines the credibility of the analysis and can harm the overall understanding of a topic.

3.4 Cyber Security

Cyber security plays a crucial role in preventing and mitigating data fudging, which involves the manipulation or alteration of data for fraudulent purposes. Effective cybersecurity measures help safeguard data integrity, confidentiality, and availability, reducing the risk of unauthorized data manipulation. Cybersecurity measures, such as data encryption and digital signatures, help ensure the integrity of data. Encryption prevents unauthorized access and modification during data transmission and storage. Digital signatures provide a way to verify the authenticity and integrity of data, making it difficult to manipulate without detection. Proper access control mechanisms limit data access to authorized individuals only. By controlling who can access and modify data, organizations can reduce the risk of data manipulation by unauthorized parties.

Intrusion Detection and Prevention Systems (IDPS): These systems monitor network traffic and system activities for signs of suspicious behavior or unauthorized access. IDPS can detect and prevent data manipulation attempts by identifying abnormal activities. Firewalls filter incoming and outgoing network traffic, helping prevent unauthorized access to sensitive data. Strong network security reduces the chances of external actors manipulating data. Robust logging and auditing mechanisms record and track all data interactions. This enables organizations to trace back any unauthorized modifications and identify potential data fudging incidents. Ensuring that software and applications are developed with security in mind helps prevent vulnerabilities that could be exploited for data manipulation. Regular security assessments and code reviews are essential. By implementing a comprehensive cybersecurity strategy, organizations can significantly reduce the risk of data fudging and maintain the integrity and trustworthiness of their data.

3.5 Data Protection in India

Data protection and addressing data fudging are important issues globally, including in India. In India, these topics are primarily governed by the Personal Data Protection Bill (PDPB) and existing regulations concerning data integrity and security. The Personal Data Protection Bill (PDPB) is a comprehensive legislation aimed at regulating the collection, processing, and storage of personal data in India. The bill outlines principles and provisions for data protection. The bill emphasizes obtaining informed and explicit consent from individuals before collecting and processing their personal data. Data collectors are required to limit the collection of personal data to what is necessary for the specified purpose. Personal data should be collected and processed only for the purpose it was collected, and any further processing must be compatible with that purpose. Organizations that collect and process personal data are considered data fiduciaries and are responsible for ensuring data protection. The bill introduces provisions for certain categories of sensitive personal data to be stored and processed only within India. Organizations are required to promptly report data breaches to the Data Protection Authority and affected individuals. Individuals have the right to request the erasure of their personal data under certain circumstances. The bill establishes an independent authority to oversee and enforce data protection regulations.

3.6 Addressing Data Fudging

Ethical Standards & Regulatory oversight : Academic and research institutions, as well as professional organizations, establish ethical standards that researchers and professionals are expected to follow. These standards emphasize accurate and transparent reporting of data. Regulatory bodies, such as the Data Protection Authority under the PDPB, can play a role in investigating cases of data manipulation, particularly if it involves personal data. Organizations and individuals found guilty of data manipulation or fudging could face penalties, including fines and legal actions, depending on the severity of the offense. Encouraging individuals to report instances of data manipulation while protecting them from retaliation is important for uncovering unethical practices. Ensuring transparency in research methodologies and making data and analysis methods available for independent audits can help detect and prevent data manipulation. Promoting awareness about the importance of data integrity and the consequences of data fudging can deter unethical practices. It's important for individuals, organizations, and regulators to collaborate

to prevent data manipulation and protect the integrity of data, especially as data-driven decision-making becomes more prevalent across various sectors in India.

Unethical practices in data fudging involve deliberately manipulating or distorting data to mislead, deceive, or gain an unfair advantage. Publishing the same or substantially similar data in multiple places without proper disclosure, which can artificially inflate the apparent significance of the findings. Presenting data as statistically significant when the significance level was not met, or misrepresenting p-values to create a false impression of validity. Including individuals who did not contribute to the research as authors or omitting individuals who did contribute, often for political or promotional reasons. These unethical practices undermine the credibility of research, analysis, and decision-making processes. They erode trust, waste resources, and can have significant negative consequences for individuals, organizations, and society as a whole. Upholding ethical standards, transparency, and accountability is essential to maintaining the integrity of data and research.

Data snooping

Data snooping, also known as data dredging, data fishing, or p-hacking, refers to the practice of repeatedly analyzing data or trying out multiple statistical tests until a desired result is obtained, without appropriately accounting for the increased likelihood of obtaining false positive results due to multiple comparisons. This can lead to the identification of false patterns or statistically significant results purely by chance, rather than due to any meaningful relationship in the data. Data snooping is a form of confirmation bias, where researchers or analysts unintentionally or intentionally focus on the results that support their hypotheses while ignoring the results that do not. This practice can lead to overestimation of the significance of findings and can result in the publication of misleading or inaccurate results. By adopting transparent and rigorous practices, researchers and analysts can avoid falling into the trap of data snooping and contribute to more reliable and credible research outcomes.

3.7 Device-based Data Protection

Device-based data protection refers to the implementation of security measures and protocols directly on the devices that store or process sensitive data. The goal is to ensure that data remains secure even if the device is lost, stolen, or compromised. This approach involves using encryption, access controls, authentication mechanisms, and other security features to safeguard the data at the device level.

For example, smartphones often utilize device-based encryption, where the data stored on the device is encrypted using keys tied to the device's hardware. This means that even if someone gains physical access to the device's storage, they would need the proper authentication credentials to decrypt and access the data as shown in Fig. 30.4

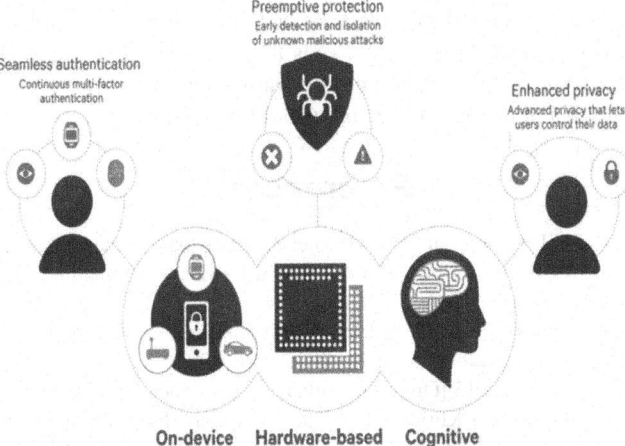

Fig. 30.4 Device based data protection at various levels [6]

Custom-based data protection

Custom-based data protection involves tailoring data protection measures to the specific needs and characteristics of an organization, system, or application. It goes beyond generic or standardized security solutions to create a custom security strategy that aligns with an organization's unique data protection requirements, risk tolerance, and technological infrastructure.

Customized access controls: Implementing access control policies and permissions that match the organization's structure and data usage patterns. Categorizing data based on its sensitivity and applying different protection measures based on these classifications. Developing security measures that address the specific vulnerabilities and threats faced by an organization's systems and applications. Customizing

data protection practices to meet specific legal or regulatory requirements that apply to the organization's industry or jurisdiction. Integrating data protection solutions seamlessly with existing IT systems and workflows. Both device-based and custom-based data protection approaches play important roles in safeguarding sensitive information. Device-based measures help secure data at the source, while custom-based strategies ensure that security practices align with an organization's unique needs and risks. Often, a combination of these approaches is used to create a robust and effective data protection framework. Unsupervised Machine learning for fraud detection as shown in Fig. 30.5

Unsupervised machine learning is indeed essential for fraud detection, particularly in scenarios where the characteristics of fraudulent activities are not well-defined or when new and previously unseen types of fraud are emerging. Unsupervised machine learning techniques play a crucial role in identifying anomalous patterns and detecting fraud without requiring labeled training data. Unsupervised machine learning algorithms are well-suited for anomaly detection, which is crucial for identifying unknown or evolving fraud patterns. These algorithms can learn the normal behavior of a system or dataset and flag instances that deviate significantly from this norm. In fraud detection, labeled training data (historical fraud instances) might be limited or outdated due to the dynamic nature of fraud. Unsupervised methods do not require labeled data; they learn from the overall dataset's patterns to identify anomalies. Fraudsters continually evolve their tactics. Unsupervised algorithms can adapt to new patterns without constant retraining, making them suitable for detecting novel and previously unseen fraud. Unsupervised methods are less likely to miss new or rare fraud patterns, as they do not rely on predefined rules or models based on historical fraud cases. Fraud can involve complex interactions and subtle variations. Unsupervised learning can capture

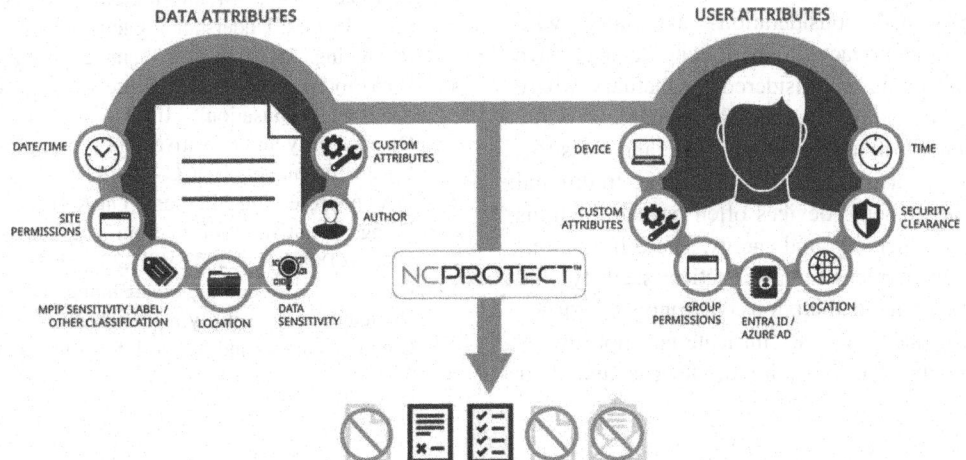

Fig. 30.5 Custom based data protection in NC project [14]

these intricate relationships that may not be easily defined in advance. Unsupervised methods base their decisions on data patterns, reducing the potential for human bias that could be present in rule-based systems. Unsupervised algorithms can quickly flag potentially fraudulent activities, allowing for timely investigation and mitigation. These algorithms can be set up to continuously monitor data streams and identify anomalies in real-time, providing rapid responses to ongoing fraudulent activities.

Popular unsupervised techniques used in fraud detection include:

- *Clustering algorithms:* These group similar data points together and can help identify clusters that deviate from the norm, potentially indicating fraud.
- *Principal component analysis (PCA):* PCA can reduce data dimensionality while retaining important information, making it useful for spotting anomalies.
- *Isolation forest:* This algorithm builds decision trees to isolate anomalies, making it particularly efficient for identifying outliers in large datasets.
- *Autoencoders:* A type of neural network, autoencoders can learn the underlying data structure and detect deviations from that structure, making them effective for anomaly detection.
- *One-class SVM:* This algorithm creates a decision boundary around the majority of the data, identifying points that fall outside this boundary as anomalies.

In conclusion, unsupervised machine learning techniques are indispensable in the field of fraud detection due to their ability to identify emerging patterns and anomalies, reduce false negatives, and adapt to evolving fraud tactics without requiring labeled training data.

4. Conclusion

Unique challenges and considerations associated with ensuring security and privacy in intelligent devices have been narrated. The home is considered a sanctuary where privacy is expected to be respected. Intelligent devices, while offering convenience, often collect personal and sensitive data, raising concerns about user privacy. Unlike traditional digital systems, smart home devices often have limitations in processing power, memory, and energy. These limitations pose challenges in implementing effective security and privacy measures within the ubiquitous environment. Privacy concerns in Intelligent devices are intricate and not always immediately apparent. The diversity of devices and their interactions can lead to unexpected privacy risks. Despite challenges, ensuring security and privacy in intelligent devices is a crucial task that must be prioritized. Comprehensive approach to security and privacy in intelligent devices, addressing challenges from device limitations to complex privacy concerns. It underscores the importance of ongoing research, collaboration, and innovative solutions to create a secure and privacy-respecting smart home environment.

References

1. D. J. Cook et al., "MavHome: An agent-based smart home," IEEE International Conference on Pervasive Computing and Communications, San Diego, CA, USA, pp. 521–524, 2003.
2. N. King, "Smart home - A Definition," Milton Keynes: Intertek Research and Testing Centre, 2003.
3. Statista, 2015 [Online]. Available: https://goo.gl/89rRIa
4. August and Xfinity, "The Safe and Smart Home: Security in the Smart Home Era," 2016 [Online]. Available: http://goo.gl/UGWb5Z
5. V. Srinivasan et al., "Protecting your daily in-home activity information from a wireless snooping attack," 10th international conference on Ubiquitous computing, pp. 202–211, 2008.
6. B. Ur et al., "The current state of access control for smart devices in homes," Workshop on Home Usable Privacy and Security, 2013.
7. S. Notra et al., "An experimental study of security and privacy risks with emerging household appliances," IEEE Conference on Communications and Network Security, pp. 79–84, 2014.
8. V. Sivaraman et al., "Network-level security and privacy control for smart-home IoT devices," Wireless and Mobile Computing, Networking and Communications, pp. 163–167, 2015.
9. T. D. P. Mendes et al., "Smart home communication technologies and applications: Wireless protocol assessment for home area network resources," Energies, vol. 8, no. 7, pp. 7279–7311, 2015.
10. C. Debes et al., "Monitoring Activities of Daily Living in Smart Homes: Understanding human behavior," IEEE Signal Processing Magazine, vol. 33, no. 2, pp. 81–94, 2016.
11. Data Protection and Privacy Preservation using Anonymisation and Pseudonamisation" – IIMT – New Delhi.
12. Data Security and Sensitive data protection using Privacy by Design technique – BDCC – 2019.
13. Privacy-preserving Microdata On A Tabular Data Publishing Using Additive Noise Approach Page No: 311-316- DOI:20.18001.GSJ.2022.V9I1.22.38503
14. C. Lee et al., "Securing smart home: Technologies, security challenges, and security requirements," IEEE Conference on Communications and Network Security, pp. 67–72, 2014.

Accident Analysis and Detection System Based on Traffic Surveillance Videos

31

Ramesh Babu Palepu[1]

Associate professor (Department of Computer Science and Engineering),
Amrita Sai Institute of Science and Technology Paritala, Andhra Pradesh, India

Shaik Asleem[2], K. Lakshmi Padmavathi[3]

M.Tech Scholor (Department of Computer Science and Engineering),
Amrita Sai Institute of Science and Technology Paritala, Andhra Pradesh, India

Abstract: Road accidents become very critical problem in India. Analyzing proper causes to accidents is very difficult and challenging task. This review will test every position on a presentation board for man-made consciousness. It will zero in on the issue of obviously and certainly finding and sorting out what traffic occasions are brought about by following cameras. Most importantly, broken vehicles do very well in the motion interaction field (MIF) way, which depends on the nun cover of various program parts that make a difference. Second, the outcomes will The Derisive v3 model is utilized to find out where the sunk boats are. A gradual packing equation is utilized to duplicate the connected courses and find out where the vehicle was going before the accident. In conclusion, a perspective change is utilized to make the course longer, which assists officials with coming to better choices. The UFIR (free limited motivation reaction) technique is utilized to sort out how quick a gadget is. This doesn't need a ton of desk work. The upward perspective on the speed and the spot of the mishap can be utilized to investigate the auto collision. Finally, a Huawei computerized reasoning conversation or music board called the HiKey970 is explored. This board was utilized to make different surveys as a whole. This test is intended to show how well and how educated the recommended strategy truly is. The board responsible for the trial of capacity gets a couple of reports with thoughts for how to fix issues. The right driving bearings are not difficult to track down, and the circumstances are interesting.

Keywords: *YOLO V5*, Speed estimation, Accident detection, Target tracking, Unbiased finite impulse response (UFIR) filter, Vehicles

1. Introduction

Over the long run, it has become more essential to utilize groundbreaking thoughts while watching traffic. The traffic management center (TMC) is most affected by what people do in the accident zone. Despite the fact that it has a few issues, human judgment is generally dependable.

Table 31.1 shows road accidents happened in India during 2021 as per Government of India (MORTH). Because it is difficult to view each traffic accident in the city from a single perspective, people may not assist the injured enough.

Table 31.1 Road accidents in India 2021 statistics

Category	Accidents	Percentage
Pedestrians	17,113	14.36
Bicycles	3,009	2.53
Two wheelers	52,416	43.99
Auto Rickshaws	5,360	4.50
Cars,Taxis,Vans	25,431	21.35
Trucks/Lorries	12,075	10.13
Buses	3,738	3.14

[1]asistithod@gmail.com, [2]shaikasleem2000@gmail.com, [3]padhunageswar@gmail.com

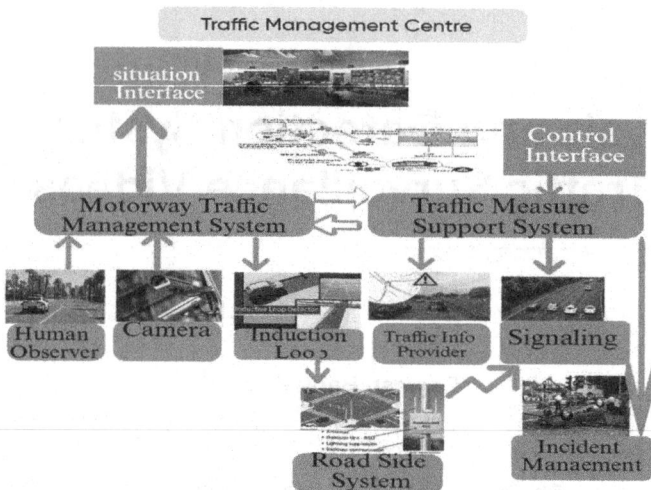

Fig. 31.1 Traffic scenario with elements

In any case, it very well may be difficult to sort out why an auto collision occurred by hand due to the fact that getting the speed and heading of the vehicle from a reconnaissance video is so difficult. Therefore, traffic event recognition and evaluation tools are crucial. Figure 31.1 shows traffic management elements.

In recent years, three different approaches have been used to build vision-based accident region structures: displaying how vehicles are linked together, concentrating on how vehicles move, and managing traffic streams. [1].The basic method uses traffic rules from various educational books to replicate standard traffic plans [2-4]. When a car doesn't go according to plan, it's a disaster [5-7]. But it's hard to tell what the effects are without crash course data. The accompanying technique checks for mishaps by setting end focuses for vehicle development, for example, speed, speed increment, and space between two vehicles [8-10]. Along these lines, every vehicle ought to be continually watched. Along these lines, the strategy is normally not exceptionally exact during busy time. The canny driver model [12] and the public limit model [11] are both utilized by the third strategy to make sense of how instruments cooperate. This innovation needs a ton of testing since it can find crashes that happen when the speed of a vehicle changes [13-14]. Yet, its capacity to engage isn't awesome.

2. Literature Review

2.1 Surveillance Using Video Analytics: Theory and Application

The TV test has gone through a lot of correction throughout the beyond couple of years, which is practically indistinguishable from the opportunity to peruse what occurs in a show in view of the quantity of cameras utilized. Actually have real

appreciation structures had the choice to isolate irksome events in isolation, and this study is at this point going on. TV handles from different cognizance cameras from one side of the planet to the other are not really investigated constantly. This makes it challenging for the public position to ponder issues like irritation, horrendous approach to acting, or people in a predicament who need help and counter, which are issues in the relationship. Thus, this is a critical issue. Understudies can rapidly find out about this present reality science behind present event programs thanks on these tracks.

2.2 Part II of the Article, "Using the Visual Intervention Influence of Pavement Marking for Rutting Mitigation," Discusses the Time of the Visual Intervention Using the Finite Element Method

As per a laborer who was tuning in, the capacity as seen with eyes mediation has major areas of strength for a toward properlingy, brings down progress wheel tracks, lessens pressure from the gathering of center pushes, and just somewhat increments (To a limited extent I). This study upholds a void out estimate technique in light of a pushed part model and uncovers a secretive system for expanding improvement rates. It similarly gives a three-step system for settling banters with the help of an expert ref and a slight drop in push-in. The make a space distortion rate bend is used to figure out measure of time its expectation for three different known chaotic top cycles to settle down. The immense news is reliably about nothing. The debilitating top of SUPERPAVE has shown ensure as another attack gadget, yet AC's savage top is especially taught. Moreover, the examination found that security from shape change is on the other hand associated with what amount of time it requires for a shape change to forge ahead toward the accompanying stage (fixed state). The attack of the flat grade cut happens more rapidly than the intrusion of the level inclination piece when a dim top happens likewise. During a discussion stage, the faint top enraged top's help past may in like manner work on by 16-31%.

2.3 Synergies Between Distributed Energy Resources and Electric Urban Transportation Systems in Smart Cities

The most energy is consumed by structures and transportation plans in urban communities. These plans (travel and the workplace) are sufficiently unique to be observed. Nevertheless, their undertakings to participate are habitually disregarded, which keeps them away from getting the benefits of composed cooperation and weight up. Taking into account both private and public sorts of energy-viable travel, for instance, energy-useful vehicles and the underground train, this study proposes one more technique for finding the most vital in rank shift and coordinating distributed energy resources (DER) in a baffling space. The vital point of convergence of this survey, which is correct, is on the critical benefits of this proficient social event. People envision that the lights used by the public vehicle office will concede the horrendous power that will be used in the batteries of electric vehicles (EVs) when it could have been used for extra trains or the real EV. Considering information from a Madrid metro line and the classified locale, several essential looks have been made. According to the data that was assembled, fundamental use saves a lot of energy across the whole design, yet the metro framework saves the most power.

2.4 In Traffic Surveillance Footage, the Motion Interaction Field can be Used to Detect Accidents

This piece shows a superior way to deal with figure out what caused a car collision by looking at the association of moving parts. Since the lengthy game plan for bringing thing exchanges was set off in one way, water waves ended up with one more exciting article on the wide experiences. With the Motion Interaction Field (MIF), Gaussian parts are used in a field construction to show how the investigation of the water surface works. By using the symmetric bits of the MIF, we can recognize traffic events and limit them without effective financial planning energy testing vehicle following. As demonstrated by fundamental news, our development works better contrasted with substitute approaches to finding and rank car crashes.

2.5 Modeling Scene Activities from Event Linkages and Global Rules to Connect the Past, Present, and Future

This study is about the main concerns that impact practices after some time and how to find them in tangled visual settings. In this way, we've thought about another point model that considers the two most convincing things that make these things happen: (1) Which things could happen that aren't absolutely for all time laid out because of how open overall scene articulations are; (2) Choices are made that blend development sprays that preceded brief pauses in light of the fact that the climate is so close. A matched irregular variable is utilized to interface these pieces during the probabilistic age interaction to figure out which of the two standards is expected for each activity. The limit of each model is proposed promoting a slash Gibbs grade conscription method. A piece of the datasets in the item direct that the model can separate between short eras at various scales: Setting-level offset interest in Markovian help and new friendship between practices that maybe used to envision the one knowledge will come to pass additional and how much rest will support entirety amount to an adequate awareness of the background's vital type.

2.6 For Video Mining Behavior, a Markov Clustering Topic Model

This test perceives how well open spot program video can be used. A staggering Markov Clustering Topic Model (MCTM) improves concerning accuracy, content, and assessment capacities than existing Surprising Bayesian Connection models (like Well) and Bayesian subject models (like Dormant Dirichlet Part). By partitioning visual times into figures out, these exercises into overall ways to deal with overseeing acting, and thereafter interacting these overall ways to deal with overseeing acting across time, our procedure shows critical areas of power for astonishing. A falling Gibbs sampler and a web based Bayesian assessment measure are made for independent learning with anonymous getting sorted out information for dynamic scene examination and ceaseless video information mining. Autonomous learning of vital background models shows the model's capacity, show approaches to acting and organized inadequacies by pursuing occurrences in three puzzling and gathered public scenes.

2.7 An Apparatus for Studying Statistical Motion Patterns

Improvement plan estimate is an inconceivable method for pregnant attitude and spot belongings that aren't average. Current approaches to looking at an improvement plan rely upon set conditions where things move in obvious ways. It's shocking that you can make game plans for object improvement that instantly offer scene information. We recommend making single information occasion plans for departure locale and direct assumption as a technique for welling support different parts in this spot thing. In the accompanying approach to acting, wool k-proposes judgment is used quickly and circumspectly to move closer assessment pixels. Each gathering centroid in the picture is associated with a drawing object if nearer view pack centroids are made and anticipated. During the most well-known approach to making a learning improvement plan, headings are consistently organized with geographical and requested information. A progression of Gaussian spreads deals with every development arrangement. Certifiable techniques are used to separate among peculiarities and speculation considering how scholastics have expected advancement previously. Our strategy was investigated using genuine picture groupings from a shut district and fake traffic. The results of the tests show that the constant assessment is reliable, that it is significant for learning development models, and that the assessments for anomaly undeniable insistence and lead question limit strong regions for are.

3. Methodology

Several looks found by deep learning are shown for seeing new boat influences. For these systems to work, the psyche ought to be facilitated in a tangled way and ready with a lot of data. Nevertheless, if you don't have even the remotest clue how to plan and have to pay a ton for affiliation, it is difficult to get these contemplations under way. Figure 2 shows system architecture followed in present study. Since there are more observational pictures of traffic, it is similarly more steadily to use a connected with framework to find and separate events all over the city. A gadget that capabilities as a wise spread accomplice should be available on each city block. Thusly, the essential device is a little structure contingent fitting contraption.

Disadvantages

1. Nonetheless, numerous establishments will before long should be rearranged in light of high treatment costs and an absence of an arranging document.
2. At the point when there is additional information from traffic following, it gets more earnestly for a solitary framework to find and break down occasions across the entire city.

In this section, we talk about how to recognize sadness and how artificial intelligence-produced voice or music recordings might help us deal with it. A model called a "motion interaction field" (MIF) can be used to grasp and circle back to traffic events right away. We utilize different degrees of get-together related to a Fundamental pull out all the stops v3 model to decide the vehicle's pre-mishap direction. We split the data appropriately before using unbiased finite impulse response (UFIR) filtering and direct shift to determine how quickly the impact occurred and where it hit. We additionally took a gander at the construction of the plan utilizing the Huawei computerized reasoning element board HiKey970.

Advantages

1. A review is finished to sort out how much the arrangement is worth and the way in which well it tends to be utilized. This is finished with a Huawei AI recording or visit board called HiKey970, which is utilized to direct the past evaluations.
2. Reality from an assortment of catastrophe films is blended in with the discourse or music recording. After a mishap is found, similar driving headings are given.

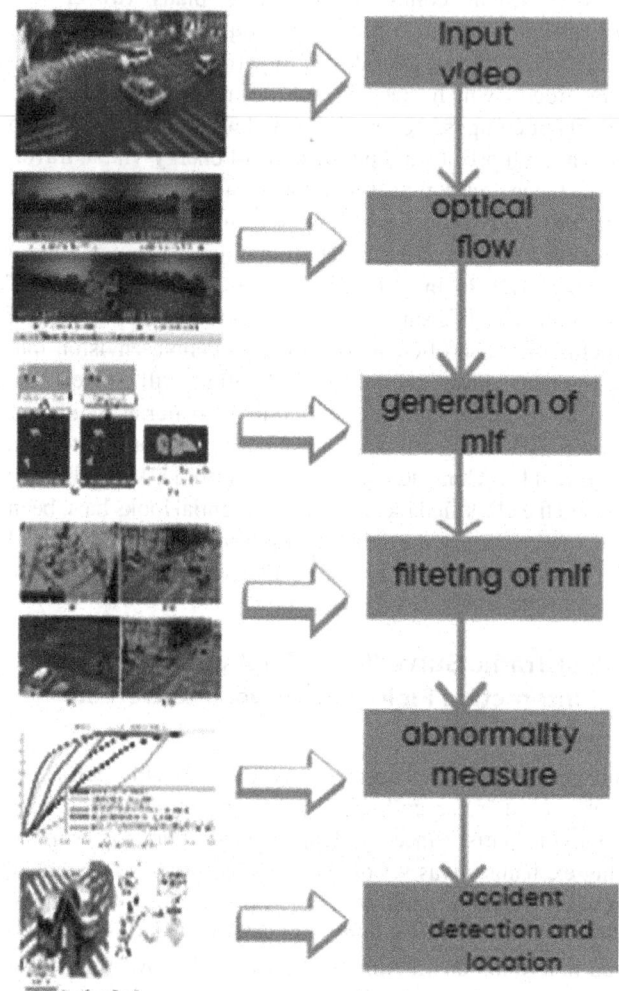

Fig. 31.2 Structure of the system

Modules

We created the modules that we really cared about in order to accomplish the project entirely.

- Examining the data: This part will be utilized to place data into the construction.
- Utilizing this device, information will be checked and watched out for.

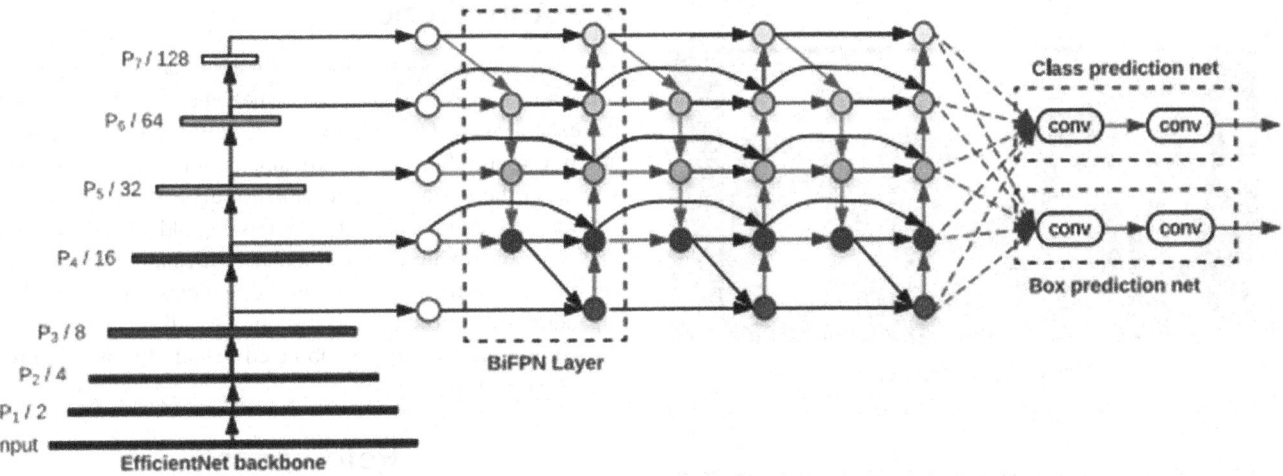

Fig. 31.3   YOLOv5 architecture

- The data in this illustration will be isolated into train and test packs.
- Step by step instructions to make a YOLOV5 rendition is shown.
- Client information exchange and login: To continue to peruse this story, you should initially enter and afterward sign in.
- At the point when this program is utilized, presumptions will be made.
- The end number that is supposed to be arrived voluntarily be shown.

4. Implementation

4.1 Algorithms

YOLOV5

"You only look once," or "YOLO," is an odd approach to placing pictures into gatherings. There are sure things that each power test attendant absolute requirement. Simply risk is one of the most widely recognized perspectives and understanding validation orders since speed and truth are so engaging. Models known as "You Only Look Once" (YOLO) are quite often utilized as excellent substance for "homicide country." Just go for it separates connections and investigates grids, and everybody understands what's inside. They make it conceivable to utilize truth streams as a solid wellspring of thought ID.

Figure 31.3 shows architecture of YOLO V5. A Convolutional Neural Network (CNN) Plan is utilized in the YOLOv5 Design. The head, neck, and spine are the most crucial

components. In the Spine, CSPNet is utilized to haul highlights out of crude pictures. To make the Pyramid part, the Neck is utilized.

Loss function

Three outputs are provided by YOLOv5, including the classes of the detected objects, their bounding boxes, and the objectless scores. To calculate the class loss and the objectness loss, BCE (Binary Cross Entropy) is used. The location loss is computed using CIoU (Complete Intersection over Union) loss. The following equation provides the final loss formula:

$$Loss = \lambda 1 Lcls + \lambda 2 Lobj + \lambda 3 Lloc$$

5. Experimental Results

By using YOLO V5 algorithm, images are run in GUI portal and output predications received as shown in Figs 31.4, 31.5 and 31.6.

Fig. 31.4   Prediction result 1

Fig. 31.5 Prediction result 2

Fig. 31.6 Prediction result 3

6. Conclusion

As a feature of this review, a method for finding and break down observational movies of vehicle crashes was made. Finding crashes on film has always been done using the MIF model method. Additionally, a YOLOv5 model was used to locate damaged autos. Thirdly, the bearings were discovered using the changing levels packing technique prior to the event. By changing the perspective, the standards were displayed on a vertical picture so the traffic police's free course could be arranged. Despite the fact that UFIR division was utilized to prevent the vehicle from moving with a specific goal in mind, it isn't completely fixed. In this way, the upstanding influence point and consistent speed were utilized to copy an accident. A Huawei PC-based mindfulness show board called HiKey970 was utilized to sort out every one of the figures for a stuff practice test. As information, a film of how individuals witnessed an occasion was displayed to the training board. The event was properly acknowledged, and the primary vehicle routes were recorded. At 2.60 GHz, the HiKey970 was 28.85 percent quicker than the Intel Center i7-9750H and 45.72 percent quicker than the Intel Center i7-9750H.

7. Future Work

Yet, there are a couple of issues that should be fixed in the following stages. At the point when the vehicle is halted, the exactness of the evidence could be further developed by attempting an alternate significant learning model. Second, whenever there is a slim likelihood of understanding reports, certain image improvement studies should be possible to work on the exhibition of mishap affirmation in various conditions. Third, the labels on the vehicles might be taken a gander at for more data. Later on, we will focus harder on controlling the way that a robotized vehicle follows and the site of an assault.

References

1. C. Regazzoni, A. Cavallaro, Y. Wu, J. Konrad, and A. Hampapur, "Video analytics for surveillance: Theory and practice," IEEE Signal Process. Mag., vol. 27, no. 5, pp. 16–17, Sep. 2010.
2. X. Zhu, Z. Dai, F. Chen, X. Pan, and M. Xu, "Using the visual intervention influence of pavement marking for rutting mitigation— Part II: Visual intervention timing based on the finite element simulation," Int. J. Pavement Eng., vol. 20, no. 5, pp. 573–584, May 2019.
3. C. F. Calvillo, A. Sánchez-Miralles, and J. Villar, "Synergies of electric urban transport systems and distributed energy resources in smart cities," IEEE Trans. Intell. Transp. Syst., vol. 19, no. 8, pp. 2445–2453, Aug. 2018.
4. K. Yun, H. Jeong, K. M. Yi, S. W. Kim, and J. Y. Choi, "Motion interaction field for accident detection in traffic surveillance video," in Proc. 22nd Int. Conf. Pattern Recognit., Aug. 2014, pp. 3062–3067.
5. J. Varadarajan, R. Emonet, and J. Odobez, "Bridging the past, present and future: Modeling scene activities from event relationships and global rules," in Proc. IEEE Conf. Comput. Vis. Pattern Recognit., Jun. 2012, pp. 2096–2103.
6. T. Hospedales, S. Gong, and T. Xiang, "A Markov clustering topic model for mining behaviour in video," in Proc. IEEE 12th Int. Conf. Comput. Vis., Sep. 2009, pp. 1165–1172.
7. W. Hu, X. Xiao, Z. Fu, D. Xie, T. Tan, and S. Maybank, "A system for learning statistical motion patterns," IEEE Trans. Pattern Anal. Mach. Intell., vol. 28, no. 9, pp. 1450–1464, Sep. 2006.
8. S. Sadeky, A. Al-Hamadiy, B. Michaelisy, and U. Sayed, "Real-time automatic traffic accident recognition using HFG," in Proc. 20th Int. Conf. Pattern Recognit., Aug. 2010, pp. 3348–3351.
9. Y.-K. Ki, "Accident detection system using image processing and MDR," Int. J. Comput. Sci. Netw. Secur., vol. 7, no. 3, pp. 35–39, 2007.

10. D. Zeng, J. Xu, and G. Xu, "Data fusion for traffic incident detector using D-S evidence theory with probabilistic SVMs," J. Comput., vol. 3, no. 10, pp. 36–43, Oct. 2008.
11. R. Mehran, A. Oyama, and M. Shah, "Abnormal crowd behavior detection using social force model," in Proc. IEEE Conf. Comput. Vis. Pattern Recognit., Jun. 2009, pp. 935–942.
12. W. Sultani and J. Y. Choi, "Abnormal traffic detection using intelligent driver model," in Proc. 20th Int. Conf. Pattern Recognit., Aug. 2010, pp. 324–327.
13. H.-N. Hu et al., "Joint monocular 3D vehicle detection and tracking," in Proc. IEEE/CVF Int. Conf. Comput. Vis. (ICCV), Oct. 2019, pp. 5389–5398.
14. A. Kuramoto, M. A. Aldibaja, R. Yanase, J. Kameyama, K. Yoneda, and N. Suganuma, "Mono-camera based 3D object tracking strategy for autonomous vehicles," in Proc. IEEE Intell. Vehicles Symp. (IV), Jun. 2018, pp. 459–464.

Note: All the figures and tables in this chapter were test data made by the authors.

A Comprehensive Survey on Brain Tumor Classification from MRI Images Using Artificial Intelligence

32

Sateesh Amarneni[1]

Research scholar, Vel Tech Rangarajan Dr. Sagunthala
R&D Institute of Science and Technology, Avadi
Assistant Professor,
VidyaJyothi Institute of Technology, Hyderabad

Valarmathi R.S.[2]

Professor, Vel Tech Rangarajan Dr. Sagunthala
R&D Institute of Science and Technology, Avadi

Abstract: Diagnosing brain tumors are a challenging and difficult process in computer-aided diagnosis (CAD), especially when it comes to abnormalities (brain tumors). Several literatures have suggested classifying brain tumors from magnetic resonance images (MRI) using ML and DL learning approaches. However, the accuracy in classification is not suitable for providing treatments. Therefore, in this paper, we analyses the techniques of brain tumor classification of various literature articles in between 2010 to 2022 based on the two categories, namely, machine learning and deep learning approach, where the articles are selected from the scientific journals: IEEE, Elsevier, Springer, and international journals. Finally, from this survey, we find that the deep learning based technique outperformed the machine learning based technique in the classification of MRI. Therefore, in future, to enhance the brain tumor classification,we shall focus on the deep learning based model.

Keywords: MRI, CAD, DWT, GLCM

1. Introduction

Nowadays, Brain tumor has been proved as a existence threatening disorder which purpose even to demise[1]. Ultimately, neuro-oncologists need to rely on the information and experience they have acquired to carry out the delicate mission of diagnosing brain tumors. [2]. One of the abnormal cell growths in the brain is a mass that is half the size of the normal one. This abnormal mass is fast and full of abnormal cells that can be heard as tumors. Brain tumors affect their shape, size, and severity [3]. Therefore correct facts aboutthe region and dimension of the tumor is essential for effective treatment [4]. However, the diagnosing of brain tumors is a challenging and difficult process in CAD for medical applications [6]. That is, identifying, segmentation, and detecting the contaminated location on MRI images of Brain tumors is a challenging and time-consuming job [7]. In the inclusion, the characteristics help classify brain tumors to distinguish tumors based on their intensity or structure. The complex structures of several MRI tumors of the brain have resulted in the extraction of useful features [8]. Super-resolution, one of the most popular subjects of modern times, improves the resolution of an important image about information (MRI) and makes the MRI image more important and clearer [9]. In addition, segmentation and subsequent quantitative assessment of lesions in clinical imaging provides valuable data for the assessment of brain disease, which is critical to treatment planning modalities, predicting disease follow-up, and the progression of patient outcomes. [10].

[1]drrsvalarmathi@veltech.edu.in, [2]Sateesh.vitae@gmail.com

2. Literature Survey

Several research papers on the classification of brain tumors with MR brain images are available in this section. However, the classification accuracy is insufficient to provide treatments. Techniques for classifying brain tumors in various literary articles from 2010 to 2020 will be explored in two categories, namely classifying brain tumors using a machine learning approach and classifying brain tumors using a deep learning approach.

Table 32.1 Chornological order of the literature survey

Author Name and Year	Classification method	Objective	Data set	Features extracted/feature extraction method	Limitations	Accuracy
El-Dahshan, et al. [11]	FP-ANN and k-neighboring neighbor (k-NN).	A hybrid approach is used to classify magnetic resonance images.	Harvard Medical School website	Discrete wavelet transformation (DWT)	Each time the image database gets bigger, new tutorials are needed	97% and 98% had gained by FP-ANN and k-NN.
Amarapur, Basavaraj [12]	Classification using the whale optimization algorithm based ANN	For the automated segmentation and classification of the brain, a cognitively based synthesis on a modified level and an adaptive ANN classification are done in the MRT.	Standard database MICCAI-BRATS	Multi-Level Bandwidth wavelet decomposition andGLCM, Gabor and Moment Unchanging Features	A proposed method for classifying tumors as edema, advanced solid core, necrotic/cystic core, progression in core classes.	The classification accuracy as 98%.
Sachdeva, et al., [13]	Genetic algorithm (GA) based totally SVM (Supported Vector Machine) and ANN (Artificial Neural Network	Interactive CAT system designed to aid radiologists in classifying brain tumors into multiple classes	Postgraduate Institute of Medical Education and Research (PGIMER), Chandigarh, India	Areas are saved as Interest Bearing Interest (SROIs). 71 depth and high-quality function are taken from this SROIs.	Hybrid 'GA-SVM' classifier and 'GA-ANN' classifier by means of amalgamating facets of GA and SVM, GA and ANN	The full accuracy of SVM elevated from 80.8% to 89% and that of ANN from 77.5% to 94.1%.
Shrot, et al.[14]	Combination of morphologic MRI, perfusion MRI, and DTI metrics	A machine learning program differentiates between different types of brain tumors based on basic and advanced MR scenarios	Basic and advanced MR sequences	Support vector machines (SVMs).	Tracing different ROIs was completed manually.	The accuracy, sensitivity, and specificity of the SVM binary classification for cleoplastoma, metastasis, meningioma, and primary central fearful device lymphoma were 95.7, 81.6, and 91.2%, respectively; 92.7, 95.1, and 93.6%;97, 90.8, and 58.3%; and 91.5, 90, and 96.9%, respectively.
Sharma et al. [15]	K-means and artificial neural network (KMANN)	Through MRI image the damaged part of brain and size of tumor are identified		They employed GLCM (Grey Level co-occurrence matrix)	A hybrid technique is used for both classification and tumor detection. The K-Means technique is used to rapid notice a brain tumor, however the records are no longer correct	A assessment was once carried out with Existing approach and proposed invented method additionally show accuracy (96.6%), Sensitivity (95.3%) and specificity (98.67%)

Author Name and Year	Classification method	Objective	Data set	Features extracted/feature extraction method	Limitations	Accuracy
Sajjad et al. [18]	A multi-grade classification system for brain tumors based on the CNN	Magnified and authentic information consider the effects are in contrast to present techniquesto display the consistent performance	The original data with NVIDIA GTX-1070 was installed on Ubuntu 16.04 with 8 GB of integrated memory and a deep learning framework Caffe	The pre-trained CNN mannequin was once specifically developed for the classification of intelligence tumors	The most standardized classification is used to balance the efficiency and accuracy of each standard with the study of lighter CNN structures.	The entire accuracy of our given approach is 87.38% for unique and it is elevated to 90.67% after information augmentation
Mittal, et al. [19]	Standard Wavelet Transform (SWT) and the Growing Convolution Neural Network (GCNN)	Was once to enhance the accuracy of the traditional gadget	BRAINIX medical images	A comparative evaluation with the SVM and the CNN	To lift up the presentation of the proposed system Grouping of most different classifiers are used	The proposed technique is based on accuracy, PSNR, MSE, and other performance parameters from SVM and CNN..
Iqbal et al [20]	Deep convolutional neural network	Have a deep CNN to divide brain tumors in MRIs	BRATS segmentation challenge	Caffe on Titan X (Pascal) GPU with 12 GB memory size.	Community shape is small, fast, and much less reminiscence disturbing	Achieved segmentation accuracy of 0.88, 0.79, and 0.73 for complete tumor, core tumor, and lively tumor, respectively, in the BRATS database
Rehman, et al. [23]	Convolutional neural networks (AlexNet, GoogLeNet, and VGGNet)	To categorize brain tumors like meningioma, glioma, and pituitary.	MRI pieces of brain tumor	Data boost approach	Another fundamentally mighty deep neural networks Structures for the temporally complex classification of brain tumors.	In terms of classification and detection highest accuracy is up to 98.69
Amin et al. [25]	Long-term memory model (LSDM) with magnetic resonance imaging (MRI)	To defeat the problems of automated brain tumor classification	BRATS 2012–15, 2018 and SISS-ISLES 2015	N4ITK and Gaussian filters having size 5 9 5	It can be brief to classify the sub-tumoral area, to measure the severity level of the tumor area, i.e., to introduce a central, complete tumor..	Achieved maximum 98% accuracy

2.1 Summary of the Survey

Classifying brain tumors is a tough and onerous task. Therefore, the classification of brain tumors is an essential lookuptopic in the clinical image diagnostic system. In the above study, we analyzed the concept of categorization of brain tumors published in the literature between December 2010 and 2020. We selected literature from academic journals from IEEE, Elsevier, Springer, and other international journals. A total of twenty documents are analyzed here, and each process has advantages and disadvantages. Therefore, there is an opportunity to continue and improve upon previous full research articles in the field of tumor classification. Some of the research breaks are.

- With quantitative approaches, the similarity error between tumor MRI and normal MRI is large and influences the detection of anomalies, even if they are classified with essential classifiers. Some of the research papers, a minimum number of features only extract for performing classification hence the accuracy is decreased.

- Some research papers, dimensionality reduction approach is not used it will increase the system complexity.
- Not able to handle the high dimensional problem.
- Computational complexity is increased in SVM based classification
- Time consumption is increased in artificial neural network-based classification
- Deep neural networks need high computational cost and high number of training data are needed

3. Analysis and Discussion

In this section, we have been analyzed and discussed the literatures which are considered as above articles. From this survey, the feature extraction is an essential stage for MRI image classification and the dimensionality reduction approach is decreased the complex problem as well as time consumption. In this survey, the DWT, GLCM, SWT, Also, several types of implementation tools are used in the literatures to analysis of the brain tumor classification. In most of the literature, MATLAB software tools are used. However, the Cafe framework are used in [18, 19] and in [26], the python tool is used for classification of brain tumor. Also, in [22], the MATLAB tool is used for the preprocessing, segmentation, dimensionality discount and the Weka 3.9 tool is used for the classification of intelligence tumor. Each article has analyzed based on the different performance measure and a different outcome. Primarily, the accuracy, sensitivity and specificity measures are used for process of classification. Therefore, from this survey, we found that the necessary to develop the accuracy of the classification system.

4. Conclusion

In this survey, totally of twenty papers have been analyzed which are from different years (2010 to 2022) and different journals. This survey has provided a grave analysis of deep learning and a machine learning algorithm based on brain tumor classification. Then the review papers are analyzed. Also, some research gaps in existing works are analyzed. This survey shows that the accuracy of the classification system needs further improvement. We have found that a through learning-based technique works better than a machine-based technique in MRI classification.. Therefore, in future, to enhance the brain tumor classification, we shall focus on the deep learning based model.

References

1. Ain, Quratul, M. ArfanJaffar, and Tae-Sun Choi. "Fuzzy anisotropic diffusion based segmentation and texture based ensemble classification of brain tumor." applied soft computing 21 (2014): 330–340.
2. Arizmendi, Carlos, Daniel A. Sierra, Alfredo Vellido, and Enrique Romero. "Automated classification of brain tumours from short echo time in vivo MRS data using Gaussian Decomposition and Bayesian Neural Networks." Expert systems with applications 41, no. 11 (2014): 5296–5307.
3. Shahzadi, Iram, Tong Boon Tang, FabriceMeriadeau, and Abdul Quyyum. "CNN-LSTM: Cascaded framework for brain Tumour classification." In 2018 IEEE-EMBS Conference on Biomedical Engineering and Sciences (IECBES), pp. 633–637. IEEE, 2018.
4. Shanthakumar, P., and P. Ganeshkumar. "Performance analysis of classifier for brain tumor detection and diagnosis." Computers & Electrical Engineering 45 (2015): 302–311.
5. Arunkumar, N., Mazin Abed Mohammed, Mohd Khanapi Abd Ghani, Dheyaa Ahmed Ibrahim, Enas Abdulhay, Gustavo Ramirez-Gonzalez, and Victor Hugo C. de Albuquerque. "K-means clustering and neural network for object detecting and identifying abnormality of brain tumor." Soft Computing 23, no. 19 (2019): 9083–9096.
6. Deepak, S., and P. M. Ameer. "Brain tumor classification using deep CNN features via transfer learning." Computers in biology and medicine 111 (2019): 103345.
7. Shree, N. Varuna, and T. N. R. Kumar. "Identification and classification of brain tumor MRI images with feature extraction using DWT and probabilistic neural network." Brain informatics 5, no. 1 (2018): 23–30.
8. Sachdeva, Jainy, Vinod Kumar, Indra Gupta, Niranjan Khandelwal, and Chirag Kamal Ahuja. "Segmentation, feature extraction, and multiclass brain tumor classification." Journal of digital imaging 26, no. 6 (2013): 1141–1150.
9. Özyurt, Fatih, Eser Sert, and Derya Avcı. "An expert system for brain tumor detection: Fuzzy C-means with super resolution and convolutional neural network with extreme learning machine." Medical hypotheses 134 (2020): 109433.
10. Amin, Javeria, Muhammad Sharif, Mussarat Yasmin, and Steven Lawrence Fernandes. "Big data analysis for brain tumor detection: Deep convolutional neural networks." Future Generation Computer Systems 87 (2018): 290–297.
11. El-Dahshan, El-Sayed Ahmed, Tamer Hosny, and Abdel-Badeeh M. Salem. "Hybrid intelligent techniques for MRI brain images classification." Digital Signal Processing 20, no. 2 (2010): 433–441.
12. Amarapur, Basavaraj. "Computer-aided diagnosis applied to MRI images of brain tumor using cognition based modified level set and optimized ANN classifier." Multimedia Tools and Applications 79, no. 5 (2020): 3571–3599.
13. Sachdeva, Jainy, Vinod Kumar, Indra Gupta, Niranjan Khandelwal, and Chirag Kamal Ahuja. "A package-SFERCB-

"Segmentation, feature extraction, reduction and classification analysis by both SVM and ANN for brain tumors"." Applied soft computing 47 (2016): 151–167.

14. Shrot, Shai, Moshe Salhov, NirDvorski, Eli Konen, Amir Averbuch, and Chen Hoffmann. "Application of MR morphologic, diffusion tensor, and perfusion imaging in the classification of brain tumors using machine learning scheme." Neuroradiology 61, no. 7 (2019): 757–765.

15. Sharma, Manorama, G. N. Purohit, and Saurabh Mukherjee. "Information retrieves from brain MRI images for tumor detection using hybrid technique K-means and artificial neural network (KMANN)." In Networking communication and data knowledge engineering, pp. 145–157.Springer, Singapore, 2018.

16. Manogaran, Gunasekaran, P. Mohamed Shakeel, Azza S. Hassanein, Priyan Malarvizhi Kumar, and Gokulnath Chandra Babu. "Machine learning approach-based gamma distribution for brain tumor detection and data sample imbalance analysis." IEEE Access 7 (2018): 12–19.

17. Kaur, Taranjit, Barjinder Singh Saini, and Savita Gupta. "An adaptive fuzzy K-nearest neighbor approach for MR brain tumor image classification using parameter free bat optimization algorithm." Multimedia Tools and Applications 78, no. 15 (2019): 21853–21890.

18. Sajjad, Muhammad, Salman Khan, Khan Muhammad, Wanqing Wu, Amin Ullah, and Sung WookBaik. "Multi-grade brain tumor classification using deep CNN with extensive data augmentation." Journal of computational science 30 (2019): 174–182.

19. Mittal, Mamta, Lalit Mohan Goyal, Sumit Kaur, Iqbaldeep Kaur, Amit Verma, and D. Jude Hemanth. "Deep learning based enhanced tumor segmentation approach for MR brain images." Applied Soft Computing 78 (2019): 346–354.

20. Iqbal, Sajid, M. UsmanGhani, Tanzila Saba, and AmjadRehman. "Brain tumor segmentation in multi-spectral MRI using convolutional neural networks (CNN)." Microscopy research and technique 81, no. 4 (2018): 419–427.

21. Kumar, Sharan, and Dattatreya P. Mankame. "Optimization driven Deep Convolution Neural Network for brain tumor classification." Biocybernetics and Biomedical Engineering 40, no. 3 (2020): 1190–1204.

22. Mohsen, Heba, El-Sayed A. El-Dahshan, El-Sayed M. El-Horbaty, and Abdel-Badeeh M. Salem."Classification using deep learning neural networks for brain tumors." Future Computing and Informatics Journal 3, no. 1 (2018): 68–71.

23. Rehman, Arshia, Saeeda Naz, Muhammad Imran Razzak, Faiza Akram, and Muhammad Imran."A deep learning-based framework for automatic brain tumors classification using transfer learning." Circuits, Systems, and Signal Processing 39, no. 2 (2020): 757–775.

24. Talo, Muhammed, Ulas Baran Baloglu, Özal Yıldırım, and U. Rajendra Acharya. "Application of deep transfer learning for automated brain abnormality classification using MR images." Cognitive Systems Research 54 (2019): 176–188.

25. Amin, Javaria, Muhammad Sharif, Mudassar Raza, Tanzila Saba, Rafiq Sial, and Shafqat Ali Shad. "Brain tumor detection: a long short-term memory (LSTM)-based learning model." Neural Computing and Applications 32, no. 20 (2020): 15965–15973.

Performance Analysis of Text Classification Machine Learning Models

33

B. Venkataramana[1]

Associate Professor, Dept. of CSE,
Holy Mary Institute of Technology & Science, TS, India

G. Venkata Koti Reddy[2]

Associate Professor & HoD, Dept. of CSE,
Holy Mary Institute of Technology & Science, TS, India

P. Bhaskarareddy[3]

Professor & Director, Dept. of ECE,
Holy Mary Institute of Technology & Science, TS, India

P. Sri Durga[4]

Assistant Professor, Dept. of CSE, VPRIT, AP, India

Abstract: Text Classification is the field of study that analyzes people's opinions, sentiments, evaluations, appraisals, attitudes, and emotions towards entities such as products, services, organizations, individuals, issues, events, topics, and their attributes. It has thus become a necessity to collect and study opinions on the Web. Of course, there are opinionated publications outside of the Web as well, and many organizations also collect consumer feedback from emails, call centers, and surveys to gauge what their customers think of their products. There are many classifier systems used for predicting the polarity of collected data while using Machine Learning (ML) algorithms. Our project is to essentially evaluate different ML models like Random Forest Classifier, Logistic Regression, and Decision Tree Classifier designed for text classification using different algorithms and analyze the performance of each model. To build our model, we use the Twitter dataset. Further data extraction, processing, and modeling are done on the dataset before using it for training the models. then models are evaluated and compared with other models based on their performance.

Keywords: Text classification, Classifier systems, Random forest classifier, Logistic regression, and Decision tree classifier

1. Introduction

In recent times, there has been a significant increase in the number of online customer reviews for products, primarily driven by the rapid growth of online reviews and the surge in social media platforms. These reviews are collected from various sources like Facebook, Twitter, and online forums, and undergo a sentiment analysis process to determine the overall sentiment polarity. This task, which involves extracting opinions or sentiments, is crucial and combines techniques from data mining and natural language processing (NLP) [1].Sentiment analysis, also known as opinion mining, is a computational study focused on understanding people's opinions, sentiments, evaluations, attitudes, moods, and emotions. It is an active area of research in fields such as NLP, data mining, information retrieval, and web mining [2]. The primary purpose of sentiment analysis is to detect whether a given text expresses a positive or negative sentiment. It is commonly employed by businesses to assess sentiment in social data, evaluate brand reputation, and gain insights into

[1]bandaruramana1@gmail.com, [2]gvkotireddy@gmail.com, [3]pbhaskarareddy@rediffmail.com, [4]psridurga2010@gmail.com

customer perspectives [3].There are various models available for performing sentiment analysis, each capable of detecting different levels of sentiment within a sentence. These models take into account different aspects of sentiments and the words used. However, each model has its own advantages and disadvantages, depending on the specific application. It is crucial to determine the efficiency of these sentiment models, particularly in terms of selecting the appropriate model based on factors such as the available data, required accuracy, and application efficiency. This research aims to provide a comprehensive analysis of the performance of various machine learning models for sentiment analysis. Different models, including logistic regression, random forest, decision tree classifiers, and support vector machine (SVM) classifiers, are constructed and evaluated to assess their performance.

1.1 Motivation

Text classification is an emerging field of study within text mining analysis. It focuses on organizing and categorizing the vast amount of unstructured text that people generate to express their ideas, opinions, and viewpoints. These opinions can originate from various sources such as the general public, consumers, social media users, athletes, the entertainment industry, and industrial organizations. With the widespread use of social networking services like Facebook, Twitter, and Google Plus, billions of users worldwide contribute to the generation of substantial textual data.

Social media platforms play a significant role in producing diverse forms of text data, including tweet IDs, status updates, reviews, author information, content, and tweet status updates. This abundance of data offers valuable insights into people's perspectives and behaviors. However, due to the unstructured nature of this data, it requires effective text classification techniques to organize, analyze, and derive meaningful information from it. Researchers are actively exploring and developing methods to classify and categorize these vast amounts of textual data for various applications and industries.

2. System Study

2.1 Related Work

Our literature review aimed to explore general online learning algorithms and determine their suitability for our specific use case. Over the past decade, sentiment analysis has been a prominent area of research, rapidly advancing into new domains. A significant portion of this research has focused on developing more accurate sentiment classifiers [4]. The emphasis in sentiment analysis research has primarily been on enhancing the precision of sentiment classifiers, often employing supervised machine learning techniques and various variables. Text classification tasks have been approached at different levels of granularity in natural language processing. Researchers such as Hu and Liu (2004), Kim and Hovy (2004), Wilson et al. (2005), Agarwal et al. (2009), and more recently, at the document level, have explored classification methods for sentiment analysis [Turney, 2002; Pang and Lee, 2004]. However, microblog data, like those found on Twitter, presents unique challenges due to the real-time nature of user comments and reactions to diverse subjects. Several studies have addressed sentiment analysis of Twitter data, including works by Go et al. (2009), Bermingham and Smeaton (2010), and Pak and Paroubek (2010). Go et al. (2009) employed distant learning techniques to gather sentiment data, distinguishing between positive and negative tweets using emoticons. They built models using Naive Bayes, MaxEnt, and Support Vector Machines (SVM), with SVM proving to be the most effective classifier. Their experimentation involved playing with features such as Unigram, Bigram, and parts-of-speech (POS) features, ultimately finding that the unigram model outperformed the others. Bigrams and POS features yielded less satisfactory results. Pak and Paroubek (2010) also utilized distant learning techniques in their data collection process. They trained classifiers using "manufactured" features to analyze Arabic sentiment. Word n-grams were the most commonly used features, employed in training SVM [5, 6, 7], Naive Bayes (NB) [8, 9], and ensemble classifiers [10]. These features are simple but lack semantic depth. To enhance accuracy, word n-grams were combined with stylistic criteria, such as letter and digit n-grams, word and document lengths, and vocabulary richness, after reducing the feature space using the Entropy-Weighted Genetic Algorithm (EWGA) [11]. Additionally, emotion lexicons were incorporated to provide deeper semantic information and improve accuracy. Various lexicons were developed for Modern Standard Arabic (MSA) and dialectal Arabic, including ArabSenti, ArSenL, and SLSA, which were used to train the models [12, 13, 14]. Other factors that could indirectly impact the system were also considered [15].

2.2 Existing System

There are many existing approaches to text classification. These approaches to sentiment analysis can be grouped into three main categories: Knowledge-based techniques classify text by affect categories based on the presence of unambiguous affect words such as happy, sad, afraid, and bored. Some In addition to listing words with obvious affect, knowledge bases often assign arbitrary words a probable"affinity" to specific emotions. Statistical techniques incorporate machine learning components including deep learning, semantic space models, support vector machines, "bag of words" for semantic orientation, and latent text categorization. Deep parsing of the text yields grammatical dependence relations. These Different approaches for analysis of lead to a conflict in the performance of the system where it is being applied. In each approach, the performance is varied based on the dataset length, amount of time for training the model, type of dataset used to train the model, and other factors that may indirectly impact the system.

2.3 Proposed System

We propose an evaluation system to analyze different Machine Learning Models for text classification, which evaluates the given models for the textual classification based on the performance evaluation factors like training accuracy, Validation accuracy, Confusion Matrix of the models, and f1 scores of each model taken for the performance analysis. The three distinct steps in the performance analysis of the models are as follows: first, the dataset, in this case the Twitter dataset, is prepared for data preparation; next, various models for sentiment analysis are developed; and finally, the models are trained using the dataset that is shared by all the models that will be used in the performance analysis. Then the next step is to evaluate the models based on the performance factors of each model.

2.4 System Architecture

See Fig. 33.1

3. Methodology

3.1 Dataset

The research utilizes the Twitter dataset, consisting of tweets from Twitter users. This dataset, along with other large datasets, provides valuable data for sentiment analysis

models. The collected data undergoes preprocessing steps, including null inspection, categorizing sentiments as positive or negative, and analyzing sentiment-related factors. The dataset is transformed into a practical format and divided into groups based on sentiment. Preprocessing includes visualizing and verifying data distribution. The data is assessed and sorted by length taking into account the limits of machine learning models. During preprocessing, unnecessary characters like hashtags are removed, and crucial characters for emotion analysis are modified. The dataset's characters are also standardized in terms of case. As an open-source dataset, user data privacy is not a concern.

3.2 Vectorization

In this methodology, machine learning models are used to classify tweets, but since tweets contain characters that cannot be directly categorized, they are converted into vectors. This vectorization process is necessary to enable the models to analyze and classify the tweets accurately. Various vectorization tools are available, such as gloVe, word2vec, and count vectorization. For this research, the count vectorization method is employed to convert tweets into vectors. The count vectorization technique is applied to the supplied dataset, which involves gathering word frequency and summing up the words in the dataset. This vectorization step allows the machine learning models to process and classify the tweet data effectively.

3.3 Collecting the Hashtags

Hashtag extraction is one of the pre-processing procedures before the data set is actually used by the models for training. The data is taken from the tweets in the data set during this preprocessing operation, and the hashtags are then divided into two categories: racist and sexiest hashtags, and non-racist and non-sexist hashtags.

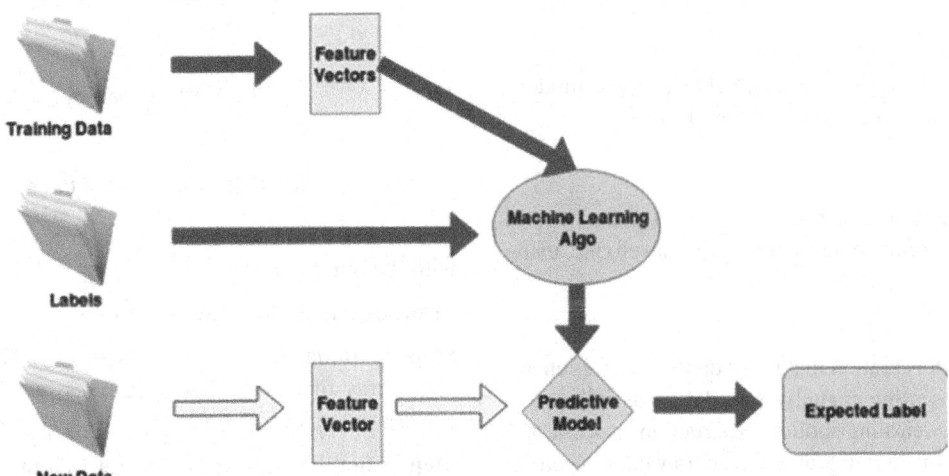

Fig. 33.1 Module design of system

3.4 Tokenising the Words

Tokenization is a technique used to break up text into "Tokens," which are words, symbols, and other significant elements. Whitespace characters can be used to divide up tokens. The normalisation procedure involves finding the abbreviations that are present in the tweets, following which the abbreviations are replaced with their full meaning, for example, "OMG" is changed to "Oh My God" [17].If a word appears more than once, its specific meaning will be eliminated. Also, remove the Stop words, Http links, and slang terms like @, RT, etc. breaking a word into tokens that can be combined to make a Unigram. Remove stop words from the list of Unigram words while constructing new words. This phrase is used to assess Chi-squared. The classifier accepted these Chi-squares. They are receiving the list of unigrams. Similar to Bigram and Trigram, the Chi-squared result is evaluated. Bigram and trigram words are being added to this list. The MPQA subjectivity lexicon includes a word list with sentiment polarity labels. We disseminated word lists from the lexicon that can be used to represent positive, neutral, and negative concepts[16].

3.5 Splitting the Dataset

Following pre-processing, the tweets are divided into two sets for training and testing. A training set contains 70% of the data, while a testing set has 30%.The next phase is the selection of the unigram, bigram, and trigram features. Using distinct training and testing data sets, we are able to identify unigrams, bigrams, and trigrams.[[links]][.1] Pre-processed data from the data set are then divided into two subdata sets called training and testing data. All of the models employed in this methodology are trained using this training data set, and after the models have classified the data, they are evaluated based on the testing data set. The testing data set evaluates the model based on how well it fits the training data set.

3.6 Training Data

Training data is utilised to fit the sentimental analysis models, and it comprises 75% to 90% of the total data set.

3.7 Testing Data

The old data, which ranges from 10% to 35%, is utilised to assess the classification models that were developed using the training data set.

Model building

The phases of model creation that make up the art of a general machine learning sentiment analysis model are as follows. In a typical machine learning model construction process, the training of a machine learning model from raw data to getting predictions from testing data requires the completion of the following phases. The following steps are carried out for each of the machine learning models that have been chosen or chosen for sentiment analysis, including SVM logistic regression, random forest classifier, decision tree classifier, and xgboost classifier.

The phases are:

1. *Model fitting:* The ML models that we are utilizing in this phase of model construction are really fitted for the training. Training data set and are afterwards used for performance evaluation.
2. *Prediction:* The trend model is tested in this stage of model construction utilizing testing data that is used to evaluate the model's performance.
3. *Training accuracy:* Training accuracy is the model's accuracy with respect to the training set of data. Shows how accurate the model is on the training data. The predictions made by the machine learning models are compared to the data set's actual outputs to establish the model's accuracy.
4. *Validation:* The process of validation involves confirming that the output in the dataset matches the output anticipated by the model.

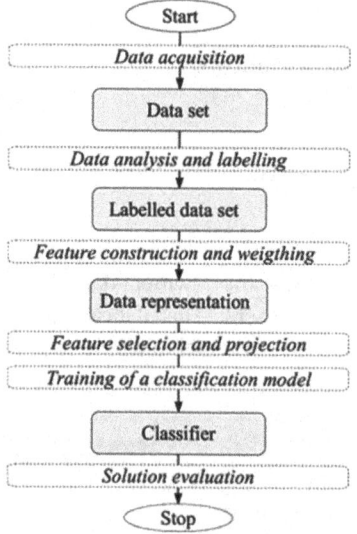

Fig. 33.2 Data flow of the system

Here is one of the machine learning models algorithms along with the other models.

Training Algorithm: Linear SVC

Step 1: Initialize the input data: D = [X, Y], where X is an array of inputs with m features, and Y is an array of corresponding class labels.

Step 2: Set the number of runs for training iterations: num_runs.

Step 3: Set the learning rate: learning_rate = random().

Step 4: Initialize the weight vector: w = zeros(m).

Step 5: Iterate for each run from 1 to num_runs:

- Set the error flag: error = 0.

- Iterate over each data point i from 1 to the length of X:

- Calculate the predicted class value: pred = dot(X[i], w).

- If (Y[i] * pred) < 1, then:

- Update the weight vector: w = w + learning_rate * ((X[i] * Y[i]) - (2 / num_runs) * w).

- Set the error flag: error = 1.

- Otherwise:

- Update the weight vector: w = w + learning_rate * (-2 / num_runs) * w.

- If error is 0 (no errors occurred during this run), then break the loop.

Step 6: Return the fitted model weights, w.

Prediction Algorithm: Linear SVC

Step 1: Take the input data, x, for which the class needs to be predicted.

Step 2: Provide the trained weight vector, w.

Step 3: Calculate the predicted value: pred = dot(x, w).

Step 4: If pred< 0, then assign class1 as the predicted class.

Step 5: Otherwise, assign class2 as the predicted class.

Step 6: Return the predicted class, c.

Evaluation

After applying all machine learning classifiers, including Naive Bayes, Random Forest, Decision Tree, Logistic Regression, Support Vector Classifier, and XGBoost, their performance is assessed through various metrics. Each classifier is evaluated based on Precision, Recall, Accuracy, and F-Measure, using a confusion matrix table. The evaluation is performed using the 10-fold cross-validation method. The entire dataset is divided into 10 equal folds or portions using this technique. Each fold is used as the test data once while the remaining nine folds serve as the training data. This process is repeated 10 times, with each fold being used as the test data exactly once. This ensures that all data points are used for both training and testing. The confusion matrix is used to analyze the performance of the classifiers. It is a table that summarizes the results of the classification task, showing the counts of true positives, true negatives, false positives, and false negatives. The confusion matrix provides a detailed breakdown of the classifier's predictions and the actual class labels. Unfortunately, the details of Figure 2, which explains the contents of the confusion matrix table, are not provided. It would typically include the true positive (TP), true negative (TN), false positive (FP), and false negative (FN) values. These values are essential for calculating performance metrics

such as Precision, Recall, Accuracy, and F-Measure. Overall, the evaluation process involves assessing the performance of each classifier using the 10-fold cross-validation method and analyzing the results using the confusion matrix to gain insights into the classifiers' performance on the given dataset

4. Results

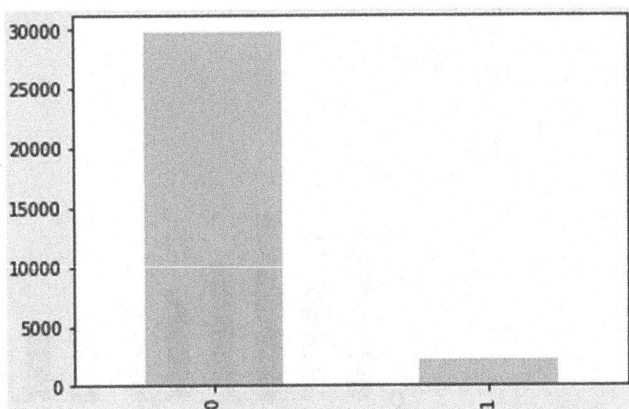

Fig. 33.3 Comparison of positive and negative tweets

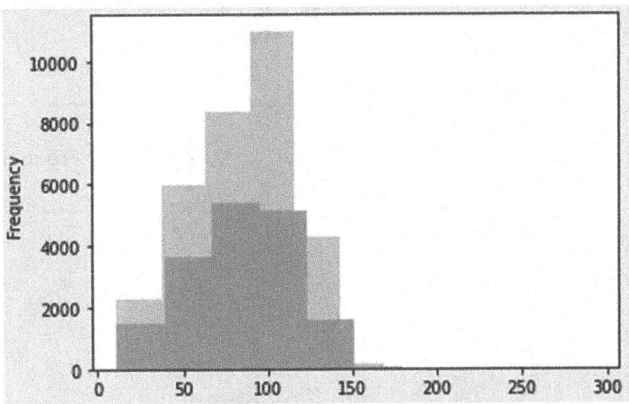

Fig. 33.4 Distribution of tweets

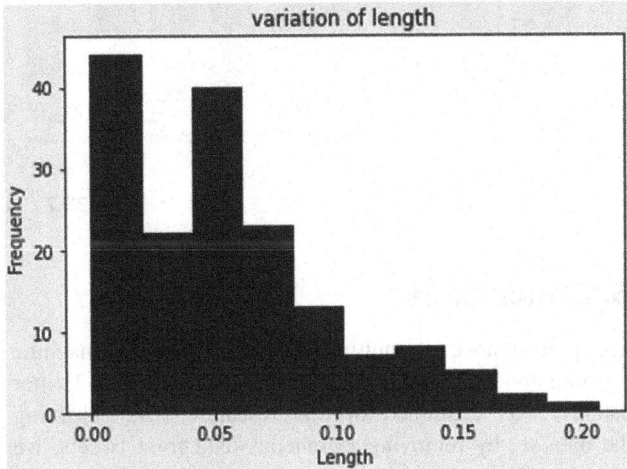

Fig. 33.5 Grouping of data based on length

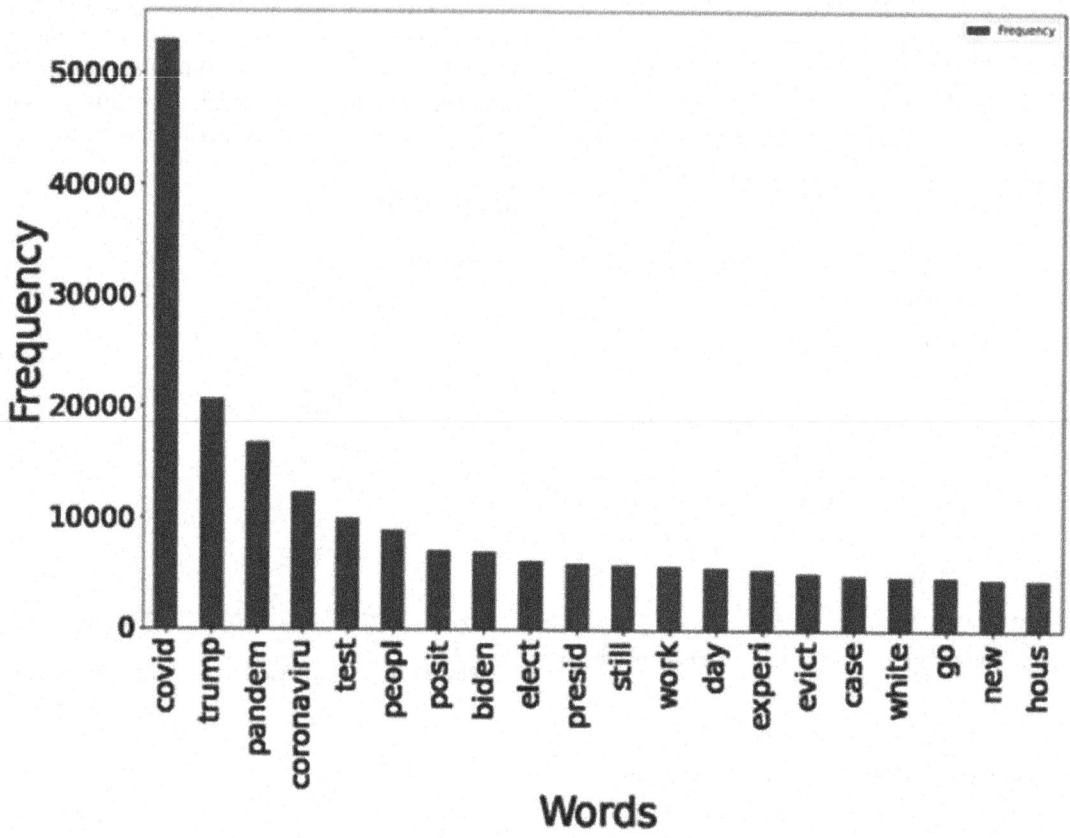

Fig. 33.6 Frequency of occurring words

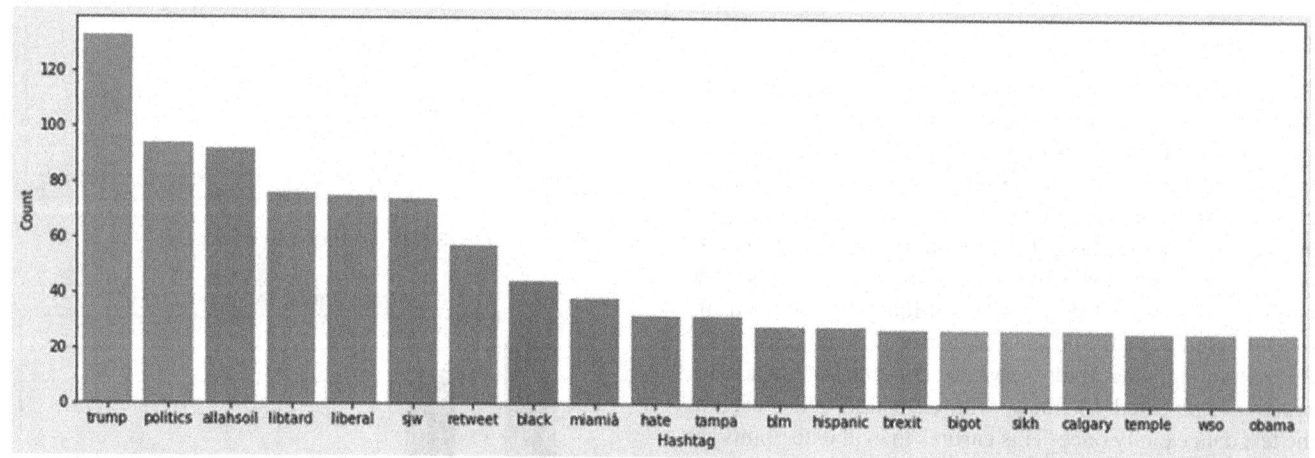

Fig. 33.7 Hastage counting

5. Conclusion

The performance of multiple sentiment analysis machine learning models, which were developed using the Twitter dataset, was examined in this research. After cleaning the data set by removing extraneous data from tweets, we preprocessed and chose the necessary features. 90% of the tweets from the Twitter dataset were used to train the sentiment classifier models, and the remaining 10% were utilized to test the trained models. Positive and negative attitudes in tweets are identified via classifiers. The classifiers are based on Machine learning algorithms like SVM classifier, logistic

```
Training Accuracy : 0.9991656585040257
Validation Accuracy : 0.9504442497810036
F1 score : 0.6000000000000001
[[7298  134]
 [ 262  297]]
```

```
Training Accuracy : 0.9851487213716574
Validation Accuracy : 0.9416843949443123
f1 score : 0.5933682373472949
[[7185  247]
 [ 219  340]]
```

```
Training Accuracy : 0.978181969880272
Validation Accuracy : 0.9521962207483419
f1 score : 0.4986876640419947
[[7419   13]
 [ 369  190]]
```

```
Training Accuracy : 0.9603687789412206
Validation Accuracy : 0.9555750218996371
f1 score : 0.5748502994011976
[[7396   36]
 [ 319  240]]
```

Fig. 33.8 Performance of the each of the models

regresion, random forest classifier, Decision tree classifier and XGboost classifier. By using the count vectorization and bagging of words the data is converted into reuired format for the utilization by the models.

6. Future Enhancement

In this study, we conducted a performance analysis of various machine learning models for sentiment analysis using the Twitter dataset. We followed a systematic approach, which included preprocessing the dataset and selecting relevant features to ensure the quality of the data used for training and testing. To begin, we cleaned the Twitter dataset by removing unnecessary information from the tweets. Next, we split the dataset into a training set (90% of the tweets) and a testing set (10% of the tweets). The training set was used to train the sentiment classifier models, while the testing set was employed to evaluate the performance of the trained models. The classifiers we employed were based on popular machine learning algorithms, such as Support Vector Machine (SVM) classifier, logistic regression, random forest classifier, decision tree classifier, and XGBoost classifier. These classifiers were designed to identify positive and negative sentiments from the tweets. To make the data compatible with the models, we employed techniques like count vectorization and bag-of-words representation. These methods transformed the textual data into a suitable format that could be utilized by the machine learning models. By evaluating the performance of these machine learning models using the provided dataset, we gained insights into their effectiveness in sentiment analysis. The study aimed to identify the strengths and weaknesses of each model, and ultimately provide valuable information for sentiment analysis tasks in the context of Twitter data.

7. Acknowledgment

We would like to express our gratitude to everyone who assisted us in writing this paper. We would like to thank Lovely Professional University for letting us work on this paper and would like to use this opportunity to thank our mentor for his guidance and support. Last but not least, we would want to express our gratitude to everyone who assisted us in finishing the paper, especially my friends and peers for their unwavering support.

References

1. Pong-Inwong, Chakrit; Songpan, Wararat; (2019) Sentiment analysis in teaching evaluations using sentiment phrase pattern matching (SPPM) based on association mining,International journal of machine learning and cybernetics (10): 2177–2186

2. Bing Liu ,"Many Facets of Sentiment Analysis"

3. Erik Cambria Dipankar Das Sivaji Bandyopadhyay Antonio Feraco;A Practical Guide to Sentiment Analysis

4. Saif M. Mohammad;"Challenges in Sentiment Analysis"

5. Rushdi-Saleh, M., Mart´ın-Valdivia, M.T., Urena-L˜opez, L. A., Perea-Ortega, J. M., 2011. Oca: Opinion corpus for arabic. Journal of the ´American Society for Information Science and Technology 62, 2045–2054.

6. Shoukry, A., Rafea, A., 2012. Sentence-level arabic sentiment analysis, in: Collaboration Technologies and Systems (CTS), 2012 International Conference on, IEEE. pp. 546–550.

7. Mountassir, A., Benbrahim, H., Berrada, I., 2012. An empirical study to address the problem of unbalanced data sets in sentiment classification, in: Systems, Man, and Cybernetics (SMC), 2012 IEEE International Conference on, IEEE. pp. 3298–3303.

8. Elawady, R.M., Barakat, S., Elrashidy, N.M., 2014. Different feature selection for sentiment classification. International Journal of Information Science and Intelligent System 3, 137–150.

9. Omar, N., Albared, M., Al-Shabi, A.Q., Al-Moslmi, T., 2013. Ensemble of classification algorithms for subjectivity and sentiment analysis of arabic customers' reviews. International Journal of Advancements in Computing Technology 5, 77.

10. Abbasi, A., Chen, H., Salem, A., 2008. Sentiment analysis in multiple languages: Feature selection for opinion classification in web forums. ACM Transactions on Information Systems (TOIS) 26, 12.

11. Abdul-Mageed, M., Diab, M.T., Korayem, M., 2011. Subjectivity and sentiment analysis of modern standard arabic, in: Proceedings of the 49th Annual Meeting of the Association for Computational Linguistics: Human Language Technologies: short papers-Volume 2, Association for Computational Linguistics. pp. 587–591.

12. Badaro, G., Baly, R., Hajj, H., Habash, N., El-Hajj, W., 2014. A large scale arabic sentiment lexicon for arabic opinion mining. ANLP 2014 165.

13. Eskander, R., Rambow, O., 2015. Slsa: A sentiment lexicon for standard arabic., in: EMNLP, pp. 2545–2550.

14. Abdul-Mageed, M., Diab, M.T., 2014. Sana: A large scale multi-genre, multi-dialect lexicon for arabic subjectivity and sentiment analysis., in: LREC, pp. 1162–1169.

15. Ganesh K. Shinde, Vaibhav N. Lokhande, Rasika T. Kalyane, Vikas B. Gore, Umesh M. Raut Sentiment Analysis on Twitter Hashtag Datasets.

16. Efthymios Kouloumpis, Theresa Wilson, Johanna Moore, "Twitter Sentiment Analysis: The Good the Bad and the OMG!". Proceedings of the Fifth International AAAI Conference on Weblogs and Social Media page no. 538–541, 2011.

Note: All the figures and tables in this chapter were made by the authors.

A Secure Routing Machine Learning Algorithms for Driverless Vehicles

34

A Divya sree[1], Chinna Archana[2], Marella Srimathi[3], Gogu Swathi[4]

Assistant professor, CSE department,
Teegala Krishna Reddy Engineering College, Hyderabad

Abstract: Many people are killed or injured in car accidents due to the increasing traffic in contemporary cities. Human drivers cause over 90% of road accidents. Autonomous vehicles (AVs), commonly known as self-driving automobiles, have made significant progress in recent years. These vehicles operate with little or no input from the driver. For AVs to function safely, they must be able to interact with other non-autonomous entities, including people, cars, and road infrastructure, without compromising their autonomy. The development of Vehicular Ad hoc Networks (VANET) is rapidly rising to maintain a smooth road network. Due to the nature of VANETs, security is still an issue. This survey article gives information on the vulnerabilities and assaults of VANETs. In this report, various current security solutions are surveyed and examined for their successes and shortcomings It is clear that VANET applications need to be secure, but several crucial issues must be addressed. As a result, these factors must be considered while building a secure system that protects privacy, productivity, and usability. As a result, further contributions to this field's study are welcome in future.

Keywords: VANET, Vehicular networks, Machine learning, Security, Privacy, Trust

1. Introduction

Electric and self-driving cars are becoming more commonplace in today's environment, and communication between the vehicles is becoming more critical. Security measures must ensure that cars and other infrastructure function smoothly in a secure vehicular network. The implementation of this infrastructure on the roadside might be costly. As a result, optimizing the roadside wifi network should be a top concern. As the number of cars in the network grows, so does this network. Developing a universal protocol for all networks and all services in a network is difficult because of the ephemerality of vehicles. Many sensors and on-board computers have been added to modern cars during the previous decade. These sensors' job is to gather information about the surrounding environment and convey it to the driver. It is vital to building an efficient routing system for this information to pass from one vehicle to another in the

form of safety messages in the vehicular network. The sheer quantity of automobiles on the road today is out of control, thanks to our frantic daily routines. Traffic congestion, road traffic accidents, and other crises have all increased due to this over-flooding of automobiles in major cities [1].

Vehicle Ad Hoc Networks (VANETs) is established by transferring the ideas of a Mobile Ad Hoc Network (MANET) to a vehicle network (MANET). MANET subsets are thus defined as such. It is a car that's equipped with the latest wireless network technology. The nature of these networks is very dynamic [2]. There is no need for infrastructure or secure connection in VANETS since they can self-organize networks [3]. On-board units (OBU), edge devices (RSU), centralized controllers, and trusted authorities compose the typical architecture of vehicular ad hoc networks (VANET), which is used to aid autonomous and non-autonomous cars (TA). A wired or wireless medium connects a vehicle network to an edge network, which connects to a backbone network. V2V,

[1]adivyareddy29@gmail.com, [2]archanachinna5491@gmail.com, [3]srimathi.marella@gmail.com, [4]goguswathi@gmail.com

V2R, infrastructure-to-infrastructure (I2I), V2I, and many more forms of communication occur between vehicles and various layers of networks [4]. Several breakthroughs have occurred due to greater connection and an expanded number of communication channels and access points. Security and data privacy are two major concerns that must be considered when developing vehicle-based solutions [5]. Different forms of assaults are possible on vehicular networks [6]. Various cryptographic techniques have been presented in the past to address various sorts of security concerns [7]. Password protection, key-based authentication, and biometric security measures are all typical forms of conventional authentication that may also be used to verify the identity of vehicles in vehicular networks. There is no way to tell whether the transferred value is bogus.

Vehicle-to-vehicle information exchange (VANET) is a potential technology that may improve traffic management, road safety, and driver and passenger information distribution. The timely delivery of notifications by VANETs may play a role in promoting road safety. That way, if anything bad happens, they will be prepared. VANETs must thus be protected by an architecture that meets all security criteria. The authors of this paper go into great length on the security needs of VANETs and provide a security architecture as an answer. For example, in [8], an overview of security and privacy considerations in vehicular applications is presented. Conditional privacy preservation and certificate revocation are two more critical concerns raised by the authors that help make the standards feasible. In VANETs, protecting the revocation of certificates and ensuring privacy are major difficulties. Because of VANET's changeable topology, maintaining network security is a difficult task. In [9], it is said that VANET requires a set of security techniques to ensure the integrity and consistency of its data. Virtual Area Networks (VANETs) provide a significant set of security challenges to address before any VANET-based systems can be put into production. For example, consider that a VANET system sends a safety message that has been altered, delayed, or completely deleted owing to an intruder or an attacker, purposefully or unintentionally. As a result, significant repercussions, including injuries, fatalities, and damage to infrastructure, are possible. Since this is the case, researchers are still trying to develop a secure VANET design [10]. [11] Addresses one of the most pressing issues in VANET security and privacy: the need for every recipient to get accurate or trustworthy information from its origin. However, the sender's privacy may be violated by this trusted information.

2. Literature Review

There is much interest in using machine learning (ML) in vehicular networks to solve various problems [12]. As a result, a number of literature reviews address a wide range of topics related to vehicle network security.

A survey on VANET security and privacy is conducted in [13]. Security issues are discussed, and possible remedies are, but no mention is made of machine learning (ML) as a means of offering security solutions. VANET security emphasizes the usage of digital signature methods.

Research on assaults in VANETs and the Internet of Vehicles is reviewed by Sakiz et al. [15]. (IoVs). The usage of ML to overcome various attack difficulties is discussed in this piece. Cryptography and machine learning as a foundation provides an overview of attack prevention strategies. Only supervised and unsupervised ML techniques are explored. There is no discussion of privacy and trust in-vehicle networks, which is an oversight in this paper. An effort to improve the security of vehicle networks is outlined in [14]. An adversary-oriented review of trust management approaches is presented in this paper. Cryptography and trust are the two categories into which the authors divide the various security frameworks. Trust and cryptography may have various relationships, as this study shows. It also provides an overview of trust-based solutions with a table comparison of current methodologies. All these methods completely ignore the use of machine learning techniques. According to this study, there is a need for new, more intelligent trust mechanisms across many VANET settings.

Locator-based services (LBS) in-vehicle networks are discussed by Asuquo et al. [16]. In this paper, we discuss outstanding issues in vehicular network location privacy. According to the literature, location privacy may be divided into two categories: cryptographic procedures (such as digital signatures and hashing), as well as privacy-enhancing strategies (such as encryption) (like mixed zones, obfuscation, silent period, and soon). Table comparisons are used to discuss these two groupings in further detail. There are also downsides to such methods in various vehicular situations that are also mentioned. No ML-based solutions are considered.

Researchers have recently begun to delve further into vehicle networks and the applications that may be found within them. High-mobility vehicle networks, such as IoT and 5G-based vehicular networks, and their derivatives, such as ML frameworks, are briefly discussed by Liang et al. [17]. (5GVN). High-mobility networks involve network architecture, channel estimation, traffic prediction, trajectory prediction, and congestion management. [17] covers these dynamics in more detail. It also discusses the use of ML to connect car intrusion detection. Network optimization in high mobility situations is the focus of most of the research featured in this study. VANET security issues are investigated by Sheikh et al. [18]. VANET architecture and associated security problems are explained in depth in this article. It summarises the most recent security and authentication techniques. This project covers some topics: symmetric cryptography, asymmetric cryptography, identity-based cryptography, and signature methods. However, ML for security issues is not included in the scope of this study. Recent developments in VANET security, privacy, and trust management are examined by Lu et al. [5]. To set the stage for this review, let us look at the architecture of VANETs and the security issues they pose. Security services and cryptography approaches are discussed in this article. In addition to security services, this section focuses on VANET's location privacy issues. Trust management in VANET is also discussed in detail, and several trust models are presented to help with this. This research does not address the significance or application of ML algorithms in VANET security. [18] describes the state-of-the-the-art technologies used to protect and maintain privacy in a VANET architecture similarly. [19] Authentication techniques are the primary focus of this study's categorization taxonomy for assaults on VANETs.

In [19], an in-depth analysis of IoV is offered, comparing it to VANET. The authors divide IoV applications into four categories: safety, comfort and entertainment, traffic efficiency, and health care, and describe their use in driving coordination and emergency warnings. An in-depth overview of assaults on IoV networks is provided. However, the ML context is hardly touched upon in this study. In [20], an the application-based poll is conducted to discuss vehicular assaults. Similar to [19], this study just touches on machine learning and other forms of learning.

According to [21], a deep reinforcement learning (DRL) survey has been conducted. Vehicle networks are not the only thing covered by this study. Aside from that, this review briefly covers network management and caching, offloading, routing and scheduling, and connectivity applications. The ML-specific survey for vehicle communication networks was later carried out by Tong et al. [22]. It examines various use cases, including vehicle platooning, navigation, security, and comfort, as well as issues with network congestion and demand and supply. There is just a very brief discussion of attack detection and prevention strategies based on machine learning in this paper. All of the ML techniques used in vehicle applications are covered in depth by Hossain et al. [24]. However, the focus of this study is confined to CR-based (CR) vehicle networks. The integration of ML in CR-VANET is a key study area in this paper. ML, VANET, and CR are all covered in this survey. An overview of CR-VANETs' usage and use of ML methodologies is provided. All of these things are included in the spectrum sharing and mobility management. The assaults in the CR environment are all discussed in terms of security risks in this paper.

Many ML-based designs are examined in Veres et al. [23] for transportation networks. [23] investigates several network dynamics, such as destination prediction, demand prediction, traffic flow prediction, travel time assessment, travel modes prediction, traffic signal control, navigation, demand serving, combinatorial optimization, and so on. [23] also investigates several network dynamics. However, this study does not address transportation security or the application of machine learning in this area.

Haydariet al. [25]For optimum traffic signal control, autonomous driving, energy management, road control, and other intelligent transportation system applications, the use of reinforcement learning (RL) and deep reinforcement learning (DRL) is examined in this study. Security is not addressed in this work. In [26], a review article focusing on security, privacy, and trust is offered. Vehicle networks are one of several mobile-IoT applications examined in this article. The use of machine learning-based solutions is also underrepresented in this paper. Another study on IoT security focusing on machine learning-based solutions is described in [27]. According to the literature, machine learning (ML) and deep learning (DL) are included in a review of IoT security. There is a short discussion of security concerns in IoT applications in [28]. "A layer-based categorization of literature on IoT security is offered with a very little discussion on transportation/vehicular applications."

Similar to [28], IoT layer-based attacks are also included in [29]. "Malware analysis, authentication, intrusion detection, and assault detection are all covered by this work's ML and DL solutions ."Many other IoT applications and a limited focus on transportation are discussed here. Security, privacy, and trust management in vehicular communication are addressed in [30], although ML-based solutions are not emphasized in this research.

Federated learning in-vehicle IoT was recently studied and reported in [31]. This paper gives a quick introduction to federated learning, as well as a comparison to other forms of education. Using FL in additional wireless IoT applications that allow vehicle use is also covered in this study. In vehicular IoT, there is a study on FL in three areas: perception, networking, and application. It examines various applications emphasizing the security and privacy of vehicle IoT networks. "However, this study sees FL as a viable research avenue for future vehicular applications in terms of security, privacy, and incentives." A recent study by Kuutti et al. [32] examines the application of intelligent processes in vehicle control systems with an emphasis on deep learning (DL). This study investigates the application of DL for various automotive network control systems. Vehicle networks might benefit from deep learning, which it sees as a promising technology. The referenced works deal with both safety and control concerns. However, there is no mention of security issues in this poll.

The authors of Rathore [33] suggest a Medium Access Control Protocol based on clusters (CMAC). This was presented as a means of handling communication between VANET vehicles. According to them, the suggested CMAC can deliver the message with little latency and high dependability. Furthermore, the protocol above is capable of resolving the issue of terminals that are either concealed or revealed. However, their idea relies heavily on the Roadside Unit's existence (RSU). As a result, the procedure will be less effective in places where RSUs are unavailable.

Researchers in Rahbari [34] have presented a strategy for detecting Sybil assaults on VANETs. Detection of such an assault is reliant on critical infrastructure. A Sybil assault will cause tremendous harm since it greatly impacts network performance. They believe that their suggested technique may quickly identify a Sybil attack since most of the actions occur at the Certification Authority, where most of the processing occurs. However, the drawback here is the algorithm's inability to identify the rogue node that initiates this sort of assault. That was left for a future effort by the authors.

Automakers have focused a lot of their emphasis on VANET security because of the danger it poses to people's lives on the road. Maintaining availability is a crucial part of security. There will be a significant impact if VANET's services are unavailable. Consequently, DoS attacks are considered one of the most serious threats to VANETs. Good security techniques should be offered to combat this sort of assault. Imagine if a DoS attack prevented a life-saving communication from reaching its intended recipient.

Soomro [35] has created a model for protecting VANETs against Denial-of-Service (DoS) assaults and a DoS severity level for use in VANETs. They also spoke about some potential remedies. As a result, they came up with a model that requires more testing before it can be implemented.

Because of the rising traffic accidents, the writers in Bitam [36] decided to work on ways to make roads safer while also providing passengers with more comfort. Using VANETs and cloud computing, they have come up with a solution. Using cloud computing to improve road safety and the travel experience is an exciting prospect. It provides flexible options like traffic light synchronization and detours. As a result, the authors developed a cloud computing strategy (where the VANET-Cloud is applied to vehicular ad hoc networks). The model they use offers a wide range of transportation options. However, as the authors have emphasized, this approach does not consider problems like security and privacy.

Li [37] says VANETs may be authenticated using a unique authentication framework with conditional privacy protection

and non-repudiation (ACPN). ID-based Online/Offline Signature (IBOOS) and ID-based Signature were utilized for authentication (IBS). The authors employed a pseudonym-based strategy while generating pseudonyms using a PKC-based mechanism to preserve anonymity. For Urban Vehicular Communications, they said ACPN met all required standards and was adequate (UVC). A large-scale network's efficiency must be shown before employing Public Key Cryptography (PKC) on a big scale.

It is still difficult to provide an acceptable authentication system in VANETs because of the need to protect user privacy at the same time. As a result, achieving a good authentication process necessitates striking this equilibrium. The authors present the asymmetric authentication technique (LESPP) in Wang [38], which employs symmetric operations to sign and verify messages. In order to maintain privacy and traceability, LESPP employs a self-generated fake identity. Message Authentication Code (MAC) generation is also used with symmetric encryption. For example, computation and communication overheads may be reduced by their approach. The suggested model performs well regarding network latency, message loss ratio, and message signing or verification in the simulation used to establish feasibility. However, using symmetric algorithms is a constraint that must be addressed. Furthermore, the outcome must be confirmed by further simulations. It is also necessary to put LESPP through its paces in real-world circumstances.

For example, the authors in Dietzel [39] highlight the value and importance of VANET aggregation and several unusual and very serious security challenges. Scalability is important because of the constraints of the wireless bandwidth medium. Acquiring and improving scalability is facilitated greatly by data aggregation. However, verifying the integrity of aggregated information is not a simple operation. As a result, an assault is not out of the question.

Several popular threats, such as Jamming, Sybil, Selfish Driver, and misbehaving nodes that provide false information and incorrect vehicle location data, are discussed by the authors in Quyoom [40]. As a serious concern, DoS assaults have been given considerable attention by researchers. DoS attacks may be identified using a new algorithm called MIPDA (Malicious and Irrelevant Packet Detection Algorithm). They contend that their technique speeds up VANET connectivity and security by reducing overhead latency. However, the authors fail to demonstrate their method's performance in a simulated or actual context.

3. Possible Attacks in VANETs

Virtual area networks (VANETs) are vulnerable to several assaults. In order to make it difficult for users to access the system or to phase out certain pieces of information, various assaults are conducted. Some definitions of assaults are derived.

3.1 Sybil Attacks

The Sybil assault involves the construction of several fictitious nodes that disseminate erroneous data. Multiple messages, each with a separate manufactured identity, are sent to the target car using an On Board Unit (OBU) installed in the target vehicle during the Sybil assault. It becomes a concern when a malevolent vehicle may assume the identity of many vehicles and so reinforce misleading data. VANET Sybil attacks may be countered using various methods, including statistical and probability analysis, signal strength measurements, and session keys [41]. "However, due to dynamic features, weather conditions, and system design, each strategy has benefits and drawbacks." For example, the signal intensity of a signal may be used as an input to a statistical and probability algorithm [41] and [42]. Positionaire claims to have determined the discrepancy between the received signal intensity and the estimated signal strength. By using statistical and probabilistic algorithms, AS analyses it.

3.2 Node Impersonation

As the name implies, node impersonation refers to the act in which one node sends out an altered message while pretending to come from the original sender for some unknown goal or purpose. A greedy approach, Detection of Malicious Vehicle (DMV) and Outlier Detection (OD), has been suggested to tackle this difficulty. RSU was used to identify and monitor anomalous behavior of nodes in the schemes. If the vehicle is trusted, the suggested technique raises the node's trust value. "If the mistrust value is more than the threshold value, the vehicle's identify (ID) will be reported to the appropriate Certificate Authority (CA)."

3.3 Sending False Information

When a node intentionally sends inaccurate or fraudulent information to another node to cause havoc, it is known as transmitting false information. To avoid misunderstanding, this circumstance should be avoided. Attackers provide false information to the vehicle for their gain. Using a fictitious traffic report, an assailant may temporarily remove a roadblock by claiming there was traffic congestion. Various techniques have been developed to identify hacked nodes that may be misbehaving [43]. A group signature technique that depends on password access is one of the options. Another vehicle receiving a message can only verify the message's authenticity using this technique when applied to a signed message. This plan, however, is impractical due to the constant turnover of group members, particularly in urban networks.

3.4 ID Disclosure

The nodes can reveal their identities and monitor the position of the target nodes in the network. A virus is sent to the target nodes' neighbors while Observer keeps tabs on them. The attacker's neighbors are infected with viruses to steal the target node ID and current position. Identity disclosure schemes have been developed to prevent vehicle tracking by identifying the keys used. Al-Hawi et al. [44] use pseudonyms to encrypt and mask a vehicle's unique identification, such as the name of the driver, the license plate number, and the location. Public Key Infrastructure (PKI) pseudonyms were used to sign the communication, making it harder to trace the sender.

4. Security Services in VANET

The security of ad hoc networks, particularly for sensitive applications, is a major concern. For an ad-hoc network to be safe, the following properties must be considered as criteria for measuring security: Authentication, confidentiality, integrity, non-repudiation, and availability.

4.1 Availability

Bandwidth and connection are two of the network services that are available to all nodes. In order to address the problem of availability, a group signature mechanism has been implemented. Vehicles and RSUs can communicate with each other utilizing this system. Using public and private keys between RSUs and cars, the suggested approach is still viable in the event of an assault on the network.

4.2 Confidentiality

Classified information on the network can never be transmitted to unknown parties because of confidentiality [45]. Unauthorized access to sensitive data, such as a person's name, license plate number, and current location, is also prevented. Using pseudonyms is the most common method for ensuring anonymity in-vehicle networks. Encrypted key pairs will be used for each vehicle node. While the vehicle node is not connected to the pseudo used to encrypt or sign messages, the appropriate authority can access this pseudo. Once an earlier pseudo expires, the vehicle must acquire a replacement from an RSU.

4.3 Authentication

Authentication is the authentication of the identification of the cars and RSUs and the certification of the integrity of the information sent between them. The proper vehicle is communicated throughout the network as a result of this. CA-based public or private keys are suggested to link automobiles, RSUs, and AS [43]. On the other hand, passwords are utilized as an authentication technique for RSUs and AS.

4.4 Integrity

Nodes, RSUs, and AS are certain that the data they get is identical to the data they created throughout the message exchanges. A password-protected digital signature is utilized to ensure the message's integrity [46].

4.5 Non-Repudiation

Ensures that neither the sender nor the recipient may claim that they never received the message, such as in the case of accident communications. Auditability refers to confirming that RSUs and vehicles have been received and dispatched in particular sectors.

5. Conclusion

Autonomous cars have great promise for enhancing human welfare by saving time, preventing accidents, and alleviating traffic congestion. The present state of technology is far from ideal since several obstacles exist. The security of this technology is a major concern. An autonomous vehicle's functioning relies on many components, all of which are prone to failure. Sensor failures, software system failures, communication failures, etc., are the most typical causes of failure. Hackers may also carry out attacks on autonomous vehicle components and networks remotely. Authenticity, integrity, availability, and secrecy are all at risk because of these assaults. Sybil attack, jamming assault, DOS attack, replay attack, etc., are some of the most popular forms of network attacks. Many features of VANETs, including protocols, coverage, and other relevant issues, are being researched extensively. As a VANET, security is vital, yet the network architecture makes it difficult to provide effective and comprehensive protection.

The deployment of secure and difficult-to-intercept routing protocols is essential to preventing network assaults and failures. VANET routing algorithms based on network density are evaluated in this paper, and a machine learning strategy is proposed to mitigate black hole attacks in VANETs.

References

1. Liang Zhao et al. "A SVM based routing scheme in VANETs ."2016 16th International Symposium on Communications and Information Technologies (ISCIT). IEEE. 2016, pp. 380–383.

2. Steven So, Prinkle Sharma, and Jonathan Petit. "Integrating plausibility checks and machine learning for misbehavior detection in VANET ."In: 2018 17th IEEE International Conference on Machine Learning and Applications (ICMLA). IEEE. 2018, pp. 564–571.

3. Jaspal Kumar, Muralidhar Kulkarni, and Daya Gupta. "Effect of Blackhole Attack on MANET routing protocols ."International Journal of Computer Network and Information Security 5.5 (2013), p. 64.

4. M. L. Sichitiu and M. Kihl, "Inter-vehicle communication systems: a survey," IEEE Communications Surveys Tutorials, vol. 10, no. 2, pp. 88–105, 2008.

5. Z. Lu, G. Qu, and Z. Liu, "A survey on recent advances in vehicular network security, trust, and privacy," IEEE Transactions on Intelligent Transportation Systems, vol. 20, no. 2, pp. 760–776, Feb 2019.

6. M. S. Al-kahtani, "Survey on security attacks in vehicular ad hoc networks (VANETs)," in 2012 6th International Conference on Signal Processing and Communication Systems, 2012, pp. 1–9.

7. Sheikh, Liang, and Wang, "A survey of security services, attacks, and applications for vehicular ad hoc networks (VANETs)," Sensors, vol. 19, no. 16, p. 3589, Aug 2019.

8. J. Guo, X. Li, Z. Liu, J. Ma, C. Yang, J. Zhang, and D. Wu, "TROVE: A context awareness trust model for VANETs using reinforcement learning," IEEE Internet of Things Journal, pp. 1–1, 2020.

9. S. A. Soleymani, S. Goudarzi, M. H. Anisi, N. Kama, S. Adli Ismail, A. Azmi, M. Zareei, and A. Hanan Abdullah, "A trust model using edge nodes and a cuckoo filter for securing VANET under the NLoS condition," Symmetry, vol. 12, no. 4, p. 609, Apr 2020.

10. E. A. Shams, A. Rizaner, and A. H. Ulusoy, "Trust aware support vector machine intrusion detection and prevention system in vehicular ad hoc networks," Computers & Security, vol. 78, pp. 245 – 254, 2018.

11. A. Le and C. Maple, "Shadows Don't Lie: n-sequence trajectory inspection for misbehaviour detection and classification in VANETs," in 2019 IEEE 90th Vehicular Technology Conference (VTC2019-Fall), 2019, pp. 1–6.

12. S. Kumar, K. Singh, S. Kumar, O. Kaiwartya, Y. Cao, and H. Zhou, "Delimitated anti jammer scheme for internet of vehicle: Machine learning based security approach," IEEE Access, vol. 7, pp. 113 311– 113 323, 2019.

13. F. Qu, Z. Wu, F. Wang, and W. Cho, "A security and privacy review of VANETs," IEEE Transactions on Intelligent Transportation Systems, vol. 16, no. 6, pp. 2985–2996, 2015

14. C. A. Kerrache, C. T. Calafate, J. Cano, N. Lagraa, and P. Manzoni, "Trust management for vehicular networks: An adversary-oriented overview," IEEE Access, vol. 4, pp. 9293–9307, 2016.

15. F. Sakiz and S. Sen, "A survey of attacks and detection mechanisms on intelligent transportation systems: VANETs and IoV," Ad Hoc Networks, vol. 61, pp. 33 – 50, 2017

16. P. Asuquo, H. Cruickshank, J. Morley, C. P. A. Ogah, A. Lei, W. Hathal, S. Bao, and Z. Sun, "Security and privacy in location-based services for vehicular and mobile communications: An overview, challenges, and countermeasures," IEEE Internet of Things Journal, vol. 5, no. 6, pp. 4778–4802, 2018.

17. L. Liang, H. Ye, and G. Y. Li, "Toward intelligent vehicular networks: A machine learning framework," IEEE Internet of Things Journal, vol. 6, no. 1, pp. 124–135, 2019

18. Sheikh, Liang, and Wang, "A survey of security services, attacks, and applications for vehicular ad hoc networks (VANETs)," Sensors, vol. 19, no. 16, p. 3589, Aug 2019

19. S. Sharma and B. Kaushik, "A survey on internet of vehicles: Applications, security issues and solutions," Vehicular Communications, vol. 20, p. 100182, 2019.

20. M. Arif, G. Wang, M. ZakirulAlamBhuiyan, T. Wang, and J. Chen, "A survey on security attacks in VANETs: Communication, applications and challenges," Vehicular Communications, vol. 19, p. 100179, 2019.

21. N. C. Luong, D. T. Hoang, S. Gong, D. Niyato, P. Wang, Y. Liang, and D. I. Kim, "Applications of deep reinforcement learning in communications and networking: A survey," IEEE Communications Surveys Tutorials, vol. 21, no. 4, pp. 3133–3174, 2019.

22. W. Tong, A. Hussain, W. X. Bo, and S. Maharjan, "Artificial intelligence for vehicle-to-everything: A survey," IEEE Access, vol. 7, pp. 10 823–10 843, 2019.

23. M. Veres and M. Moussa, "Deep learning for intelligent transportation systems: A survey of emerging trends," IEEE Transactions on Intelligent Transportation Systems, pp. 1–17, 2019.

24. M. A. Hossain, R. M. Noor, K. A. Yau, S. R. Azzuhri, M. R. Zaba, and I. Ahmed, "Comprehensive survey of machine learning approaches in cognitive radio-based vehicular ad hoc networks," IEEE Access, vol. 8, pp. 78 054–78 108, 2020.

25. A. Haydari and Y. Yilmaz, "Deep reinforcement learning for intelligent transportation systems: A survey," IEEE Transactions on Intelligent Transportation Systems, pp. 1–22, 2020.

26. V. Sharma, I. You, K. Andersson, F. Palmieri, M. H. Rehmani, and J. Lim, "Security, privacy and trust for smart mobile-internet of things (M-IoT): A survey," IEEE Access, vol. 8, pp. 167 123–167 163, 2020. [

27. M. Stoyanova, Y. Nikoloudakis, S. Panagiotakis, E. Pallis, and E. K. Markakis, "A survey on the internet of things (IoT) forensics: Challenges, approaches, and open issues," IEEE Communications Surveys Tutorials, vol. 22, no. 2, pp. 1191–1221, 2020.

28. M. A. Al-Garage, A. Mohamed, A. K. Al-Ali, X. Du, I. Ali, and M. Guizani, "A survey of machine and deep learning methods for internet of things (IoT) security," IEEE Communications Surveys Tutorials, vol. 22, no. 3, pp. 1646–1685, 2020.

29. F. Hussain, R. Hussain, S. A. Hassan, and E. Hossain, "Machine learning in IoT security: Current solutions and future challenges," IEEE Communications Surveys Tutorials, vol. 22, no. 3, pp. 1686–1721, 2020.

30. J. Huang, D. Fang, Y. Qian, and R. Q. Hu, "Recent advances and challenges in security and privacy for V2X communications," IEEE Open Journal of Vehicular Technology, vol. 1, pp. 244–266, 2020.

31. Z. Du, C. Wu, T. Yoshinaga, K. L. A. Yau, Y. Ji, and J. Li, "Federated learning for vehicular internet of things: Recent advances and open issues," IEEE Open Journal of the Computer Society, vol. 1, pp. 45–61, 2020.

32. S. Kuutti, R. Bowden, Y. Jin, P. Barber, and S. Fallah, "A survey of deep learning applications to autonomous vehicle control," IEEE Transactions on Intelligent Transportation Systems, pp. 1–22, 2021.

33. Rathore NC, Verma S, Tomar GS, editors. CMAC: A cluster-based MAC protocol for VANETs. International Conference on Computer Information Systems and Industrial Management Applications (CISIM); 2010 Oct 8–10.

34. Rahbari M, Jamali MAJ. Efficient detection of Sybil attack based on cryptography in vanet. arXiv preprint arXiv:11122257; 2011.

35. Soomro IA, Hasbullah H, Ab Manan J-l. Denial of Service (DOS) attack and its possible solution in VANET; 2010

36. Bitam S, Mellouk A, Zeadally S. VANET-cloud: A generic cloud computing model for vehicular Ad Hoc networks. IEEE Wireless Communications. 2015; 22(1):96–102.

37. Li J, Lu H, Guizani M. ACPN: A novel authentication framework with conditional privacy-preservation and non-repudiation for VANETs. IEEE Transactions on Parallel and Distributed Systems. 2015; 26(4):938–48.

38. Wang M, Liu D, Zhu L, Xu Y, Wang F. LESPP: Lightweight and efficient strong privacy-preserving authentication scheme for secure VANET communication. Computing. 2014:1–24

39. Dietzel S, Schoch E, Konings B, Weber M, Kargl F. Resilient secure aggregation for vehicular networks. IEEE Network. 2010; 24(1):26–31.

40. Quyoom A, Ali R, Gouttam DN, Sharma H, editors. A novel mechanism of detection of Denial of Service attack (DoS) in VANET using Malicious and Irrelevant Packet Detection Algorithm (MIPDA). 2015 International Conference on Computing, Communication, and Automation (ICCCA); 2015, May 15–16

41. B. Xiao, B. Yu, and C. Gao, "Detection and localization of sybil nodes in VANETs," B. Proceedings of the 2006 Workshop on Dependability Issues in Wireless ad Hoc Networks and Sensor Networks, 2006, pp. 1–8

42. G. Jyoti, S. G. Manoj, and L. Vijay, "A novel defense mechanism against sybil attacks in VANET," in Proceedings of the 3rd International Conference on Security of Information and Networks (SIN '10). ACM, New York, NY, USA, pp. 249–255. 2010

GUI Based Market Movement Using Digital and Binary

35

G.koushik[1]

Department of Computer Science and Engineering,
Prasad V Potluri Siddhartha Institute of Technology, Vijayawada, India

G. Krishna Mohan[2]

Professor, Department of Computer Science and Engineering,
Koneru Lakshmaiah Education Foundation, Vijayawada, India.

J. Satish Babu[3]

Asst. Professor, Department of Computer Science and Engineering,
Koneru Lakshmaiah Education Foundation, Vijayawada, India

Abstract: An emerging method of vision-based input is the visual user interface, which computes to complete the task based on the present and prior vision and selects the form by using what you see as the input. It takes on the role of a user and uses the supplied data to do the task at hand. The main objective of this is to make it easy to do analyses on data that is displayed visually rather than on extensive tables of data with numerical values in order to increase the success of forex trading. by the use of the pixels that are present in each of the provided photos and a likelihood match. This suggests that data is transferred to a built-in bot, which automates the necessary analysis and makes decisions with or without user interaction. This makes it possible to explain future market circumstances across several countries, regardless of the subject or subset of data the bot is intended to predict.

Keywords: GUI (Graphical User Interface), Currencies, PYAUTOGUI, GDP, HLOC, OTC (Over the Counter)

1. Introduction

Currencies represent the financial worth of the respective nations. Any currency's value changes in response to supply and demand for all imports and exports. The only way to calculate the value of the national currency to an exact digit is by comparing it to the values of other nations. The exchange rate with respect to each country is extremely varied when determining a currency's value.

The state of the economy may be good or bad depending on the market price, political sway, Gross Domestic Product(GDP) and investors. Because the situation is uncertain, many people started trading on changes in currency prices. The forex market data which is considered can be either linear or nonlinear, the data is in the form of tables and the same may be represented in graphical format.

By using machine learning, computer models may analyse data without human intervention and come to their own conclusions. In this area of artificial intelligence, models self-train using historical data. They have the ability to recognize various patterns after training. They may be able to make their own decisions through certain procedures.

In the financial markets, machine learning techniques are well-known and often employed. A daily trade volume estimate of the financial market shows that foreign exchange makes up a sizable portion as well. The factors influencing price changes on the FX market have interacted in intricate

[1]Koushikgonuguntla03@gmail.com, [2]gvlkm@kluniversity.in, [3]jampanisatishbabu@kluniversity.in

ways. The Forex market forecast is important because of this. Forecasts for the financial markets are essential for traders, researchers, economists, and analysts.

Options trading has piqued the curiosity of many individuals. The reason for this is that it is easier than traditional trading approaches. We don't require time to determine whether to use your favourite broker application's High or Low button. Some people consider options trading is a gamble, while others argue that it is not since it takes an innovative approach to predict whether it will be higher or lower. The main difference between Options and traditional trading is time. The time is fixed in Options trading, but traders in traditional trading can select when to end a position. We must decide if the value is greater than or less than the existing Position.

In options trading, the investor chooses between two options: call (long) or put (short) depending on their forecast of whether the price would be greater or lower than the strike price. If the prediction is true, the investor gets a positive payout %(depending on the market condition) on their investment; otherwise, they get nothing.

2. Literature Survey

There are some academician who are continuously working on various methods of gathering information from speech and computing the necessary tasks for each unique purpose [1]-[3]. If speech can be captured, then the screen that displays information to the user may also be interpreted by the bot. Perhaps some of the different type of concepts have been computed using the vision-based information provided by device[4]-[6]. Now the input is being used as the basic for visualisation, it may be observed by the user without the aid of additional devices, with the complexity or effort level and even the degree of variation increasing the deviation's ability to be altered. When compared to the field of currencies, there are some researchers on object detection using YOLO (You Only Look Once)[7], algorithmic trading bot [8] using the similar ideology, there are some other researches where linear regression[11], multiple regression perhaps termed as nonlinear regression[15] in other researches takes the data in the form of tables and replicated to the graphical format is a huge task for the user and machine.

There have been studies conducted in the past on the storage of data in tabular and numerical formats [9]. Utilising the basic data, some alternative models have been developed using CNN [10], machine learning [11], reinforcement learning [12] and at the binary options[17]. There are several papers that predate all of these that have the same idea, but their time frames are lengthy, and as it takes time to book a decent amount of profit, a shorter time period may result in a greater profit booking [13]. Researchers are also looking

at merging techniques, and they frequently employ neural network models to build a bot and use the neural network bot to complete the task. Because bots are produced by users, each of whom has a unique approach to competing in the market, it is preferable to keep some fundamental information as opposed to teaching them, and then have the bot act on the basis of each user's unique observations.

The bot may therefore dynamically inform itself on whether or not to do the essential action by utilising certain predetermined indicator variants. Reviewing all of the earlier trials and attempts on the idea shows that there is still a chance to maximise the processing or memory capacity, even if erroneous or inaccurate information is presented.

3. Methodology

A programme by employing data sets in the form of a GUI interface model and using the available GUI Libraries, is capable of correctly predicting outcomes precisely when compared to humans without making any mistakes. A programming language is employed that fulfils its capabilities based on the current trend in data estimation and analysis.

PYAUTOGUI, a library that is already inbuilt for the usage of any GUI interface task loaded into Python, can be used to trade currencies as we are going to provide the information or required data for its input in the form of the Graphical model form. It results in the automation of all keyboard and mouse operations. This automation's main objectives are to do the work effectively and promptly assess the trend. If a human computation takes 30 seconds and there is a 30-second time delay before an order is placed. In contrast, a bot that has been trained to perform the same activity is not required to check its results. Checking the upcoming trend is the only way to set orders that will allow you to manage market hatred that has already occurred and produce successful results at the end of each market open day. It is feasible to evaluate the model's efficacy using a precise projection of price movement based on the previously trained collection of all photos.

The functioning of the model is calculated based on whether the provided image matches the market situation or not. A job with only two outputs has a 50% chance of occurring, according to the probability formula. There is a change in the likelihood of performing an event that may be determined by analysing the provided inputs. There is a strong chance that the selected sequence will be accurate if testing and study are conducted to establish the best feasible task based on the previously occurred moves. Concentrating on both goals might result in a lot of work, memory use, and wasted time. It is considerably simpler to evaluate which of the two possibilities is more likely to occur in the future when only one of the two possibilities is considered.

To train the bot dynamically and then use it for a clear and perfect choice, as demonstrated above, some input must be provided to observe the clear way of finding the outcome. To optimise the algorithm of the find approach, only one dependent variable must be taken into consideration and its likelihood must be checked.

If something doesn't happen, the outcome will automatically be the different option. Therefore, in order to provide a clearer understanding of the concept, we will only take into account the method that establishes the SELL condition.

If the condition has a larger possibility of occurring, the result is "SELL," which may even be carried out by mouse using an automated job. The sole alternative is to choose to BUY, which is the only option if the prerequisites are not met or the current circumstances does not warrant it.

There are numerous currency alternatives available on the market, some of which include

Digital

1. EUR/USD
2. EUR/JPY

Binary

1. EUR/USD(OTC)
2. EUR/JPY(OTC)

3.1 Candlestick Pattern Behaviour

When estimating the optimum result from a calculation, some required data must be present. The calculation uses just the data that are required for processing. To fulfil our essential tasks, such as determining the prices' opening and closing prices as well as upper and lower deviations and ranges, we need information from the candlestick patterns. Since the findings from the second time period might potentially be drastically different, it is necessary to tabulate and compute all the data in order to address the issues of greater data storage and execution latency. The only data type saved is the pixel format, and the accuracy of the image is assessed using a probability scale ranging from 0 to condition. The shortest feasible value turn in the candle behaviour may be calculated using this straightforward approach based on the movement of the image candles.

The fundamental guidelines of the candlestick view are shown in Fig. 35.1. Even though the candlestick is a data representation form and has open, close, low, and high data ranges present, it is also instantly feasible to compare an image with its pixel view. While candlestick charts are easier to comprehend, traders still prefer them even though HLOC (High Low Open Close) charts offer almost unique amount of data. It's helpful to experiment with both to see which works best for you.

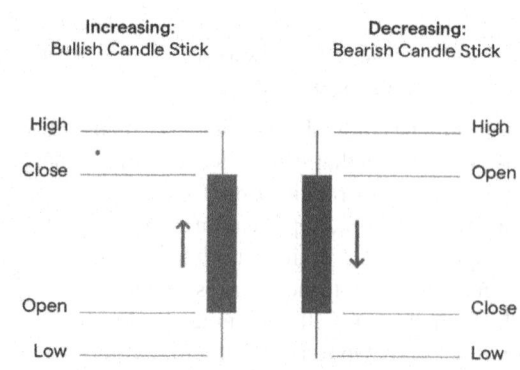

Fig. 35.1 Data from candlesticks[14]

3.2 SELL Condition Possibilities

Here are some basic conditions taken into consideration from the observed market condition and the present time frames then by considering these we have computed the whole task and even the results are also be done using these data only.

Table 35.1 Various input defined fields

All possible SELL conditions	Each trend meaning	
	1st candle meaning (1st check)	2nd candle meaning (2nd conformation)
	Higher fall with a upper move next.	Continuous to go more down.
	A normal downtrend candle.	More downtrend.
	Down with rebounded back.	More completely down trend.
	Down with rebounded back.	Comes up until the previous high and then goes down.
	Downtrend candle.	Reboundedhigh upper trend iswith minimal probability.
	Uptrend hammer candle.	Normal candle of down trend.
	Downtrend hammer candle.	Downtrend candle leads to more down

Only a probability greater than 50% is necessary. It is a clear trend probable move after a clear match with the previously set input ranges.

(A positive condition results in a good outcome, while a negative condition, if it fails, likewise produces a positive outcome.)

Note: The exact same input may not work for the others there are some of the probability values which are been observed from the behaviour of market and given to bot with those inputs.

If there is any influence from other factor such as a market meeting, then the market is already stabilising at that moment, but after the event it may show some of variant behaviour and sets back to normal. In the cases that are performed, a series of tests are computed to check the particular feasible value, and then out of those tests, an optimal value is taken into consideration.

4. Results

When the programme is started using the coded programme, it runs a set by set of code and provides outputs in accordance with the present situations. This is because the model or bot that is built is entirely a dynamically running bot with the predetermined value using the inputs. The outcomes in my instance is as follows.

Fig. 35.2 Results for particular tested bot

Figure 35.2 shows the possible outcomes when the bot is used for testing. A minimal basic amount is first used to compute and test bot after a higher value of capital is used and higher profits are booked.

The deals listed above were carried out on separate days and led to the outcomes. When you look at the accompanying chart, you can see that even after several checks and trials, there are still some uncommon occurrences. In this scenario, the likelihood of the market being absolutely correct is relatively low. However, there is a lot of danger to be faced in those circumstances if a significant amount of market cash is involved can be observed in Fig. 35.3. If multiple orders are being place in a single time frame there is a chance of at least 1 order being correct.

A user can also be able to halt the bot when a good point of chance is gained by the user and can put a guarantee order with either a large amount. The bot itself stabilised the maximum and minimum obtained by maintaining an order

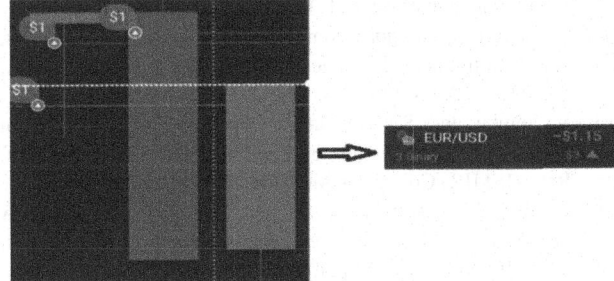

Fig. 35.3 Result with failure case with minimal loss

in the possible limits. It allows for the estimation of the following candle's frame size. As long as only multivariate dependent true situations are taken into account, there are alternate and potentially effective approaches to provide the exact right result. The sole goal at this time is to do a single check and provide an effective result. The examples and outcomes mentioned above demonstrate that it is feasible.

5. Conclusion

Financial estimates cannot be predicted very effectively since they are markets and human-emotion-dependent, non-linear, multi-variable dependant functions. The proposed methodology doesn't need to be trained like machine learning algorithms based on the graphical data using the probabilistic analysis irrespective of the data trend with in a time frames of 30 seconds which leads to a quick decision making, the bot makes the prediction easily. The proposed method is a generic since the model is based probabilistic analysis.

When short ranges are taken and multiple orders are placed with a small amount of capital size, the market is on a long windy day and there is only a small loss. Perhaps instead of counting the profit on the maximum capital size, one should consider the smallest amount of incorrect predictions. If that amount is manageable, then additional small frequent orders can bring in the profit and balance the day only on those market bad days.

Finally, it is feasible to make a forecast, but there are very few instances in which the prediction will be incorrect. If the market has any unsustainable elements, such as governmental interference or the influence of powerful industries, etc. The main objective of the assignment is to maximise profit since placing repeated orders for little amounts of cash in short time intervals will always result in a steady and large profit.

References

1. C.-L. Hsu, D. Wang, J.-S. R. Jang, and K. Hu, "A tandem algorithm for singing pitch extractionand voice separation from music accompaniment," IEEE Trans. Audio, Speech, Language Process., vol. 20, no. 5, pp. 1482–1491, Jul. 2012.

2. J. Salamon, E. Gomez, D. P. W. Ellis, and G. Richard, "Melody extraction from polyphonic music signals: Approaches, applications, and challenges," IEEE Signal Process. Mag., vol. 31, no. 2, pp. 118–134, Mar. 2014.

3. S. Vembu and S. Baumann, "Separation of vocals from polyphonic audio recordings," in Proc. Int. Soc. Music Inf. Retr. (ISMIR) Conf., London, U.K., 2005, pp. 1–8.

4. Schroff. F, Kalenichenko. D and Philbin.J, "FaceNet: A unified embedding for face recognition and clustering," in Proc. IEEE Conf. Comput. Vis. Pattern Recognit. (CVPR), Boston, MA, USA, Jun. 2015, pp. 7–12.

5. Y. Sun, X. Wang, and X. Tang, "Deep learning face representation from predicting 10,000 classes," in Proc. IEEE Conf. Comput. Vis. Pattern Recognit., Columbus, OH, USA, Jun. 2014, pp. 23–28.

6. Y. Taigman, M. Yang, M. Ranzato, and L. Wolf, "DeepFace: Closing the gap to human-level performance in face verification," in Proc. IEEE Conf. Comput. Vis. Pattern Recognit., Columbus, OH, USA, Jun. 2014, pp. 23–28.

7. Birogul, Serdar, Gunay Temur, and Utku Kose. "YOLO Object Recognition Algorithm and 'Buy-Sell Decision' Model Over 2D Candlestick Charts." IEEE Access 8 (2020): 91894–91915. Web.

8. Mathur, Medha & Mhadalekar, Satyam & Mhatre, Sahil & Mane, Vanita. (2021). Algorithmic Trading Bot. ITM Web of Conferences. 40. 03041. 10.1051/itmconf/20214003041.

9. Refet Gürkaynak & Justin Wolfers, 2005. "Macroeconomic Derivatives: An Initial Analysis of Market-Based Macro Forecasts, Uncertainty, and Risk," NBER Chapters, in: NBER International Seminar on Macroeconomics 2005, pages 11–50, National Bureau of Economic Research, Inc.

10. Dunis, Christian & Williams, Mark. (2011). Modelling and Trading the EUR/USD Exchange Rate: Do Neural Network Models Perform Better?. Derivatives Use, Trading and Regulation. 8.

11. Sarkar, Md & Ali, U A Md Ehsan. (2022). EUR/USD Exchange Rate Prediction Using Machine Learning. 8. 44–48. 10.5815/ijmsc.2022.01.05.

12. Thibaut Théate, Damien Ernst,"An application of deep reinforcement learning to algorithmic trading",Expert Systems with Applications,ISSN 0957-4174.

13. Lantana Dioren Rumpa et al 2021 IOP Conf. Ser.: Mater. Sci. Eng. 1088 012107.

14. IG, Patrick Foot, 15 June 2020, https://www.ig.com/en/trading-strategies/japanese-candlestick-trading-guide-200615.

15. Martha Flores-Sosa, Ernesto León-Castro, José M. Merigó, Ronald R. Yager, Forecasting the exchange rate with multiple linear regression and heavy ordered weighted average operators, Knowledge-Based Systems, Volume 248, May 2022, 108863, ISSN 0950-7051.

16. Ertek G, Al-Kaabi A, Maghyereh AI. Analytical Modeling and Empirical Analysis of Binary Options Strategies. Future Internet. 2022; 14(7):208.

17. Sharma, Dinesh Kumar. "Prediction of Foreign exchange rate using regression techniques." Review of Business and Technology Research, Vol. 14, No. 1, 2017, ISSN 1941-941.

Note: All the figures and table in this chapter were made by the authors.

Extracting Trending Topics from Social Media Posts Using Word Embedding Based Clustering

36

P. S. Mahalle[1], P. N. Chatur[2],
A. V. Deorankar[3], K. A. Waghmare[4]
Department of Computer Science and Engineering,
Government College of Engineering, Amravati, Maharashtra

A. S. Jumde[5]
Department of Computer Engineering,
Bajaj Institute of Technology, Wardha

Abstract: With the ever-increasing volume of online textual data, the need to automatically identify and track trending topics has become crucial for various applications, such as social media monitoring, news analysis, and market research. In this study, we propose a methodology for detecting trending topics by combining the power of Word2Vec, t-SNE, TF-IDF, and clustering algorithms. We utilize the Word2Vec algorithm to generate high-dimensional word embeddings from a large dataset of text documents. These embeddings capture the semantic relationships between words, enabling us to measure their similarity and represent their meanings in a continuous vector space. Next, we apply the t-SNE (t-Distributed Stochastic Neighbor Embedding) algorithm to drop off the dimensionality of the word embeddings while preserving their inherent structures. This step helps visualize the relationships between words in a two-dimensional or three-dimensional space, making it easier to identify clusters and patterns. To further enhance the trend detection process, we leverage the TF-IDF (Term Frequency-Inverse Document Frequency) approach. TF-IDF measures the importance of a word within a document and across a corpus, allowing us to prioritize terms that are both frequently occurring and specific to a particular document. We analyze the clusters formed by the t-SNE visualization and identify keywords that exhibit high TF-IDF scores within those clusters. Our study contributes to the field of trend detection by providing a comprehensive methodology that combines Word2Vec, t-SNE, and TF IDF algorithms.

Keywords: Trending topic detection, Word2Vec, t-SNE, TF-IDF, Dimensionality reduction, Clustering

1. Introduction

The proposed work on the detection of trending topics from social media posts using clustering builds upon the growing importance of social media as a source of real-time information and public sentiment analysis [1-2]. With the rise of platforms like Twitter, Facebook, Instagram, and YouTube, social media has become a commanding medium for individuals and organizations to express their opinions, share content, and engage in discussions. Social media platforms generate an enormous volume of data, with millions of posts, comments, and interactions being generated every day [3].

Extracting valuable insights from this vast amount of data is a challenging task due to its dynamic nature, noise, and the need to filter out irrelevant or spam content. The proposed work aims to address the challenges associated with trend detection in social media by leveraging clustering algorithms. It focuses on data collection, pre-processing, feature extraction, and the application of clustering techniques to identify trending topics. The objective is to present a thorough comprehension

[1]mahallepriyanka2@gmail.com, [2]prashant_chatur@rediffmail.com, [3]avdeorankar@gmail.com, [4]waghmare.kamlesh@gmail.com,
[5]amol.jumde@bitwardha.ac.in

of the procedure and its potential applications. By developing effective techniques for detecting trending topics from social media posts, this project contributes to the field of social media analytics and offers practical insights for decision-making, marketing strategies, and understanding public sentiment.

Here we summarize our main contributions:

- Proposed an effective and efficient method for detecting trending topics from social media posts using clustering techniques
- Proposed a comprehensive methodology that combines Word2Vec, t-SNE, and TF-IDF algorithms
- Prepared a new dataset by merging social media posts from different avenues.

The organization of the paper is as follows. The existing work carried out in this field is explained in Section 2. Section 3 describes the proposed approach for trending topics detection. The experimental results and analysis are given in Section 5. Section 4 discusses the implementation and experimentation. Finally, we draw conclusions and suggest future scope in Section 5.

2. Related Work

Social media platforms have revolutionized communication and information sharing, leading to the generation of massive amounts of data. Analyzing this data to extract valuable insights and detect topics of interest has become crucial for various applications. Traditional methods of topic detection in social media often face challenges in capturing the semantic relationships among words. Word embedding-based clustering techniques [15, 19, 21] have emerged as a powerful approach to overcome this limitation.

World embedding techniques, such as Word2Vec, GloVe, FastText, ELMo, and BERT, have been widely adopted in natural language processing tasks [4]. These techniques map words to high-dimensional vectors, capturing their semantic meanings and contextual relationships. By encoding words into continuous vector representations, world embeddings enable clustering algorithms to measure the similarity between words based on their vector distances. This, in turn, facilitates the identification of coherent topic clusters [1, 3, 14, 15].

Clustering algorithms play a vital role in grouping similar words and identifying topics in social media data. K-means clustering, agglomerative clustering, DBSCAN, and hierarchical clustering are commonly employed algorithms in this context [22]. These algorithms partition the word embeddings into clusters based on their proximity, enabling the detection of topics with related terms grouped together. The choice of clustering algorithm depends on factors such

as the size of the dataset, the preferred number of clusters, and the density of the data points. World embedding-based topic detection approaches have been explored extensively in recent research. Traditional clustering methods that utilize world embeddings often involve applying clustering algorithms directly to the word embeddings. Enhanced clustering algorithms integrate world embeddings as additional features, enriching the clustering process with semantic information. Hybrid approaches combine world embeddings with other techniques, such as graph-based algorithms or topic modeling, to improve the accuracy and robustness of topic detection.

To evaluate the performance of world embedding-based clustering for topic detection, various evaluation metrics are utilized. Purity measures the extent to which each cluster contains instances of a single true topic. Normalized Mutual Information (NMI) quantifies the mutual dependence between the true and detected topics. F1-score assesses the balance between precision and recall in topic detection. The silhouette coefficient measures the coherence within clusters by considering the compactness of data points within a cluster and the separation between different clusters [9-10].

Case studies and applications have demonstrated the effectiveness of world embedding-based clustering in topic detection for social media data. Twitter data, with its real-time nature and limited text length, has been extensively studied for topic detection using world embeddings [16-18, 20]. Sentiment analysis, which aims to detect the sentiment associated with specific topics in social media, has also benefited from world embedding-based clustering. Additionally, real-time topic detection in social media streams has been explored, enabling the identification of emerging topics as they unfold [23].

3. Proposed Methodology

The proposed trending topic detection system is depicted in Fig. 36.1. First, we collect the data from different social media platforms like Twitter, Instagram, Kaggle, etc. We apply the preprocessing steps on the data like merging all the data, removing punctuation, tokenization, stopwords removal, converting into lowercase. We convert the words into the word vectors using the word2vec tool. Now, the word vectors are grouped into the clusters by applying the K-Means clustering algorithm. Finally, the labels for each of the clusters is determined using the tf-idf algorithm. We discuss each of these components here.

3.1 Tokenization

The process that breaks a set of sentences into tokens, individual words, symbols, terms, sentences, or other important elements is known as tokenization. It is the process

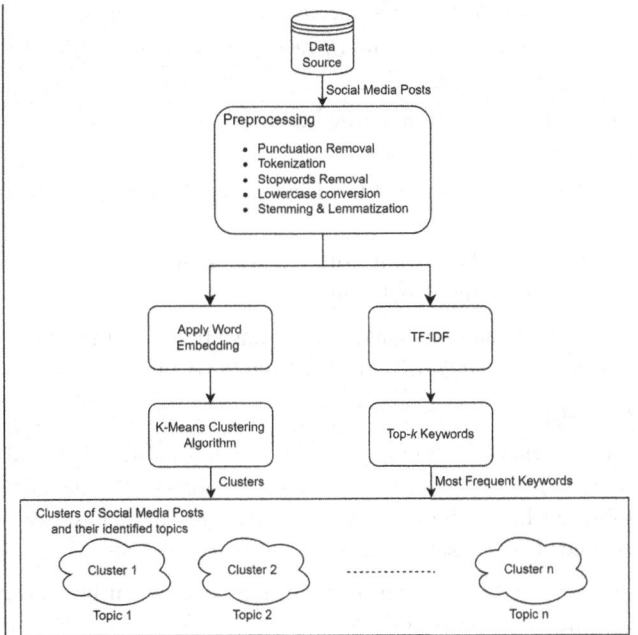

Fig. 36.1 Block diagram of the proposed system

Source: Author

of splitting natural language text into data parts that can be viewed as discrete units.

Example:

```
word_tokenize("The sole meaning of life is to serve humanity")
['The', 'sole', 'meaning', 'of', 'life', 'is', 'to', 'serve', 'humanity']
```

3.2 Stop Words Removal

Stop words are the words that do not contribute much in the NLP task and therefore, generally skipped in the pre-processing stages. Such terms, which are among the most prevalent in any language (together with articles, conjunctions, pronouns, prepositions, etc.), do not provide great information for the text. In English, stop words include "the," "a," "an," "so," and "what."

Product review: "The dress quality was very poor."

After stop-word removal: "dress quality poor"

3.3 Word2Vec

The word2vec [24] word embedding model takes text corpus to produce the word vectors. These vectors are the vector representations of the corpus words. The word vectors capture the semantic relationship between the words. Therefore, similar words are placed close to each other in the vector space. Many applications from machine learning and natural language processing fields can use the generated word vectors file as features to find similar or dissimilar words.

3.4 Clustering

Clustering falls under unsupervised type of machine learning. Clustering algorithms receive abundant amount of unlabeled data and they iteratively group data into disjoint set of clusters. Elements inside one cluster are similar to each other as per the similarity criteria considered for clustering. Clustering finds its application in domain like pattern discovery, feature engineering, or it is utilized in an initial stages of the other algorithms. It is a useful tool to get the insights from the unknown data.

Comparison schemes

Comparisons with the following clustering approaches [3,5] are made to determine how effective the proposed technique is compared to current techniques.

1. *K-Means*: The well-known partitioning-based clustering algorithm is k-means. Although it was initially developed in 1967 [6], it is still commonly used today due to its simplicity and low computing cost.

2. *Agglomerative clustering*: This clustering approach [10] is hierarchical. In which data points are gathered into clusters to produce a dendogram [9], a structure resembling a tree.

3. *Density-based*: This approach is a reliable variation of the density-based clustering algorithm DBSCAN [11]. It is called "based spatial clustering of applications." It is a relatively new clustering technique that outperformed a number of other algorithms [7-8].

4. *Genie*: A relatively new hierarchical clustering technique [12] uses Gini index [13], a well-liked statistical index for assessing spreading in a given set of frequency values. Genie makes sure that the Gini index value stays below a predetermined limit. The smallest cluster is combined with its closest neighbor if it exceeds.

5. *Disambiguated Core Semantics* (DCS): This method [14] employed lexical chains—groups of semantically connected words formed from WordNet—to identify the most crucial ideas in a document. K-means is then used to extract the document clusters from this smaller set of topics.

6. *Stamantic Clustering* (STC): By combining statistical and semantic data with TF-IDF as a scoring system and utilizing semantic relationships with WordNet, this method [3] achieves document clustering.

3.5 Trend Detection

Here the objectives are:

(a) Monitor the generated clusters and analyze their characteristics to identify trending topics.

(b) Develop algorithms to detect and track the emergence and evolution of trending topics over time.

(c) Utilize metrics such as cluster size, activity spikes, or time-based decay functions to determine the significance of trends.

3.6 Data Visualization

Data visualization is a technique to plot the data to interpret the relationship, pattern, and trends present between the data. Libraries like matplotlib, plotly, seaborn provide ready tools to visualize the data easily.

4. Experimentation

4.1 Datasets

Twitter

The dataset is titled "Cricket Tweets" and it contains tweets about cricket, which is a popular team sport worldwide. The dataset is in the form of a CSV file with 16 columns. Here is a brief description of the columns in the dataset:

- *id*: tweet identifier.
- *user_name*: tweeter user
- *user_location*: tweeter user location
- *user_description*: user's description.
- *user_created*: profile creation date
- *user_followers*: count of followers

- *user_friends*: count of acounts followed by the user
- *user_favourites*: count of tweets marked as favorites by the user
- *user_verified*: is a verified user?
- *date*: date of tweet
- *Label*: A categorical label indicating the range of tweet counts.
- *Count*: The count of tweets falling within the corresponding label range.

To perform additional analysis, we have retrieved the "*user_description*" and "*text*" columns from the dataset.

Instagram

The provided dataset is named "most_followed_ig" and sourced from Kaggle. It contains information about the most followed Instagram accounts as of the time the dataset was created. The dataset includes the following columns:

- *RANK*: The ranking of the Instagram account based on their number of followers.
- *BRAND*: The name of the brand or individual associated with the Instagram account.
- *CATEGORIES 1*: The primary category associated with the account.
- *CATEGORIES 2*: The secondary category associated with the account.
- *FOLLOWERS*: The number of followers the account has on Instagram.

```
In [3]:    1  data.head()
```

Out[3]:

	id	user_name	user_location	user_description	user_created	user_followers	user_friends	user_favourites	user_verified	dat
49729	1458460493979389956	Rishi	Earth	interested in Data Science | Formula 1| AutoSe...	2012-12-20 02:12:22	385	2592	18000	False	2021 11-1 15:44:1
47427	1451591381206507539	Ankit Shah ð□□®ð□□³	Ahmedabad, Gujarat, à¤- à¤¾à¤°à¤¤ ð□□®ð□□³	Be an #Encourager the world has plenty of #Cri...	2012-03-03 20:47:03	173	50	603	False	2021 10-2 16:48:4
12845	1435100881771499521	Ibrar ali Khokhar	Sindh, Pakistan	IAM Not ð□□« BoRn To ImPRessS UH ð□□□	2021-09-07 02:03:11	51	381	80	False	2021 09-0 04:41:2
51355	1458285796989947910	CricketTimes.com	Lucknow, Uttar Pradesh	Latest Cricket News, Cricket Facts, Live Score...	2015-01-29 05:18:34	39237	311	188	False	2021 11-1 04:10:0
60209	1468972961000394761	Ø³Ù□اØ§Ù□ اا§Ù□ Ø´ااØ±Ú©Ø²Ø¡	Azad Kashmir Province	Member of Barakzai welfare Youth Moment.\nBara...	2019-11-16 13:13:24	1416	1037	11589	False	2021 12-0 15:57:0

Fig. 36.2 Electronic media text data

Source: Author

```
In [8]: df.head()

Out[8]:
```

	RANK	BRAND	CATEGORIES 1	CATEGORIES 2	FOLLOWERS	ER	iPOSTS ON HASHTAG	MEDIA POSTED
0	1	Selena Gomez	celebrities	musicians	105.4Mæ(=)	2.62%æ(1342)	14.5Mæ(48)	1.2kæ(2135)
1	2	Taylor Swift	celebrities	musicians	95.2Mæ(=)	1.96%æ(2040)	10.5Mæ(66)	958æ(2669)
2	3	Ariana Grande	celebrities	musicians	92.3Mæ(=)	1.43%æ(2759)	16.9Mæ(41)	2.8kæ(824)
3	4	Beyonce	celebrities	musicians	90.6Mæ(=)	2.53%æ(1427)	9.2Mæ(70)	1.4kæ(1897)
4	5	Kim Kardashian West	celebrities	tv	89.3Mæ(=)	1.39%æ(2812)	5.1Mæ(130)	3.6kæ(550)

Fig. 36.3 Experiment data

Source: Author

- *ER*: Engagement rate, expressed as a percentage.
- *iPOSTS ON HASHTAG*: The number of posts that include the specified hashtag.
- *MEDIA POSTED*: The total number of media posts on the account.

Each row in the dataset represents an Instagram account, and the values in each column provide relevant information about that account. The dataset is likely intended to be used for analysis and insights regarding popular Instagram accounts and their followership.

We had fetched the "BRAND" and "CATEGORIES" columns from the dataset for further analysis. We intentionally kept social media posts of three categories *i.e.* Instagram, Governments & Policymaking, and Sports mainly Cricket, for cross-validation. Therefore, at the end, we expect three clusters and these categories as their topics names.

4.2 Libraries Used

1. *Pandas:* An open-source Python package for data manipulation and analysis is pandas. For managing structured data, it offers efficient, powerful, easy-to-use APIs and data interpretation capabilities.

2. *Nltk:* Working with human language data is made easy with the NLTK (Natural Language Toolkit) toolkit, an effective open-source Python package. It offers a variety of tools and resources for conducting NLP tasks like tokenization, semantic reasoning stemming, parsing, tagging.

3. *Numpy:* A robust open-source Python library called NumPy (sometimes known as "Numerical Python") supports high, multi-dimensional arrays, tensors and matrices and offers several mathematical functions to effectively work with them. It serves as the foundational library for scientific computing in Python.

4. *sklearn.manifold:* The sklearn.manifold module in scikit-learn (sklearn) is a sub-module that provides methods for manifold learning and dimensionality reduction. Manifold learning algorithms aim to uncover the underlying structure and relationships within high-dimensional data by mapping it to a lower-dimensional space.

5. *Matplotlib:* It is one of the most widely used visualization libraries. It provides easy-to-use tools to create various types of plots, including bar plots, line plots, histograms, scatter plots, 3D plots, and more. It is highly customizable, giving you full control over the appearance and style of your plots.

6. *Gensim:* Gensim is an open-source Python toolkit for natural language processing (NLP), document similarity analysis, and topic modelling. It provides a simple and efficient way to analyze and process large text corpora.

7. *sklearn.feature.extraction.text:* The sklearn.feature _extraction.text module in sklearn provides tools for feature extraction from text data. It provides numerous techniques for transforming text documents into numerical feature representations that may be fed into machine learning algorithms. Key classes and functions available in sklearn.feature _extraction.text include:

   - *CountVectorizer:* This class turns a matrix of token counts into a collection of text documents. It tokenizes the text, creates a vocabulary of recognised terms, and tracks how many times each word appears in each document.
   - *Tf idf Vectorizer:* With the help of this class, a group of text documents can be transformed into a matrix of TF-IDF (Term Frequency-Inverse Document Frequency) properties. It determines how significant each word is in relation to the total corpus of a document.

5. Results and Discussion

In K-Means clustering technique, *k* clusters are initially generated at random, and then they are iteratively adjusted until their *k*-centroids converge. However, we need to figure out the suitable value of *k*. We employ the Elbow approach to calculate k. In this method, we constantly iterate from $k = 1$ to $k = n$. We determine the within-cluster sum of squares (WCSS) value for each value of k. The WCSS is the square sum of the distances between the centroids and each point. Now, we draw a graph of k vs their WCSS value to determine the ideal number of clusters (k). Surprisingly, the graph resembles an elbow as shown in Fig. 36.4. Additionally, the WCSS value is maximum when $k = 1$, but it starts to fall as k increases. When the graph begins to resemble a straight line, which is the value of k that we chose. For our merged dataset, from the Fig. 36.4, it is evident that the $k = 3$ is the best value of the number of clusters which was our anticipated value.

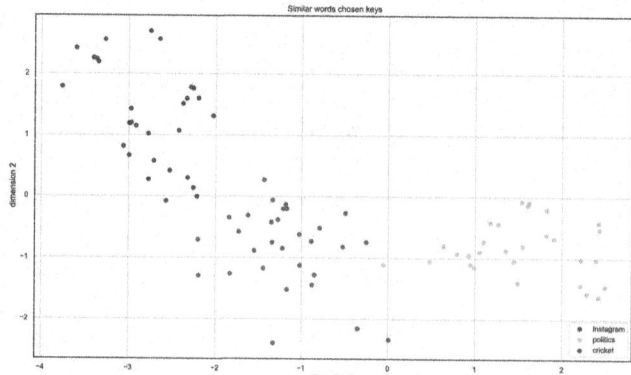

Fig. 36.5 Visualization of the results of the proposed system. Trending topics in the identified clusters are Instagram, politics, and cricket

Source: Author

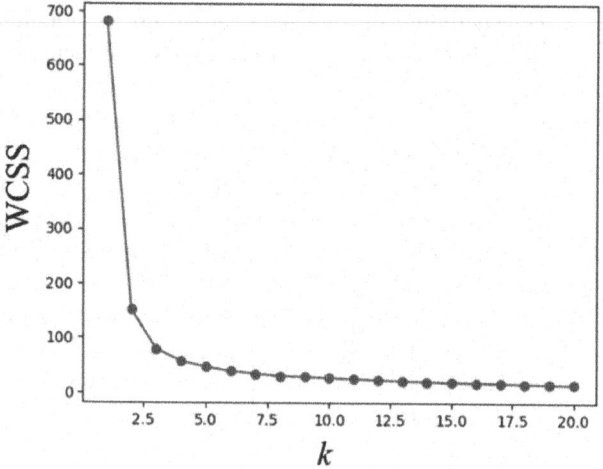

Fig. 36.4 Elbow method to determine k value [10]

After clustering, t-SNE algorithm is used to reduce the dimension of the clustered data. t-SNE is a popular dimensionality reduction technique and ideal for a visualization purpose. We reduced the 150-dimensional data into 2-dimensional data using t-SNE and plotted the scatter plot (Fig. 36.5) for clear understanding of the performance of the proposed system. Proposed topic extraction is system is able to find out the trending (most frequent) keyword from the clusters and which are matching with the topics that we kept purposefully.

6. Conclusion

In conclusion, this study proposes a comprehensive methodology for detecting trending topics in large text datasets. By integrating the power of Word2Vec, K-Means, t-SNE, and TF-IDF algorithms, we are able to capture semantic relationships, visualize word clusters, and prioritize important terms. Our approach leverages Word2Vec to generate word embeddings, t-SNE to reduce dimensionality and visualize relationships, and TF-IDF to prioritize terms based on frequency and specificity. By analyzing clusters formed by t-SNE and identifying high TF-IDF scoring

keywords, we can identify current trending topics. The proposed system for detecting trending topics through social media posts using clustering techniques has the potential to provide valuable insights for social media managers, digital marketers, or researchers who are interested in monitoring public sentiment, tracking brand mentions, and identifying emerging trends in real-time. The obtained results also complements this observation.

References

1. Comito, Carmela, Agostino Forestiero, and Clara Pizzuti. "Word embedding based clustering to detect topics in social media." IEEE/WIC/ACM International Conference on Web Intelligence. 2019.
2. Onan, Aytuğ. "Two-stage topic extraction model for bibliometric data analysis based on word embeddings and clustering." *IEEE Access* 7 (2019): 145614–145633.
3. Mehta, Vivek, Seema Bawa, and Jasmeet Singh. "WEClustering: word embeddings based text clustering technique for large datasets." *Complex & intelligent systems* 7 (2021): 3211–3224.
4. Manning CD, Schütze H (1999) Foundations of statistical natural language processing. MIT press, New York
5. Jain AK (2010) Data clustering: 50 years beyond k-means. Pattern Recogn Lett 31(8): 651–666
6. Fränti P, Sieranoja S (2018) K-means properties on six clustering benchmark datasets. Appl Intell 48(12):4743–4759
7. McInnes L, Healy J (2017) Accelerated hierarchical density based clustering. In: 2017 IEEE international conference on data mining workshops (ICDMW). IEEE, pp 33–42
8. McInnes L, Healy J, Astels S (2017) hdbscan: Hierarchical density based clustering. J Open Sour Softw 2(11): 205
9. Saxena A, Prasad M, Gupta A, Bharill N, Patel OP, Tiwari A, Er MJ, DingW, Lin CT (2017) A review of clustering techniques and developments. Neurocomputing 267:664–681
10. Xu D, Tian Y (2015) A comprehensive survey of clustering algorithms. Ann Data Sci 2(2): 165–193

11. Ester M, Kriegel HP, Sander J, Xu X et al (1996) A density-based algorithm for discovering clusters in large spatial databases with noise. Kdd 96:226–231
12. Gagolewski M, Bartoszuk M, CenaA(2016) Genie: a new, fast, and outlier-resistant hierarchical clustering algorithm. Inf Sci 363: 8–23
13. Ceriani L, Verme P (2012) The origins of the gini index: extracts from variabilità emutabilità (1912) by corrado gini. J Econ Inequal 10(3): 421–443
14. Wei T, Lu Y, Chang H, Zhou Q, Bao X (2015) A semantic approach for text clustering using wordnet and lexical chains. Expert Syst Appl 42(4):2264–2275
15. Carmela Comito, Word Embedding-based Clustering to Detect Topics insocial-media, Nat. Research Council of Italy (CNR), Institute for HighPerformance, Rende, Italy, 2022.
16. Md Shoaib Ahmed, Online Topical Clusters Detection for Top-k Trending Topics in Twitter, Jahangirnagar University, Dhaka, Bangladesh, 2020.
17. Ma. Shiela C. Sapul, Trending Topic Discovery of Twitter Tweets Using Clustering and Topic Modeling Algorithms, Assumption University, Bangkok, Thailand, 2017.
18. W. Ahmed, S. Hameed, A. Anwar, A. Tariq, A. Raza, "Efficient Trending Topic Detection and Tracking on Twitter," in Proceedings of the 2019 International Conference on Frontiers of Information Technology (FIT), Islamabad, Pakistan, 2019.
19. A. Abbasi, S. Chen, A. Salem, "Topic Detection and Tracking in Social Media: A Survey," ACM Computing Surveys (CSUR), vol. 51, no. 4, article no. 82, 2018.
20. G. Li, J. Yang, "Trending Topic Detection in Twitter Using Machine Learning Techniques," in Proceedings of the 2018 IEEE International Conference on Big Data (Big Data), Seattle, WA, USA, 2018.
21. N. Hosseinzadeh, A. SiamiNamin, "Trending Topic Detection in Social Media: A Review," Information Processing & Management, vol. 53, no. 5, pp. 994–1015, 2017.
22. S. Boorboor, P. El Khoury, K. Toutanova, C. De Vleeschouwer, "Joint Clustering of Users and Hashtags in Social Media," in Proceedings of the 2016 IEEE/ACM International Conference on Advances in Social Networks Analysis and Mining (ASONAM), San Francisco, CA, USA, 2016.
23. J. Yang, T. Huang, Y. Wu, "Trending Topic Detection on Twitter Based on Temporal and Social Analysis," in Proceedings of the 2015 IEEE International Conference on Data Science and Data Intensive Systems (DSDIS), Sydney, Australia, 2015
24. Mikolov, Tomas, Kai Chen, Greg Corrado, and Jeffrey Dean. "Efficient estimation of word representations in vector space." *arXiv preprint arXiv:1301.3781* (2013).

Heart Disease Prediction System Using Machine Learning

37

**Vaishnavi Bhuyar[1], Anil Deorankar[2],
Kamlesh Waghmare[3]**

Department of Computer Science and Engineering,
Government College of Engineering Amravati, India

Abstract: Heart disease (HD) is a leading cause of death worldwide. It is therefore vital that health care providers work together to protect patient health and save lives. In this paper, we compared the performance of various classifiers to accurately classify a heart disease dataset and predict heart disease cases using several features.

The healthcare sectors amassed massive amounts of data containing sensitive information. This data collection helps us make informed decisions. To achieve reliable results and make educated data judgements, several advanced data mining techniques are used. In this case, the "HDPS" system for predicting heart disease was developed by means of logistic regression, decision tree, K-nearest neighbor, support vector machine and random forest algorithm to calculate the risk level for heart illness.

The end results show that Logistic Regression has the highest accuracy at 88.15%, while decision tree, K-neighbor classifier, support vector classifier and random forest classifier have 75%, 63.15%, 60.52% and 82.89% corresponding.

Keywords: Logistic Regression, Machine learning, heart disease, accuracy, random forest classification

1. Introduction

Data mining is a method of discovering previously unknown themes, patterns and recent trends in databases and using this knowledge to construct prediction models. In order to find patterns and create correlations within vast databases, data mining technology uses numerical study, machine learning algorithms and database technology management systems.

World health statistics from 2012 highlight a problem: one in three adults of all ages was susceptible toward high blood pressure, a condition responsible for partly of demises from heart disease as well as stroke. Heart disease, well-known as cardiovascular disease (CVD), is not just heart illness, but a variety of conditions that affect the heart. This moment has proven fatal to a human each 34 seconds in the United States.

Coronary artery illness, cardiovascular and cardiomyopathy health problems are specific sub-areas where the body's blood is pumped and circulated.

Diagnosis is an main task that must carry out capably. This can be generally done beneath the supervision of a specialist. This leads to unsuitable comes about and expensive medical expenses for patient care. Therefore, this prediction system turns out to be very advantageous.

Many scientists have tried to develop models that can predict early heart disease, but no model is perfect. Each proposed framework has its claim disadvantages. In existing systems, Chen et al. came up with the thought of predicting heart disease. For classification and prediction purposes, he used vector quantization techniques, one of the technique is artificial intelligence. A neural network is trained using back-propagation to calculate the prediction system. We achieved about 80% accuracy on the test suite during the tesing phase. It takes time to actually use the data gathered from historical records. Low accuracy rate.

To clarify this problem, we use a logistic regression algorithm to obtain precise results quickly. Machine learning

---
[1]vaishnavibhuyar1998@gmail.com, [2]avdeorankar@gmail.com, [3]waghmare.kamlesh@gmail.com

has become a main concern in many new bioapplications and medicine. In the field of prediction, Machine learning will play an important role. Our subject is cardiac disease prediction through patient records and patient data processing i.e., number of users whose probability of developing heart disease needs to be predicted.

2. Related Work

Many studies have been prepared focusing on top of the analysis of heart illness. They applied dissimilar data mining techniques to diagnose as well as obtained different probabilities for different techniques.

These parameters are evaluated by the system utilising data mining classification algorithms. Python evaluates datasets using two major machine learning methods. In conditions of cardiac accuracy, the naive bayes algorithm and decision tree algorithm were considered the better of the two techniques [1].

Aditi Gavhane et al. Predicted myocardial infarction intended for before time determination toward decrease mortality. For this, machine learning is heavily emphasized in this article. This prophecy takes people out of life's danger zone. In this article, we use the random forest and KNN algorithms to predict myocardial infarction [2].

Senthil Kumar et al. Introduced predictive models using various feature combinations and several well-known classification techniques. A hybrid random forest (HRFM) heart disease prediction model with a linear model [3] achieved a performance level improvement of 88.7% accuracy.

Pabitra Kumar Bhunia et al. Introduced Predictive models using various feature combinations and several well-known machine learning algorithms. Of all the algorithms, the most accurate test was the logistic regression and SVM method, with 92.32% accuracy [4].

Himanshu Sharma et al. Described as well as demonstrated how deep learning plus machine learning algorithms present new opportunities for precisely predicting heart attacks. This article provides a wealth of information on the most advanced deep learning and machine learning techniques. Analytical comparisons are offered to support forward-thinking researchers in this area [5].

Mr. Nikhil Kumar and associates, used eight algorithms for heart disease prediction, including Logistic Model Trees Decision Trees J48 Algorithms, Random Forest Algorithms, ANNs, Naive Bayes, Support Vector Machines and Near Neighbors algorithms best. The more attributes you use, the more accurate your prediction class will be [6].

Amandeep Kaur et al. He described data mining as an essential step in the KDD method with the intention of to predict, treat, and diagnose disease in healthcare. This article gives a summary of the various data mining techniques and strategies used toward identify heart disease [7].

Pahulpreet Singh Kohli. To foresee the early stages of heart disease, the ENDDP (Advanced New Dynamic Data Processing) algorithm was developed. The outcomes show how well the proposed system performs [8].

3. Data Set Information

The title of dataset is heart_disease_data.csv. Figure 37.3 shows the description of dataset. This dataset contains 303 cases of heart disease or healthy people. Of the 303 cases, 165 (54.45%) involved a person who has heart disease and 138 (45.54%) involved a person with no heart disease. There are 14 attributes in total as shown in Table 37.1. There are no values in the dataset that are lost or invalid.

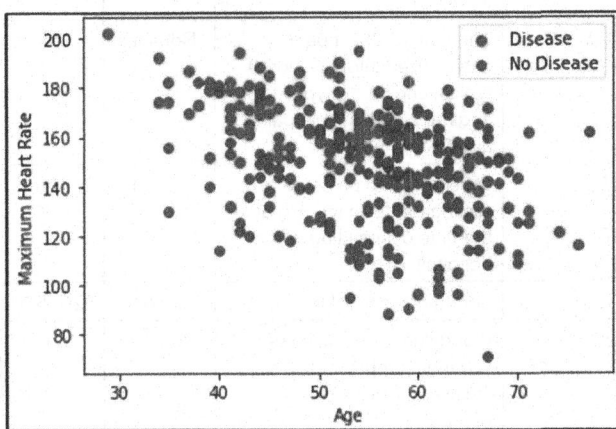

Fig. 37.1   Positive and negative cases

Source: Authors

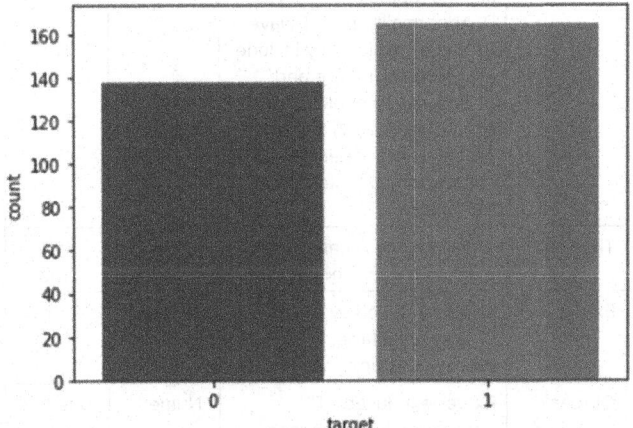

Fig. 37.2   Explore gender dataset

Source: Authors

```
      age  sex  cp  trestbps  chol  fbs  ...  exang  oldpeak  slope  ca  thal  target
0     63    1   3      145    233    1   ...     0      2.3      0    0    1      1
1     37    1   2      130    250    0   ...     0      3.5      0    0    2      1
2     41    0   1      130    204    0   ...     0      1.4      2    0    2      1
3     56    1   1      120    236    0   ...     0      0.8      2    0    2      1
4     57    0   0      120    354    0   ...     1      0.6      2    0    2      1
..    ...  ...  ..     ...    ...   ...  ...    ...     ...     ...  ..  ...    ...
298   57    0   0      140    241    0   ...     1      0.2      1    0    3      0
299   45    1   3      110    264    0   ...     0      1.2      1    0    3      0
300   68    1   0      144    193    1   ...     0      3.4      1    2    3      0
301   57    1   0      130    131    0   ...     1      1.2      1    1    3      0
302   57    0   1      130    236    0   ...     0      0.0      1    1    2      0
```

Fig. 37.3 Description of dataset

Source: Analyticsvindhya.com

Table 37.1 Attributes of UCI heart disease data set

Attributes	Value		
	Description	Type	Ranges
Age	Patient's age in completed years	Numeric	29 to 77
Sex	Patient's Gender (male represented as 1 and female as 0)	Nominal	0 or 1
Cp	The type of Chest pain categorized into 4 values: 0. typical angina, 1. atypical angina, 2.non-anginal pain and 3. asymptomatic	Nominal	0 to 3
Trestbps	Level of blood pressure at resting mode (in mm/Hg at the time of admitting in the hospital)	Numeric	94 to 200
Chol	Serum cholesterol in mg/dl	Numeric	126–564
FBS	Blood sugar levels on fasting> 120 mg/dl; represented as I in case of true, and 0 in case of false	Nominal	0 or 1
Restecg	Results of electrocardiogram while at rest are represented in 3 distinct values: Normal state is represented as Value 0. Abnormality in ST-T wave as Value I (which may include inversions of T-wave and/ or depression or elevation of ST of >0.05 mV) andany probability or certainty of LV hypertrophy by Estes' criteria as Value 2	Nominal	0 to 2
Thalach	The accomplishment of the maximum rate of heart	Numeric	71 to 202
Exang	Angina induced by exercise. 0 represents false, I represents true.	Nominal	0 or 1
Oldpeak	Exercise-induced ST depression in comparison with the state of rest	Numeric	0 to 6.2

Attributes	Value		
	Description	Type	Ranges
Slope	ST segment measured in terms of the slope during peak exercise depicted in three values: 1. unsloping, 2. flat and 3. downsloping	Numeric	0 to 2
Ca	Fluoroscopy colored major vessels numbered from 0 to 3	Numeric	0 to 3
Thal	Status of the heart illustrated through three distinctly numbered values. Normal numbered as 3, fixed defect as 6 and reversible defect as 7.	Nominal	3, 6, 7
Target	The label column, those have heart disease are represented by 1, and those do not have heart disease are represented by 0	Nominal	0 or 1

Source: Authors

Characteristics include age, sex, cp (chest pain category), trestbps (resting blood pressure in mm/Hg), chol (serum cholesterolim in mg/dl), fbs (fasting blood sugar>120 mg/dl), restecg (resting electrocardiographic), thalach (maximal heart rate achieved), exang (exercise induced angina), oldpeak (measure of ST depression induced by exercise), slope (measure of slope in favor of peak exercise ST segment), ca (numeral of large vessels), thal (thallium stress test) and target. The bar graph shows positive and negative cases (1- positive, 0- negative) (Fig. 37.1). Scatter plot showing explore gender dataset (Fig. 37.2).

4. Proposed Methodology

4.1 Data Set Details

The main purpose of this reading was to develop a mechanism for predicting cardiac disease. The system can discover and extract secret information about illnesses from historical heart data sets. To help predict heart disease, the Cardiac prediction system uses data mining techniques from medical datasets. (Fig. 37.4) shows the distribution of tags on each feature.

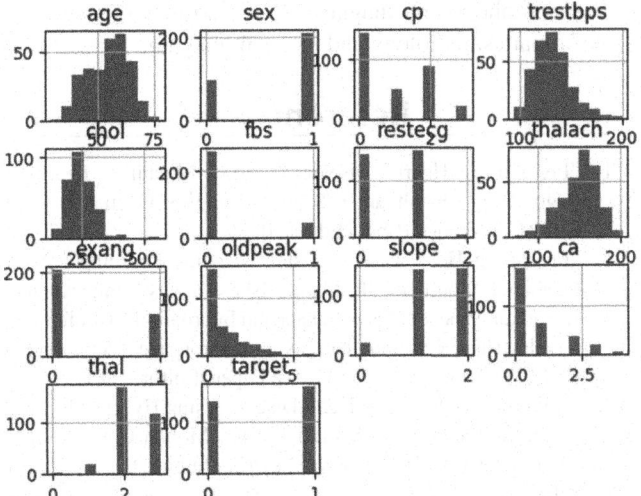

Fig. 37.4 Histogram of the distribution of target on each attribute

Source: Authors

4.2 Training and Testing

In the training phase, features (independent variables) are extracted from the data set as well as in the testing phase, the dependent variables are used to compute the performance of the model in conditions of predictions. The data set has been divided into two phases. These are the phases of training and testing. We divide the data set into a 90% training period also a 10% testing period and assume a random state of 1. A fixed internal random number generator is initialized with a random state parameter that defines how the records is separated into training and testing indices. Setting the random state guarantees a constant value to creates the similar order of random integers each point in time the code is executed. The data was then scaled using normal scatter plots and the transition between test and training data was corrected.

```
from sklearn.preprocessing import StandardScaler
sc=StandardScaler()
X=df.drop('target',axis=1)
Y=df['target']
df=sc.fit(X).transform(X)
from sklearn.model_selection import train_test_split
X_train,X_test,y_train,y_test=train_test_split(X,Y,test_size=0.25,random_st
ate=3)
```

4.3 Classification Used

Logistic regression (LR)

Logistic regression is one of the mainly well-liked machine learning algorithms and is classified as a supervised learning technique. It is used to examine data sets in which one or more independent variables determine the outcome. Logistic regression is entered through a random state of 0. After that the training model is fixed. The test's accuracy is 88.15%.

KNN classifier

For pattern matching and clustering, the K-nearest neighbor algorithm is used. In predictive analytics, it is often used. As new data arrives, the K-nearest neighbor [8] algorithm determine the closest existing data point to it. I imported the K-Neighbors classifier from sklearn.neighbors through n_neighbors= 5. After that the training model was fixed. The test's accuracy is 63.15%.

Support vector classifier

Support Vector Machine (SVM) is a supervised machine learning algorithm used for regression plus classification. An advantage of this algorithm is that it generates best fit decision lines or boundaries that can divide the n-dimensional space into classes, in order that newly added data points can be easily tested in future and be classified as appropriate . The SVM was imported from sklearn and the kernel was kept linear with gamma=0.1 and c=1.0. And the training model has also been adjusted. The accuracy of the test is 60.52%.

Random forest classifier

The Random Forest Classifier is a powerful supervised classification algorithm that comprises of numerous decision trees cooperating as a group. Random Forest generates the entire structure of the classification tree from a specified dataset instead of a particular classifier tree. Each one of these trees generates a taxonomy in favor of a particular set of attributes. I imported a Random Forest Classifier from sklearn.ensemble and kept the random state on 0. After that the training model was fixed. The test's accuracy is 82.89%.

Decision tree classifier

The accuracy of the test is 75.00%. Decision trees are treelike structures in which the inner nodes correspond to attribute tests, each branch correspond to the result of the test, and every leaf node correspond to a class label. I imported decision tree as kept the random state 42 and the training model was adjusted. The accuracy of the test is 75.00%.

5. Result

Of all the classification methods (Table 37.2), the most accurate test was the logistic regression method, with 88.15% accuracy. (Fig. 37.5) shows the results obtained by applying algorithms.

Table 37.2 Compare different performance classifiers

	Model	Train_Accuracy	Test_Accuracy
0	LogisticRegression	0.854626	0.881579
1	DecisionTreeClassifier	1.000000	0.750000
2	SVC	1.000000	0.605263
3	KNeighborsClassifier	0.775330	0.631579
4	RandomForestClassifier	1.000000	0.828947

Source: Authors

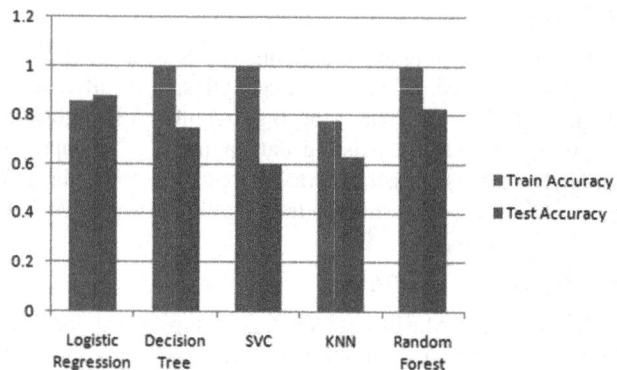

Fig. 37.5 Accuracy comparison of ML algorithms using the split train/test technique

Source: Authors

6. Conclusion

With an accuracy of 88.15%, this heart disease prediction model helps people, particularly medical professionals, react to unlike situations. They are well-versed in human health and can rapidly recognise age-related health issues, thereby alerting patients in advance. Alternatively, the person can confer with a doctor first and get tested to avoid the development of heart disease. Therefore, this representation help out to build belief and create a intelligence of protection among public.

7. Future Scope

In the upcoming, this application may be improved by updating some features. For example, if the user has a heart condition, all family members will be notified in advance by message. Also, the information should be sent to the nearby hospital. Another aspect is the online discussion with the nearest doctor.

In this context, it should be noted that ML applications with various powerful algorithms are used not only in disease prophecy and diagnosis, but also in fields such as bioinformatics, radiology and medical imaging.

References

1. "Predicting Heart Disease Using Machine Learning Algorithms" Krishnan J Santhana and S Geetha ICIICT |2019|Conference Papers|Editor: IEEE.
2. "Predicting Heart Disease Using Machine Learning". Aditi Gavhane, Gouthami Kokkula, Isha Panday, and Kailash Devadkar, Proceedings of the Second International Conference on Electronic Communications and Aerospace Technologies (ICECA) |Year: 2018|Conference Papers|Editor: IEEE.
3. "Effective Prediction of Heart Disease Using Hybrid Machine Learning Techniques" Senthil Kumar, Mohan Chandrasegar Thirumalai, Gautam Srivastva |2019|Conference Papers|Editor: IEEE.
4. "Heart Disease Prediction using Machine Learning" International Journal of Engineering Research & Technology (IJERT), Pabitra Kumar Bhunia and Arijit Debnath |2021|Conference Papers|Editor: IEEE.
5. "Predicting heart disease using machine learning algorithms:"A Survey" International Journal of Recent Innovation Trends in Computing and Communications Vol. 5 No. 8, Himanshu Sharma and M A Rizvi |2019|Conference Papers|Editor: IEEE.
6. "Predicting Heart Disease Using Algorithms and Tools for Data Mining and Machine Learning," International Journal of Scientific Research in Computer Science Engineering and Information Technology, IJSRCSEIT, M. Nikhil Kumar KV S Koushik and K. Deepak|Year:2019|Conference Papers|Editor:IEEE.
7. "Prediction of heart disease using data mining techniques: A Survey" Amandeep Kaur and Jyoti Arora International Journal of Advanced Research in Computer Science IJARCS |Year:2019|Conference Materials|Editor:IEEE.
8. "Application of Machine Learning in Disease Prediction", Pahulpreet Singh Kohli and Shriya Arora, 4th International Conference on Computer Communication and Automation (ICCCA). |Year:2018|Conference Papers|Editor:IEEE.

A Systematic Review on Predicting Skin Cancer Disease Using Deep Learning

38

K. Srilatha[1]

Research Scholar, Dept. of CSE, Chaitanya Deemed To Be University,
Hanamkonda, Warangal, Telangana, India

N. Satheesh Kumar[2]

Professor, Dept. of CSE, Chaitanya Deemed To Be University,
Hanamkonda, Warangal, Telangana, India.

Abstract: As an outcome of the latest advances in technology, a significant amount of healthcare data is generated every day, and this information includes significant and helpful patient information. For some diseases, particularly skin conditions, machine learning based on images is becoming more common. The primary factors affecting how accurately a computer-based system can diagnose a patient are the selection of the proper feature, the classifier employed, the accessibility of the dataset, and the volume of images utilized to train the model. These days, Convolution Neural Networks (CNN) is frequently used for pattern recognition and classification tasks.

Image-based skin disease diagnosis is a challenging issue because there are many different skin diseases. The following challenges to classifying skin diseases have been identified by researchers: 1) Different types of lesions can occur in a disease. 2) Because many diseases can have similar visual symptoms. Doctors may find it difficult to recognize the problem through visual examination. 3) The differing skin tones and skin types (due to ageing) complicate computer-based diagnosis. Computer-based diagnostics must choose the most pertinent features in order to accurately diagnose these disorders. In the field of medicine, there are numerous techniques available for the diagnosis of skin issues. However, computer-based automated diagnosis speeds up the process and is far more valuable for assisting with medical decisions. For instance, patients will not suffer unnecessarily from a lack of expertise if such an automated system is utilised at healthcare facilities.

Keywords: ANN,CNN,SVM,KMC, GLCM, ABCDE rule, ACO-GA

1. Introduction

The largest and most delicate part of the human body is the skin, which shields our innermost vital organs and body parts from the outside world and keeps them from coming into contact with germs and viruses. The regulation of body temperature also benefits from skin. Cells, pigmentation, blood vessels, and other substances make up the skin. It has three basic layers: epidermis, dermis, and hypodermis. A waterproof and protective covering is created around the body's surface by the epidermis, the skin's outermost layer.

Under the epidermis, there is a layer of connective tissue called the dermis that shields the body from strain and stress. The dermis and epidermis are connected by a basement membrane. The hypodermis, which is also referred to as subcutaneous tissue and is situated below the dermis, is not a component of the skin. It not only provides blood vessels and nerves, but it also connects the skin to the underlying bone and muscle. People of all ages are susceptible to skin ailments, which are quite common. Modern skin disease detection still relies on manual surveillance and straightforward clinician observation.

[1]csephdscholar2021@gmail.com, [2]meet.nskumar@gmail.com

Around 7 percent of all newly diagnosed cancer cases worldwide have a skin cancer incidence [1], which cost the Medicare programme in the United States more than $8 billion in 2011. According to clinical evidence, there may be racial disparities in skin cancer outcomes Despite the fact that individuals with darker skin tones are around 20-30 times more likely than those with lighter skin tones to get melanoma, it has been discovered that depending on their skin tone, those with darker skin tones have either a higher or lower mortality risk for specific forms of melanoma. A skin lesion must be correctly identified in order to administer the appropriate treatment. This method improves survival by detecting melanoma early in dermoscopy photos and photographs. Dermatologists who have undergone considerable training in the numerous skin lesions caused by melanomas are the best prepared to provide an accurate diagnosis. As a result of the lack of a defined boundary between skin lesions and the skin itself, the visual resemblance of malignant and non-melanoma skin lesions, and other factors, diagnosing melanoma can be challenging. Thus, pathologists will greatly benefit from the development of a reliable automated method for detecting skin tumours, such as a system that can automatically analyse skin lesions. This is especially important in a time when knowledge is in short supply. According to the findings of this study, the classification algorithms of K-nearest Neighbors, Support Vector Machines, and Decision Trees all provided results that were not precise and accurate enough. Further investigation into the mathematics underlying classification revealed that using Deep Learning models—also known as deep learning models—was the most sophisticated way to achieve the desired results.

We came to the conclusion that the depth and quality of activation that pre-trained models provided did not match. In order to create a model known as a Dense Convolutional Network, we combined our mathematical expertise. We used Deep Learning, a subfield of artificial intelligence that is incredibly strong and potent in its capabilities, to achieve effective and reliable picture categorization.

The main goal is to develop a method of skin cancer screening that is easy for the general public to use and reasonably quick. In most cases, the sooner it is identified, the greater the chance that the person will fully recover. According to the American Academy of Dermatology Association, the majority of dermatological malignancies are curable if detected early. A skin biopsy is only the first step after using the trained model. If you want to be certain whether you have skin cancer or not, go to a dermatologist and have a skin biopsy done. The only way to make an accurate diagnosis is through this method.

2. Literature Survey

Numerous studies using CNN models have been conducted on the classification of skin disorders and some of these models have demonstrated excellent classification performance. The relevant academic articles of a few researchers on the topic of classifying skin disease images are summarised below. Qiwei Wu, Mingjun Wei, and others 2023 [1] Reliable multi-class CNN models have been presented by a number of researchers. Mobiny et al.'s [1] The Bayesian DenseNet-169 model, which calculates model uncertainty without requiring any new parameters or any changes to the underlying network design, is an approximation risk-aware deep Bayesian model. In comparison to the baseline DenseNet169 model, its classification accuracy on the HAM10000 dataset increased from 81.35% to 83.59%. Wang et al. suggest incorporating human interpretability into their CNN model. The inputs to this multi-class classification model are images of skin lesions and information about the patient. On the HAM10000 dataset, it received scores of 95.1% accuracy and 83.5% sensitivity. Allugunti et al. created a multi-class CNN model for detecting skin cancer. The suggested model distinguishes between lesion maligna, nodular melanoma, and superficial spreading melanoma. (Electronics 2023, 12, 438, pg. 3 of 19). By doing this, the virus can be quickly identified, and the necessary actions to contain it and stop its spread can be taken right away. The inclusion of new layers by Anand et al. [2], including a pooling layer, two thick layers, and a dropout layer, improved the original Xception model. The seven categories of skin illnesses in the initial FC layer have been expanded to include a new FC layer. On the HAM10000 dataset, it had a classification accuracy of 96.40 percent. Ensemble learning is also useful for enhancing the categorization accuracy of the model. For the classification of skin lesions, Thurnhofer-Hemsi et al. [3] proposed merging an ensemble of enhanced CNNs with a regularly spaced test-time-shifting strategy.. It creates a large number of test input images using the shift technique, which are then distributed to each classifier in the ensemble and used in a combined classification. It classified objects with an accuracy of 83.6% using the HAM10000 dataset. Classification efficiency can be improved by strengthening the model's feature extraction capabilities with an attention module. Karthik et al. [4] reduced the number of training parameters in the EfficientNetV2 [4] model by replacing an Efficient Channel Attention block for the Squeeze-and-Excite block. When tested on four separate datasets pertaining to skin illnesses (acne, actinic keratosis, melanoma, and psoriasis), the model's overall accuracy was 84.70. Image classification accuracy can be enhanced by employing a variety of image processing techniques such as picture conversion, equalization, enhancement, and segmentation.

Abayomi-Alli et al. [5] offer an improved data augmentation model to aid in the correct diagnosis of melanoma skin cancer. The method produced bogus melanoma images by oversampling data from a nonlinear low-dimensional manifold. On the PH2 [6] dataset, it had an accuracy of

92.18 percent, a sensitivity of 80.77%, a specificity of 95.11 percent, and a f1-score of 80.84.4 percent. Wide-ShuffleNet and a fresh segmentation technique are used in Hoang et al.'s [7] unique strategy to classify skin lesionsInitially, an entropy-based weighted sum first-order cumulative moment (EW-FCM) is used to segregate the lesion from the rest of the skin image. The segmented data is sent into a cutting-edge deep learning architecture dubbed wide-ShuffleNet, which then classifies the data. It has a 96.03% specificity on the HAM10000 data set, a 70.71% sensitivity, a 75.15% precision, a 72.61% accuracy, and a f1-score of 84.80%. Malibari et al. [7] suggested a computer-aided diagnosis approach powered by an ideal deep neural network to improve their skin cancer detection and classification model. The pre-processing stage using Wiener filtering and a U-Net segmentation method forms the basis of the model. The maximum accuracy of the model was 99.90%.

Nawaz et al. [8] proposed the DenseNet77-based UNET model, which is an improved Deep-Learning-based method.. Their findings proved the model's validity and showed that it could correctly categorize a variety of skin lesions, regardless of size or color. On the ISIC2017 dataset and the ISIC2018 [9] dataset, respectively, its accuracy was 99.21% and 99.51%.

In [10], scientists develop a model employing the K-NN classifier for image classification that can recognize and categorize a range of skin diseases. The HSV and lightness red-green-yellow (L*a*b) color models are two examples of the models used to extract attributes. The HSV color model outperformed the L*a*b color model, according to their research (91.80% accuracy vs. 81.60% accuracy).

The classification of 24 different skin illnesses could be further automated by applying the Transductive SVM (TSVM), according to Ahmed, Ema, and Islam [11]. For picture segmentation, the proposed system employs a hybrid genetic algorithm. Ant colony optimization (ACO-GA) and generalized linear discriminant analysis (GLCM) were also employed to determine its differentiating characteristics. They accomplished their work with a 95% accuracy rate. utilizing the technique specified by Hajgude et al. [12], there

are 408 pictures of eczema, impetigo, melanoma, and other ailments. The Otsu method for segmenting lesions, the 2D Wavelet transform for extracting statistical features (such as entropy and standard deviation), and the generalized linear discriminant analysis (GLCM) for extracting texture features (such as contrast and correlation) are all used in the model's construction. The SVM and CNN classifiers' respective classification accuracy for the diseases was 90.7% and 99.1%.. The most well-known image processing libraries and the CNN classifier are both described in [13]. Then, using CNN, they attempt to diagnose skin conditions with a 70% success rate, despite the fact that a sizable dataset could raise that percentage to over 90%. The authors of [14] suggest a web-based method for assessing skin conditions in Ghana (Africa). In their study, they use a convolutional neural network (CNN) to categorize 254 photos of disorders such atopic dermatitis, acne vulgaris, and scabies. Finally, they were able to reach accuracy of 88%, 85%, and 84.7% for each condition. With the suggested method, more patients might be diagnosed more quickly given that it only needs 0.0001 seconds to complete. In several earlier studies, the effectiveness of SVM and CNN has been demonstrated. The study also showed how crucial image processing is in aiding in the classification of skin conditions. Increasing the number of photos in the training set may improve classification accuracy as well.

3. Proposed Work

3.1 Considering and Contrasting CNN with Various Skin Diseases:

HaofuLiao [14] has presented a technique that makes use of CNN to automate the diagnosis of skin conditions. They trained CNN's architecture using the Dermnet database and a dataset of 20,000 images of skin conditions. Additionally, the effectiveness of the system in comparison to an OLE database has been assessed. The system performed better when trained with fewer images than it did when trained with more images. Due to the addition of time complexity, this represents the

Table 38.1 National institute of health

References	Type of skin Disease	Objectives	Classifier	Database	Evolution metrics	Drawback
1	Melanoma	To identify Melanoma skin disease	CNN	ISIC 2018	Accuracy	Classification output is low
2	Melanoma	To classify the skin lesion as malignant or benign	CNN	ISIC Archive Dataset	Sensitivity, Specificity, and Accuracy	Skin lesion classification problem
3	Melanoma	To classify melanoma skin disease	CNN	ISIC Archive Dataset	Accuracy	Accuracy varies for different types of images
4	Melanoma	To detect Melanoma skin disease	CNN	ISIC Archive Dataset	Accuracy	Poor accuracy

Source: https://www.ncbi.nlm.nih.gov/pmc/articles/PMC9327733/

system's main limitation. They should increase accuracy as a result while training on a bigger dataset.

CNN has been advocated for use in the diagnosis of skin diseases by Jainesh Rathod et al. [28]. The aforementioned classification method is originally used to anticipate the features by eliminating background noise from the database's skin images. CNN and SoftMax classifiers are used in the feature extraction and classification process for better outcomes. The system performed better when trained with fewer images than it did when trained with more images. The time complexity this method introduces is one of its main drawbacks.

3.2 Data sets and Machine Learning Algorithms

The development of a method for identifying skin disorders Prepare the data set first, which is the fundamental component of the project and can be obtained from a number of sources, including but not limited to hospitals and clinics, electronic websites on the Internet, official websites, and authorized repositories that specialize in offering data sets to students and researchers According to prior studies and their own review of the data, ISIC, OLE, DermNet, PH2, DermIs, DermQuest, and other sources rank among the most reliable for the data used in the research team's study[15]. Choosing the appropriate supervised machine learning classification algorithm is a second factor. In machine learning, supervised algorithms concentrate on datasets with labels and a function that already maps inputs to desired outputs. Some of the most important algorithms are Artificial Neural Networks (ANN), Support Vector Machines (SVM), Convolutional Neural Networks (CNN), Decision Trees (DT), Boosted Trees (BT), K-Nearest Neighbors (KNN), K-Means Clustering (KMC), Random Forests (RF), Naive Bayes (NB), Genetic Algorithms (GA), Fuzzy Algorithms (FA), and others. Modeling comes after data processing and feature extraction. Any machine learning algorithm-based categorization process must begin with the selection of the features that will be utilized.. Different types of criteria were applied to different studies as required by the various classification systems. Dermatological pictures may be identified with respectable levels of accuracy using ANN, SVM, and KNN algorithms as well as a range of texture and color features. The early diagnosis of skin lesions was recommended by Murugan, A., and Murugan, A., et al. (2019). One thousand photos from the ISIC dataset were used to assess the system's efficacy. The watershed technique was used to isolate the lesion area after the input images had been cleansed of noise. The lesion region contained a wide range of properties, and several methods, including the ABCD rule, GLCM, and form features, were employed to extract features from it. The retrieved traits allowed for the accurate identification of the tumour. Support Vector Machine (SVM),

K-Nearest Neighbor (KNN), and Random Forest classifiers were used to diagnose the skin lesions. With remarkable 87.68% sensitivity, 83.76% specificity, and 85.72% accuracy, the SVM classifier performed best [16].

The ISIC 2017 dataset was used to evaluate two deep learning frameworks, the Lesion Indexing Network (LIN) and the Lesion Feature Network (LFN). LIN's JA and AUC for lesion segmentation and classification are significantly higher than cutting-edge deep learning systems, at 0.753 and 0.912, respectively. The suggested LFN demonstrates good task handling ability by achieving the best average precision and sensitivity for dermoscopic feature extraction, or 0.422 and 0.693, respectively.

4. Datasets

Datasets name	Images	Classification
BCN2000 [M Combalia et al 2019]	19424	8
HAM10000[P Tschand et al 2018]	10015	7
ISIC Archive[ISIC Project-ISIC Archive. Accessed: May 23, 2021]	23906	7
ISIC 2016[D Gutman et al 2016]	1279	2
ISIC 2017[N Codella et al 2017]	2000	3
ASan[S S Han et al 2018]	17125	12
Pigmentary dermatos[Y Yang et al 2020]	12816	6

5. Conclusion

Non-dermoscopic digital camera images have been utilized as a telemedicine melanoma detection method. In this work, clinical photographs were combined using a deep learning-based computationally advanced approach. This technique was able to identify between benign and malignant melanoma instances. We were able to improve the system's ability to discriminate by passing images through lighting correction, which raised the accuracy of the system. For training, we chose a tiny dataset that was easily accessible. To make more images, image cropping, scaling, and rotation were employed. Our suggested method gave CNN the duty of extracting features from the data while traditional learning approaches try to do so. Our superior accuracy in comparison to other detection algorithms was demonstrated by experimental results.

6. Acknowledgement

The authors would like to thank their University for providing facilities to complete this work.

References

1. Upadhyay, A., Chauhan, A., &Kudtarkar, D. (2020). LDA-and QDA-Based Skin Lesion Melanoma Detection" Computing and Smart Communication 2019 (pp. 1057–1064). Springer, Singapore.

2. Mingjun Wei, Qiwei Wu, Hongyu Ji, Jingkun Wang, Tao Lyu, Jinyun Liu, and Li Zhao, "A Skin Disease Classification Model Based on DenseNet and ConvNeXt Fusion", Electronics 2023, 12, 438. https://doi.org/10.3390/electronics12020438

3. Aryan Mobiny, Aditi Singh and Hien Van Nguyen, "Risk-Aware Machine Learning Classifier for Skin Lesion Diagnosis", Journal of Clinical Medicine, August 2019.

4. EyePACS-1. The World of Eyepacs. 2015. Available online: http://www.eyepacs.com (accessed on 10 February 2022).

5. Wang, S.; Yin, Y.; Wang, D.; Wang, Y.; Jin, Y. Interpretability-based multimodal convolutional neural networks for skin lesion diagnosis. IEEE Trans. Cybern. 2021, 52, 12623–12637.

6. Allugunti, V.R. A machine learning model for skin disease classification using convolution neural network. Int. J. Comput. Program. Database Manag. 2022, 3, 141–147

7. Anand, V.; Gupta, S.; Koundal, D.; Nayak, S.R.; Nayak, J.; Vimal, S. Multi-class Skin Disease Classification Using Transfer Learning Model. Int. J. Artif. Intell. Tools 2022, 31, 2250029.

8. Chollet, F. Xception: Deep learning with depthwise separable convolutions. In Proceedings of the Proceedings of the IEEE Conference on Computer Vision and Pattern Recognition, Honolulu, HI, USA, 21–26 July 2017; pp. 1251–1258.

9. Thurnhofer-Hemsi, K.; López-Rubio, E.; Domínguez, E.; Elizondo, D.A. Skin lesion classification by ensembles of deep convolutional networks and regularly spaced shifting. IEEE Access 2021, 9, 112193–112205.

10. Karthik, R.; Vaichole, T.S.; Kulkarni, S.K.; Yadav, O.; Khan, F. Eff2Net: An efficient channel attention-based convolutional neural network for skin disease classification. Biomed. Signal Process. Control 2022, 73, 103406.

11. Tan, M.; Le, Q. Efficientnetv2: Smaller models and faster training. In Proceedings of the International Conference on Machine Learning, Shenzhen, China, 26 February–1 March 2021; pp. 10096–10106.

12. Abayomi-Alli, O.O.; Damasevicius, R.; Misra, S.; Maskeliunas, R.; Abayomi-Alli, A. Malignant skin melanoma detection using image augmentation by oversampling in nonlinear lower-dimensional embedding manifold. Turk. J. Electr. Eng. Comput. Sci. 2021, 29, 2600–2614.

13. Mendonça, T.; Ferreira, P.M.; Marques, J.S.; Marcal, A.R.; Rozeira, J. PH 2-A dermoscopic image database for research and benchmarking. In Proceedings of the 2013 35th Annual International Conference of the IEEE Engineering in Medicine and Biology Society (EMBC), Osaka, Japan, 3–7 July 2013, pp. 5437–5440

14. Hoang, L.; Lee, S.-H.; Lee, E.-J.; Kwon, K.-R. Multiclass Skin Lesion Classification Using a Novel Lightweight Deep Learning Framework for Smart Healthcare. Appl. Sci. 2022, 12, 2677.

15. Malibari, A.A.; Alzahrani, J.S.; Eltahir, M.M.; Malik, V.; Obayya, M.; Al Duhayyim, M.; Neto, A.V.L.; de Albuquerque, V.H.C. Optimal deep neural network-driven computer aided diagnosis model for skin cancer. Comput. Electr. Eng. 2022, 103, 108318

16. Nawaz, M.; Nazir, T.; Masood, M.; Ali, F.; Khan, M.A.; Tariq, U.; Sahar, N.; Damaševičius, R. Melanoma segmentation: A framework of improved DenseNet77 and UNET convolutional neural network. Int. J. Imaging Syst. Technol. 2022, 32, 2137–2153.

Privacy Preservation in Dynamic Data Through Synonymous Linkage on Micro Aggregation

39

M. Suresh Babu[1]

Professor, Teegala Krishna Reddy Engineering College,
Hyderabad, India

G. Kathyayini[2]

Professor, Department of Applied Mathematics,
Yogi Vemana University, YSR Kadapa

B. Md Irfan[3]

Research Scholar, Department of CST,
Sri Krishnadevaraya university, Anantapur, A.P India

Mohammed Raziuddin[4]

Research Scholar, Osmania University, Hyderabad, India

Abstract: The rise of the big data age and the growth of the mobile Internet and intelligent gadgets are undoubtedly responsible for the digitalization of personal information. However, the dissemination of such information raises the possibility of privacy violations. To address this issue, privacy preserving data publishing methods have been proposed. However, current methods based on anonymous models may not be effective in protecting non-numerical sensitive information that may contain synonymous linkages leading to privacy breaches. This paper proposes a microaggregation-based dynamic data publishing strategy to get around this restriction. The suggested approach adds a number of indicators to assess the relationships between values that are not numerically sensitive, which enhances the clustering impact of the microaggregation anonymous approach. In order to support the dynamic release and update of data, a dynamic update programme is also included. In comparison to current state-of-the-art methodologies, experimental research reveals that the suggested method offers greater privacy protection and publishable data availability. Therefore, the suggested microaggregation-based privacy-preserving dynamic data publishing strategy may provide a practical way to safeguard sensitive data while facilitating the sharing and publication of huge data.

Keywords: Privacy preserving, Microaggregation, Big data, Clustering

1. Introduction

The value of big data has attracted significant attention from governments, industries, and research departments worldwide, leading to the development of numerous technological innovations and applications. However, the publication of big data without adequate privacy protection can lead to the leakage of sensitive information, endangering personal safety and property, damaging reputation, and leading to discriminatory treatments. Traditional privacy preserving data publishing methods, such as deleting identifying attributes, may not provide sufficient protection from linking attacks. The K-anonymity and l-diversity models were put out as solutions to this problem, however they might not be successful in securing naturally semantically relevant non-numerical sensitive information.

[1]deanstudentaffairstkr@gmail.com, [2]kathyagk@gmail.com, [3]irfan@nalsar.ac.in, [4]raziuddin5807@gmail.com

This problem is addressed by the research, which suggests a microaggregation-based dynamic data release technique that protects privacy by avoiding the synonymous linking of sensitive variables. The proposed method offers a number of indicators to evaluate synonymous relationships between non-numerically sensitive variables, which strengthens the clustering effect of the microaggregation algorithm. The approach also includes a dynamic update plan that enables the dynamic release and refresh of data. The work is divided into various sections, including a summary of relevant research, an introduction to the fundamental indicators utilised in the conventional microaggregation approach, and a description of the design indicators for the suggested algorithm. The study then finishes with a summary of its contributions and the presentation of experimental findings. Overall, the suggested privacy-preserving dynamic data release technique based on microaggregation offers a practical method for safeguarding naturally semantically relevant non-numerical sensitive information while facilitating the sharing and publication of huge data. The paper's contributions include the introduction of new indicators, an improved microaggregation algorithm, and a dynamic update program. These contributions provide a significant step forward in the development of privacy-preserving data publishing methods in dynamic and real-time big data scenarios.

1.1 Related Work

The topic of privacy-preserving data publishing has gained a lot of attention lately, with the K-anonymity model being the most often used tactic. The K-anonymity model generalises the quasi-identifier properties of records in a dataset to a certain value range in order to divide records into comparable classes with at least K records having the same quasi-identifier values. Other more effective anonymous models, such as l-diversity and t-closeness, have been proposed to enhance the K-anonymity model's capacity to safeguard user privacy. The generalisation operation on quasi-identifiers is used by several existing anonymization approaches, which can be computationally costly and lead to considerable information loss. However, microaggregation and clustering techniques can also be used to produce the split of analogous classes based on quasi-identifiers. Various researchers have proposed different microaggregation and clustering methods to achieve K-anonymity and improve the privacy protection of static and dynamic data sets. These methods include knowledge-based numerical mapping, hierarchical clustering, genetic algorithms, distance metrics, information entropy, fuzzy possibilistic clustering, linear discriminant analysis, optimized prepartitioning strategy, efficient clustering method, weighted K-member clustering, K-center clustering approach, particle swarm optimization algorithm,

and equi-cardinal clustering. Overall, the literature shows a growing interest in privacy-preserving data publishing, and researchers continue to propose new and improved methods for achieving K-anonymity and other anonymous models. Top of Form

1.2 Prior Knowledge

Data records are divided into equivalent classes using the microaggregation approach based on the highest intra-class and lowest inter-class similarity. To evaluate similarity, a distance metric is used. Attributes in a relational database can be continuous or discrete, with discrete attributes further classified as nominal or ordinal. While ordinal characteristics have a meaningful order or ranking among values, nominal qualities might have semantic connections or no associations between values. Pincode is a discrete nominal property with semantic correlations, Sex is a discrete nominal attribute with semantic correlations, and Religion is a discrete nominal attribute in the micro data table in Table 39.1.

Table 39.1 Table of mixed attributes

ID	Age	Pincode	Sex	Religion	Capital gain
1	34	515001	F	Hinduism	fair
2	36	515401	M	Christian	excellent
3	48	515230	F	Sikhism	fair
4	52	515430	M	Christian	moderate

Capital gain is a discrete ordinal property having three values—moderate, fair, and excellent—and non-semantic relationships. In order to evaluate the relationships between records with numerous attributes, a number of distance metrics have been established.

Definition 1 (Distance for a characteristic that is continuous). The distance between two values vi, vj C for any continuous attribute C in data table T may be defined as:

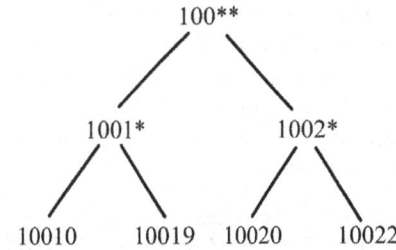

Fig. 39.1 Tree of taxonomy for attribute pincode

$$d_C(v_i, v_j) = \frac{|v_i - v_j|}{\max(C) - \min(C)} \quad (1)$$

where max(C) and min(C) denote the highest and lowest values of a continuous attribute C, respectively

Definition 2 (Distance for nominal property with semantic association). The distance between two values vi, vj Ns for any semantic correlation nominal property Ns in data table T may be written as:

$$d_{N_s}(v_i, v_j) = \begin{cases} 0, & v_i = v_j \\ \dfrac{|\,\text{Parent}(v_i, v_j)\,|}{|\text{Tree } N_s|} & v_i \neq v_j \end{cases} \quad (2)$$

Where |TreeNs| is the total number of leaf nodes for TreeNs, where TreeNs is the taxonomy tree for the semantic correlation nominal property Ns. Parent(vi, vj) is the parent node that both vi and vj share, and |Parent(vi, vj)| represents the total number of leaf nodes that have Parent(vi, vj) as their root, according to TreeNs.

Definition 3 (Distance for nominal characteristic with non-semantic association). The distance between two values vi, vj N for any non-semantic correlation nominal property N in data table T may be written as follows:

$$d_N(v_i, v_j) = \frac{p - \text{match}(v_i, v_j)}{p} \quad (3)$$

Where p is the total number of non-semantic correlation nominal values contained in N and match(vi, vj) is the number of matches between vi and vj.

Definition 4 (Distance for the ordinal characteristic). Any ordinal property O in data table T may have a distance between two values vi, vj that may be defined as:

$$d_O(v_i, v_j) = |\phi(v_i) - \phi(v_j)| \quad (4)$$

$$\phi(v) = \frac{\text{rank}(v) - 1}{|O| - 1} \quad (5)$$

where |O| is the number of different values in the ordinal attribute O, and rank(v) is the rank of value v in ascendant order.

Definition 5 (The separation of two recordings). The distance between two records r1, r2 T is defined as follows for a data table T with continuous characteristics Ci(i = 1,..., m), semantic correlation nominal attribute Nsj(j = 1,..., n), non-semantic correlation nominal attribute Ng(g = 1,..., x), and ordinal attributes Oh(h = 1,..., y):

$$d(r_1, r_2) = \frac{1}{|QIA|} \left(\sum_{i=1}^{m} d_C(r_1(C_i), r_2(C_i)) \right.$$
$$+ \sum_{j=1}^{n} d_{N_s}(r_1(N_{sj}), r_2(N_{sj}))$$
$$+ \sum_{g=1}^{x} d_N(r_1(N_g), r_2(N_g))$$
$$\left. + \sum_{h=1}^{y} d_O(r_1(O_h), r_2(O_j)) \right) \quad (6)$$

where the related continuous, nominal, and ordinal distance functions identified in Definitions 1-4 are denoted by the letters dC(r1,r2), dNs(r1,r2), dN (r1,r2), and dO(r1,r2). The quantity of quasi-identifiers in data table T is given by |QIA|.

2. Microaggregation for Privacy Protection in Opposition to Synonymous Linking

This approach is achieved by introducing two indicators: the semantic correlation index and the semantic similarity index. The semantic correlation index evaluates the degree of semantic correlation between two non-numerical sensitive values, while the semantic similarity index measures the degree of semantic similarity between two sensitive values. These indicators are used to supplement the traditional microaggregation metrics, such as Euclidean distance, which only consider the numerical attributes of data records. By incorporating these indicators into the microaggregation method, the proposed approach can effectively prevent the linkage of non-numerical sensitive information while ensuring the privacy protection of numerical attributes. Additionally, the proposed method can dynamically update data records, ensuring the long-term effectiveness of privacy protection measures. According to experimental findings, the suggested technique performs better in terms of data accessibility and the efficiency of privacy protection than current state-of-the-art solutions. Therefore, it is a potential strategy for disseminating data while protecting privacy, especially in big data scenarios that are dynamic and real-time. The privacy of people in the dataset is better secured by countering synonymous assaults in this manner.

2.1 Microaggregation-based Dynamic Data Release with Privacy Protection

The proposed privacy-preserving microaggregation approach addresses the challenge of privacy protection in the context of data dissemination, especially for non-numerical sensitive information having natural semantic value. It aims to optimise the trade-off between disclosure risk and information loss by contrasting a unique microaggregation metric versus synonymous linkage. The first two components of this metric aim to minimise information loss by reducing the distance between quasi-identifier qualities within each equivalent group, while the third component aims to broaden the semantic range of sensitive attributes to prevent synonymous linkage. The approach also includes a dynamic updating programme that allows for the insertion, deletion, and modification of data while using dynamic adjustment to prevent synonymous linkage. Overall, the proposed algorithm is effective in mitigating privacy leakage caused by synonymous linkage in both static and dynamic data publishing scenarios.

3. Microaggregation Publishing Algorithm

Algorithm Based on what you have provided, it seems that the DRASL algorithm utilizes the K-anonymity model to obscure specific values into a range to safeguard personal data. To balance the trade-off between disclosure risk and knowledge loss, a unique microaggregation metric versus synonymous linkage is used. The third and final component of this measure maximises the semantic variance of sensitive features in each comparable group in an effort to prevent synonymous coupling. The input data table T and the privacy protection parameter K are used in the publishing process of the DRASL algorithm. Based on a predetermined set of sensitive variables, the approach delivers the anonymous data table T and the clustered equivalent groups GID. The best record to join the comparable group is selected using the enhanced microaggregation metric against synonymous linkage

3.1 Dynamic Insertion of Record

To address these issues, the proposed DRASL algorithm includes a dynamic adjustment process for insertion of record. The procedure works as follows:

Input:

D: the current dataset

r: the record to be inserted

K: the desired anonymity level

L: the number of nearest neighbors to consider

Output:

D': the updated dataset with the new record r inserted

Steps:

Input: Data table T; parameter K; predefined catalogue of sensitive values;
Output: Clustered equivalent groups GID; anonymous data table T^*
1: **if** ($|T| \leq K$) **then**
2:   Return
    end if
    Let $GID = \emptyset$ //Create an empty list of equivalent groups
5: Let $T^* = \emptyset$ //Create an empty anonymous table of T
6: **while** ($|T| > K$) **do**
7:   Select a record r from T randomly
8:   $T = T - \{r\}$
9:   $gid = \{r\}$
10:   **while** ($|gid| < K$) **do**
11:     Find a record $r' \in T$ s.t. $\max\{f_{link_{SA}}(gid, (gid \cup \{r'\}))\}$
12:     Find a group $gid_j \in GID$ s.t. $\max\{f_{link_{SA}}(gid, (gid \cup \{gid_j\}))\}$
13:     **if** ($f_{link_{SA}}(gid, (gid \cup \{r'\})) > f_{link_{SA}}(gid, (gid \cup gid_j))$) **then**
14:       $T = T - \{r'\}$
15:       $gid = gid \cup \{r'\}$
16:     **else**
        $GID = GID - gid_j$
18:       $gid = gid \cup gid_j$

19:       **end if**
        end while
        $GID = GID \cup gid$
22: **end while**
    Create an empty list Q
24: **while** ($|T| \neq 0$) **do**
25:   Select a record r from T randomly
26:   $T = T - \{r\}$
27:   **for** each group $gid_j \in GID$ **do**
28:     $Q \longleftarrow f_{link_{SA}}(gid_j, (gid_j \cup \{r\}))$
29:   **end for**
    $j =$ the sequence number of the element with the maximal value in Q
31:   $gid_j = gid_j \cup \{r\}$
32: **end while**
    $T^* \longleftarrow generalization(GID)$
34: **return** GID, T^*

Compute the microaggregation metric for each equivalent group in D, based on Definition 11.

Find the L nearest neighbors of r in D, based on the quasi-identifier attributes.

For each of the L nearest neighbors, compute the microaggregation metric for the equivalent group that would result if r were added to that neighbor's group.

Choose the equivalent group with the best microaggregation metric among the L+1 options (i.e., the L nearest neighbors and the group without any nearest neighbors).

If the chosen group has fewer than K-1 records, add forged records to it until it has at least K-1 records.

Add r to the chosen group.

Perform dynamic adjustment to prevent synonymous linkage as follows: a. If the chosen group already has a forged record, update the sensitive value of the forged record randomly. b. Otherwise, generate a new forged record with a different sensitive value from that of r and add it to the chosen group.

Output the updated dataset D'.

The proposed dynamic adjustment algorithm in situation 1 uses a two-step process to prevent synonymous linkage. Firstly, it generates a set of candidate forged records with different sensitive values, and then selects the best candidate based on the semantic diversity metric (Definition 10) to minimize the probability of synonymous linkage. In situation 2, the algorithm also generates multiple candidate forged records to prevent synonymous linkage. By doing so, the

Table 39.2 Similar groupings from a micro patient table clustered together

GID	Age	Pincode	Disease
1	[21-22]	[12***–14***]	Gastro
1	[21-22]	[12***–14***]	Cancer
2	[27-29]	[19***–26***]	Thyriod
2	[27-29]	[19***–26***]	Cardio

Table 39.3 Equivalent groupings were updated once a new record was added

GID	Age	Pincode	Disease
1	[21–23]	[13***–16***]	Gastro
1	[21–23]	[13***–16***]	Cancer
1	[21–23]	[13***–16***]	Thyriod
1	[21–23]	[13***–16***]	Cholera
2	[26–28]	[19***–26***]	Thyriod
2	[26–28]	[19***–26***]	Cardio

algorithm can effectively protect privacy and maintain data utility even in the face of dynamic data updates.

The suggested approach (approach 3) looks to deal with the problem of sensitive values being exposed during dynamic record insertion by either establishing a new value for the forged record or a new fabricated record entirely. In particular, lines handle the situation where there is already a forgery in the group and the algorithm changes its value to a new value that is not identical to the sensitive value of the new record being inserted. On the other hand, lines handle the situation where there isn't a forgery in the group and the algorithm creates a new forgery with a random sensitive value that is different from the sensitive value of the tuple and has no connection.

Table 39.4 Equivalent groupings were updated once a record is deleted

GID	Age	Pincode	Disease
1	[22-24]	[13***–16***]	Gastro
1	[22-24]	[13***–16***]	Cancer
1	[22-24]	[13***–16***]	Thyriod
1	[22-24]	[13***–16***]	Asma
2	[27-29]	[19***–26***]	Thyriod
2	[27-29]	[19***–26***]	Cardio

The dynamic process for deletion in the suggested DRASL algorithm is designed to ensure that sensitive information is protected even after record deletion. Algorithm 4 provides a step-by-step guide on how this process works. When a record r is deleted from an equivalent group Gidj and the size of the group is less than K, the algorithm first removes Gidj from the clustered equivalent group GID (Line 7). Next, a random record r is selected from gidj and removed (Lines 9-10). The algorithm then identifies the most suitable equivalent group for r to join using the improved microaggregation metric (Lines 11-13). Once the best group Gi has been identified.

If the size of Gidj is still greater than or equal to K after the record deletion, then Algorithm 3 is called to update the group Gidj (Lines 3-5 and 16-18). This ensures that the equivalent group continues to be properly protected and that sensitive information is not exposed. Overall, the dynamic adjustment process for record deletion in DRASL helps to stabilize secrecy of the data even after records have been removed.

Input: Clustered equivalent groups *GID*; deleted record *r*; parameter *K*; predefined catalogue of sensitive values;
Output: Updated clustered equivalent groups *GID*
1: Let $gid_j \in GID$ be the equivalent group currently containing the deleted record r
2: $gid_j = gid_j - \{r\}$
3: **if** ($|gid_j| \geq K$) **then**
4:     $gid_j \leftarrow recall\_Algorithm\ 3(gid_j, r)$
5:     $GID = GID \cup gid_j$
6: **else**
7:     $GID = GID - gid_j$ //retrieve the group gid_j from the set GID
8:     **while** ($|gid_j| \neq 0$) **do**
9:         Select a record r from gid_j randomly
10:        $gid_j = gid_j - \{r\}$
11:        **for** each group $G_i \in GID$ **do**
12:            $Q \leftarrow f_{link_{SA}}(G_i, (G_i \cup \{r\})))$
13:        **end for**
14:        i = the sequence number of the element with the maximal value in Q
15:        $G_i = G_i \cup \{r\}$
16:        $G_i \leftarrow recall\_Algorithm\ 3(G_i, r)$
17:        $GID = GID \cup G_i$
18:    **end while**
19: **end if**
20: **return** *GID*

Algorithm: Microaggregation

Input: modified record r_mod and the original record r_ori

(a) Call Algorithm 4 to remove the original record r_ori from its equivalent group if the alteration is on quasi-identifiers.

(b) Invoke Algorithm 2 to add the altered record r_mod to the most appropriate equivalent group.

If the modification is on sensitive value:

(a) Call Algorithm 3 to update the sensitive value of the equivalent group containing the original record r_ori

(b) If the updated sensitive value is different from the original one, call Algorithm 4 to delete the original record r_ori from its equivalent group and add the new record r_mod into the equivalent group containing the updated sensitive value

(c) If the updated sensitive value is synonymous linked with the modified sensitive value, generate a new value and repeat Step 3a.

The algorithm handles the modification of records either by deleting the old record and inserting the modified record into a suitable equivalent group (if the modification is on quasi-identifiers), or by updating the sensitive value and checking whether it is still protected from synonymous linkage. If the updated sensitive value is not protected, the algorithm generates a new value and repeats the process until the new value is suitable for insertion.

Input: Clustered equivalent groups *GID*; modify record *r*; parameter *K*; predefined catalogue of sensitive values;
Output: Updated clustered equivalent groups *GID*
1: Let $gid_j \in GID$ be the equivalent group currently containing the modify record r
2: **while** (record change $r \rightarrow r'$) **do**
3:     **while** (modification includes only quasi-identifiers $r[QID]$) **do**
4:         $gid_j \leftarrow recall\_Algorithm\ 4(gid_j, r)$
5:         $GID = GID \cup gid_j$
6:         $GID \leftarrow recall\_Algorithm\ 2(GID, r')$
7:     **end while**
8:     **while** (modification includes only sensitive value SA_r) **do**
9:         **if** (Cash table H contains the modify record r) **then**
10:            $H = H \cup H\{r \rightarrow SA_r = SA_{r'}\}$
11:        **end if**
12:        $gid_j = gid_j \cup gid_j\{r \rightarrow SA_r = SA_{r'}\}$
13:        $gid_j \leftarrow recall\_Algorithm\ 3(gid_j, r')$
14:        $GID = GID \cup gid_j$
15:    **end while**
16: **end while**
17: **return** *GID*

The privacy protection effect of the proposed DRASL algorithm can be evaluated by measuring the reduction in probability mass synonymous linkage between sensitive values in the published data. The DRASL algorithm introduces dynamic adjustments to the microaggregation process, which can improve the protection of privacy in published data by reducing the possibility of synonymous linkage between sensitive values. Additionally, the paper highlights the trade-off between privacy protection and data utility. The suggested approach optimises the microaggregation measure against synonymous linkage in an effort to strike a compromise between these two criteria. The DRASL method efficiently protects privacy while maintaining the usefulness of the released data by minimising information loss during microaggregation and maximising the semantic variety of sensitive qualities within each analogous group. This compromise is crucial because too stringent privacy protection methods may result in substantial information loss and lower the usefulness of public data for analysis and decision-making.

4. Conclusion

Overall, the suggested solution solves the shortcomings of previous methods in preventing privacy leakage caused by semantic links between non-numerically sensitive variables, which is a potential contribution to the field of privacy preserving data publication. The use of microaggregation and updates enables more efficient clustering of comparable groups and improved privacy protection. It is crucial to remember that the suggested approach might not be appropriate for all large data publishing scenarios, including those involving unstructured data and graph data. To create strategies for privacy preservation in these situations, more study is required. All things considered, the suggested approach has the potential to be used in a variety of publication scenarios and can support ongoing attempts to achieve privacy preservation in voluminous data.

References

1. Ge, M., Bangui, H. & Buhnova, B. Big data for internet of things: a survey. *Fut. Gen. Comput. Syst.* **87**, 601–614 (2018).
2. Zhu, L., Yu, F. R., Wang, Y., Ning, B. & Tang, T. Big data analytics in intelligent transportation systems: a survey. *IEEE Trans. Intell. Transp. Syst.* **20**(1), 383–398 (2019).
3. Qi, C. Big data management in the mining industry. *Int. J. Miner. Metall. Mater.* **27**, 131–139 (2020).
4. Shamsi, J. A. & Ali, K. M. Understanding privacy violations in big data systems. *IT Professional.* **20**(3), 73–81 (2018).
5. Lv, Z. & Qiao, L. Analysis of healthcare big data. *Fut. Gen. Comput. Syst.* **109**, 103–110 (2020).
6. Anupam, D., Sarma, K. & Deka, S. Data security with DNA cryptography. *Proceedings of the World Congress on Engineering* **2019**, 246–252 (2019).
7. Samarati, P. & Sweeney, L. Protecting privacy when disclosing information: k-anonymity and its enforcement through generaliza- tion and suppression. *SRI Computer Science Laboratory.* 1–19 (1998).
8. Sweeney, L. K-anonymity: a model for protecting privacy. *Internat. J. Uncertain. Fuzziness Knowl.-Based Syst.* **10**(5), 557–570 (2002).
9. Samarati, P. Protecting respondents' identities in microdata release. *IEEE Trans. Knowl. Data Eng.* **13**(6), 1010–1027 (2001).
10. Machanavajjhala, A., Gehrke, J., Kifer, D. & Venkitasubramaniam, M. l-diversity: Privacy beyond k-anonymity. *ACM Trans. Knowl. Discov. Data* **1**(1), 3 (2007).
11. Li, N., Li, T., Venkatasubramanian S. & CSMDL. t-closeness: Privacy beyond k-anonymity and l-diversity. *IEEE 23rd International Conference on Data Engineering.* 106–115 (2007).
12. Palanisamy, B., Liu, L., Zhou, Y. & Wang, Q. Privacy-preserving publishing of multilevel utility-controlled graph datasets. *ACM Trans. Internet Tech.* **18**, 1–21 (2018).
13. Temuujin, O., Ahn, J. & Im, D. H. Efficient l-diversity algorithm for preserving privacy of dynamically published datasets. *IEEE Access.* **7**, 122878–122888 (2019).
14. Xiao, Y. & Li, H. Privacy Preserving data publishing for multiple sensitive attributes based on security level. *Inf. (Switzerland).* **11**, https://doi.org/10.3480/info11030166 (2020).
15. Domingo-Ferrer, J. & Torra, V. Ordinal, continuous and heterogeneous k-anonymity through microaggregation. *Data Min. Knowl. Disc.* **11**(2), 195–212 (2005).
16. Domingo-Ferrer, J. & Mateo-Sanz, J. M. Practical data-oriented microaggregation for statistical disclosure control. *IEEE Trans. Knowl. Data Eng.* **14**(1), 189–201 (2002).
17. Domingo-Ferrer, J., Sanchez, D. & Rufian-Torrell, G. Anonymization of nominal data based on semantic marginality. *Inf. Sci.* **242**, 36–48 (2013).

Note: All the figures and tables in this chapter were self test data prepared by the authors.

Design a Secure Authentication Protocol for Blockchain Based Sharing of Electronic Health Records

40

Etikala Aruna[1]

Research Scholar of Computer Science and Eng ineering,
VELS Institute of Science, Technology & Advanced Studies, Pallavaram, Chennai,

Arun Sahayadhas[2]

Professor of Computer Science and Engineering, VELS Institute of Science,
Technology & Advanced Studies, Pallavaram, Chennai

Abstract: Over the last ten years, the Internet performed a more important part in our everyday lives. The performance of online services including the e-banking, e-rail, and e-health has increased due to modern communication technologies. The Electronic Health Record (EHR) not only has great value, but it also implies a great deal of privacy. The exchange of health data is a key technique for improving medical care quality while decreasing costs. Health information must be encrypted in order to remain private, but doing so reduces the flexibility and usefulness of searches. In this study, a secure authentication protocol for blockchain-based sharing of electronic health records is presented. Through utilizing blockchain technology, the presented solution offers two significant benefits. Given that it began by being decentralized and absence of a single point of failure, it is indeed decentralized. The workload is distributed throughout the consensus nodes of the blockchain network, making it computationally efficient. The described method produced a results: constant key generation execution time and a linearly increasing file size-dependent encryption time.

Keywords: Blockchain, Electronic health record (EHR), Personal privacy, Health data, Cloud computing, ABSE

I. Introduction

Due to the advancement of information technology, the healthcare industry has experienced a paradigm shift, moving from remote access to medical records towards the real-time exchange of information from several patients internal sensors [1].

EHRs are a directory of data about medical procedures, treatments, and patient data as well as information on diseases, medications and medical images (like name, race, age, identification number, background information, medical prescription, physical examination, or laboratory test) [2]. It is accessible to authorized medical professionals, facilitating their evaluation of the patient's state and clinical diagnosis. As a result, sharing of EHR data has significant research

implications for nations, EHR organizations and people [3]. EHR data sharing among institutions, doctors, and patients can help them accurately diagnose patients and assess their conditions, and having complete EHR data at the individual level can help to enhance the quality of EHR services. EHR data is a great index of the population's general health at the national level. As a result, the question of how to implement secure data exchange based on the confidentiality of patient information and the security of EHR data storage has become increasingly important [4].

Being a centralized system, the cloud server's database could be a prime target for an attacker [5]. Patients may have major problems if an attacker intrudes the cloud server's database and alters, falsifies, or deletes recorded data. The centralized issue with cloud servers can be addressed by blockchain

[1]arunae.phd@velsuniv.ac.in, [2]arun.se@velsuniv.ac.in

technology, which functions as a distributed ledger. Every transaction is tracked by this technology in a ledger, which is then chained together to create a blockchain using hash values [6]. An attacker cannot alter the transactions on the blockchain as each user of the network maintains the ledgers.

They use blockchain and cloud storage technologies to effectively store EHR data, ensuring that it is stored in both of these formats. Blockchain innovation has been extensively employed in elections, supply chains, healthcare, the Internet of Things, and other applications because of its appealing qualities of accessibility, autonomy, and tamper proof resistance. It keeps the encrypted EHR data in the cloud as well as the EHR data key words in context on the blockchain due to the restricted storage capability of the blockchain. This considerably reduces the storage space required by the blockchain and guaranteeing data integrity and confidentiality.

The blockchain prevents arbitrary data modification by requiring all transactions to be confirmed by a consensus procedure in an untrusted environment. A distributed topology of computing nodes is used to build blockchain, making it resistant to errors and attacks. The interoperability of medical information can also be improved by blockchain technology and smarter contracts. Blockchain therefore has a lot of prospective for usage in the medical industry. They use the consortium blockchain, which is run by a number of carefully chosen healthcare clinics, to provide a secured framework for exchanging EHRs in view of data security and patient confidentiality concerns in the healthcare industry. The consortium blockchain offers better privacy protection since it lets us manage which user nodes become part of the network in comparison to the public blockchain.

Although encryption guarantees security, it significantly hinders data accessibility. Even the simplest action, like searching, becomes extremely difficult as a result. This issue is resolved by the Attribute Based Searchable Encryption (ABSE) method. With ABSE, the cloud server can seek for sensitive information without revealing any specifics about the content being looked for [7]. Additionally, users from different domains interact in a cloud environment; access control needs to be put in place in order to provide capabilities like fine-grained searches. The request for cloud-assisted blockchain-based Cyber-Physical System (CPS) for healthcare is consistent by all the characteristics listed in encryption algorithm like access control and search functionality. Following is how the rest of this analysis is structured: Associated works are covered in Section II, Section III elaborates the described methodology based on Blockchain, result analysis is described in Section IV. The conclusion is described in section V.

2. Literature Survey

Ramani V, Bracken A, Kumar T, Liyanage M, Ylianttila M et al. [8] demonstrated a secure and efficient access to medical data solution based on blockchain. Smart contracts are used by the system to control how medical data is used by doctors, and blockchain technology ensures that consumers may access EHR in a secure and reliable manner.

Liu J, Li X, Ye L et al. [9] An electronic medical data privacy protection solution referred to BPDS (Blockchain-based Privacy Preserving Data Sharing) was introduced. To guarantee that a consensus is achieved for each transaction, the system utilizes Delegated Proof Of Stake (DPoS) resulting in suitable and adequate transaction verification. Only a small number of patients medical data are taken into account by this technique, and it is difficult to tell whether information has been altered using BPDS.

Zuobin Ying, Ximeng Liu, Qi Li, Lu Wei, Jie Cui, et. al. [10] on the basis of CP-ABE, a policy-preserving EHR system was presented. One-to-many encryption and accurate access control are both possible with the Ciphertext-Policy Attribute-Based Encryption (CP-ABE) cryptography prototype. Although the ciphertext in CP-ABE is linked to the access policy, it is not secured, which will also result in certain privacy leakage. This created an algorithm that can both hide the whole access policy and retrieve the hidden properties from the access matrix. The suggested technique only introduces minimal overhead costs, according to the assessment of element insert, lookup, and recovery that follows.

Jianhong Zhang, Hao Xiao, et. al. [11] suggests a unique public auditing mechanism for the exchange of dynamic information. It can make it such that the designated data users can edit a particular section of the shared information file. As far as it is aware, it is the initial strategy to accomplish this purpose. The auditor's constant contact and computing costs are another remarkable feature of the method. The results of a detailed simulation and comparisons with the most recent scheme demonstrate that the described method has a number of benefits in regards to user revocation, data block updating as well as integrity verification time. Lastly, it explicitly establishes the described scheme's security and assesses the effectiveness of the auditing process.

Xia, Q., B., Asamoah, Sifah, E. K. O., Gao, J., and Du, X. et al. [12] developed a plan for exchanging medical data to supply data sources, audit and monitor confidential information in clinical data. The issue of sharing medical data between sizable clinical databases without trust can be resolved by this plan. The system is built on the blockchain, tracks data efficiently using smart contracts and access control mechanisms and can identify when data is permitted improperly. However, sharing data is a complicated task.

X. Liang, J. Zhao, S. Shetty, J. Liu and D. Li, et al. [13] presented a decentralized and permissioned blockchain-based system for user-centric health data exchange. In this investigation, smartphone application is used to gather clinical information from individual wearable technology. Healthcare providers as well as health insurance firms received the data after it had been synchronized to the cloud. G. Zyskind, O. Nathan, et. al. [14] utilizes blockchain to safeguard the confidentiality of personal data. In order to ensure that users own and control their information, the authors created a method that turns a blockchain into an automatic access-control manager without the requirement for third-party trust.

Teresa C. Piliouras, Robert J. Suss, Pui Lam Yu, et. al. [15] presents methods for integrating digital imaging technology with clinical workflows and EHR systems. A case analysis using the open source EHR program, OpenEMR (Electronic Medical Record), is provided as an illustration of the difficulties in combining imaging and the EHR innovation within and among medical practices. Vendors of Radiology Information Systems (RIS), Picture Archiving and Communication Systems (PACS), and EHR systems must improve the system interoperability and image sharing features of their product offerings. To make this possible, it needs standards that are better and more comprehensive. This would lessen the pressures placed on medical personnel and Information Technology personnel who should utilize and build device architecture while maintaining adherence to the Health Insurance Portability and Accountability Act (HIPAA) and Meaningful Use (MU).

3. Block Chain Based Sharing of Electronic Health Records

Figure 40.1 demonstrates the architecture of the Secure Authentication Protocol for Block Chain Based Sharing of EHR.

An study of the system's overall algorithm's architecture is provided in this section of the analysis. It additionally provided a definition of security for the specified system and a model for game-based security. The Data Owner (DO) entity is responsible for hospitals, patients, clinical data, electronic health records, clinical trials, and disease registries. The DO (in this case, health data) is the owner of certain data, as represented by the name, and is the one who decides whether to share it with a third-party cloud server. With the help of the GenIndex algorithm, the data owners create the encrypted text, it is subsequently sent to a cloud server for secure transmission. The keywords are encrypted initially, and the clinical data file and related data are extremely outsourced. The owner of the information in the system is

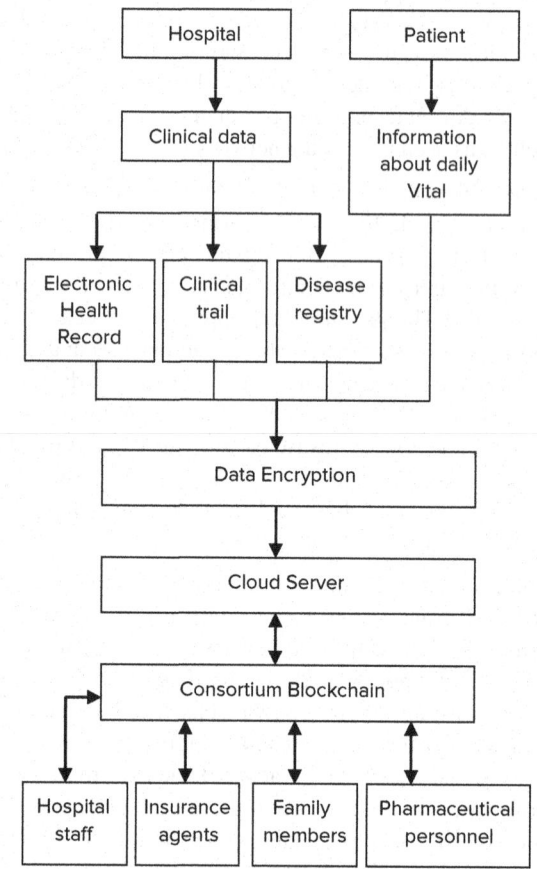

Fig. 40.1   Architechture of the secure authentication protocol for block chain based sharing of EHR

the patient who encrypts their health information with any widely used encryption technique, most likely the symmetric key (rapid computation) approach. The method provided uses the GenIndex algorithm to encrypt the keywords present in the information additionally the associated symmetric key used to perform encrypting data.

The monotonic access architecture, often known as the attribute-based searchable encryption (ABSE) approach, is represented as: The attribute of the world is represented by 'μ'. A non-empty subset of is included in the monotonic access framework so that if a subset is a member of A, thus its superset is necessary to be included in A, i.e., if $\forall$ B is a member of $C \in \mathbb{A}$, if $B \in \mathbb{A}$ is a member of $B \subseteq C$, then $C \in \mathbb{A}$ is a member of A. They use the following linear secret sharing scheme (LSSS), over Z_p, to distribute secrets throughout the access architecture:

1. 'Z_p' is the made up of secret shares $S \in Z_p$ is the vector over.

2. A share generating matrix, $\times n$, whose elements develop from Z_p, exists for each access structure created from the attribute universe.

Cloud Server (CS): The entity in charge of keeping encrypted health data, serving as the data user's agent during keyword searches, and providing the results is known as CS. The matter in consideration is often handled by a healthcare provider.

Consortium Blockchain (CB): The system will be initialised and global parameters will be created by this entity. The user's public key connected to their unique Global ID (GID) is also registered. Although, a user contacts the CB with the qualities that best characterize the search they want to perform, the user will receive a partial search token from the CB. The user's qualities will be verified by the CB group of consensus nodes, and depending on the characteristics, distribute the client that submitted the request with a partial search token for the user.

Role of CB in the suggested scheme:

1. *System Setup:* Shamir's secret sharing mechanism is used by the consensus nodes to begin operating the system and set the general public settings.

2. Through keeping the public key that corresponds to their particular global identity, users can sign up for CB. This process also results in the creation of a partial search token. Users who want to recover encrypted data from the cloud can connect with the CB to have a partial search token produced that is personalized. The search token must be produced and maintained

3. *No longer controlled by the Master Secret:* explained the system management for the CB-assisted plan is not handled by a single organization.

Data User (DU): DU wants to access the data that is stored in a cloud-based servers. In order to obtain the file containing the search term, he or she will first develop a thorough search token, which is then submitted to the cloud-based servers. Data in a healthcare system may be used by physicians, researchers, hospital staff, insurance agents, relatives, and other people. Utilizing the user's entire search token, CS performs the search algorithm; the results are then passed back to DU. Furthermore, blockchain consensus nodes handle user characteristics, in contrast to previous ABSE methods which require for a central trusted authority. Increasing the number of consensus nodes improves the security and dependability of a system. Since more nodes must now recognize in order for consensus to be reached. Regarding the amount of transactions that it can process, the system also becomes very adaptable. The consensus process also consumes less processing resources since only consensus nodes in a consortium blockchain are permitted to contribute new blocks to the blockchain. Consensus nodes are chosen by associations with a high degree of confidence.

4. Result Analysis

The given method is implemented in Java on a 64-bit Windows 10 device with an Intel Core i3 CPU operating at 2.00 GHz and 4 GB of RAM using the Netbeans 8.1 integrated development environment (IDE) and the Java Pairing-Based Cryptography (JPBC) Library. The baseline field is configured to be 512 bits in size, providing security identical to 1024 bits, while the origin group and destination group orders are configured to be 160 bits each. It primarily focuses on the routine tasks carried out by end users and also how the blockchain innovation helped to lessen the load of those end users and cloud, compares it to current multi-authority searchable encryption systems. The size and number of the group components are used to calculate the storage cost. The computational cost, in contrast, is calculated based on the quantity and variety of operations used to produce the results of each approach. The consensus technique and computing power of such consensus nodes, that are not primary subject of the analysis, determine how well blockchain operations perform. In order to demonstrate performance, it changed the collection's access policy from 8 to 40 with a step duration of 8. In the attribute universe, it has changed a wide range of characteristics.

Figure 40.2 shows how long the key generation technique for Attribute-Based Searchable encryption with BlockChain (ABSE+BC) and comparable systems techniques Multiauthority Attribute-Based Keyword Search (MAAKS), decentralized attribute-based conjunctive keyword search (DACKS). The given scheme's key generation time is persistent and does not vary with the amount of characteristics, unlike MAAKS and DACKS, where a linear graph may be seen.

Table 40.1 Average execution time for key generation

No. of Attributes	ABSE+BC	Maaks	Dacks
8	0.0046	0.1	0.12
16	0.0046	0.2	0.23
24	0.0049	0.25	0.29
32	0.0046	0.29	0.35
40	0.0047	0.36	0.4

The properties are not contained in the secret key that is generated for the user in the described approach. The consensus nodes perform it directly at the same moment, during the production of the search token. In this case, to register themselves on the blockchain network, a public and private key pair that reflects their own worldwide identity will be created by the user for themselves. As a result, the

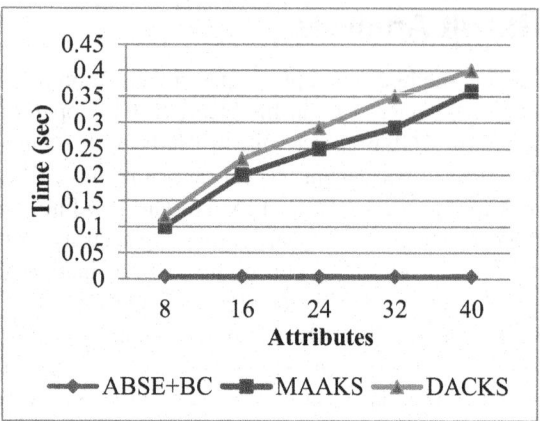

Fig. 40.2 Average execution time for key generation

described approach does away with the requirement for a central authority to control user attributes. When a user requests the generation of a search token, the consensus nodes carry it out.

The smart contracts are responsible for enforcing Access Control regulations. It constructed a large number of arbitrary access policy rules for this test to see how the number of access control rules affected the speed at which transactions were processed. In order for the smart contract to be deployed, a minimum of three administrative privilege rules are necessary. The transactions per second (tps) is the unit used to measure the throughput of a system. Figure 40.3 demonstrates that when records are submitted to the blockchain, the proportion of access control regulations has very little of an impact just on latency and throughput of blockchain network. A blockchain network's performance is comparable to a network with fewer limitations, while having 300 access control rules. They may not have accurately reflected the complicated access control rules that must be implemented in real-world situations because it has only constructed arbitrary rules of relatively modest complexity.

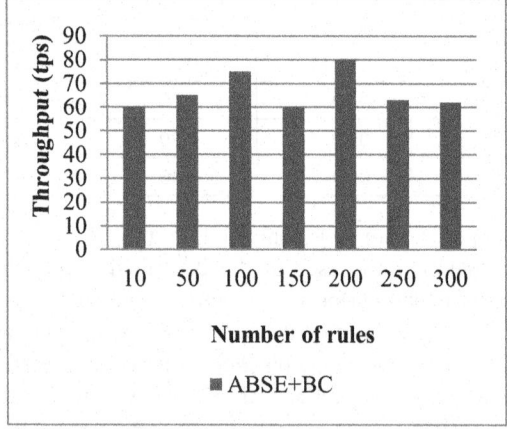

Fig. 40.3 Throughput of Secure authentication protocol ABSE +BC

When taking into account that healthcare providers must have quick access to medical information at the point of care, the time it takes to encrypt documents is crucial. The time required for ABSE encryption (per record) is enhanced linearly with file size, as seen in Fig. 40.4.

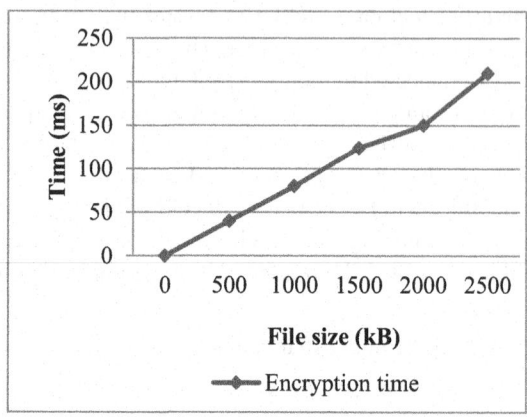

Fig. 40.4 Encryption time

Therefore from results it is clear that, described model or scheme for sharing of EHRs through cloud by using blockchain is very efficient and secured in terms of average execution of key generation, throughput and Encryption time.

5. Conclusion

Secure Authentication Protocol for Block Chain Based Sharing of Electronic Health Records is described in this paper. Healthcare professionals and patients may easily access and exchange medical records due to the sharing EHR system's dedicated client service. The consensus nodes, who are in charge of setting up the system's initial state, are the ones that produce users' partial search tokens. As a result, there is less need for an worldwide central authority to establish a system, and the computational burden on data consumers is reduced. Additionally, ABSE schemes work like blockchain's consensus nodes, which are a decentralized, trusted authority that controls user features. The given technique has a fixed key generation time that is unaffected by the implementation of additional attributes. According to a performance investigation, file size (in kB) affects how quickly data gets encrypted. The blockchain enables health care providers to share records that are securely preserved and encrypted while additionally requesting approval.

References

1. Jeong Hyeon Han, Joo Yeoun Lee, "Digital Healthcare Industry and Technology Trends", 2021 IEEE International Conference on Big Data and Smart Computing (BigComp), Year: 2021.

2. Tejashwa Kumar Tiwari, Apoorva Tyagi, Gunseerat Kaur, Sumit Badotra, "Design of cloud-based interoperable electronic health record with advanced security for Indian healthcare industry", 2021 6th International Conference on Signal Processing, Computing and Control (ISPCC), Year: 2021.

3. Shelly Sachdeva, Disha Batra, Shivani Batra, "Storage Efficient Implementation of Standardized Electronic Health Records Data", 2020 IEEE International Conference on Bioinformatics and Biomedicine (BIBM), Year: 2020.

4. Nan Liu, Chuan Wang, Xinyu Miao, Hua Bai, Yunan Wang, Limin Yang, Yiming Lei, Wei Zhang, Hong Wang, "A New Data Visualization and Digitization Method for Building Electronic Health Record", 2020 IEEE International Conference on Bioinformatics and Biomedicine (BIBM), Year: 2020.

5. Yong Wang, Aiqing Zhang, Peiyun Zhang, Huaqun Wang, "Cloud-Assisted EHR Sharing With Security and Privacy Preservation via Consortium Blockchain", IEEE Access, Volume: 7, Year: 2019.

6. Xiaodong YANG, Ting LI, Rui LIU, Meiding WANG, "Blockchain-Based Secure and Searchable EHR Sharing Scheme", 2019 4th International Conference on Mechanical, Control and Computer Engineering (ICMCCE), Year: 2019.

7. Deepthi Rao, D.V.N. Siva Kumar, P. Santhi Thilagam, "An Efficient Multi-User Searchable Encryption Scheme without Query Transformation over Outsourced Encrypted Data", 2018 9th IFIP International Conference on New Technologies, Mobility and Security (NTMS), Year: 2018.

8. Ramani V, Kumar T, Bracken A, Liyanage M, Ylianttila M, "Secure and efficient data accessibility in blockchain based healthcare systems" in Proc. GLOBECOM, Dec. 2018, pp 206–212.

9. Liu J, Li X, Ye L, "BPDS: A blockchain based privacy-preserving data sharing for electronic medical records", Proceedings of the 2018 IEEE Global Communications Conference (GLOBECOM) IEEE, 2018, pp 1–6.

10. Zuobin Ying, Lu Wei, Qi Li, Ximeng Liu, Jie Cui, "A Lightweight Policy Preserving EHR Sharing Scheme in the Cloud", IEEE Access, Volume: 6, Year: 2018.

11. Jianhong Zhang, Hao Xiao, "Public Auditing Scheme of Dynamic Data Sharing Suiting for Cloud-Based EHR System", 2017 International Conference on Cyber-Enabled Distributed Computing and Knowledge Discovery (CyberC), Year: 2017.

12. Xia, Q., Sifah, E. B., Asamoah, K. O., Gao, J., and Du, X., "MeDShare: Trust-less medical data sharing among cloud service providers via blockchain", IEEE Access, 2017.

13. X. Liang, J. Zhao, S. Shetty, J. Liu and D. Li, "Integrating blockchain for data sharing and collaboration in mobile healthcare applications," 2017 IEEE 28th Annual International Symposium on Personal, Indoor, and Mobile Radio Communications (PIMRC), Montreal, QC, 2017, pp. 1–5.

14. G. Zyskind, O. Nathan, "Decentralizing privacy: Using blockchain to protect personal data," in Security and Privacy Workshops (SPW), 2015 IEEE. IEEE, 2015, pp. 180–184.

15. Teresa C. Piliouras, Robert J. Suss, Pui Lam Yu, "Digital imaging & electronic health record systems: Implementation and regulatory challenges faced by healthcare providers", 2015 Long Island Systems, Applications and Technology, Year: 2015.

Note: All the figures and table in this chapter were made by the authors.

Enhanced Feature Based Attribute Encryption for Data Privacy and Access Control

41

Somireddy Pavani[1]

Research Scholar of Computer Science and Engineering,
VELS Institute of Science, Technology & Advanced Studies, Pallavaram, Chennai

Arun Sahayadhas[2]

Professor of Computer Science and Engineering, VELS Institute of Science,
Technology & Advanced Studies, Pallavaram, Chennai

Abstract: The healthcare organization collects an enormous amount of complicated healthcare data, but it is not mined for the private information required to make wise decisions. Personal health records are an emergent concept in healthcare that centers on the patient and collects their health information on the cloud. It enables anyone to create, manage, and share data over the cloud. However, necessitates consideration of security and privacy issues. This paper explains Enhanced Feature based Attribute Encryption for Data Privacy and Access Control, which is provided as a compromise to these demanding requirements. Manage a huge amount of data by implementing a flexible and fine-grained access control system. On unreliable servers, it is predicted to be possible to acquire and change the entry with fine-grained information. According to these methods, before uploading or downloading files to the cloud, owners of the data must encrypt. They must also re-encrypt their files anytime a client's login information changes. The results show as effectively and consistently data access and sharing are handled in cloud computing. The result analysis describes the comparison of access control and computing costs of different methods. This approach lowers the cost of computing for customers by outsourcing the procedure.

Keywords: Feature based attribute encryption, Access control, Data privacy, Fine grained authentication

1. Introduction

For modern information outsourcing systems, flexible systems for controlling access are required [1]. Differentiated access services are frequently preferred because they enable data access restrictions to be determined by user characteristics or responsibilities. In traditional access control systems like reference, a trusted server is responsible for establishing and carrying out access control procedures.

Because users desire the ability to create an access policy and impose it on the contents, as well as share confidential content with a set of people they choose, modern data outsourcing technologies no longer operate under this presumption [2]. Consequently, it is preferable to provide data owners control over access policy decisions.

The providing of high services at reasonable prices is among the most difficult problems facing the healthcare industry (hospitals, medical facilities) [3]. The level of service reflects how well patients are diagnosed and how well they receive effective treatment. Poor clinical decision-making can have disastrous results that are unacceptable. The cost of clinical tests must also be decreased by hospitals. They can achieve these outcomes by using the proper computer-based information and/or decision support tools. Users and owners utilizing the Physical Health Records (PHR) service are multiplying immensely as the use of cloud computing rises tremendously [4].

Access control policies are determined by various aspects of the requester, environment, or data item in newly presented access control methods like attribute-based access control

[1]somi.phd@velsuniv.ac.in, [2]arun.se@velsuniv.ac.in

[5]. Additionally, present phenomenon of storage outsourcing necessitates greater data protection, including access control techniques with cryptographical enforcement. A promising strategy that satisfies these conditions is the invention of Feature Based Attribute Encryption (FBAE). ABE has a method that allows access policies and assigned properties between private keys and ciphertexts to be used to control access to encrypted material. In particular, ciphertext-policy ABE (CP-ABE) offers a scalable data encryption approach in which encryptor specifies the collection of properties that the decryptor must possess to decrypt the ciphertext [6]. According to security policy, certain users are therefore permitted to decode various pieces of data. As a result of this, it is no longer necessary to depend on the storage server to stop illegal data access.

Encrypting the data prior to outsourcing would be a workable and viable strategy. In essence, Encryption methods and who may access each file should be decided by the PHR owner. PHR files ought to be kept private and only accessible to people who have been provided the necessary decryption keys [7]. Additionally, access privileges can always be granted and revoked by the patient as they see it necessary. But sustainability in a PHR system frequently clashes with the objective of patient-centric privacy. Either for personal or professional usage, the PHR may need to be accessed by the authenticated persons.

Some disease predictions are not automated. Public key encryption-based techniques were initially employed to encrypt the data from personal health records. However, the encryption was one-way and there was a lot of key management work required. A solution for effective key management and encryption is described despite this one-way approach, such as FBAE which is a one-to-many encryption methodology. It was described that the data may be encrypted using a set of qualities that would allow many users to decrypt the encrypted data using the provided key. The distinguishing feature of attribute-based encryption is that it stops multiple users from cooperating.

The system's significant features include, discovery of PHR record done by the revoked customers collaborating with current users in these approaches, it concentrate on developing an FBAE method with effective client revocation for cloud infrastructure and furthermore, this approach would minimize significant computation cost for users. The techniques may be used for client revocation capabilities-required fine-grained improved attribute access control in cloud storage systems [8]. With a diverse range of customers, including relatives, friends, professionals, specialists, and security offices, now, a user can have accessibility to his medical records in an appropriate manner.

2. Literature Survey

Zhang Y., Zheng D., and Deng R. H., et al. [9] a framework for privacy-aware s-health access control has been established, and it suggests a wide range of CP-ABE with partially hidden access rules. ABE systems have an issue in that the ciphertext size and decryption time cost increase together with the size of the access policy.

Zheng Yan, Mingjun Wang, Yuxiang Li, Athanasios V. Vasilakos, et. al. [10] presents a method based on ABE to allow safe data access control and de-duplicate encrypted information stored in cloud. Employment rate for storage facilities, particularly for massive data, is significantly reduced by the possibility of storing duplicated data that is encrypted using various encryption algorithms on the cloud. Recently, a number of data de-duplication strategies have been suggested. The performance of the scheme is assessed by the authors through analysis and implementation. Results demonstrate the scheme's scalability, effectiveness, and efficiency for prospective adoption in practice.

Zheng Yan, Xueyun Li, Mingjun wang and Athanasios V. Vasilakos, et. al. [11] discusses methods to manage flexibility data access in cloud computing on the basis of the data owner's assessment of the trust and/or reputations created by a number of reputation centers by using proxy re-encryption and ABE. In order to support multiple control situations and methods, it integrates reputation evaluation and context-aware trust into a cryptographic system. Through in-depth research, security proof, comparison, and application, the performance and security of the method are assessed and justified. The outcomes describe, adaptability, effectiveness and efficiency of data access control approach for cloud computing.

Seo J. H.and K. Emura, et. al. [12] shown that the identity-based encryption (IBE) prior revocation techniques are susceptible to decryption key disclosure. Revocable Identity-Based Signatures (RIBS) are reviewed from perspectives of security methods and structures. Initially, it demonstrates that, with the exception of the Boneh-Franklin design, all prior RIBE constructs are susceptible to the actual risk of decryption key leak. Second, through integrating the (selectively secure) Boneh-Boyen IBE technique with the (adaptively secure) Waters IBE scheme, it develops the first scalable RIBE approach with resistance to decryption key disclosure. Then it demonstrate that RIBE approaches is highly effective than remaining prior flexibly secure scalable RIBE approaches. In order to demonstrate the viability of the solutions, it present implementation outcomes.

K. Yang and X. Jia, et. al. [13] presents, a dynamic auditing system that is effective and secure for cloud computing data storage. In the beginning, it develops a cloud storage system auditing architecture and suggests a successful and confidential auditing protocol. Then, because data dynamic operations are effective and secure in the random oracle paradigm, it modifies the auditing guidelines to support them. Without the aid of a reliable organizer, it further expand the auditing guidelines to accommodate auditing batch for the numerous owners and multiple clouds. Analysis as well as simulation findings demonstrate that presented auditing methods are effective, secure, particularly in lowering the auditor's computing costs.

M. Bellare, S. Keelveedhi, and T.Ristenpart, et al. [14] Deduplication of the encrypted data was achieved using Message-Locked Encryption (MLE), a novel encryption method. Clients must encrypt their data using the object data's hash value in MLE-based schemes to guarantee that distinct clients can access the same encrypted copies of the identical plaintext information. Ibraimi L., Hartel P., Nikova S., Petkovic M., and Jonker W., et al. [15] The data server's semi-trusted proxy is used to offer CP-ABE schemes that provide quick attribute revocation instead of periodic or planned revocation. In the context of outsourcing data, additionally, they have failed to create a refined user access control system.

N. Attrapadung and H. Imai, et. al. [16] Another user-revocable ABE strategy that combines ABE techniques with broadcast encryption systems is presented as a solution to the problem. For direct user revocation to be possible with this method, the data owner must fully take over management of all membership lists for each attribute group. This plan cannot be used with the architecture for data outsourcing because, after data owners have outsourced their data to an external data server, they won't be able to directly affect how the data is dispersed longer.

3. Enhanced Feature Based Attribute Encryption

Figure 41.1 shows the architecture of the Enhanced Feature Based Attribute Encryption (FBAE) system for Data Privacy and Access Control.

A specific cryptographic method is required for feature-based attribute encryption. In instance, in an FBAE plot, customers obtain their premium private keys from a real estate expert. When the endorser's features are shown to the verifier as qualifying the marking predicate, the mark is considered important. With open access keys, cryptography systems are relatively well-known. Using an online middlemen for each trade is a requirement for intervening cryptography. In order to represent the Scanning Electron Microscope (SEM) cryptography's property-based version, Security Mediated Certificate (SMC) cryptography has also recently undergone

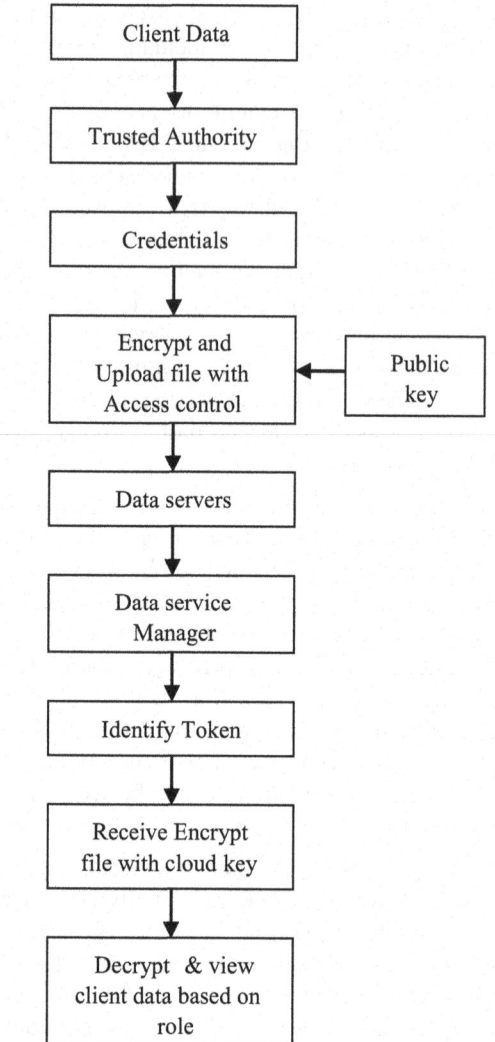

Fig. 41.1 Overview of the architecture system [4]

modifications. These different PHR clients and access holders use the PHR application procedure. A public health record owner is someone who is in charge of the patient information included in a PHR. Although they are regarded as authorized users in order to carry out PHR sharing, PHR owners have access to each patient's PHR records.

An important source for the collection of attributes is a reputable source. For the system, it produces both open and closed variables. Its responsibilities include providing attribute keys to users and changing those keys as required. It provides each user a different set of access rights based on their qualities. Each and every organization utilizing the data outsourcing system is totally dependent on this one party.

A client is referred to be a data owner if they have information and desire to move it to one of the service provider's external data servers. A data owner may implement (attribute-based) by encrypting the data that is relevant to the access policy before providing it to a third party. An organization that wants access to the data that has been outsourced is known as a client. If users follow the requirements of the access policy issued by the data owner for this encrypted content, they will be permitted to decode the ciphertext and obtain the data and do not have their rights in either of the attribute groups removed. Owners of the data are responsible for encrypting it before uploading or downloading it to or from the cloud. If a customer's login information change, they are also in responsible for re-encrypting the data.

The term "service provider" refers to a person who provides data outsourcing services. Data servers and a data service manager are the system's components. On data servers are kept the data that the data owners have outsourced. Access to the servers holding the outsourced data must be restricted for outside users, and the data service manager is in responsibility of offering the necessary content solutions. Like the older designs, it takes into account the curiosity and integrity of the data service manager. As a result, it will faithfully do the tasks that have been assigned by the right system participants. However, it would want to learn as much as it can about contents that are encrypted. Every attribute group's keys for that attribute are under the control of the data service manager.

The data service manager receives attribute groups Gj from the reliable authority for each "$\_\lambda_j \in \Lambda$ ". For instance, if u_1, u_2, u_3 are connected to $\{\lambda_1, \lambda_2, \lambda_3\}$, $\{\lambda_2, \lambda_3\}$, $\{\lambda_1, \lambda_3\}$, respectively, trusted authority gives data service manager $G_1 = \{u_1, u_3\}$, $G_2 = \{u_1, u_2\}$, $G_3 = \{u_1, u_2, u_3\}$.

Key Encryption Key (KEK) Generation: After running KEKGen(U), the data service manager will generate KEKs for clients in U. A binary KEK tree for the user universe is developed initially by the data service manager, which is then used to assign keys of the attribute group to the members of U $\subseteq \mu$. The 'v_j' represents the each node of the tree, denoted by the phrase "the KEK_j," includes KEK. Path keys are collection of KEKs on nodes that create the path between a leaf and a root. The following phase involves identifying various tokens or properties.

During data decryption, the attribute group key is extracted from the header Hdr, and the message is then decrypted from the encrypted CT^l format. CipherText Every time a data service manager sends ciphertext (Hdr, "CT") to a user, He initially collects keys from a Hdr for each attribute group in Λ possessed by the user. If a user's u_t has the viable attribute λ_j (i.e., $u_t \in G_j$), the attribute group key K_{λ_j} is decrypted by Hdr using KEK, which is common in both KEK (G_j) and PK_t (that is, $KEK \in KEK (G_j) \cap PK_t$). Since there can only be one such KEK, the user can only be a part of one subset that is rooted by one KEK in KEK (G_j).

The Multiple User Access FBAE (MA-FBAE) was updated with the following revoking attributes.

- Public key components for the affected attribute have been modified.
- Substances that are updated in a secret key to represent each client's status as revoked. The Cipher text property had furthermore been modified on that server.

As a result, the framework provides advantages that are extremely strict and are maintained by highly demanding limited management.

4. Result Analysis

The effectiveness of the presented approach in terms of access control granularity is examined in this section. The efficiency of suggested scheme is then analyzed and contrasted with that of the earlier CP-ABE strategies in terms of theory. In the network simulation, the described scheme's effectiveness is shown in terms of communication costs. It also discuss about how effective it is when used with certain parameters and contrast the outcomes with those of the other systems.

The level of access control granularity and rekeying for each method are shown in Table 1. In the suggested approach, the data service manager rather than CP-ABE can perform the rekeying right away. As a result, a user or an attribute may be removed at any moment, even before any possible attribute expiration period. By closing the windows of susceptibility that permit unwanted access to the data, this enhances the confidentiality of the outsourced data's backward and forward secrecy. The suggested system also implements more precise user access control for every attribute.

Table 41.1 Access control comparison

Used model	Access Control Granularity	Rekeying
CP-ABE	Attribute revocation	Timed rekeying
CP-ABE with efficient revocation	System level user revocation	Immediate rekeying
FBAE	Attribute level user revocation	Immediate rekeying

Source: Authors

Figure 41.2 compares the cost of calculation for various levels of privacy derived models as CP-ABE with efficient Revocation, CP-ABE and Feature based Attribute Encryption. The CP-ABE approach consumes 210.645 milliseconds and 232.869 milliseconds, respectively, when there are 400 and 1000 files, as can be shown. The computation time for CP-ABE along with effective Revocation for the same number of files is 26.029 ms and 25.987 ms, respectively. FBAE for Data Privacy and Access Control model, which is more effective than other models, uses 10.012 milliseconds and 9.011 milliseconds, respectively, almost a negligible constant.

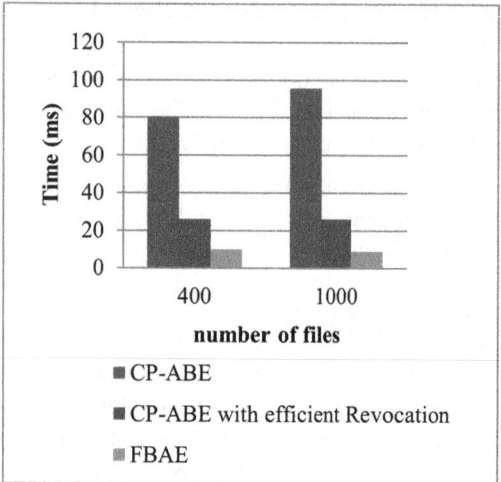

Fig. 41.2 The comparison of computation cost for different models

Source: Authors

5. Conclusion

Enhanced FBAE for Data Privacy and Access Control is described in this paper. For the protection of someone's health information, the most important system for managing health-related information is the PHR. These data are stored on unreliable servers in an effort to develop a secure information exchange system. A data owner can set the access control policy and then apply it to data that has been outsourced using the method provided. Access to and manipulation of fine-grained data is guaranteed on un-trusted servers. In addition, it features a technique that enables efficient attribute and user revocation, providing more accurate access control. The result analysis describes the comparison of access control and computing costs of different methods. The results show how successful and effective the method is for accessing and sharing data on the cloud.

References

1. Sultan Badran, Nabil Arman, Mousa Farajallah, "An Efficient Approach for Secure data outsourcing using Hybrid data Partitioning", 2021 International Conference on Information Technology (ICIT), Year: 2021
2. Qinlong Huang, Yixian Yang, Wei Yue, Yue He, "Secure data Group Sharing and Conditional Dissemination with Multi-owner in Cloud Computing", IEEE Transactions on Cloud Computing, Year: 2021
3. Mohini Bhardwaj, Nitin Pandey, Vinod Kumar Shukla, Ajay Vikram Singh, Neetu Gupta, "Review and Analysis of Security Model in health care System", 2021 9th International Conference on Reliability, Infocom Technologies and Optimization (Trends and Future Directions) (ICRITO), Year: 2021
4. S. Kanaga Suba Raja, A. Sathya, L. Priya, "A Hybrid data Access Control Using AES and RSA for Ensuring Privacy in Electronic Healthcare Records", 2020 International Conference on Power, Energy, Control and Transmission Systems (ICPECTS), Year: 2020
5. Rohit Ahuja, Sraban Kumar Mohanty, "A Scalable attribute based access control Scheme with Flexible Delegation cum Sharing of access Privileges for Cloud Storage", IEEE Transactions on Cloud Computing, Volume: 8, Issue: 1, Year: 2020
6. Guangli Xiang, Beilei Li, Xiannong Fu, Mengsen Xia, Weiyi Ke, "An Attribute Revocable CP-ABE Schem", 2019 Seventh International Conference on Advanced Cloud and Big Data (CBD), Year: 2019
7. Alex Roehrs, Cristiano André da Costa, Rodrigo da Rosa Righi, Sandro José Rigo, Matheus Henrique Wichman, "Toward a Model for Personal Health record Interoperability", IEEE Journal of Biomedical and Health Informatics, Year: 2019
8. Wei Li, Bonnie M. Liu, Dongxi Liu, Ren Ping Liu, Peishun Wang, Shoushan Luo, Wei Ni, "Unified Fine-Grained Access Control for Personal Health records in Cloud Computing", IEEE Journal of Biomedical and Health Informatics, Volume: 23, Issue: 3, Year: 2019
9. Y. Zhang, D. Zheng, and R. H. Deng, "Security and privacy in smart health: Efficient policy-hiding attribute-based access control," IEEE Internet Things J., vol. 5, no. 3, pp. 2130–2145, Jun. 2018.
10. Zheng Yan, Mingjun Wang, Yuxiang Li, Athanasios V. Vasilakos, "Encrypted Data Management with Deduplication in Cloud Computing", IEEE Cloud Computing, Volume: 3, Issue: 2, Year: 2016
11. Zheng Yan, Xueyun Li, Mingjun wang and Athanasios V. Vasilakos, "Flexible Data Access Control based on Trust and Reputation in Cloud Computing" 2015, IEEE transactions on cloud Computing, DOI 10.1109/TCC 2015.2469662.
12. J. H. Seo and K. Emura, "Revocable identity-based cryptosystem revisited: Security models and constructions," IEEE Trans. Inf. Forensics Security, vol. 9, no. 7, pp. 1193–1205, Jul. 2014.
13. K. Yang and X. Jia, "An efficient and secure dynamic auditing protocol for data storage in Cloud computing," Parallel and Distributed Systems, IEEE Transactions on, vol. 24, no. 9, pp. 1717–1726, 2013
14. M. Bellare, S. Keelveedhi, and T. Ristenpart, "Message-locked encryption and secure deduplication," in Proc. IACR Cryptol. ePrint Archive, P. Q. Nguyen, Ed., 2012, pp. 296–312.
15. L. Ibraimi, M. Petkovic, S. Nikova, P. Hartel, and W. Jonker, "Mediated Ciphertext-Policy Attribute-Based Encryption and Its Application," Proc. Int'l Workshop Information Security Applications (WISA '09), pp. 309–323, 2009.
16. N. Attrapadung and H. Imai, "Conjunctive Broadcast and Attribute-Based Encryption," Pairing '09: Proc. Int'l Conf. Palo Alto on Pairing-Based Cryptography, pp. 248–265, 2009.

Mahammad Shabana[1]

Associate Professor, Department of Computer Science & Engineering,
Neil Gogte Institute of Technology, Hyderabad, India

J. Praveen Kumar[2]

Associate Professor, Teegala Krishna Reddy Engineering College,
Hyderabad, India.

Kaja Masthan[3]

Associate Professor,
Department of CSE Sphoorthy Engineering College, Hyderabad

Abstract: In spite of the fact that duplicate move fraud, otherwise called cloning, is a sort of picture control in which one piece of an image is reordered into one more piece of similar picture, the two sections appear to be indistinguishable. The availability of high-quality picture editing software has made malicious image alteration, editing, and the production of phoney images quite straightforward. Therefore, in order to validate the veracity of digital photos, reliable PBIF (Passive-Blind Image Forensics) techniques are required. CMFD (Copy-Move Forgery Detection) employs the DoG (Difference of Gaussian) blob detector to detect regions in images and combines it with an ORB (Oriented Fast and Rotated Brief) rotation-invariant and noise-resistant feature detection).

Keywords: Blobs, CMFD, DoG, ORB, PBIF

1. Introduction

Many applications rely heavily on visual features, and digital pictures are among the most versatile forms of data due to their distinct representation and ease of modification. There are primarily three types of picture tampering, all of which involve making changes or omissions to an image for harmful intentions without leaving a clear signature of fabrication. Contraband copying or When cloning, a picture is cut out of one area and put into another. Plagiarism via cutting and pasting By combining or splicing together many photos, a forger might create a fake that bears little resemblance to the original. Image In the process of retouching, images may have their characteristics enhanced or reduced, but the image's topic will always be the same. Forensics analysis of

digital photos focuses on finding signs of manipulation or alteration. The term "Passive-Blind Image Forensics" (PBIF) is used to describe methods of assessing picture authenticity that do not need any special knowledge or security codes in order to detect subtle signs of manipulation in an image. Copy-Move Forgery Detection (CMFD) is the main topic of this work. You may use either a block-based or a keypoint-based approach to CMFD. The image is broken up into tiny chunks, some of which may overlap while others don't. The blocks are compared to each other to find the pairs that are a match. Best for spotting fakes when exposed to JPEG compression and Gaussian noise. Using feature vector matching within the image to locate the duplicated regions, the keypoint-based method computes highlight vectors for high entropy regions in an image without utilizing picture

[1]mdshabana5824@gmail.com, [2]praveenkrecit@gmail.com, [3]drkhaja.cse@gmail.com

development. Best method for spotting fakes when they've been distorted. Among the many difficulties inherent in detecting copy-move forgeries is maintaining accuracy in the face of standard picture handling strategies like pressure, commotion expansion, and processing time.

2. Literature Survey

Copy move image forgery by Dr. Archana and B. Patankar

Copy move fabrication is a strategy for controlling computerized pictures where a particular area of interest is reordered into a similar advanced picture before it is covered. Since the forged target zone retains many of the picture's essential properties, it is difficult to tell that the image has been altered. As a result, spotting this sort of fake is challenging. The authors of this work propose a method for locating these fabricated areas that makes use of the hybrid wavelet transform. To achieve the same result, the Kekre Wavelet Transform and the Hadamard Transform are combined. The resultant Mixture Change is then applied to try and request covering blocks, and elements got from each block are contrasted and attributes of any remaining blocks. With the same picture, varying the size of the overlapping blocks produces different outputs, which are then selected subjectively.

Digital picture copy-move forgery detection by Jan Lukas, David Soukal, and Jessica Fridrich

Digital photographs can be easily altered thanks to the availability of sophisticated tools for image processing and editing. There are no longer any telltale signs of manipulation when a picture has key details added or removed. Authenticating digital photographs, confirming their information, and identifying forgeries will become more important as digital cameras and videos gradually replace their analogue predecessors. This research focuses on methods for identifying cases of malicious alteration in digital photographs. This study focuses on the detection of a particular type of digital fraud known as a copy-move assault. In this type of fraud, a significant visual feature is obscured by copying and pasting a portion of the image elsewhere in the image. In this study, we go into the challenge of spotting copy-move forgeries and provide a strategy that works well and can be relied upon. Even if the fake picture is saved in a lossy organization like JPEG and thecloned portion is enhanced or retouched to blend in with the background, approach may still be able to identify the forged component. The suggested method's efficacy is proved on a variety of fake pictures.

Digital picture copy-move forgery detection by Loai Alamro and Nooraini Yusoff

One of the most widely recognized types of advanced picture treating is known as duplicate move, which involves moving parts of a digital image around. Copy-move forgeries of digital images may be difficult to detect due to geometric modification. In this kind of picture fraud, the pieces that have been copied and pasted are either rotated or re-scaled. As a result, we suggest in this research that DWT and SURF be used together to identify copy-move behaviour. With an accuracy rate of 95%, the aftereffects of the analysis show that the suggested strategy is superior. Digital picture copy-move attacks have been discovered, and the approach may also identify fraudulent components that have been compromised by a rotation or scale problem.

An effective approach for copy-move forgery detection using discrete wavelet transform by Saiqa Khan and Arun Kulkarni

Picture fabrication might be separated into numerous subtypes, one of which is designated "Duplicate Move phony," in which one part of a computerized picture is reordered onto one more segment of a similar picture. In this examination, we give a technique for identifying Copy-Move fraud using a blind forensics methodology. Our method use DWT to provide a lower-dimensional representation of the input picture. Next, overlapping blocks are created from the compressed picture. Phase Correlation is then used as a similarity criteria to rank the blocks and find any duplicates. Due to the use of DWT, detection is first performed on the most primitive form of the picture representation. This method significantly speeds up the detecting procedure.

Discovering duplicated image regions to reveal digital forgeries by Alin C Popescu and Hany Farid

Here, we detail a reliable method for automatically identifying instances of picture duplication. This method reduces the number of dimensions by first conducting principal component analysis to picture blocks of a defined size. This representation can withstand modest shifts in the original picture brought on by things like lossy compression or additive noise. After that, we do a lexicographic sort on all the picture blocks to look for duplicate areas. We demonstrate the technique's effectiveness on convincing fakes and evaluate its resistance to and aversion to added substance commotion and lossy JPEG pressure.

3. Existing System

Several transform-domain based methods exist for detecting copy-move image forgery (CMFD). A lexicographic order and a blocking strategy based on DCT coefficients were proposed by Fridrich et al. Utilizing a blend of DWT and DCT, Khizar et al. conceived CMFD. Toqeer et al. proposed CMFD utilizing fixed wavelet change (SWT) and discrete cosine change (DCT).Tetrolet transform was proposed for CMFD by Kunj et al. Some of these techniques are not resilient to post-processing processes like blurring, while others have a high computational cost. Diaa et al. proposed the CMFD method, which makes use of Hessian characteristics and the CSLBP. Using AKAZE characteristics and nonlinear scale space, Guzin et al. proposed CMFD. Fan et al. proposed a CMFD that utilizes a blend of KAZE and Filter highlights. By melding customary block-based and keypoint-based strategies, Priya et al. proposed CMFD. The CMFD proposed by Vaishnavi et al., 2019 makes use of symmetry-based local characteristics. Therefore,

3.1 Existing System Disadvantages

Complexity in terms of computing is high. Not all of them can withstand effects like blurring in post-production.

4. Proposed System

Sobel edge detection, feature extraction (using DoG and ORB), and feature matching are the three primary processes involved. In an effort to combat Copy-Move forgeries, To defeat these hindrances, we need to coordinate block-based and keypoint-based recognition strategies into a solitary model. There are certain drawbacks to using block-based methods; for example, it may be difficult to decide on the appropriate block size, the matching process can become computationally demanding with tiny blocks, huge blocks can't be used to spot forgeries in small places, and uniform sections in the original picture will appear more than once. To go around these restrictions, we use blobs in place of traditional picture blocks. After keypoints have been matched, several keypoint-based algorithms use the RANSAC (Random Sample Consensus) algorithm to reduce spurious matches. Our method does not involve using RANSAC to filter out irrelevant matches.

After the coloured picture is grayscaled, it is separated into covering or non-covering blocks of the same size. Some attributes or statistical information are gathered for each block of this generated picture. These characteristics were compared to those of the other building blocks to see whether a match could be made.

Last but not least, finding the match block makes any copy_ move forgeries in the picture immediately obvious. Copy_ move forgery is detected by first processing each block with the feature-based technique, which extracts picture characteristics like SIFT, SURF, etc.

4.1 Proposed System Advantages

Efficient detection of copied regions. Comparatively our system is simple and feasible

5. Architecture Diagram

The actual execution of a product framework's parts might be visualised with the use of an architectural diagram. It lays forth the software system's overall layout, together with all of the interconnections, constraints, and boundaries that exist inside it. There is a lot of complexity and change in software environments.

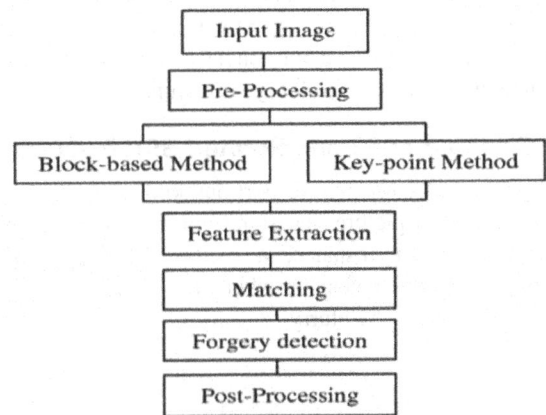

Fig. 42.1 Architecture diagram

6. Methodologies

6.1 Sobel Edge

Detection Using the Sobel edge detector, we can calculate the 2D spatial gradient of an image. In the end, we get a 2D guide addressing the inclination at each point (the Sobel picture), with the districts of high slope represented by white lines. To improve blob localisation, we employ a Sobel picture as the blob detector's input.

6.2 Feature Extraction

At this point, the blob detection and ORB feature detection processes that constitute feature extraction are executed. 3.2.1 Blobs are areas in a digital picture that have different qualities, such brightness or colour, compared to adjacent regions, and DoG (Difference of Gaussian) is a blob detector that is designed to find such regions. The DoG technique

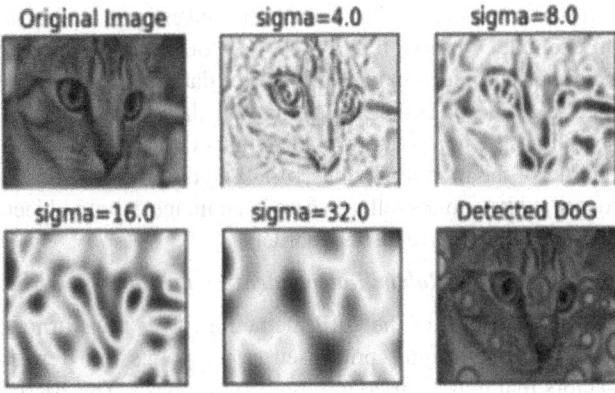

Fig. 42.2 Screenshot: Detecting dog

[12] is being used to identify Blobs. In addition, Laplacian of Gaussian (LoG) was employed in the model's assessment. DoG is the distinction between two pictures convolved with two Gaussian channels g(x, y, σk) and g(x, y, σ) as displayed in Figure g(x, y, σ) = 1 2πσ2 e(−x2 + y2 2σ2) (1) Canine = g(x, y, σk) ∗ I(x, y) − −g(x, y, σ) ∗ I(x, y) (2) g(x, y, σ) is Gaussian channel, ∗ is convolution head, k is a scale variable, σ is standard deviation and I(x, y) is picture.

6.3 Oriented FAST and Rotated BRIEF (ORB)

ORB, first, seeks for picture portions known as key spots. Those parts of a picture that stand out the most are called "key points." For instance, a picture's borders (where the brightness fluctuates, or where the pixel values shift dramatically) might be blurry.

Fig. 42.3   Screenshot: Keypoint detection

Feature Detection using FAST (Features from Accelerated Segments Test)

Pick a pixel p in the picture that has to be classified as either an intriguing landmark or a background feature. Intensity should rest with me. Find the right value for the threshold, t. Take a 16-pixel radius around the investigated pixel as an example. Observe the Fig. 42.4. If there are at least two neighboring pixels that are brighter or darker than pixel 'p,' then 'p' is chosen as the pivot point. With these enhancements, searching a whole picture for landmarks may be done in a quarter of the time it previously took. In this way, the FAST-identified key points tell us where in an image certain objects are located that have distinct borders.

BRIEF (Binary Robust Independent Elementary Features)

The second stage of the ORB method involves transforming the FAST algorithm's prioritised set of points into feature vectors that may be used to represent an object. The BRIEF

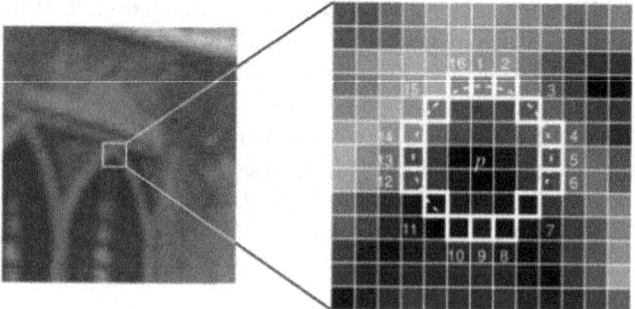

Fig. 42.4   Screenshot: Feature detection using FAST

algorithm is used by ORB to generate feature vectors. From a small number of identifiers, binary feature vectors are generated. As is common knowledge, a binary feature vector is just a feature vector that consists of only ones and zeros; this is also known as a binary descriptor. Using BRIEF, you may skip the tedious search and get straight to the binary strings. It takes a patch of a smoothed picture and picks a certain nd(x,y) locations to compare the intensities of their pixels. In a nutshell, BRIEF can calculate and match feature descriptors more quickly than other methods. Unless there is a significant in-plane rotation, it also offers a good recognition rate.

7. Algorithm

7.1 DoG (Difference of Gaussian)

The difference of gaussians may be used as a technique for working on the nature of computerized pictures by making finer distinctions between edges and other picture features more apparent. Edge detection using the Difference of Gaussians (DoG) is a popular technique. The procedure involves subtracting two Gaussians, one of which has a lower standard deviation than the other. Edge detection in this picture is the outcome of a convolution between the subtracted kernels and the input image.

7.2 ORB (Oriented FAST and Rotated BRIEF)

Fast and robust local feature detector for application in computer vision tasks like object identification and 3D reconstruction, Oriented FAST and rotated BRIEF (ORB) was initially reported by Ethan Rublee et al. in 2011. It competes well with SIFT and SURF and is an effective alternative to them. The fact that SIFT and SURF are protected by patents was a major inspiration for developing ORB. You may utilize ORB without cost however. ORB is over two orders of magnitude quicker than SIFT while achieving comparable performance on these tasks. ORB is based off of two other popular tools: the FAST key point detector and the BRIEF description. Both of these methods are enticing due to their high efficiency and inexpensive price.

8. Output Results

Select "upload image" and then "scan" to upload the picture.

Fig. 42.5 Screenshot: GUI page

Here we see the outcome, which is

Fig. 42.6 Screenshot: Output

9. Conclusion

DoG blobs detector and ORB feature detection are the basis of a novel model presented for Copy-Move Forgery Detection. Multiple copy-move forgeries inside a single picture, as well as geometric manipulations like rotation and scaling, are all within the scope of the proposed method's efficacy. We just paired Circle highlight descriptors from isolated masses, which significantly diminished figuring time, how much elements to coordinate, and bogus matches since the first region and the moved locale generally brought about various masses. The upsides of both block-based and key point-based fabrication discovery approaches are joined in a solitary model when Canine and Sphere highlights are consolidated. To assess the adequacy of Duplicate Move Imitation Discovery methods on very big datasets, researchers want to focus their efforts on removing the need for human interpretation of the techniques' output.

References

1. Dhiman, N., Kumar, R.: Classification of copy move forgery and normal images by ORB features and SVM classifier. In: ICITSEM 2017, pp. 146–155 (2017)
2. Malviya, A.V., Ladhake, S.A.: Pixel based image forensic technique for copy-move forgery detection using auto color correlogram. Procedia Comput. Sci. 79, 383–390 (2016)
3. Lee, J.-C.: Copy-move image forgery detection based on Gabor magnitude. J. Vis. Commun. Image Representation 31, 320–334 (2015)
4. Amerini, I., Ballan, L., Caldelli, R., Del Bimbo, A., Serra, G.: A SIFT-based forensic method forcopy-move attack detection and transformation recovery. IEEE Trans. Inf. Forensics Secur. 6(3),1099–1110 (2011)
5. Christlein, V., Riess, C., Jordan, J., Riess, C., Angelopoulou, E.: An evaluation of popular copy-move forgery detection approaches. IEEE Trans. Inf. Forensics Secur. 7(6), 1841–1854 (2012)
6. Rublee, E., Rabaud, V., Konolige, K., Bradski, G.: ORB: an efficient alternative to SIFT orSURF. In: ICCV, pp. 2564–2571 (2011)
7. AlSawadi, M., Ghulam, M., Hussain, M., Bebis, G.: Copy-move image forgery detection usinglocal binary pattern and neighborhood clustering. In: EMS (2013)
8. Popescu, A., Farid, H.: Exposing digital forgeries by detecting duplicated image regions. Dartmouth College, Computer Science, Technical report, TR 2004–515 (2004)
9. Fridrich, J., Soukal, D., Lukas, J.: Detection of copy-move forgery in digital images. In: Digital Forensic Research Workshop, Cleveland, OH, pp. 19–23 (2003)
10. Jing, L., Shao, C.: Image copy-move forgery detecting based on local invariant feature. J. Multimed. 7(1), 90–97 (2012)
11. Shivakumar, B.L., Santhosh Baboo, S.: Detection of region duplication forgery in digital images using SURF. IJCSI Int. J. Comput. Sci. Issues 8(4), 199–205 (2011)
12. Lowe, D.G.: Distinctive image features from scale-invariant keypoints. Int. J. Comput. Vis.60(2), 91–110 (2004)
13. Ng, T.-T., Chang, S.-F., Lin, C.-Y., Sun, Q.: Passive blind image forensics. In: Multimedia Security Technologies for Digital Rights Management, pp. 383–412 (2006) Digital Image Forensics Technique for Copy-Move Forgery Detection 483
14. Gupta, C.S.: A review on splicing image forgery detection techniques. IJCSITS, 6(2), (2016)
15. Mushtaq, S., Hussain, A.: Digital image forgeries and passive image authentication techniques: asurvey. Int. J. Adv. Sci. Technol. 73, 15–32 (2014)
16. Rathod, G., Chodankar, S., Deshmukh, R., Shinde, P., Pattanaik, S.P.: Image forgery detection on cut-paste and copy-move forgeries. Int. J. Adv. Electron. Comput. Sci. 3(6) (2016). ISSN: 2393–2835
17. Redi, J., Taktak, W., Dugelay, J.: Digital image forensics: a booklet for beginners. Multimed. Tools Appl. 51(1), 133–162 (2011)

18. Lindeberg, T.: Detecting salient blob-like image structures and their scales with a scale-spaceprimal sketch: a method for focus-of-attention. Int. J. Comput. Vis. 11(3), 283–318 (1993)

19. Calonder, M., Lepetit, V., ¨ozuysal, M., Trzcinski, T., Strecha, C., Fua, P.: BRIEF: computing a local binary descriptor very fast. IEEE Trans. Pattern Anal. Mach. Intell. 34(7), 1281–1298 (2012)

20. Hassaballah, M., Abdelmgeid, A.A., Alshazly, H.A.: Image features detection, description andmatching. In: Awad, A.I., Hassaballah, M. (eds.) Image Feature Detectors and Descriptors. SCI, vol.630, pp. 11–45. Springer, Cham (2016). https://doi.org/10.1007/978-3-319-28854-3 2

21. Audi, A., Pierrot-Deseilligny, M., Meynard and, C., Thom, C.: Implementation of an IMU aided image stacking algorithm in a digital camera for unmanned aerial vehicles. Sensors 17(7), 1646(2017)

22. Calonder, M., Lepetit, V., Strecha, C., Fua, P.: BRIEF: binary robust independent elementary features. In: Daniilidis, K., Maragos, P., Paragios, N. (eds.) ECCV 2010. LNCS, vol. 6314, pp.778–792. Springer, Heidelberg (2010). https://doi.org/ 10.1007/978-3-642-15561-1 56

23. Kong, H., Akakin, H.C., Sarma, S.E.: A generalized Laplacian of Gaussian filter for blob detection and its applications. IEEE Trans. Cybern. 43(6), 1719–1733 (2013)

24. Tralic, D., Zupancic, I., Grgic, S., Grgic, M.: CoMoFoD - new database for copy move forgery detection. In: Proceedings of 55th International Symposium ELMAR2013, pp. 49–54 (2013)25. Irwin, S., Gary, F.: A 3 x 3 Isotropic Gradient Operator for Image Processing. The Stanford Artificial Intelligence Project, pp. 271–272 (1968)

25. Vincent, O.R., Folorunso, O.: A descriptive algorithm for Sobel image edge detection. In: Proceedings of Informing Science & IT Education Conference (InSITE) (2009)

Note: Author self data and the graphs/images are part of the research work

K Nearest Neighbour: A Dynamic Model

43

Tanya Tripathi[1]

UG Student, Faculty of Science and Technology, IFHE, Hyderabad

B. Deevena Raju[2]

Faculty of Science and Technology, IFHE, Hyderabad

M. Suresh Babu[3]

Professor, Department of CSE,
Teegala Krishna Reddy Engineering College, Hyderabad

Abstract: This study proposes a novel dynamic model to allocate the value of 'K' in the K Nearest Neighbour algorithm based on geometrical calculations. K-nearest neighbour algorithm is a well-known machine-learning algorithm widely used to classify data. It faces certain disadvantages when working on a large dataset or data with several dimensions.

One of the significant factors that affect the accuracy of the model is the value of K. Fundamentally; the value of K is fixed for all data points. In this study, the value is calculated at the local level for all test points. This study performs a comparative study between suggested model and the traditional KNN model.

Keywords: K-Nearest neighbor, Dynamic model, Classify data

1. Introduction

The value of k is a hyperparameter that needs to be tuned for each dataset. It is the same principle as "one size does not fit all," meaning that we are looking forward to find localized value of k for each data points and evaluating its impact on the accuracy of the model by dynamically allocating the value of K. To improve the prediction of unknown values, we hypothesize that removing the dependency on K and shifting the voting criteria to the density of data points surrounding the test subject will improve the prediction. In this model, we are constructing geometric constraints which are flexible as per the location of each data point and also act as barriers to avoid taking votes from irrelevant data points that reduces the impact of noise in the decision-making process. The datasets will first be pre-processed, and then the predictions of both models will be compare to find the more accurate model.

2. Literature Review

The traditional K Nearest Neighbour (KNN)

Was first proposed by Cover and Hart in 1967 [1] works to find the k nearest neighbours to the unknown data point. It takes the majority of votes from all k data points and assigns the majority voted value to test point.

Algorithm for KNN is as follows:

Step1: Choose the value of 'K', K stands for number of nearest neighbours that we consider for voting.

Step 2: Find the distance between all the data points and unknown test point.

Step 3: Identify k number of nearest neighbours

Step 4: Take votes from those K neighbours

[1]tanyatripathi2001@gmail.com, [2]deevenaraju@ifheindia.org, [3]sureshcse@tkrec.ac.in

Step 5: Predict the data point as outcome of majority votes from those K neighbours.

However, KNN model was improved in different ways over the years, resulting in variations in the original model. This model has a number of notable variations, including weighted KNN by Saha et al in 1991 [2]. In weighted KNN, all votes are not treated equally. The closer neighbour influences more through its voting power

Another approach includes Radius Neighbour Classification. In this approach, a certain radius is fixed for a particular dataset. All the points within the circular area are considered for voting and the majority vote is applied to an unknown data point which was proposed by Leo Breiman and Jerome Friedman in 1984 [3].

More often than not, data is available in steady and sustained form as values not categories. In such cases, we take the mean sum of values from all the nearest voters to evaluate the value of the test data point as suggested by Altman in 1992 [4].

Inducing other major geometrical findings are often fruitful for the accuracy and efficiency of the model as in the case of dynamic nearest neighbour queries in Euclidean space by Mohammed Einus Ali [5] as it revolves around the construction of Voronoi diagrams and classifying the points as per the boundaries laid by Voronoi diagram.

Another method is the centroid method by MacQueen, J. B. [6] in which each cluster is identified by centroids of data points of a labelled class. When we need to classify a new unknown data point the algorithm calculates the distance between the unknown data point and the centroids using a distance matrix. The unknown data point is classified as the label of closest centroid.

This method has another variation which was proposed by Tibshirani, R., Walther, G., & Hastie, T [7] which involves scaling down the features. Scaling down is performed by dividing the values of one feature by the variance of that feature this method is called shrunken centroid method and it is useful to reduce the influence of noisy data and help to make the model more accurate.

We have another improvised model named as Neighbourhood Components Analysis by Goldberger, J., Roweis, S., & Hinton, G. E [8], this method takes labelled training set as input and makes a matrix that maps higher dimension inputs to lower dimension values. NCA employs an optimization algorithm, such as stochastic gradient descent, to iteratively update the transformation matrix to maximize the objective function and then it follows the majority voting from the nearest neighbours and classifies the unknown test points.

Another major outbreak was achieved by Hassant, M. A [9] by using Hassanat distance. Hassant distance is useful as it

is resistant to variations in values for different features along with this it is also sturdy against noise in data.

Hassant distance is calculated using the following formula

$$d_h(x, y) = \|x - y\|_2 / (1 + \|x\|_2 + \|y\|_2)$$

where,

$\|x\|2$ is the Euclidean norm of x.

It is the square root of the sum of the squares of the vector's components. Given below is mathematical formula for Euclidean norm of x.

$$\|x\|_2 = \sqrt{\{x_1^2 + x_2^2 + \ldots + x_n^2\}}$$

Where x_1, x_2 and so on are values of a particular feature.

Another advancement in improving KNN was

Fuzzy KNN it was introduced by Dubois, D., Prade, H., & Yager, R. R [10]. In classical KNN the unknown test point is strictly classified into one of the known categories where as in fuzzy KNN the unknown data point is given a degree of membership for each know category as it allows partial membership in classes.

Another variations in KNN in Mutual KNN by Ding, C., He, X., & Simon, H [11] which is an unsupervised learning algorithm in which points are merged and linked together. Here clustering of data points happen due to calculated mutual similarities between data points. This algorithm is very useful in anomaly detection for malfunctioning equipment or for illegal/scam transactions.

Another method was proposed by Gongde Guo1 and Hui Wang [12] to overcome the limitation of being dependent on Fixed value of k, we outline certain regions to represent classified data points set. In the selection of each representative set, we use the optimal but different k allocated by the dataset itself to abolish dependency on k without the user's intervention.

Let us look at a distance calculation method which is often used in KNN called Hamming distance by R.W. Hamming [13]. It is used to compute the similarities between two data points that are conventionally represented as binary vectors. Binary vectors are often used to represent data that can be classified into two categories like "yes" or "no", "pass" or "fail" and so on.

$$V1 = [1, 1, 0, 1, 0] \quad V2 = [0, 1, 1, 1, 1]$$

The following two vectors have the hamming distance of 3. The more the similar the data points are lesser is the hamming distance.

In this variation of KNN the unknown test point is compared with all the points present in dataset.

If the value of K is 3. The algorithm takes majority voting from K nearest neighbours and classifies the unknown data points into category that has majority votes. One of the major drawback is it cannot be used for non-binary data which acts as short coming because everything in this world cannot be classified as "black" or "white", world consist of wide spectrum of grey colour too.

Another simple method of calculating the distance called as Jaccard similarity by Jaccard, P [14]. Let us understand Jaccard similarity using an example S1 and S2 are the two sets

$$Set\ S1 = \{1,6,7,5\}$$
$$Set\ S2 = \{1,5,3,4\}$$

Jaccard similarity (S1, S2) = |S1 ∩ S2| / |S1 ∪ S2|

|S1 ∩ S2| is the size of the intersection of S1 and S2

|S1 ∪ S2| is the size of the union of S1 and S2

Here the intersection consists of {1,5}

The union consist of {1,3,4,5,6,7}

Jaccard similarity (S1, S2) = 2/6

On simplification we get 1/3 or .333

The larger the value the closer Jaccard similarity.

If K is equal to 3, then the algorithm groups unknown data items into categories with a majority of votes using the votes of the K closest neighbours.

3. Methodolgy

As one shoe cannot fit all, similarly, one value of K is not optimal for all unknown test points. In this model, we calculate the value of K of each data point by using mathematical calculations.

First step, we calculate the distance of all points to test point. Second step, we find the index of the nearest point and the distance between both the test point and the nearest data point. Let us call this distance as "d".

Now we calculate K as half of the value of "d." We do this because we require to set the value of K to a flexible value that should be yielding and adaptable to the close neighbourhood of the test point. If we declare a fixed value of K, it might be troublesome where the local density of data points varies drastically at that point of time over fitting can occur when noisy or isolated data points dominate the classification choice. Further, if k is set too small than under fitting can occur, conversely, if k is set too high, the classification judgments are influenced by the data points that would be too far from the test point.

Therefore, if the data points are more densely packed, the closest neighbours would be nearer and the value of K will be less, and if the data points are more sparsely packed, the

nearest neighbour would be further away and we require a bigger value of K. Hence, assume K as half of the value of "r," which helps in running the model smoothly and more flexibly for each test point.

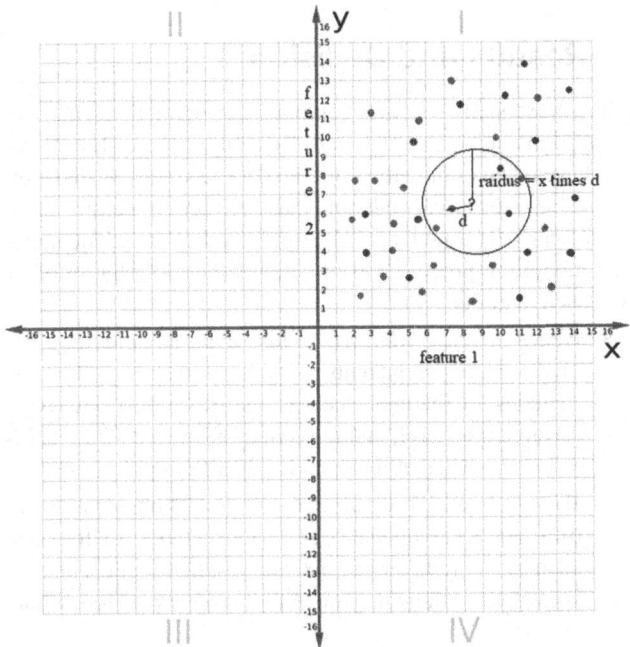

Fig. 43.1 Feature 1

Where d represents distance between the test point and the nearest data point.

Third step is to find all the training points within a radius equivalent to twice/ "x times" the distance from the nearest data point. We decide the value of x by feature engineering and find the optimal value of x for particular dataset. The majority voting is taken from all points in that particular area. The unknown data point takes the value of the majority vote. We repeat this process for all the test data points.

The following procedure is discussed as step by step algorithm for dynamic KNN model:

Step 1: Load the data set

Step 2: Split the dataset into test and train

Step 3: Create a function to predict the values

Step 4: Calculate distance from test point to all data point

Step 5: Find the index of the nearest neighbour and its distance

Step 6: Calculate K as half of the distance to nearest neighbour

Step 7: Find all the training points within a radius of let us say 2d

Step 8: Take a majority vote of the classes of the training points with in the radius

Make predictions for the test points

Step 9: Calculate the accuracy of the model.

3. Results and Decision

As a next step, we will compare the predictions of both algorithms on various data sets. To analyse these algorithms, we have taken publically available data sets, such as diabetes, wine, heart disease, Seeds, and glass. All pre-processing steps have been performed on all datasets and they are adequate for the modelling process. Given below is the description of all the used datasets

- *Heart disease dataset:* This dataset contains 303 observations of patients with heart disease. The data includes 14 features, such as age, sex, chest pain type, serum cholesterol, and maximum heart rate. The goal is to predict whether a patient has heart disease or not.
- *Wine dataset:* This dataset contains 178 observations of different types of wine. The data includes 13 features, such as alcohol content, pH, and residual sugar. The goal is to classify the wine into one of three types: red, white, or rose.
- *Diabetes dataset:* This dataset contains 768 observations of patients with diabetes. The data includes 10 features, such as age, sex, body mass index (BMI), and fasting plasma glucose. The goal is to predict whether a patient will develop diabetes within 5 years.
- *Seeds dataset:* This dataset contains 351 observations of different types of seeds. The data includes 7 features, such as seed length, width, and area. The goal is to classify the seeds into three types: Kama, Pinto, and Wheat.
- *Glass dataset:* This dataset contains 214 observations of different types of glass. The data includes 9 features, such as refractive index, sodium content, and silicon content. The goal is to classify the glass into one of six types: window, bottle, tableware, etc.

Given below is the table for comparison between KNN and Dynamic KNN.

Table 43.1 This table consists of comparison in terms of accuracy of both KNN and dynamic KNN.

Dataset	No. of Observations	KNN Accuracy	Dynamic KNN Accuracy
Heart Disease	303	66.7	73.3
Wine	178	88.9	88.9
Diabetes	768	74	74.7
Seeds	351	90.5	95.2
Glass	214	100	100

Given below is line graph for comparison between KNN and Dynamic KNN.

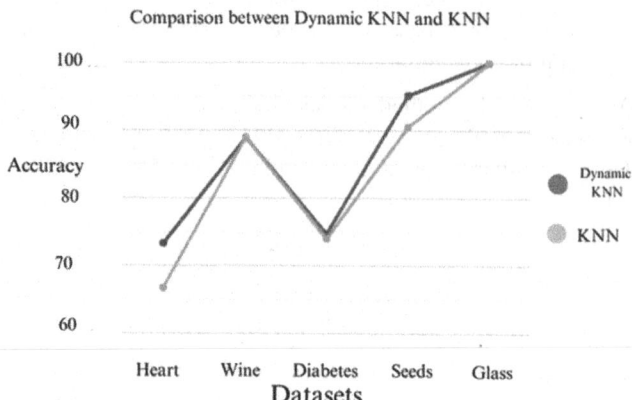

Fig. 43.2 Line graph showing comparison between both KNN and dynamic KNN

5. Conclusion

We have tested our model on various data sets like heart diseases, wine, diabetes, seed and glass. In all the tested cases above the dynamic KNN has provided accuracy equal to or greater than classical KNN model. One of the main reasons for greater accuracy is overcoming the fixed nature of value of K and locally determining the value of K for each test point and while voting it also restricts the influence of data points that are beyond certain range.

6. Future Scope

The model can be extended to regression and clustering algorithms in the future, and it can be tested on various datasets and the accuracy patterns of both algorithms will be compared and also model can be further extended to 10 fold cross validation.

References

1. T. M. Cover and P. E. Hart, "Nearest neighbour pattern classification," in IEEE Transactions on Information Theory, vol. 13, no. 1, pp. 21–27, January 1967. doi: 10.1109/TIT.1967.1053964.
2. Saha, D. W., Kibler, D., & Albert, M. K. (1991). Instance-based learning algorithms. Machine learning, 6(1), 37-66. doi: 10.1023/A:1022683801
3. Breiman, L., & Friedman, J. H. (1984). Classification and regression trees. CRC press.
4. Altman, N. S. (1992). An introduction to kernel and nearest-neighbor nonparametric regression. The American Statistician, 46(3), 175-185.
5. Ali, M. E. (2016). Dynamic nearest neighbor queries in euclidean space using voronoi diagrams. IEEE Access, 4, 2996-3010. doi: 10.1109/ACCESS.2016.2566778

6. MacQueen, J. B. (1967). Some methods for classification and analysis of multivariate observations. In Proceedings of the fifth Berkeley symposium on mathematical statistics and probability (pp. 281–297). Berkeley, CA: University of California Press.

7. Tibshirani, R., Walther, G., & Hastie, T. (2001). Estimating the number of clusters in a data set via Gaussian mixture models. Journal of the Royal Statistical Society: Series B (Statistical Methodology), 63(2), 411–423.

8. Goldberger, J., Roweis, S., & Hinton, G. E. (2004). Dimensionality reduction by learning an invariant mapping. Neural Computation, 16(6), 1205–1226.

9. Hassant, M. A. (2004). A new distance measure for pattern recognition. Pattern Recognition, 37(1), 155–162.

10. Dubois, D., Prade, H., & Yager, R. R. (1986). A fuzzy k-nearest neighbor algorithm for pattern IEEE Transactions on Pattern Analysis and Machine Intelligence, 8(6), 675–685.

11. Ding, C., He, X., & Simon, H. (2005). Mining anomalous data using mutual k-nearest neighbor graphs. In Proceedings of the 2005 ACM SIGMOD international conference on Management of data (pp. 503-514). New York, NY: ACM.

12. Guo, G., & Wang, H. (2004). A fuzzy k-nearest neighbor algorithm. IEEE Transactions on Systems, Man, and Cybernetics, Part B (Cybernetics), 34(4), 2021–2033.

13. R. W. Hamming, "Error Detecting and Error Correcting Codes," Bell System Technical Journal, Vol. 29, No. 2, April 1950, pp. 147–160.

14. Jaccard, P. (1901). Étude comparative de la distribution florale dans une portion des Alpes et des Jura. Bulletin de la Société Vaudoise des Sciences Naturelles, 37, 547–579.

Note: All the figures and tables in this chapter were authors' original test data and experimental output.

A Survey on Efficient Memory Management in Large-Scale Deep Learning Optimization

44

Jitendra Alaparthi[1]

ReseasrchScholar,Department of CSE, SRK University,
Jhatkedi,Misrod, Bhopal, India

Varsha Namdeo[2]

Professor, Department of CSE,
SRK University, Jhatkedi, Misrod, Bhopal, India

Abstract: Deep learning has achieved remarkable success in various domains, powered by the availability of vast amounts of data and advanced computational resources. However, training large-scale deep neural networks (DNNs) poses significant memory challenges due to the massive number of parameters and complex model architectures. Efficient memory management in deep learning optimization becomes crucial to overcome memory limitations, reduce computational overhead, and enable the training of large-scale DNNs. This survey paper comprehensively reviews and analyses the state-of-the-art memory management techniques applied to tackle memory issues in large-scale deep learning optimization. The survey explores memory optimization methods, distributed computing, parameter storage techniques, and other memory-efficient strategies that play a critical role in making deep learning more accessible and effective in real-world applications.

Keywords: Deep learning optimization, Memory management, Large-scale DNNs, Memory-efficient training, Distributed computing, Gradient accumulation, Mixed precision training, Model parallelism, Survey

1. Introduction

1.1 Background

Deep learning has emerged as a revolutionary technology in the field of artificial intelligence and machine learning. It has demonstrated unprecedented performance in various tasks, such as image recognition, natural language processing, and speech synthesis. Deep neural networks, a core component of deep learning, consist of multiple layers of interconnected neurons that can automatically learn hierarchical representations from data. However, the success of deep neural networks comes at the cost of computational complexity, memory requirements, and long training times. As the size and complexity of deep neural networks continue to grow to accommodate more challenging tasks and datasets, efficient memory management becomes a critical concern in deep learning optimization.

1.2 Motivation

The increasing popularity of deep learning applications in diverse domains, including healthcare, autonomous vehicles, finance, and robotics, has intensified the need for efficient memory management techniques. Training large-scale deep neural networks demands substantial memory resources, leading to challenges in both research and practical implementation. The motivation behind this study is to explore and analyze the various memory management strategies and techniques employed in deep learning optimization to overcome memory limitations, reduce computational overhead, and enhance the scalability of deep learning models.

[1]jittu28@gmail.com, [2]varsha_namdeo@yahoo.com

1.3 Objectives

The primary objective of this research is to conduct a comprehensive study on efficient memory management in large-scale deep learning optimization. This study aims to:

(a) Identify and review the state-of-the-art memory management techniques used in deep learning optimization.

(b) Analyse the advantages and limitations of each memory management approach concerning memory usage, computational efficiency, and model performance.

(c) Investigate the impact of memory management on model training time, convergence, and generalization.

(d) Explore the trade-offs between memory optimization and model complexity, accuracy, and resource utilization.

(e) Provide insights into the practical implementation and applicability of memory management techniques in real-world deep learning scenarios.

1.4 Scope and Organization

This research focuses on memory management techniques for large-scale deep learning optimization, specifically addressing the challenges posed by memory limitations and computational overhead. The study covers a wide range of memory management methods, including parameter pruning, quantization, model compression, and distributed computing. It explores techniques to reduce the memory footprint of deep neural networks during training while maintaining or improving their performance.

The organization of this thesis consists of several chapters, each dedicated to exploring different memory management strategies. Chapter 2 provides an overview of deep learning optimization, highlighting the significance of memory management. Subsequent chapters delve into specific memory management techniques, such as data parallelism, gradient accumulation, mixed precision training, and model parallelism. The evaluation and comparative analysis of memory optimization approaches are presented in a dedicated chapter. Real-world applications and challenges are discussed in separate chapters, followed by concluding remarks and future research directions in the final chapter.

2. Deep Learning Optimization: An Overviw

2.1 Gradient-Based Optimization Algorithms

Gradient-based optimization algorithms form the backbone of deep learning model training. These algorithms aim to find the optimal set of model parameters that minimize the error or loss function. Common gradient-based optimization methods include:

Stochastic Gradient Descent (SGD): The basic form of gradient descent that updates model parameters after each individual data sample, making it computationally efficient but noisy.

Mini-Batch Gradient Descent: Similar to SGD, but updates model parameters after processing a mini-batch of data samples. It strikes a balance between the efficiency of SGD and stability of full-batch gradient descent.

Adam (Adaptive Moment Estimation): A popular adaptive optimization algorithm that computes adaptive learning rates for each model parameter. It combines the advantages of both Adagrad and RMSprop, providing faster convergence and robustness.

2.2 Memory Requirements in Deep Learning Training

Deep learning training requires a considerable amount of memory due to the large number of parameters and intermediate activations involved. The main components contributing to memory requirements are:

Model Parameters: Deep neural networks can have millions or even billions of parameters, and storing them in memory during training can be memory-intensive.

Activation Tensors: The intermediate activation tensors generated during forward and backward passes need to be stored for computing gradients. As the depth and width of the network increase, so does the memory footprint of activation tensors.

Batch Size: A larger batch size requires more memory to store gradients, leading to increased memory consumption during training.

2.3 Challenges in Large-Scale Deep Learning Optimization

Training large-scale deep neural networks poses several challenges related to memory management:

Memory Limitations: Deep learning models may exceed the available memory capacity, leading to out-of-memory errors during training.

Computational Overhead: Memory-intensive operations, such as gradient computations, can lead to longer training times and increased computational overhead.

Training Efficiency: Memory bottlenecks can hinder efficient utilization of computational resources like GPUs, resulting in suboptimal training efficiency.

Generalization and Overfitting: Memory optimization techniques may affect the model's capacity to generalize well to unseen data or increase the risk of overfitting.

Scalability: As deep learning models scale up, the memory demands grow significantly, making it challenging to handle large-scale datasets and architectures.

Addressing these challenges requires innovative memory management strategies that strike a balance between memory optimization and model performance. Researchers and practitioners are actively exploring various techniques to efficiently manage memory during large-scale deep learning optimization, thereby enabling the training of more complex models and facilitating broader applications of deep learning technology.

3. Memory Management Techniques

Memory management techniques in large-scale deep learning optimization aim to reduce memory consumption without compromising model performance. This section explores various memory optimization approaches:

3.1 Parameter Pruning and Sparsity

Parameter pruning techniques involve removing redundant or less important parameters from deep neural networks. By identifying and eliminating unnecessary connections, the model's memory footprint is significantly reduced. Sparse models, which contain a high proportion of zero-valued parameters, can be efficiently stored and processed, leading to memory savings during training and inference.

3.2 Low-Rank Approximation

Low-rank approximation techniques aim to approximate the weight matrices of deep neural networks with lower-dimensional matrices. By compressing the parameters using low-rank matrices, memory usage is reduced while retaining most of the model's representational capacity. This approach is particularly effective for models with large fully-connected layers.

3.3 Quantization Techniques

Quantization involves reducing the precision of model parameters and intermediate activations. By representing numerical values with fewer bits, memory requirements are reduced. Techniques like weight quantization and activation quantization have been explored to achieve memory-efficient deep learning models. For example, 8-bit quantization can significantly reduce memory usage compared to the standard 32-bit representation.

3.4 Model Compression and Distillation

Model compression techniques aim to reduce the size of deep neural networks without sacrificing their performance. Techniques like knowledge distillation involve training a smaller student network to mimic the behavior of a larger, more complex teacher network. The student network can be trained with fewer parameters, resulting in reduced memory consumption while maintaining performance comparable to the original larger model.

These memory management techniques play a crucial role in enabling the training and deployment of large-scale deep learning models on resource-constrained devices and platforms. They offer a balance between memory optimization and model accuracy, paving the way for the application of deep learning in various real-world scenarios. Researchers continue to explore and innovate in memory management strategies, striving to overcome the memory challenges posed by ever-growing deep learning models and datasets.

4. Data Parallelism and Distributed Computing

Data parallelism and distributed computing techniques are essential for training large-scale deep learning models efficiently. These approaches allow the distribution of computational workload across multiple devices or nodes, enabling parallel processing of data and model updates. This section explores various data parallelism and distributed computing techniques used in deep learning optimization:

4.1 Synchronous and Asynchronous Training

In synchronous training, all devices or nodes update their model parameters simultaneously after processing a mini-batch of data. Synchronous training ensures consistency among all devices, but it may suffer from communication overhead, especially when devices have varying processing speeds. Asynchronous training, on the other hand, allows devices to update their parameters independently, leading to faster training times. However, asynchrony can result in parameter inconsistency and affect model convergence. Striking the right balance between synchronous and asynchronous training is crucial for efficient distributed deep learning optimization.

4.2 Model Parallelism

Model parallelism involves partitioning a deep neural network across multiple devices or nodes, allowing each device to handle a specific portion of the model. This approach is particularly useful for training extremely large models that cannot fit entirely in the memory of a single device. Model parallelism enables training complex architectures with limited memory resources, although it may introduce communication overhead for synchronizing model parameters between devices.

4.3 Data Parallelism and Gradient Accumulation

Data parallelism is a popular technique where each device or node processes a unique mini-batch of data and computes gradients independently. After gradient computation, the gradients are averaged or accumulated across all devices, and the model parameters are updated accordingly. Data parallelism enables efficient utilization of GPUs and accelerates training by distributing the workload. Gradient accumulation further reduces the communication overhead, as model updates can be performed less frequently, improving training efficiency.

4.4 Communication Overhead Reduction

Communication overhead refers to the time and resources consumed in exchanging data between devices or nodes during distributed training. To reduce communication overhead, researchers have explored various techniques, such as gradient compression, to transmit gradients more efficiently. Additionally, gradient quantization and sparsification can reduce the amount of data transferred during communication. Strategies like hierarchical communication and pipelined training can further mitigate communication bottlenecks in distributed deep learning optimization.

Data parallelism and distributed computing are indispensable in addressing memory and computational challenges in large-scale deep learning optimization. These techniques empower researchers and practitioners to train complex models on massive datasets, enabling breakthroughs in artificial intelligence and facilitating advancements in various industries. As deep learning models continue to scale, innovative approaches to data parallelism and distributed computing will remain at the forefront of large-scale deep learning optimization research.

5. Efficient Gradient Accumulation Strategies

Efficient gradient accumulation strategies are essential for managing memory usage and improving training efficiency in large-scale deep learning optimization. This section explores several techniques that aim to optimize the gradient accumulation process:

5.1 Adaptive Batch Size Selection

In adaptive batch size selection, the batch size used for gradient accumulation is dynamically adjusted during the training process. Instead of using a fixed batch size for all iterations, adaptive methods increase or decrease the batch size based on various criteria. For example, the batch size could be increased during early training stages to benefit from larger batch parallelism, and then decreased later to improve generalization. Adaptive batch size selection helps strike a balance between training speed and memory consumption, optimizing the trade-off between batch size and convergence performance.

5.2 Gradient Accumulation with Stochastic Batch Sampling

In traditional gradient accumulation, each device processes a mini-batch of data independently and accumulates gradients across multiple iterations before updating the model parameters. However, instead of using the same mini-batch data for gradient accumulation in each iteration, stochastic batch sampling randomly selects different mini-batches for each gradient update. This technique introduces randomness in gradient accumulation and can lead to a more diverse set of gradients, potentially improving the exploration capability during training. Stochastic batch sampling can also help avoid overfitting in scenarios with limited data.

5.3 Trade-offs between Batch Size, Convergence Speed, and Memory Usage

The choice of batch size in gradient accumulation involves trade-offs between several factors. Larger batch sizes can lead to faster convergence due to reduced variance in gradient estimates and improved parallelism. However, larger batch sizes also consume more memory, which may be limiting for large-scale models. Smaller batch sizes require less memory but may lead to noisier gradient estimates and slower convergence. Striking the right balance is crucial, and researchers often perform hyperparameter tuning to find an optimal batch size that maximizes convergence speed while staying within memory constraints.

Memory-efficient gradient accumulation strategies play a pivotal role in making deep learning optimization more accessible and practical for a wide range of applications. These techniques enable researchers and practitioners to train large-scale deep learning models on resource-constrained devices and platforms. Efficient gradient accumulation, coupled with other memory management and optimization techniques, empowers the advancement of artificial intelligence and fosters innovation in diverse domains. As deep learning continues to evolve, further advancements in gradient accumulation strategies are likely to contribute to more efficient and scalable deep learning optimization.

6. Reducing Activation Memory Footprint

Reducing the memory footprint of intermediate activation tensors is crucial for efficient memory management in large-scale deep learning optimization. This section explores several techniques to achieve this goal:

6.1 Activation Quantization

Activation quantization involves reducing the precision of activation tensors during training and inference. Instead of using standard 32-bit floating-point representations, lower precision data types (e.g., 16-bit or 8-bit fixed-point or integer representations) are employed. By using fewer bits to represent activations, memory usage is significantly reduced. Quantization techniques can be combined with specialized hardware support (e.g., Tensor Processing Units) to accelerate inference while maintaining acceptable model accuracy. However, aggressive quantization may lead to information loss and affect model performance, necessitating a careful balance between precision reduction and model accuracy.

6.2 Activation Pruning

Activation pruning techniques identify and remove inactive or low-impact neurons or channels from intermediate activation tensors. Neurons or channels that consistently produce close-to-zero activations or have minimal impact on the model's output can be pruned without significantly affecting the model's performance. Pruning reduces the size of activation tensors, leading to memory savings during training and inference. Activation pruning is particularly useful in convolutional neural networks where individual filters or channels may have varying levels of importance in the model's representation. Careful pruning strategies and retraining are employed to ensure that pruning does not compromise model performance.

6.3 Low-Memory Activation Allocation Strategies

Low-memory activation allocation strategies focus on optimizing the allocation and management of activation tensors during forward and backward passes. Instead of storing all activation tensors in memory simultaneously, memory-efficient allocation techniques only retain the necessary tensors required for gradient computation. Techniques like gradient checkpointing, where intermediate activations are recomputed during the backward pass instead of storing them, can effectively reduce the memory footprint. However, this approach introduces additional computation overhead, and the trade-off between memory reduction and computation time must be considered.

Reducing the memory footprint of activation tensors is critical for handling large-scale deep learning models with complex architectures and large batch sizes. These techniques complement other memory management strategies, such as parameter pruning and low-rank approximation, to achieve efficient memory usage throughout the training and inference processes. By optimizing activation memory usage, researchers and practitioners can enable the deployment of deep learning models on resource-constrained devices, paving the way for widespread adoption of deep learning in various real-world applications. As the field of deep learning progresses, further advancements in activation memory reduction techniques are expected to contribute to more memory-efficient and scalable deep learning optimization.

7. Adaptive Learning Rate Scheduling

Adaptive learning rate scheduling techniques dynamically adjust the learning rate during the training process to improve convergence speed and model performance. These approaches aim to strike a balance between rapid progress in the early training stages and fine-tuning towards convergence. This section explores several adaptive learning rate scheduling techniques:

7.1 Memory-Aware Learning Rate Decay

Memory-aware learning rate decay adjusts the learning rate based on the available memory resources during training. As the memory usage increases, the learning rate is automatically adjusted to prevent out-of-memory errors. This technique is particularly useful in scenarios where the model's memory requirements vary throughout the training process, such as when dealing with changing batch sizes or different layers' memory demands. Memory-aware learning rate decay ensures stable training in resource-constrained environments, optimizing both memory usage and convergence performance.

7.2 Dynamic Learning Rate Adjustment

Dynamic learning rate adjustment techniques adapt the learning rate based on the model's training progress. Common approaches include cyclic learning rates, where the learning rate oscillates between minimum and maximum values over a predefined number of iterations. Another technique is the one-cycle learning rate policy, where the learning rate starts at a minimum value, increases linearly to a maximum value, and then decreases to a final minimum value. Dynamic learning rate adjustment helps accelerate convergence by finding optimal learning rates for different stages of training.

7.3 Adaptive Gradient Clipping

Gradient clipping is a technique used to prevent the gradients from becoming too large during training, which can lead to unstable optimization and exploding gradients. Adaptive gradient clipping dynamically adjusts the clipping threshold based on the training progress or the gradients' statistical properties. By adapting the clipping threshold, the model can handle varying gradients more effectively, improving convergence stability and avoiding divergence.

Adaptive learning rate scheduling techniques play a critical role in deep learning optimization, allowing models to converge efficiently and effectively. These methods enable researchers and practitioners to fine-tune learning rates automatically, without manual tuning or fixed schedules. By dynamically adjusting the learning rate based on various criteria, adaptive scheduling ensures that the optimization process is both memory-efficient and robust, leading to improved model performance and reduced training time. As deep learning models continue to grow in complexity, adaptive learning rate scheduling will remain an important area of research, contributing to the advancement of large-scale deep learning optimization.

8. Memory-Efficient Backpropagation

Memory-efficient backpropagation techniques focus on reducing the memory overhead during the backward pass of deep learning optimization. The backward pass involves computing gradients with respect to model parameters, which requires storing intermediate activations and gradients, contributing to significant memory consumption. This section explores several memory-efficient backpropagation techniques:

8.1 Backpropagation Through Time Truncation

Backpropagation Through Time (BPTT) is a common algorithm used for training recurrent neural networks (RNNs). In BPTT, the computation graph is unrolled over time to handle sequential data. However, for long sequences, BPTT can lead to an explosion of memory usage due to the accumulation of activations and gradients over time steps. Backpropagation Through Time Truncation involves truncating the sequence length during backpropagation. Instead of considering the entire sequence, the computation is truncated after a certain number of time steps. This reduces memory usage at the cost of losing some information from earlier time steps. Careful selection of the truncation length is essential to balance memory efficiency and model performance.

8.2 Gradient Checkpointing Techniques

Gradient checkpointing is a technique that selectively recomputes intermediate activations during the backward pass instead of storing them in memory. In traditional backpropagation, all intermediate activations are retained to compute gradients efficiently. However, gradient checkpointing allows recomputing only the necessary activations on-the-fly, significantly reducing memory usage. This trade-off introduces additional computation overhead, as some activations may be recomputed multiple times, but it can be a valuable memory optimization strategy, especially for models with a large number of layers.

8.3 Memory-Optimized Backpropagation Algorithms

Researchers have explored memory-optimized backpropagation algorithms that efficiently manage memory usage during gradient computations. These algorithms aim to minimize the storage of intermediate gradients by directly computing the gradients without explicitly storing them in memory. Techniques like the Krylov subspace method and Hessian-free optimization are examples of memory-optimized backpropagation algorithms that can handle large-scale models with reduced memory overhead.

Memory-efficient backpropagation is essential for training deep learning models with limited memory resources, especially when dealing with large batch sizes and complex architectures. By optimizing memory usage during the backward pass, researchers and practitioners can train deeper and wider models, leading to improved performance in various tasks. Efficient backpropagation techniques contribute to the scalability and practicality of deep learning in real-world applications, allowing models to be trained on a wide range of devices and platforms. As the field of deep learning continues to progress, further advancements in memory-efficient backpropagation algorithms are expected to enhance the training process and foster innovation in the deep learning community.

9. Mixed Precision Training

Mixed precision training is a memory-efficient technique that utilizes lower-precision data types for certain parts of the deep learning model while maintaining higher precision for critical components. By combining different numerical precisions, mixed precision training reduces memory usage without significantly compromising model accuracy. This section explores various aspects of mixed precision training:

9.1 Precision Trade-offs and Numerical Stability Concerns:

Mixed precision training involves a trade-off between memory savings and numerical stability. Lower precision data types (e.g., 16-bit floating-point) consume less memory, but they are more susceptible to numerical overflow and underflow, which can affect training stability and model performance. It is essential to carefully choose which parts of the model can use lower precision while ensuring that critical components, like the loss function and batch normalization layers, retain higher precision for numerical stability. Advanced optimization techniques, such as gradient scaling and loss scaling, are often employed to maintain numerical stability during mixed precision training.

9.2 Automatic Mixed Precision Techniques

Manually implementing mixed precision training can be challenging and time-consuming. Automatic mixed precision techniques use specialized libraries and hardware support (e.g., NVIDIA's Automatic Mixed Precision or Google's TensorFlow Automatic Mixed Precision) to automatically manage the numerical precision of operations in the model. These libraries enable seamless conversion of operations to lower precision and handle scaling factors to mitigate numerical stability issues. Automatic mixed precision simplifies the adoption of mixed precision training and allows researchers and practitioners to experiment with lower-precision data types without extensive code modifications.

9.3 Mixed Precision Training for Large-Scale DNNs

Mixed precision training is particularly beneficial for large-scale deep neural networks, where memory limitations can be a significant concern. By using lower precision for non-critical parts of the model, mixed precision training allows researchers to train deeper and wider models on GPUs with limited memory. Additionally, mixed precision training can accelerate the training process, as lower precision arithmetic operations are faster to compute on modern hardware.

Mixed precision training has become an indispensable technique in large-scale deep learning optimization, enabling memory-efficient training of complex models. By leveraging lower-precision data types while maintaining numerical stability, mixed precision training strikes a balance between memory optimization and model accuracy. As hardware capabilities and software libraries continue to improve, mixed precision training is expected to play an increasingly vital role in deep learning optimization, facilitating the training of even larger and more powerful models in the future.

10. Model Parallelism for Extremely Large Models

Model parallelism is a technique used to train extremely large deep learning models that cannot fit entirely in the memory of a single device or GPU. It involves partitioning the model across multiple devices or nodes, allowing each device to handle a specific portion of the model. This section explores various aspects of model parallelism for handling extremely large models:

10.1 Model Partitioning Strategies

Model partitioning involves dividing the layers or components of the deep learning model across multiple devices. There are several model partitioning strategies:

Layer-wise Partitioning: In this approach, each device handles a specific layer of the model. For example, one device may handle the convolutional layers, while another handles the fully connected layers. Layer-wise partitioning is relatively straightforward to implement, but it may introduce communication overhead for synchronizing the intermediate results between devices.

Module-wise Partitioning: In module-wise partitioning, each device handles a group of related layers or modules. This approach strikes a balance between granularity and communication overhead, as related layers are kept together on the same device.

Data Parallelism Model Parallelism Hybrid: In some cases, a combination of data parallelism and model parallelism is used to distribute the workload. Data parallelism is employed across multiple devices, each handling a subset of the data, while model parallelism is applied within each device for extremely large models.

10.2 Parameter Synchronization Techniques

In model parallelism, devices work on different parts of the model, and their computations need to be synchronized to ensure consistent model updates. Parameter synchronization techniques are used to manage the communication between devices during the training process. Some common techniques include:

All-reduce: All-reduce is a collective communication operation that synchronizes gradients across all devices. It is often used in distributed training settings to ensure that each device has access to the latest gradient updates from other devices.

Gradient Accumulation: In gradient accumulation, devices accumulate gradients over several iterations before performing the parameter update. This reduces the frequency of communication, which can help reduce communication overhead.

10.3 Handling Computational Overhead in Model Parallel Settings

Model parallelism introduces computational overhead due to the need for communication and synchronization between devices. Handling computational overhead is crucial to ensure that the training process remains efficient. Some techniques to address computational overhead include:

Overlap Communication and Computation: Devices can overlap communication and computation to minimize the waiting time during parameter synchronization. For example, while gradients are being communicated, the device can continue performing computations for the next iteration.

Pipeline Parallelism: In pipeline parallelism, multiple devices work on different parts of the forward and backward pass simultaneously, creating a pipeline-like structure. This can help in reducing idle time and maximizing device utilization.

Asynchronous Model Parallelism: Asynchronous model parallelism allows devices to update their model parameters independently without waiting for the synchronization with other devices. This can improve the training speed, but it may lead to parameter inconsistency and affect model convergence.

Model parallelism for extremely large models is essential for pushing the boundaries of deep learning research and applications. By efficiently partitioning and synchronizing the model across multiple devices, researchers and practitioners can train models with unprecedented size and complexity. Advanced model parallelism techniques, along with improvements in communication protocols and hardware capabilities, are expected to drive the training of even larger models and enable breakthroughs in artificial intelligence.

11. Challenges and Open Research Directions

11.1 Limitations and Trade-offs

Despite the significant advancements in deep learning optimization and memory management, several challenges and trade-offs still exist:

Performance vs. Memory Efficiency Trade-off: Many memory optimization techniques come with a trade-off between memory efficiency and model performance. Aggressive memory reduction may lead to degraded model accuracy or slower convergence. Striking the right balance remains an ongoing challenge.

Model Complexity: As deep learning models continue to grow in complexity, memory management and optimization become even more critical. Handling extremely large models with billions of parameters poses unique challenges in terms of memory consumption, training efficiency, and communication overhead.

Hardware and Platform Heterogeneity: Deep learning models are often deployed across various hardware and platforms, such as GPUs, TPUs, and edge devices. Memory optimization techniques need to adapt to different hardware architectures and efficiently utilize the available resources.

Generalization: Memory-efficient optimization techniques must not compromise model generalization. Techniques that reduce memory usage may lead to overfitting or reduced model capacity to generalize to unseen data.

11.2 Potential Extensions and Future Research

Future research in deep learning optimization and memory management should focus on addressing the following areas:

AutoML for Memory Optimization: Developing automated tools that can automatically select and configure memory optimization techniques based on the model architecture and hardware environment. AutoML-driven memory management can simplify the process of choosing the most appropriate memory optimization strategies.

Hardware-Optimized Memory Management: Investigating memory management techniques that are tailored to specific hardware architectures, taking advantage of specialized hardware features for memory-efficient deep learning.

Novel Memory Reduction Techniques: Exploring innovative approaches to further reduce memory requirements, such as more effective quantization methods, novel parameter pruning techniques, and memory-efficient activation allocation strategies.

Scalable Distributed Training: Advancing distributed training techniques to handle even larger-scale models with greater efficiency, reducing communication overhead and synchronization bottlenecks.

Memory-Efficient Transfer Learning: Studying how memory-efficient optimization techniques can be applied to transfer learning scenarios, where pre-trained models are fine-tuned on new tasks with limited resources.

Robustness and Stability: Ensuring that memory optimization techniques maintain model stability, numerical precision, and robustness across various optimization settings and architectures.

Real-World Applications: Conducting empirical studies and case studies to evaluate the practicality and effectiveness of memory optimization techniques in real-world applications, such as healthcare, finance, robotics, and autonomous vehicles.

As the field of deep learning continues to evolve, addressing these challenges and exploring new research directions will be crucial in unlocking the full potential of large-scale deep learning optimization, making deep learning more accessible, efficient, and applicable to a wide range of domains and applications.

12. Conclusion

12.1 Summary of Survey Findings

In this survey, we explored various topics related to deep learning optimization and memory management. We began by understanding the significance of memory management in large-scale deep learning optimization, where the memory requirements of models and data play a crucial role in determining the feasibility and efficiency of training. We discussed the challenges posed by extremely large models, computational overhead, and the trade-offs between memory optimization and model performance.

Next, we delved into several memory management techniques, such as parameter pruning, quantization, and model compression, which aim to reduce memory usage while maintaining model accuracy. We explored data parallelism and distributed computing as strategies to handle the computational demands of training large-scale models efficiently. Additionally, we investigated memory-efficient backpropagation and activation memory reduction techniques to optimize memory usage during the training process.

12.2 Implications and Significance

The survey findings have significant implications for the field of deep learning optimization and memory management. Efficient memory management techniques enable the training of larger and more complex models on resource-constrained devices, expanding the applicability of deep learning technology. They also open up new opportunities for innovation in various domains, including healthcare, finance, robotics, and autonomous vehicles, where deep learning models can solve complex problems and improve decision-making processes.

The survey highlights the importance of striking a balance between memory optimization and model accuracy. While memory-efficient techniques can lead to memory savings,

it is crucial to consider the trade-offs and ensure that model performance and generalization are not compromised. Advanced hardware support, automated tools, and AutoML-driven memory optimization can aid researchers and practitioners in selecting the most appropriate memory management strategies for their specific scenarios.

12.3 Concluding Remarks

In conclusion, the survey provides a comprehensive overview of the latest trends and techniques in deep learning optimization and memory management. Memory-efficient optimization strategies play a critical role in overcoming the challenges posed by large-scale deep learning models and datasets, making deep learning more accessible, efficient, and scalable. As deep learning continues to advance and tackle even more complex tasks, memory management will remain a vital area of research and innovation.

Researchers and practitioners are encouraged to explore and adopt these memory management techniques to unlock the full potential of deep learning and address real-world challenges. By optimizing memory usage, we can accelerate model training, enhance model capacity, and pave the way for ground breaking applications of deep learning in diverse industries. As the field evolves, continuous exploration and advancements in memory-efficient deep learning optimization will shape the future of artificial intelligence and drive the next wave of innovation.

References

1. Optimization Methods for Large-Scale Machine Learning: https://arxiv.org/abs/1606.04838 by Shamir et al.
2. Large-Scale Deep Learning Optimization Techniques Tutorial: https://cvpr2023.thecvf.com/virtual/2023/tutorial/18575 by Ameet Talwalkar and others.
3. A Survey of Large-Scale Deep Learning Serving System Optimization: Challenges and Opportunities: https://arxiv.org/abs/2111.14247 by Wang et al.
4. Large-Scale Deep Learning Optimizations: A Comprehensive Survey: https://arxiv.org/abs/2111.00856 by Zhang et al.
5. A Comprehensive Guide on Optimizers in Deep Learning: https://www.analyticsvidhya.com/blog/2021/10/a-comprehensive-guide-on-deep-learning-optimizers/ by Analytics Vidhya.
6. Optimization in Deep Learning- Learn with examples: https://www.e2enetworks.com/blog/optimization-in-deep-learning-learn-with-examples by E2E Networks.

Firmware Security: Challenges, Vulnerabilities, and Mitigation Strategies 45

Firmware Security: Challenges, Vulnerabilities, and Mitigation Strategies

Devi Gujjula[1]

Assistant Professor, Department of CSE-IoT,
Holy Mary Institute of Technology and Science, Bogaram,Telangana, India

G. V. Koti Reddy

Professor & HoD, Department of CSE-IoT,
Holy Mary Institute of Technology and Science, Bogaram,, Telangana, India

P. Bhaskar Reddy

Director, Holy Mary Institute of Technology and Science,
Bogaram, Telangana, India

Abstract: Firmware security is a critical aspect of modern computing, as firmware serves as a crucial bridge between hardware and higher-level software. However, firmware is often overlooked and can be susceptible to various challenges and vulnerabilities that threaten the security of embedded systems. This survey paper presents an in-depth exploration of firmware security, encompassing the challenges faced in securing firmware, common vulnerabilities prevalent in firmware code, and effective mitigation strategies to safeguard against potential threats. We delve into the intricacies of firmware analysis techniques, secure boot mechanisms, hardware-based security, and firmware update processes to address the vulnerabilities present in firmware. Furthermore, we discuss the importance of continuous monitoring, threat modelling, and best practices to enhance the overall security posture of firmware-based systems. The paper aims to serve as a comprehensive resource for researchers, practitioners, and industry professionals interested in comprehending the multifaceted landscape of firmware security and taking proactive measures to safeguard their embedded systems.

Keywords: Firmware security, Embedded systems, Firmware analysis, Secure boot, Hardware-based security, Firmware update, Vulnerability mitigation, Continuous monitoring, Threat modelling, Secure coding practices, Security standards

1. Introduction

Embedded systems play an increasingly vital role in modern technological landscapes, powering a wide array of devices and critical infrastructure. Firmware, the embedded software that facilitates communication between hardware and higher-level software, forms the foundational layer of these systems.

However, firmware security is often overlooked, leaving devices vulnerable to potential cyberattacks and unauthorized access. As attackers increasingly target firmware for exploitation, it becomes imperative to comprehensively address the challenges and vulnerabilities inherent in this critical component.

I.A. Background and Significance of Firmware Security

Firmware security has gained prominence due to the rising number of cyber threats targeting embedded systems. The potential consequences of firmware breaches are far-reaching, including device malfunction, data theft, system compromise, and even physical harm in certain sectors like healthcare and industrial automation.

[1]dgujjula28236@gmail.com, [2]gvkotireddyhit@gmail.com, [3]pbhaskarareddy@rediffmail.com

1.1 Objectives of the Survey

The primary objective of this survey paper is to provide a comprehensive analysis of firmware security, outlining the challenges, vulnerabilities, and potential threats faced by embedded systems. By examining the various attack vectors and prevalent firmware vulnerabilities, this survey aims to raise awareness of the critical need for robust firmware security measures.

2. Firmware Analysis Techniques

Firmware analysis plays a pivotal role in identifying potential vulnerabilities and security weaknesses within embedded systems. This section explores various firmware analysis techniques, each with its distinct advantages and limitations.

2.1 Static Analysis of Firmware

Static analysis involves examining firmware code without executing it. By analysing the firmware's source code or binary, static analysis aims to detect potential security flaws, unsafe coding practices, and vulnerabilities. This technique utilizes advanced algorithms to analyse the code's structure, control flow, and data dependencies, identifying potential buffer overflows, code injection points, and other vulnerabilities.

2.2 Dynamic Analysis of Firmware

Dynamic analysis, in contrast, involves executing firmware in a controlled environment to observe its behavior at runtime. By monitoring the firmware's actions, interactions with hardware, and network communications, dynamic analysis can identify runtime vulnerabilities, memory corruption issues, and unintended behaviors

2.3 Hybrid Approaches for Comprehensive Analysis

To address the limitations of individual analysis techniques, hybrid approaches combine static and dynamic analysis. By integrating the results from both methods, researchers can gain a more comprehensive understanding of firmware security. Static analysis can be used to guide dynamic analysis, identifying interesting code paths to explore during runtime analysis.

2.4 Challenges and Limitations of Firmware Analysis

Despite the benefits of firmware analysis, several challenges and limitations persist. Obfuscation techniques, commonly employed to hinder analysis, can obscure critical aspects of firmware code, making it challenging to identify potential vulnerabilities.

3. Common Firmware Vulnerabilities

Firmware vulnerabilities are potential weaknesses that can be exploited by attackers to compromise the security and integrity of embedded systems. This section examines some common firmware vulnerabilities that pose significant risks to the overall security of devices.

3.1 Buffer Overflows and Memory Corruption

Buffer overflows and memory corruption vulnerabilities occur when a program writes data beyond the boundaries of allocated memory buffers. Attackers can exploit this vulnerability to overwrite adjacent memory regions, potentially leading to arbitrary code execution or system crashes. Firmware lacking proper input validation and bounds checking is susceptible to buffer overflows, making it crucial for developers to implement secure coding practices and ensure robust memory management.

3.2 Insecure Authentication and Authorization

Firmware often includes authentication and authorization mechanisms to control access to privileged functionalities and sensitive data. Insecure authentication practices, such as hardcoded or weak credentials, can allow unauthorized users to gain unauthorized access. Similarly, improper authorization checks may grant excessive privileges, leading to privilege escalation attacks.

3.3 Injection Attacks and Command Injection

Injection attacks occur when untrusted data is improperly processed, leading to unintended execution of malicious code. Firmware is vulnerable to injection attacks, such as SQL injection and command injection, if it fails to validate and sanitize user input properly.

3.4 Insecure Firmware Updates and Boot Procedures

Insecure firmware update and boot procedures pose critical risks to the integrity of embedded systems. Attackers can compromise the firmware update process to inject malicious code, leading to unauthorized modifications and potential backdoor installation. Similarly, insecure boot procedures may allow attackers to bypass security measures and gain unauthorized access to the system.

3.5 Backdoors and Unauthorized Access

Backdoors are intentionally inserted by developers or attackers to provide covert access to a system. Unauthorized access through backdoors can be exploited to execute malicious code, exfiltrate sensitive data, or launch further attacks.

4. Secure Boot and Code Signing

Securing the boot process is a fundamental step in ensuring the integrity and authenticity of firmware and preventing unauthorized code execution. This section delves into the importance of secure boot, the components involved in the secure boot process, code signing techniques for firmware authentication, and the limitations and best practices associated with secure boot implementation.

4.1 Importance of Secure Boot

Secure boot is a critical security mechanism that verifies the authenticity and integrity of firmware during the system's boot-up process. By establishing a chain of trust from the initial firmware to the operating system, secure boot ensures that only authorized and properly signed firmware is loaded and executed. This mitigates the risk of firmware tampering, malicious code injection, and unauthorized modifications, safeguarding the embedded system against firmware-based attacks. Secure boot is a foundational security measure that establishes a trusted computing environment, making it challenging for attackers to compromise the system's firmware and overall security.

4.2 Secure Boot Process and Components

The secure boot process involves several key components that collectively establish a secure chain of trust.

4.3 Code Signing Techniques for Firmware Authentication

Code signing is a key aspect of secure boot that ensures the authenticity and integrity of firmware images. Firmware is signed using cryptographic algorithms, such as RSA or ECDSA, with a private key held by the firmware developer or manufacturer.

4.4 Limitations and Best Practices for Secure Boot

While secure boot is an essential security measure, it is not without limitations and challenges. For instance:

Dependency on Secure Hardware: Secure boot relies on the integrity of hardware components involved in the boot process. An attacker with physical access to the hardware or with the ability to compromise it may circumvent secure boot protections.

Key Management: Proper key management is crucial for secure boot. Safeguarding private keys and ensuring the authenticity of public keys are critical aspects of a secure boot implementation.

Firmware Update Mechanisms: Secure boot should be complemented with secure firmware update mechanisms. Otherwise, an attacker might exploit vulnerabilities during the update process to compromise the system.

5. Hardware-Based Security for Firmware

Hardware-based security solutions offer a robust and tamper-resistant foundation for protecting firmware and ensuring the integrity of embedded systems. This section explores various hardware security features, trusted platform modules (TPMs), and unique hardware-based identities, emphasizing their significance in safeguarding firmware and enhancing overall system security.

5.1 Hardware Security Features for Embedded Systems

Modern embedded systems often integrate dedicated hardware security features to fortify their defense against attacks. These hardware security features encompass a range of capabilities, including secure boot mechanisms, hardware-based encryption, secure storage, and cryptographic acceleration. Hardware security modules (HSMs) and cryptographic co-processors provide dedicated hardware for cryptographic operations, ensuring faster and more secure encryption and decryption. These hardware-based security features are essential in establishing a trusted foundation for firmware execution and secure communication.

5.2 Trusted Platform Modules (TPMs) and Hardware Roots of Trust

Trusted Platform Modules (TPMs) are specialized hardware components that serve as hardware roots of trust. They provide a secure environment for cryptographic operations, secure key generation and storage, and secure measurements of the firmware and system state. TPMs play a crucial role in establishing the authenticity and integrity of firmware during the boot process and throughout system operation. By attesting to the system's trustworthiness, TPMs enable secure remote attestation and are a fundamental building block inestablishing a chain of trust for firmware-based systems.

5.3 Leveraging Hardware-Based Security for Firmware Protection

Hardware-based security measures offer significant advantages in protecting firmware from various threats, including malware injection, tampering, and unauthorized access. By employing hardware-based secure boot mechanisms, firmware integrity can be verified during the boot process, ensuring that only authenticated and verified firmware images are executed

6. Secure Firmware Updates

Secure firmware updates are essential for maintaining the security and functionality of embedded systems throughout their lifecycle. This section explores the challenges in ensuring secure firmware updates, the importance of secure over-the-air (OTA) update mechanisms, techniques for preventing firmware rollback attacks, and best practices for implementing secure firmware updates.

6.1 Challenges in Secure Firmware Updates

Securely updating firmware in embedded systems presents several challenges:

Authentication and Integrity: Ensuring the authenticity and integrity of firmware updates is crucial to prevent unauthorized and tampered updates.

Secure Delivery: Firmware updates must be securely delivered to prevent interception and modification during transmission.

Resource Constraints: Embedded systems often have limited resources, making it challenging to implement robust cryptographic mechanisms and secure storage for updates.

Rollback Attacks: Firmware rollback attacks can exploit vulnerabilities in the update process to revert to older, potentially vulnerable firmware versions.

Continuous Monitoring: Continuous monitoring of firmware updates and validation mechanisms is necessary to detect and respond to potential attacks or update failures.

6.2 Secure Over-the-Air (OTA) Update Mechanisms

Secure OTA update mechanisms are designed to address the challenges in securely delivering and installing firmware updates over wireless networks. These mechanisms typically include:

Secure Protocols: Utilizing secure communication protocols, such as TLS/SSL, to encrypt the firmware update data during transmission and authenticate the update server.

Code Signing: Cryptographically signing the firmware update to verify its authenticity and integrity during installation.

Update Validation: Performing integrity checks on the update before installation to prevent the installation of tampered or corrupted firmware.

Secure Boot Integration: Integrating secure boot mechanisms to verify the authenticity and integrity of the updated firmware before executing it.

Rollback Protection: Implementing mechanisms to prevent firmware rollback attacks, such as version checks and secure counters.

6.3 Secure Firmware Rollback Prevention

Firmware rollback prevention is crucial to ensure that once a firmware update is installed, the system cannot revert to an older, potentially vulnerable version. Techniques to prevent firmware rollback attacks include:

Version Checks: Storing the version number of the currently installed firmware and refusing to install older versions.

Cryptographic Checksums: Storing cryptographic checksums of the installed firmware and using them to validate the firmware during boot.

Digital Signatures: Including digital signatures in the firmware update metadata to prevent installation of unsigned or incorrectly signed updates.

6.4 Firmware Update Best Practices

To ensure the security and effectiveness of firmware updates, the following best practices should be adopted:

Secure Update Channels: Establish secure channels for delivering firmware updates, such as encrypted connections and trusted update servers.

Code Signing and Validation: Digitally sign firmware updates and perform validation checks to ensure authenticity and integrity.

Secure Boot Integration: Integrate secure boot mechanisms to verify the authenticity of the updated firmware during the boot process.

Rollback Protection: Implement techniques to prevent firmware rollback attacks, ensuring that the system stays up-to-date with the latest secure firmware.

Error Handling: Include robust error handling in the update process to handle update failures and prevent system instability.

7. Continuous Monitoring and Threat Modelling

Continuous monitoring and threat modelling are essential components of a proactive firmware security strategy. This section discusses the importance of continuous firmware monitoring, the process of threat modelling for firmware security, and the implementation of incident response plans to effectively address firmware attacks.

7.1 Importance of Continuous Firmware Monitoring

Continuous monitoring of firmware is crucial to detect and respond to potential security threats and anomalous behaviour promptly. Unlike traditional software, firmware operates at a lower level and is often more challenging to inspect for security vulnerabilities and unauthorized changes.

7.2 Threat Modelling for Firmware Security

Threat modelling is a structured approach to identify potential security threats and vulnerabilities in the firmware. The process includes the following steps:

Assess Vulnerabilities: Evaluate the potential vulnerabilities that could be exploited by the identified threats. Consider the impact and likelihood of successful exploitation for each vulnerability.

Mitigation Strategies: Develop mitigation strategies to address the identified threats and vulnerabilities. These strategies may include secure coding practices, access controls, cryptographic protections, and secure update mechanisms.

7.3 Implementing Incident Response Plans for Firmware Attacks

Despite robust preventive measures, the possibility of firmware attacks cannot be entirely eliminated. Therefore, organizations must have well-defined incident response plans to handle firmware-related security incidents effectively. Key components of incident response plans for firmware attacks include:

Early Detection: Implement monitoring tools and anomaly detection mechanisms to identify potential firmware security incidents as early as possible.

Forensic Analysis: Conduct a thorough forensic analysis of the compromised firmware and affected systems to understand the nature and extent of the attack.

Firmware Restoration: If necessary, restore firmware from trusted backups or use secure update mechanisms to reinstall the firmware.

8. Best Practices for Firmware Security

To establish a strong Défense against potential cyber threats and ensure the resilience of embedded systems, adhering to best practices for firmware security is paramount.

8.1 Secure Coding Practices for Firmware Development

Secure coding practices are foundational to building firmware that is robust against security vulnerabilities and attacks. Some essential secure coding practices for firmware development include:

Input Validation: Validate all user input and external data to prevent buffer overflows, injection attacks, and other forms of code injection.

Memory Safety: Implement safe memory management techniques to prevent memory corruption vulnerabilities, such as buffer overflows and null pointer dereferences.

Least Privilege: Adhere to the principle of least privilege by granting firmware components only the necessary privileges and access rights to perform their designated functions.

Code Reviews: Conduct regular code reviews to identify and address potential security issues, logical flaws, and coding errors.

Error Handling: Implement robust error handling to prevent information leakage and ensure that the firmware gracefully handles unexpected situations.

Secure Communication: Utilize secure communication protocols, such as TLS/SSL, to protect sensitive data transmitted between firmware components and external systems.

8.2 Secure Configuration and Hardening of Firmware

Secure configuration and hardening involve configuring firmware settings and features in a way that minimizes security risks and potential attack surfaces. Key practices for secure configuration and hardening of firmware include:

Default Passwords: Eliminate default passwords and enforce the use of strong, unique passwords during initial setup.

Access Controls: Implement granular access controls to restrict privileged functions and limit access to sensitive areas of the firmware.

Secure Boot and Update Mechanisms: Integrate secure boot mechanisms and ensure that firmware updates are cryptographically signed and authenticated.

Secure Protocols: Use secure communication protocols for management interfaces and remote access.

Secure Key Management: Safeguard cryptographic keys and certificates used for firmware authentication and encryption.

8.3 Security Testing and Firmware Validation

Security testing and validation are critical to assessing the effectiveness of firmware security measures and identifying potential vulnerabilities. Some essential practices for security testing and firmware validation include:

Penetration Testing: Conduct regular penetration testing to simulate real-world attack scenarios and identify potential weaknesses in the firmware.

Code Signing Verification: Validate the authenticity and integrity of firmware using code signing verification during the boot process.

Update Validation: Validate firmware updates to ensure they come from trusted sources and are not tampered during transmission.

9. Industry Standards and Frameworks

To promote consistency and best practices in firmware security, various industry standards and frameworks have been developed. This section examines firmware security standards and guidelines, as well as industry frameworks that provide a structured approach to firmware security assurance.

9.1 Firmware Security Standards and Guidelines

Several organizations have established standards and guidelines specifically focused on firmware security. Some notable ones include:

ISO/IEC 15408 (Common Criteria): This international standard provides a framework for evaluating the security properties of IT products, including firmware. It helps establish the trustworthiness of firmware through rigorous evaluation and certification processes.

CIS Benchmarks The Center for Internet Security (CIS) publishes benchmarks for secure configuration and hardening of various systems, including firmware. These benchmarks offer prescriptive guidance on securely configuring firmware settings.

Firmware Security Best Practices by OWASP: The Open Web Application Security Project (OWASP) provides a comprehensive list of best practices and guidance for securing firmware, covering areas like secure boot, cryptography, and secure communication.

9.2 Industry Frameworks for Firmware Security Assurance

In addition to specific standards and guidelines, some industry frameworks offer comprehensive approaches to firmware security assurance:

Trusted Computing Group (TCG): TCG provides open standards for hardware-based security technologies, such as Trusted Platform Modules (TPMs) and secure boot

mechanisms. These standards help establish a chain of trust and ensure the integrity of firmware.

Platform Security Architecture (PSA): Developed by Arm, PSA is a comprehensive framework for building secure connected devices. It includes guidelines, threat models, and security analysis tools to help device manufacturers implement robust firmware security measures.

Firmware Security Framework (FSF): The FSF is an open-source framework that aims to provide a standardized and structured approach to firmware security.

Integration of these industry standards and frameworks into firmware development processes can greatly enhance the security of embedded systems and ensure compliance with widely recognized best practices.

10. Future Directions and Emerging Trends

The field of firmware security is continuously evolving to keep pace with the ever-changing threat landscape and technological advancements. This section explores some future directions and emerging trends in firmware security, including advancements in firmware analysis techniques, quantum-resistant cryptography, and the integration of security automation and AI-driven firmware protection.

10.1 Advancements in Firmware Analysis Techniques

Firmware analysis techniques are expected to undergo significant advancements to better detect and mitigate firmware vulnerabilities.

10.2 Quantum-Resistant Cryptography for Firmware

As quantum computing capabilities advance, traditional cryptographic algorithms may become vulnerable to attacks. Quantum-resistant cryptography, also known as post-quantum cryptography, is expected to gain importance in firmware security to ensure the longevity of secure communication and authentication.

10.3 Security Automation and AI-Driven Firmware Protection

Automation and artificial intelligence (AI) are likely to play a key role in enhancing firmware security. Some emerging trends include:

AI-Driven Threat Intelligence: AI-powered threat intelligence platforms will continuously analyze and correlate security data to identify emerging threats and improve firmware protection strategies.

AI-Based Firmware Behavior Analysis: AI algorithms that monitor firmware behavior in real-time can detect and respond to anomalous activities, mitigating potential threats.

11. Conclusion

In conclusion, firmware security is a multifaceted discipline that requires a comprehensive approach encompassing analysis, coding practices, hardware-based protection, and continuous monitoring. Through constant innovation, collaboration, and proactive measures, firmware security can be effectively strengthened to safeguard the integrity and functionality of embedded systems in an ever-evolving cybersecurity landscape.

References

1. Top 7 Vulnerability Mitigation Strategies: https://reciprocity.com/blog/top-7-vulnerability-mitigation-strategies/ byReciprocity
2. Security Vulnerability and Mitigation in Photovoltaic Systems: https://ieeexplore.ieee.org/document/9494252 by IEEE
3. Impact, Vulnerabilities, and Mitigation Strategies for Cyber-Secure Critical Infrastructure: https://www.mdpi.com/1424-8220/23/8/4060 by MDPI
4. How to Mitigate Firmware Security Risks in Data Centers, and Public and Private Clouds: https://www.gartner.com/en/documents/3947141 by Gartner
5. Mitigations for Security Vulnerabilities in Control System Networks: https://www.cisa.gov/sites/default/files/2023-01/MitigationsForVulnerabilitiesCSNetsISA_S508C.pdf by CISA

Designing a Multi-Sensor Based Wireless Sensor Network System for Monitoring the Wellness of Elderly Individuals: Experimental Studies

46

K. Srinivasa Reddy[1], K. Venkata Murali Mohan[2]
Associate Professor,
Department of Electronics and Communication Engineering

Geetha Reddy Evuri[3]
Department of Electrical and Electronics Engineering,
TKREC, Hyderabad, Telangana, India

B. Padmini[4]
Associate Professor, Asst Professor, TKR Engineering College

Abstract: There is a rising need for geriatric care as the percentage of aged persons in the population rises quickly. In the foreseeable future, it is anticipated that this trend will continue, putting a substantial strain on national resources and driving up the price of elder care. Many older people choose to live independently rather than in retirement communities or nursing homes, but they still need ongoing care and access to emergency medical care. In-home monitoring systems for the elderly have been the subject of much study, frequently utilizing wireless LANs. For older people and their families, wireless sensing technology has significantly improved the quality of life. For senior people, a wireless sensor-based smart home monitoring system provides a secure, pleasant, and safe living environment. These systems are made up of several wireless sensors that offer useful data. Such systems are able to spot unanticipated problems and facilitate early medical intervention by keeping track of the old person's everyday activities and identifying any odd trends. This project aims to construct an intelligent wireless sensor-based smart home system that can identify a person's daily activities based on their interactions with various gadgets. The technology can detect early indicators of behavioral abnormalities in the elderly by comparing everyday patterns, enabling prompt medical intervention. Although there are many types of off-the-shelf sensors available, it is difficult to make them "intelligent" when monitoring the elderly. A framework has been created to manage the implementation problems and design complexities of innovative sensors that are especially aimed at an older person's digital home environment in order to meet these concerns. The monitoring system created using this framework takes into account the everyday activities and way of life of older people who live alone. The system successfully records the person's everyday behavior while just employing a few sensors. The sensors are intended to be non-invasive, assuring human approval, and it is simple to install in domestic settings. These sensors are essentially invisible to the user, which increases their comfort and usability in contrast to cameras and microphones, which might be considered as obtrusive.

Keywords: WSN, Zigbee, Sensor, Heartbeat sensor

1. Introduction

The system for determining the wellness of elderly individuals through a wireless sensor network involves assessing their daily activities and monitoring their health The existing system discussed in the research work aims to enhance home monitoring activities for elderly individuals using wireless sensor networks. The motivation behind this system is to

[1]drksreddytkrec@gmail.com, [2]tkrec.principal@gmail.com, [3]gre.413@gmail.com, 4padmin@tkrec.ac.in

overcome the limitations and acceptability issues associated with camera-based home monitoring solutions. Instead of relying on cameras, the system incorporates various sensor technologies within the household environment to monitor human activity behavior.

These sensor technologies include:

- Infrared small motion detectors
- Passing sensors
- Operation detectors
- IR motion sensors

The combination of these sensors allows the system to monitor and interpret human activity in real-time..In the broader context of personal wellness monitoring and safety, RFID technology allows for individual identification through radio waves, enabling positional information to be collected by interacting with RFID tags. Additionally, some individuals may be hesitant to continuously wear a monitoring system on their body, making such solutions less viable for healthy elderly individuals who desire a certain level of independence and privacy. To address these challenges, the research work proposes the use of wireless sensor networks (WSNs) as an alternative to traditional RFID implementations. WSNs offer several advantages over RFID technology:

Rapid Deployment: WSNs allow for the swift deployment of multi-hop networks, enabling ad hoc and temporary usage.

Multiple Interconnected Nodes: Unlike RFID, which is limited to single-hop communication, WSNs can create networks with multiple interconnected nodes, providing better coverage and reliability.

Local Data Storage: WSNs have the capability to store data locally in the sensor nodes when they are out of the network range and then upload that data when they re-enter the network. This feature ensures data integrity and resilience even in the case of temporary disconnections.

By leveraging the benefits of WSNs, the proposed intelligent home monitoring system can enhance the monitoring and evaluation of the well-being of elderly individuals living alone. The system's use of different sensor technologies and the integration of wireless sensor networks offer a comprehensive approach to gather real-time data on the elderly person's activities and behavior. This data can then be analyzed at various levels of abstraction, considering time and the sequence of sensor usage, to provide valuable insights into their daily routines and ensure their safety and well-being. This technology allows for the collection of data on the person's activities and well-being, ensuring their safety and providing valuable insights for caregivers or healthcare professionals.

2. Proposed System

The deployment of multiple sensors, strategically placed throughout the home environment, allows the system to monitor a wide range of activities and interactions of the elderly individuals. The sensors are designed to detect the usage of various electrical devices, beds, chairs, and other relevant elements within the home. By capturing data based on the interactions with these objects, the system can gather information about the status (active or inactive) and identity of each sensor. For example, sensors placed near electrical appliances can detect when these appliances are being used, sensors near beds and chairs can determine if they are occupied or unoccupied, and other sensors may track movement patterns of the elderly individual throughout the home. As the sensors continuously collect data on these activities, they transmit this data wirelessly through the ZIGBEE network to the smart sensor coordinator, which acts as the central hub for data collection and transmission. The collected sensor data serves as the fundamental source of information about the elderly individual's activities and behavior within their home. By analyzing this data, the subsequent software module (at the higher level of the system) can gain insights into the real-time activity behavior of the elderly individual. Indeed, the intelligent software module in the home monitoring system plays a crucial role in assessing the well-being of elderly individuals living alone. By employing intelligent mechanisms and algorithms, the module processes the collected sensor data at various levels of abstraction. Factors like time and the sequence of sensor usage are taken into consideration during this data analysis.

The system's ability to continuously monitor and assess the elderly individual's activities and interactions with their environment provides valuable support to caregivers and family members. They can proactively respond to any potential risks or emergency situations, ensuring that the elderly person receives the necessary support and care when needed. The intelligent home monitoring system described here is an essential tool for ensuring the safety and well-being of elderly individuals living alone. By continuously monitoring and analyzing their daily activities using intelligent algorithms, the system can detect unsafe situations and predict abnormal behavior, empowering caregivers, healthcare providers, or family members to provide timely and necessary support to the elderly person. The system's real-time monitoring and alerting capabilities enhance the overall quality of care and contribute to improving the elderly person's independence and comfort in their home environment.

3. Need for Early Detection of Ageing Changes/Problems of Elderly People

As the world's population ages, the number of people with impairments and those in need of assistance also rises. There is an urgent need for an in-home ubiquitous network that can help residents by controlling household appliances, granting access to medical information, and allowing contact during crises. Sadly, disturbing headlines like "Elderly man lay dead for days" or "Woman found starving in flat" are commonly seen. These incidents highlight the lack of support and care for these individuals that might come from within their own families. On the other hand, as a society, we highly respect our independence and our capacity to make our own decisions. Many individuals worry about being compelled to live in a care facility or with their adult children. Any unusual occurrences or mishaps that an older person has at home can frequently result in more serious diseases or even death. To deal with these circumstances, efficient monitoring methods should be taken into consideration, especially when diminished mobility or other conditions are present. Technology of today provides a more effective solution to these problems. It is feasible to offer technological help and keep an eye on people's well-being in their homes by using a centralized and dispersed network of wireless sensors that are positioned strategically all throughout the house. The creation and use of smart homes mark a significant advancement in the provision of all-encompassing care and assistance for the ageing population.

4. Our Approach and Solution

The ability to evaluate each person's health state and transmit timely updates to carers and employers is an additional advantage of in-home monitoring. This enables the creation of customised, focused preventative interventions [13]. Smart houses with wireless sensor networks are advantageous for healthcare practitioners and their patients. The technology reduces the demand for human work, which results in decreased costs and greater efficiency. In order to accomplish this, it continuously monitors patients and promptly notifies physicians or other healthcare professionals of any aberrant situations. Furthermore, by addressing crucial issues like privacy, independence, dignity, and convenience through the provision of services within the comfort of their own homes, smart homes assist and improve the quality of life for the elderly. It is important to note that various applications use wireless sensor networks in very diverse ways. It is essential to conduct a detailed analysis of performance metrics. Off-the-shelf sensors are easily accessible, but it is difficult to make

them "intelligent" for a particular application setting, like geriatric monitoring. An intelligent wireless sensor system has the capacity to gather and analyse data in order to spot any anomalies, going beyond just identifying consumption patterns of everyday appliances. The system uses a limited number of wireless sensors and a controller that takes input from these sensors to handle these issues. It proves the created approach's viability, dependability, practicability, and scalability. Future developments in intelligent monitoring systems for the well-being of the elderly are made possible by this proof-of-concept system.

5. Results

The integration of all sensors with ARM7 is depicted in the image below. The sensors are used to monitor and control people using Zigbee and GSM modules.

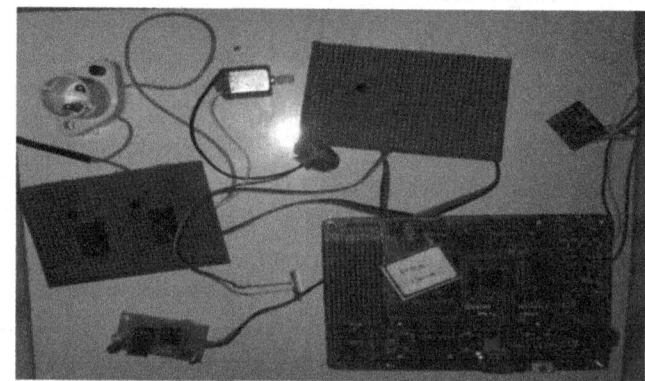

Fig. 46.1 Integration of all sensors with ARM7

The additional information you provided about the intelligent home monitoring system sounds even more comprehensive and advanced. By connecting the sensors to an ARM7 board through GSM and ZIGBEE modules, the system gains enhanced capabilities for monitoring and controlling various aspects of the home environment to ensure the safety and well-being of the elderly individual. Here's a breakdown of the system's features based on the new information:

Sensor Connectivity: The sensors strategically placed throughout the home environment are now connected to an ARM7 board. This board serves as the central processing unit for the system, enabling it to collect data from the various sensors.

GSM Module Integration: The inclusion of a GSM module enables the system to send messages or alerts directly to mobile phones. This feature allows caregivers or family members to receive real-time notifications about critical events or abnormal conditions at the elderly individual's home.

ZIGBEE Module Integration: The ZIGBEE modules continue to facilitate wireless communication between the interconnected sensors. This mesh network structure ensures efficient data transmission and communication within the system.

Remote Control Operations: The system can now perform remote control operations based on the data collected from the sensors. For example:

High Temperature: If the temperature in the home environment exceeds a certain threshold, the system can trigger a message alert to the caregivers' mobile phones through the GSM module. The caregivers can remotely respond by turning on the fan using a mobile application or by sending specific commands to the ARM7 board, which will control the fan.

Low Light Intensity: Similarly, if the light intensity is insufficient, the system can alert the caregivers, who can remotely turn on the lights through the ARM7 board using their mobile phones.

Abnormal Heartbeat Rate: In case the system detects an unusually high heart rate in the elderly individual, it can immediately notify the caregivers via SMS using the GSM module. The caregivers can then decide whether to take the person to the hospital or seek medical attention.

Figure 46.4 shows display of temperature and heart beat rate of elder person who is alone at home.

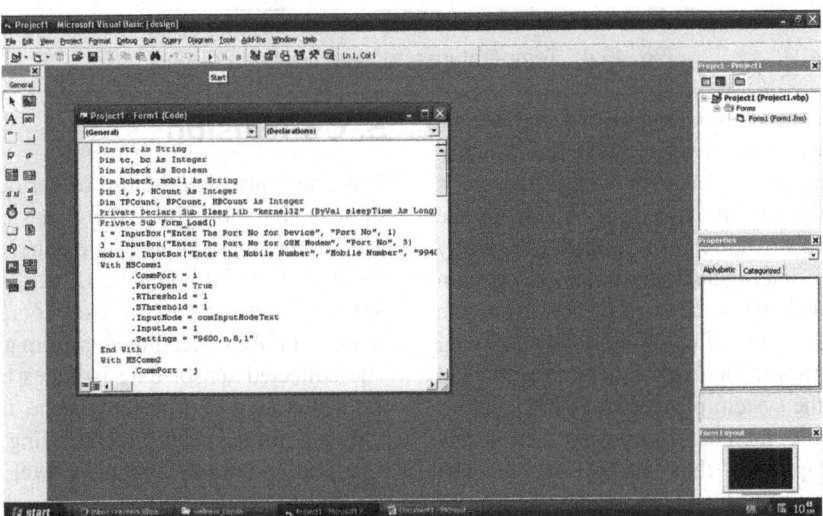

Fig. 46.2 Code execution using visual basic

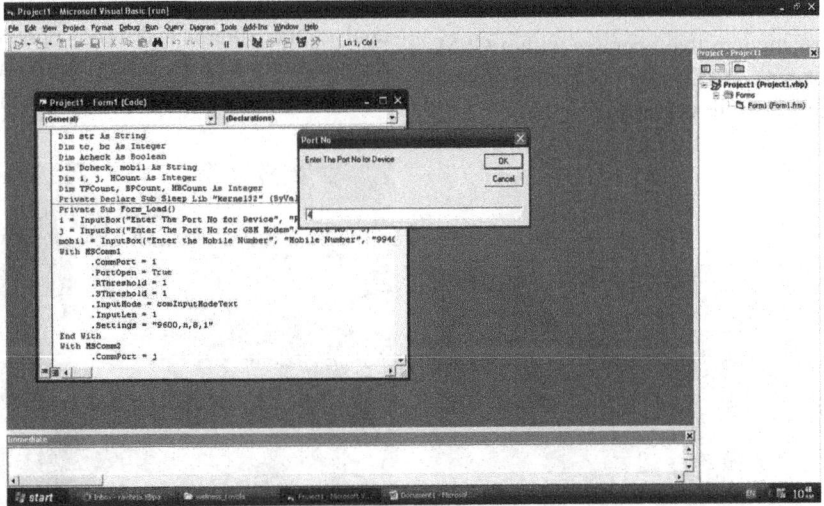

Fig. 46.3 Entering the port number for execution of code

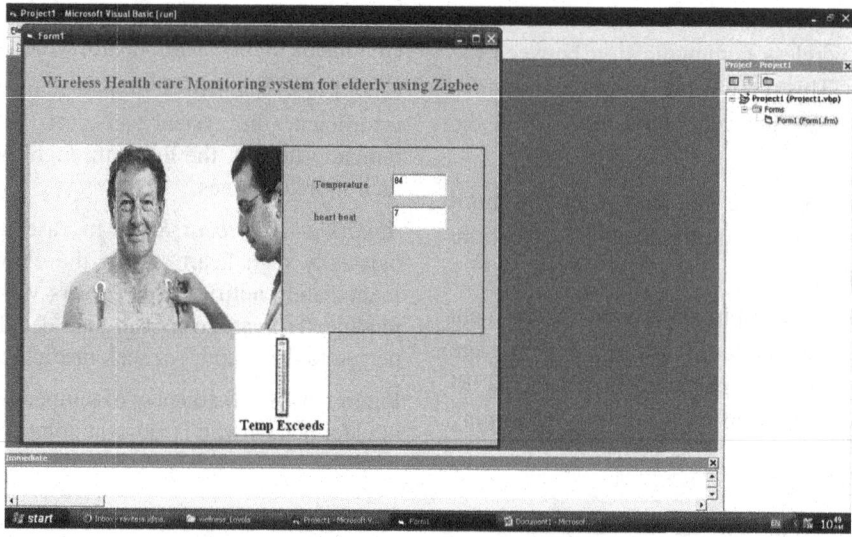

Fig. 46.4 Temperature and heart beat rate display

The system's major objective was to monitor and evaluate the health of older people who were living alone in their homes. The system was designed to continuously monitor the user's behaviors and interactions with numerous objects in the home environment in order to identify and foresee potentially dangerous scenarios. The intelligent home monitoring system presented in this research work effectively addressed the safety and well-being concerns of elderly individuals living alone. By combining sensor technology, data processing, and intelligent algorithms, the system provided valuable support to caregivers and family members, ensuring the elderly person's safety and improving their quality of life. The successful implementation of this system holds significant potential for supporting the aging population and promoting independent living while maintaining a sense of security and care.

GSM Modem: A wireless modem that operates with a GSM wireless network is known as a GSM modem. Similar like a dial-up modem, a wireless modem operates.

5. Conclusion

This paper involves the use of various smart sensors placed strategically throughout the home environment to capture data on the interactions of the elderly with different objects, such as electrical appliances, beds, chairs, etc. At the core of the system is a smart sensor coordinator, which acts as a central hub for collecting data from all the deployed sensors. The collected sensor data includes basic information about the status (active or inactive) and identity of each sensor, allowing for a clear understanding of which sensors are triggered and when. The next level of the system involves a software module that processes the collected data using intelligent mechanisms. This module operates at various levels of data abstraction, considering factors such as time and the sequence of sensor usage. The intelligent algorithms analyze the patterns and trends observed over time, allowing the system to understand the daily routines and behaviors of the elderly occupants. This capability enables continuous

Fig. 46.5 GSM

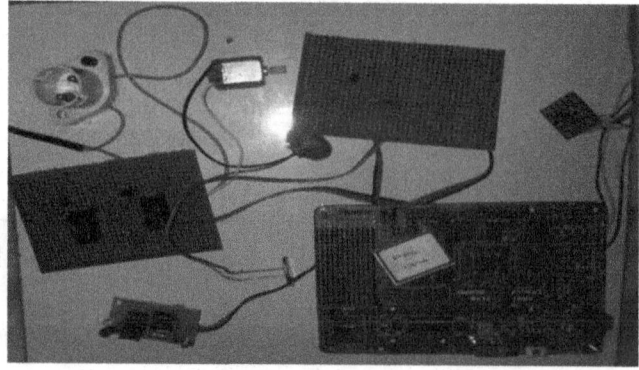

Fig. 46.6 Integration of all sensors with ARM7

monitoring and analysis of their activities, making it possible to detect any abnormal or concerning behavior promptly. The information provided by the system can be invaluable for caregivers, healthcare providers, or family members. It empowers them to take appropriate actions and provide timely intervention or support if any unsafe or unusual behavior is detected. For example, if the system identifies a prolonged inactivity period, it could alert caregivers, enabling them to check on the individual's well-being and provide assistance if needed. The data collected by the smart sensors, though low-level in nature, serves as the foundation for understanding the overall activity status of the elderly individual. However, the subsequent software module enhances the interpretation of this data by applying intelligent techniques. This augmentation of the data analysis significantly improves the system's ability to detect patterns and changes in behavior, which are critical for assessing the well-being of the elderly. By combining sensor technology, data processing, and intelligent algorithms, the system provides an effective means of real-time monitoring and assessment of the elderly individuals' well-being. It offers peace of mind to caregivers and family members, knowing that they can promptly respond to any issues or emergencies that may arise. Overall, the system contributes to enhancing the safety and quality of life for elderly individuals living alone.

References

1. H. Nasution and S. Emmanuel, Intelligent video surveillance for monitoring elderly in home environments, in Proc. IEEE 9th Workshop Multimedia Signal Process., Oct. 2007, pp. 203–206.

2. Z. Zhongna, D. Wenqing, J. Eggert, J. T. Giger, J. Keller, M. Rantz,and H. Zhihai, A real-time system for in-home activity monitoring of elders,in Proc. Annu. Int. Conf. IEEE Eng. Med. Biol. Soc., Sep. 2009, pp. 6115–6118. 1972 IEEE SENSORS JOURNAL, VOL. 12, NO. 6, JUNE 2012

3. S. J. Hyuk, L. Boreom, and S. P. Kwang, Detection of abnormal living patterns for elderly living alone using support vector data description,IEEE Trans. Inf. Technol. Biomed., vol. 15, no. 3, pp. 438–448, May 2011.

4. Wood, J. Stankovic, G. Virone, L. Selavo, H. Zhimin, C. Qiuhua, D. Thao, W. Yafeng, F. Lei, and R. Stoleru, —Context-aware wireless sensor networks for assisted living and residential monitoring, IEEE Netw., vol. 22, no. 4, pp. 26–33, Jul.–Aug. 2008.

5. J. K. Wu, L. Dong, and W. Xiao, —Real-time physical activity classification and tracking using wearble sensors, in Proc. 6th Int. Conf. Inf.,Commun. Signal Process., Dec. 2007, pp. 1–6.

6. Z. Bing, Health care applications based on ZigBee standard, in Proc. Int. Conf. Comput. Design Appl., vol. 1. Jun. 2010, pp. V1-605–V1-608.

7. K. P. Hung, G. Tao, X. Wenwei, P. P. Palmes, Z. Jian, W. L. Ng, W.T. Chee, and H. C. Nguyen, —Context-aware middleware for pervasive elderly homecare, IEEE J. Sel. Areas Commun., vol. 27, no. 4, pp.510–524, May 2009.

8. H. Yu-Jin, K. Ig-Jae, C. A. Sang, and K. Hyoung-Gon, —Activity recognition using wearable sensors for elder care, in Proc. 2nd Int. Conf. Future Generat. Commun. Netw., vol. 2. Dec. 2008, pp. 302–305.

9. A. Moshaddique and K. Kyung-Sup, —Social issues in wireless sensor networks with healthcare perspective,Int. Arab J. Inf. Technol., vol. 8,no. 1, pp. 34–39, Jan. 2011.

10. K. Hara, T. Omori, and R. Ueno, —Detection of unusual human behaviour in intelligent house, in Proc. 12th IEEE Workshop Neural Netw. Signal Process., Nov. 2002, pp. 697–706.

11. S.-W. Lee, Y.-J. Kim, G.-S. Lee, B.-O. Cho, and N.-H. Lee, —A remote behavioral monitoring system for elders living alone, in Proc. Int. Conf.Control, Autom. Syst., Oct. 2007, pp. 2725–2730.

Note: All the figures in this chapter were authors' original test data and experimental output.

Transfer Learning Based Kidney Stone Detection in Patients Using ResNet50 with Medical Images

47

P. Naresh

Assistant Professor, Department of Information Technology,
Vignan Institue of Technology and Science, Hyderabad, India

**A. Jahnavi Reddy, S. Prem Kumar,
CH. Nikhil, T. Chandu**

UG Scholar, Department of Information Technology,
Vignan Institue of Technology and Science Hyderabad, India

Abstract: The kidneys play a crucial role in human health by filtering the blood. Maintaining normal amounts of sodium, potassium, and blood pH depends on the kidneys functioning normally. Humans are increasingly susceptible to renal failure as a result of modern living, diet, and illnesses like diabetes. Timely therapy of renal stones requires accurate early prognosis. The success rate of image processing-based diagnostic methods is higher than that of other methods of identification. A DL model of a cross-residual network (XResNet-50) was presented for renal stone categorization by Yildirim et al. 2021. For precise diagnostics, the suggested XResNet-50 uses four layers of computing. ResNet layers at every step of the process boost the model's ability to identify features. According to the experiments, the suggested cross-layered model has a better precision (96.23

Keywords: Kidney stone, ResNet50, Deep learning, Machine learning

1. Introduction

Individuals with chronic kidney disease (CKD) experience progressive loss of renal capability north of a while to years. At first, patients might encounter no signs;however, they may develop limb edema, fatigue, sickness, lack of hunger, and disorientation later on. The endocrine dysfunction of the kidneys can lead to bone disease, anemia, and high blood pressure, all of which are often caused by stimulation of the renin-angiotensin system. Increased mortality and hospitalisation rates are associated with the circulatory problems that CKD patients face. Diabetes, hypertension, glomerulonephritis, and polycystic kidney infection are potential main drivers of ongoing kidney disappointment. [1]A history of severe renal illness in the family is one risk factor. For diagnosis, albumin tests and the blood's estimated glomerular filtration rate (eGFR) are used. An ultrasound or renal sample can help find the root of the problem. There are multiple severity-based queuing methods in use today, . People who are at risk should be screened. Medication to reduce blood pressure, blood sugar, and lipids may be used as an initial therapy. Angiotensin converting enzyme inhibitors (ACEIs) or angiotensin II receptor blockers (ARBs) are normally utilized as the first-line treatment for hypertension in light of their capacity to forestall renal illness and lower the gamble of cardiovascular sickness. Circle diuretics can be utilized to lessen liquid maintenance and hypertension. You ought to avoid NSAIDs .In addition to these medications, doctors advise their patients to lead busy lives and make adjustments to their diets, such as eating less sodium and more protein.

[1]nareshintell4@gmail.com, [2]janvireddy1601@gmail.com, [3]siliverupremkumar03@gmail.com, [4]nikhil.chidhirala@gmail.com, [5]chanduthanneru8@gmail.com

Therapies for iron deficiency and bone disease may likewise be required.[2]Patients with end-stage renal disappointment might require either hemodialysis or peritoneal dialysis or a kidney relocate. In 2016, 753 million individuals worldwide were afflicted with chronic renal disease; 417 million women and 336 million men. The number of fatalities it caused in 2015 was 1.2 million, up from 409,000 in 1990. Hypertension represents 550,000 fatalities yearly, followed by diabetes at 418,000 and glomerulonephritis at 238,000.

2. Literature Survey

In the last decade, chronic renal disease has been responsible for roughly 58 million fatalities around the globe, according to the globe Health Organization. Moreover, more than two million people around the globe require dialysis or a kidney donation to stay alive due to renal failure. However, this figure may only reflect 10The banking organization, for example, has used it widely for fraud prediction and credit rating, marketing, quality control, and repair plans up-selling and cross-selling. The discovery of oil spills is just one of many potential uses. Other applications include medical diagnostics, abnormality detection, flaw diagnosis, and e-mail screening. Stakeholders can use data mining's classification, clustering, association rule mining, regression analysis, and anomaly detection tools to examine the data from a variety of perspectives in light of the numerous issues they have raised. In particular,[4] classification is a data mining method that is employed to aid in the analysis of data and the prediction of possible outcomes by teaching a set of characteristics for the purpose of evaluating a set of attributes with which the practitioner is unfamiliar. Chronic Kidney Disease (CKD) is being diagnosed with more and more people, but there is currently no widely recognised prognostic model for CKD, as reported in. Prediction of CKD is the primary focus of this study. Patients' renal diseases are categorised using a supervised categorization method, such as a Two-Class Neural Network or a Two-Class Decision Forest. For the purpose of creating categorization models for distinguishing between people with chronic kidney disease (CKD) and those with non-chronic kidney disease (NOCKD),[5] Microsoft Azure Machine Learning Studio and the Two-Class Decision Forest and Two-Class Neural Network methods were utilized. Other algorithms that have been discussed in a survey by include linear discriminant, decision trees, linear support vector machines, quadratic support vector machines, k-nearest neighbors, weighted k In 2016, four chronic kidney diseases were predicted, including Nephritic Syndrome, Chronic Glomerulonephritis, and Acute Renal Failure. SVM and ANN, both guided algorithms, were used to make the renal illness prediction. When compared to SVM,[6] the findings demonstrated that

ANN was the superior predictor due to its higher precision and shorter processing time. Patients with renal illness have been classified as either chronic or non-chronic using Naive Bayes and KNN algorithms in recent research. Best First Search and the Wrapper approach in WEKA were used to narrow down candidate features for inclusion in the final model.(BFS). The findings demonstrated that the algorithms' efficiency improved with fewer characteristics. On a smaller sample consisting of hand-picked characteristics,[7] WEKA's KNN classifier implementation, the IBK classifier, outperformed the competition.[8] The forecast method has also advanced the use of data pre-processing and data mining methods to determine whether or not there is a connection between the observed factors or parameters and the outcome of the patients.[9] Two dynamic calculations were utilized to reenact the choice rules to estimate the death rate in view of explicit elements that were made sense of by clinical staff at four dialysis offices.[10] On renal dialysis information, we analyzed the exhibition of three particular characterization calculations—ANN, DT, and Coherent Relapse (LR)—in. Compared to the DT and LR algorithms, the experimental findings demonstrated that ANN achieved the greatest precision suggested a different renal failure prediction model using Apriori and[11] k-meansalgorithms on a total of 42 characteristics. For the purpose of the analysis, the calibration graphs and the Receiver Operating Characteristic (ROC) curve were utilized. [12]Back Propagation Algorithm (BPA), Support Vector Machine (SVM), and Radial BasisFunction (RBF) were some of the additional algorithms for renal stone diagnosis that were taught and evaluated in.[13] The primary objective of this study was to improve doctor productivity and speed up the diagnostic process. The trial findings showed that BPA greatly enhanced the categorization technique for medicinal applications.[14] Finally,utilised Support Vector Machines (SVM) and Random Forest (RF) with varying kernel values to categorise additional illnesses like malignancy, liver disease, and heart disease.[15] It was determined that different kernel functions yielded various outcomes, and that these differences could be optimised for with careful parameter selection.

3. Problem Statement

In most cases, medical imaging techniques provide an accurate diagnosis of a variety of illnesses. Many machine learning techniques and deep learning techniques are used to develop kidney stone detection using CNN, which reduces the doctor's mental fatigue. A convolutional brain organization (CNN) is a sort of brain network that orders highlights separated from an info picture by using another brain organization. The brain network performs arrangement in view of the recovered component signals. Consequences be damned performed well in both the preparation and test sets for 2D location. The principal trouble in adjusting Just go for it to 3D clinical pictures is that it requires 2D photographs as info. The model has difficulty detecting the kidney's extremities, either because of its reduced size or the attenuation of distinctive characteristics. These are more prone to go undetected, as was the case with the kidneys' bottom thirds.

4. Architecture and Methodology

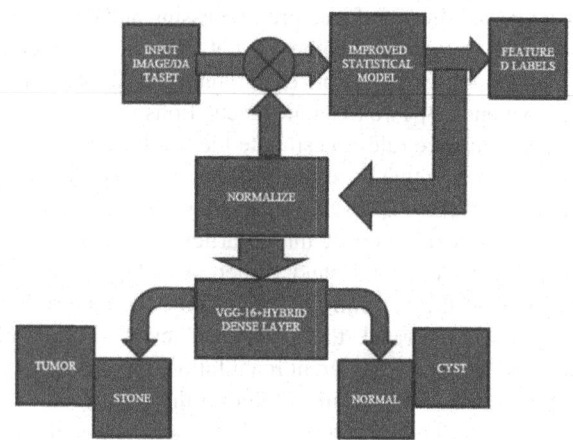

Fig. 47.1 System architecture

Our study seeks to solve these problems by developing an adaptable, automatically predictive system for kidney stones in real time. Using RESNET50 and CNN, we suggested a technique for extracting meaningful characteristics from photos. Additionally, we will create a scalable system that employs transfer learning to detect kidney stones in real time. We will develop a framework using deep learning that aims to diagnose the kidney stones using images . RESNET50 increases the feature extraction capability enhancing the accuracy and efficiency.

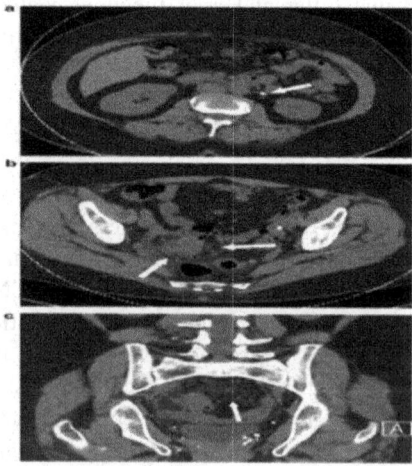

Fig. 47.2 Kidney images

5. Implementation

The process of machine learning begins with the construction of a model, which is then educated using a set of training data, after which it is able to analyse more data and provide predictions. Machine learning systems have made use of, and conducted research on, a wide variety of different models. Artificial neural networks.

An artificial neural network is similar to the vast network of neurons in the brain in that it is made up of connected nodes. Artificial neurons are represented by the spherical nodes in this picture, and connections between them are shown by the arrows. Biological neural networks form the basis of animal brains, and artificial neural networks (ANNs), also known as connectionist systems, are a kind of computer system that is looselymodelled after these biological neural networks. These kinds of systems quot learn quot to carry out tasks by thinking about examples, often without being programmed with any rules that are relevant to the jobs themselves. An ANN is a kind of model that is constructed using a network of interconnected units, or nodes, artificial neurons quot; These neurons are intended to generally imitate the neurons that are tracked down in a human mind. Each connection, which is analogous to the synapses in a real brain,has the ability to convey information in the form of a quot, signal quot from one artificial neuron to another. When it receives a signal, an artificial neuron has the ability to process it and then pass on the information to other artificial neurons that are linked to it. In the majority of ANN implementations, the message at a linkage among both biological neuron is a true number, and indeed the result from each perceptron is tabulated by some quasi function of the sum of its inputs. To put it another way, the signal at a neuronal linkage is a real number. Edges are a term used to describe the connections that are made between artificial neurons. The weight of an artificial neuron or edge will normally change as learning progresses whether it is an artificial neuron or edge. The intensity of the signal at a connection may be increased or decreased depending on the weight. The signal may only be sent if the overall signal strength is greater than a certain threshold, which may be present in artificial neurons. Layered organisation is a common method for artificial neurons. The inputs to the various levels may be subjected to a variety of changes, depending on the layer. Signals stream from the first layer, which is known as the info layer, the whole way to the last layer, which is known as the output layer. This may include travelling through the layers more than once. The purpose of the ANN technique, the original idea was to try to solve problems in the same way that the human brain would. However, as time passed, the emphasis shifted from biology to completing particular tasks,which resulted in departures from biological norms. There are many applications for artificial neural networks, such as computer vision, voice recognition, machine translation, filtering in social networks, playing board games and video games, and medical diagnosis. The use of numerous hidden layers inside an artificial neural network is what constitutes deep learning. This method works by simulating the manner in which the human brain converts light and sound into the senses of vision and hearing. Computer vision and voice recognition are two areas that have seen significant progress because to deep learning. Decision trees In decision tree learning, the nodes represent observations, and the branches provide judgements about the goal value of the item based on those observations (represented in the leaves). It may be used as a kind of predictive modelling in the areas of statistics, datamining, and machine learning. The objective variable in classification trees may take on an infinite number of values. The nodes of such trees reflect individual traits, while the leaves represent class designations. When the dependent variable may take on a continuous range of values (often represented by real numbers), a decision tree style known as a regression tree is utilised. A decision tree is a useful tool for decision analysis because it may graphically and clearly reflect choices and the decision-making process. A decision tree is used to visualise data in the area of data mining, and the resulting classification tree may be utilised as a guiding factor in making choices. Support vector machines For the reasons for order and relapse, support vector machines (SVMs), otherwise called help vector organizations, are a bunch of interconnected regulated learning calculations. When given a set of training examples labelled with one of two classes, a support vector machine (SVM) training process will generate a model that can determine which class a new example belongs to. During training, a support vector machine (SVM) is a binary, non-probabilistic, linear classifier, it may be employed in a probabilistic classification situation with the help of techniques like Platt scaling. Support vector machines (SVMs) are not limited to linear classification; by the use of the kernel method, which translates their inputs implicitly into high-dimensional feature spaces, they are also capable of doing non-linear classification efficiently. Regression analysis When it comes to estimating the nature of the connection that exists between the variables that are fed into the model and the characteristics that are connected with those models, regression analysis makes use of a wide range of statistical techniques. Linear regression, in which a mathematical criterion like ordinary least squares is used to fit a straight line to the data, is the most prevalent type of regression analysis. This type of linear regression is the most prevalent. The latter is sometimes supplemented by regularization (mathematics)approaches, such as ridge

regression, in an effort to reduce instances of overfitting and bias. Microsoft Excel's trendline fitting feature uses polynomial regression, when dealing with quasi problems, some examples of go-to modeling techniques include logistic regression, which is used for statistical classification, and kernel regression, which reveals non-linearity by mapping input parameters to a higher-dimensional space using the kernel function. Genetic algorithms The term quot; genetic algorithm quot; refers to both a search algorithm and a heuristic technique. A evolutionary algorithms (GA) are search algorithms that use pattern recognition methods like mutations and crossovers to produce new genetic traits in the hope of finding effective solutions to a particular issue. This mimics the process of natural selection. Throughout the1980s and 1990s, genetic algorithms were used in the field of machine learning. On the other hand, methods from machine learning have been applied to genetic and adaptive methods in order to enhance their current effectiveness. Training models Machine learning models, in general, call need a substantial amount of input data in order to function well. While training a machine learning model, it is often necessary to gather a sizable and representative subset of the data included in the model;s training set. A lot of text, a lot of pictures, or data from individual service users are all examples of data that could be included in the training set. While training a machine learning model, one thing to keep an eye out for is something called overfitting. Federated learning is a novel approach to decentralizing the training of machine learning algorithms. This removes the need that users provide their data to a centralised server, therefore preserving the users right to privacy. This also boosts efficiency since it distributes the process of training over many different devices. For instance, Gboard makes use of federation deep learning to train querying forecasting model locally on users; cellular telephones. This eliminates the need for users to submit Google their personal search queries.

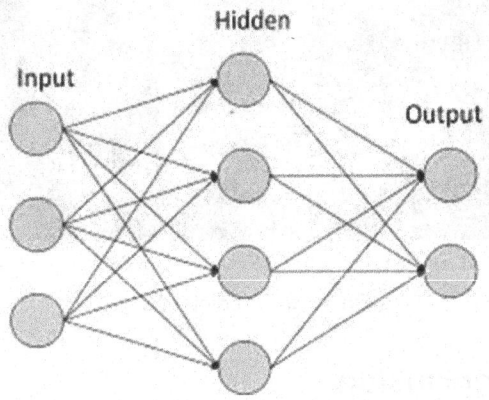

Fig. 47.3 ANN [2]

6. Results

```
Model: "sequential"

Layer (type)                    Output Shape             Param #
=================================================================
conv2d (Conv2D)                 (None, 198, 198, 32)     896

max_pooling2d (MaxPooling2D     (None, 99, 99, 32)       0
)

conv2d_1 (Conv2D)               (None, 97, 97, 32)       9248

max_pooling2d_1 (MaxPooling     (None, 48, 48, 32)       0
2D)

conv2d_2 (Conv2D)               (None, 46, 46, 64)       18496

max_pooling2d_2 (MaxPooling     (None, 23, 23, 64)       0
2D)

conv2d_3 (Conv2D)               (None, 21, 21, 64)       36928

max_pooling2d_3 (MaxPooling     (None, 10, 10, 64)       0
2D)

conv2d_4 (Conv2D)               (None, 8, 8, 128)        73856

max_pooling2d_4 (MaxPooling     (None, 4, 4, 128)        0
2D)

conv2d_5 (Conv2D)               (None, 2, 2, 128)        147584

max_pooling2d_5 (MaxPooling     (None, 1, 1, 128)        0
2D)

flatten (Flatten)               (None, 128)              0

dense (Dense)                   (None, 512)              66048

dense_1 (Dense)                 (None, 4)                2052

-----------------------------------------------------------------
Total params: 355,108
Trainable params: 355,108
Non-trainable params: 0
```

Fig. 47.4 Model Summary

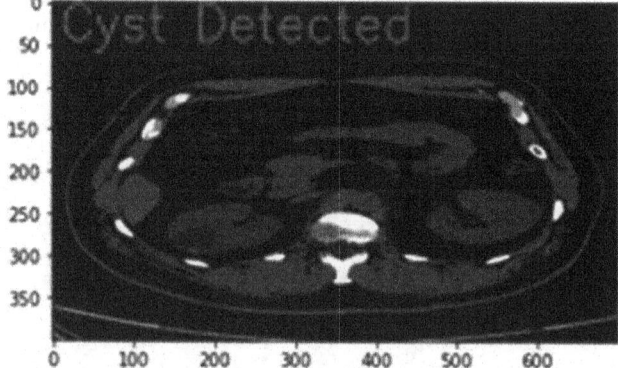

Fig. 47.5 Predicted: Cyst

7. Conclusion

The proposed research shed light on how CKD patients are diagnosed so that they can address their condition and receive

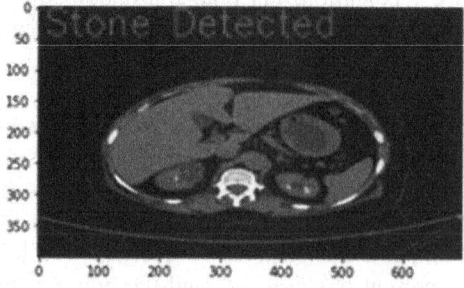

Fig. 47.6 Predicted: Stone

treatment early on. A total of 400 cases were analyzed, and 24 features were discovered. The dataset was split in half, with one quarter utilised for testing and validation and the other half for training. Mean and mode statistical measures were used to replace missing nominal and numerical values and eliminate outliers from the dataset, respectively. The most strongly representative CKD characteristics were chosen using the VGG Hybrid Model indicating the positive cases. The scope of the work is to improvise a solution with loss estimation and its improved accuracy for all different dataset chosen in real time.

8. Acknowledgement

We would like to convey our heartfelt gratitude towards the Department of Information Technology at Vignan Institute of Technology and Science, Hyderabad, for providing our team with every tool and resource possible,assistance,and guidance necessary to complete this research.

References

1. Fadil Iqbal1, Aruna S. Pallewatte2, Janaka P. Wansapura, "Texture Analysis of Ultrasound Images of Chronic Kidney Disease", 2017 International Conference on Advances in ICT for Emerging Regions (ICTer): 299–303.
2. Chi Hu1 ,Xiaojun Yu1*, Qianshan Ding2 , Zeming Fan1, Zhaohui Yuan1 ,Juan Wu1and Linbo Liu3, "Cellular-Level Structure Imaging with Micro-optical Coherence Tomography (μOCT) for Kidney Disease Diagnosis", 2019 the 4th Opto-electronics Global Conference.
3. Ahmad Amni Johari Mohd Helmy Abd Wahab Aida Mustapha, "Two-Class Classification: Comparative Experiments for Chronic Kidney Disease", 2019 4th International Conference on Information Systems and Computer Networks (ISCON) GLA University, Mathura, UP, India. Nov 21–22, 2019.
4. Rahul Gupta1, Nidhi Koli2, Niharika Mahor3, N Te-jashri4, "Performance Analysis of Machine Learning Classifier for Predicting Chronic Kidney Disease", 2020 International Conference for Emerging Technology (INCET) Belgaum, India. Jun 5–7, 2020.

5. Akash Maurya, Rahul Wable, Rasika Shinde, Sebin John, Rahul Jadhav, Dakshayani R., "Chronic Kidney Disease Prediction and Recommendation of Suitable Diet plan by using Machine Learning", 2019 International Conference on Nascent Technologies in Engineering (ICNTE 2019).

6. Dr. Uma N Dulhare Professor, CSED, MJCET Hyderabad, India Uma.dulhare@mjcollege.ac.in Mohammad Ayesha PG Student, CSED, MJCET Hyderabad, India mohammadayesha8993@gmail.com, "Extraction of Action Rules for Chronic Kidney Disease using Naïve Bayes Classifier", 978-1-5090-0612-0/16/31.002016*IEEE.*

7. Yedilkhan Amirgaliyev Institute of Information and Computing Technologies (IICT), Almaty, Kazakhstan amir$_e$d@mail.ru

8. Mubarik Ahmad, Vitri Tundjungsari, Dini Widianti, Peny Amalia, Ummi Azizah Rachmawati, "Diagnostic Decision Support System of Chronic Kidney Disease Using Support Vector Machine".

9. Sheng-Min Chiu1*, Feng-Jung Yang2, Yi-Chung Chen3, Chiang Lee1, "Deep learning for Etiology of Chronic Kidney Disease in Taiwan", 2nd IEEE Eurasia Conference on IOT, Communication and Engineering 2020.

10. S. Ramya and Dr. N. Radha, "Diagnosis of Chronic Kidney Disease Using Machine Learning Algorithms", International Journal of Innovative Research in Computer and Communication Engineering, Volume 4, Issue 1, January 2016, pp 813–820.

11. Brugnara C, Eckardt KU. Hematologic aspects of kidney disease. In: Taal MW, ed. Brenner and Rector's The Kidney. 9th ed. Philadelphia: Saunders; 2011: 2081–2120.

12. Chaurasia, V., Pal, S., Tiwari, B. B. (2018). Prediction of benign and malignant breast cancer using data mining techniques. Journal of Algorithms Computational Technology, 12(2), 119–126.

13. Chaurasia, V. and Pal S. Performance analysis of data mining algorithms for diagnosis and prediction of heart and breast cancer disease Rev Res 2014; 3: 1–13

14. Bhavya Gudeti, Shashvi Mishra, Shaveta Malik, Terrance Frederick Fernandez, Amit Kumar Tyagi, Shabnam Kumari. A Novel Approach to Predict Chronic Kidney Disease using Machine Learning Algorithms, 2020.

15. AKM Shahariar Azad Rabby, Rezwana Mamata, Monira Akter Laboni, Ohidujjaman, Sheikh Abujar, Machine Learning Applied to Kidney Disease Prediction: Comparison Study, IIT – Kanpur, Kanpur, India, 2019. Vikas Chaurasia, Saurabh Pal, B.B. Tiwari, Chronic Kidney Disease: A Predictive model using Decision Tree, Jaunpur, UP, India, 2018.

Note: All the figures except Fig. 47.3 in this chapter were authors' sample test data and researchgate.

System for Managing Batteries in Electric Vehicles

48

Srujana Athimamula[1], Srilatha Attaluri[2]

Department of Electrical and Electronics Engineering,
Vidya Jyothi Institute of Technology, Hyderabad, India

Abstract: The battery is a crucial component in Electric Vehicles, which are a progress in sustainable transportation. An essential part of Electric and Hybrid Vehicles is the Battery Management System (BMS). The BMS goal is to provide reliable and secure battery use. Functionalities including state monitoring,evaluation,charge regulation and cell balancing have been introduced to BMS to protect battery safety. Battery functions varies depending on the operating and environmental circumstances because it is an electrochemical product. The implementation of these functions presents a challenge because a battery's performance is unpredictable. A crucial responsibility for a BMS is the examination of a battery's state including its charge, health and life.

Keywords: Battery management system, Lithium-ion battery, Battery monitoring and management, State of charge, State of health, State of life, Electric vehicles

1. Introduction

Batteries are frequently employed as the primary energy source in a variety of applications from mobile devices to Electric Vehicles (EV's). EV's can lower fuel usage by up to 75%. EV batteries are returning in the automotive industry. Safety and dependability are users' key concerns in order to expand EV's share of the market. However, in addition to battery technology, they rely on the battery management system. In order to optimize vehicle operation and improve battery performance, a battery management system which serves as an interface between the vehicle and the battery. The need for a complete and advanced BMS is crucial with the EV market's rapid development. Similar to a gasoline vehicle's engine control system, the BMS in EV's should have an indicator meter. BMS indications display the battery's condition with regard to safety, use, performance and longevity. An overcharged lithium-ion battery may catch fire due to variations in flammability, volatility and entropy. This is a serious problem; an explosion might cause an awful accident [3]. Due to chemical reactions that are irreversible,

over discharge often lowers cell capacity. As a result, the BMS must monitor and regulate the battery in accordance with the safety circuitry. The BMS should notify the user and initiate corrective action, when inappropriate conditions, such as over voltage or overheating occur. BMS also communicates with specific components and operators, examines the system temperature and provides a more efficient power consumption plan[2].

2. Electricvehicle Batteries

Batteries for Electric Vehicles are distinguished by their high specific energy, power-to-weight ratio and energy density. The performance of vehicles can be improved by the use of smaller, lighter batteries. Modern Electric Vehicles most frequently use lithium-ion and lithium-polymer batteries because of their high energy density per unit of weight [5]. Since they have a high discharge volt and high-power density and widespread application in Electric Vehicles (EV's) lithium-ion batteries are a common component [1]. To ensure proper performance, a precise temperature range

---

[1]eeehod@vjit.ac.in, [2]asrilathaeee@vjit.ac.in

must be maintained for lithium-ion batteries. Their power and life cycle will be substantially compromised among other things, by overheating. A lithium-ion cell's lifespan has been observed to decrease when working in the range of 30 to 40 °C[4].

3. Proposed System for Managing Batteries

A careful examination of the present methods identifies the shortcomings of the current BMS. The complete and developed BMS have the elements and basic functionalities shown in Fig. 48.1 in order to address these limitations.

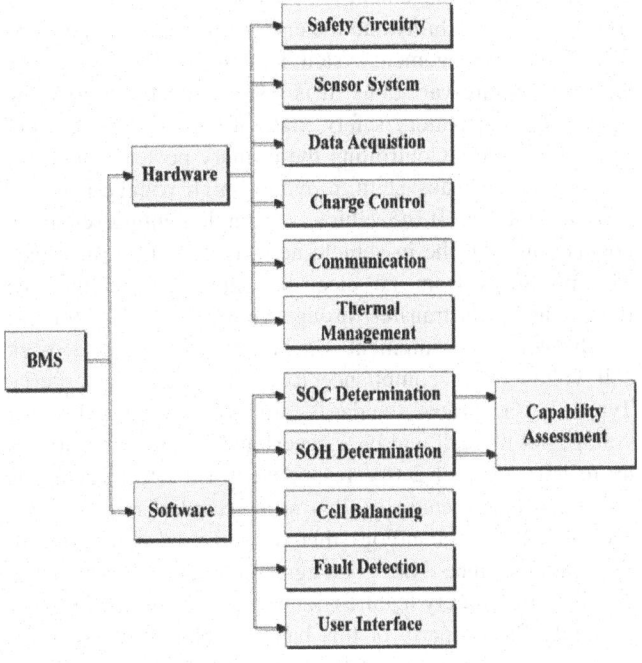

Fig. 48.1 Components of BMS

4. Hardware

In BMS security circuitry has been employed & more sensors are added to the proposed BMS, the safety circuitry designs are improved, with the addition of precise alarms and controls to prevent overcharges/ discharges and overheating. A variety of sensors make up the sensor system, which tracks and measures battery measurements such as battery temperature, battery current and cell voltage. The viability of conducting measures outside the laboratory environment is hampered by both a lack of available space and the high cost of the instrument. Therefore, it is necessary to measure current, voltage and temperature to enhance state tracking in practical applications. BMS contains vital components for databases that are analyzed and built for data collection and modeling

of data storage systems. The component that regulates the discharge protocol is charge control. A potentiostat and a galvanostat are necessary components of the constant current/constant voltage (CC/CV) approach, which is frequently used to charge batteries. To assist in cell balancing or measure internal resistance, a variable inhibitor might be needed. To properly balance the battery pack and determine the battery's condition, a crucial design element needs to be improved. Since the majority of the BMS subsystems are standalone modules, data transfer between them is necessary. One of the main ways of exchanging data within the BMS is communication over the CAN bus. As smart batteries get more advanced, more information may be gathered to allow the battery's embedded microchips to communicate with the user and charger. In order to improve communication between batteries and chargers, wireless and telecommunications technologies are also rapidly being implemented into charging systems. To properly balance the battery pack and determine the battery's condition reliability and performance, a module is essential for thermal management. According to reports, the temperature difference between the cells must be kept as little as possible, and these cells must be observed and maintained at the correct temperature.

5. Software

In order to make decisions and determine the system's state, all hardware activities are controlled by the BMS software, which also analyzes sensor data. The software of the BMS must be used to regulate the switch control, sampling rate monitoring, cell balance control and even the dynamic protection circuit design in the sensor system. In order to continuously update and regulate battery functions, online data processing and analysis is also necessary. Because the analysis evaluates the status and identifies the defect, reliable and robust automated data analysis is crucial to success. The user will be presented with this information via a user-friendly interface with the necessary suggestions. A capacity evaluation that also provides the battery life situation and determines operational restrictions using modern algorithms like fuzzy logic, neural networks and so on will include the determination of SOC and SOH [5]. Without overcharging or over-discharging the cell, the battery's performance should be maximized balancing is done. Because of this, SOC levels tend to be close to one another. The controller will regulate the charging procedure using a thorough plan that is dependent on the SOC of each cell. In order to improve equilibrium, accurate SOC calculation of each cell is essential. Online data processing will find the majority of soft fault issues. To alert a battery issue and identify a state of out-of-tolerance, intelligent data analysis is necessary. Historical data will be saved and will show a status prior to potential issues. The user interface should provide users with access to the necessary BMS data. Based on the battery's SOC, the dashboard should show the remaining range. Users must also be informed about unexpected hazards and replacement recommendations relating to battery predict and forecasts [2].

6. BMS Program Structure

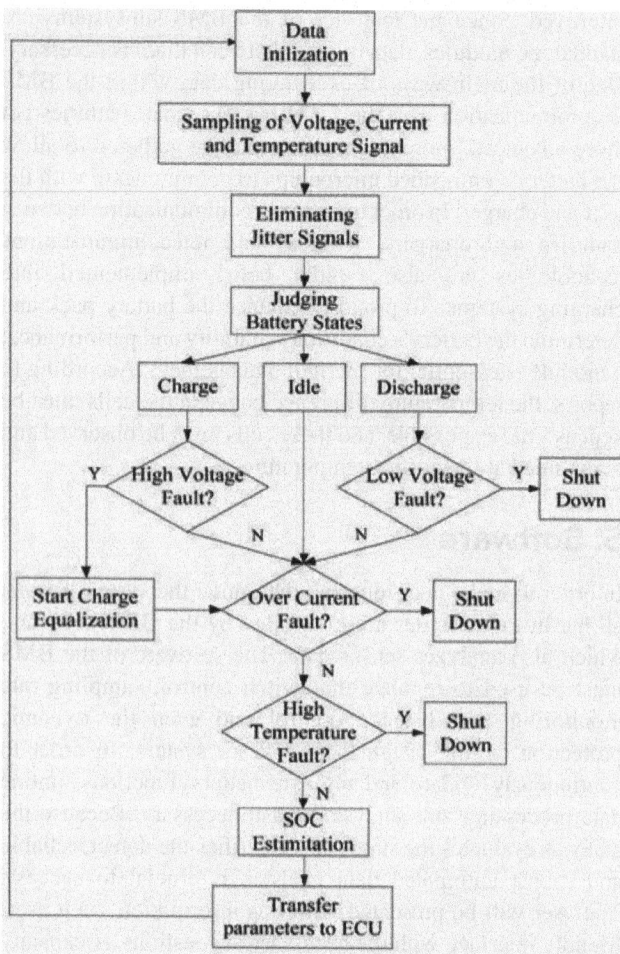

Fig. 48.2 BMS software flow pattern

The flowchart for the C-programmed BMS in this paper is illustrated in Fig. 48.2. The expert system of the program "Code Warrior" automatically generates hardware interface programs. The MCU program implements procedures for reading voltage, current, temperature signals, anti-jittering the current signal, over-current, over-charge/over-discharge, high/low temperature protection, charge equalization and SOC estimation. The current signal's anti-jitter technique involves picking out signals that are outside of the threshold that was previously set, removing them and computing the signal at that precise moment by smoothing the signals from earlier. We use the referenced charge equalization method.

The MCU would shut down the entire circuit if one of the faults of low voltage, over current or high temperature occurred to prevent long-term damage to the battery stack or even explosion events. One of the hardest aspects of BMS design is SOC estimate, which will be detailed in the following section. The Fig. 48.2 software is recycled once the BMS begins operating until abnormal circumstances are detected. To safeguard the battery stack, the MCU would shut down the entire circuit [3].

7. Battery Monitoring and Management

Li-ion battery management is necessary for ensuring battery life, energy availability and the energy storage system safety. The battery management system's inputs are battery voltage and temperature variations. It is responsible for conducting assessments of battery safety, state-of-charge (SOC), state-of-health (SOH). Controlling the primary power switch, the heating/cooling subsystem, providing high voltage isolation from the interior of the vehicle and implementing separated connection with the in-vehicle network are other functions. The problems from an electronic approach include the precise and data transfer through various voltage domains, synchronous measurement of battery current and pack cell voltage, and compliance to ASIL-C security criteria. Typically, accuracy standards are set around 0.1% for voltages at the cell and pack levels and 0.5-1% for currents up to 450 A and 1-2 mV respectively. LiFePO4 chemicals are the primary cause of this strict standards for voltage accuracy. It is a technology that is often used in automotive applications, since it has good agreements in terms of energy density, price, safety features, longevity and cycle resistance. A highly flat property of this battery technology exists for open circuit voltage vs state charge stands. Due to this, it is exceedingly challenging to precisely place charge using voltage measurements, especially between 20% and 80% state-charge. Other automotive Li-ion chemistry, such as Li-Titanate or Li-Manganese are less demanding in terms of voltage monitoring precision than LiFePO4 cells. From the viewpoint of the semiconductor component, it is necessary to design for integrated error compensation techniques. For accurate determination of product parameters, modern product qualification and high-precision manufacturing testing equipment are also required [5].

8. Integration of Batteries in Electric Vehicles

Numerous factors affect how a battery system is integrated into an Electric Vehicle. As a result, it is impossible to implement a rule that applies to every possible integration instance. Instead, a concrete example will be used to highlight the most important points. The illustration is based on knowledge obtained from the E3 Car project. For the battery system to achieve the needed 400 Vof voltage, 96 series-connected 50 Ah Li-ion cells are used. Modules of four cells are used to divide the cells. There are 24 modules that make up the battery pack, each of which has a specially developed circuit for monitoring and balancing cells. The battery system also has a pack management unit in addition to the module management unit-equipped battery module. Its duties include gathering information from current sensors and module management units, computing battery measurements including charge levels and state-of-health, connecting to vehicle control systems, operating power switches, managing batteries, heating and cooling system. A CAN bus can be used for communication between battery module management units and pack management units. It is vital to ensure galvanic separation of the power supply and the communication cables of the monitoring circuit from the rest of the vehicle since MMUs are required to monitor voltages at levels up to roughly 400 V, describes in more detail the battery monitoring module and its monitoring circuit. The battery monitoring IC serves as the monitoring circuits brain. This monitoring IC interacts with a microcontroller through the SPI bus and numerous digital I/O lines. The microcontrollers responsibilities include managing cell data collecting, keeping track of the balancing procedure and enabling communication between the IC and the PMU over the CAN bus. Therefore, using an 8-bit microcontroller in this instance, an ATMEL AT90CAN128 is sufficient [5].

9. BMS Solutions

Progostics and Health Management (PHM)is an enabling strategy for BMS, consisting of techniques and methodologies. Estimates of battery positions, including SOC, SOH and SOL can be made by tracking sensor signals and processing data from a BMS in real-time and to provide the end user with an accurate "gauge meter" in an EV. The BMS chooses the appropriate maintenance techniques based on the data gathered. While waiting for prediction results to be updated, signals for abnormality detection can be employed to ensure the dependability and safety of batteries. Battery internal responses that are difficult to access and variable external loads affect the battery's precise performance. It is important to build modeling that considers the imposed factors. A competitive strategy for modeling battery degradation has been suggested using regression technologies and a state-space model. In order to decrease the rate of degradation based on particular battery materials, regression algorithms

use data training. Empirical degradation characteristics can be integrated with real-time state information after the Markov process employs the fitted parameters as its initial input to produce reliable forecast results. In order to increase viability and save design costs, our strategy is to measure and gather current, voltage and temperature as the major operating characteristics [2].

10. Conclusion

Since batteries are an Electric Vehicle's main source of electricity, how powerful they are is greatly influenced by how well they work. Manufacturers are therefore striving for improvements in BMS and battery technology. Battery degradation may vary depending on the environment because operating conditions have an effect on the chemical reactions in batteries. Manufacturers who wish to increase their product's market share must own a fully developed BMS. The major BMS challenges are addressed in this paper includes cell balancing, modeling and battery state evaluation, where techniques for assessing battery status were seen as challenges. In order to compare the corresponding functions for SOC, SOH and SOL batteries were examined. A BMS framework was created to solve the deficiencies of the current BMS in research and commercial items. The specific issues that BMS face and potential answers were outlined based on prior work as a foundation for future research. A standard solution was not necessary because real-world applications conditions are always changing. Various strategies should be used, depending on the circumstances, to enhance and optimize the performance of BMS in upcoming EVs.

References

1. Vikas Gupta, Aamani Ravada, 2015 "Model Based Battery Management System for Electric Vehicles", Indo American Institutions Technical Campus, Visakhapatnam, India.
2. Yinjiao Xing, Eden W. M. Ma, Kwork L. Tsui and Michael Pecht, 2011, "Battery Management Systems in Electric and Hybrid Vehicles".
3. Xueqing Yuan, Lin Zhao, BoLi, Naiming Liu, 2015 "Battery Management System for Electric Vehicle and the Study of SOC Estimation", China.
4. Jiwen Cen, Zhibin Li, Fangming Jiang, 2018 "Experimental investigation on using the electric vehicle airconditioning system for lithium-ion battery thermal management" Chinese Academy of Science (CAS), China.
5. M. Brandl, M. Wenger, F. Baronti, A. Thaler, 2012, "Batteries and battery management systems for electric vehicles".

Note: Authors' self data and the graphs/images are part of the research work.

Improved Light Weight Crypto System for Secure Image Sharing Using IoT

49

**Harini Saraswathi[1], Pavani Prathyusha Meenige[2],
Nithish Kumar Reddy Mulageri[3], C. N. Sujatha[4],
Syed Jahangir Badashah[5]**

Electronics and Communication Engineering,
Sreenidhi institute of Science and Technology, Hyderabad, India

Abstract: The Internet of Things (IoT), which has been the subject of this paper, integrates physical and also digital worlds to provide a networked environment for communication. The transfer of medical data has become regular with the introduction of IoT-based remote digital healthcare systems. In order to secure the confidentiality and reliability of the patient's diagnostic information is transported to the IoT ecosystem and received there,it is important to build an effective model. Steganography methods and system encryption algorithms are used to accomplish this purpose by concealing digital data in a picture. On the other hand, The security and integrity of the medical data have become crucial challenges for healthcare services applications as a result of the rapid rise of IoT in the healthcare industry. A hybrid and minimal security paradigm is presented in this study to protect the diagnostic text data in medical photographs. The suggested hybrid encryption technique is used with a 2-D discrete wavelet transform steganography technique to produce the suggested model. The proposed hybrid encryption technique combines the Feistel and Advanced Encryption Standard (AES) encryption algorithms.

Keywords: Steganography, Encryption algorithms, Hybrid encryption, Discrete wavelet transform, Advanced encryption standard, Feistel encryption algorithm

1. Introduction

The healthcare industry has grown dramatically in recent years, which has had a large positive impact on both employment and revenue. Prior to a few years ago, identifying illnesses and other bodily anomalies needed a physical examination at the hospital. The majority of patients were required to remain in the hospital for their treatment. This raised the cost of healthcare while also putting strain on institutions in remote and rural areas. Using tiny devices like smart watches, it is now feasible to diagnose a wide range of illnesses and monitor one's health thanks to advancements in technology throughout time. Technology has also transformed the healthcare system, moving it away from hospitals and toward

people. It is possible to assess blood pressure, blood sugar, pO2, and other without the aid of a medical expert, at home. Furthermore, advanced telecommunications technologies enable the transmission of clinical data from rural areas to healthcare facilities. Because of utilization of these services for communication and swiftly evolving technologies (similar as machine literacy, big data analysis, Internet of effects(IoT), wireless seeing, mobile computing, and cloud computing), healthcare facilities are now more accessible.

IoT creates an integrated communication ecosystem of networked platforms and devices by fusing the physical and digital worlds. The transfer of medical data has become regular with the introduction of IoT-based remote digital healthcare systems. Steganography methods and system

[1]saraswathiharini000@gmail.com, [2]prathyushameenige@gmail.com, [3]nithishreddy0504@gmail.com, [4]cnsujatha@sreenidhi.edu.in, [5]sydjahangir@sreenidhi.edu.in

encryption algorithms are used to accomplish this purpose by concealing digital data in a picture. The Internet of Things has increased human freedom while increasing opportunities for contact with the outside world. Modern protocols and algorithms have greatly impacted IoT's impact on worldwide communication. Numerous devices, wireless sensors, home appliances, and electrical apparatus are all connected to the Internet through this. IoT is employed in industries including agriculture, transportation, construction, and healthcare. The benefits of enhanced accuracy, lower costs, and better event prediction make the Internet of Things (IoT) a growingly popular technology. The rapid development of IoT has also been aided by the development of wireless technology, advancements in computer and mobile technologies, and the expansion of the digital economy.

The benefits of enhanced accuracy, lower costs, and better event prediction make the Internet of Things (IoT) a growingly popular technology. The rapid development of IoT has also been aided by the development of wireless technology, advancements in computer and mobile technologies, and the expansion of the digital economy.

In order to monitor and share data, IoT devices (sensors, actuators, etc.) have been coupled with other physical devices utilising a variety of communication protocols, including Bluetooth, Zigbee, IEEE 802.11 (Wi-Fi), and others. In applications to healthcare, embedded or sensors that are worn, utilised to gather physiological information from patients' bodies, such as their body's temperature, blood pressure, electrocardiogram (ECG), electroencephalogram (EEG), and more. Temperature, humidity, time, and date are just a few examples of environmental data that may be recorded. These data enable making precise and pertinent judgments regarding the patients' state of health. The Internet of Things system (sensors, mobile phones, e-mail, software, and apps) is reliant on the availability and storage of a sizable amount of information that is gathered and collected from a variety of sources.

1.1 Healthcare IoT (HIoT) Architecture

The HIoT topology is composed of several IoT medical system/network configurations that in a healthcare setting make sense to be connected. Publisher, broker, and subscriber make up the three essential parts of the basic HIoT system. The publisher depicts an interconnected network of sensors and other medical equipment that may simultaneously/sequentially gather vital data on the patient.

Due to the IoT's rapid expansion,sensors that are worn, medical technology, portable gadgets, and other items have all improved in cost and use. Patient data may be gathered using these systems, spot ailments, observe the patient's health, and communicate with them in the case of a medical emergency.

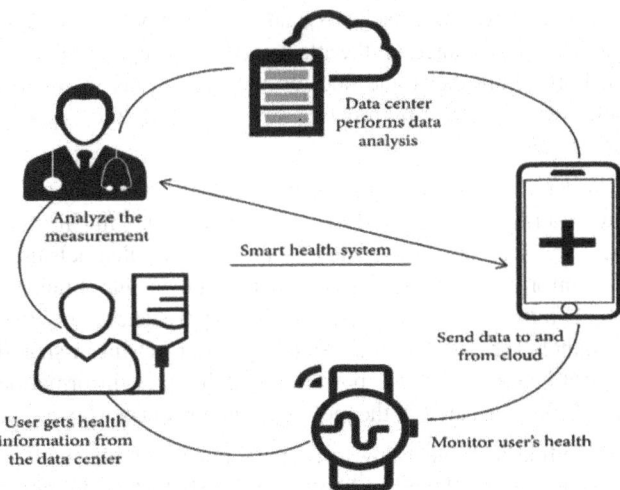

Fig. 49.1 Smart health systems

2. Literature Survey

A secure and resource-efficient technique for sending patient information sent from wearable IoT sensors to the base station (BS) was presented by Humayun, M., Jhanjhi, and Alamri in 2020. To gather and send real-time data, IoT sensors are extensively used in the healthcare sector. However, because to their computational and storage resource limitations, these sensors are more susceptible to security breaches and other dangers. Also, the energy quality of IoT sensors deteriorates over time, which might occasionally result in the loss of important medical data[1].

Lightweight group-based authentication that is secure was recommended by Almulhim, M., and Zaman, N. (2020) for Internet of Things (IoT)-based E-health applications. The recommended paradigm will offer energy-efficient computation and mutual authentication for IoT-based healthcare applications. The recommended model's aforementioned qualities are provided by elliptic curve cryptography (ECC) ideas, which will be used in this[2].

The method was put forth by Mallikarjuna, B., Kiranmayee, D., Saritha, and Krishna, P. V. (2021), and tested using the NodeJs software and ApacheJmeter open source environment for JMeter Cloud Testing. This demonstrated that the outcomes of the BEHR simulation were superior to those of the current conventional system in terms of reaction time, file storage, and EHR transmission[3].

In contrast to the standard RSA cryptographic algorithm, Amare and Vuda (2021) developed an upgrade in which two public keys are generated throughout the key creation process and used simultaneously instead of only one. Unlike to the conventional RSA technique, which only sends the public key once, in this case the public key is sent twice independently. Because to his lack of knowledge of the encryption key, the attacker is unable to decrypt the message. In rare circumstances, if the attacker can intercept the sending of both public keys and does so with malicious intent, the attacker can utilise both public keys to decode the encoded message[4].

Three colour image steganography methods were suggested by Bairagi et al. (2019)to safeguard data in an IoT architecture. Red, green, and blue are the three channels used for information transmission whereas green and blue are the colours used in the first and third approaches, only two channels are used in the second approach. Using a shared secret key and dynamic positioning techniques, information has been concealed in the picture channels' deeper layer[5].

A method to protect any form of pictures, notably medical photos, was developed by Anwar et al. in 2020. They attempted to preserve the accuracy of electronic medical records by guaranteeing their accessibility and authenticating the data to ensure that only those with the proper authorization may access it. First, the first component was encrypted using the AES method. This work includes the ear print as well; from the ear image,seven values were extracted to create a feature vector. By transferring medical photos over the internet, the suggested method enhanced their security and protected them from being accessed by any unauthorised individuals[6].

A combined security strategy based on encryption, steganography, and watermarking techniques was proposed by Razzaq et al.in2021. It broke down into three phases:

1. Utilising the XOR method to encrypt the cover picture
2. The stego-image is produced via an embedding method employing least significant bits (LSBs).
3. Applying a spatial and frequency watermark to the stego-image. Experimental findings demonstrated that the suggested strategy was highly effective and secure[7].

By applying the idea of a decision tree to mask the data, Jain et al. (2019) suggested a novel method for transferring the patient's medical data into the medical cover image. Coding is completed using several blocks that are evenly dispersed. When data has to be concealed, a mapping procedure based on breadth-first search allocates cover image secret code blocks in order to insert the data. The RSA encryption method was initially used to encrypt the data[8].

A transformation technique is used to arrange the blocks in the original image's group into turns. This algorithm was presented by Zawand Phyoin 2019. Following that, the Blowfish method is used to encrypt the altered image. By employing smaller block sizes and more blocks, it was discovered that the correlation declines and the entropy rises[9].

Pushkar Kishore, Kulamala Vinod Kumar, Swadhin Kumar Barisal, and Durga Prasad Mohapatra A new method using a time stamp is suggested for handling a replay attack at IEEE International Conference on Communications, ICC 2021, 1-6, 2021. Using the Elliptic Curve Discrete Logarithm Problem (ECDLP), we guarantee robust forward security, making it difficult for an attacker to decipher the security settings. Last but not least, it is made sure that the bits of the hash function preserve the entropy of the key utilised in the security model[10].

As a result, the suggested approach enhances the security of the E-Health model while also protecting privacy. A system employing block chain and IPFS (Inter Planetary File System) was proposed by Sarath Sabu, Swaraj Hegde, et al. in Global Transitions Proceedings 2 (2), 429-433, 2021 to offer a solution to all issues. It also includes limitations and safety measures on what can be done with your personal information and what can't be done in certain situations. The IPFS data will be divided up across the nodes. Interplanetary File System (IPFS), which has the benefit of being distributed and making data immune to alteration, may be used to store health records[11].

3. Methodology of Proposed Work

In this paper, in order to protect the transmission of medical data in Internet of Things (IoT) environments, we offer a paradigm for healthcare security. The suggested model includes four separate processes:

1. Utilising RDWT, ciphertext data is cloaked to make a stego-image in a cover image;
2. The ciphertext data is encrypted using are commended hybrid light weight encryption approach that combines AES and FBC encryption methods.

The original data is then obtained by decrypting the extracted data

3.1 Scheme for Data Encryption

The recommended model incorporates the cryptography technique. Encryption and decryption processes make up the cryptographic system. Odd and even parts of the simple text T are separated throughout the encryption process. Using a private public key and the AES encryption algorithm, data is encrypted. The secret public key m is used to encrypt data using the AES algorithm. In order to strengthen security, the private key x utilised in the recipient side decryption process is encrypted using the FBC technique and provided to the receiver in an encrypted form.

3.2 Embedding Technique

RDWT was used in this process. Similar to RDWT, 2D-DWT-2L is a sequential transformation that applies low-pass and high-pass filters along the picture's rows and columns, respectively. The suggested paradigm uses steganographic method. The steganographic consists of the embedding and extraction procedures. A cover photo Canda secret text message Tare combined by the embedding technique to produce a stego-picture S. The embedded message is extracted while the message is extracted backwards. These equations may be used to mathematically explain it. During the hidden text is converted to ASCII representation and separated into even and odd numbers throughout the embedding procedure. Vertical coefficients, which cover up the odd values, are mentioned in LH2. The diagonal coefficients that disguise the even values are specified by the HH2.

3.3 Extraction Procedure

The hidden message is retrieved and the cover photo is recovered using the RDWT approach after the text has been combined with the cover photo. The cover image is delivered to steganography and then it is extracted using RDWT. After that the decryption process takes place through AES and FBC algorithms.

3.4 Data Decryption Scheme

Returning the user ciphertext data in a format they are comfortable with is known as decryption; it is the opposite of the encryption process. During the encryption procedure, the cipher-text must be protected with the same key that was used by the sender.

3.5 Redundant Discrete Wavelet Transform

This part describes complex encoding and extraction technique of the suggested research. The recommended method used the texture of the photos by computing the entropy levels for each picture block for both cover and secret photos. The utilisation of entropy in the embedding process is the proposed work's main contribution. The cover image and the hidden image will both be divided into equal-sized, non-overlapping pieces before embedding. For each picture block, an entropy value will be computed. Next,the values will be sorted from highest to lowest entropy value in descending order. This applies to both the cover and the hidden picture. The blocks will next be subjected to RDWT, starting with the building block having the highest entropy rating. The cover picture block's LL sub-band will then undergo QR decomposition. The hidden image information is implanted by altering the LL sub-band's R value in the cover image.

The LL sub-band is then obtained by performing an inverse QR decomposition. The updated picture block will then be obtained using inverse RDWT. The embedding of all hidden image information will then commence. All updated image blocks will then be combined to create a stego image. The benefit of using RDWT over other transform techniques is that it does away with the up- and down-sampling of coefficients that DWT uses. Moreover, RDWT has a greater capacity for embedding and can boost robustness. As opposed to SVD, QR decomposition offers more imperceptibility and solves the false positive problem. One can use entropy to determine an electronic image's texture.

An image block includes more information the greater its entropy is. The cover image block with the highest entropy value will be mixed with the hidden image block with the highest entropy value throughout the embedding phase. The entropy values won't be significantly affected by the minor adjustments made to the specific picture block. Imagine that the cover picture block with the lowest entropy is inserted within the hidden image block with the highest entropy value, which carries more information. This kind of approach will alter the visual texture more than the previous example, which will indirectly impact how undetectable the scheme is.

4. Results and Discussions

The robustness and invisibility of the recommended approach are examined in this section. In order to determine the best adaptive scaling factor for watermarks of different sizes, the scaling factor across PSNR and MSE is first looked at. In the trials, watermarks of different sizes are scaled using adaptive optimum scaling factors. Utilizing both objective quantitative analysis and subjective eye observation, the suggested technique's invisibility and durability are assessed. The resistance is also examined by means of a variety of assaults with various characteristics. The strength and conceal ability of the suggested technique are then assessed in light of prior studies. Double-clicking the 'run.bat' file will start the project and display the screen. Enter some message in 'Secret Message' field in the obtained screen. We have input some text on the screen, and when we click the "Hybrid Encryption" button, the text is encrypted using both RSA and AES. We can see the encrypted odd and even messages on the screen above thanks to the hybrid AES-RS Atechnique. To encrypt the message using AES and Feistel encryption, click the "Extension" button now.

The whole message is seen above with ODD and EVEN sections before both parts are encrypted using AES and Feistel. Now that the message is prepared, use the "Embedding Algorithm" button to submit a picture before hiding the encrypted message choosing a photo from the collection that has both colour and grayscale variations, and then clicking "Open" to see the outcome.

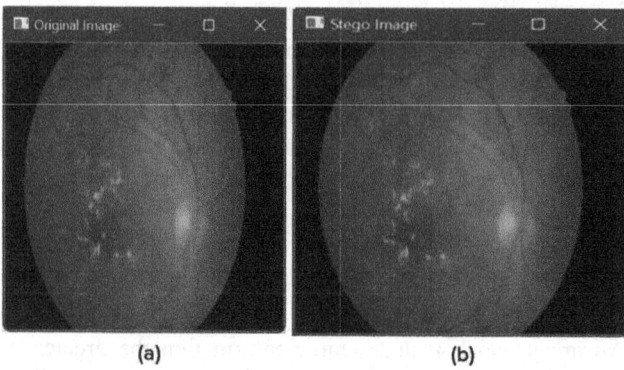

Fig. 49.2 (a) Actual Image, (b) Stego Image

The first picture on the screen above is the actual image, while these cond image is a hidden message created using steganography. Both messages have a comparable visual quality. Close both the photos. If not we will get an error.

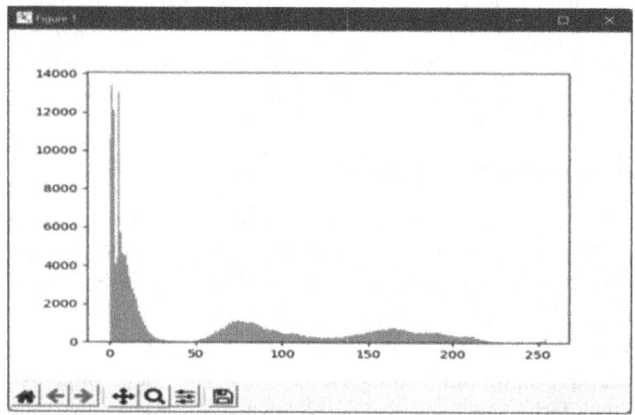

Fig. 49.3 Histogram

In the histogram shown above, both pictures have equal-sized bars that are visible after the message has been hidden. In the Stego image and the text area shown, the PSNR and MSE values are shown. You may submit other photographs and test in a similar way. To extract and decode the message from the picture, click the "Extraction Algorithm" button. We grabbed the encrypted message from the text area of the mentioned screen and then decrypted it to get the original information.

Conclusion

Grayscale images have been created as a cover carrier for a secure patient diagnostic data transmission paradigm that uses both colour and for an IoT-based healthcare setting. The proposed approach combined AES and FBC encryption with DWT steganography. Future studies may need to focus further on deterring other assaults like rotation and cropping attacks using the recommended steganography technique. Additionally, the performance of steganography may be

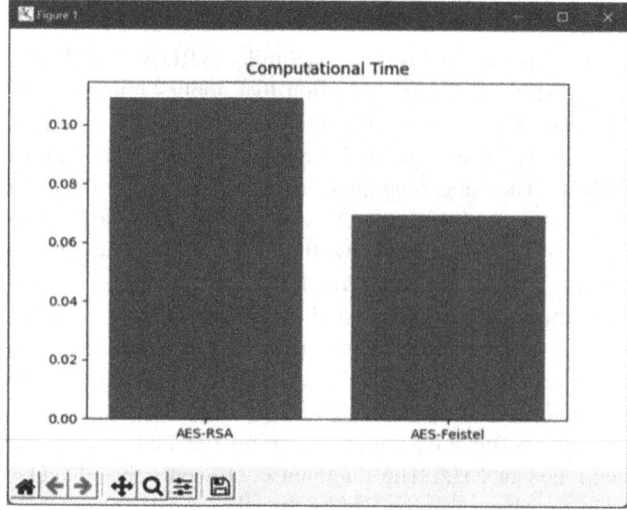

Fig. 49.4 Computational time

improved even further if the enhanced FOA approach is use IoT safe picture sharing requires the use of light weight crypto systems, which are crucial to its success. These systems are meant to provide reliable encryption and decryption techniques while using the least amount of processing power and storage space possible. This is crucial for IoT devices since they have limited processing power and battery life. The privacy and confidentiality of sensitive data maybe helped by the use of portable crypto systems in safe picture sharing over IoT. It can stop cyber attacks, data breaches, and illegal access.

Moreover, the IoT may utilize light weight crypto systems to increase data transmission and storage efficiency while lessening the stress on the network and devices. Overall, effective and safe methods of transferring photos over IoT devices are provided by lightweight crypto systems. The use of these systems may enable the advantages of IoT in a various sectors, including the health care, manufacturing, and transportation, while reducing the dangers related to data breaches and cyber attacks.

Light weight crypto systems for safe picture sharing with the Internet of Things have a bright future ahead of them, with plenty of room for expansion and improvement. First, it's probable that research into creating ever more effective and lightweight cryptographic algorithms will continue. This will make it possible for IoT devices to complete difficult encryption and decryption operations with even less resource use, which will increase the uptake of secure picture sharing through IoT. Second, the security and privacy of picture sharing in the Internet of Things may be further improved by the combination of blockchain technology with lightweight crypto systems.

Blockchain technology may provide a decentralized, tamper-proof way to store and share data,which might help guard against illegal access and alteration of private photographs. Lastly, more standardization and interoperability of lightweight crypto systems will be required as more IoT devices are deployed. The broad adoption of safe picture sharing via IoT will be made possible by the seamless integration of security measures across multiple devices and platforms. The introduction of 5G networks will provide Internet of Things (IoT) devices quicker and more dependable communication, enabling real-time picture sharing in a variety of sectors like healthcare, transportation, and manufacturing.

Inorder to ensure that this data is transferred safely and effectively, light weight crypto systems will be essential. In conclusion, the development of lightweight cryptosystems for safe picture sharing through IoT has a bright future, and we can anticipate more growth and advancement in this area, which will result in more efficient and secure image sharing across a range of sectors.

References

1. Humayun, M., Jhanjhi, N. and Alamri, M. (2020). IoT-based Secure and Energy Efficient scheme for E-health applications. Indian JSci Technol,13(28), 2833–2848.

2. Almulhim, M., and Zaman, N. (2020, February). Proposing secure and light weight authentication scheme for IoT based E-health applications. In 2018 20th International Conference on advanced communication technology (ICACT)(pp.481–487).

3. Mallikarjuna, B., Kiranmayee, D., Saritha, V., and Krishna, P. V. (2021, June).Development of efficient e-health records using iot and block chain technology. In ICC2021-IEEEIntern ationalConferenceonCommunications (pp. 1–7). IEEE.

4. Amare and Vuda(2021). Edge Devices for Internet of Medical Thing Technologies,Techniques, and Implementation, eddah22246-48.

5. Sarath Sabu, Swaraj Hegde et.al Global Transitions Proceedings 2 (2), 429-433, 2021[6]. Bairagi et al. (2019), IoT-Based Healthcare-Monitoring System towards ImprovingQualityofLife.hulna9208, Bangladesh

6. Shahzadi, R., Niaz, A., Ali,M., Naeem,M., Rodrigues, J.J., Qamar, F., and Anwar, S.

7. Hussain, A., Ali, T., Adeelaziz, F., Draz, U., Irfan, M., Yasin, S., ... and Alqhtani, S.(2021). Security framework for IoT based real-time health applications. Electronics, 10(6),719.

8. Karolak, M., Razzaque, A., and Al-Sartawi, A. (2021). E-services and M-services using IoT: an assessment of the Kingdom of Bahrain. In Artificial Intelligence Systems and the Internet of Things in the Digital Era: Proceedings of EAMMIS 2021 (pp. 523-533). Cham: Springer International Publishing.

9. Dhatterwal, Jagjit Singh, Kuldeep Singh Kaswan, Anupam Baliyan, and Vishal Jain. "Integration of Cloud and IoT for Smart e-Healthcare." In Connectede-Health: Integrated IoT and Cloud Computing, pp. 1–31. Cham: Springer International Publishing, 2022.

10. Soni, M., & Singh, D. K., seyyedi (2021). LAKA: Lightweight authentication and key agreement protocol for internet of things based wireless body area network. Wireless Personal Communications, 1–18.

11. Pushkar Kishore, Swadhin Kumar Barisal, Kulamala Vinod Kumar, Durga Prasad Mohapatra ICC 2021-IEEE International Conference on Communications, 1–6, 2021.

12. Kaur, M., Singh, D., Kumar, V., Gupta, B. B., and AbdEl-Latif, A.A. (2021). Secure and energy efficient-based E-health care framework for green internet of things. IEEET ransaction son Green Communications and Networking, 5(3), 1223–1231.

13. Farahat, AS Tolba, Mohamed Elhoseny, Waleed Eladrosy Security in smart cities: models, applications,and challenges, 117–142, 2019.

14. Momin, Md Sarfaraz, Abu Sufian, Debaditya Barman, Paramartha Dutta, Mianxiong Dong, and Marco Leoand Zawand Phyo. "In-home older adults' activity pattern monitoring using depth sensors: Areview." Sensors 22, no. 23(2022): 9067.

15. Marco Leo and Zawand Phyo. "In-home older adults' activity pattern monitoring using depth sensors: Areview." Sensors22, no. 23(2022): 9067.

16. T Venkat Narayana Rao, A Govardhan and S J Badashah "Improved Lossless Embedding and Extraction-A Data Hiding Mechanism" International Journal of Computer Science & Information Technology,(IJCSIT)ISSN:0975-4660, www. airccse.org , April, Volume-2,number-2 Pp: 77–88.

17. T V N Rao, Dr A Govardhan and S J Badashah, "Statistical Analysis For Performance Of Image Symentation Quality Using Edge Detection Algorithms" International Journal of Advanced networking and Application, (IJANA) ISSN:0975-0282,www.ijana.in, Nov/Dec 2011, Volume-3, issue-3, pp1184-1193.

Note: Authors' self data and the graphs/images are part of the research work.

An Approach to Recognize Hand Gestures Using Convolution Neural Networks and Recurrent Neural Networks

50

Rajesh Bhaskarla[1], Srujana Athimamula[2], Ravi C. N.[3]
Department of Electrical and Electronics Engineering,
Vidya Jyothi Institute of Technology Hyderabad, India

Abstract: The objective of this work is to build a good model for identifying gestures to control a video in a smart Television. This feature enables any user to watch a video just by using gestures to control the television without using a remote. Two model configurations using Neural Networks are implemented and the best model is selected for this application. The results and observations are discussed in this paper.

Keywords: Neural networks, Deep learning, Convolution neural network, Recurrent neural network, Python programming, AI techniques and gesture recognition

1. Introduction

The aim of gesture recognition is to control appliances with less effort and to spend the least possible amount of energy by avoiding battery based remote controls. Gesture recognition is not an easy task in computer vision or computer-based applications. Latest improvements in computing and image processing have brought in the possible to build automated systems for human interaction. Computer systems can detect hands with these capabilities of hand gesture recognition, identify them, and follow various movements of hand. Natural Image contains many technical and digital specifics that can be used in various areas of computer vision [1]. With the advance in Deep Leaning, gesture recognition can be identified with the help of various types of convolution neural networks. This can be used for multiple applications especially for computer vision and decision taking applications. Gesture recognition has numerous applications in various domains, that enable natural and insightful interactions between humans and machines. Few applications of them are Human Computer Interaction, Education and Training, Gaming, Virtual and Augmented Reality, Health care and Rehabilitation, Automotive Industry and Assistive Technology. As technology advances and research progresses, we can expect to see even more innovative and impactful applications of gesture recognition in the future. Traditional TV remote controls have been the primary means of navigating and interacting with televisions for decades. However, these devices often come with a myriad of buttons and complex interfaces, leading to user frustration and reduced usability. In contrast, hand gesture recognition offers an elegant and user-friendly alternative by enabling users to control TV functions through simple hand movements and gestures. This paper presents an in-depth exploration of hand gesture recognition techniques for TV video control. The goal is to develop a seamless and intuitive interaction paradigm, enhancing the user experience and accessibility of modern television systems. By leveraging computer vision and machine learning algorithms, we aim to create a gesture recognition system capable of accurately interpreting users' hand gestures and translating them into meaningful commands for TV video control. The potential impact of implementing gesture recognition for TV control is substantial. Firstly, it eliminates the need for traditional remote controls, streamlining the user experience and reducing clutter in the living room. Moreover, gesture-based TV control holds significant promise for individuals with mobility impairments, providing them with a more inclusive means of interacting with their entertainment devices. The

[1]raajesh6@gmail.com, [2]eeehod@vjit.ac.in, [3]dr.ravicn@gmail.com

challenges in this field lie in developing robust and accurate algorithms capable of handling varying lighting conditions, diverse hand shapes, and natural variations in gesture patterns. Additionally, real-time performance is crucial to ensure seamless interaction and prevent any perceivable delays in TV command execution. In the following sections, review of two Neural Networks applications in hand gesture recognition, exploring different methodologies and techniques employed in this domain are presented. The dataset used for training and evaluating the proposed gesture recognition system is also made available. Furthermore, discussion on Python programming used, evaluation metrics, and results obtained through rigorous testing.

2. Literature Survey

Hand gesture recognition using neural networks has gained significant traction in recent years due to the remarkable advancements in deep learning. Neural networks offer the ability to automatically learn complex features from raw image data, making them well-suited for recognizing intricate hand gestures. Here, a comprehensive literature review on the topic, highlighting key studies and methodologies is presented:

2.1 Convolutional Neural Networks (CNNs)

CNNs have become the backbone of many computer vision tasks, including hand gesture recognition. They excel at learning hierarchical features from images through a series of convolutional and pooling layers. CNNs can capture spatial patterns and local features in hand gesture images, making them well-suited for this task [2].

2.2 Recurrent Neural Networks (RNNs)

RNNs are generally used to handle sequential data and have most applications in dynamic hand gesture recognition. RNNs, especially variants like Long Short - Term Memory (LSTM) as shown in Fig. 50.2 and Gated Recurrent Unit (GRU), can capture temporal dependencies in gesture sequences, making them suitable for recognizing gestures involving motion or sign language [3].Convolution LSTM Neural Network architecture can be observed from Fig. 50.2.

2.3 Three Dimensional Convolutional Neural Networks (3D CNNs):

3D CNNs enhance the concept of 2D CNNs for processing spatiotemporal information from video or depth sequences. They are effective for recognizing dynamic hand gestures, where temporal information is crucial for accurate classification [8][9]. The 3D CNN architecture can be witnessed from Fig. 15.3.

2.4 Capsule Networks (CapsNets)

CapsNets are a more recent development that aims to overcome some limitations of traditional CNNs. They focus on learning spatial hierarchies and rotations in images and can be useful for capturing the fine-grained details in hand gesture images [4].

2.5 Spiking Neural Networks (SNNs)

SNNs are inspired by the spiking behavior of neurons in the brain. They are event-driven and particularly well-suited for real-time gesture recognition applications due to their low power consumption and efficient event processing[5].

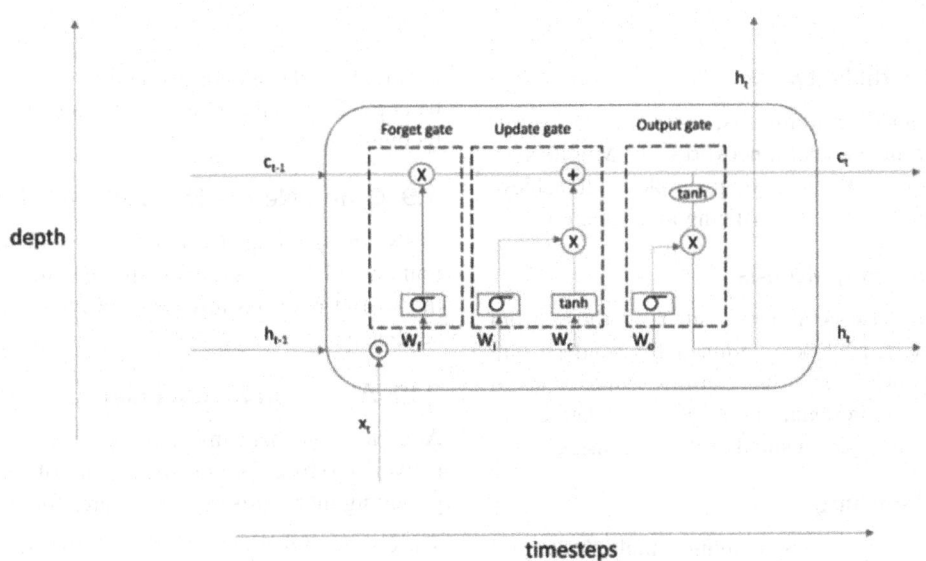

Fig. 50.1   LSTM cell

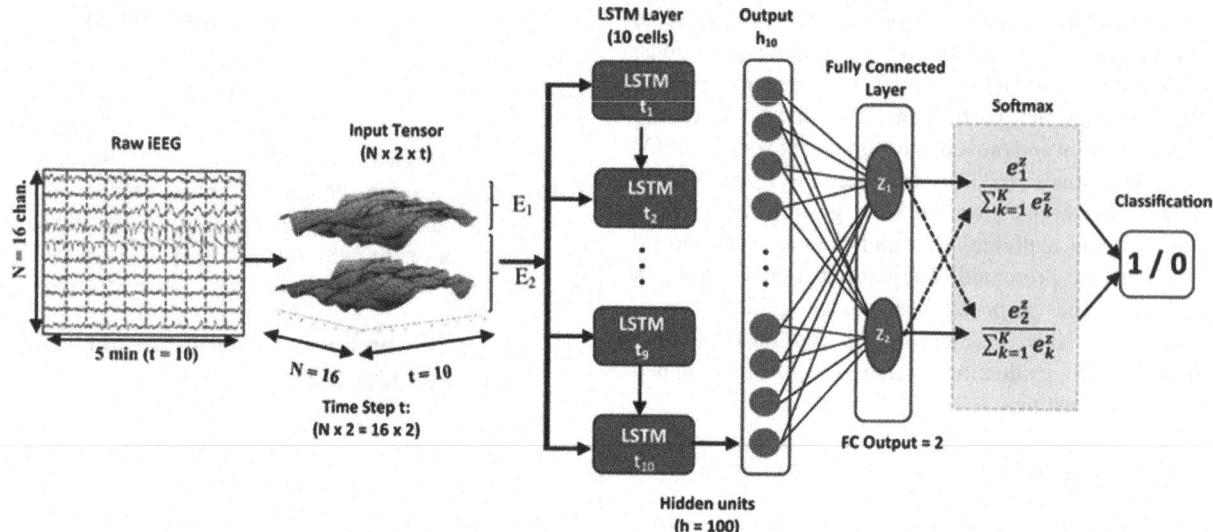

Fig. 50.2 Convolution LSTM neural network architecture

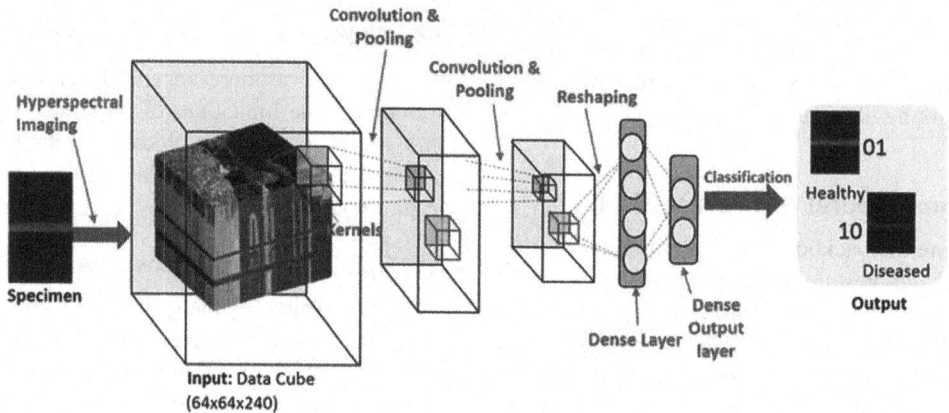

Fig. 15.3 3D Convolution neural network architecure

2.6 Efficient Neural Networks

In resource-constrained environments, such as edge devices and smartphones, efficient neural networks like MobileNets, ShuffleNets, and SqueezeNets are often employed to achieve real-time performance without sacrificing much accuracy.

2.7 Transfer Learning Models

Pre-trained neural network models, such as those from the ImageNet dataset, may be fine-tuned for hand gesture recognition applications. Transfer learning contributes to leverage the trained representations from larger datasets to increase their performances on small gestures datasets.

2.8 Ensemble Learning

Ensemble learning techniques combine multiple neural network models to enhance classification accuracy. Techniques like model averaging, voting, and stacking can be applied to create a more robust and accurate hand gesture recognition system.

2.9 Graph Neural Networks (GNNs)

GNNs are gaining traction for recognizing gestures in 3D point cloud data or skeletal representations. They can capture the spatial relationships between joints or points in gesture sequences.

2.10 Attention Mechanisms

Attention mechanisms can be incorporated into neural networks to focus on specific regions of hand gesture images, providing more informative features for classification.

The choice of neural network architecture depends on factors such as the nature of the hand gesture data, computational resources, real-time requirements, and available labeled data for training[6]. Researchers continue to explore novel architectures and techniques to further improve hand gesture recognition performance using deep learning methods[7].

3. Methodology and Implementation

Gesture recognition using generator models of 3D CNN, CNN+RNN, and Convolutional LSTM architectures involves a multi-step process that combines the strengths of these models. Here's a high-level methodology for implementing such an approach: Python Programming is used and the following libraries are imported for the same and can be seen in Fig. 50.4.

```
In [1]: from skimage.transform import resize
        from imageio import imread
```

```
In [2]: # Importing the necessary libraries

        import numpy as np
        import os

        import datetime
        import os
        import warnings
        warnings.filterwarnings("ignore")
        import abc
        from sys import getsizeof
```

```
In [5]: # importing some other libraries which will be needed for model
        import cv2
        import matplotlib.pyplot as plt
        %matplotlib inline
        from keras.models import Sequential, Model
        from keras.layers import Dense, GRU, Flatten, TimeDistributed,
        from keras.layers.convolutional import Conv3D, MaxPooling3D, Cc
        from keras.layers.recurrent import LSTM
        from keras.callbacks import ModelCheckpoint, ReduceLROnPlateau,
        from keras import optimizers
        from keras.layers import Dropout
```

```
In [6]: nroject folder="/home/datasets/Project data"
```

Fig. 50.4 Libraries and modules used in Python programming

Step 1: Data Collection and Preprocessing

A dataset of hand gesture videos or depth sequences with corresponding labels is acquired. The data is appropriately labeled to represent different gestures accurately. The TV webcam continuously monitors the gestures given out by the operator. Every gesture represents a specific command represented in Table 50.1:

Table 50.1 Gestures used

Gesture Used	Stop	Swipe Right	Swipe Left	Thumbs Down	Thumbs Up
Action Expected	Pause Video	Move forward 10 seconds	Move backward 10 seconds	Decrease Volume	Increase Volume

Every video maintains a bunch of thirty frames (or images).

Step 2: Generator is built

This is one of the most important part of the code.. In the generator, images are preprocessed as there are images of two different dimensions and a group of video frames are created. It is experimented with `y`,`z` `img_dx` and normalization is done to attain high accuracy.

Step 3: Feature Extraction

For 3D CNN and CNN+RNN models:

The model is made using different features that Keras delivers. Convolution3D and MaxPooling3D for a 3D convolution model are used. Time Distributed while developing a Convolution2D + RNN model is done. The final layer is softmax. The network is designed carefully in a way that the model gives good accuracy on the smallest number of parameters in order to fit in internal memory of the Television webcam.

Step 4: Sequence Generation

1. For 3D CNN and CNN+RNN models:

   The feature vectors obtained from the CNN layers are used as input to a generator model, such as a GAN or VAE. This generator model generates realistic and diverse sequences of feature vectors representing different hand gestures.

2. For Convolutional LSTM models:

   The Convolutional LSTM architecture inherently generates sequences of feature vectors by considering the temporal dependencies in the input sequences.

Step 5: Sample Cropping

Sample cropping can be observed from the following Fig. 50.5.

Step 6: Training

The generator model is trained to generate realistic sequences of feature vectors that correspond to various hand gestures.

Step 7: Gesture Recognition

1. After training, the generator model generates sequences of feature vectors for unseen hand gesture videos or depth sequences.

2. For 3D CNN and CNN+RNN models:

   The generated feature sequences are classified using an RNN (LSTM) to predict the corresponding hand gesture labels.

3. For Convolutional LSTM models:

   The generated feature sequences are directed into the Convolutional LSTM model for gesture recognition.

Step 8: Evaluation and Fine-tuning

1. The performance of the combined architecture on a separate validation is evaluated and its accuracy, precision, recall, are assessed.

Sample Cropping

```
In [11]: test_generator=ModelConv3D1()
         test_generator.initialize_path("/home/datasets/Project_data")
         test_generator.initialize_image_properties(image_height=160,image_width=160)
         test_generator.initialize_hyperparams(frames_to_sample=16,batch_size=3,num_epochs=20)

         g=test_generator.generator(test_generator.val_path,test_generator.val_doc,augment=True)
         batch_data, batch_labels=next(g)
         fig, axes = plt.subplots(nrows=1, ncols=2)
         axes[0].imshow(batch_data[0,15,:,:,:])
         axes[1].imshow(batch_data[3,15,:,:,:])
         plt.show()
```

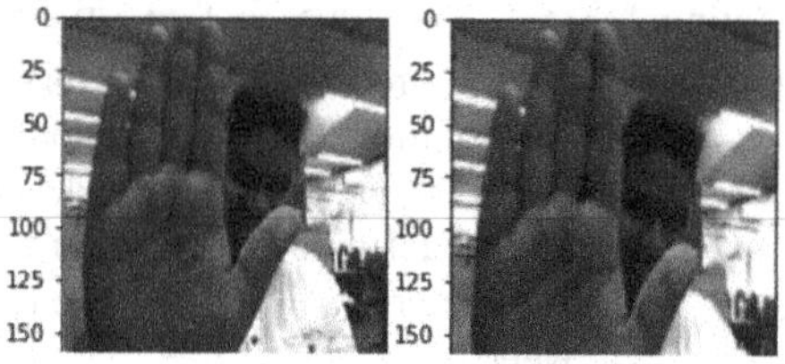

Fig. 50.5 Sample cropping

2. The model is fine tuned to improve performance or generalize better to different hand gestures.

The gesture recognition implementation using generator models of 3D CNN, CNN+RNN, and Convolutional LSTM requires careful architectural design, data preprocessing, and hyperparameter tuning. Proper training and evaluation are crucial to achieving accurate and robust recognition performance for various hand gestures.

Two types of models were built to analyze and solve the problem:

1. 3D Convolution Neural Networks (Conv3D)
2. CNN + RNN architecture

Neural Network Architecture development and training:

Two model configurations and hyper-parameters and number of iterations, sequences of batch sizes, size of filter, dimensions of image, stride length and padding are experimented with. Metrics (val_loss) remain unchanged in between epochs.

Adam () optimizer is used as it leads to improvement in model's accuracy by rectifying high variance in the model's parameters.

Batch Normalization, pooling and dropout layers are also used when our model started to overfit, this could be easily witnessed when our model started giving poor validation accuracy inspite of having good training accuracy.

4. Results and Discussion

It was observed that as the number of trainable parameters increase, the model takes much more time for training. The computation was done on Jarvis Lab AI as the models require high GPU and computational power for training. Increasing the batch size reduced the training time but this also has a negative impact on the model accuracy. This conveys that there is always a trade-off here on basis of priority.

The model is loaded and tested and the following are the losses and accuracy and it is observed that the losses got decreased and improvement in accuracy is observed form Fig. 50.6 and Fig. 50.7.

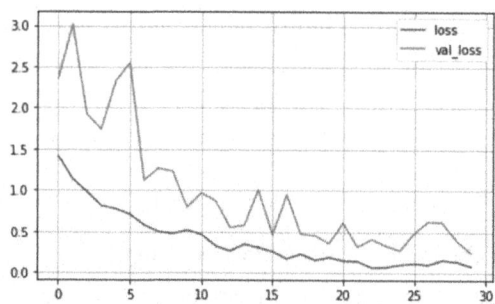

Fig. 50.6 Loss and variable loss

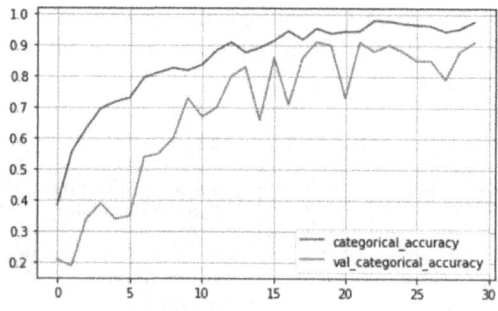

Fig. 50.7 Accuracy

Data Augmentation and Early stopping helped in solving the problem of overfitting.

CNN+LSTM based model with GRU cells had better performance than Conv3D. As per our understanding, this is something which depends on data used, the architecture developed and the hyper-parameters chosen.

The observations with respect to the models can be observed from Table 50.2.

Table 50.2 Results and observations from the models

Exp. No.	Model	Result	Decision & Explanation
1	Conv3D	Total params: 1,736,389 Trainable params: 1,735,525 Non-trainable params: 864 Warning message was thrown	To reduce the batch size and reduce the number of neurons in Dense layer
2	ConvLSTM	categorical_accuracy: 0.9759 val_loss:0.2395 val_categorical_ accuracy: 0.9100 after 30/30 epochs. Accuracy improved over epochs and it remained constant in the last epochs.	This result is good as the model is not overfitting
Final Model	ConvLSTM	It is the best model as its accuracy is near to validation accuracy	This model is suitable

5. Conclusion

3D Convolution Neural Networks and CNN+RNN with LSTM models are built and after validation the CNN+RNN model performs better for this set of data, for gesture recognition based on the accuracy and loss of variables. 3D Convolution Networks also can give better results for a different set of layers and parameters. Based on the results obtained CNN+RNN model is the best model for gesture recognition.

6. Acknowledgment

We would like to express our sincere gratitude to the management of Vidya Jyothi Institute of Technology which provided access to computing resources, data repositories, and research facilities, enabling us to conduct experiments and analysis efficiently. We are also grateful to Kaggle Data Sets, which provided Data Sets of Hand Gestures and insights during various stages of this project. Their diverse perspectives and expertise significantly enriched this work process.

References

1. Cao, Z., Simon, T., Wei, S. E., & Sheikh, Y. (2017). Realtime multi-person 2D pose estimation using part affinity fields. In CVPR.
2. Molchanov, P., Gupta, S., Kim, K., Kautz, J., & Kim, V. G. (2016). Hand gesture recognition with 3D convolutional neural networks. In CVPR.
3. Hochreiter, S., & Schmidhuber, J. (1997). Long short-term memory. Neural computation, 9(8), 1735–1780.
4. Sabour, S., Frosst, N., & Hinton, G. E. (2017). Dynamic routing between capsules. In NeurIPS.
5. Wu, J., Zhang, H., & Huang, K. (2016). Learning a recurrent visual representation for gesture recognition. In ECCV.
6. Li, Z., Zhang, Z., & Liu, Z. (2018). Hand gesture recognition with spiking neural networks. In ICANN.
7. Huang, J., & LeCun, Y. (2006). Large-scale learning with SVM and convolutional nets for generic object categorization. In CVPR.
8. Zhang, C., & Konrad, J. (2017). Towards privacy-preserving hand gesture recognition for wearable devices using convolutional neural network. In ICME.
9. Pu, J., Song, X., Li, W., Zhao, J., & He, L. (2018). Hand gesture recognition based on CNN and LSTM. In ICMLC.
10. Wang, C., Liu, W., & Chen, S. (2020). Real-time hand gesture recognition system with lightweight 3D CNN. In ICCV.

Note: Authors' self data and the graphs/images are part of the research work.

Comparative Study of Spatial and Frequency DomainImage Steganography Techniques on Grayscale Images

51

Pydimarri Padmaja[3], T. Rajeshwari[2]

Department of ECE, Teegala Krishna Reddy Engineering College,
Hyderabad, India

Abstract: Data Security is a major concern in digital communications. The Various ways of ensuring security of the transmitted data include water marking, Cryptography and Stegano graphy. Stegano graphy is a technique used to hide the presence of data using a cover so that the intruder will not be able to figure out the message that is kept inside as to him it is only the cover that appears. The different media that can be used as covers include text, audio, video and image. The most widely used cover is an image because of its high redundancy. It also serves as an effective media to communicate one's thoughts. The basic classification of stegano graphic techniques include spatial domain Techniques and Transform domain Techniques. This paper ensures a detailed analysis and comparison of spatial domain technique, transform domain technique and Hybrid technique. Their performances are evaluated using the Mean Square Error (MSE) and Peak Signal to Noise Ratio (PSNR).

Keywords: Water marking, Cryptography, Stegano graphy, Spatial domain, Mean square error (MSE), Peak signal to noise ratio (PSNR)

1. Introduction

Transmitting data over the noisy channel is a problem of high concern today. One cannot assure the security and authenticity of the data over the insecure channel during its transmission. Hackers concentrate on the sensitive information like the credit card details, passwords and other personal information. Especially military information requires secure transmission of data as if something is corrupted or accessed without permission, the whole country is at risk. Cryptography is a leading technique to encrypt the information.

The original message or Plain Text is converted into an unreadable format called cipher text so that the intruder has no idea of what is being sent as the message. This process of converting the a leading technique to encrypt the information.

The original message or Plain Text is converted into an unreadable format called cipher text so that the intruder has no idea of what is being sent as the message. This process of converting the Plain text to cipher text is called encryption.

The process of re-converting the cipher text to plain text is called decryption. This technique has been used since ancient times to prevent any unwanted person from gaining intelligence that might result in severe loss [2]. Watermarking is a technique where a unique signature of the owner is embedded for copyright protection. To detect illegal copies or modifications in digital media, Watermarking is widely employed.

If a person tries to remove the watermark, the data gets distorted [4].Stegano graphy is the technique of hiding, where the presence of information is kept secret. The information is hidden inside a cover object. Secret message can be accommodated in an image, audio, video, and text. Usually, images are chosen as covers because they have large storage space and can be altered easily. Steg analysis is the art of analyzing the information and detecting the presence of information. It has originated from Greek words: steganos" meaning "hidden", and "graphy" means "writing" [1].

[1]padmajavattem@tkrec.ac.in, [2]raji3061.iiit@gmail.com

2. Stegano Graphy Techniques

2.1 Types of Stegano Graphy

Image Stegano graphy: The binary equivalent of them essageis taken and the bits are sequentially embedded in the pixels of the image. Usually, the least significant bits are chosen as they contribute very less to the intensity of a pixel. In Audio Stegano graphy the audio is broken and messages can be hidden in those gaps.

The other technique used is the low frequency components are often dominated by the high frequency components. Human ear is susceptible to such low frequencies. So, those low frequency signals occurring just before a high frequency can be used to store them message [5].Video Stegano graphy: Video as a cover has huge capacity to hold data and minor changes are not noticeable due to continuous flow of image spere very frame.[3].

Text stegano graphy is a methods in volve hiding the message by changing the format of the text file into the cover. Applications include Confidential communications and secret data storing, protection of data from alteration, access control systems for digital content distribution, media database systems.

2.2 Least Significant Bit (LSB) Technique

Direct Least Significant Bit Replacement (DLSBR) is one of the oldest spatial domain techniques, where large secret data can be easily hidden in the cover pixels without much distortion. It is highly prone to attack due to its simplicity. If the presence of data is known one can easily get access by accessing the LSB bits of cover pixels [1].

Suppose the message bits are 10101101. These are hidden in the LSB of cover pixels. When the number of message bits hidden in each pixel increases more to than 3, then noticeable changes are observed in the cover image.

Table 51.1 LSB's of cover pixels are being modified

10001000	10001001
10001011	10001010
10001011	10001011
10101100	10101100
10101100	10101101
10101010	10101011
10111101	10111100
10110001	10110001

Thus, cover gets distorted. There are several adaptive techniques employed to increase the visual quality of the stego image at the cost of reduced embedding capacity.[7]

2.3 Discrete Wave letTransform (DWT) Technique

Discrete Wavelet Transform decomposes the image pixel into four wavelet sub-bands: LL, HL, LH, and HH. The LL sub-band holds the approximate coefficients (averages)of the image, the HL holds the horizontal details, the LH holds the vertical details, and the HH holds the diagonal details of transform image. Figure 51.1 shows the different processing levels of DWT. At the level 1 decomposition, the LL sub band is at the top leftpart of the wavelet sub band which contains the mean value of the image. The most important parts of the image lie in the LH, HL and HH sub bands. The information relating to edges is contained in them. As edges are high frequency components, any changes in these edges can be easily identified [1]. Hence only the LL (approximate sub band) can be used to hide the information. Multi-level DWT can be used to increase the security. But at the same time, the capacity of information that can be hidden reduces as the order increases.

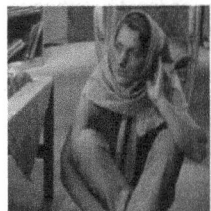 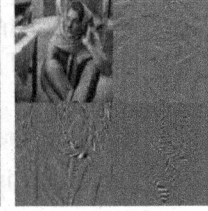 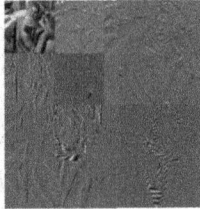

(a) Original image   (b) level 1 DWT   (c) level 2 DWT

Fig. 51.1 Different levels of DWT

2.4 Integer Wavelet Transform (IWT) Technique

The major short coming of DWT transform is the use of waveletalters that have floating point coefficients. Data being hidden in their coefficients, any truncations of the floating-point values of the pixels may cause the loss of the hidden information which may lead to the failure of perfect recovery of the data. To avoid these problems of floating-point precision of the waveletalters Integer Wavelet Transform ensures that there will be no loss of information through forward and inverse transform. IWT coefficients are integers.

The IWT transform also divides the entire image into sub bands LL, LH, HL, HH. Lifting scheme can be used to perform the IWT. IWT coefficients of an image (M x N):

$$LL = [(I)/2]$$

$$HL_{i,j} = [(I_{2i,2j+1} - I_{2i,2j})]$$

$$LH_{i,j} = [(I_{2i+1,2j} - I_{2i,2j})] HH_{i,j} = [(I_{2i+1,2j+1} - I_{2i,2j})]$$

Inverse IWT coefficients of the image:

$$I_{2i,2j} = LL_{i,j} - floor(HL_{i,j}/2)$$

$$I_{2i,2j+1} = I_{2i+1,2j} + floor(HL_{i,j+1}/2)$$

$$I_{2i+1,2j} = I_{2i,2j+1} + (LH_{i,j}) - (HL_{i,j}) \, I_{2i+1,2j+1}$$
$$= I_{2i+1,2j} + (HH_{i,j}) - (LH_{i,j})$$

Where $1 <= i <= M/2$ and $1 <= j <= N/2$

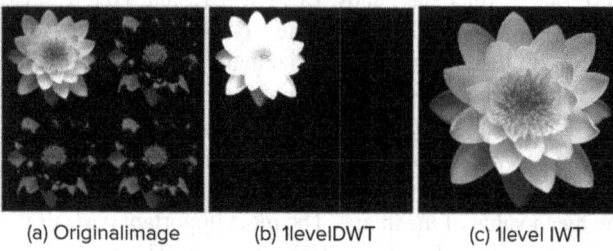

(a) Originalimage   (b) 1levelDWT   (c) 1level IWT

Fig. 51.2 Comparison of LL sub-band of DWT and IWT

2.5 Hybrid DWT-DCT Technique

Discrete Cosine Transform is widely used in JPEG compression. These algorithms are strong in opposition to easy image processing operations like adjustment, blurring, brightness, compare and low pass filtering.

But it is computationally expensive. In order to enhance the security and increase robustness of transform domain techniques, a hybrid technique of DWT along with DCT is used.

Input: Cover image and secret messageoutput:Stego image

Step 1:Apply 1 level DWT to the cover image

Step 2:Apply DCT to the LLband

Step 3:Embed the secret message

Step 4:ApplytheinverseDCT

Step 5: Apply the inverse DWT to obtain stego image.

Input: stego image

Output: retrieved message

Step1: Apply1 level DWT to the stego image

Step2: Apply DCT to theLLband

Step3: Extract the secret message

3. Problem Definition

The Information sent or received through the internet requires a high level of security. Implementing the existing Techniques like LSB, DWT, IWT for hiding text inside images and a hybrid Technique DCT. DWT.Their performances are evaluated using parameters like BER, PSNR and NCC

4. Parameters

4.1 Mean Square Error (MSE)

Mean Square Error is calculated by the square of difference between the original image pixels and stego image pixels.

$$MSE = \frac{\sum\limits_{i=0}^{M-1}\sum\limits_{j=0}^{N-1}(I_{s(i,j)} - I_{c(i,j)})^2}{MN}$$

4.2 Peak Signal to Noise Ratio (PSNR)

PSNR is the ratio of maximum intensity in the image to the standard deviation of the noise.

$$PSNR = 10\log_{10}\left(\frac{c_{max}^2}{MSE}\right)$$

4.3 Normalized Cross Correlation (NCC)

NCC is a metric used to evaluate the degree of similarity between two images. Normalization is done to limit the value obtained from cross correlation of stego and original cover image Where $I_{c(i,j)}$ is the cover image pixel and $I_{s(i,j)}$ is the stego image pixel, M,N are the width and height of the cover image

$$NCC = \frac{\sum\limits_{i=0}^{M-1}\sum\limits_{j=0}^{N-1}I_{c(i,j)}I_{s(i,j)}}{\sum\limits_{i=0}^{M-1}\sum\limits_{j=0}^{N-1}I_{c(i,j)}^2}$$

5. Simulation Results

5.1 LSB Technique

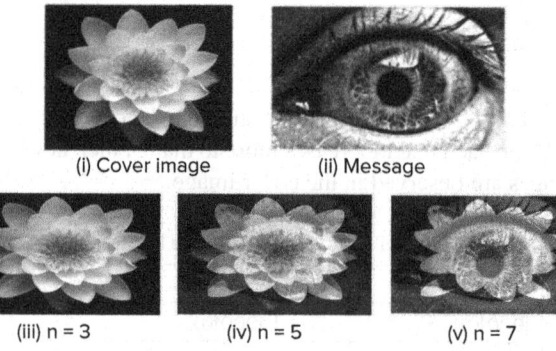

(i) Cover image   (ii) Message

(iii) n = 3   (iv) n = 5   (v) n = 7

Fig. 51.3 LSB stego-image for different number of bits in cover pixel being replaced by message bits

Table 51.2 DWT and IWT

Cover Image	Stego Image (DWT)	Stego Image (IWT)
(a) Lady		

Cover Image	Stego Image (DWT)	Stego Image (IWT)
(b) Lena 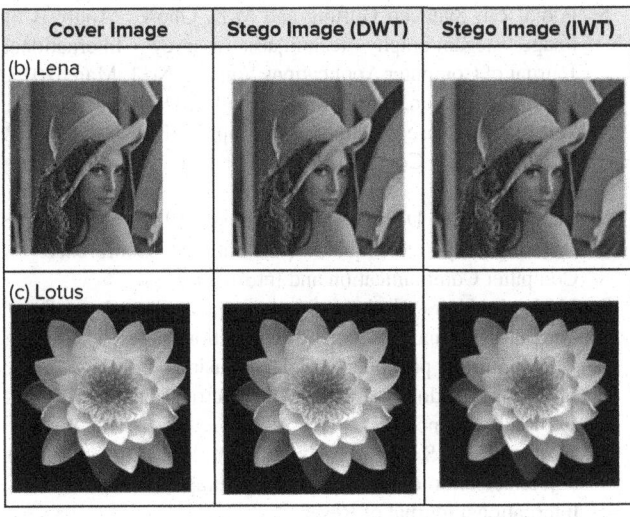		
(c) Lotus		

5.2 Hybrid DWT-DCT

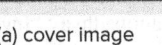

(a) cover image	(b) stego image

Fig. 51.4 Image Steganography using DWT-DCT

5.3 Comparison

The below Table 51.3 is a comparative table using all techniques for different message bits.

(i) Mean-square error (MSE)

Table 51.3 Comparison of MSE for lotus

Length of msg (bits)	LSB	DWT	IWT	DWT-DCT
5	1.904e-0 5	0.0056	7.616e-05	2.5012e-05
10	3.808e-05	0.0115	1.523e-04	3.656e -05
15	5.7124e-05	0.0173	2.248e-04	4.4161e-05
20	7.6165e-05	0.0232	3.046e-04	5.517e-05
25	9.5206e-05	0.0290	3.808e-04	7.1345e-05
30	1.4125e-04	0.0348	4.569e-04	8.2687e-05
35	1.3329e-04	0.0355	4.956e-04	8.907e-05
40	1.5233e-04	0.0370	5.324e-04	1.008e-04

(ii) Peak signal-to-noise ratio (PSNR)

Table 51.4 Comparison of PSNR for lotus

Length of Msg (bits)	LSB	DWT	IWT	DWT-DCT
5	95.3676	70.6829	89.3475	94.1833
10	92.3576	67.5578	86.3369	92.5347
15	90.5966	65.7843	84.5761	91.7144
20	89.3472	64.55099	83.3266	90.7477
25	88.3782	63.5408	82.3576	89.6312
30	87.5862	62.7490	81.5657	88.9904
35	86.9168	62.6625	81.2130	88.6671
40	86.3369	62.4828	80.9021	88.1613

(iii) Normalized cross correlation

Table 51.5 Comparison of NCC

Length of Msg	LSB	DWT	IWT	DWT-DCT
5	1	1	1	0.9989
10	1	1	1	0.9989
15	1	1	1	0.9989
20	1	1	1	0.9989
25	1	1	1	0.9989
30	1	1	1	0.9989
35	1	1	1	0.9989
40	1	1	1	0.9989

6. Result

The below graphs shown are the comparison of MSE measurement for different data sequence using different techniques.

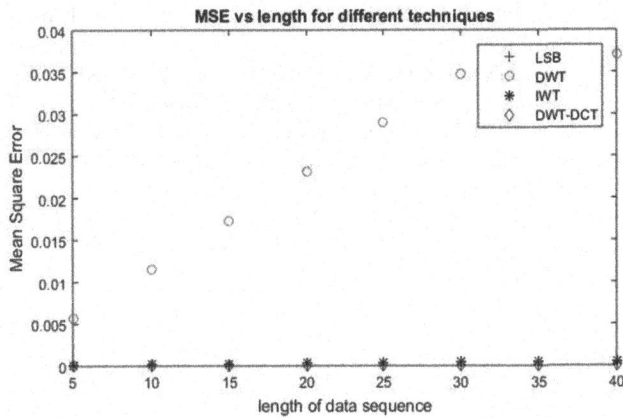

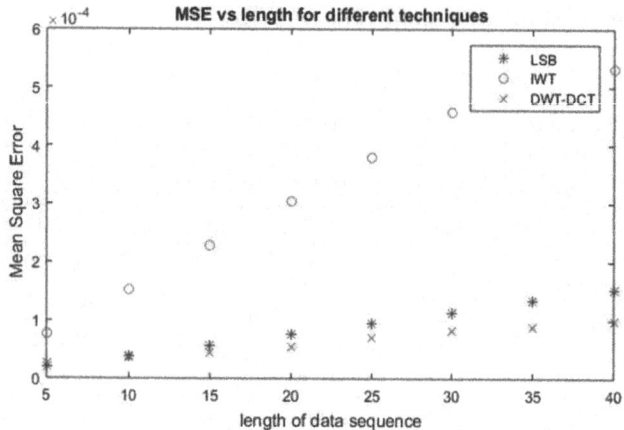

Hybrid algorithms increase payload capacity with acceptable PSNR and MSE values. Use of Dct-Dwt ensures better results. It ensures a higher level of security along with reasonable embedding capacity and high correlation.

7. Future Scope

Analyze the effect of various noises on Steganography techniques and analyse their performance over AWGN channel. Study the effect of Gaussian noise on data being transmitted.

References

1. Elshazly Emad, Abdel Wahab Safety, "A Secure Image steganography algorithm based on least significant bit and Integer wavelet transform journal of System Engineering and Electronics, Vol. 29, No. 3, June 2018.

2. Unik Lokhande, "An effective way of using LSB steganography inimages along with cryptography" International Journal of Computer Applications,Vol.88–No.12,February2014.

3. Stuti Goel, Arun Rana and Manpreet Kaur,Comparison of image steganography techniques" International Journal of Computers and Distributed Systems, Vol.No.3,IssueI,April-May2013.

4. Aayushi Verma,Rajshree Nolkha,Aishwarya Singh and Garima Jaiswal, "Implementation of Image steganography using 2 level DWT Technique" International Journal of Computer Science and Business Informatics, Vol.1,No.1.MAY2013.

5. Pooja Rai, Sandeep Gurung and M.K. Ghose, "Analysis of image steganography techniques: asurvey" International Journal of Computer Applications,Vol.114–No.1, March2015.

6. T. Kartheeswaran, V. Senthooran and T D D L Pemadasa, "Multi Agent based Audio steganography" IEEE International Conference on Computational Intelligence and Computing Research.

7. Thangadurai K, Devi G S, "An analysis of LSB based image steganography techniques" (International Conference on Computer Communication and Informatics.

8. Muthyala Veera Venkata Satyanarayana Chowdary, Dr T Venkata Ramana, "Automatic recognition of color sensation with controlled phosphene brightness using pre-trained CNNs framework" Indonesian Journal of Electrical Engineering and Computer Science.

9. Muthyala V V S Chowdary, Ch Shekar, "An Efficient salient object detection of video with spatiotemporal deep features" International journal of Research.

10. Muthyala V V S Chowdary, Ch Shekar, "Enhancing methodology for Video resolution for 3D TV" International journal of Research.

11. Jan Odstrcilik, Radim Kolar, Jiri Jan, Jiri Gazarek, Zdenek Kuna, Martina Vodakova, "Analysis of retinal nerve fiber layer via Markov random fields in color fundus images",2012 19th International Conference on Systems,Signals and Image Processing (IWSSIP),Year: 2012.

12. N.V. Sibirev, Y.S. Berdnikov, V.N. Sibirev, V.G. Dubrovskii, "Stabilization of wurtzite crystal phase in arsenidenanowires via elastic stress", 2020 International Conference Laser Optics (ICLO), Year: 2020.

13. Sandip Bhattacharya, Rajib Saha, Subhrajit Sikdar, Subrata Mandal,Chirantan Das, Sanatan Chattopadhyay, "Investigation of density and alignment of ZnO- Nano wires grown by double-step chemical bath deposition (CBD/CBD) technique on metallic, insulating and semiconducting substrates", 2020 International Symposium on Devices, Circuits and Systems (ISDCS), Year: 2020.

14. Xu Shoulong, Zou Shuliang, Huang Youjun, "γ-Ray Detection Using Commercial Off-the-Shelf CMOS and CCD Image Sensors", IEEE Sensors

15. Yumei Zhou, Jun Luo, Juqi Hu, Hengyu Li, Shaorong Xie, "Bionic eye system based on fuzzy adaptive PID control", 2012 IEEE International Conference on Robotics and Biomimetics (ROBIO), Year: 2012.

Note: Authors' self data and the graphs/images are part of the research work.

Molecular Interactions in Binary Liquid Mixtures

52

Nagarjuna A.[1]

PDF Scholar, Department of Physics,
Srinivas University, Mangalore, Karnataka-574146, India

Sunitha G.[2]

Department of Physics, T K R Engineering College (Autonomous),
Medbowli, Meerpet, Hyderabad, Telangana, India

VSN Raju K.[3]

Department of Physics, Adikavi Nannaya University, Srinivas University,
Tadepalligudem, West godavari, Andhra Pradesh, India

Praveen. B. M.[4]

Director, Research and Innovation Council,
Srinivas University, Mangalore, Karnataka, India

Abstract: Using the accepted methods at 303.15, 308.15, and 313.15K, viscosity, density, and speed of sound of binary mixture of ethyl-4-hydroxybenzoic acid with various mole fractions of aniline (AN), o-chloroaniline (OCA), and o-toluidine (OT) were measured. Many thermo-acoustic characteristics, including compression that is isentropic (K_s^E), inter-molecule free length (L_f^E), molar volume (V_m^E), acoustic impedance (Z^E), and their excess properties, were derived using these observe values by using least squares approach, the obtained values have fitted to a polynomial equation of Redlich-Kister type to predict, binary co-efficient and normal deviance. Further examine hydrogen bonding creation in terms of inter-actions between dissimilar substances of current binary system, FT-IR spectroscopic studies of these mixes were also conducted.

Keywords: Binary liquid mixtures, Redlich-Kister polynomial, Ethyl-4-hydroxy benzoic acid FTIR spectroscopy and standard techniques

1. Introduction

Understanding the changes in thermodynamic behavior in a given mixture requires a study of Physico-chemical characteristics of fluid mixtures. One of the most significant uses of these features is their potential to reveal details of system structure effects and intermolecular interactions. Anilines, also known as amino benzene or phenyl amine, are organic compounds that belong to the category of substances in organic chemistry. These substances are classified as aromatic amines and are thought to be poisonous in nature. An oily liquid with a musty and fishy smell, aniline is yellowish and somewhat brownish in color. It has the stench of rotting fish. On the other hand Ethyl-4-hydroxy-benzoate,($C_9H_{10}O_3$) is ethyl ester of p-hydroxy benzoic acid, or Ethyl parable's. It is also utilized as an antifungal preservative and has a high melting point ranging from 115^0-118 °C. The antibacterial properties of ethyl-4-hydroxy benzoate are used extensively in the food, cosmetic, and textile industries. It serves as an ingredient in cleaners, hair dyes, beverages, lotions, shampoos, and other products.

[1]dranagarjunaphysics@tkrec.ac.in, [2]gsunithahs@tkrec.ac.in, [3]dr.kvsnr@gmail.com, [4]researchdirector@srinivasuniversity.edu.in

The excess characteristics were calculate by using observe principles of viscosity (η), density (ρ),& sound speed (U), and these values were then got to simplified Redlich-Kister (R-K) equations using Legendre polynomials. This result was then interpreted in molecular inter-actions and impacts on structures. The non-bonding interactions like hydrogen bonding and additional forces present, such as charge transfer, dispersive force, and dipole-dipole interactions that happen in the investigated liquid mixes are explained in this work. Similar sorts of findings have been published by earlier researchers[1-4] according to a quick review of the literature, however in this study, a thorough and methodical approach is being used to learn more about the molecular behavior of liquid mixes.

2. Experimental

Just before the experiment began, the binary liquid mixes were created using mass variation for several samples. The final mole fraction's degree of uncertainty was calculated to be less than 0.0001 percent. Table 52.1 displays the purity analysis of pure liquids.

Measurements of the speed of sound ranged from 303.15 K to 313.15K by using ultrasonic inter-ferometer (M/s Mittal Enterprises, India) operating at a fixed frequency of 2 MHz with an accuracy of ±0.1m/s. Regarding binary systems, E-4-HB with aniline, O-Chloroaniline, O-toluidine. The unknown for speed of sound was ±0.7 m s^{-1}.

The density of pure component and its multiple components was determined with an accuracy of ± 0.02mg using 10ml specific gravity bottles on a high-precision electronic digital balance (Baijnath Premnath SF 400A, Kanpur, USA). Viscosity was measure with an Viscometer by Ostwald. Various temperatures were used to calibrate the viscometer, by using double-distiller water. The experimentally measured data on density, viscosity and sound velocity are compared with research values and are shown in Table 52.2.

3. Results and Discussions

The excess possessions of research binary system E-4-HB containing aniline, O-chloro aniline and O-toluidine at, T = 303.15 K, 308.15 K and 313.15K was calculate by using an excel spread sheet and are shown in Figs 52.1–52.3.[9]

Table 52.1 Analysis

Comp.	CAS number	Source	First-order mole fraction	Mole fraction towards the end	Analysis
Ethyl-4-Hydroxy benzoate	120-47-8	Hi Media Lab Pvt. Ltd. Mumbai, India	0.992	0.995	Vacuum drying
Aniline	62-53-3		0.993	0.997	
O-Chloroaniline	95-51-2	Sigma Aldrich, India	0.995	0.997	Distillation in gas and liquid
O-toluidine	95-53-4		0.993	0.996	

Table 52.2 Comparison of measurements for pure fluid densities, viscosities, and sound speeds with published values.

Liq.	Temp T (K)	Density (ρ)kg m^{-3}		Viscosity (η) mPaS		Speed of sound (U) m.s^{-1}		C_P J K^{-1} mol^{-1}
		lit	Expt	lit	Expt	lit	Expt	
E-4-HB [a]	303.15	1.1035	1.1038	1.283	1.291	1523.0	1527.4	304.03
	308.15	1.0313	1.0318	1.198	1.192	1541.4	1543.6	317.64
	313.15	0.9973	0.9982	1.175	1.171	1572.6	1578.2	338.27
Aniline [b]	303.15	0.8019	0.8026	2.224	2.236	1228.2	1223.4	191.01
	308.15	0.7982	0.7988	1.958	1.941	1217.8	1212.6	192.05
	313.15	0.7946	0.7861	1.700	1.722	1198.8	1194.2	193.70
O-Chloro Aniline [c]	303.15	1.2026	1.2022	3.825	3.812	1469.6	1462.4	218.41^e
	308.15	1.1980	1.1968	3.412	3.477	1453.3	1459.2	222.37^e
	313.15	1.1930	1.1944	3.060	3.103	1435.2	1436.8	231.82^e
O-toluidine [d]	303.15	0.9900	0.9932	4.433	4.466	1572.2	1576.4	211.30^e
	308.15	0.9859	0.9875	5.666	5.681	1553.2	1558.2	213.41^e
	313.15	-----	0.9743	-----	6.124	-----	1532.6	218.66^e

[a] Reference[5] [b] Reference[6] [c] Reference[7] [d] Reference[8] ^{e}Estimated from NIST web book

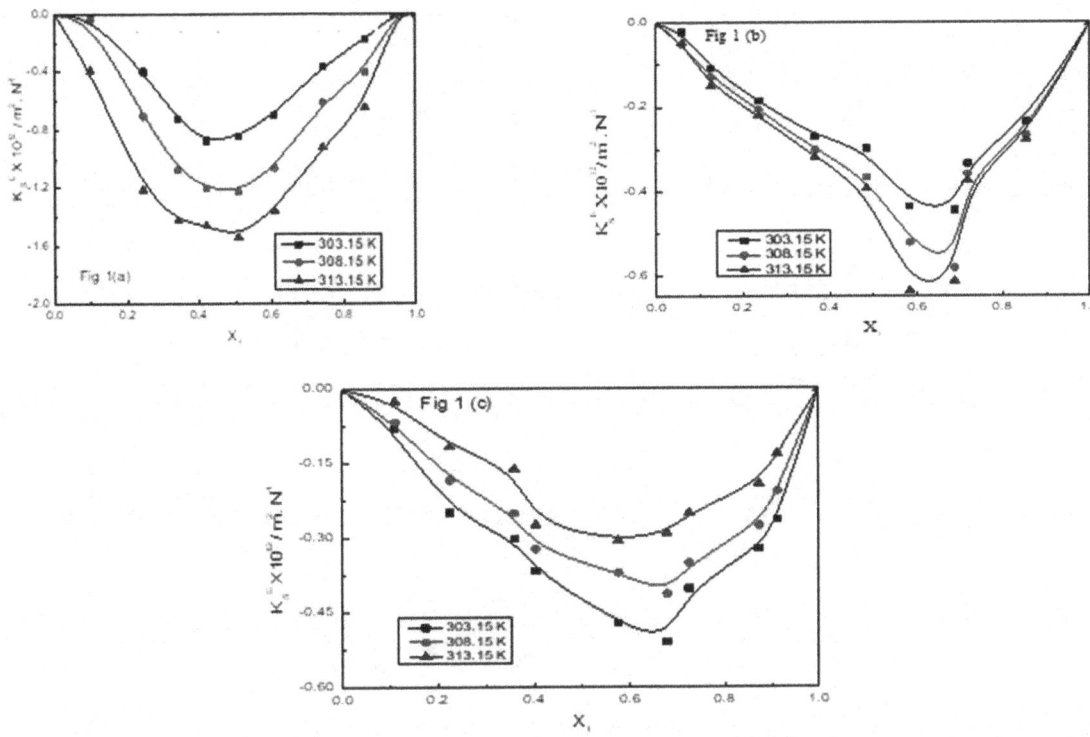

Fig. 52.1 Excess isentropic compressibility with respect to mole fraction of E-4-HB + Aniline/OCA/OTatthree different temperatures

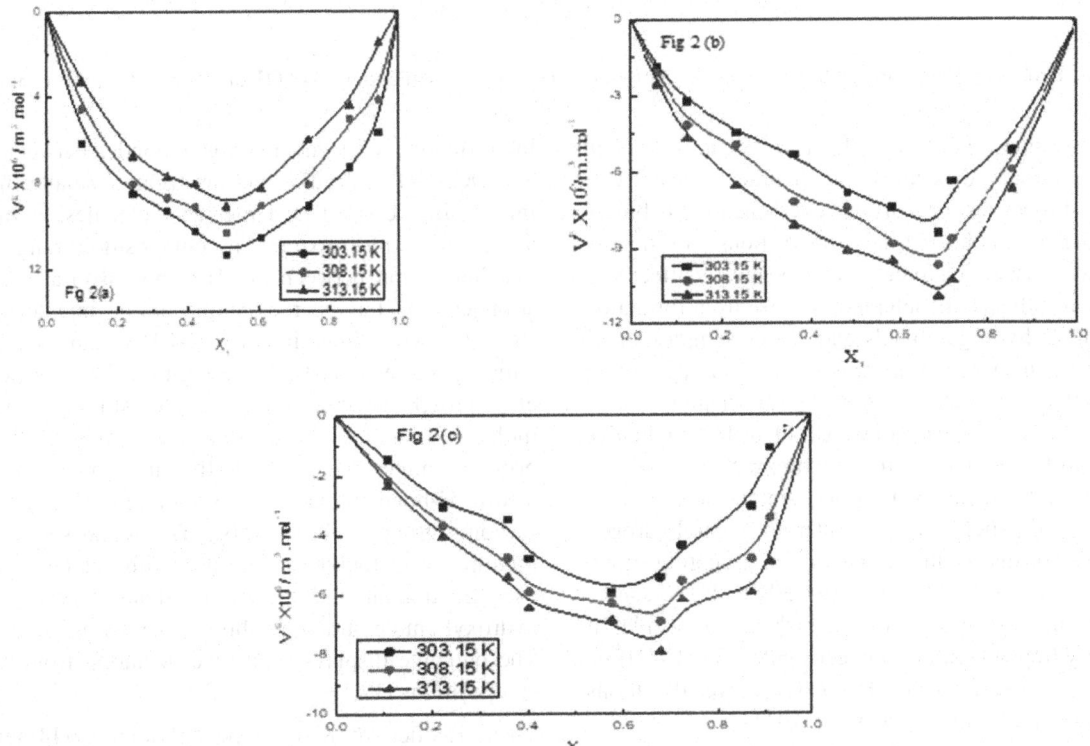

Fig. 52.2 Excess molar volume changes depending on mole fraction of E-4-HB + Aniline/OCA /OT at three various temperatures

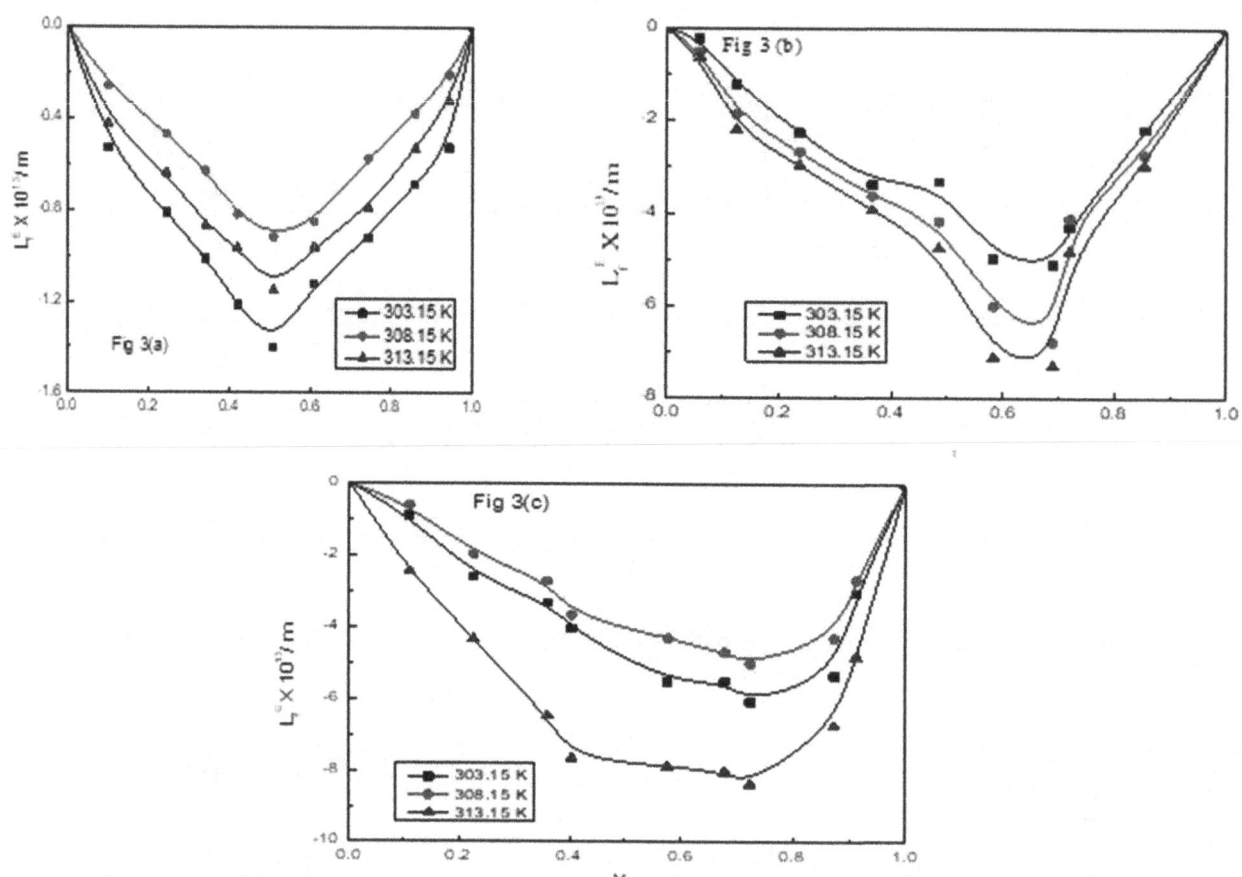

Fig. 52.3 Extra free length in relation to mole fraction varies of E-4-HB + Aniline/ OCA / OT at three different temperatures

The observed values of K_s^E, V_m^E, L_f^E and Z^E can be explain by different effects depending on chemical, physical,& geometrical properties of current components of binary system under research. Physical interactions are mainly concerned with dispersion forces, which positively affect V_m^E and L_f^E. Chemical/ specific interactions arise from formation and breaking of hydrogen bonds and anther complex inter-actions, ending in negative upshots of L_f^E and K_s^E due to geometric effects that allow molecules of smaller size to enter the structure of other molecules. Plausible qualitative interpretations for behavior of this mixture with composition have proposed. As mentioned earlier, amine molecule has an extensive intra molecular and intermolecular hydrogen bonding network due to the strong proton accepting group C=O in the molecule. Addition of E-4-HB with OT causes slow hydrogen bond dissociation in hydroxyl molecule, at same time, hydrogen bonding between Aniline and E-4HB is also possible because it has three H-bond moreover, donations to the C=O group of proton acceptors. [10, 11]

The positive V_m^E values over entire blend composition range may be due to disruption of predominant hydrogen structures

by formation of weak hydrogen bonds, between E-4-HB and OCA/OT molecules and insertion of smaller molecules into cavities created by larger an molecules, Z values that are positive for mixtures in composition range indicate that interaction between E-4-HB and substituted amines is stronger than the ester or amine–amine interactions. How strongly two things interact E-4-HB and was replaced amino particles appears to be highly reliant on position of ethyl groups in amino molecules. Substituted amine, when molecules are pure, they have a smaller more H-bonds than primary amines. Mixing E-4-HB with replaced amines can lead to H-bond formation between amide H-atom and ester carbonyl group, and vice versa. The sound speed of amine mixtures was found to decrease with increasing E-4-HB mole fraction and that of substituted amines with increasing hydroxyl mole fraction three contrasting temperatures. The outcome displays a positive deviation from Z^E for all compositions [12,13]

Then, Redlich & Kister type polynomial [14]anon-linear regression method and values of co-efficient, aired determine by least quires and are reported in Table 52.3. The out comes of Redlich & Kister, concluding that redundancy quantities $(K_s^E, V_m^E \text{ and } L_f^E)$ give a general view of springs that are not ideal in mixtures because of the strong correlations but can still be quite difficult to capture specially large systems with overflows when (σ) is smaller.

4. FT-IR Spectra

FT-IR data of fewer cleaners(Figs 52.4-52.7) and their mix with other compositions were performed in the wave number range of 500-3500 cm⁻¹. A partial assignment of solvent mixtures and pure solvent bands has been performed and discuses accordingly. The main absorption bands of esters, are due to C=O (extending) & N-H (extending) motions. The C=O shift (extending) frequency of various amines in E-4-HB also results from the breaking of intra- and intermolecular hydrogen bonds in amides and their formation between different molecules in liquid mixtures. The single-shot FT-IR spectra of this three blends with different compositions show that there are some differences in peaks that include

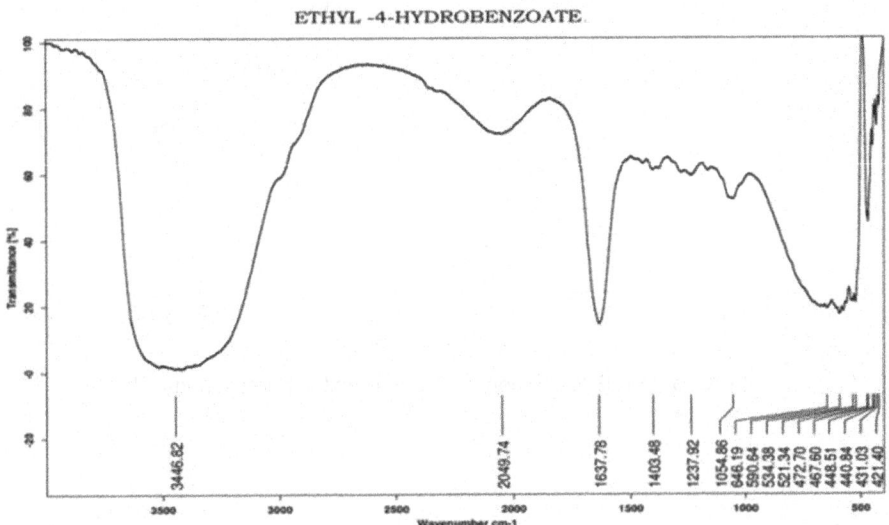

Fig. 52.4   FT-IR analysis of Ethyl-4-hydroxybenzoate pure component

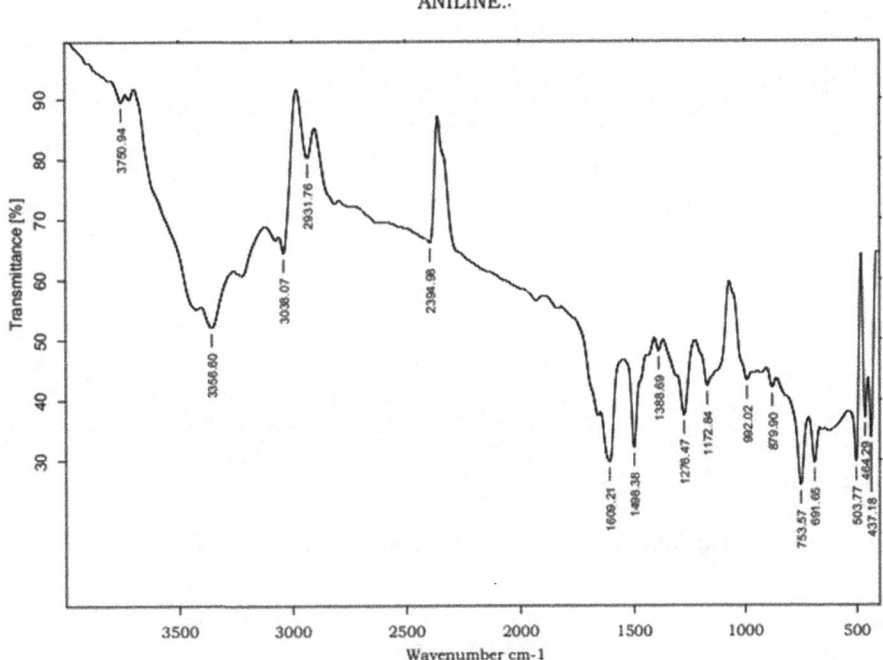

Fig. 52.5   FTIR analysis of aniline pure component

O-CHLORO ANILINE.0

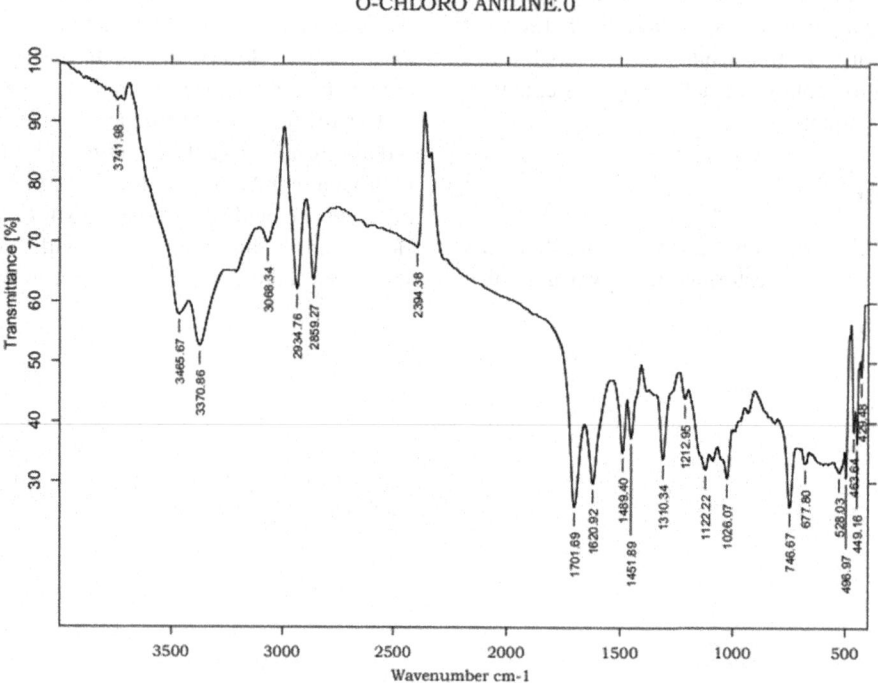

Fig. 52.6 FT-IR evaluation of O-Chloroanilinepurely chemical

O-TOLUIDENE

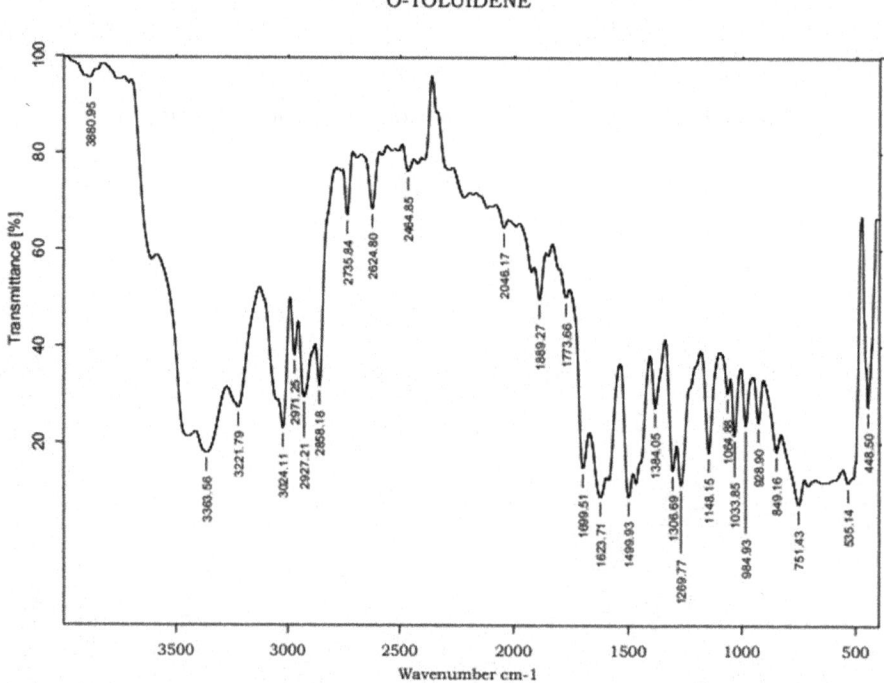

Fig. 52.7 FT-IR analysis of O-Toluidinepurely chemical

Table 52.3 Parameters (A k) and typical Deviation (σ) for E-4-HB with aniline, O-CA and OT at T= 303.15 K, 308.15K and 313.15K

Parameter	Temp.(K)	A0	A1	A2	σ
E-4-HB + Aniline					
$10^{12}K_s^E/$ $(m^2.N^{-1})$	303.15	2.3704	-1.3748	-0.9249	0.2007
	308.15	2.0288	-1.0836	-0.7129	0.3504
	313.15	1.3350	-0.9372	-0.4082	0.5701
$10^6 V^E/$ $(m^3.mol^{-1})$	303.15	17.3747	6.8536	-3.2657	0.6440
	308.15	19.9782	9.2488	-7.0781	0.3010
	313.15	23.4971	14.1826	-9.1038	0.1494
$10^{13}L_f^E/(m)$	303.15	7.1156	-5.8063	-1.9931	0.1275
	308.15	9.1964	-8.6581	-4.5350	0.4114
	313.15	12.5114	-10.3484	-7.1696	0.8385
E-4-HB+O-Chloroaniline					
$10^{12}K_s^E/$ $(m^2.N^{-1})$	303.15	1.4614	-0.8296	-0.1058	0.2518
	308.15	1.7546	-1.0317	-0.1870	0.3048
	313.15	1.9477	-1.1420	-0.2124	0.5396
$10^6 V^E/$ $(m^3.mol^{-1})$	303.15	18.2028	-9.0543	-6.9865	0.3216
	308.15	25.7451	-13.3676	-9.9245	0.6305
	313.15	36.9254	-17.5164	-13.8109	0.9648
$10^{13}L_f^E/(m)$	303.15	7.3817	-1.7284	-2.0534	0.2144
	308.15	10.5263	-5.0221	-3.4669	0.5187
	313.15	13.2294	-8.5134	-5.3038	0.9614
E-4-HB +O-Toluidine					
$10^{12}K_s^E/$ $(m^2.N^{-1})$	303.15	1.6911	-1.2104	-0.5379	0.3161
	308.15	1.1416	-1.0316	-0.3752	0.2664
	313.15	1.1011	-0.8025	0.2283	0.1979
$10^6 V^E/(m^3.$ $mol^{-1})$	303.15	11.5881	-6.0457	-5.6157	0.4402
	308.15	19.9257	-9.9456	-11.7846	0.7486
	313.15	25.6942	-12.9867	-20.8105	0.9134
$10^{13}L_f^E/(m)$	303.15	4.5685	-9.6651	-1.9716	0.5762
	308.15	7.6253	-11.8947	-4.7564	0.7234
	313.15	10.7254	-14.2574	-6.1755	0.9166

components, specifically 30% E-4-HB with 70% aniline and two more compositions for E-4-HB+OCA/OT blends, but there is no difference in compositional peaks of other two blends. This is due to intra-inter molecular hydrogen-bond dissociation of amine molecules or its formation between molecules. [15, 16]

5. Conclusions

First, the experimental the sound speed, densities, and binary mixture viscosities of E-4-HB with amines at 303.15 K, 308.15 K, and 313.15 K was determined. In addition, thermo acoustic variables of various types were determined by comparing gathered info by using standard relations taken from research. The excess molar volume values are discovered to be positive and low with relation to all binary systems across compositional range at all temperatures. The extra attributes K_s^E and L_f^E are negative for mixture binary over entire range of moles per mole at three degrees. The outcomes are explicable by breakdown of intra- intermolecular H-bonds of amines and its creation between different components of liquid mixtures. In addition, certain interactions between dipoles may exist in all blends, particularly between OCA and OT, which is confirmed by the higher value of acoustic impedance of the thermo-acoustic function. In addition, the value of the redundant The Redlich & Kister polynomial equation is fitted using parameters, and the results are shown. Analysis of the pure solvent and its mixes using FT-IR, assigning the main ester and amine bands, is also reported. The shift of main bands supports both the breaking and formation of new H-bonds between the components of mix.

6. Acknowledgement

Authors Dr. Nagarjuna. A, is thankful to his PDF supervisor Dr. B. M. Praveen, Director, Research and innovation Cell, Srinivas University, Mangalore for his valuable suggestions and discussions

References

1. S.S. Sastry, S.M. Ibrahim, L.T. Kumar, S. Babu, and H.S. Tiong, "Excess thermodynamic and acoustic properties for equimolar mixture of ethyl benzoate and 1-alkanols with benzene at 303.15 K", Int. J. Eng. Res. Technol, 4, pp. 315–324, 2015.
2. S.S. Sastry, S. Babu , T. Vishwam,and H.S Tiong, "Excess thermodynamic and acoustic properties for the binary mixtures of Methyl Benzoate at T = (303, 308, 313, 318 and 323) K", Phys. Chem. Liq(Taylor &Francis), 52,pp. 272–286,2014. doi:10.1080/00319104.2013.820302.
3. A.N. Kannapan, and V. Rajendran, "Excess transport properties of the binary mixtures at different temperatures", Indian. J. Phys. 68, pp. 131–135, 1994.
4. B. Garcia, R. Alcalde, and J.M. Leal, " Excess properties for binary liquid mixtures of propionic acid with aniline derivatives", Can. J. Chem, 69, pp. 369–372, 1991. doi:10.1139/v91-056.
5. A. Nagarjuna, and Shaik. Babu,"Spectroscopic and volumetric study of binary liquid mixtures containing Ethyl –4-hydroxy benzoate and alkanols" Chem. data collections, 26, pp. 100347-100357, 2020 https://doi.org/10.1016/j.cdc.2020.100347
6. A. Nagarjuna, Shaik Babu,and M. Gowri sankar," Thermo-physical and spectroscopic studies of liquid mixtures

containing p-methoxy benzoic acid"Int. J. Amb. Energy (Taylor &Francis), 43(1), pp. 1974–1985, 2020

7. M. Chandra Sekhar, A. Venkatesulu, M. Gowrisankar, and T.Srinivasa Krishna," Thermo-dynamic study of interactions in binary liquid mixtures of 2-Chloroaniline with some carboxylic acids",Phys. Chem. Liq (Taylor & Francis), 55, pp.196-217, 2017

8. Neeti Saini, J.S. Yadav, Sunil K. Jangra, Dimple Sharma, and V. K. Sharma, "Thermodynamic studies of molecular interactions in mixtures of o-toulidine with pyridine and picolines: Excess molar volumes, excess molar enthalpies, and excess is entropic compressibilities," J. Chem. Therm, 43, pp. 782–795, 2011

9. S. Suriya Shihab, K. G Rao, M. Gnana Kiran, S. Babu, and S.S Sastry, "Excess thermodynamic and acoustic properties for equimolar mixture of methyl benzoate and alkanols with benzene at 303.15 K", Rasayan J. Chem, 10, pp. 59–63, 2017 doi:10.7324/RJC.2017.1011552.

10. R.C Reid, J.M Prausnitz, and B.E Poling, "The Properties of Gases and Liquids", 4th ed., (McGraw Hill book company, New York) 1987.

11. P. Venkateswara Rao, L. Venkatramana, and M. Gowrisankar, "Volumetric, acoustic and spectroscopic properties of 3-chloroaniline with substituted ethanols at various temperature", J. Chem. Therm, 94, pp. 186–196, 2016.

12. D.S. Kumar,and D.K. Rao, "Study of molecular interactions and ultrasonic velocity in mixtures of some alkanols with aqueous propylene glycol", Indian J. Pure Appl. Phys, 45 pp.210–220,2007

13. C. Narasimharao, S. Ramnaiah, A. Chowdappa, and P. Venkateswarlu," Thermodynamic and transport properties of binary liquid mixtures of cyclohexanol with isomeric chlorotoluenes", Phys. chem. liqs (Taylor & Francis), 68, pp. 264–274, 2014

14. O. Redlich and A. T Kister, Algebraic representation of thermodynamic properties and the classification of solutions, "Ind Eng Chem". 40, pp. 345–348, 1948.

15. Awasthi, M. Rastogi, and J.P. Shukla, "Ultrasonic and IR study of molecular association process through hydrogen bonding in ternary liquid mixtures", Fluid Phase Equilib. 215, pp. 119–127, 2014 doi: 10.1016/j.fluid.2003.08.017.

16. T. Karunakar, and C. H. Srinivasu, "Thermo acoustic and infrared study of molecular interactions in binary mixture aniline+ 1-butanol", Res. Rev. Pure Appl. Phys. 1, pp. 5–10, 2012.

Note: Authors' self data and the graphs/images are part of the research work.

S. P. V. Subba Rao[1]

HOD ECE Dept, Sreenidhi Institute of Science and Technology,
Hyderabad, India

Ramaswamy T.[2]

Associate Professor-ECE, Sreenidhi Institute of Science and Technology,
Hyderabad, India.

T. Supriya[3]

ECE Dept, Sreenidhi Institute of Science and Technology,
Hyderabad, India

Abstract: To address this shortcoming, this paper first presents a new modeling strategy to generate scale- free network topologies, which considers the constraints in WSNs, such as the communication range and the threshold on the maximum node degree. Then, Rose a novel robustness enhancing algorithm for scale-free WSNs, is proposed. The extensive experimental results verify that our new modeling strategy indeed generates scale-free network topologies for WSNs,and ROSE can significantly improve the robustness of the network topologies generated by our modeling strategy. Moreover, we compare with two existing robustness enhancing algorithm, showing that ROSE outperforms.

Keywords: Scale-free IoT networks, Centrally measures, Rewiring, Malicious attacks, Networks optimization, Random attacks, Robustness

1. Introduction

An important sort of network for sensing the environment and gathering data is a wireless sensor network (WSN). A vast number of sensor nodes, such as sink nodes and sensor nodes, are installed in different locations in order to perceive environmental characteristics. These nodes come together to form a multi-hop ad hoc network system, which carries out given tasks in accordance with application specifications. WSNs have been installed in residences, structures, mountains, woods, etc. The creation of multiple routing and communication protocols for networks are essential for network properties like network lifetime, energy consumption, reliability, and data latency, is based on the sensor network topology, which outlines wireless communication between various sensor nodes in WSNs. Increasing the resilience of the WSNs has become a crucial topic in recent years due to the rise in cyber attacks. The resilience of the network topologies for WSNs was the main goal of this study. In other words, our goal was to develop a technique that would allow the connection of as many nodes as possible to be maintained in the event of certain node failures brought on by cyber attacks. There are two different sorts of assaults in terms of target choice: malevolent and arbitrary. In random assaults, the target nodes are chosen at random from the network topology, but in malevolent attacks, the target nodes are those with a high node degree. It is well known that some network topologies are resilient to intentional assaults, whereas others are resilient to random attacks.

---

[1]spvsubbarao@sreenidhi.edu.in, [2]ramaswamyt@sreenidhi.edu.in, [3]supriyavk2018@gmail.com

2. Ease of Use

(i) Towards prolonged lifetime for deployed WSNs in outdoor environment monitoring

For unattended Outdoor Environment Monitoring (OEM) applications, Wireless Sensor Networks (WSNs) have recently become a potent and affordable alternative. These applications provide a number of difficulties for WSN deployment, such as 3-D (3-D) locations, difficult operational circumstances, and constrained energy resources. We suggest using Relay Nodes (RNs) in addition to Sensor Nodes (SNs) in a distributed way to extend the lifespan of the deployed WSN while reducing the consequences of these difficulties. While RNs aid in travelling to far-off places, SNs can save their meagre energy supplies for sensing and data collection. Additionally, Mobile RNs (MRNs), a group of RNs that may be reassigned (i.e.mobilized) at any moment throughout the network lifespan, can be employed to avoid any potential link or node failure brought on by the challenging circumstances.

The deployment method suggested in this article for heterogeneous WSNs (composed of SNs, RNs, and MRNs) is based on a 3-D grid. The goal of the issue, which is to maximize network longevity while preserving predetermined levels of fault-tolerance and cost-efficiency, is to be solved using a Mixed Integer Linear Program me (MILP). Furthermore, an Upper Bound (UB) on the deployed WSN lifespan has been driven, assuming no unforeseen node/link failures. In-depth calculations based on real-world/demanding OEM trial scenarios demonstrate that the suggested grid-based deployment approach can average out the projected UB.

(ii) Energy efficient trust based routing and attack detection in WSN

Increasingly, wireless sensor networks are being seen as a looming technology due to their wide range of applications in several industries. The sensors saving proficiency limit, the range of the transportation frequency, and the evaluation proficiency all have restrictions in WSN.

They are extremely vulnerable to all forms of attacks because of the restricted assets and their disregarded environment. Active Trust, a living system, analyses black hole attacks and contributes routing proposal that depends on trust. The most crucial aspect of active trust is that it effectively creates a number of illumination channels or routes to instantly disclose and evaluate nodes trust readings and so increase the information route security. This prevents black holes from forming. The trust measurements for each node in the current system were calculated randomly, which is not a fair way to determine how important a node is to the network. On the ground, several parameters like error rate, latency, and lingering energy are accepted in the predicted proposal.

The trust management node will estimate trust readings based on these parameters, which will improve the accuracy and dependability of data packet delivery in WSN. A hashing method called SHA is used to generate hash values in order to secure the data. Other attack methods, such as DOS, Eavesdropping, Selective Forwarding, Message Modification, Grayhole, etc., are also considered in the anticipated proposal in addition to black hole assault. Working with both of the aforementioned assaults will be rewarding under the anticipated plan.

After conducting a review of the methods already in use for safe and economically viable data transmission in WSN, we discovered that the Active Trust methodology was more successful. Consequently, it adds to improving the active trust approach now in use. Based on how often the proper data is sent favorably through that node, node trust levels are assessed. As a result, the current method is more successful and precise in revealing a node's reputation in a certain network zone. Due to the fact that we are sending the data through the shortest path possible, which is determined by a number of factors like nodes' trust levels, latency, error rate, etc., our system will be more energy efficient, dependable, and secure. We have also considered the possibility of other assaults, such as denial of service attacks and black hole attacks, on WSN, and in response, we will offer a strategy for dealing with such attacks. In our upcoming research, we'll examinethe strategy for further WSN assaults.

3. Implementation

1. Degree of difference module
2. Angle sum operation module

In the scale-free network structure, the higher degree nodes are coupled and surrounded by lower degree nodes. A hierarchical structure is shown by the node degree from the centre to the boundary.

The nodes that have the same or comparable degrees are connected to one another in an onion-like structure. A high degree node has high degrees in all of its neighbors. When a node fails, its neighbors might take up its original duties and maintain the network's connectivity. In WSNs, the ability to neutralize harmful assaults is significantly reduced as a result. The scale-free network characteristic is still retained by the onion-like structure. Most nodes have modest degrees.

Commercial and opensource simulators

Some network simulators are commercial; hence they won't give free access to the source code for their programme or any packages that are associated with it. All users must pay to obtain individual packages for their unique use needs or to get the license to utilize their programme. The OPNET is one example of this. Commercial simulators have benefits and drawbacks. The benefit is that it typically has thorough and current documentation, and that company's specialized team can continuously update them. The opensource network simulator has a drawback in this regard, and generally speaking, there aren't enough qualified individuals working on the documentation. When new features are added to the various versions, it might be challenging to track down or comprehend the older programmes without the proper documentation.

Table 53.1 Network simulator types

Type	Name
Commercial	OPNET, QuaNet
Open Source	NS2, NS3, OMNeT++, SSFNet, J-Sim

Implementation and testing:

One of the most crucial project jobs is implementation, which is also the phase in which one must exercise caution because all of the efforts made will be very interactive. The most important step in creating a successful system and providing users trust that the new system is practical and efficient is implementation. Utilizing sample data, each programme is independently tested at the time of development to ensure that it links with other programmes in the manner outlined in the programme specification. The user-satisfying testing of the computer system and its surroundings.

Implementation

System design is more innovative than implementation. User education and file conversion are its main concerns. It's possible that the system needs substantial user training. As a result of programming, the system's initial settings should be changed. A straightforward working technique is offered to help the user quickly and easily comprehend the many capabilities.

The user has access to both dot matrix and inkjet printers, which may be used to print the various reports.

The recommended approach is rather easy to put into practice. The act of converting a freshly developed or updated system design into a functional one is referred to as "implementation" in general.

4. Screenshots

I'm running the sims Hill and Rose here. When harmful nodes arise, Hill will locate the MAX degree node and take no action, but ROSE will take the following node and make it the MAX degree node.

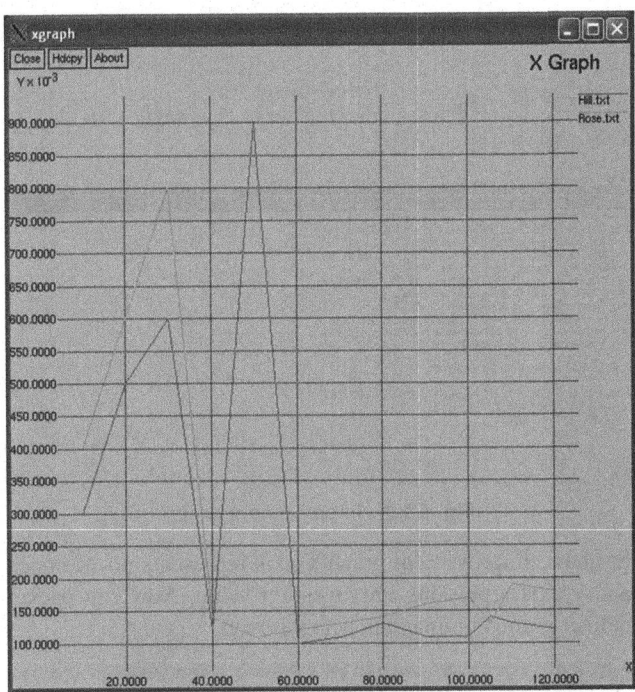

Fig. 53.1 Calculating using hill technique

Fig. 53.2 Simulation screen

In above screen we are calculating using hill technique no of max degree nodes and their neighbor coverage. Below is the simulation screen.

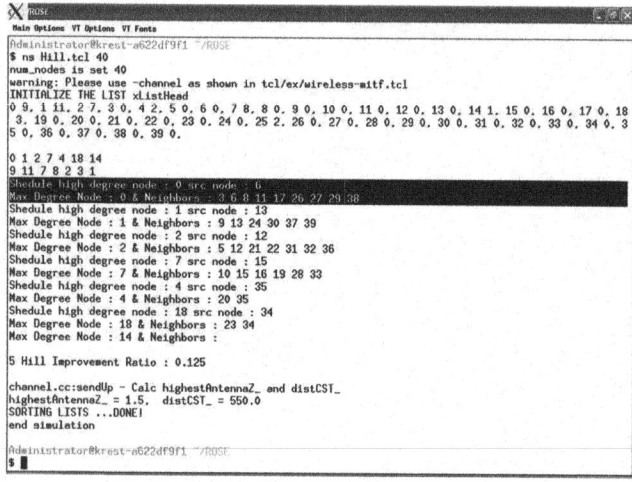

Fig. 53.3 Improvement graph

In above Improvement graph x-axis represents no of nodes and y-axis represents improvement value. Now run packet delivery ratio command in below screen

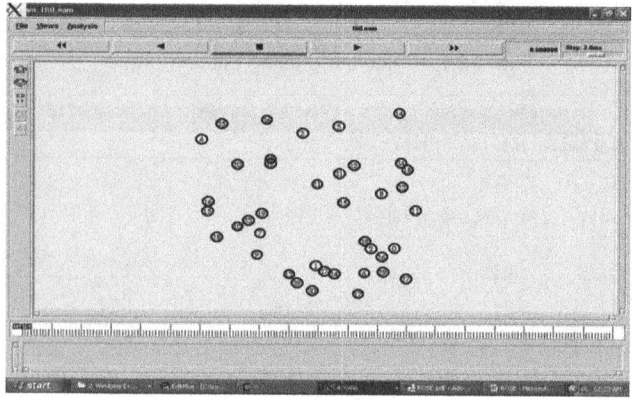

Fig. 53.4 Packet delivery ratio command

References

1. F. M. Al-Turjman, H. S. Hassanein, and M. Ibnkahla, "Towards prolonged lifetime for deployed WSNs in outdoor environment monitoring," Ad Hoc Netw., vol. 24, pp. 172–185, Jan. 2015.

2. S. Ji, R. Beyah, and Z. Cai, "Snapshot and continuous data collection in probabilistic wireless sensor networks," IEEE Trans. Mobile Comput., vol. 13, no. 3, pp. 626–637, Mar. 2014.

3. J. Long, A. Liu, M. Dong, and Z. Li, "An energy-efficient and sinklocation privacy enhanced scheme for WSNs through ring based routing," J. Parallel Distrib. Comput., vols. 81–82, pp. 47–65, Jul. 2015.

4. A. Munir, A. Gordon-Ross, and S. Ranka, "Multi-core embedded wireless sensor networks: Architecture and applications," IEEE Trans. Parallel Distrib. Syst., vol. 25, no. 6, pp. 1553–1562, Jun. 2014.

5. P. Eugster, V. Sundaram, and X. Zhang, "Debugging the Internet of Things: The case of wireless sensor networks," IEEE Softw., vol. 32, no. 1, pp. 38–49, Jan. 2015.

6. T. Qiu, R. Qiao, and D. Wu, "EABS: An event-aware backpressure scheduling scheme for emergency Internet of Things," IEEE Trans. Mobile Comput., to be published, doi: 10.1109/TMC.2017.2702670.

7. Z. Li and H. Shen, "A QoS-oriented distributed routing protocol for hybrid wireless networks," IEEE Trans. Mobile Comput., vol. 13, no. 3, pp. 693–708, Mar. 2014.

8. T. Tanizawa, G. Paul, R. Cohen, S. Havlin, and H. E. Stanley, "Optimization of network robustness to waves of targeted and random attacks," Phys. Rev. E, Stat. Phys. Plasmas Fluids Relat. Interdiscip. Top., vol. 71, no. 4, p. 047101, 2005.

9. Y. Lou and L. Zhang, "Defending transportation networks against random and targeted attacks," Transp. Res. Rec., J. Transp. Res. Board, vol. 2011, no. 2234, pp. 31–40, Dec. 2011.

10. R. Albert, H. Jeong, and A.-L. Barabasi, "Error and attack tolerance of complex networks," Nature, vol. 406, no. 6794, pp. 378–382, 2000.

11. S. Scellato, I. Leontiadis, C. Mascolo, P. Basu, and M. Zafer, "Evaluating temporal robustness of mobile networks," IEEE Trans. Mobile Comput., vol. 12, no. 1, pp. 105–117, Jan. 2013.

12. X. Huang, J. Gao, S. V. Buldyrev, S. Havlin, and H. E. Stanley, "Robustness of interdependent networks under targeted attack," Phys. Rev. E, Stat. Phys. Plasmas Fluids Relat. Interdiscip. Top., vol. 83, no. 6, p. 065101(R), 2011.

Note: All the figures and table in this chapter were authors' original test data and experimental output.

Enhanced Power Quality Management with DSTATCOM and Model Predictive Control

54

Kalagotla Chenchireddy[1]

Assistant Professor, Departmet of EEE,
Teegala Krishna Reddy Engineering College, Telangana, India

N. Rajasekhar Varma[2]

Professor, Departmet of EEE,
Teegala Krishna Reddy Engineering College, Telangana, India

Khammampati R. Sreejyothi[3]

Assistant Professor, Departmet of EEE,
Teegala Krishna Reddy Engineering College, Telangana, India

K. Santhosh[4]

Assistant Professor, Departmet of EEE,
Teegala Krishna Reddy Engineering College, Telangana, India

Abstract: the paper focuses on using Model Predictive Control to operate a DSTATCOM in a distribution system. By considering discrete-time current converter outputs and selecting the switching state that brings the real currents closer to their references, the DSTATCOM can effectively reduce harmonics on the source side and improve overall power quality in the distribution system. This approach aims to enhance the efficiency and reliability of power delivery to consumers.

Keywords: MPC, DSTATCOM, Power quality, Reactive power

1. Introduction

In Paper [1] presents MPC with DSTATCOM to improve the Power quality in distribution systems. This paper compared Model Predictive control with sliding mode controller. The conventional control method delivers switching states; those are actual currents closure to their references. The Total harmonic distortion in load side compensation maintained IEEE 519 standards. The implementation of a model predictive controller for a three-level H-bridge DSTATCOM is shown in this study [2]. The suggested method efficiently balances the DC capacitor voltage even when the load varies by using a discrete-time model of the DSTATCOM. The D-STATCOM effectively corrects for reactive power and current harmonics by using the real reference current. To further address power

quality issues, [3] introduces a Four-leg distribution static compensator. A cost function is used to determine how effective the control variable is after it is obtained from the reference variable. The differences between the reference and actual currents determine the cost function variables.

A grid-connected DSTATCOM is also given a Finite Control Set Model Predictive Control (FCS-MPC) implementation in [4], which offers enhanced performance and grid integration capabilities.

This algorithm directly enables the direct current (DC) and alternating Current (AC) side voltage control to reduce the switching losses and harmonic current. In this paper [5] Proposed model predictive control associated with LMS control for power quality improvement. The control algorithm

[1]chenchireddy.kalagotla@gmail.com, [2]drnrsvarma@gmail.com, [3]krs.jyothi@gmail.com, [4]Santhosh.btech245@gmail.com

is developed to gain the basic quantity of load currents in a three-phase four wire DSTATCOM. The extracted load currents are used to divergences the proposed work to improve the power quality.

This study [6] presents a novel method for efficiently switching DSTATCOM between the Current Control Mode and the Voltage Control Mode. Utilizing a single DSTATCOM to effectively address power quality concerns affecting both current and voltage is the major goal. A Distribution Static Compensator is also presented in article [7], which was created specifically for load compensation using a single DC source-based cascaded H-bridge multilevel inverter (SDCHBMLI). The capacitor voltage, which is related to diode-clamped flying capacitor multilevel inverters, is reduced by using a single DC source. A model predictive control approach is also used in paper [8] to develop a four-level Nested Neutral Point Clamped (NNPC) MLI-based DSTATCOM. This strategy ensures accurate voltage management while successfully eradicating issues with current-related power quality. A thorough modeling and simulation study of MPC algorithm-based DSTATCOM for power quality enhancement is also presented in paper [9]. In order to achieve unit power factor and overall harmonic distortion levels that are acceptable with IEEE standards, the control approach optimizes the switching states to closely follow reference signals. The applications of Nonlinear Model Predictive Control (NMPC) in various sectors are also covered in article [10]-[11]. It offers a thorough analysis of NMPC theory, which is necessary for developing efficient NMPC applications. This paper presents power quality enhancement using model predictive control technique. The simulation is verified by using MATLAB/simulink software.

2. Proposed Topology

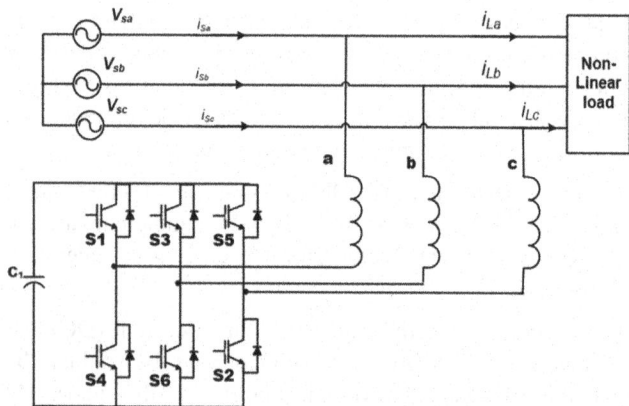

Fig. 54.1 Basic circuit diagram of DSTATCOM

Figure 54.1 displays the A Power electronics-based DSTATCOM (Distribution Static Compensator) is a tool

for enhancing power quality in distribution networks. It is intended to make up for different power quality problems while keeping the necessary voltage and current levels within allowable bounds. In the distribution network, DSTATCOMs are often located at key locations such distribution substations.

The following are a few power quality concerns that DSTATCOM can resolve:

Voltage Regulation: To provide a steady and dependable supply of electricity to consumers, DSTATCOM can regulate the voltage levels in the distribution system and assist in mitigating voltage sags (dips) and swells (increases).

Reactive Power Compensation: To keep the system's power factor close to unity, DSTATCOM can offer reactive power support, whether capacitive or inductive. This aids in reducing losses and maximizing the distribution network's use.

Harmonic Mitigation: In the distribution system, non-linear loads may create harmonics that distort the waveform. In order to eliminate the distortion and provide a cleaner power supply, DSTATCOM can inject equal and diametrically opposed harmonics [12]-[13].

Flicker Mitigation: Certain loads, such as arc furnaces or welding equipment, can cause rapid voltage changes, or flicker. These variations can be tamed with DSTATCOM, which can also cut flicker to a manageable level. Load Balancing: By shifting electricity between phases, DSTATCOM can assist in balancing three-phase loads, lowering imbalance and increasing overall system effectiveness.

Compensation for Voltage Flicker: DSTATCOM can help reduce voltage flicker brought on by fluctuating loads or sporadic renewable energy sources, improving the quality of the electricity for delicate equipment.

DSTATCOM device implementation in distribution systems can, all things considered, result in higher power quality, less downtime, and greater utilization of the distribution infrastructure. DSTATCOM is a crucial component of contemporary power distribution systems because it can balance out variations and maintain voltage levels by dynamically injecting or absorbing reactive power.

3. Model Predictive Control

Model predictive control (MPC), a potent advanced control approach, is employed in distribution systems to improve power quality shown in Fig. 54.2 It offers efficient real-time correction and regulation of various power quality concerns when used with DSTATCOMs or other power electronic devices. The following is an example of how MPC can be used to enhance power quality in a distribution system: The DSTATCOM can dynamically modify its output to account for voltage variations, harmonics, and reactive

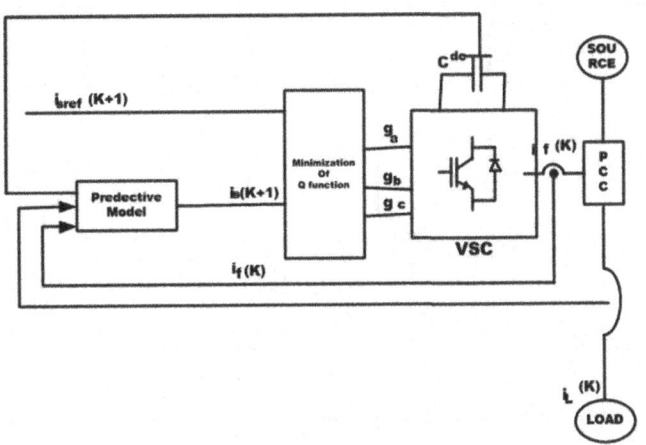

Fig. 54.2 Diagram of model predictive control

power imbalances in the distribution system. It is an MPC for improving power quality. This control strategy improves power quality and stabilizes the distribution network by enabling the DSTATCOM to react promptly to changing circumstances and disruptions.

The following are some of the main benefits of MPC for enhancing power quality:

Flexibility: MPC may manage numerous control targets concurrently, enabling a thorough and flexible method for power quality improvement.

Robustness: Because MPC is predictive, it can successfully manage changes and ambiguities in the distribution system.

Rapid Reaction: MPC is capable of reacting quickly to system changes, making it appropriate for improving power quality in real time. Overall, MPC provides a complex and intelligent approach to enhancing power quality in distribution systems, resulting in a more dependable and effective power supply for customers. The anticipated variances in source and load current are equal to the reference compensating current.

$$I_{sm1} = \frac{P_{3\phi}}{3V_m^+} \tag{1}$$

The Sum of reference currents is equal to zero

$$i_a + i_b + i_c = 0 \tag{2}$$

The Equations S0, P0, and q0 represent the Zero sequence, Positive sequence and Negative sequence Power components.

$$S_0 = 3v_0^* i_0^* = p_0 + jq_0 \tag{3}$$

$$p_0 = \frac{1}{3}(v_a + v_b + v_c)(i_a + i_b + i_c) \tag{4}$$

$$q_0 = 0 \tag{5}$$

To calculate the positive sequence power we represented S+, P+, and q+.

$$S_+ = 3v_+ i_+^* = p_+ + jq_+ \tag{6}$$

$$p_+ = \frac{1}{2}[(v_a i_a + v_b i_b + v_c i_c) - \frac{1}{3}(v_a + v_b + v_c) \tag{7}$$

$$(i_a + i_b + i_c)]$$

$$q_+ = -\frac{1}{2\sqrt{3}}[v_a(i_b - i_c) + v_b(i_c - i_a) + v_c(i_a - i_b)] \tag{8}$$

To calculate the negative sequence power we represented S-, P-, and q-.

$$S_- = 3v_- i_-^* = p_- + jq_- \tag{9}$$

$$p_- = \frac{1}{2}[(v_a i_a + v_b i_b + v_c i_c) - \frac{1}{3}(v_a + v_b + v_c)(i_a + i_b + i_c)] \tag{10}$$

$$q_- = \frac{1}{2\sqrt{3}}[v_a(i_b - i_c) + v_b(i_c - i_a) + v_c(i_a - i_b)] \tag{11}$$

The total Instantaneous active power is

$$S = (S_0 + S_+ + S_-)$$
$$= (p_0 + p_+ + p_-)$$
$$= p_{3\phi} = (v_a i_a + v_b i_b + v_c i_c) \tag{12}$$

$$\frac{P_{3\phi}}{2} = p_+ = \frac{3V_{m1}^+}{\sqrt{2}} \frac{I_{sm1}^+}{\sqrt{2}} \tag{13}$$

$$I_{sm1}^+ = I_{sm} = \frac{P_{3\phi}}{3V_{m1}^+} \tag{14}$$

So

$$i_{sx}^* = i_{sxref} = I_{sm} U_{xl} \tag{15}$$

$$i_{fx}^* = i_{fxref} = i_{Lx} - i_{sxref} \tag{16}$$

$$\dot{S}_x = \dot{i}_{Lx} + \frac{v_{sx}}{L_{fx}} + (g_x)\left[\frac{V_{dc}}{2L_{fx}}\right] - I_{sm} * \dot{U}_x \tag{17}$$

$$g_x = m1[g_a + ag_b + a^2 g_c] \tag{18}$$

Where $a = e^{j\frac{2\pi}{3}}$ and $m1 = \frac{2}{3}$

4. Simulation Results and Discussion

Figure 54.3 shows the three-phase supply voltages. The circuits have three voltage and current measurement blocks. The measurement blocks are used for measuring current and voltages from supply side. Three voltage measurement blocks outputs are connected to three goto blocks.

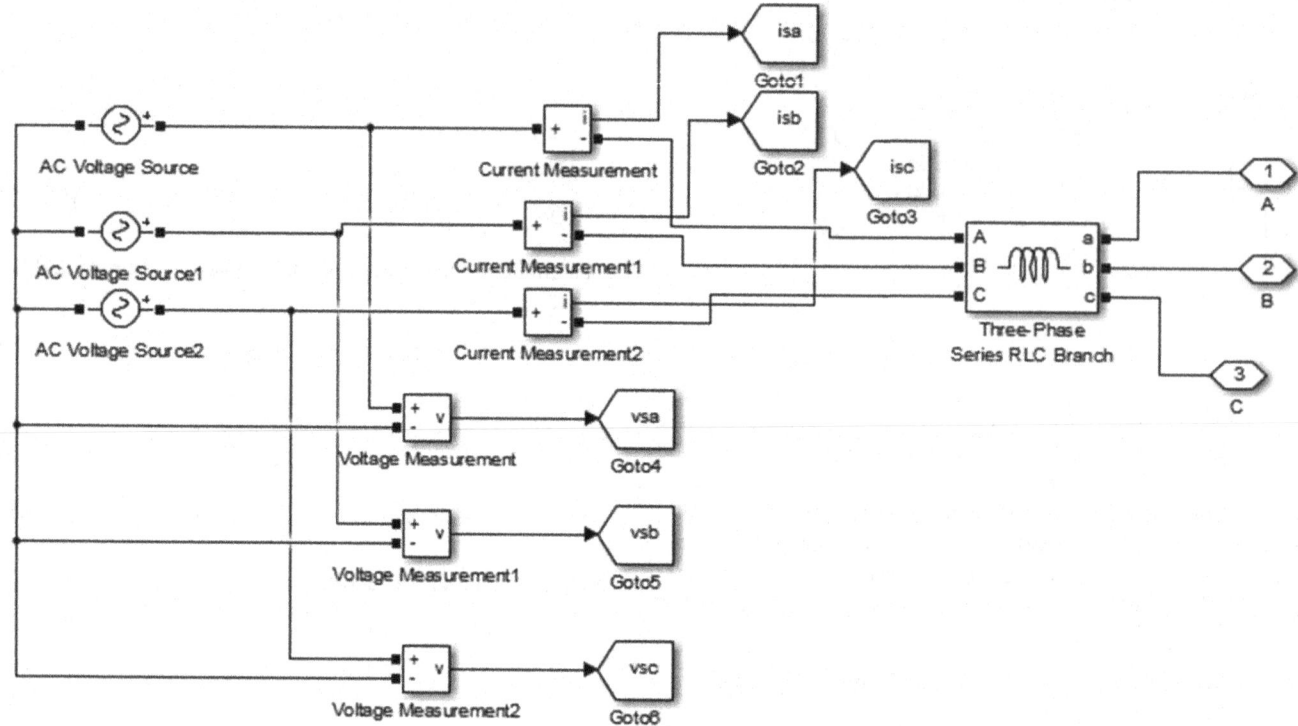

Fig. 54.3 3-phase, 415V input voltage

Figure 54.4 shows the SPWM technique, the specialty of this control scheme very easy implemented and easily controlled and generated pulses for inverter. This diagram have three reference voltages, one triangular block, six relational operators, the logic operator compared input signal and generated pulse for inverter.

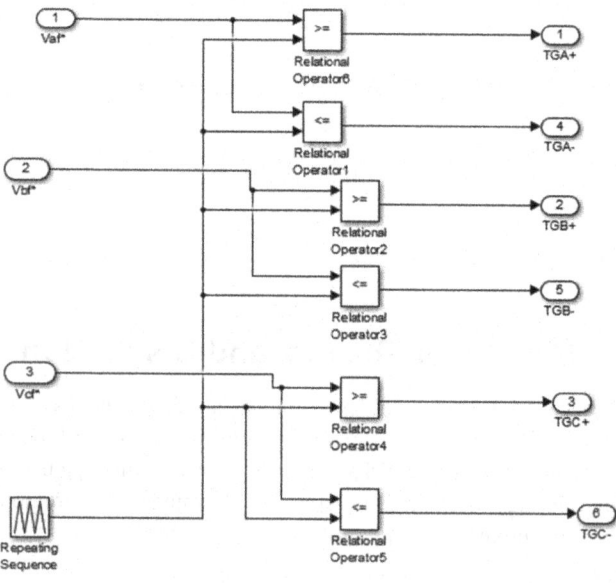

Fig. 54.4 SPWM pulse generator

Figure 54.5 shows the two-level inverter, this inverter acts as a DSTATCOM. This MATLAB circuit six IGBT switches, three-phase resistive and inductive branch, DC link capacitor and three current measurements. The main function of DC link capacitor is it compensating reactive power. The main cause for using IGBT switches it operating high current and high power rating.

Figure 54.6 shows the three-phase non-linear load. The non-linear load is inductive. The diode bridge rectifier is acts as DC load as well as non-linear load. These circuits have three-current measurement blocks and three-goto blocks. The measurement blocks are measuring three-phase current. Three goto blocks are connected to scope.

Figure 54.7 shows the three-phase supply voltage Vsabc and load current Isabc. The amplitude of supply voltages are equal and balanced waveform. The second waveform is current waveform; the current waveform is three different variations. These are first variation is without compensation, the time duration is without compensation 0 to 0.38 sec. the DSTATCOM operated at 0.38 sec, this instant the supply current raised suddenly and twice the supply current. The duration of this transient current is 0.38 to 0.42 sec.

Figure 54.8 shows the load voltages and current waveforms. The circuits have three-phase voltage is equal magnitude. The load current is 2A, the load is non-linear load. The simulated

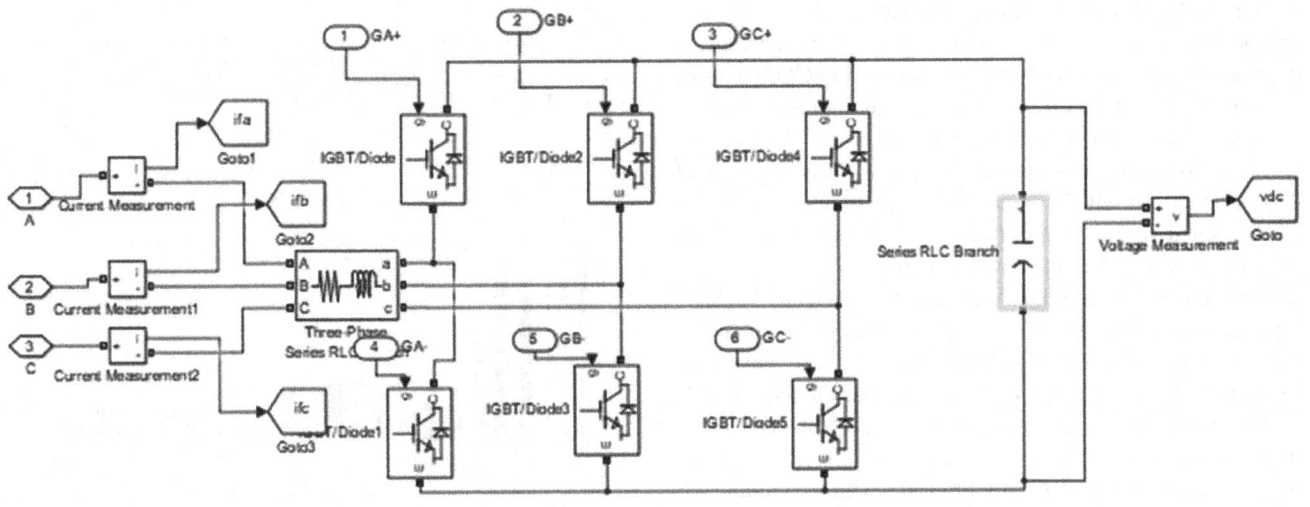

Fig. 54.5 DSTATCOM block diagram

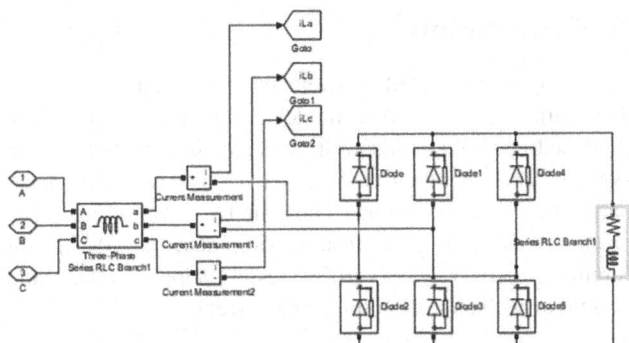

Fig. 54.6 Non-linear load

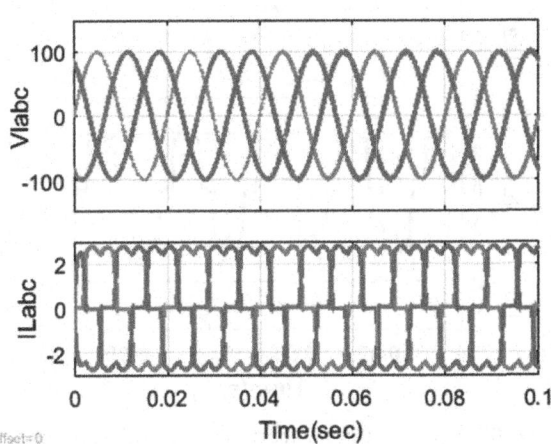

Fig. 54.8 Three-phase load voltage and current

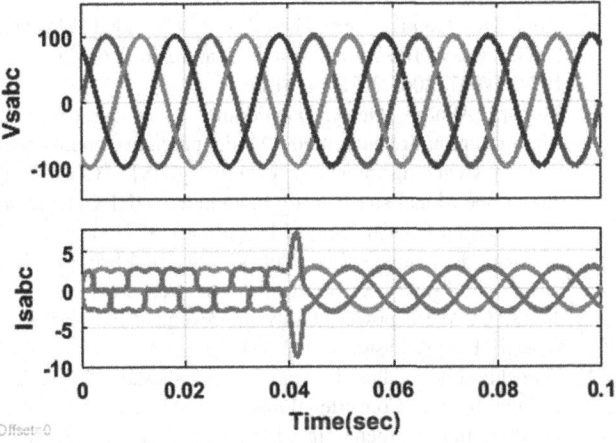

Fig. 54.7 Three-phase source voltage and current

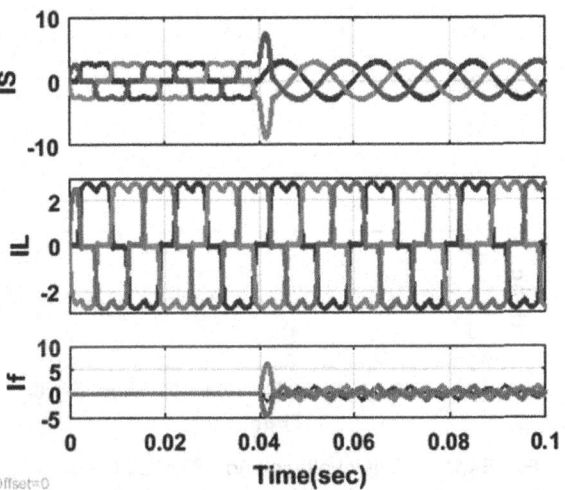

Fig. 54.9 Three-phase supply current, load current and compensation current

waveform is 0.1 sec. Fig. 54.9 shows the supply current, load current and compensating current. The Fig. 54.9 without compensation time duration is 0 to 0.38 sec. This waveform the load current is maintaining constant from 0 to 0.01 sec.

Figure 54.10 shows the source current, load current and compensation current waveforms. The supply current is distorted at 0 to 0.04sec duration. The DTSTACOM is turn on at 0.04 sec, at this time on words the source current is maintained constant current without distortion. The source current is presents Isa. The second waveform is ILa is load current, this current distorted due to nonlinear load. This compensated source side current only. The third waveform is compensating current waveform this waveform is zero amplitude current upto 0.04 sec. after that the DSTATCOM turn on it compensating source current. Figure 54.10 shows the DC link voltage and compensating current waveform relation. This waveform 0 to 0.04 sec the amplitude of DC link voltage is zero and compensating current also zero. After 0.04 sec the DSTATCOM is working. Figure 54.11 shows the source current waveform with THD. The THD value of source current is 7.02%.

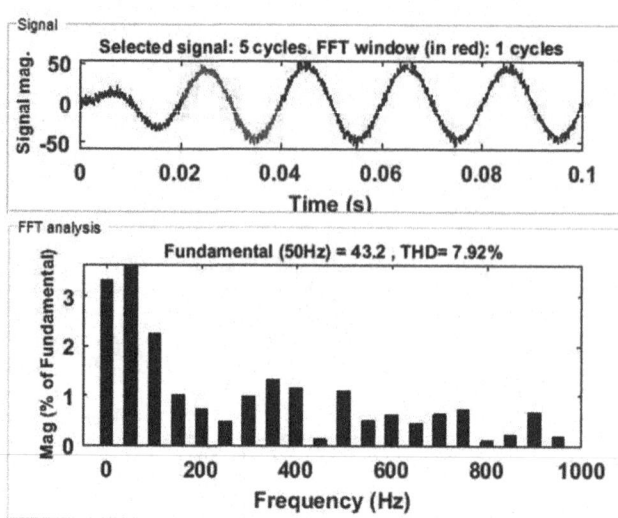

Fig. 54.12 Source current waveform with THD

5. Conclusion

This paper presented power quality improvement in distribution system with model predictive control. The proposed controller improved power quality in distribution system and compensated reactive power, eliminated harmonics, and maintained constant DC link voltage. The proposed DSTATCOM results simulated in MATLAB/Simulink software, the verified results supply voltage and current as well as load voltage and current.

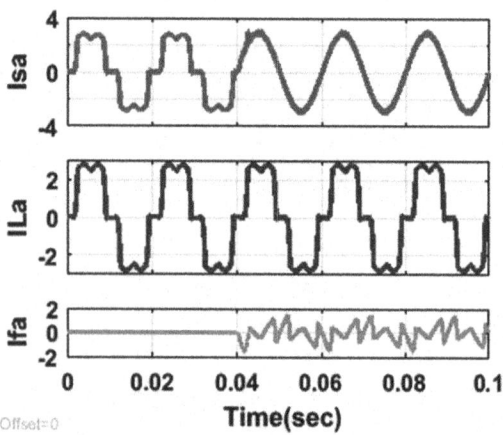

Fig. 54.10 Source current, load current and compensating current wavform

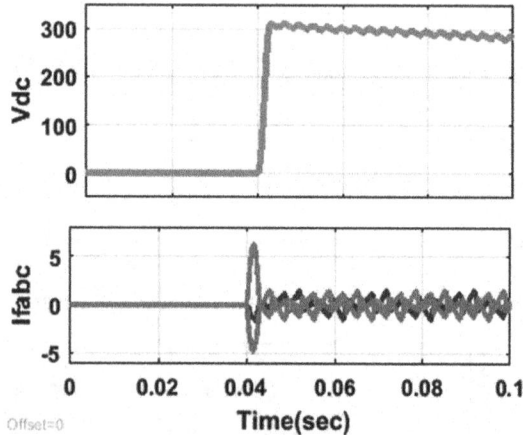

Fig. 54.11 DC link voltage and DSTATCOM currents

References

1. Y. Srinivasa Rao & M. K. Pathak (2020) Model Predictive Control for Three-Level Cascaded H-bridge D-STATCOM, IETE Journal of Research, 66:1, 65–76, DOI: 10.1080/03772063.2018.1476189
2. Kumar, Alladi Pranay, Ganjikunta Siva Kumar, and Dharmavarapu Sreenivasarao. "Model predictive control with constant switching frequency for four-leg DSTATCOM using three-dimensional space vector modulation." IET Generation, Transmission & Distribution 14.17 (2020): 3571–3581.
3. Kampara, Ravisankar, Kamaraju Viriyala, and Srinivasa Rao Rayapudi. "Finite control set model predictive control based d-STATCOM for power quality improvement in distribution system." Energy Systems (2023): 1–17.
4. Chenchireddy, Kalagotla, and V. Jegathesan. "Three-Leg Voltage Source Converter-Based D-STATCOM for Power Quality Improvement in Electrical Vehicle Charging Station." AI Enabled IoT for Electrification and Connected Transportation. Singapore: Springer Nature Singapore, 2022. 235–250.
5. Patel, Dhairya A., et al. "A dual functional DSTATCOM for power quality improvement." Journal of The Institution of Engineers (India): Series B 102 (2021): 881–893.

6. R. Chakrabarty and R. Adda, "Model predictive control of DSTATCOM employing a single DC source cascaded H-bridge multilevel inverter in a weak distribution system," 2018 20th National Power Systems Conference (NPSC), Tiruchirappalli, India, 2018, pp. 1–6, doi: 10.1109/NPSC.2018.8771824.

7. A. P. Kumar, G. S. Kumar and D. Sreenivasarao, "Model Predictive Control of Four Level NNPC DSTATCOM for Power Quality Improvement in Distribution System," IECON 2019 - 45th Annual Conference of the IEEE Industrial Electronics Society, Lisbon, Portugal, 2019, pp. 7063–7068, doi: 10.1109/IECON.2019.8926955.

8. Holkar, K. S., and Laxman M. Waghmare. "An overview of model predictive control." International Journal of control and automation 3.4 (2010): 47–63.

9. Qin, S. Joe, and Thomas A. Badgwell. "An overview of nonlinear model predictive control applications." Nonlinear model predictive control (2000): 369–392.

10. Chenchireddy, Kalagotla, and V. Jegathesan. "ANFIS Based Reduce Device Count DSTATCOM." Journal of Applied Science and Engineering 26.11 (2023): 1657–1666.

11. Santhosh, K., et al. "Time-Domain Control Algorithms of DSTATCOM in a 3-Phase, 3-Wire Distribution System." 2023 International Conference on Intelligent Data Communication Technologies and Internet of Things (IDCIoT). IEEE, 2023.

12. Chenchireddy, Kalagotla, and V. Jegathesan. "ANFIS Based Reduce Device Count DSTATCOM." Journal of Applied Science and Engineering 26.11 (2023): 1657–1666.

13. Chenchireddy, Kalagotla, et al. "Grid-Connected 3L-NPC Inverter with PI Controller Based on Space Vector Modulation." 2023 9th International Conference on Advanced Computing and Communication Systems (ICACCS). Vol. 1. IEEE, 2023.

Note: All the figures in this chapter were authors' test data as part of the research work - MATLAB experimental analysis.

Design and Implementation of Area Efficient MAC Unit Using Reversible Logic Gates

55

Ch. Sravana Lakshmi[1]

Electronics and Communication Engineering,
Anurag University, Hyderabad, Telangana, India

Rajesh Thumma

Electronics and Communication Engineering,
Anurag University, Hyderabad, Telangana, India

Abstract: A MAC is a basic block for digital signal processing application such as digital filters, Image processing and neutral networks. By leveraging the unique properties of reversible logic gates, we proposed methodology to design an area efficient MAC unit that maintains the reversibility of operations while minimizing physical area requirements. The proposed design starts by decomposing the MAC operation into its constituent logic gates, including Feynman gate and Peres gate. These gates form the basics for generating partial products and performing accumulation. To optimise the area efficiency of MAC unit, you can employ technique such as gate sharing, gate re-use and gate level optimisation. This technique helps to decreases the overall number of gates and minimise the physical area required for implementation. The final results show's the potency of the proposed design in means of area efficiency, the consumption of power reduces and performance improvement and this can be finally implemented in FPGA with less number of LUT's (97) and the delay of 1.224 ns. These contribute the development of computing systems and can be valuable for researchers and engineers working in the field of DSP and low power digital design.

Keywords: Multiply and accumulate (MAC) unit, Reversible logic gates, Area efficiency, Digital signal processing, Gate level optimisation RAM chips, Row hammering, Reliability, Probability, High speed, Energy overhead, Area overhead

1. Introduction

The Multiply-Accumulate (MAC) operation is a fundamental addition operation widely used in many signal processing algorithms and mathematic operations. It combines a multiplication operation and an addition operation in a single step, thereby reducing the number of required instructions and improving computational efficiency. The MAC unit is a basic component in processors and digital signal processing (DSP) systems, and optimizing its performance is of great importance.

Traditionally, MAC units employ a multi-stage process involving multiplication, partial product generation, and reduction. In this process, firstly the products should be multiplied the operand and then they added up by series of additions and accumulations to get final result. However, this sequential nature of the process restricts the performance of the MAC unit[1].

To address this restrict, this paper proposes a high level-performance MAC unit design that additions and accumulations integrate the reduction process of partial product which is shown in Fig. 55.1. By these operations, the proposed design reduces the number of stages, thereby improving the productiveness and reducing the potential of the MAC unit

The proposed high-performance MAC unit design introduces novel architectural enhancements to optimize the partial

[1]sravanib1517@gmail.com, [2]rajesh.thumma88@gmil.com

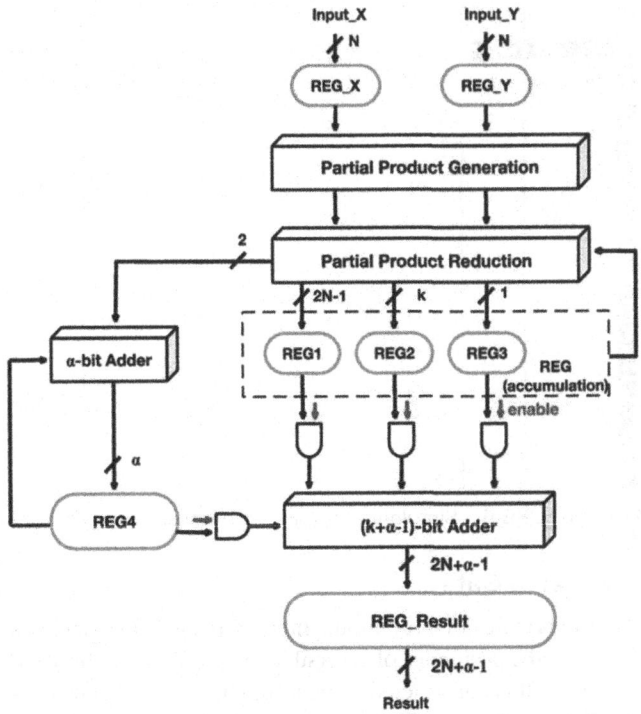

Fig. 55.1 The existing MAC architecture [2]

enable computation without loss of information. These gates are capable of performing both forward and backward computations, allowing there storation of input values from out put values [3] [4]. These logic gates have emerged as an approach for reducing in power consumption and dissipating less heat, which is especially crucial for energy-constrained systems.

In this paper, we propose a familiar design of MAC unit which is shown in Fig. 55.2, that exploits the advantages of reversible logic gates to achieve area efficiency. The objective is to minimize the hardware foot print without compromising the performance and accuracy of MAC operations.

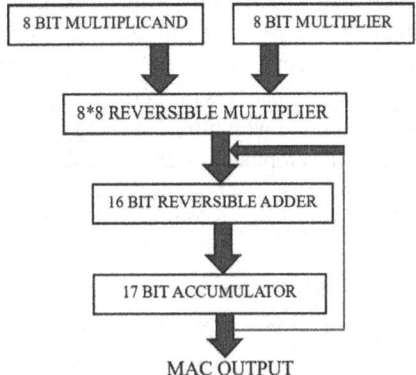

Fig. 55.2 The Proposed MAC Architecture

Source: Made by author

The proposed design not only contributes to reducing power but also facilitates the integration of MAC units into compact and resource-constrained systems.

The proposed approach leverages the properties of reversible logic gates to design an area-efficient MAC unit. The design process involves decomposing the MAC operation in to its constituent reversible logic gates and optimizing their arrangement to minimize the overall area requirements[5].

To achieve area efficiency, techniques such as gate reuse, gate sharing, and gate merging are employed. These techniques aim to be that gates number has been decreases and the complexity of the MAC unit while preserving the accuracy and precision of the computations. Furthermore, the design incorporates reversible adders and accumulators that ensure the reversibility of the MAC unit's operations while minimizing the gate count and optimizing the area utilization.

Reversible computing has come out as a promising paradigm for energy-efficient and information-preserving computation. Unlike traditional irreversible logic gates, reversible logic gates allow for computation where input scan be especially reconstructed from the outputs, results in no loss of information[6].These gates have found applications in diverse areas, including low-power electronics, quantum computing, and nano technology. The two terminal input, named after the

product reduction process. The key idea is to leverage parallelism and exploit the inherent mathematical properties of multiplication addition, and accumulation operations to achieve higher performance.

The MAC unit incorporates parallel addition and accumulation circuits, which operate concurrently with the partial product generation stage[2]. This parallel is mallows for a more efficient utilization of hardware resources and enables a reduction in the path delay.

Furthermore, the proposed design employs advanced techniques such as carry-save addition and tree-based accumulation to further enhance the execution of the MAC unit. These techniques minimize the propagation delay and maximize the utilization of available hard ware resources, resulting in improved overall performance.

2. Proposed System

The Multiply-Accumulate (MAC)operation is a fundamental arithmetic functioning widely utilized in various computational tasks, including digital signal processing, scientific simulations, and neural networks. As the demand for high-performance processors continues to grow, there is a pressing need to develop MAC units that not only deliver superior computational capabilities but also minimize the area requirements. Reversible logic gates, unlike their conventional counterparts, exhibit unique properties that renowned physicist Richard Feynman, is a reversible logic gate that has garnered considerable interest in the sector of reversible computing.

Inspired by the principles of quantum mechanics,Feynman proposed this gate as a means to perform computation without dissipating energy or losing information.

2.1 Feynman Gate

The Feynman Gate is a two-terminal input, two terminal output gate is shown in Fig. 55.3, and the logic diagram is shown in Fig. 55.4, that operates based on quantum mechanical principles. It is controlled of a series of controlled-NOT (CNOT) gates and Toffoli gates, which are generally used reversible logic gates. By arranging these gates in a specific configuration, the Feynman Gate can perform various computational operations. This property allows for reversible logic circuits to operate with minimal energy dissipation and is crucial for the development of energy-efficient computing systems.

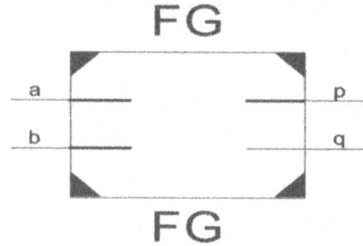

Fig. 55.3 Logic symbol of Feynman gate [2]

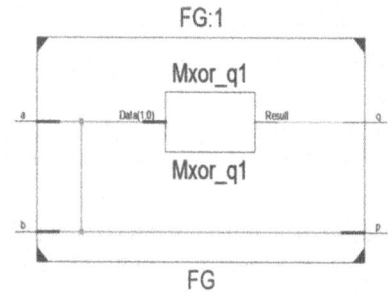

Fig. 55.4 Logic diagram of Feynman gate [4]

Reversible computing has come out as a promising paradigm for energy-efficient and information-preserving computation. The simulation output is shown in Fig. 55.5. Reversible logic gates, unlike traditional irreversible gates, allow for computations where the input scan be taken or reconstructed from the outputs[7] [8]. These gates find uses in various areas, such as low-power electronics, quantum computing. The Peres Gate is a reversible logic gate that has gained attention for its unique properties and applications in reversible computing.

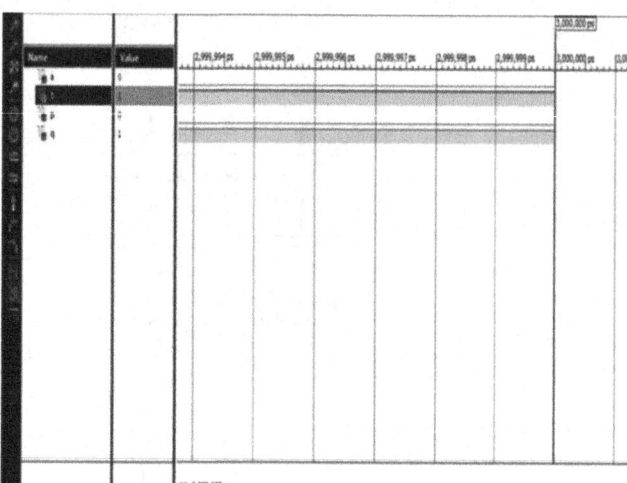

Fig. 55.5 Simulation output of Feynman gate [4]

2.2 Peres Gate

The Peres Gate is a three-input, three-output gate that operates based on the principles of reversible logic symbol is shown in Fig. 55.6. It is constructed using a combination of controlled-NOT (CNOT) gates, Toffoli gates, and ancillary inputs. The order of these gates number of operations allows to perform various operations and logic diagram is shown in Fig. 55.7.

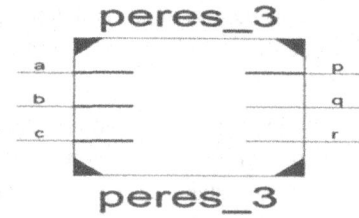

Fig. 55.6 Logic symbol of Peres gate [6]

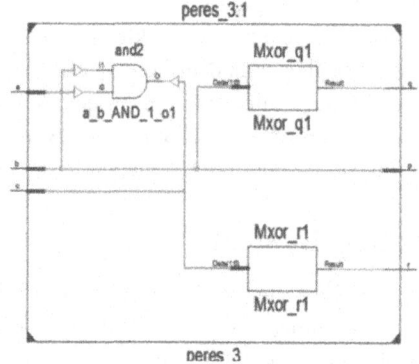

Fig. 55.7 Logic diagram of Peres gate [6]

The study of gate mention its ability to execute both front and back computations, performing that the taken inputs can be determined from the outputs, and vice versa [9]. This property enables reversible logic circuits to operate without information loss and is essential for the development of energy-efficient computing systems[10]. The simulation output is shown in Fig. 55.8.

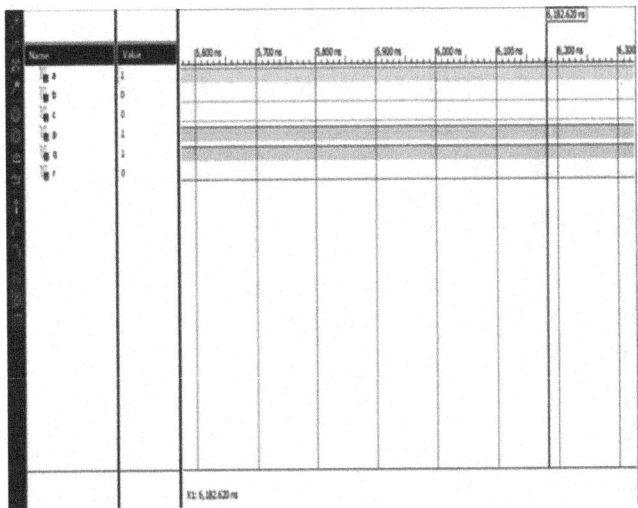

Fig. 55.8 Simulation output of Peres gate [6]

2.3 Reversible Full-adder and Full-Subtractor

Reversible computing has a promising paradigm for energy-efficient computation that preserves information. These circuits are such as Full Adders and Full Subtractors is shown in Fig. 55.9, play a vital role in performing essential arithmetic operations while maintaining reversibility. These circuits find applications in various fields, including quantum computing, low-power electronics, and error correction [10][11]. The model and implementation of Adder and Subtractor circuits shown in Fig. 55.10, aim to achieve arithmetic functionality while ensuring that the inputs can be determined from the outputs, resulting in information-preserving computation. The reversible Full Adder circuit combines the inputs of two

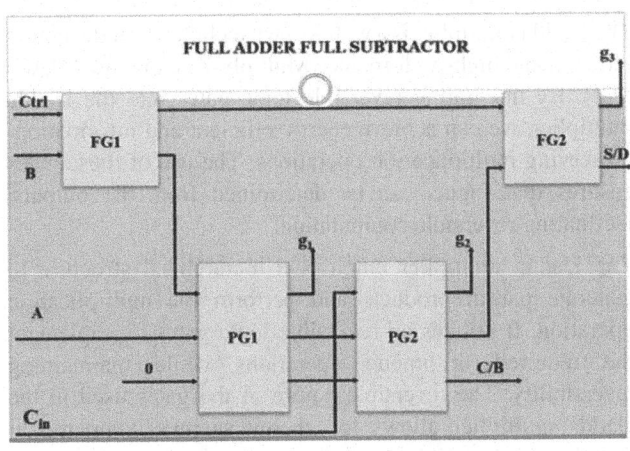

Fig. 55.9 Full adder full subtractor [14]

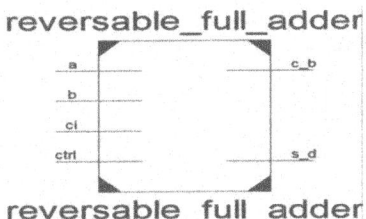

Fig. 55.10 Logic symbol of full adder and full subtractor [14]

binary numbers the logic diagram is shown in Fig. 55.11 and also with a carry input and generates a sum output and a carry output. The circuit employs reversible gates can achieve reversibility. Additional ancillary inputs are utilized to ensure the uniqueness of input recovery[12][13].In the same way, these circuit performs subtraction by combining the inputs of two binary numbers along with a borrow input, producing a difference output and a borrow output is shown in Fig. 55.12 and Fig. 55.13. The circuit incorporates reversible logic gates and ancillary inputs to achieve reversibility while preserving computational accuracy

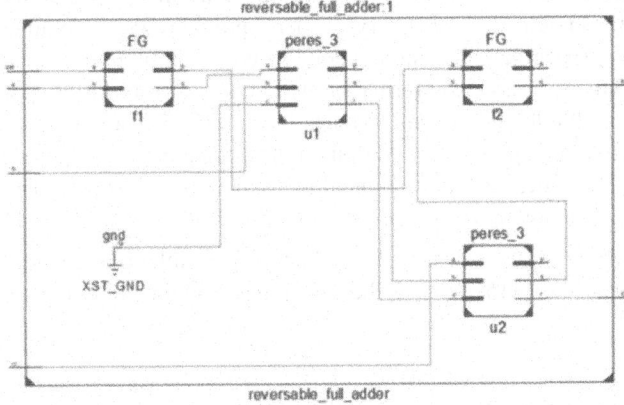

Fig. 55.11 Logic diagram of full adder and full subtractor [14]

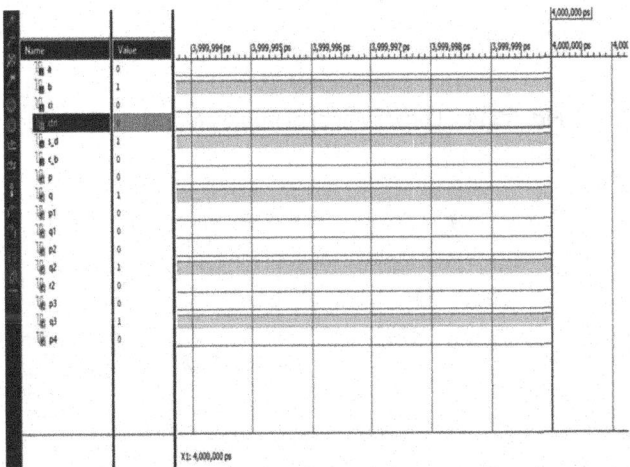

Fig. 55.12 Simulation output of reversible full adder [14]

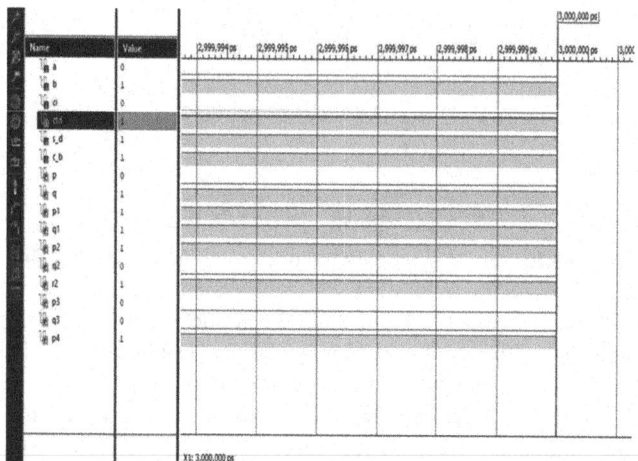

Fig. 55.13   Simulation output of reversible full Subtractor [13]

2.4 Reversible Ripple Carry Adder

In traditional computing,the three input Adder is widely used in binary addition. However, due to its irreversible nature, it dissipates energy and loses information during computation. Reversible Ripple Carry Adders provide an alternative approach that allows for energy-efficient addition operations while preserving information which is shown in Fig. 55.14 and the logic diagram is shown in Fig. 55.15. These circuits find applications in various fields, including low-power electronics, quantum computing, and error correction. The simulation output is shown in Fig. 55.16.

These Adder is designed based on the principles of reversible gates [14] [15], such as the Toffoli gate and the Fredkin gate.

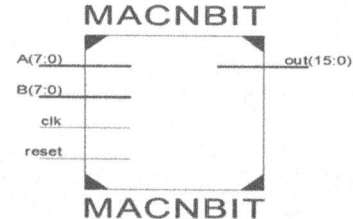

Fig. 55.14   Logic symbol of ripple carry adder [13]

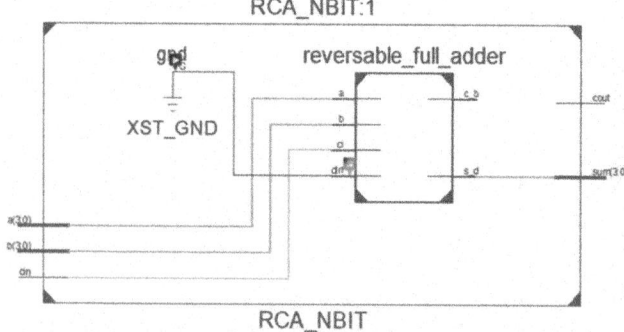

Fig. 55.15   Logic diagram of ripple carry adder [13]

Fig. 55.16   Simulation output of RCA [13]

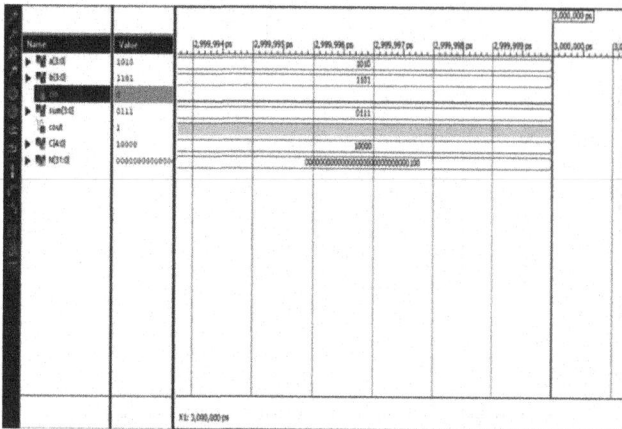

Fig. 55.17   Logic symbol of MACNBIT [13]

By utilizing these gates, the carry propagation mechanism in the Ripple Carry Adder is modified to achieve reversibility. The circuit employs ancillary inputs and additional reversible gates to ensure that the inputs can be determined from the outputs. The design and implementation of the Reversible Ripple Carry Adder focus on minimizing gate count, reducing power consumption, and preserving computational accuracy. Various optimization techniques, including gate decomposition and gate count reduction, can be applied to enhance the efficiency of the circuit

3. Results

The Dadda multiplier is a well-known technique for designing efficient and high-performance Multiply-Accumulate (MAC) units. By integrating reversible logic gates into the Dadda multiplier, we can achieve energy-efficient and information-preserving multiplication operations. The use of these gates ensures that inputs can be determined from the outputs, facilitating reversible computation.

The Dadda multiplier employs a hierarchical structure to generate partial products and perform the multiplication operation. It utilizes all reversible logic gates to implement the required arithmetic operations while maintaining reversibility. The reversible nature of the gates used in the Dadda multiplier allows for energy savings compared to traditional irreversible designs.

The design of the Dadda multiplier using reversible logic gates is shown in Fig. 55.18, offers several advantages. Firstly, it reduces power consumption, making it suitable for low-power applications. Secondly, it enables precise and reversible computation, which is essential for error correction techniques in computing systems. Additionally, the Dadda multiplier using reversible logic gates is compatible with quantum computing architectures, where reversible operations are fundamental. The simulation output of MACNBIT is shown in Fig. 55.19.

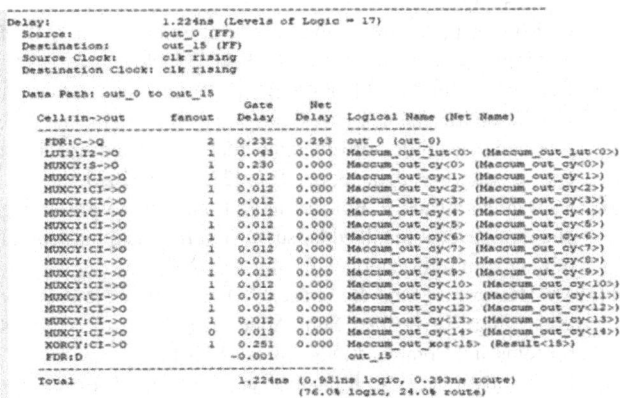

Fig. 55.18 Logic diagram of MACNBIT

Source: Made by author

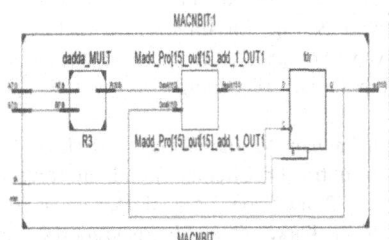

Fig. 55.19 Simulation output of MACNBIT

Source: Made by author

The results obtained from using the Dadda multiplier with reversible logic gates demonstrate its effectiveness that can be in performance and power efficiency. The integration of reversible logic gates into the Dadda multiplier improves the overall efficiency of the MAC unit, allowing for faster and more accurate multiplication and accumulation operations.

Further research and optimization of the Dadda multiplier using reversible logic gates can lead to even better results in terms of performance is shown in Fig. 55.20, power consumption, and error correction capabilities, delay which is shown in Fig. 55.21. The final comparison table of both existing and proposed method is shown in Table 55.1. By harnessing the advantages of reversible logic gates in the design of MAC units, we can contribute to the development of energy-efficient and reliable computing systems

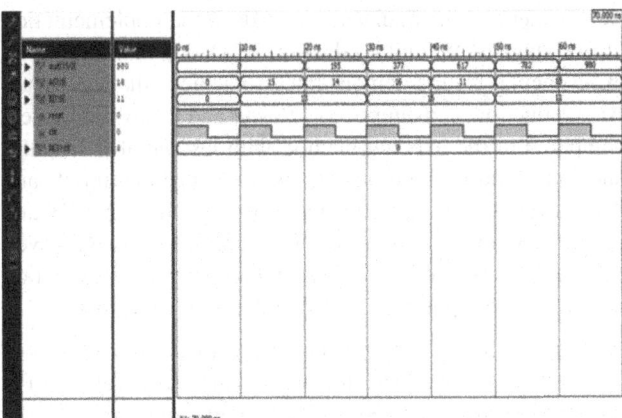

Fig. 55.20 Area utilization of MAC Unit

Source: Made by author

Table 55.1 Comparison of existing and proposed method

RESULTS		
PARAMETER	**EXISTING METHOD**	**PROPOSED METHOD**
No. of LUT'S	167 LUTS	97 LUTS
DELAY	3.185 ns	1.224 ns

Source: Made by author

Device Utilization Summary			
Slice Logic Utilization	**Used**	**Available**	**Utilization**
Number of Slice Registers	16	35,200	1%
Number Used as Flipflops	16		
Number Used as Latches	0		
Number Used as Latch - thrus	0		
Number Used as AND/OR Logics	0		
Number of Slice LUT's	97	17,600	1%
Number Used as Logic	97	17,600	1%
Number Using O6 Output Only	74		
Number Using O5 Output Only	0		
Number Using O% and O6	23		
Number Used ROM	0		
Number Used as Memory	0	6,000	0%
Number Used Exclusively Route-thrus	0		
Number of Occupied Slices	47	4,400	1%
Number of LUT Flipflop Pairs Used	97		

Fig. 55.21 Delay of MAC unit

Source: Made by author

4. Conclusion

In this paper, we developed an approach to design an area efficient MAC unit using Reversible logic gates. In this article aimed to address the challenges of Physical area, Power consumption, Delay requirements. Through a systematic design methodology, we decomposed the MAC operation into a Reversible logic gates, and utilised the gate sharing, gate reuse and gate level optimizations to optimize area efficiency. The contributions of proposed design has implemented by using with the stages of Feynman and Peres gate, Reversible full adder and subtractor, Reversible ripple carry adder and the final stage is leads to the implementation of two stage MAC unit by the type of Dadda multiplier. The experimental results demonstrated to achieve the significant Area reduction, consumption of Power and Delay parameters compared to the existing method with less number of gates and can produce eccentric outputs from specified inputs and this Reversible circuits do not loose any information and finally this can be evaluated in DSP applications. Moreover, the area efficient MAC unit executed competitive execution in terms of Computation, Speed and Energy consumption.

As future work, it would be beneficial to explore additional optimization techniques, investigate the scalability of the proposed design for larger MAC unit, and evaluate the performance of the area efficient MAC unit in real-world applications. Additionally further research can focus on developing specialised synthesis and optimization tools to facilitate the design and implementation of Reversible logic-based MAC units.

References

1. Che-Wei Tung, Shih-Hsu Huang, Senior member, IEEE 'A high performance Multilply-Accumulate Unit by integrating additions and accumulations into partial product reduction process' VOLUME XX 2020.
2. Mahmoud Masadeh, Osman Hasan, and Sofiene Tahar. "Input-Conscious Approximate Multiply- Accumulate (MAC) unit for Energy-Efficiency." IEEE access, volume 7, 2019.
3. Pavel Lyakhov, Maria Valueva, Georgii Valuev, Nikolai Nagornov "High performance digital filtering on truncated multiply-accumulate units in the residue number system." IEEE access, volume8, 2020.
4. Bin Zhou, Guangsen Wang, Guisheng Jie, Qing Liu, andZhiwei Wang " A High-Speed Floating- Point Multiply-Accumulator Based on FPGA's". IEEE Transactions on very large-scale integration (VLSI) systems, vol_29, NO. 10, OCT 2021.
5. Abdelgawad, A. (2013). Low power multiply accumulate unit (MAC) for future wireless sensor networks. In Proceedings of the IEEE sensors applications symposium (pp. 129–132), Galveston, TX, USA.
6. Abdelgawad, A., and Bayoumi, M. (2007). High speed and area-efficient multiply accumulate (MAC) unit for digital signal processing applications. In Proceedings of the IEEE international symposium on circuits and system (pp. 3199–3202), New Orleans, LA, USA
7. R. Landauer, "Irreversibility and heat generation in the computing process", IBM J. Research and Development, vol. 5, no. 3, pp. 183–191, 1961
8. C.H. Bennett, "Logical Reversibility of computations", IBM J. Research and development, pp. 525–532, November 1973.
9. Garipelly Raghava, P. Madhu Kiran and A. Santhosh Kumar, "A Review on Reversible Logic Gates and their Implementation", International Journal of Emerging Technology and Advanced Engineering, vol. 3, no. 3, pp. 417–423, March 2023.
10. M. Haghparast and, A. Bolhassani, Optimized parity preserving quantum reversible full adder/subtractor. International Journal of Quantum Information, vol. 14, no. 3, 2016.
11. M. Haghparast, and S. Shoaei, "Design of a New Parity Preserving Reversible Full Adder", Journal of Circuits Systems and Computers, vol. 24, no. 1, 2015
12. H Thapliyal and N. Ranganathan, "Design of Reversible Latches Optimized for Quantum Cost Delay and Garbage Outputs", Centre for VLSI and Embedded System Technologies International Institute of Information Technology
13. Gupta Aakash, Singla Pradeep, Gupta Jitendra and Maheshwari Nitin, "An Improved Structure of Reversible Adder And Subtractor", International Journal of Electronics and Computer Science Engineering, vol. 2, no. 2, pp. 712–718.
14. V. Kamalakannan, V. Shilpakala , and H. N. Ravi, "Design of Adder/Subtractor Circuits Based on Reversible Gates", Ijareeie, vol. 2, no. 8, pp. 3796–3804, August 2013.
15. P. K. Lala, J.P. Parkerson and P. Chakraborty, "Adder Designs using Reversible Logic Gates", WSEAS Transactions on Circuits and Systems, vol. 9, no. 6, June 2010.

A Novel Bionic Eye with a Nanowire Array Artificial Retinal Regeneration

56

**Muthyala V. V. S. Chowdary[1], N. Aravind[2],
R. Suneel Kumar[3]**
Research scholar (Ph. D) Department of EECE,
GITAM Deemed to be University Visakhapatnam, India

T. Venkata Ramana[4]
Senior Associate Professor Department of EECE,
GITAM Deemed to be University Visakhapatnam, India

Abstract: Bionic eye is an artificial electronic eye. One of the new emerging revolutionary techniques is artificial vision which is very useful for blind people. By the implantation of the photoreceptors array or camera in eye gives the solution to this blind problem. A small chip is implanted at the back the retina in the eye ball by using surgical methods. The people who are suffering from age related blindness, it can be restoring the eye sight. The most important part in visual nervous system is 'retina with complicated structure'. The blindness occurred from any disorder inside retina due to neural signals are transferred irregularly to the brain. In present years, the nanostructures emergence and nanowires are providing a viable means that to enhance the regeneration of retina. The light sensing nature acquired by the nanowires and converting them into electrical signals, therefore additional cellular electrical properties are simulated. For the applications of retina these are represented as newest nanostructures. This system represents a novel integrating concept of bionic eye with the nanowire array artificial retinal regeneration which is latest approach. Gold nanoparticles with core-shell nanowires are used for retinal neural system regeneration and the functionalized nanowires are attracted to attention. Retinal regeneration uses the applications of nanowires and therefore complete blindness is eliminated.

Keywords: Artificial electronic eye, Regeneration of retinal, Nanowires, Eye replacement

I. Introduction

Bio Electronic eye is generally referred as Bionic eye which is an electronic device. The functionality of the eye is replaced with this device [1]. An external camera is worn on dark glasses pair that send the images in the digital form to the receiver of radio which is placed in eye. To the visionless eyes, technology adds life to them[2]. For cell microenvironment the nanostructures are most widely used from nanowires. The properties of electrodynamic having permanently affected cellular functions namely adhesion, proliferation, differentiation, morphology [3]. Moreover, for better cell adhesion, biocompatibility, and electro-

conductivity, and the researchers are developing the new structures[4]. The nanowires can represent the nerve signal simulation in retina also for improving the vision loss by the damaged retina and it transfers among layers. To explain the vision recovery by nanowires, where lost by degeneration of retinal, which is starts with describing the anatomy of retinal and how the blindness occurred by various disorders of retinal [5]. Therefore nanowires are investigated which can determine the organization of retina with light sensing, then converted as electrochemical signals with various properties, structures, materials. Hence,for vision recovery the nanowires application and challenges are discussed [6].

[1]mvvsvenkatchowdary716@gmail.com, [2]aravind.nalika@gmail.com, [3]suneelkumar.rebaka@gmail.com, [4]vteppala@gitam.edu

For retinal applications, the advanced structures are developed by the nanostructures and its functionality can be depends on properties and materials. Therefore required functions are improved by the scaffold combinations with available nano materials. It is observed that, for retina regeneration the nanowires with different structures are used. The nanowires fabrication and synthesis with subtracting narrow range like TiO2 and gold nanoparticles are combined with poly and polymers are assigned to the anodized aluminum oxide template and utilized for applications of regeneration of retina[7].

Generally human brain can remember the information more than 80% about surroundings through eyes. The human eye with light management component and concavely hemispherical retina is noticed particularly for its excellent adaptively to the optical environment and exceptional characteristics such as high resolution of one arcmin per line pair at the fovea, wide FOV up to1500[8]. The retina dome shape has the optical system merit reduction complexity by direct compensating the aberration from curved focal plane particularly. According to principle, the retina of the human can reach this goal by a hemispherical image sensor design [9]. Planar device structures are used by the CCD(commercial charge-coupled device) and CMOS (complementary-metal-oxide-semiconductor) and these are designed by using the process of mainstream planar micro-fabrication. Therefore the hemispherical device fabrication is possible [10].

An innermost multilayered eye structure is retina with approximate 0.50 mm thickness. At first, the light translates into biochemical message by retina therefore the biochemical messages are prepared and converted as electrical messages, it may be cause for the visual information with transforming to primary brain visual cortex through an optic nerve. Therefore retina is combination of neuro-retina and RPE (retinal pigment epithelium)where it again classified into 9 layers[12]. The layer of neuro-retina contains photoreceptors (PL) inner and outer segments, outer limiting membrane, nerve fiber layer (NFL), outer plexiform layer (OPL), outer nuclear layer (ONL), inner plexiform layer (IPL) [13], inner nuclear layer, ganglion cell layer (GCL). From nearest layer to choroid up to nearest layer to vitreous contains the inner limiting membrane (ILM). This part contains 5 neuron types are amacrine cells, horizontal cells, bipolar cells,retinal ganglion cells,visual receptors cells (cones & rods) respectively.

2. Human Eye and Bionic Eye

2.1 The Human Eye

The eye of human is an organ which can react to light for various applications. The human eye ball seems like spherical shape. The retina is most crucial part of eye is very responsible for vision. Moreover, light sensitive tissue lining the inner eye surface is also called retina. An eye is focused on the light sensitive cell sheet when light falling on eye. In retina the photosensitive ganglion cells are receiving the signals of light and affect the adjustment of pupil size. In retina, ganglion cells are connected to cone cells and rods. The viewed image color recognition can be responsible by the cone cells and movements are distinguished by rod cells and image contrast placed on retina. The optic nerve is defined as 'retina is connected to a nerve' that connects eye and brain. In skull, a special cone shaped region is designed for eyeball for it best protection and it is called as socket or origin. The measurement of this cone shape diameter is approximately one inch.

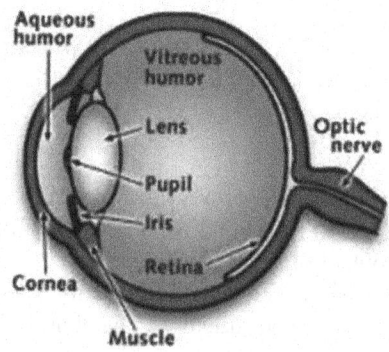

Fig. 56.1 Human eye structure [2]

2.2 Bionic Eye

By stimulating nerves the bionic eye is worked, that are activated by electrical impulses [14]. Here, small equipment is placed in patient body which can receive the signals of radio then that signals are transmitted and interpret the image. The artificial eyes creation is the one of most dramatic bionics applications. At first it uses silicon-based photo detectors, but in human body the silicon is toxic then unfavorably reacted

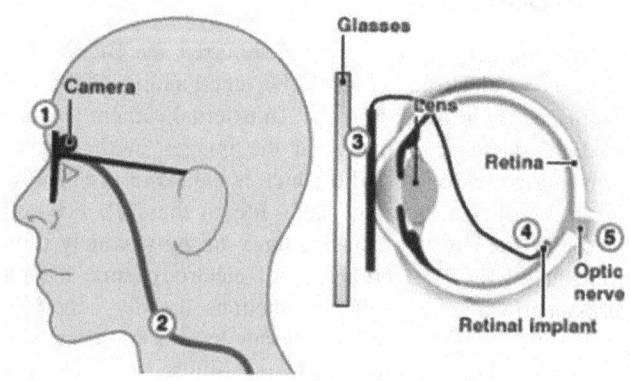

Fig. 56.2 Bionic eye system [2]

by fluids in eye. Hence, a new material is developed and used by the University of Houston at Space Vacuum Epitaxy Centre (SVEC) Texas. The incoming light can be detected by tiny ceramic photocells and malfunctioning human eyes are repaired. Bionic eye system is represented in Fig. 56.2.

2.3 Eye Disorders

The eye disorders representation is below [15],

- Macular Degeneration
- Retinitis Pigmentosa

Retinitis Pigmentosa

Retina of the eye hereditary diseases group gives the name of Retinitis Pigmentosa (RP). Neurons degeneration is involved in this disease which is commonly called as outer retina progressive blinding disorder. Damaging of some portion functionality of retina cones or rods is caused with this Retinitis Pigmentosa (RP).

Macular Degeneration

The one of the medical condition is macular degeneration where it can affect the old, adults commonly. Due to cones breakdown in retina the macular degeneration was occurred.

3. Bionic Eye with Nanowire Array Artificial Retinal Regeneration

The component schematic to be implanted contains stimulation-current driver,power as well as signal transceiver, the placing of secondary receiving is located very close to the cornea, processing chip, polyimide or thin silicon materials are used in the fabrication of system electrode array, silicon rubber by cables of ribbon to connect the devices. Therefore polyimide biocompatibility is studied and their lightweight, thin consistencies suggest its possible utilization like non-intrusive material for array of electrode. To keep an array of electrode in place acyanoacrylate glue or Titanium tack is used.

In following manner, the system is operated. The external camera can acquire the image then images are preprocessed regarding as certain parameters. Through RF telemetry, encoded data stream is transmitted to an intraocular transceiver. By modulating carrier signal of high frequency the data signal can be transmitted. Therefore this signal is filtered, rectified and diagnostics have the capability of extracting the clock signal, data and power. On the retina of patients,derived image can be stimulated. The system can have 2 parts are placed separately interior and exterior to the eyeball. The receiver and transmitter are equipped by the every part.

The configuration of stimulating electrodes is settled by providing the primary coil which is accompanied by frequency modulated signal 10 kHz with carrier signal 0.5-10 MHz. By an incoming RF signal rectification a DC power supply is attained. On other side the receiver extracts the four data bits for every pixel from incoming RF signal and produces amplification, demodulation, filtering. By electrode signal driver an extracted data is interpreted. At final, appropriate currents are generated for stimulating the electrodes in terms of frequency, pulse width, magnitude respectively.

4. Working Strategy

In brain the artificial eye provokes visual sensations by stimulating various optic nerve parts directly. With the stimulating nerves the bionic eye is worked and activated with

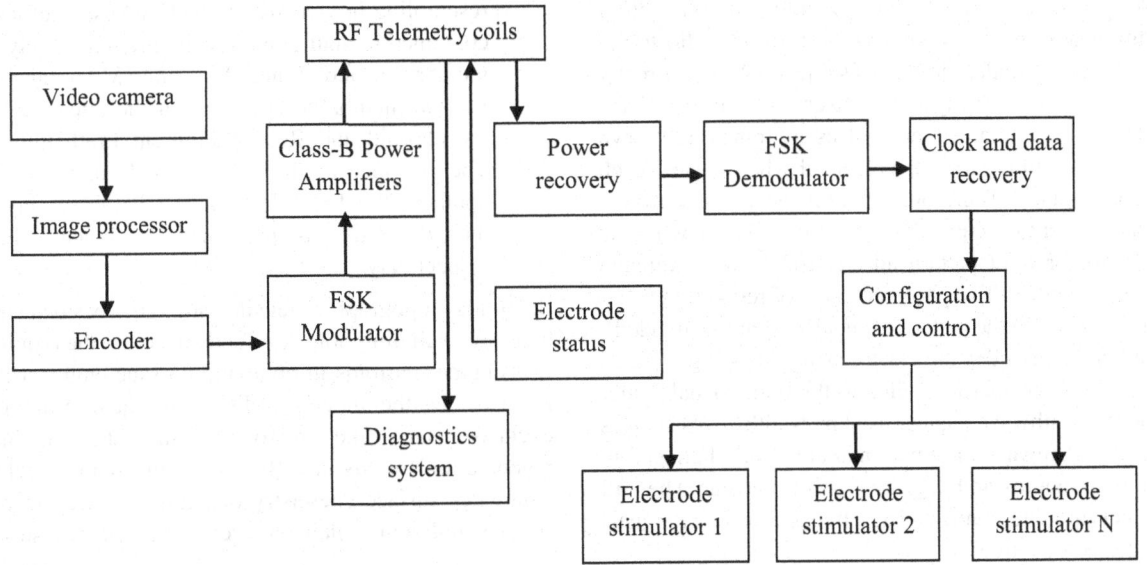

Fig. 56.3 Functional block diagram

Source: Made by author

the help of electrical impulses. Here, small device is placed in patient body then signals of radio are received and transmitted to nerves respectively. The Argus II implant contains the electrodes array that are attached to the retina also further used for video processing system and conjunction with an external camera for providing a rudimentary sight form to the subjects implantation. The light detection and sight are provided by the Argus II Retinal Prosthesis System for people who are not have blind from degenerative eye disease. The eye photoreceptors are damaged by this disease. Behind the retina, light patterns are perceived and transfer to brain in the form of nerve impulses, which the patterns of impulse pulses are interpreted like images. These photoreceptors are taken by Argus II system[16]. The Second Sight's retinal prosthesis second incarnation contains 5 major parts:

- *Radio transmitter:* The pulses are transmitted wirelessly to the receiver implanted under the eye or above the ear.
- *Video processing microchip:* It is established into handheld unit, in the form of electrical pulses, images are processed which are represents the light or dark patterns. Then in glasses by using radio transmitter, pulses are sending.
- *Digital Camera:* The pair of capture images and glasses are developed in real-time and send the images to microchip.
- *Retinal implant:* The 60 electrodes array on a chip measuring 1 mm by 1 mm.
- *Radio receiver:* The receiver send the pulses to retina implantation by implanted wire, a thin hair.

When the image is captured by camera, means a tree, an image is in dark or light pixel form. Therefore it send image to the video processor, it convert tree-shaped pixel pattern to electrical pulses series that represent "dark" as well as "light". In glasses a processor send these pulses into radio transmitter and again pulses are transmitted in the form of radio to the implanted receiver under the skin of subjects. Wire is directly connects the electrode array and receiver, and at the eye back electrode array is implanted then pulses are transferred down to wire. The electrode array is presented when pulses reach the retinal implant. Therefore an array acts like artificial retina photoreceptors equivalence. In accordance with the encoded dark or light pattern an electrodes are stimulated that representing tree, like photoreceptors of retina if they are working (except that the pattern cannot be digitally encoded). By stimulated electrodes the electrical signals are generated then as neural signals are travelled to the brain visual center by the normal pathways manner used by healthy eyes such as optic nerves. Pathways of optical neural are not damaged in the retinitis pimentos and degeneration of macular. Then all these signals are interconnected as a tree and send the signals as "You're seeing a tree".

4.1 Nanogenerator Power Supply

With the Zinc oxide nanowires the nano generator devices are made where when released and bent the electric charges are produced. For small scale devices, power supply is produced by this generator with establishment of interconnected arrays which are having the number of such nano wires. The nano-generator internal structure can be shown in Fig. 56.4, where it contains PZT nanofibers placed on substrate of silicon.

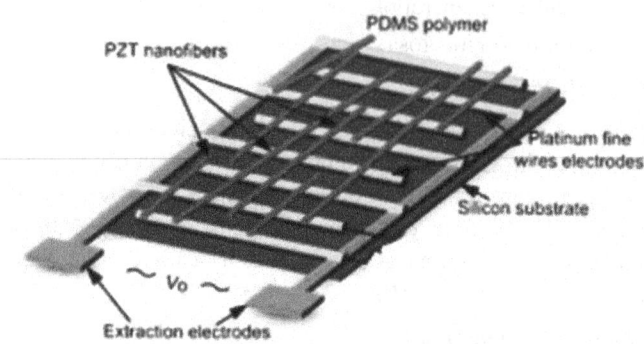

Fig. 56.4 Nanogenerator [6]

4.2 Extracellular Matrix Simulation and Cell Adhesion

To redevelop the organs or injured tissues the tissue engineering is specified. It has two foremost policies are:

1. *Cell-based:* The place modification before transplanted to the host body when cells are critical substance.
2. *Scaffold-based:* In vivo structures an Extra Cellular Matrix (ECM) from biomaterials, are simulated and designed efficient architecture is required for recreation of retina which is replica or imitates the physiological and morphological features with extracellular matrix resembling in vivo structure. Two basement layers are contained in mature mammalian retina namely, brunch membrane and Inner Limiting Membrane (ILM). Brunch membrane is present at the interface between the choroid and Retinal Pigment Epithelium (RPE). The interface of the vitreous body and neural retina contains the LIM. Retina two basement layers and their respective morphogenesis is shown in Fig. 56.5 respectively.

Scaffolds topography, tensile strength besides the cell adhesion, cell morphology effectiveness are the requirements of usingthe scaffolds in ophthalmic tissue engineering. This report shows the underlying ECM significance in endorsing exclusive nano and micro environments that improves regeneration of tissue. But, the information related to nanowires surface chemistry or nanowires are not reported ECM simulation which represents the cell adhesion and it

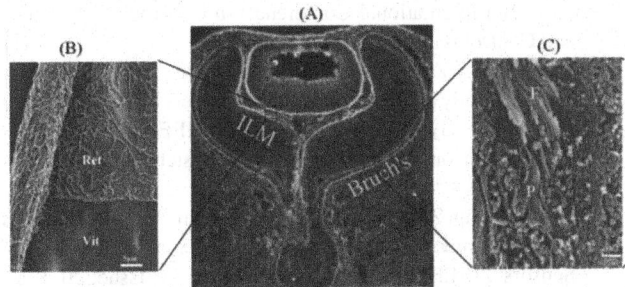

Fig. 56.5 Basement layers in the retina [15]

is used in retina regeneration. However, all reports are using single nanowire for retina regeneration.

4.3 Extracellular Electrical Simulation

As mentioned earlier, the light is converted into neural signals then neural signals are transferred to the brain with retina which makes visual perception. In nervous systems like retina, an extracellular electrical involving ion channels that plays an important role. The silk is one of the photovoltaic polymers which is having high potential for connecting the layers of retina for vision restoration with electrical signals transferring. One of unknown metrics is Intracellular voltage which is used by the number of researchers for utilization of recording signal strength with nanowires as nano electrodes. The tremendous potential can be shown by nanowires based on extracellular electrical cell stimulation to promote the differentiation, adhesion, cell growth respectively. Extracellular electric simulation with nanowires is represented in Fig. 56.6 and the effect of the

vision recovery after implantation in mice eye and nanowires cell adhesion is implemented.

4.4 Light Sensation

Extensively the nanowires are explored like a photovoltaic component for improving the sensing light efficiency for applications of retina. The single nanowire utilization, the nanostructures of photovoltaic represents various important advantages, where it can be a cause to produce compatibility, high efficiency, robust with cells.

5. Conclusion

In biomedical engineering, the bionic eye (Bio-Electronic Eye) is one of the breakthrough that give the vision to people who are with total or partial visual loss. In this case, the biomimetic eye is demonstrated with hemispherical retina made by lightsensitive and high-density NWs. It has a structure with high similarity degree to human eye when achieving greater imaging resolution with potency if implementing the proper contact strategy. The bionic eye changes the visually challenged people world. Surely it is observed that,better resolution, higher quality even color is possible in upcoming years. Nowadays the sight restoration for the blind not a dream. But it is possible with "Bionic Eyes". For controlling neural signals and directing the cells to polarization of nanowires and are developed and more useful for complex architecture of engineering also retina. Already the researchers have a plan that is third version which is 1,000 electrodes on the implant of retina, they have hope as allowed

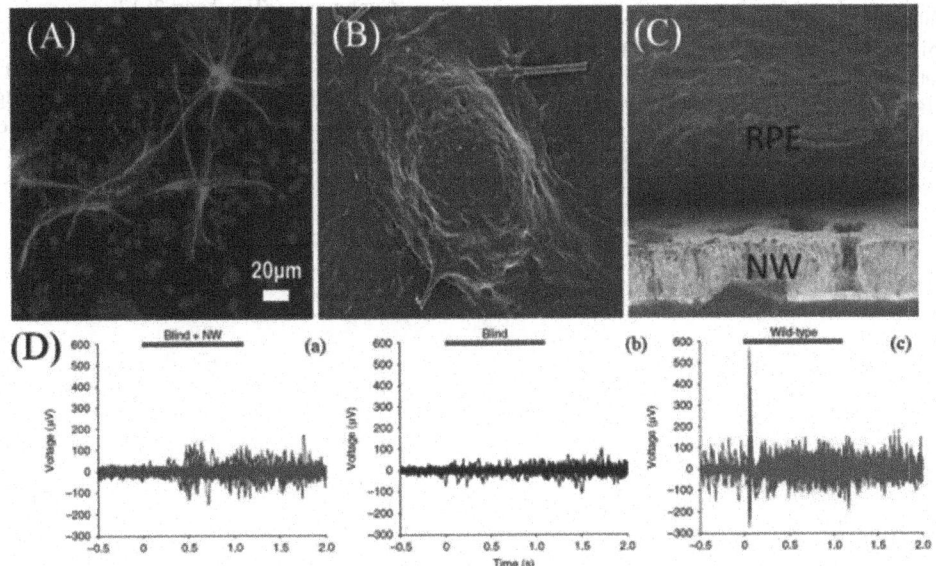

Fig. 56.6 Nanowires simulation

Source: Made by author

for capabilities offacial-recognition and have believe for allowing user to seen the colorful images possibly.

References

1. Yueqi Zhai, Jiaqi Niu, Jingquan Liu, Bin Yang,Bionic Artificial Compound Eyes Imaging System Based on Precision Engraving", 2021 IEEE 34th International Conference on Micro Electro Mechanical Systems (MEMS), Year: 2021

2. Avali Banerjee, S.B. Amreen Saba, Sangeeta Rana, Samayita Chakraborty, "Bionic Eye - A Review", 2020 8th International Conference on Reliability, Infocom Technologies and Optimization (Trends and Future Directions) (ICRITO), Year: 2020

3. N. V. Sibirev, Y. S. Berdnikov, V. N. Sibirev, V. G. Dubrovskii, "Stabilization of wurtzite crystal phase in arsenide nanowires via elastic stress", 2020 International Conference Laser Optics (ICLO), Year: 2020

4. Sandip Bhattacharya, Rajib Saha, Subhrajit Sikdar, Subrata Mandal, Chirantan Das, Sanatan Chattopadhyay, "Investigation of density and alignment of ZnO-nanowires grown by double-step chemical bath deposition (CBD/CBD) technique on metallic, insulating and semiconducting substrates", 2020 International Symposium on Devices, Circuits and Systems (ISDCS), Year: 2020

5. Jacque L. Duncan, "Adaptive Optics Scanning Laser Ophthalmoscopy in Retinal Degenerations: New Insights in Structure and Function", 2018 Conference on Lasers and Electro-Optics (CLEO), Year: 2018

6. Chennupati Jagadish, "Semiconductor Nanowires for Optoelectronics Applications", 2018 IEEE Research and Applications of Photonics In Defense Conference (RAPID), Year: 2018

7. Enrique Quiroga-González, A. Jesús Arzola Flores, Enrique Soto Eguibar, M. Audrey Ortega Ramírez, Octavio González Petlacalco, "Fabrication and testing of multielectrode matrix of disordered Si nanowires for brain tissue sensing", 2018 40th Annual International Conference of the IEEE Engineering in Medicine and Biology Society (EMBC), Year: 2018

8. Lintao Liu, Jun Liu, Ziqiang Huang, "Human-Eye Tracking and Location Algorithm Based on AdaBoost-STC and RF", 2017 2nd International Conference on Cybernetics, Robotics and Control (CRC), Year: 2017

9. Tian Zhang, Zhihui Lin, Mingliang Song, Bin Zhou, Rong Zhang, "Study on the thermoforming process of Hemispherical Resonator Gyros (HRGs)", 2017 IEEE International Symposium on Inertial Sensors and Systems (INERTIAL), Year: 2017

10. Xu Shoulong, Zou Shuliang, Huang Youjun, "γ-Ray Detection Using Commercial Off-the-Shelf CMOS and CCD Image Sensors", IEEE Sensors Journal, Volume: 17, Issue: 20, Year: 2017

11. Sanket Mehta, Arpita Patel, Jagrat Mehta, "CCD or CMOS Image sensorfor photography", 2015 International Conference on Communications and Signal Processing (ICCSP), Year: 2015

12. Weiguang Ding, Mei Young, Serge Bourgault, Sieun Lee, David A. Albiani, Andrew W. Kirker, Farzin Forooghian, Marinko V. Sarunic, Andrew B. Merkur, Mirza Faisal Beg, "Automatic detection of subretinal fluid and sub-retinal pigment epithelium fluid in optical coherence tomography images", 2013 35th Annual International Conference of the IEEE Engineering in Medicine and Biology Society (EMBC), Year: 2013

13. Jan Odstrcilik, Radim Kolar, Jiri Jan, Jiri Gazarek, Zdenek Kuna, Martina Vodakova, "Analysis of retinal nerve fiber layer via Markov random fields in color fundus images",2012 19th International Conference on Systems, Signals and Image Processing (IWSSIP), Year: 2012

14. Yumei Zhou, Jun Luo, Juqi Hu, Hengyu Li, Shaorong Xie, "Bionic eye system based on fuzzy adaptive PID control", 2012 IEEE International Conference on Robotics and Biomimetics (ROBIO), Year: 2012

15. Gökay Akinci, Ediz Polat, Orhan Murat Koçak, "A video based eye detection system for bipolar disorder diagnosis", 2012 20th Signal Processing and Communications Applications Conference (SIU), Year: 2012.

16. Muthyala Veera Venkata Satyanarayana Chowdary, Teppala Venkata Ramana "Automatic recognition of color sensation with controlled phosphene brightness using pre-trained CNNs framework", Indonesian Journal of Electrical Engineering and Computer Science (IJEECS),Year: 2023.

ICU Monitoring System Using Zynq Architecture Based SoC

57

Dharmavaram Asha Devi[1]

Dept. of ECE, Sreenidhi Institute of Science & Technology,
Hyderabad, India

M. Suresh Babu[2]

Dept. of CSE, Teegala Krishna Reddy Engineering College,
Hyderabad, India

G.Sirisha[3], Muthyala V. V. S. Chowdary[4]

Dept of ECE,Teegala Krishna Reddy Engineering College,
Hyderabad, India

Abstract: Monitoring the health status of the ICU patients is a very important aspect because a little delay in treating the patient can result in mortality. In the case of babies, the situation gets worse when the baby is delivered prematurely. A baby may be at high risk and have higher possibility of being admitted to the intensive care unitdue to any of the reasons like breathing difficulties, heart issues, infections, or birth abnormalities. At those conditions, the ICU provides care for infants who have medical disorders and health issues. For these issues we developed an ICU (Intensive Care Unit) Monitoring System Using ZYNQ Architecture based SoC design. It is a semi custom ASIC design meant for ICU application. Here, the required soft core IPs: Zynq Processing Unit, AXI interface Unit, PMOD sensor IPs, and the PMOD OLEDrgb IPs from Xilinx and Digilent, are integrated at the block level using Xilinx Vivado tool. The application software is developed in C at the SDK environment and finally verified on Zed board. The advantage of the proposed Zynq architecture based design is enhanced flexibly by adding some more IPs and the corresponding software with the same target.

Keywords: ASIC, AXI, ICU, IP, PMOD, Zynq architecture.

1. Introduction

ICU is an essential hospital unit that provides life support to the patients with critical illnesses. The most crucial ICU task is patient monitoring because even a slight delay in a patient's condition deteriorating can cause fatality. Failures in monitoring that are directly tied to individuals frequently cause ICU patients' deteriorating to go unnoticed. Numerous studies show that ICU healthcare staff has tremendous workload, which may lead to the improper monitoring of the patients.

The HVAC system, also referred to as the ICU's heating, ventilation, and air conditioning system, is used in accordance with the environmental requirements. It must have full air conditioning with humidity, temperature, and air flow controls. It is recommended that the air be 99% efficiently filtered to a size of 5 microns. In critical care units with enclosed patient modules, each patient module shall have a thermostat that can be set to a temperature of 16 to 25 degrees Celsius.

Although there are many embedded systems-based monitoring and controlling solutions for ICUs, more

[1]ashadevi.d@rediff.com, [2]sureshcse@tkrec.ac.in, [3]gshirisha4@gmail.com, [4]mvvsvenkatchowdary716@gmail.com

reconfigurable monitoring units with higher precision are required for such applications. So, we are implementing an architecture-based ICU environmental status monitoring. In this, we are monitoring the temperature, humidity and light in the ICU.

2. Literature Review

Sanyadwia Ghinasni Zen et al. 2021[1], A slight delay in recognizing a patient's deterioration in the intensive care unit (ICU) might result in a lasting disability or even death. With the use of IoT, this project aims to enhance the operational procedure of ICU monitoring. The Complex Proportional Assessment (COPRAS) method is used to rank the risks related to the current monitoring procedure in order to select the optimum IoT. Through the integration of IoT and the Business Process Reengineering methodology, this study aims to optimize the operational processes for ICU patient monitoring.

D. A. Devi et al. 2021 [2] demonstrated an FPGA development board based on the Zynq architecture. They created a Wi-Fi networked ICU control and monitoring system in this project. The functionality of this system is to analyse and show the ICU environmental data on an OLED screen while remotely managing several appliances close to a Wi-Fi network. In order to ensure the adaptability and efficient monitoring of ICUs, such systems are helpful in both business and public hospitals.

Vaibhavi Bhelkar et al. 2016 in [3] designed implemented the health monitoring device using FPGA. In this project they used Spartan 3 FPGA board and it is used to calculate various parameters like temperature, heartbeat etc. The used heartbeat sensor to calculate the heartbeat and accelerometer to find the acceleration which may be due to result of motion. The patient's temperature is determined using the LM35 sensor. They used different software's like Xilinx and visual basic to implement the project. It also sends an emergency alert message to their relatives and doctors to take care of the patient in case of any emergency.

D. A. Devi et al. 2020 in [4] had shown the project design and implementation of reconfigurable System on Chip based data acquisition system with high performance. It is a semi-custom design that uses Zynq processing system IP, programmable logic on a reconfigurable 7-series FPGA, hygro, ambient light sensor, and OLEDrgb accessory module IPs. The approach suggested utilizes a reconfigurable SoC. with a 100MHz operating frequency intended for fast data acquisition systems. The system works well for fast and affordable real-time embedded systems.

Yu Wang et al. 2019in [5], implemented a low-cost, reconfigurable FPGA-based IoT a multi-sensor healthcare

infrastructure can be used to connect a pulse sensor interface intended for monitoring a user's pulse indicators and tell whether their heart rate is normal or abnormal. They used FPGA because it can be reconfigurable . A low-cost early validation platform design can be produced using the FPGA-based platform. They implemented a pulse sensor interface using VHDL programming and FPGA technology.

S. M. Hussain et al. 2021 in [6], implemented a project on an FPGA implementation of a health monitoring system using IoT. They designed the project using the Spartan3AN FPGA board. They interfaced different sensors like temperature sensor, pulse sensor, A/D converter, Wifi module, and LCD to the FPGA board. They used Xilinx ISE software for the project simulation. In this project, they used the advantages of FPGA and IoT technology. This project continuously monitors the temperature and heart rate of a person. This project also used the concept of IP address to monitor the health of the patient.

Anjali Chindham et al. 2019 in [7], discussed on a project, FPGA-based health monitoring system so that medical practitioners may keep an eye on their patients. This technique is useful for keeping track of regular physical examinations for regular people as well. An FPGA-based health monitoring system can offer actual time evaluation of data regarding the physiological state of a human body or patient. Body temperature, blood pressure, respiration rate, and pulse rate information from sensors are used as input parameters by this system. Verilog HDL and the Xilinx ISE EDA tool were used to create this system with the Spartan-3 FPGA in mind. This system will accurately track the patient's health state by considering the aforementioned factors.

While performing any ICU monitoring and control of operations, ubiquitous, sensitive and secured data is to be maintained [8],[9]. Scheduled routing algorithms are used in this communication networks. Smart monitoring is to be implemented in Ambulances also[10],[11]. For the time dependent routine operations, real time clock[12] is very much essential in the monitoring and control system.

3. Methodology

The proposed work is a reconfigurable SoC design. It is a block level design using Zynq Processing Unit and the required PMOD IPs interconnected via AXI. Each block is composed of soft IP cores that are integrated into a design. Once the design has been correctly connected, create output ports, and then verify the validity of the design. After validation is finished, create design's associated HDL wrapper. The next three processes are synthesis, implementation, and bit stream creation. Xilinx Vivado Systems Design Suite is the software platform that is utilized throughout all of these processing

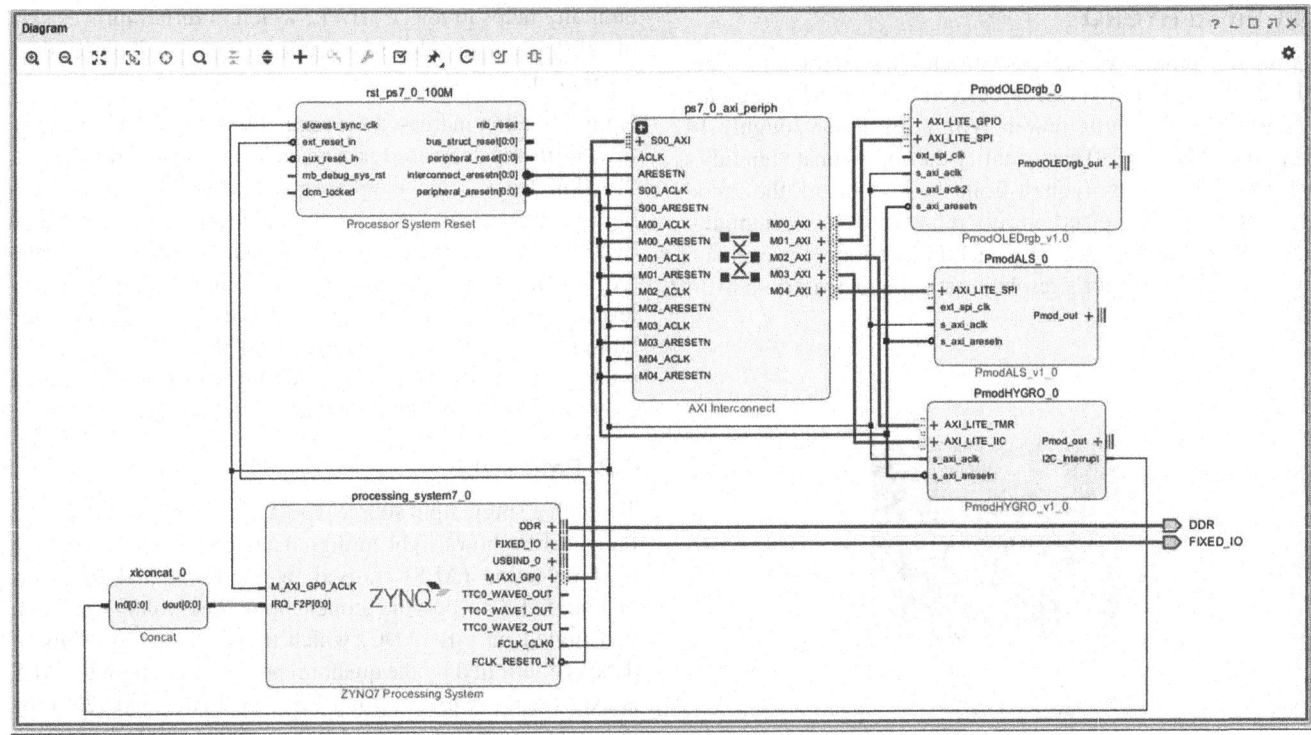

Fig. 57.1 Architectural block diagram of ICU monitoring system

phases. Following the bitstream has been created,export the device and launch the Software Development Kit(SDK). This integrated development environment is used to create embedded programmes.

The necessary application logic is constructed using C/C+ and HTML for the website's page in an application project that is built on the SDK platform. Build the project, and then use programming to load the intricate layout into the programmable logic of the FPGA. It processes application software theARM9 dualcore processor architecture. Following the hardware debugging option's execution, keep an eye on the outcomes both at the hardware level and on the results-monitoring webpage.

The components of the ICU Monitoring system's block diagram as illustrated in Fig. 57.1 are the ZYNQ7 Processing System, processor system reset, PmodOLED, PmodALS, PmodHYGRO, and Concat. Programmable logic (FPGA) and an ARM7 dual-core CPU make up the ZYNQ7 processing system. All of the P mods are connected to the ZYNQ7 processing system using the AXI interface. When necessary, the system can be reset using the processor system reset. Temperature and humidity sensors are built into PmodHYGRO. Concat is used to relatively flip between the two sensors.

4. Hardware Components

Zed Board: The Zed Board illustrated in Fig. 57.2 automatically establishes a connection with the host PC with proper configuration setup of hardware and software. Over USB-UART, a new connection can be established. If using the Terminal to facilitate simple board-PC communication. Keep in mind that a third micro-USB port (USB- OTG) is also provided for connecting USB accessories.

Fig. 57.2 Zed board

4.1 Pmod HYGRO

A relative humidity detector with an integrated temperature detector, the Pmod HYGRO delivers substantially accurate measurements at little power. With a resolution roughly 14 bits, the TI HDC1080 can quantify the approximate humidity of the terrain. Upon request from the host board, the Pmod HYGRO is built to electronically report the relative humidity and ambient temperature. By enabling for longer conversion periods, each detector's resolution can be increased up to 14 bits.

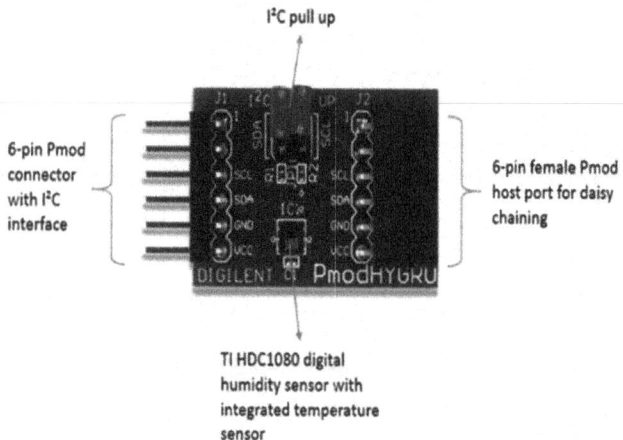

I²C pull up

6-pin Pmod connector with I²C interface

6-pin female Pmod host port for daisy chaining

TI HDC1080 digital humidity sensor with integrated temperature sensor

Fig. 57.3 Pmod HYGRO

Importance of Pmod HYGRO

A temperature sensor is necessary for every process heating application because it provides temperature information about the process that can be employed to track or manage the operation. The ideal temperature for intensive care units is between 21 and 24 C. Monitoring newborn babies in ICU is ineffective. Due to the serious morbidities and mortality brought on by neonatal hypothermia, it is essential to accurately monitor newborns' body temperatures. The main goal of taking a newborn's temperature is to identify any signs of cold stress because fever is a rare sign of sickness and is frequently influenced by environmental variables.

It is possible to stop the transmission of disease and maintain a secure and healthy environment by using sensors to monitor humidity levels. The function of a humidity sensor is to detect, gauge, and report the air's relative humidity (RH) or assess how much water vapour is present in a combination of gases (air) or a pure gas. Australia's recommendations for burns ICU advised a higher humidity range (30-95%) likely to speed up wound healing compared to the USA's prescription of 40-60%.Use of humidity in newborns Trans epidermal water loss (TEWL) is more likely to occur in newborns who were born before 30 weeks of gestation because their stratum corneum and epidermis are still developing. Utilising ambient

humidity helps to lower TEWL, which in turn improves skin integrity, fluid and electrolyte management, and temperature regulation.

Humidification increased the regulation of body temperature from birth and decreased skin water loss, but it did not prevent the skin from developing normally after birth. Despite a high incubator air temperature, babies frequently became hypothermic when not adequately humidified when being fed. Babies thrive in environments with greater humidity levels. Even some medical professionals advise keeping the humidity in baby rooms between 50 and 70 percent. The issue with it is that excessive humidity encourages bacterial growth in the air, which can make our infant sick.

4.2 Pmod ALS

Through a single light source detector, the Digi sophisticated Pmod ALS shows light-to-digital vision. A single ambient light detection (ALS) is used in the functional overview of PmodALS to provide stoner input. The voltage position transmitted into the ADC, which transforms it to 8 bits of data, is controlled by the quantum properties of light the ALS is exposed to. A place with a value of 0 is considered to be in low light, while a value of 255 is considered to be in high light.

Fig. 57.4 Pmod ALS

Importance of ALS

One of the most common stressors experienced with a severe illness is sleep deprivation. Sleep loss and sleeplessness have been named by survivors as important sources of stress and nervousness while receiving treatment in the ICU. Melatonin secretion's disrupted circadian rhythm has been seen in ICU patients who were also being intensively ventilated and on drugs. According to a theory, sleepiness in critically sick patients undergoing mechanical ventilation is brought on by a decline in bloodstream melatonin levels linked to the interruption of circadian rhythms. In addition to delirium associated with critical care, neuropsychological impairment, the duration of mechanical breathing, and compromised immune system, sleep deprivation in the critically ill may also contribute to emotional distress. Both immediate and long-term sleep disruption are brought due to light, which also has an impact on circadian function, a critical mechanism supporting restful sleep. The circadian rhythm is disrupted by brief bursts of light at night and a lack of bright daylight, which impairs sleep. According to several researches, bright blue light, as opposed to bright red or conventional white fluorescent light, lessens acute kidney impairment in sepsis-exposed rats.

Therefore, the monitoring light intensity is crucial for ICU patients.

4.3 Pmod OLED rgb

An organic RGB LED module from Digilent called the Pmod OLED rgb has a 96 x 64-pixel display and supports 16-bit colour resolution. Designed to work in confluence with a host board, communicating via the SPI protocol, the Pmod OLED rgb is a graphical display that allows druggies the option to light up any individual pixel on the OLED screen. The Solomon Systech SSD1331 present controllers are used by Pmod OLED RGB to receive data from the host board and show the necessary data on the OLED screen.

Fig. 57.5   Pmod OLED rgb

5. Implementation

The SDK is a tool for designing, constructing, and programming an FPGA. The target board must be created before creating an application project must be connected to the computer system where it is already developed. There are two distinct types of relationships present here. The first is used for programming with USB to JTAG, whereas the additional one is for asynchronous connection with UART.

6. Results

As illustrated in Fig. 57.6, the PMOD OLED rgb, PMOD ALS and PMODHYGRO are connected to the Zed board as illustrated in the Fig. 57.6.

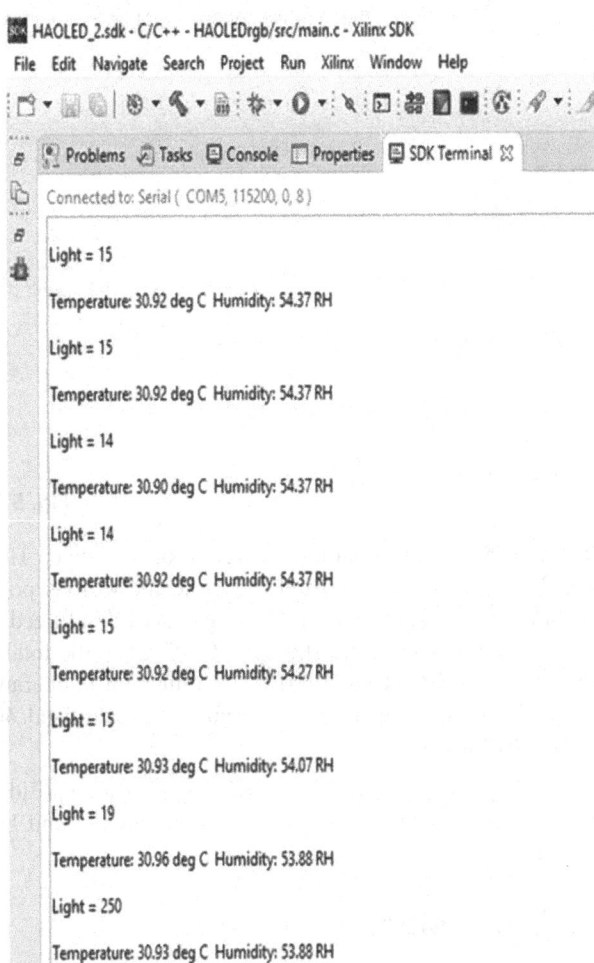

Fig. 57.6   Zed board Interfacing with PMODs

The Fig. 57.7 illustrated the monitoring parameters of the ICU. We can see the output at the SDK Terminal of various parameters values of the ICU. We are measuring the various parameters like temperature, humidity and light.

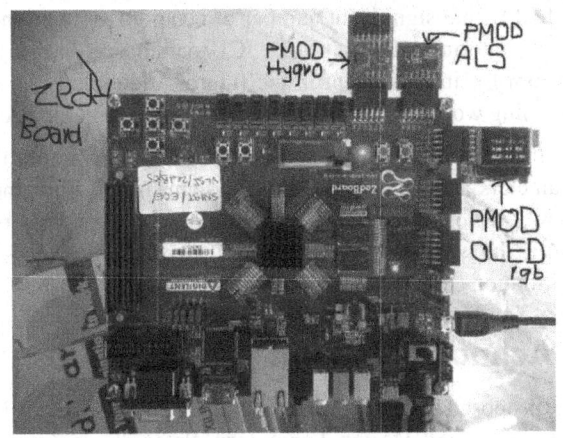

Fig. 57.7   SDK terminal

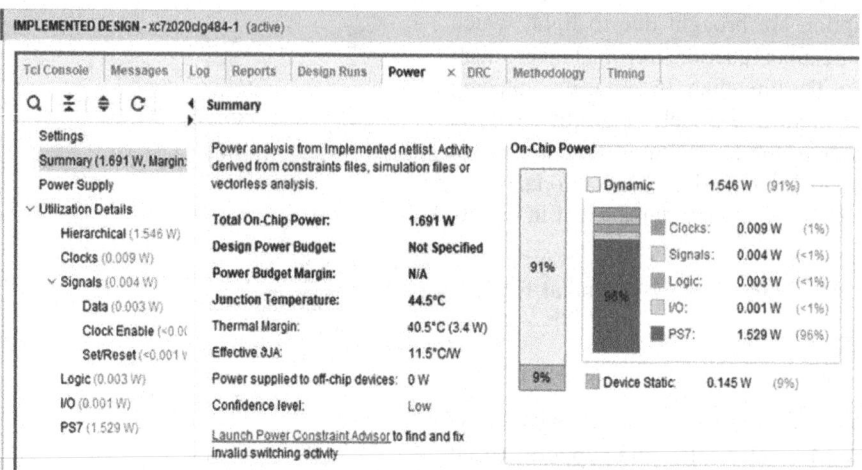

Fig. 57.8 Power report

The Fig. 57.8 tells about the utilization of the power. Here, the dynamic power is utilized 91%.And the various power parameters like clock, signal, IO, Logic and PS7 utilized the remaining 9%.The device static has 0.145w and the total on chip power is about the 1.691W. The junction temperature utilized 44.5⁰C. The thermal margin power isutilized 40.5 degree Celsius.

The advantages of the proposed system: Patients/Elder's safety can be assured, low power, reconfigurable and high speed.

7. Conclusion

The suggested study has a solid track record of using an embedded programme to monitor and regulate the ambient light, humidity, and temperature parameters. For real-time ICU, a comparable technique can be applied. Microprocessors and microcontrollers can both implement the same kind of application. Flexible Reconfigurability yet is lacking. The Zynq architecture-based SoC design is executed in this method with a significant number of benefits, as mentioned. The monitoring and control of ICU parameters have a lot of room for improvement in the future. Additionally, in the upcoming work, the patient monitoring can be established.

For the upcoming work, we plan to construct a multiple parameters accessing and control system. It will implement a power-efficient medium access control (MAC) convention within the smaller measured sensor organization for the information consistency.

8. Acknowledgment

The proposed research work is implemented in the VLSI Technology and Design Laboratory, Sreenidhi Institute of Science and Technology, Hyderabad. The authors of this paper express their gratitude to the management of this institute for the support and encouragement given.

References

1. Sanyadwia Ghinasni Zen, M. Dachyar. (2015). The Intensive Care Unit (ICU) Patient Monitoring Improvement Using Internet of Things (IOT) Based on BPR Approach, 11th Annual International Conference on Industrial Engineering and Operations Management Singapore, March 7–11, 2021.
2. Devi, D.A., Savithri, T.S., Babu, M.S. (2021). Monitoring and Controlling of ICU Environmental Status with WiFi Network Implementation on Zynq SoC. In: Suma, V., Chen, J.IZ., Baig, Z., Wang, H. (eds) Inventive Systems and Control. Lecture Notes in Networks and Systems,s vol 204. Springer, Singapore. https://doi.org/10.1007/978-981-16-1395-1_48
3. V. Bhelkar and D. K. Shedge, "Different types of wearable sensors and health monitoring systems: A survey," 2016 2nd International Conference on Applied and Theoretical Computing and Communication Technology (iCATccT), Bangalore, India, 2016, pp. 43–48, doi: 10.1109/ICATCCT.2016.7911963.
4. Dharmavaram Asha Devi, Tirumala Satya Savithri and Sai Sugun.L, "Design and Implementation of Real Time Data Acquisition System using Reconfigurable SoC" International Journal of Advanced Computer Science and Applications(IJACSA), 11(9), pp. 325–331, 2020.
5. Yu Wang, Sunghoon Jang A pulse sensor interface design for FPGA based multisensor health monitoring platform, International Journal of Biosensors & Bioelectronics. 2019; 5(1): 23–27 at DOI: 10.15406/ijbsbe.2019.05.00147.
6. S. Munavvar Hussain, A. Jhansi Naga Sai Surekha, D. Archana, N. Hannah Priyanka, An FPGA Implementation Of Health Monitoring System Using Iot, International Journal of Creative Research Thoughts, Vol 6, Issue 2,l pp. 75–80, April 2018.

7. Anjali Chindham, Donthagani Rakesh, Sabavath Virisha, Racha Ganesh, Design of Health Monitoring System using FPGA, International Journal of Research Publication and Reviews, Vol 3, no 12, pp 2106–2110 December 2022.

8. M. Suresh Babu, K. Bhavana Raj, D.Asha Devi.: Future Trends of Business Intelligence and Big Data Analytics in Ubiquitous Environment, International Journal of Engineering and Advanced Technology, Vol.8, Issue-3S, pp. 773–778, (2019)

9. Babu, M.S., Raj, K.B., Devi, D.A. (2021). Data Security and Sensitive Data Protection using Privacy by Design Technique. In: Haldorai, A., Ramu, A., Mohanram, S., Chen, MY. (eds) 2nd EAI International Conference on Big Data Innovation for Sustainable Cognitive Computing. EAI/Springer Innovations in Communication and Computing. Springer, Cham. https://doi.org/10.1007/978-3-030-47560-4_14

10. Devi, D.A., Jaga, S., "Analysis of scheduled routing algorithms on 5-port router for network on chip application", International Journal of Scientific and Technology Research, 8(9), pp. 2148–2153, 2019.

11. M. Bhavani and D. A. Devi, "Design of smart Monitor for automobiles using FPGA based Data Logger," *2019 International Conference on Communication and Electronics Systems (ICCES)*, Coimbatore, India, 2019, pp. 1940–1945, doi: 10.1109/ICCES45898.2019.9002034.

12. D. A. Devi and N. S. Rani, "Design and Implementation of custom IP for Real Time Clock on Reconfigurable Device," 2019 Third International Conference on Inventive Systems and Control (ICISC), 2019, pp. 414–418, doi: 10.1109/ICISC44355.2019.9036428.

13. Dharmavaram Asha Devi, Niharika Reddy Kathula, Gopinath Kalluri, and Leela Sai Bondalapati, "Design and Implementation of Image Processing Application with Zynq SoC, ISSN (2210-142X) Int. J. Com. Dig. Sys.14, No.1 (Jul-23).

Note: All the figures in this chapter were author self test data and experiment results were performed as part of their research work.

Noise Cancellation of Electrocardiogram Signal Using Moving Average Filter

58

M. Thomas Chinmai Chowdary[1], Ch. Lekhaz Kumar[2],
N. R. Kavitha[3], Dharmavaram Asha Devi[4]
Dept. of Electronics and Communication Engineering,
Sreenidhi Institute of Science and Technology, Hyderabad, India.

Abstract: The electrocardiogram (ECG) is a vital tool for monitoring heart conditions in the field of cardiac medicine. The raw ECG signal often comes with white noise; hence, it is really important to consider signal filtering. As part of the pre-processing procedure, we must apply a filter to the signal to lower the noise. The Moving Average filter, a type of smoothing filter, eliminates noise and short-term overshoots from a signal while preserving the original signal representation. The major goal of this study is to evaluate the efficacy of a moving average filter by examining how well it works to minimize noise in an ECG signal. Using the moving average filter on ECG data produced promising noise reduction results. When processing ECG data, moving average filters demonstrated good noise cancellation by successfully attenuating several kinds of noise. In conclusion, by implementing the filter using tools like MATLAB and VIVADO in our paper, we get to finalize that this filter indeed helps in noise-free ECG signal analysis. Despite its simplicity, this filter is anticipated to have a significant and broad impact on the signal processing industry.

Keywords: Electrocardiogram, Moving average filter, White noise, Effective noise cancellation, Signal analysis

1. Introduction

Monitoring the overall condition of the heart is an important step that should be handled with careful attention to detail and must be accurate. The Electrocardiogram (ECG) is a device that offers specific details regarding the cardiac pulse, rhythm, and general condition of the heart. By carefully examining the wave-forms it presents, it helps maintain heart health and identify potentially dangerous cardiac diseases. Electrodes are usually fastened to a person's body during an ECG. These electrodes record a variety of wave-forms when the device collects the electrical impulses from the heart's contractions. These wave-forms, which are frequently referred to as ECG tracings, show the depolarization and re-polarization of the heart's chambers. ECG's are also used to monitor patients taking medications to screen for any cardio-toxic effects. In the field of medical research and development, an ECG may be used to assess the safety of recently discovered

medications as well as any unexpected cardiac activity that may occur when taking that particular medication.

ECG signals can easily be impacted by noise and interference, which can significantly alter the tracings and prevent us from perceiving the information. Electrical interference from a variety of sources, including power lines, electronic gadgets, and muscle movements, may have an influence on ECG readings. Due to the possibility of noise in the ECG signal caused by these outside influences, it could be challenging to accurately assess the heart's activity.

Hence Noise cancellation is considered the most significant step in signal analysis and is done during the pre-processing stage. Many noise cancellation techniques can help in removing the said noise, but the most significant point here is that there must be no signal loss, which is achieved by very few filters. We can eventually agree that the trade-off between noise reduction and signal preservation is a crucial factor to

[1]chinmaichowdarymaddineni @gmail.com, [2]chlekhazkumar@gmail.com [3]kavithanr04@gmail.com, [4]ashadevi.d@rediff.com

take into account when choosing the best noise cancellation technique.

2. Moving Average Filter

The Moving Average filter is an uncomplicated yet useful filter that is employed in pre-processing to help remove the noise from an ECG signal. It is a filter whose impulse response happens to become zero in a finite period of time. It is a basic and extensively used digital signal processing method for reducing noise and smoothing time-series data.

Its versatility in the area of signal processing is a result of the fact that it not only attenuates noise but also maintains an accurate step response and preserves the originality of the signals. The two foremost benefits of a moving average filter are the simplicity of the conceptual model and the ease of implementation. Because it is easy to use, efficient at eliminating high-frequency noise, and maintains crucial cardiac information, the moving average filter is commonly used to de-noise ECG data. By averaging adjacent samples, it smooths the signal and minimizes noise spikes without changing the fundamental waveform. The filter is simple to operate, exhibits little phase distortion, and is efficient computationally. This makes it seem more appealing and applicable.

This filter acts as a low pass filter in signal processing. It attenuates the high frequency components and smooths the data.

A moving average filter computes the average of a certain number of successive data points and incorporates the result as the output value for each data point. When given an array of data and a specific subset size (window size), the average of the first subset in the series is calculated to form the first component of a moving average filter. To get the subsequent member, the subset is then advanced. It skips the first number and continues on to the next.

Figure 58.1 describes the step-by-step operation of a Moving average filter In the form of a flowchart.

2.1 Operation of a Moving Average Filter

- The buffer is initialized once the filtering process has begun. The buffer, which is also the main window where the averaging action takes place, is utilized to store n data points. This indicates that the window size is n.
- In step 2, new input samples are collected. These are the inputs for data that will eventually be kept in the buffer.
- On the basis of one specification, the samples are added to the buffer.
- According to the flowchart, if the buffer size after adding the input sample is larger than the window size, the oldest

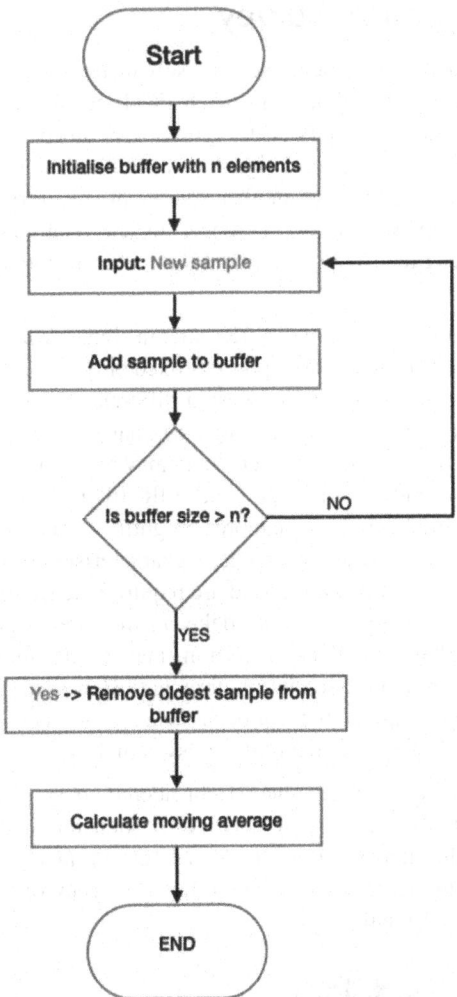

Fig. 58.1 Flowchart of moving average filter

sample is deleted from the buffer; otherwise, the input sample is added without removing any other samples.

- After completing the window's averaging, the output data points are calculated.

The output data for this filter of order N and of window size L for an input signal x(n) is specified in the below equation.

$$y[n] = \frac{1}{L}\sum_{k=0}^{L-1} X[n-k] \qquad (1)$$

Electrocardiogram signal when applied to a moving average filter removes the baseline wander which is caused by the patient's sudden movements. Power-line interference (PLI), which results from the 50/60 Hz electrical impulses present in the surroundings, is a typical cause of noise in ECG data. The moving average filter can reduce PLI, particularly if a large enough window size is selected.

3. Literature Survey

There are many cases where noise can be produced in an ECG signal. The denoising of an ECG signal can be done with the help of filters at the time of pre-processing. It has been noted that the ECG signal's amplitude contains 50% of the baseline noise and another significant category of noise is due to power line interference, muscle contractions also known as Electromyography (EMG) and instrumental noises [1].

A finite impulse response (FIR) filter in signal processing is a filter where its impulse response is fixed only for limited time. Many researchers have suggested different filters to reduce noise in ECG signals. This signal is hard to decipher. The original information after denoising may be obtained through low-pass, moving average, and FIR filters. The moving average filter is one of the more straightforward digital filters [2]. In most cases, linear phase characteristics throughout the filter's specified pass band are required, hence FIR filters are often utilized. The only delay in the input signals, but the avoidance of phase distortion, makes this linear phase characteristic an essential requirement.[3] We can see from the filtered signal that noise has been eliminated since its power is lower than that of the noisy signal [4].

The moving average filter is considered to be one of the most widely used digital filter. An n-point moving average (MA) filter may be used to attenuate or eliminate, the EMG (Electromyography noise which has frequency of >100Hz), from ECG signals [5].

4. Methodology

4.1 Existing Method

Noise reduction is not usually required since the electrocardiogram (ECG) signal is often clear and does not experience severe noise interference. However, there are a few circumstances in which noise cancellation methods could be beneficial for enhancing the ECG signal's quality. There are delays and an increase in chip size as power consumption increases when utilizing complex adaptive filtering techniques like Wavelet transform, Kalman filtering, Independent Component Analysis, and Principal Component Analysis [6]. It may be quite difficult to remove noise from a signal. The original data might be lost, which could even result in the signal components being lost. In addition, it is important to keep in mind that the adaptive filtering strategy can only be effective if the reference input is significantly linked with the motion artifact overlaid on the ECG [8]. As a result, it is believed that these filters are less effective at processing noisy signals.

4.2 Proposed Method

Random white noise has the ability to corrupt modern ECG. The moving average filter is used in the suggested methods to remove the noise from ECG data signals. Our study's major goal is to demonstrate how noise may be successfully eliminated from ECG data using a moving average filter. Utilizing two different yet potent tools, MATLAB and VIVADO [11], we implement this filter.

The proposed Verilog implementation of the computationally efficient and hardware-friendly Moving Average Filter may be used for real-time ECG signal analysis. In order to improve the accuracy and dependability of cardiac diagnosis and monitoring, it may be easily integrated into portable medical equipment and ECG monitoring systems. With a minimal length of code, Verilog has been used as a building block to create several large-scale circuits. After designing the circuit and testing its functionality, further implementation is also done using Artix-7 Field Programmable Gate Array (FPGA) which gives us the precise report and is adaptable [13].

The circuit is framed using the moving average filter equations. Unit delays are implemented using D flip-flops, and the intermediate stages involve multiplying and adding the inputs at each separate point, which are then averaged [14], to produce the de-noised output.

A 4th order Moving Average FIR filter (Fig. 58.2) typically averages the input data from the current sample as well as the three samples before it. The average is calculated by adding together the four samples and dividing the total by four.

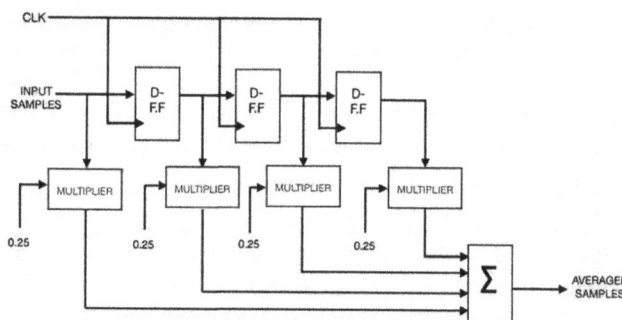

Fig. 58.2 4th order moving average filter

The Moving Average filter equations are:

$$y[n] = 1/4 \ (x[n] + x[n-1] + x[n-2] + x[n-3]) \quad (2)$$

Which is equal to:

$$y[n] = 0.25(x[n] + x[n-1] + x[n-2] + x[n-3]) \quad (3)$$

The methodology implementation first starts with generation of an ideal and noise-free ECG signal. To create the noisy ECG signal, this signal is used as a reference, to which varying amounts of noise are then added. This completes the formation of the noisy ECG signal. These signals are further converted to binary representation, which makes sure that they are suitable and prepared for Verilog implementation. MATLAB has been used throughout the entire process thus far.

We next proceed by using VIVADO to implement the Moving Average filter. Movable parameters, such as the filter length and average window size, are used in the filter design. This increases its adaptability and flexibility to meet numerous requirements. The noise cancellation system is tested using ECG signals produced with all the noise characteristics once the filter design has been completed.

The system's functionality is verified by writing the test bench, where we dump the MATLAB-generated ECG signals as a single file, which is eventually taken as the input signals. The developed Moving Average filter's performance is assessed, and the simulation outcome is witnessed when the filtered ECG signal is created. We may contrast the filtered ECG signal with the unfiltered reference signal to determine the extent of noise reduction the Moving Average filter achieved. Results from the simulation demonstrate how the suggested noise cancellation approach efficiently lowers noise and improves the quality of the ECG signal.

While the Moving Average Filter successfully eliminates undesired artifacts, baseline drift, and high-frequency noise, the crucial QRS complexes and other important ECG waveform components are also maintained. The input reference signal, noisy ECG signal, and scaled signals are illustrated in Fig. 58.3, Fig. 58.4 and Fig. 58.5 respectively.

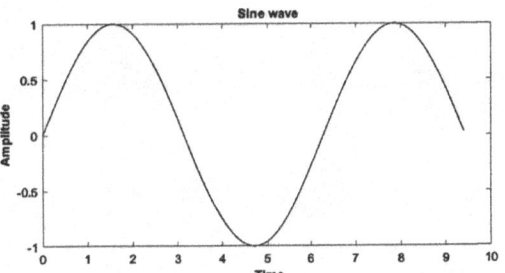

Fig. 58.3   Input reference signal

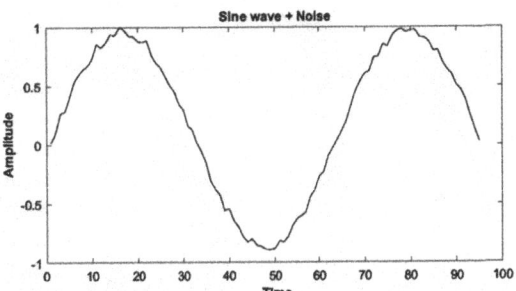

Fig. 58.4   Noisy ECG signal

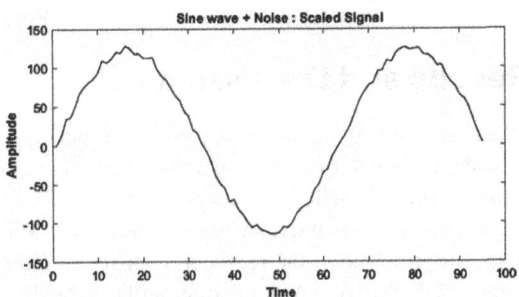

Fig. 58.5   Scaled signal

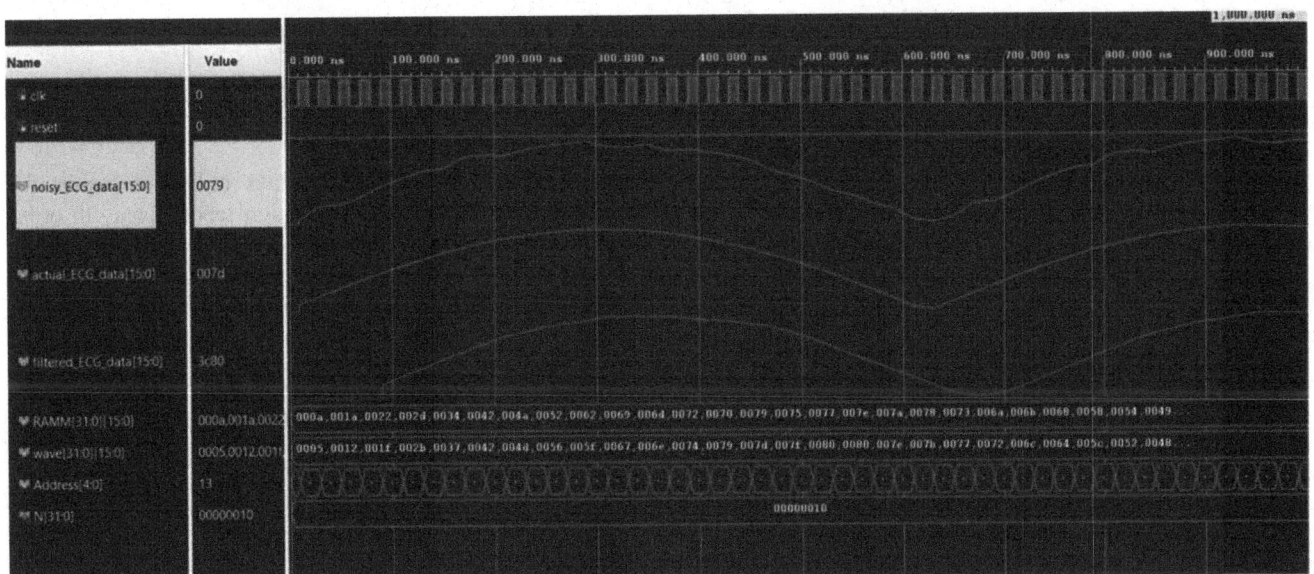

Fig. 58.6   Vivado simulation output

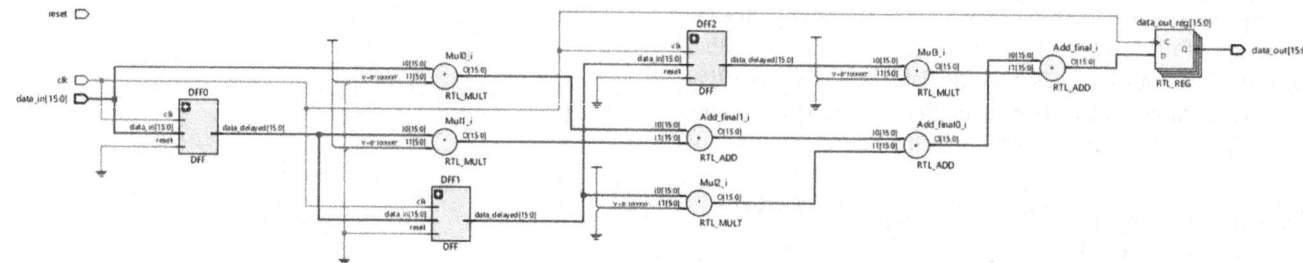

Fig. 58.7   RTL schematic

Name	Slack ^1	Levels	Routes	High Fanout	From	To	Total Delay	Logic Delay	Net Delay	Requirement	Source Clock
Path 1	∞	9	10	2	data_in[0]	data_out_reg[15]/D	5.517	4.090	1.427	∞	input port clock
Path 2	∞	9	10	2	data_in[0]	data_out_reg[13]/D	5.497	3.934	1.563	∞	input port clock
Path 3	∞	9	10	2	data_in[0]	data_out_reg[14]/D	5.452	4.025	1.427	∞	input port clock
Path 4	∞	9	10	2	data_in[0]	data_out_reg[12]/D	5.392	3.829	1.563	∞	input port clock
Path 5	∞	8	9	2	data_in[0]	data_out_reg[11]/D	5.270	3.707	1.563	∞	input port clock
Path 6	∞	8	9	2	data_in[0]	data_out_reg[10]/D	5.205	3.642	1.563	∞	input port clock
Path 7	∞	8	9	2	data_in[0]	data_out_reg[9]/D	5.049	3.346	1.703	∞	input port clock
Path 8	∞	8	9	2	data_in[0]	data_out_reg[8]/D	4.944	3.241	1.703	∞	input port clock
Path 9	∞	7	8	2	data_in[0]	data_out_reg[7]/D	4.591	2.888	1.703	∞	input port clock
Path 10	∞	7	8	2	data_in[0]	data_out_reg[6]/D	4.482	2.775	1.707	∞	input port clock

Fig. 58.8   Timing report

5. Results and Discussion

Following the design's simulation, which is accomplished with the development of a test bench for it, the functionality is evaluated and a conclusion is reached. It shows us the denoised signal from the provided sample noisy ECG data as the output. Furthermore, the Register Transfer Logic (RTL) Schematic of the Moving Average filter will also be displayed after the RTL analysis. The interconnection of various components in the design is shown in the schematic diagram (Fig. 58.7) which helps us to easily understand the flow of the data.

Additionally, we can inspect the Moving Average Filter's timing, power, and usage data, which is found by implementation using Artix-7 FPGA.

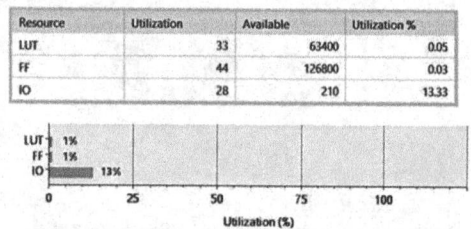

Fig. 58.9   Utilization report

From the above results we can see the worst path delay is 5.517n seconds along with Total On-Chip Power is 0.128 W where Dynamic power covers 24% of the total power and device static power is 76% as illustrated in Fig. 58.10.

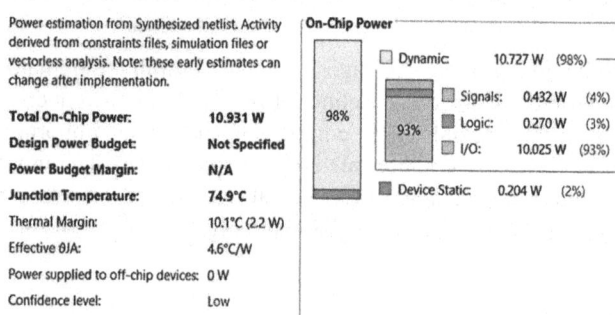

Power estimation from Synthesized netlist. Activity derived from constraints files, simulation files or vectorless analysis. Note: these early estimates can change after implementation.

Total On-Chip Power:	10.931 W
Design Power Budget:	Not Specified
Power Budget Margin:	N/A
Junction Temperature:	74.9°C
Thermal Margin:	10.1°C (2.2 W)
Effective θJA:	4.6°C/W
Power supplied to off-chip devices:	0 W
Confidence level:	Low

Fig. 58.10   Power report

We can also observe the utilization report where it utilizes 33 LUT's of 63400 which gives 0.05% and Utilization of flip-flops doesn't cross 0.03%.

Fig. 58.11 and Fig. 58.12 displays utilization report for a 32 tap LMS filter and a utilization report for a 4th order moving average filter, respectively. These complicated signal processing systems are implemented using Virtex-II FPGA chips. According to the above-mentioned findings, Adaptive Filter implementation requires more LUT's, flip-flops, and

Arch. Length	AC97 Control	FIR Filter			AC97 Initi.	Error Control
		8	16	32		
Slices (10752)	181	542	1073	2124	127	36
LUT4 (21504)	188	524	1032	2023	199	63
FF (21504)	137	682	1337	2649	76	55
MUL18 (56)	0	9	17	33	0	0
Freq.(MHz)	146	66	67	61	97	96

Fig. 58.11   Utilization report of 32 Tap LMS filter [7]

Resource	Utilization	Available	Utilization %
LUT	33	63400	0.05
FF	44	126800	0.03
IO	28	210	13.33

Fig. 58.12 Utilization report of 4ᵗʰ order moving average filter

other units than Moving Average Filter. Adopting complex adaptive filtering techniques results in delays, a rise in chip use, and an increase in power consumption. Therefore, it is believed that these filters perform less well while processing noisy signals.

Despite how straightforward the Moving Average filter may be to use, it has certain drawbacks. The trade-off between removal of noise and smoothing the signal will always exist. Additionally, the Moving Average filter cannot be used directly in the frequency domain. It is possible to change this issue by utilizing a variety of strategies, but additional research is necessary.

Moving Average Filter's future potential lies in the development of progressive filtering methods or in the study of other applications through integration with other signal processing techniques. Finding the most advantageous option is the first thing to perform because every solution includes trade-offs.

6. Conclusion

Overall, this study shows that using a Moving Average Filter to reduce noise in ECG data is both practicable and efficient. For scientists and engineers working on the creation of ECG signal processing algorithms and hardware designs, the suggested Verilog implementation is a useful resource. The employed methodology is cost-effective, secure, and advantageous to the environment. Moving average filters are also proved to be useful in many circumstances. Before applying the Moving Average filter, it's vital to take the application into account because it performs well in the time-domain but the opposite in the frequency domain. As a result, it is advised against using moving average filters in the spatial or temporal domain while processing spectral data in the frequency domain.

7. Acknowledgment

We extend our heartfelt thanks to Dr. D. ASHA DEVI, Professor, Electronics and Communication Engineering Department of Sreenidhi Institute of Science and Technology, Senior Member IEEE, FIE, FIETE, who was involved as our project guide.

References

1. S. Subbiah, R. Patro, and K. Rajendran, "Reduction of noises in ECG signal by various filters," in International Journal of Engineering Research & Technology (IJERT), vol. 3, no. 1, pp. 1–6, January 2014.
2. H. Magsi, A. Sodhro, F. Chachar, and S. Abro, "Analysis of signal noise reduction by using filters," in 2018 International Conference on Computing, Mathematics and Engineering Technologies (iCoMET), Sukkur, Pakistan, 2018, pp. 1–6.
3. S. Patil, P. Patil, I. Patil, and S. Jadhav, "Implementation of FIR filter using VLSI," AESS Journal, October 2017.
4. N. Singh, S. Ayub, and J. Saini, "Design of digital IIR filter for noise reduction in ECG signal," in 2013 5th International Conference on Computational Intelligence and Communication Networks (CICN), Mathura, India.
5. R. Kher, "Signal processing techniques for removing noise from ECG signals," Journal of Biomedical Engineering, vol. 1, no. 1, pp. 1–9, 2019.
6. J. Zhu and X. Li, "Electrocardiograph signal denoising based on sparse decomposition," Healthcare Technology Letters, vol. 4, no. 4, pp. 134–137, 2017.
7. A. Elhossini, S. Areibi, and R. Dony, "An FPGA implementation of the LMS adaptive filter for audio processing," in 2006 IEEE International Conference on Reconfigurable Computing and FPGAs (ReConFig 2006), San Luis Potosi, Mexico, pp. 1–8, 2006.
8. M. Milanesi, et al., "Multichannel techniques for motion artifacts removal from electrocardiographic signals," in Proceedings of the 28th Annual International Conference of the IEEE Engineering in Medicine and Biology Society, pp. 3391–3394, 2006.
9. S. Reza and R. Sameni, "Writing efficient MATLAB® codes," in V. K. Ingle and J. G. Proakis (Eds.), Digital Signal Processing Using Matlab, pp. 1.2, August, PWS Publishing Company, 2006.
10. D. Preethi and R. Valarmathi, "Classification and suppression of noises in fetal heart rate monitoring: A survey," Microelectronics, Electromagnetics and Telecommunications, vol. 521, pp. 607, 2019.
11. A. Devi Dharmavaram, S. Babu M and S. Vunnisa Sayyad, "Design and Analysis of Encoding Decoding Methods used for 5G Communications," 2022 IEEE Delhi Section Conference (DELCON), pp. 1–7, 2022.
12. Devi, D.A., Jaga, S., "Analysis of scheduled routing algorithms on 5-port router for network on chip application", International Journal of Scientific and Technology Research, 8(9), pp. 2148–2153, 2019.
13. Dharmavaram Asha Devi, Tirumala Satya Savithri and Sai Sugun.L, "Design and Implementation of Real Time Data Acquisition System using Reconfigurable SoC" International Journal of Advanced Computer Science and Applications (IJACSA), 11(9), pp. 325–331, 2020.
14. Devi, D.A., Chandana, Y.S., Akshitha, M., "Analysis of Walsh-Hadamard Transformation with CFBMC Communication", Proceedings - International Conference on Artificial Intelligence and Smart Systems, pp. 1246–1252, 2021.
15. Babu, M.S., Raj, K.B., Devi, D.A., "Data Security and Sensitive Data Protection using Privacy by Design Technique", EAI/Springer Innovations in Communication and Computing, pp. 177–189, 2021.

Note: All the figures in this chapter were author self test data and experiment results were performed as part of their research work.

The Smart Home Technology by WSN and GSM Using Mobile Applications MAN

59

C. Anna Palagan[1]

Professor in ECE, Teegala Krishna Reddy Engineering College,
Meerpet, Hyderabad, Telangana, India

**K. Renuka[2], Sangeeta Jawar[3],
J. Rachana[4], G. Rani[5]**

Assistant Professor in AIML, Teegala Krishna Reddy Engineering College,
Meerpet, Hyderabad, Telangana, India

Abstract: The smart home technology plays a substantial role in present day way of life due to its progress in applying at dissimilar places with more quality. It will be minimizing the human work consequently. The homebased computerization is rechargeable devices remain retiring, it is programmatic, for example, devices inaccessible controller and correspondence agenda. Homebased computerization uses for electrical controller devices anywhere. This modernization is focus on controller the electronic devices such as light, fan, TV, AC automatically and also detect the gas and fire, in this case administrator send the alarm notification to the user. In difficult situation, In this applicable for disabled human beings. We have planned to a household apparatus monitor in computerization operating Global System for Mobile communication. In generally structure of intense Homebased Computerization technology by minimal effort and remote technology. The switches ON/OFF procedure of electronic devices remain possible at all.

Keywords: Microcontroller, Arduino, Mobile phones, Smart home, Electrical controller devices

I. Introduction

The Smart homebased technology perception was a maximum number of the years with enhanced highlights. It is correspondence resolution for electric devices. It fundamentally uses rechargeable power communication for signals and controller, where the signal includes wireless recurrence of computerized information and persist the most generally extended range available. Homebased robotization is the private expansion of construction devices and includes the control and robotization of light, fan, cooling-machines, and security of these systems[1]. Because of the forced remote modernization, there are a few distinctive of association are presented in GSM, WIFI, and Bluetooth. Every one association takes their remarkable details, highlights, application[2]. This proposed technology shows the general plan assembly of Household Robotics Scheme with low-cost remote framework.

The Smart home technology [3] recovers the household devices. That innovation was monitor and controlled the electronic devices like light, fan, AC, and so on[4]. It is used to unhelpful condition, for disabled humans. We are proposed a homebased device are controlled in robotization utilizing GSM is included[5]. In this paper identifies with different undertakings from multiple points of view. We are using Bluetooth because it can give short range. However, the different activities are going to give GSM unit through which the scope of the framework is comfort. Here exist numerous methodologies which don't consume any calculated connection with each other[6].

[1]annapalagan@tkrec.ac.in, [2]renukacse@tkrec.ac.in, [3]sangeetajawar@tkrec.ac.in, [4]rachanajakkula@tkrec.ac.in, [5]ranicse@tkrec.ac.in

GSM system has a favorable circumstance to develop innovation, wide covering zone, long correspondence separation, and sound correspondence impact etc. The remote home security framework exhibited in this paper unites such huge numbers of points of interest of WSN and GSM any place the clients are, when some hazardous occasion occurs in home like, gas releasing or criminal interfering, this framework sends alert message to the clients over GSM arrange quickly, illuminating individuals the conceivable risky conditions in home. Also, the remote sensor system set up in homebased has the highlights of simplicity foundation, consumption of connection, and low-control consumption.

The WSN information gathering Centre components are associated with pyroclastic flow, temperature device, smog alarm and gas sensor independently. At that point when the pyroclastic flow finds that a few persons interfere into the home strangely, the smog sensor recognizes smoke, when the gas radar recognizes the flammable gas fixation, the sensors will send message sign to the homebased controller focus the remote device system built up in house. When the remote-control focus gets alert sign to the clients over the GSM segment.

2. Literature Review

The perception of communicate consumes useful to the improvement of the GSM constructed Home robotization technology. The house holder has the option toward get return criticism status of any household devices controlled and monitor whether the switches are ON/OFF automatically from the user device. ATmega328 microcircuit with the combination of GSM gives the homebased technology by the ideal image. [1] This exploration effort examines on the capability of smart home machine controlled and Saves the time, which is the point of the Homebased Computerization framework the and Execution of the Homebased Computerization Innovation via GSM kit to interface Home devices like Light, Fan, AC, etc.[2] The alert message introduced in this exploration description effort concentrated on capacities of the GSM allows client to control goal framework anywhere, applied recurrence transmission capacity range.[3] Here many frameworks are Homebased Computerization that can control household devices using mobile. Homebased safety alarm system [4] was established by WSN and GSM. The single chip C5081F310 hardware was used in that system. The software system developed by C51 languages are used.at The result of this technique easily to notice the leaking gas, theft, fire and send alarm notification through mobile phones. Visual programming frame work [5] was managed the complexity of increasing smart phone devices. In this system easy to enable the configuration and upgrade of WSN.IOT enable sensor are used and codes ae disposition in python

script. [6] Mesh-based design was used toward appreciate the homebased perception. In this application Checksum method are used the error uncovering of data packet. The routing procedure is based on Digi Mesh topology.

Kinect WSN video sensor [7] was used to control the old person in smart homes. In this application they are used Wi-Fi with mesh topology. According to this paper The Kinect data using USB over IP encapsulation. The essential of development Zigbee [8] was short range and low power consumption. In this system not only increase the distance of the devices, it also used 3G smart phones. The system can suitable of existing environment for user. The interactive GUI (Graphical User Interface) [9] was used to average user. In this paper WSN are used to hand-held devices with android application. This application is easy to control the home devices used in power monitoring. Security and privacy for WSN [10].

3. Proposed System

Our proposed system consists of two modules: User and Admin. The User install the application. If the user is the new user first register into the application. On registration time user will get the Login id and password. The login id should be user name or mobile number. The user can use the Login id and password for login the application time. Once user login the application then user monitor and control the current home status. The two parameters are used to control the home status they are Switch ON/OFF and control the fire, gas and smog.

In this application, first page is the user login /Register page. In Register page user can enter the user details and next page is designed User select the location of the home room position or room names. Once a user selects any room switches User and administrator can view the current status. If the user forgets to switch off the light or fan or any other devices in this case administrator send the alert message to the user mobile phones also detect the gas and fire, in this case administrator send the alarm notification to the user. This application is very useful for old age and disabled human beings, when the user is not in the house.

3.1 User Module

In this module, User can login the app and enters the login id and password, once user login the app then user can see the home situation which is the electronic devices switch ON/OFF. User can control the current home status and also change the temperature in anywhere through mobile phones. After user registration user login id and password are stored in the data base.

3.2 Arduino

Arduino is used to manage the present situation of the household devices. In GSM input, using microcircuit Arduino changes the device states. The sensors are maintained by microcircuit.

3.1 Server

In this module, server displays the message among the Arduino and GSM. All the users Request and Response are maintained by the server.

3.4 Data Base Module

In this data modules data base are used, All the user data are stored in this module. Get the data from the user. The data encryption and decryption are done by this module.

3.5 Hardware Module

We set up an example model framework in room. In place of referenced overhead, chip C8051F310 MCU as the information preparing component of WSN focus Centre knob and information gathering Centre knob. Here, C8051F310 has a temperature radar enclosed to identify the room temperature. Later equipment association, announce the proper programming created C51 on MCU-based covered remote-control focus. At that point, we can start the trial in this model framework by changes present temperature power. When the real room temperature exceeds this present temperature limit, then controller focus on time activate TC35 GSM segment to send an alarm message to user mobile. This model framework worked effectively and successfully with solid and well correspondence.

In this framework, first, the application sends an analysis to GSM tool kit by means of the signal medium. At that point GSM component refers the inquiry to Arduino microcircuit. The microcircuit controls a position and then Send to the Application. It is the activity station and it is refreshed outcome. Also, that linked data SQL database is place away and update as each time.

The data encryption and decryption we are used AES algorithm. AES algorithm is very protected, when relates to the other encryption and decryption algorithm.

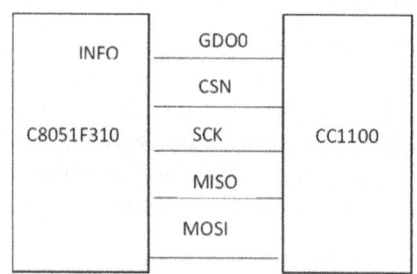

Fig. 59.1 Hardware design

4. Smart Home Architecture Diagram

Remote Organization and controller device, are handle effectively. In this, the DTMF innovation utilized in this approach a client needs to call the mobile phones associated with the framework with the assistance of receiver at that point using the devices automatically answer the client. In awaken the composing cipher, the procedure of the client necessity completes by utilizing a microcircuit. This framework gives the security is disturbing when an interruption happens. The framework is isolated monitoring device like GSM tool kit. It is used to controller, checking to the execution remotely by utilizing limitations like Temperature, Pressure and so on. At that point when the estimations of the x value or on the other hand predefine value, at that point warning SMS is sent to the client.

The system specification of the proposed system. The Graphical user interface offers the entering the Input data for the user. The input of the system the user submits the query. In the output system the user enters the query into the application. The output screen displays the relevant data. The processor P contain three parameters they are input, output and subordinates. The deterministic data represented the ON/ OFF function trigger values.

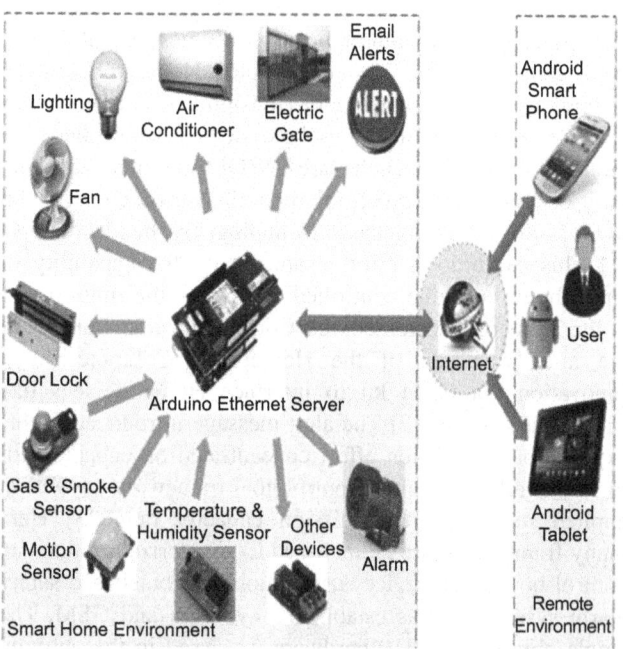

Fig. 59.2 Architecture diagram

4.1 Function

Step 1: Identified the processor P

HAS= {In, Op, P}

P= {TD, AD}

Step 2:

TD [C, MAX, LS]

C = Switch ON/OFF

MAX = {0,1},

If MAX=0;

Theswitch is OFF

If MAX=1;

The switch is ON

LS= Output based on C where Light ON/OFF

Step 3:

AD= {LS, Arduino, Status}

LS=Input of AD

Arduino= It is used for Light ON/OFF

Status= It is used for the light mode. This status is visible on the users.

4.2 Algorithm

In this algorithm, First the user login the application. Then user activate the trigger value. Call the Arduino function and get the input value to the trigger data. The max value of trigger data is 0 is denoted as the switch is OFF state and the trigger data 1 is denoted as the switch is ON. In this trigger data is generated by the user. Then processing the trigger data. At last display the result on the user mobile screen.

Step 1: User login the application.

Using login id and password.

Step 2: Activate value access by user

Step 3: Call the Arduino functions then get input to Trigger Data.

Step 4: for i=0, It is created by user

Step 5: Get Trigger Data as a input.

Step 6: Call the Arduino function.

Step 7: Processing the Trigger data.

Step 8: Exhibition the Result if the switch is ON/OFF.

Step 9: Stop.

4.3 Design and Implementation of Prototype Model

Homebased Automation frameworks remain confronting four primary challenges; they are significant expense of proprietorship, adaptability, reasonability, and troublesome making security. The principle goals of the projected framework are plan and to actualize a modest open source homebased computerization framework that is monitoring and programming the majority of the household machines through a simple method to track and keep up to verified home mechanization framework essential. The projected framework utilizes message administration gave by GSM system to confirmed client and remotely checking, and monitoring his home apparatus.

4.4 Experimental Analysis

The experimental validation of the system has been carried out considering all the loads. The results are presented here considering various combination of three different loads; Living room light of 10 W (L1), Kitchen CFL of 15 W (L2) and Neighboring CFL load of 20 W (L3). room2 light load of 5(L4) only under emergency and socket modes of operation. Each of the loads is controlled (turning ON or OFF) by an individual relay assigned for the load. The relays receive signals generated by Arduino after getting command from Smart Home Automation App in personal mobile via Bluetooth.

Table 59.1   Power calculation table

Load	Total expected power=load+30W	Theoretical current (A)	Practical current (A)	Practical power (w)
L1	10+30=40	0.200	0.19	40.2
L2	15+30=45	0.214	0.22	46.2
L3	20+30=50	0.250	0.25	50.6
L4	5+30=35	0.195	0.20	35.8

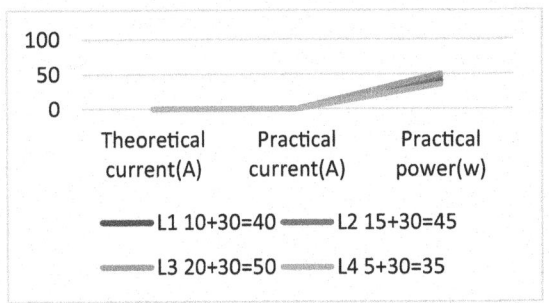

Fig. 59.3   Power variation graph

5. Conclusion

In this paper monitor and control the homebased electronic devices over mobile phones. In this framework based on the WSN and GSM. This application is easy to handle the home electronic devices like TV, AC, Light, Fan, etc. And send notification to the user. The smart home technology also detects the smog, gas, theft and fire and send the alarm notification automatically toward the user mobile phones. The single chip C5081F310 and CC110 are used to receiving and sending messages. This application is successfully control and monitoring the home devices through mobile phone. It is mainly used by old age and disabled person. The system hardware and software are described.

References

1. Rozita tey mourzadeh, Salah addin Ahamed et al. Smart GSM based home automation, IEEE, 2013.
2. M.Angeles serna, Cormac j.sreenan, Szymon fedor et al. A visual programming framework for WSN in smart home application, IEEE, 2015.
3. [3] Huiping Huang, Shide Xiao, Xiangyin Meng, Ying Xiong, A Remote Home Security System Based on Wireless Sensor Network and GSM Technology, pp 535–538, 2010.
4. Maulana Yusuf Fathany, Trio Adiono, Wireless Protocol Design for Smart Home on Mesh Wireless Sensor Network, pp 462–466, 2015.
5. A.W.Ahamed,N.jan et al. Implementation of Zigbee-GSM based home security monitoring an remote Control system, IEEE, pp 1–4, 2011.
6. Asma BEN HADJ MOHAMED, Thierry VAL, Laurent ANDRIEUX, Using a Kinect WSN for home monitoring: principle, network and application evaluation, 2013.
7. Mrs. Bhagyashri R. Wankar, Prof. Vidya Dhamdhere, Application of WSN to Intelligent Home Automation and Power Monitoring Using Android Smart Phone, IJCSMC, vol. 4, pp 242–245, 2015.
8. Rancisco J. Bellido Outeiriño, CE Soc. Member, IEEE, José Flores Arias CE Soc. Member, IEEE, Matías Liñán-Reyes and Emilio Palacios-Garcia, In-Home Power Management System Based on WSN, IEEE International Conference on Consumer Electronics (ICCE), pp 546–547, 2013.
9. Andreas Kamilaris, Vlad Trifa and Andreas Pitsillides, HomeWeb: An Application Framework for Web-based Smart Homes, pp 134–139, 2011.
10. Shubham Magar, Varsha Saste, Ashwini Lahane, Sangram Konde, Supriya Madne,Smart home automation by GSM using android application, IEEE, 2017.

Note: All the figures and table in this chapter were the orginal work done by the author - test data.

Grid-connected Single-stage PV System

60

Ankathi Manjula[1]

Research Scholar, Dept of Electrical and Electronics Engineering,
Sathyabama Instituteof Science and Technology, Tamil Nadu, India

A. Ramesh Babu[2]

Professor, Dept of Electrical and Electronics Engineering,
Sathyabama Institute of Science and Technology, Tamil Nadu, India

Abstract: This paper shows the perfection of the PV system connected with the grid. The reduced harmonic output of the PV system can be connected to the grid without any mismatching and switching losses. The proposed topology has simple in design, low in cost, and easy to operate. This model doesn't require any boost converter, or single-stage operation, and requires less time to operate. Maximum power will be tracked by a PV module, the inverter with IGBTs can give the controlled ac output with the help of the PWM technique and lower switching losses.

Keywords: PV cell, Grid, IGBT inverter, PWM technique

1. Introduction

Solar light can be converted into electrical energy. Solar energy is a natural and nonconventional energy. Even though if we can use millions of years there is no chance of exhaust. A solar cell converts light energy to dc from electrical energy [1,2]. To receive an alternate source of electricity, an inverter must be placed between the solar cell and the network. That AC form of electrical energy must be controlled by using PWM technology in the inverter [3, 4]. The new inverter topology's key characteristic in this study is that it produces a controlled and properlymodulated AC output voltage from the DC input. Based on the immediate duty cycle [12, 13]. Controlled ac output from the inverter can be filtered with the filters, and finally, this proper AC output sends to the grid. This study covers operation, analysis, modulation, control scheme, and experimental results.

2. PV Cell

2.1 Design of PV Cell

When the sunlight falls on the solar cell then that light energy can convert into electrical energy due to the photovoltaic effect. Photons can help move electrons movement on the solar cell. We can collect that electrical energy by using two wires. Fig. 60.1 shows the design of solar cells with semiconductors. PV cell consists of two materials one is a P-type thick layer of boron-doped silicon and the other is an N-type thin layer of phosphorous-doped silicon and it is placed on a boron-doped silicon layer. Electrons can move from P-typeto N-type when the photons fall on it. If a solar cell connected with load current can flow through the load, otherwise generates only voltage like DC0.5V to 0.6V, when the load is not present. These solar cells are usedfor

[1]manju.ankathi708@gmail.com, [2]rameshbabueee@sathyabama.ac.in

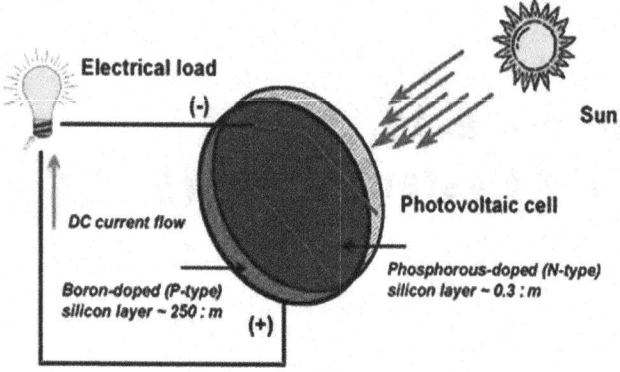

Fig. 60.1 PV cell

solar water heating, solar room heating, solar lighting, and electrical energy generation purposes [9,14].

2.2 PV Cell Interconnection and Design of Module

The combination of several cells is called solar arrays, combination solar arrays are called solar modules.

Fig. 60.2 PV module

Figure 60.2 represents the design of a solar modulethat is made up of 36 solar cells.one PV module can give 12V DCoutput. PV modules can be protected from air and moisture by using special materials in design. It is made up of a back sheet as an insulator, two EVA sheets before and after pv cells for protection from air and moisture, transparent glass to transmit heat through the glass, an EVA sheet to PV cells, and finally aluminum frame for strong support to the PV module components.

3. The Inverter Topology

Most regular, voltage source inverter is used and here also voltage source bridge type inverter can be used. By using IGBT switches in the design of the inverter, switching losses are reduced due to these switches. Here sinusoidal PWM technique can be used to control ac output of the inverter so that the controlled output is harmonic less output. With the help of the control scheme, we can get a proper supply of grid-matched output. We can get more information [4, 9,

and 10] onthe analysis, control, and experimentationof this topology of the inverter.

4. Principle and Operation of Inverter

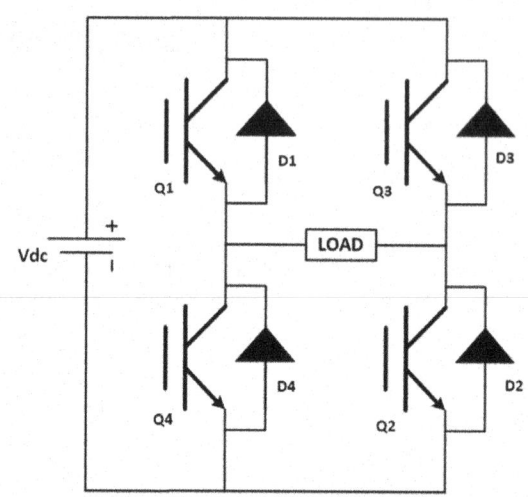

Fig. 60.3 Single-phase full bridge inverter

Figure 60.3 consists of four IGBTs and 4 diodes. Obtaining the output ac voltage at the load terminals by the operating of 4 IGBTs with input DC supply to the circuit. After giving gate signals to the Q1 and Q2, load currents start flow dc source to Q1, load,Q2 then finally reach to source terminals. So due to the conduction of Q1 and Q2 positive half cycle of the ac output will come. When gate signals are given to the Q3 and Q4, load current will start to flow from source to Q3, load, Q4, and reaches source terminals. But in this case load,the current direction is opposite to the first case. So,the output voltage is in a negative half cycle like we can do ac output. Here D1, D2, D3 &D4 are feedback diodes. When the load is inductive, then these diodes are feedback on the energy from the load to source through them. Figure 60.4 represents the DC input and AC output voltage waveforms of the inverter

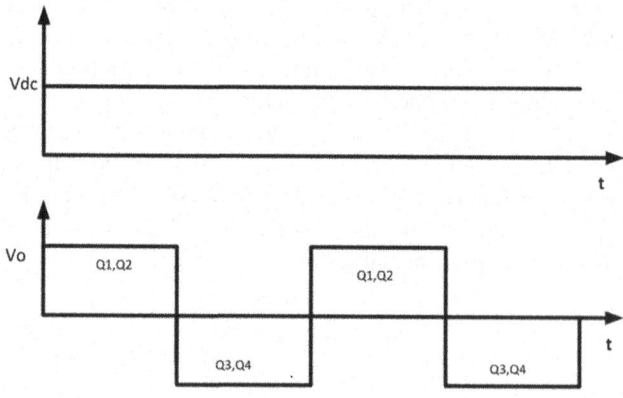

Fig. 60.4 Input and output waveforms of inverter

conduction of Q1 and Q2 from the positive half cycle, Q3, and Q4. Single-phase inverter gives single-phase AC output and, a three-phase inverter can give three-phase AC output.

5. Description of the Circuit

In the design of this circuit, I used the main devices' PV module, three-phase IGBT inverter, LC filter, and controllers. Whenever the solar light falls on the PV module then electrical energy will come from that PV module due to the photovoltaic effect. And that energy is in dc form. The main intention of this paper is the proper connection of the PV system with the grid without any mismatching conditions and switching losses. The order of connections of devices in the proposed circuit PV modules,inverter, LC filter, and grid. From the PV module, we can get DC output. Here no need for any boost converter, in a single stage only we can get the required DC output, due to this advantage the cost and size of this circuit are reduced. That DC output has given as input to the inverter, this inverter can convert dc to ac output That ac output is controlled output that can be controlled by the inverter with a sinusoidal PWM technique, next, the LC filter can filter the AC output of the inverter, then harmonic less and suitable ac output will come from the filter. From that filter ac output can send to the grid. There are no output voltage and output current mismatching problems with the

grid. And the output can be controlled by input with the help of feedback which is the control scheme. Switching losses are also reduced by the inverter with IGBT switches. Figure 60.5 represents the proposed circuit and which can be created in MATLAB and Simulink software. Figure 60.6 represents the controlling scheme circuit which controls the gate pulses of the three-phase inverter by sinusoidal pulse width modulation.

6. Simulation of the Circuit

6.1 Control scheme

See Fig. 60.6.

7. Simulation Results

Figure 60.7 represents the output voltage of the PV module. Which is in DC form and this dc voltage can be converted as ac by the 3-phase inverter.

Figure 60.8 represents the three-phase output voltage of the inverter, which is properly matched with the grid voltage.

Figure 60.9 represents the three-phase output current of the inverter, that current is properly matched with the Grid current.

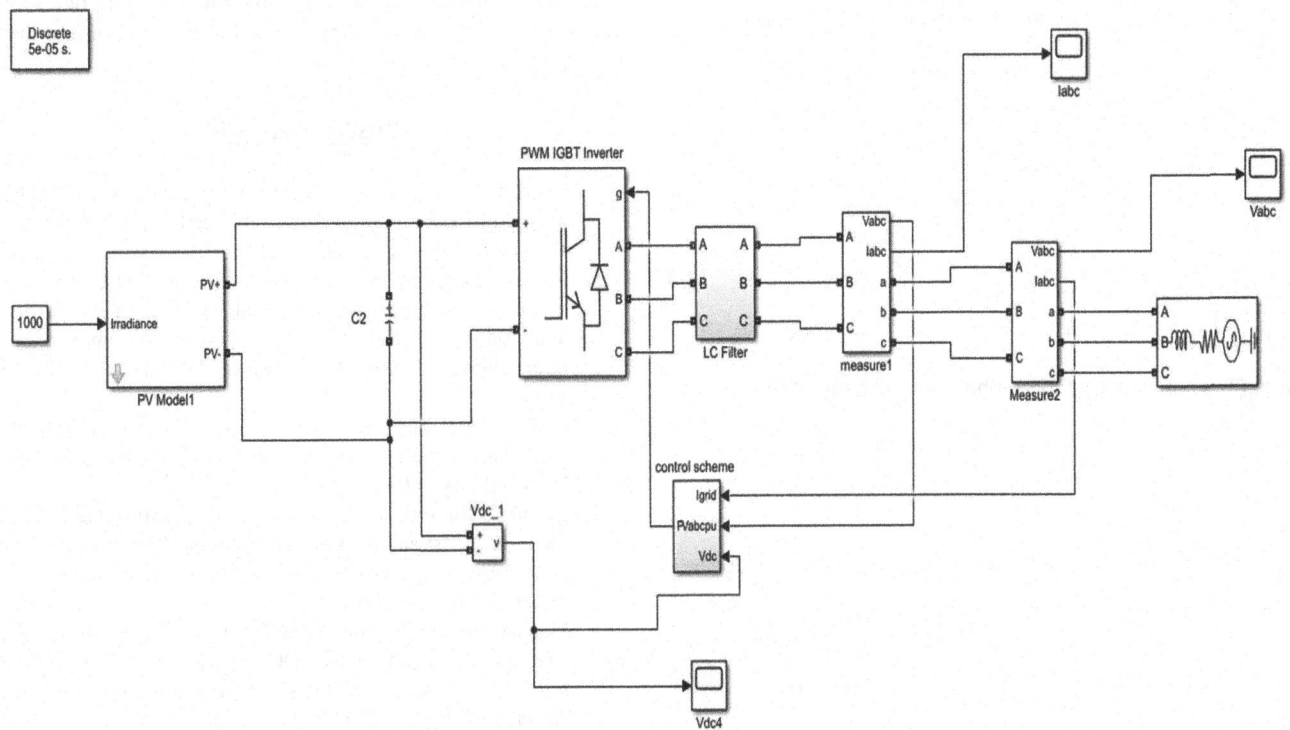

Fig. 60.5 Simulation circuit

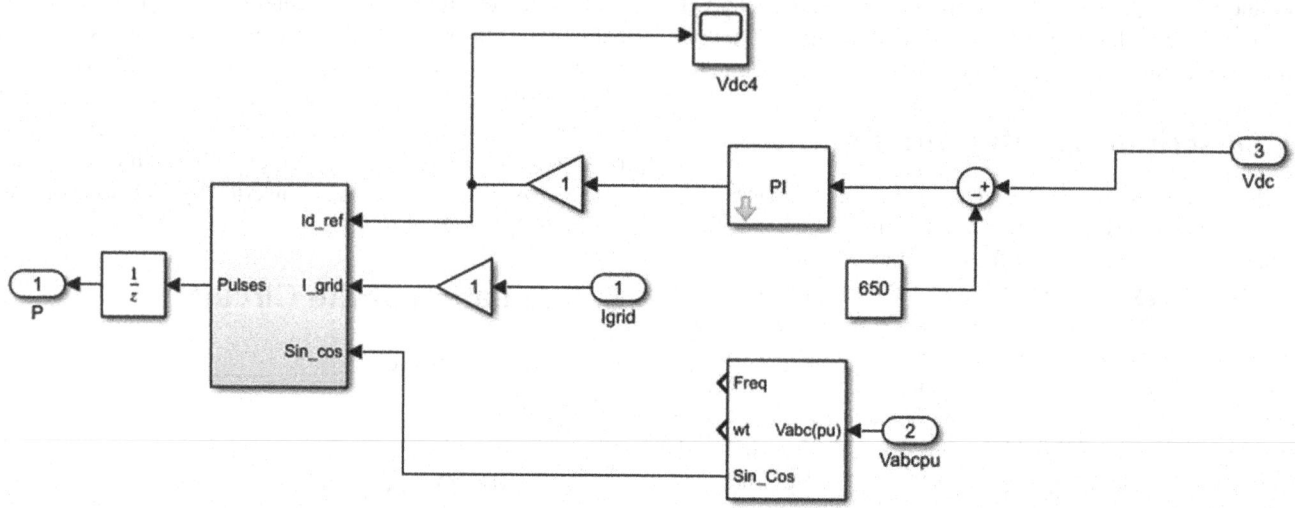

Fig. 60.6 The control scheme of the simulation circuit

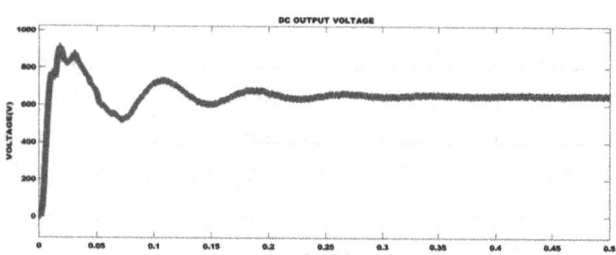

Fig. 60.7 The output dc voltage of thePV module

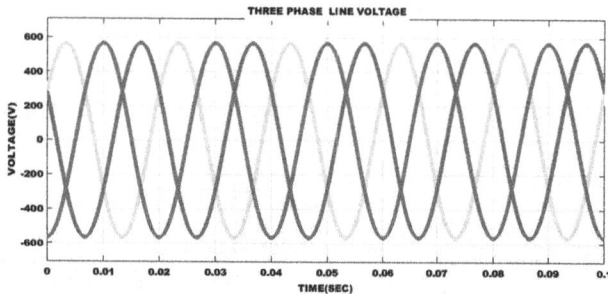

Fig. 60.8 The output three-phase line voltage of the inverter

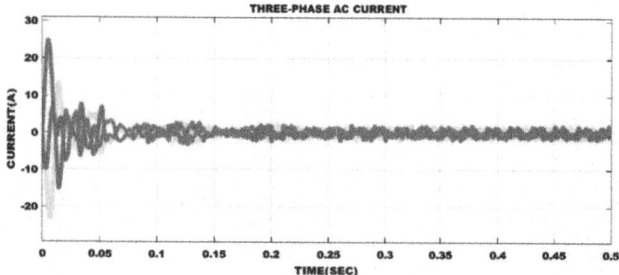

Fig. 60.9 The output three-phase current of the inverter

8. Conclusion

This paper concludes that proper matching of The PV system with the grid. We can take that supply from the grid to our loads without any distortions, so we can use our loads as usual because of the proper AC output collected from this proposed model. Size and cost are very less due to the absence of a boost converter. We can set up this circuit very easily, and we can use this model for our home electrical applications, stand-alone applications. And also sellto the government like a grid.

References

1. Rivas, C., and A. Rufer. "PWM Current converter for electric energy production systems from fuel-cells." *Proc. Eur. Conf. Power Electron. Appl. EPE.* 2001.
2. Ramesh Babu, A., et al. "Novel cascaded H-bridge sub-multilevel inverter with reduced switches towards low total harmonic distortion for photovoltaic application." *International Journal of Ambient Energy* 39.2 (2018): 117–121.
3. Mohan, Ned, Tore M. Undeland, and William P. Robbins. *Power electronics: converters, applications, and design.* John Wiley &Sons, 2003.
4. Vázquez, Nimrod, et al. "Analysis and experimental study of the buck, boost and buck-boost inverters." *30th Annual IEEE Power Electronics Specialists Conference. Record.(Cat. No. 99CH36321).* Vol. 2. IEEE, 1999.
5. Menaka, S., and S. Muralidharan. "Design and performance analysis of novel boost DC-AC converter." *2011 3rd International Conference on Electronics Computer Technology.* Vol. 2. IEEE, 2011.

6. Barzegar, Farhad, and Slobodan Cuk. "Solid-state drives for induction motors: Early technology to current research." *Proc. IEEE Region.* Vol. 6. 1982.

7. Barzegar, Farhad, and Slobodan Cuk. "A new switched-mode amplifier produces clean three-phase power." *Proceedings Ninth International Solid-State Power Conversion Conference.* 1982.

8. Bilbao, Julia, et al. "Test reference year generation and evaluation methods in the continental Mediterranean area." *Journal of Applied Meteorology and Climatology* 43.2 (2004): 390–400.

9. Vorperian, V. "Simplify your PWM converter analysis using the model of the PWM switch PART II: Discontinuous conduction mode." *Current (Virginia Polytech Newsletter)* (1989): 6–12.

10. Tymerski, Richard, et al. "Nonlinear modeling of the PWM switch." *IEEE Transactions on Power Electronics* 4.2 (1989): 225–233.

11. Orosco, R., and N. Vazquez. "Discrete sliding mode control for DC/DC converters." *7th IEEE International Power Electronics Congress. Technical Proceedings. CIEP 2000 (Cat. No. 00TH8529).* IEEE, 2000.

12. Caceres, Ramon O., and Ivo Barbi. "A boost DC-AC converter: analysis, design, and experimentation." *IEEE Transactions on power electronics* 14.1 (1999): 134–141.

13. Menaka, S., and S. Muralidharan. "Design and performance analysis of novel boost DC-AC converter." *2011 3rd International Conference on Electronics Computer Technology.* Vol. 2. IEEE, 2011.

14. Chenchireddy, Kalagotla, and V. Jegathesan. "A Review Paper on the Elimination of Low-Order Harmonics in Multilevel Inverters Using Different Modulation Techniques." *Inventive Communication and Computational Technologies: Proceedings of ICICCT 2020* (2021): 961–971.

Note: All the figures in this chapter were author's self data and the work was part of the research program.

Reliability Evaluation of Distribution System Using Micro-Grid Model

61

Raju Kaduru[1]

Associate Professor of EEE, TKR College of Engineering and
Technology Hyderabad (autonomous), Telangana, India

G. N Srinivas[2]

Professor of EEE, Department of Electrical and Electronics Engineering, JNTUH UCEST,
Jawaharlal Nehru Technological University Hyderabad

Abstract: This paper describes a reliability of micro-grid (M-G) model on the IEEE RBTS Bus 6 distribution network. The reliability indices are enhanced by M-G accumulating to distribution network. The reliability indices are SAIFI, SAIDI, CAIDI and ASAI. Modelling technique of M-G depends on state transitions of components in the feeder. In this paper, develop new analytical technique based on algorithm of FMEA. The main advantage of this method based on FMEA is continuously identified along with the changing position of M-G in the distribution network. This is the because of power source changed with the main grid to micro –grid. This method here is tested on Feeder 4 of IEEE RBTS Bus 6 and results are show that this method is efficient and suitable for practical distribution system applications. A MATLAB program is developed to verify this method.

Keywords: Distribution Network, Micro-Grid, Modified FMEA, Reliability Indices

I. Introduction

The distribution network (DSN) operates at a low voltage and is divided into sub-transmission and radial/meshed segments. Within this DSN, approximately 80% of outages occur due to a combination of faults and the natural end of component lifetimes. The failure rate (FR) and outage (OTT) count can be considered as reliability indicators at various load points, which tend to reflect the overall system's reliability. In practice, conventional electricity, supplemented by a limited capacity of distributed generation (DG) and renewable sources, forms the backbone of the energy supply, transmitting power from one location to the loads of end consumers. This evolution may take on a competitive structure, influenced by changing environmental conditions, maintenance requirements for power lines, and the economic evaluation of energy policies. Given the circumstances outlined above, the CDP are seeking the opportunity to receive their power supply from alternative sources. In this scenario, a well-suited micro-grid (M-G) incorporating distributed generation (DG) is providing the required electricity to the customer load points.. M-G is a self-reliant gadget, which encompass the distributed generation (DGs), strength storages gadgets, tracking and shield gadgets. The modelling of RDN and its working is relies upon at the power glide, that is improved via delivered to DG [1]. The problem is identified, which includes thing failures within the distribution feeder; it's far indicated the unavailable of supply close to LPs. It confirmed the exchange to electric industry in aggressive surroundings conditions. The reliability of the network by way of the use of the analytical approach and MC simulation has been effects addressed in [2]. In analytical technique, network reliability parameters are used in mathematically modelling; it's far frequently simplified, and evaluates indices. The reliability in the distribution network may be found by using simulating the real technique and random behavior of the gadget has considered. This

[1]kadururaju@tkrcet.com, [2]gnsgns.srinivas785@gmail.com

process has been executed using a simulation technique. The reliability modeling approach, which involves integrating DG into the distribution network, is detailed in reference [3]. However, this approach does not thoroughly investigate reliability enhancement but focuses on backup systems within the network, aiming to reduce both the frequency and duration of interruptions. The RDN, which includes Wind Turbine Generators (WTG), was studied using a sequential simulation method (SSM) as detailed in reference [5]. This study took into account variables like fluctuating wind speeds, unpredictable WTG failures, and repair costs within the network. However, it did not significantly contribute to enhancing RE, particularly in relation to REI. IEEE RBTS Bus-2 has 4 exceptional feeders and it is evaluate reliability via Monte Carlo simulation technique [6]. Furthermore, radial networks with exceptional running situations are included and their impacts of indices that aren't explored. With the effect of nature of climate situations, the time various failure price and time required for recuperation of the useful resource, the radial network reliability indices evaluated has been showed in [7]. Alternatively, such as chronological troubles and machine random conduct for figuring out the reliability indices of the network with assist of sequential simulation technique are taken into consideration. In this situation, reliability improvement in the network does not explore. The MCS manner has used to evaluate reliability of the network, in which the system in random nature has discussed in [10]. On this discussion, examine network reliability blanketed distinct reliability parameters and its purchaser statistics and cargo information are taken into consideration.

The reliability of the network is determined based totally on idea hourly based totally time various traits with appreciate to machine parameters has provided in [12]. In which, assessment of distribution network reliability considering with DG primarily based on working conditions for the feeder and its failure of reliability parameters of these sections. Reference [13] delves into a comprehensive exploration of network RE, leveraging an AN Method to meticulously assess the network's performance in the context of its integration with DG. This rigorous analysis not only encompasses the fundamental aspects of FR and OTD but also meticulously scrutinizes various critical system indices, delving deep into the intricate web of factors that influence these RE metrics. These factors, indispensable to the RE evaluation process, include failure rates, load levels, the strategic placement of DG units within the network, and the nuanced parameters governing DG generation.. This impact does no longer confirmed in improvement of reliability network. The network with big scale DG and small scale DG, the assessment of reliability using an analytical method has mentioned in [14]. Then again, decided reliability of the network in association of DG unit at diverse distances from the main deliver, and

DG is positioned at the feeder for every section, it is able to be confirmed that reliability improvement of the network have much less as in line with decided element on range and period of interruptions. The calculation of network RE is executed through a meticulous assessment based on FR and duration metrics, as elucidated in reference [15]. Notably, it is observed that enhancements in RE do not exert a discernible influence on the overall system performance. Furthermore, reference [16] introduces an optimization strategy for determining the most advantageous placement of DG within the distribution network. This strategic approach is designed to maximize improvements in reliability, particularly with regard to reducing both the frequency and duration of customer interruptions, a vital consideration in the pursuit of enhanced network performance In this evaluation, with the exception of the DG reliability parameters are required to assess the reliability of the network. Alternatively, development in reliability at each load point via this DG not explored. The paper showed the reliability of active distribution networks primarily based on SMC technique with admire to the customer overall performance factor of view [17]. Further, a powerful operation of the network and its optimization strategies do not affect reliability development within the network. The author addressed to the reliability of DG on feeder in each region of load point primarily based on healing time algorithm has been mentioned in [18]. The paper addressed to distributed technology in phrases of waste warmth, which is based totally at the plug-and-play version [19]. The impact of micro generation on reliability of the network, consisting of the LV and the MV network has been addressed through author in [20]. In this concerned, as results suggests that micro generation to micro loads is protected improvement of reliability in the device. In this case, impact on micro generation on reliability of the network that protected a small wide variety of hundreds. On the other hand, mathematically version is advanced and it's going to relies upon on component primarily based on the PV device which includes intermittent effect and uncertain working conditions are the evaluation reliability of the distribution network has been presented in [21]. Furthermore, effects of reliability at the distribution network via connecting the PV network. The author presented the reliability modelling in accordance from growing older additives in systems and such as the records of components and its failure in structures has been provided in [22]. The paper highlights the applicability of the MCS method in modeling various aspects of system operation, including failures and stochastic factors. Furthermore, in reference [23], the authors delve into the discussion of achieving maximum RE enhancement, emphasizing that placing Distributed DG units in close proximity to customer premises yields the most substantial improvement, ensuring optimal accessibility in terms of serving a greater number of customer. So many analytical techniques can be found in the literature. These have been implemented to test systems under DG at different load points. But the analytical technique during the micro-grid at the load points has not been found in the literature including component failures. In order to overcome the problem associated with the traditional method, algorithm based on modified FMEA is proposed in this paper, which chooses the search component failure from the complete feeder of the system.

In this paper, an attempt has been made to analyze modified FMEA analytical method for an examine reliability distribution network with connection of M-Gs with a 100 % dependable. The proposed technique is implemented on modified IEEE RBTS Bus 6, blanketed foremost feeder four, the sub feeder F5, F6 and F7 and 23 load points respectively. The take a look at network has studied according to exclusive M-G models. The M-G place on the load points is based totally on the component failure and repair fee of the section in the distribution feeder. The remainder of the paper is structured as follows: Section II provides an explanation of the DSN reliability indices, while Section III elaborates on the development of the M-G model. Section 4 delves into the discussion of case studies, while Section 5 presents and discusses the results. Finally, Section 6 provides a summarization of the conclusions reached in this study.

2. Distribution Network Reliability Indices

Contemplate a radial DSN test network, as illustrated in Fig. 61.1, comprising various sections denoted as SE1, SE2, SEj, along with associated laterals, breakers, and relatedcomponents.

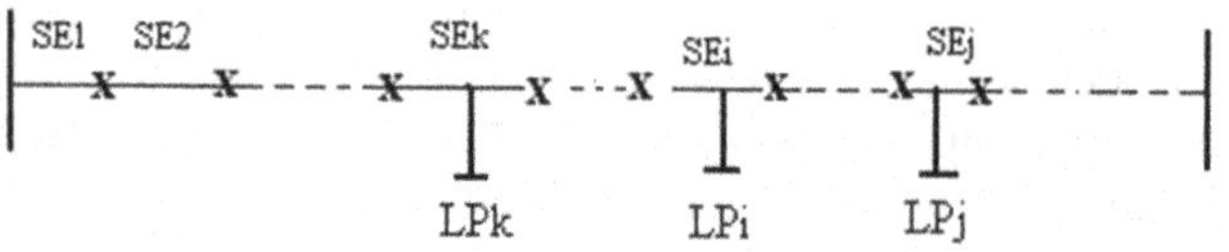

$$
\begin{array}{c}
\text{SE1 \quad SE2} \qquad \text{SEk} \qquad \text{SEi} \qquad \text{SEj} \\
\text{LPk} \qquad \text{LPi} \qquad \text{LPj}
\end{array}
$$

Fig. 61.1 The radial DSN

Employing the classical methodology to evaluate the DSN, we obtain three pivotal reliability metrics: the AFR, symbolized as λs, the AOD (rs), and the AAOT, denoted as Us, all of which are analyzed across a range of load factors. These RE indices are derived through the application of equations (1), (2), and (3). Furthermore, the comprehensive system performance indicators for both individual feeders and the entire network are computed utilizing equations (4) to (7), as extensively elucidated in references [4] and [9].

$$\lambda_s = \sum\nolimits_{i=1}^{N} \lambda_i \, (f/yr) \tag{1}$$

$$U_s = \sum\nolimits_{i=1}^{N} \lambda_i r_i \, (hr/yr) \tag{2}$$

$$r_s = \frac{U_s}{\lambda_s}(hrs) \tag{3}$$

$$SAIFI = \frac{\sum \lambda_i N_i}{\sum N_i}(\text{int}./cust.yr) \tag{4}$$

$$SAIDI = \frac{\sum U_i N_i}{\sum N_i}(hrs./cust.yr) \tag{5}$$

$$CAIDI = \frac{\sum U_i N_i}{\sum \lambda_i N_i}(hrs./cust.yr) \tag{6}$$

Here, λ_s be the load point failure and U_s is the average annual outage time, N_i is the number of customers to the load point i (LP_i)

Average service availability index (ASAI)

$$= \frac{\sum N_i \times 8760 - \sum U_i N_i}{\sum N_i \times 8760} \tag{7}$$

Where, 8760 is the number of hours in a calendar year

3. Development Micro-grid Model for Distribution System

In Figure 61.2, the M-G configuration is depicted, featuring a collaborative operation between a DG system and a battery that is interconnected with the main grid. In this arrangement, energy is provided to customers from both the primary grid and the DG system within the M-G, which also integrates a battery for enhanced reliability and efficiency. The M-G model is hooked up to the main –grid and works collectively, in which the fault does now not regarded inside the predominant grid. This operation is called regular operation.

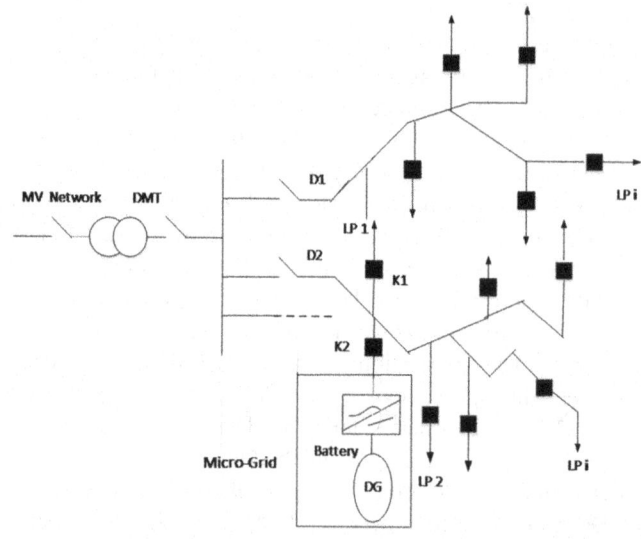

Fig. 61.2 Micro-grid model

If fault is passed off at the primary-grid, the circuit breaker (CB) acts as an open role. The M-G is formed within the state of an island. Each M-G and the main grid operated wherein fault is repaired and then the circuit breaker (K2) acts as closed (see Fig. 61.2).

The M-G model is separated from the main-grid, if fault is regarded at the main-grid. At some point of this circumstance, the circuit breaker K2 is opened. The M-G is change into an operation mode, where fault is isolated and the CB acts as closed. In this way, hooked up a FMEA table and confirm the scope affected by each failure, after which determine the RE indices of every load point of the DSN with M-G. On this phase, the following assumptions are taken to assessment of the distribution system reliability with M-G as

- CB is hooked up in between the main grid and M-G, which is separated from the main-grid.
- M-G functioning operation is split into two approaches.
- Both DG and the load random behaviour is neglected.
- The requirement of load is fulfilled by using the DG and electricity storage device.

The M-G can't affect the loads inside M-G, wherein the fault is acting on the outdoor of the M-G. If the fault seemed in the M-G, the main-grid is separated from M-G with the assist of CB and isolated from the primary grid. Based on this, M-G model is developed on distribution network for improving the distribution DSRE indices. The M-G in radial DSN is to identify an best location which could provide a minimal failure rate and outage time without violating the operational constraints. The running constraints right here are the load capacity and feeder capability. The interruption rate and outage time of load point i after M-G is connected load point, the AFR and AOT of load point is calculated through Eq. (8)- (9),

$$\lambda_s' = \sum_{i=1}^{A-B} \lambda_i + \sum_{l=1}^{A-B} \lambda_l \qquad (8)$$

$$U_s' = \sum_{i=1}^{A-B} \lambda_i r_i + \sum_{l=1}^{A-B} \lambda_l s \qquad (9)$$

Where,

λ_s' - Failure rate of load point i with micro-grid

λ_i - Component failure rate of section i within the micro-grid

λ_l - Lateral component failure rate of section i within the micro-grid

U_s' - Annual average outage time of load point i with M-G

r_i - Component repair time of section i within the M-G

s - Switching time

3.1 Algorithm for Calculating the Reliability Indices of IEEE Roy Billinton Test System Bus 6- Feeder 4

Step-1: Consider each load point at the feeder within the distribution system;

Step-2: Apply failure mode analysis to each LPs;

Step-3: Check the load points within the M-G or not

Step-4: Select load points inside the M-G within a circuit to shape an island, that's referred to as a micro – grid;

Step-5: consider micro- grid is an autonomous operation from the main grid;

Step-6: Due to change of failure rate and average annual outage time of load points within the M-G, the micro- grid place can shape within the sub feeder;

Step-7: Find the failure rate and average annual outage time of load points within the M-G and evaluate with & without micro-grid;

Step-8: Repeat the step 5 to step 6, for all sub feeders in the distribution system;

Step-9: Using step-7, find the SAIFI and SAIDI for all feeders inside the distribution network including different M-Gs.

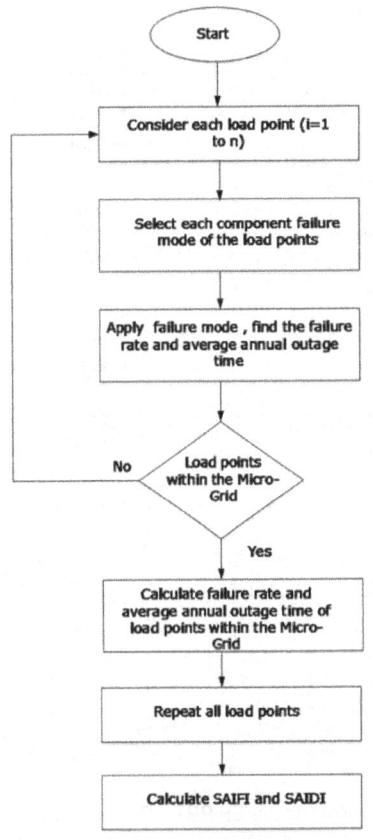

Fig. 61.3 The flow chart of RE evaluation

4. Case Studies

In this section, examine the impact of various M-Gs on the IEEE RBTS Bus 6-F4, which includes sub-feeders F5, F6, and F7 within the DSN, as illustrated in Fig. 61.4. To assess system RE, apply a modified Failure Modes and Effects Analysis (FMEA) method and perform a series of case studies on the test system. These investigations encompass RE parameters of system components, customer data, total average load, peak load, and the count of load points and of IEEE RBTS Bus-6-F4 are provided in the Table 61.1 and Tab. 2 [8]. In these case studies, 2nd order overlapping outages, inclusive of outage time is omitted. Similarly, open points are removed. Consider M-G is 100% dependable for modelling and evaluating the reliability indices of distribution system. Case study-1: RE evaluation of IEEE RBTS Bus-6 without M-G. Case study- 2: RE evaluation of M-G location A-B in

Table 61.1   Feeder data

S. No	Components	
1	No. of Lines	30
2	Load points	23
3	No. of Fuses	23
4	No. of Transformer	23
5	No. of Breakers	4
6	Disconnect switch	1
7	Failure rate of the transformer	0.05 f/yr
8	The repair time	48 hrs/yr
9	Loop switch	0.5 hrs/f

Table 61.2   Customer data

Load points	Customer type	Number of customers
2	Residential	126
1,6	Residential	147
5	Residential	132
8, 11, 14, 19	Residential	79
10, 12,16,19	Residential	76
15, 20	Farm	1
3, 13, 17	Farm	1
4, 18	Farm	1
7, 23	Farm	1
9,21	Farm	1

Feeder 4 (see Fig. 61.4), Case study-3: RE evaluation of M-G location C-D in Feeder 4 (see Fig. 61.4) Case study-4: RE evaluation of M-G location E-F (see Fig. 61.4). By means of undertaking all of the instances, we have to be aware that the RE parameter of M-G which include its failure rate and restore is 100% reliable may be taken in to account.

4.1 Case Study-1: Reliability Evaluation of Distribution Network without M-G for F4-IEEE RBTS BUS-6

The line length is proportional to the failure fee of line duration. The feeder four as proven in Fig. 61.4 and in this most important feeder sections and laterals sections, whose FR, RT, the FR of the transformer, RT, the entire isolation and switching time and consumer load information shown in the Table 61.1 and Table 61.2. The failure rate of the load point i(where i = LP 1 to LP 23) in the feeder four is determined through Eq. (10) as

$$\lambda_s = \sum_{i=1}^{N} \lambda_i + \lambda_{li} \ (\text{f/yr}) \tag{10}$$

Where, λ_i and λ_{li} are the FR of the main feeder section and lateral components at load point i respectively.

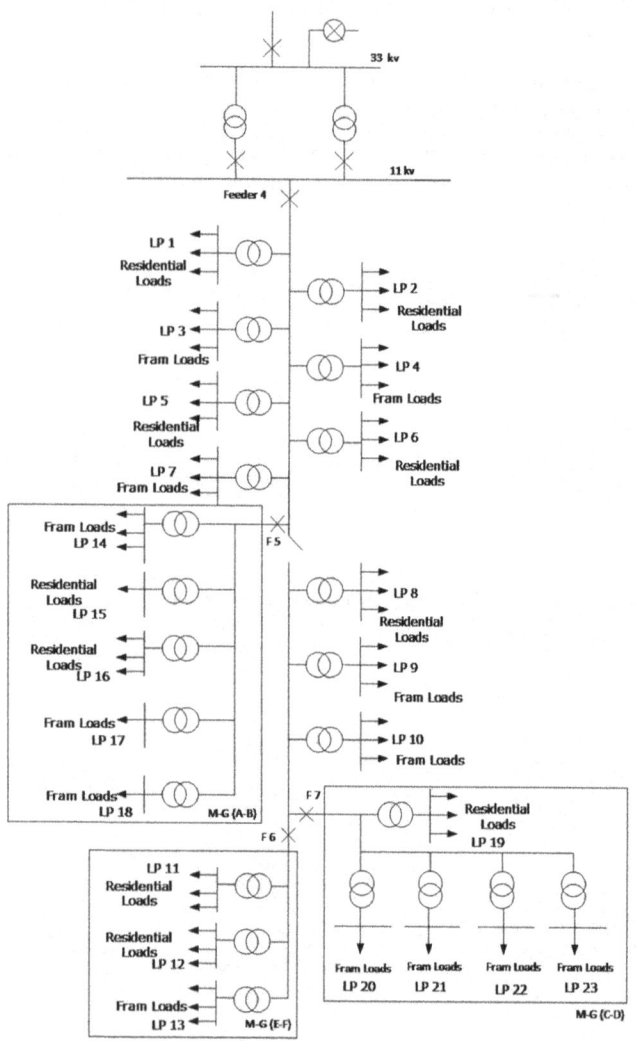

Fig. 61.4   Modified IEEE RBTS Bus -6, F4 DSN with M-G

The AAOT of (U_{LPi}) of load point i is determined by Eq. (11);

$$U_s = \sum_{i=1}^{N} \lambda_i r_i + \lambda_{li} S \qquad (11)$$

Where, the FR of the main feeder section and are lateral components at load point i. is repair time and 'S' is the switching time. Case-1, obtained results are tabulated inside the Table 61.3.

4.2 Case Study- 2: Reliability Evaluation of M-G Location A-B in F4-IEEE RBTS BUS-6

In this case study, radial feeder of IEEE RBTS Bus 6 with M-G (A-B) is proven in Fig. 61.4. In this situation examine, the reliability of all LPs from LP 14 to LP 18 at the feeder F4 has been calculated. The test system is used to calculate to reliability indices of distribution network without DG is mentioned in the case examine-1. Those outcomes are shown that the system higher interruption rate and its duration at LP 14 to LP 18, consequently, it could lead to system performance indices are imply that higher value. At this point, we must recollect that the system of minimization of reliability indices and development of radial feeder 4, to include the M-G at LP

14 to LP18 within the distribution feeder. On the subject of Fig. 61.4 (see M-G), apply modified FMEA method to test system for calculating the average failure rate and average annual outage time from LP 14 to LP 18 of feeder 4 through Eq. (12) and Eq. (13) as

$$\lambda_s' = \sum_{i=1}^{A-B} \lambda_i + \sum_{l=1}^{A-B} \lambda_l \qquad (12)$$

λ_s' - Average failure rate at LPs within M-G.

A-B - M-G location

i = 1 - Load point 1

$$U_s' = \sum_{i=1}^{A-B} \lambda_i r_i + \sum_{l=1}^{A-B} \lambda_l s \qquad (13)$$

U_s' - Average annual outage time at LPs within M-G.

r_i - Average repair time at LPs within M-G

S - Switching Time at LPs within M-G

The case study-2 results are tabulated inside the Tab 61.4.

Table 61.3 Results of load point indices of case-1

LP	λ(f/yr)	r (hrs)	U(hrs/yr)
LP1	1.188	5.2822	6.2753
LP2	1.188	5.2822	6.2753
LP3	1.188	5.2822	6.2753
LP4	1.188	5.2822	6.2753
LP5	1.188	5.2822	6.2753
LP6	1.215	5.3459	6.4953
LP7	1.224	5.8830	7.2009
LP8	1.188	8.7676	10.416
LP9	1.215	8.7540	10.6362
LP10	1.188	8.7676	10.416
LP11	1.579	8.3728	13.2208
LP12	1.579	8.3728	13.2208
LP13	1.579	8.3728	13.2208
LP14	1.836	6.0771	11.1577
LP15	1.873	6.1143	11.4521
LP16	1.836	6.0771	11.1577
LP17	1.836	6.0771	11.1577
LP18	1.836	6.0771	11.1577
LP19	1.781	8.3341	14.8432
LP20	1.816	8.3303	15.128
LP21	1.781	8.3341	14.8432
LP22	1.781	8.3341	14.8432

Table 61.4 Results of load point indices of case-2

LP	λ (f/yr)	r (hrr)	U (hrs/yr)
LP1	1.188	5.2822	6.2753
LP2	1.188	5.2822	6.2753
LP3	1.188	5.2822	6.2753
LP4	1.188	5.2822	6.2753
LP5	1.188	5.2822	6.2753
LP6	1.215	5.3459	6.4953
LP7	1.224	5.8830	7.2009
LP8	1.188	8.7676	10.416
LP9	1.215	8.7540	10.6362
LP10	1.188	8.7676	10.416
LP11	1.579	8.3728	13.2208
LP12	1.579	8.3728	13.2208
LP13	1.579	8.3728	13.2208
LP14	0.6268	8.9572	5.6144
LP15	0.6636	8.9041	5.9088
LP16	0.6268	8.9572	5.6144
LP17	0.6268	8.9572	5.6144
LP18	0.6268	8.9572	5.6144
LP19	1.781	8.3341	14.8432
LP20	1.816	8.3303	8.3303
LP21	1.781	8.3341	14.8432
LP22	1.781	8.3341	14.8432
LP23	1.781	8.3341	14.8432

Table 61.5 Results of load point indices of case-3

LP	λ (f/yr)	r (hrr)	U (hrs/yr)
LP1	1.188	5.2822	6.2753
LP2	1.188	5.2822	6.2753
LP3	1.188	5.2822	6.2753
LP4	1.188	5.2822	6.2753
LP5	1.188	5.2822	6.2753
LP6	1.215	5.3459	6.4953
LP7	1.224	5.8830	7.2009
LP8	1.188	8.7676	10.416
LP9	1.215	8.7540	10.6362
LP10	1.188	8.7676	10.416
LP11	0.406	9.7733	3.968
LP12	0.406	9.7733	3.968
LP13	0.406	9.7733	3.968
LP14	0.6268	8.9572	5.6144
LP15	0.6636	8.9041	5.9088
LP16	0.6268	8.9572	5.6144
LP17	0.6268	8.9572	5.6144
LP18	0.6268	8.9572	5.6144
LP19	0.6084	8.9861	5.4672
LP20	0.6429	8.9332	5.7432
LP21	0.6084	8.9861	5.4672
LP22	0.6084	8.9861	5.4672
LP23	0.6084	8.9861	5.4672

4.3 Case Study-3: Reliability Evaluation of M-G Location C-D in F4- IEEE RBTS BUS-6

In case study-3, radial feeder 4 of IEEE RBTS Bus-6 with M-G is positioned from LP 19 to LP 23 shown in Fig. 61.4. Fundamental reliability of feeder-4 of an IEEE RBTS Bus -6 has been decided and method for calculating to load point indices from LP 19 to LP 23. This analysis is just like case study-1. The received values of failure rate and average annual outage time from LP 19 to LP 23 give higher value. On the way to decrease of failure rate and average annual outage time from LP 19 to LP 23, the M-G (C-D) is hooked up to LP 19 to LP 23 within the feeder 4 of Fig. 61.4. In keeping with the M-G location C-D, apply FMEA method to test system for calculating the average failure rate and average annual outage time from LP19 to LP23 with the help of Eq. (14) to Eq. (15) as

$$\lambda_s' = \sum_{i=1}^{A-B, C-D} \lambda_i + \sum_{l=1}^{A-B,C-D} \lambda_l \tag{14}$$

λ_s' - Average failure rate at LPs within M-G.

A-B, C-D means M-G locations

i = 1- Load point 1

$$U_s' = \sum_{i=1}^{A-B,C-D} \lambda_i r_i + \sum_{l=1}^{A-B,C-D} \lambda_l s \tag{15}$$

U_s' - Average annual outage time at LPs within M-G.

r_i - Average repair time at LPs within M-G location A-B, C-D.

S - Switching Time at LPs within M-G location A-B and C-D.

The calculated results are tabulated inside the Table 61.5.

4.4 Case Study-4: Reliability Evaluation of M-G Location E-F in F4- IEEE RBTS BUS-6

In this case study, the M-G (E-F) is hooked up from LP 11 to LP 13 in the feeder 4 of Fig.4. Assessment of the outcomes of reliability indices from LP11 to LP13 including failure rate and average annual outage time of feeder 4 of the IEEE RBTS Bus-2 is similar to case study-1. As a way to reduce the load point indices and improve system indices of feeder 4, M-G (E-D) is brought from LP 11 to LP 13 within the feeder. The AFR and AAOT of load point LP 11 to LP 13 are calculated with admire to M-G (E-F) consistent with Fig. 61.4 and the use of those Eq. (16) and (17) respectively.

$$\lambda_s' = \sum_{i=1}^{A-B, C-D,E-F} \lambda_i + \sum_{l=1}^{A-B,C-D,E-F} \lambda_l \tag{16}$$

λ_s' - Average failure rate at LPs within M-G.

A-B, C-D and E-F means M-G locations

i = 1 - Load point 1

$$U_s' = \sum_{i=1}^{A-B,C-D,E-F} \lambda_i r_i + \sum_{l=1}^{A-B,C-D,E-F} \lambda_l s \tag{17}$$

Us' - Average annual outage time at LPs within M-G.

r_i - Average repair time at LPs within M-G location A-B, C-D and E-F

S- Switching time at LP within M-G location A-B,C-D and E-F.

The calculated results of case study-2 are tabulated inside the Table 61.6.

Table 61.6 Results of load point indices of case-4

LP	λ (f/yr)	r (hrr)	U (hrs/yr)
LP1	1.188	5.2822	6.2753
LP2	1.188	5.2822	6.2753
LP3	1.188	5.2822	6.2753
LP4	1.188	5.2822	6.2753

LP	λ (f/yr)	r (hrr)	U (hrs/yr)
LP5	1.188	5.2822	6.2753
LP6	1.215	5.3459	6.4953
LP7	1.224	5.8830	7.2009
LP8	1.188	8.7676	10.416
LP9	1.215	8.7540	10.6362
LP10	1.188	8.7676	10.416
LP11	0.406	9.7733	3.968
LP12	0.406	9.7733	3.968
LP13	0.406	9.7733	3.968
LP14	0.6268	8.9572	5.6144
LP15	0.6636	8.9041	5.9088
LP16	0.6268	8.9572	5.6144
LP17	0.6268	8.9572	5.6144
LP18	0.6268	8.9572	5.6144
LP19	0.6084	8.9861	5.4672
LP20	0.6429	8.9332	5.7432
LP21	0.6084	8.9861	5.4672
LP22	0.6084	8.9861	5.4672
LP23	0.6084	8.9861	5.4672

5. Results and Discussions

In this section, an effect of different M-G places to give unique reliability indices of the IEEE RBTS Bus 2 and these results are compared to without M-G. A MATLAB software program programme is advanced in R2016a, 2GHz, and a personal computer based totally on the mathematical version. AMATLAB programme has been developed based on the modified method and it is showed in Appendix A. The reliability indices of distribution feeder with all case studies is shown in Table 61.3 Table 61.4, Table 61.5, Table 61.6 and Table 61.7, consequences show that of system performance indices in all cases.

Table 61.7 Results of system performance indices for all cases

Indices	Case-1	Case-2	Case-3	Case-4
SAIFI (f/yr)	1.4076	1.2459	1.091	0.9368
SAIDI (hrs/yr)	9.5489	8.8067	7.575	6.3519
CAIDI (hrs/f yr)	6.7838	7.0685	6.943	6.7804
ASAI	0.99891	0.9989	0.9991	0.99927

Table 61.7, showed that system performance indices for all cases including without M-G. In the case 2, SAIFI and SAIDI are reduced as brought of M-G. All through these cases, the results of SAIFI and SAIDI for all cases as confirmed in Fig. 61.5 and Fig. 61.6. Further to CAIDI and ASAI are

increased because these indices depend on number of load pints. . In case-3, SAIFI and SAIDI are reduced as modified with the M-G position. Because SAIFI is depends on the total component failure rate and SAIDI is depends on the repair rate of components. In this situation, CAIDI and ASAI are also increased as compared to preceding instances along with case 1 and case 2. Inside the case -4, SAIFI and SAIDI are a whole lot greater decreased as compared to the case 2 and case 3. Also, CAIDI and ASAI are expanded. Therefore, the system performance indices of IEEE RBTS Bus 6-F4 radial distribution network are improved with the effect of M-G. The results of CAIDI and ASAI for all case as showed Fig. 61.7 and Fig. 61.8.

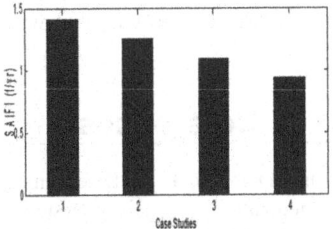

Fig. 61.5 The results of SAIFI for all cases

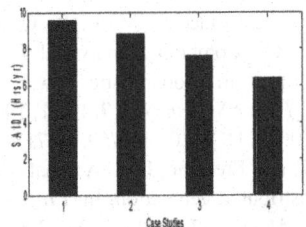

Fig. 61.6 The results of SAIDI for cases

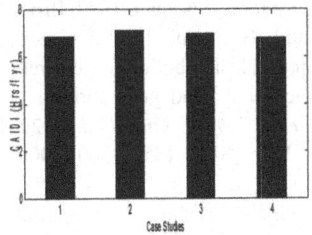

Fig. 61.7 The results of CAIDI for all cases

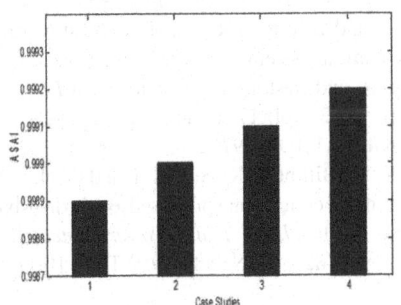

Fig. 61.8 The results of ASAI for all cases

6. Conclusion

This paper described an analytical technique of various Micro-Grid models to assess the reliability indices of the distribution network. The approach has to improve the reliability indices such as SAIFI, SAIDI, CAIDI and ASAI with impact of build-up of Micro-Grid position within the group of load points inside the feeder community. Reliability evaluation algorithm has developed for individual Micro-Grid and this algorithm depends on the modified FMEA method. The consequences show that the interruption rate and outage time of load points within Micro-Grid has decreased efficaciously, however reliability indices of load points outside to Micro-Grids does not change. The reliability indices of the distribution network without Micro-Grid are used to examine with the proposed approach.

References

1. Barke, R. and Demello, R.W. Determining the impact of distributed generation on power systems: Par 1-Radial distribution systems. In: *IEEE Summer Power Meeting*: 2000, pp. 1645–1656. ISBN. 0-7803-6420-1. DOI: 10.1109/PESS.2000.868775.

2. Pereira, M. V. F., Maceira, M. E. P., Oliveira, G. C. and Pinto, M. V. G.. Combining analytical models and Monte Carlo techniques in probabilistic power system analysis. *IEEE Trans. Power System*. vol. 7, Iss. 1, 1992, pp. 265–272. ISBN.1558-0679. DOI: 10.1109/59.141713.

3. Brown, R. E. and Freeman, L. A. A. Analyzing the reliability impact of distributed generation. In: *Proc. IEEE Power Eng. Soc. Summer Meeting*. 2001, 15-19 July2001, pp. 1013–1018. ISBN.0-7803-7173-9. DOI: 10.1109/PESS.2001.970197.

4. Billinton, R. and Allan, R.N. Reliability Evaluation of Engineering Systems. Springer. 1992. ISBN: 0306440636.

5. Wand, P. and Billinton, R. Time-sequential simulation technique for rural distribution system reliability cost/worth evaluation including wind generation as alternative supply. *IEE proc. Genr. Transm. Distribution*, 2001, Vol. 148, Iss. 4, pp. 355–360. ISSN 1350-2360. DOI: 10.1049/ip-gtd: 20010406.

6. Ou, Y and Goel, L. Monte-Carlo simulation for overall distribution system worth assessment, *IEE proc. Gener Transm. Distribution*, 1999, Vol.146 Iss. 5, pp. 535–540. ISSN 1350-2360. DOI: 10.1049/ip-gtd: 19990542.

7. Wang, P. and Billinton, R. Reliability cost/worth assessment of distribution systems incorporating time-varying weather conditions and restoration resources. *IEEE Trans. Power Delivery*, 2003. Vol. 17, Issue. 1, pp. 260–265. ISSN 0885-8977 DOI: 10.1109/61.974216.

8. Allan, R. N., Billinton, R., Sjarief, I. and Goel, L A. reliability test system for educational purposed-Basic distribution system data and results. *IEEE Trans. Power Systems*, 1991.Vol. 6, Iss.2, pp. 813–820. ISSN 0885-8950. DOI: 10.1109/59.76730.

9. Billinton, N. and R. N. Allan. Reliability Evaluation of Power Systems. 2nd ed. New York and London: Plenum Press, 1996. ISBN 0-30645259-6.

10. Billinton, R. and Li, W. Reliability Assessment of Electric Power Systems Using Monte Carlo Methods, 1994. ISBN 03064-47819.

11. Billinton, R. and Jonnavithula. A test system for teaching overall power system reliability assessment power system. *IEEE Transactions on Power Systems*, 1996. Vol. 11 Iss.. 4, pp. 1670–1676. ISSN 0885-8950. DOI: 10.1109/59.544626.

12. Subae, I., Kim, J., Chul Kim, J.and SINGH, C. Optimal Operating Strategy for Distributed Generation Considering Hourly Reliability Worth. *IEEE transactions on power systems*, 2004. Vol. 19, Iss. 1. ISSN 0885-8950. DOI: 10.1109/TPWRS.2003.818738.

13. Fotuhi-Firuzabad, Md. and Ghahnavie Abbas, R. An Analytical Method to Consider DG Impacts on Distribution System Reliability. In: *IEEE/PES Transmission and Distribution Conference & Exhibition*, 2005. ISBN 0-7803-9114-4. DOI: 10.1109/TDC.2005.1547168.

14. Waseem, I., Pipattanasomporn, M. and Rahman, S. Reliability benefits of distributed generation as a backup source. *IEEE power & energy society general meeting*, 2009, pp. 1–8. ISSN 1932-5517. DOI: 10.1109/PES.2009.5275233.

15. Imankhonakdarr, T., Sheikholeslami, A., Barforoushi, T. and Seyed Mohammad B. S. Investigating Impacts of Distributed Generation on Distribution Networks Reliability: A Mathematical Model. In: *IEEE Electric power quality and Supply Reliability Conference*, 2010. pp. 117–124. ISSN 978-1-4244-6978-9. DOI:10.1109/PQ.2010.5550010.

16. BOONTHIENTHONG, B., RUGTHAICH AROENCHEEP, N. and AUCHARIYAMET, S. Service Restoration of Distribution System with Distributed Generation for Reliability Worth. In: *47th International Universities Power Engineering Conference* (UPEC), 2012, Sept 4-7, ISBN. 978-1-4673-2856-2. DOI: 10.1109/UPEC.2012.6398620.

17. GENGFENG, L., ZHAOHONG, B., HAIPENG, X. X. and WANGI FAN, W. Reliability Evaluation of Active Distribution Networks Considering Customer Satisfaction. In: 6 *International Conferences on Applied Energy*, 2014, pp. 591 – 594.ISSN1876-6102DOI:10.1016/j.egypro 2014.11.1177.

18. RAJU, K.., and SRINIVAS, G.N. Distribution System reliability with Distributed Generation based on Customer Scattering. *Journal of Advances in Electrical and Electronics Engineering*, 2015, Vol. 13, Iss. .2, pp. 64-73. ISSN 1336-1376. DOI: 10.15598/aeee.v13i2.1025.

19. LASSETER, R.H., and PAIGI, P. Micro grid a conceptual solution. In: *IEEE Power Electronic Specialists Conference Madison*, 2004, pp. 4285-4290.ISSN.0275-9306. DOI:10.1109/PESC. 2004.1354758

20. COSTA, P.M., and MANUEL A.M. Assessing the contribution of micro grids to the reliability of distribution network. *Journal of Electric Power Systems Research*, 2009. Vol. 79, Issue. 2, pp. 382–389. ISSN 0378-7796. DOI: 10.1016/j.epsr.2008.07.009.

21. ESAU, Z., and JAYAWEERA, D. Reliability assessment in active distribution networks with detailed effects of PV systems. *Journal of Modern power systems Clean Energy*, 2014, Vol. 2 Iss.. 1, and pp. 59-68. ISSN 2196-5420. DOI: 10.1007/s40565-014-0046-2.

22. KIM, H., and SINGH, C. Power system reliability modeling with aging using thinning algorithm. In: *IEEE Conf. Bucharest Power Tech*, 2009. pp. ISBN 978-1-4244-2234-0. DOI: 10.1109/ PTC.2009. 5281858.

23. AHMAD, S., SARDA, S., NOOR.B and ASAR. UL. A. Analysing distributed generation impact on the reliability of electric distribution network. *International Journal of Advanced Computer Science & Applications,* 2016. vol. 1, no. 7, pp. 217–221, 2016. ISSN. 2156-5570. DOI: 10.14569/ issn.2156-5570.

Note: All the figures and tables in this chapter were made by the authors.

Appendix A

Abbreviations

The following abbreviations are used in this manuscript:

Case-1: Reliability evaluation of distribution network without M-G for F4- IEEE RBTS Bus-6

```
% A program to find the reliability indices of IEEE RBTS
BUS 6 Feeder 4

clc

% customer data
Customer data = [1 147
                 9 1
                 13 1
                 16 76
                 20 1];
Load point = customer data (:,1);
noc = customer data(:,2);
% data of RBTS BUS 6 FEEDER 4
%function [sec ln fr rt swt]
data = [1 2.80 0.0460 8 0.5
         2 2.50 0.0460 8 0.5
         3 1.60 0.0460 8 0.5
         4 0.90 0.0460 8 0.5
         5 1.60 0.0460 8 0.5
         6 2.50 0.0460 8 0.5
         7 0.60 0.0460 8 0.5
         8 1.60 0.0460 8 0.5
         9 0.80 0.0460 8 0.5
        10 0.90 0.0460 8 0.5
        11 3.20 0.0460 8 0.5
        12 2.80 0.0460 8 0.5
        13 0.60 0.0460 8 0.5
        14 3.50 0.0460 8 0.5
        15 1.60 0.0460 8 0.5
        16 2.80 0.0460 8 0.5
        17 3.20 0.0460 8 0.5
        18 2.50 0.0460 8 0.5
        19 3.20 0.0460 8 0.5
        20 1.60 0.0460 8 0.5
        21 0.80 0.0460 8 0.5
        22 2.80 0.0460 8 0.5
        23 2.50 0.0460 8 0.5
        24 3.20 0.0460 8 0.5
        25 2.80 0.0460 8 0.5
        26 2.50 0.0460 8 0.5
        27 0.75 0.0460 8 0.5
        28 1.60 0.0460 8 0.5
        29 3.20 0.0460 8 0.5
        30 2.80 0.0460 8 0.5
        31 1 0.0150 48 1];
%finding basic indices
for i=1:31
 data(i,6)=data(i,2)*data(i,3);
end
%failure rate data(1,7)=data(1,6)+data(2,6)+data(3,6)+
data(4,6)+data(5,6)+data(6,6)
+data(8,6)+data(10,6)+data(11,6)+data(12,6)+data(14,6)+
data(15,6)+data(31,6); data(2,7)=data(1,7)+data(13,6);
data(3,7)=(data(1,7)-data(15,6))+data(16,6)+data(17,6)+
data(18,6);
data(4,7)=data(1,7)+data(19,6)+data(20,6)+data(22,6)+
data(23,6) +data(24,6);
data(5,7)=data(1,7)+data(19,6)+data(26,6)+data(27,6)+
data(28,6)+data(29,6)+data(30,6);
%average annual outage
timedata(1,8)=((data(1,6)+data(2,6)+data(3,6)+data(4,6)+
data(5,6)+data(6,6)+data(8,6)+data(10,6))*8)+((data(11,6)+
```

```
data(12,6)+data(14,6)+data(15,6))*0.5)+(data(31,6)*48);
data(2,8)=((data(1,6)+data(2,6)+data(3,6)+data(
result(:,1)=data(:,1);
 result(:,2)=data(:,7);
 result(:,3)=data(:,8)
%finding performance indices
 T1=0;NC=0;T2=0;
for i=1:5
T1=T1+data(i,7)*noc(i);
NC=NC+noc(i);
T2=T2+data(i,8)*noc(i);
End
SAIFI(1)=T1/NC
SAIDI(1)=T2/NC
CAIDI(1)=SAIDI/SAIFI
ASAI(1)=(NC*8760-T2)/(NC*8760)
 ASUI(1)=1-ASAI(1)
```

Case-2: Reliability evaluation of M-G location A-B in F4-IEEE RBTS Bus-6

```
% A program to find the reliability indices with Micro-grind
in location A-B
clc
%customer data
Customer data = [1 147
                 9 1
                 13 1
                 16 76
                 20 1];
Load point=customer data(:,1);
noc=customer data(:,2);
% data of RBTS BUS 6 Feeder 4
%function [sec ln fr rt swt]
data = [1 2.80 0.0460 8 0.5
        2 2.50 0.0460 8 0.5
        3 1.60 0.0460 8 0.5
        4 0.90 0.0460 8 0.5
        5 1.60 0.0460 8 0.5
        6 2.50 0.0460 8 0.5
        7 0.60 0.0460 8 0.5
        8 1.60 0.0460 8 0.5
        9 0.80 0.0460 8 0.5
        10 0.90 0.0460 8 0.5
        11 3.20 0.0460 8 0.5
        12 2.80 0.0460 8 0.5
        13 0.60 0.0460 8 0.5
        14 3.50 0.0460 8 0.5
        15 1.60 0.0460 8 0.5
        16 2.80 0.0460 8 0.5
        17 3.20 0.0460 8 0.5
        18 2.50 0.0460 8 0.5
        19 3.20 0.0460 8 0.5
        20 1.60 0.0460 8 0.5
        21 0.80 0.0460 8 0.5
        22 2.80 0.0460 8 0.5
        23 2.50 0.0460 8 0.5
        24 3.20 0.0460 8 0.5
        25 2.80 0.0460 8 0.5
        26 2.50 0.0460 8 0.5
        27 0.75 0.0460 8 0.5
        28 1.60 0.0460 8 0.5
        29 3.20 0.0460 8 0.5
        30 2.80 0.0460 8 0.5
        31 1 0.0150 48 1];
%finding basic indices
for i=1:31
data(i,6)=data(i,2)*data(i,3);
end
%failure rate
data(1,7)=data(1,6)+data(2,6)+data(3,6)+data(4,6)+
data(5,6)+data(6,6)+data(8,6)+data(10,6)+data(11,6)+
data(12,6)+ data(14,6)+data(15,6)+data(31,6);
data(2,7)=data(1,7)+data(13,6);
%average annual outage time
data(1,8)=((data(1,6)+data(2,6)+data(3,6)+data(4,6)+
data(5,6)+data(6,6)+data(8,6)+data(10,6))*8)+((data(11,6)+
data(12,6)+data(14,6)+data(15,6))*0.5)+(data(31,6)*48);
result(:,1)=data(:,1);
 result(:,2)=data(:,7);
 result(:,3)=data(:,8)
%finding performance indices
 T1=0;NC=0;T2=0;
for i=1:5
T1=T1+data(i,7)*noc(i);
 NC=NC+noc(i);
 T2=T2+data(i,8)*noc(i);
End
SAIFI(1)=T1/NC
SAIDI(1)=T2/NC
CAIDI(1)=SAIDI/SAIFI
ASAI(1)=(NC*8760-T2)/(NC*8760)
ASUI(1)=1-ASAI(1)
```

Design of Electrical Vehicle Lithium Battery Charger Using MATLAB Simulink

62

B. Vidyasagar[1], N. Rajasekhar Varma[2]

Professor, Teegala Krishna Reddy Engineering College,
Hyderabad, India

Dhasharatha G[3], Kalagotla Chenchi Reddy[4]

Assistant Professor, Teegala Krishna Reddy Engineering College,
Hyderabad, India

Abstract: Due to their reduced environmental effect and lower operating costs when compared to conventional gasoline-powered vehicles, electric vehicles (EVs) are becoming more and more popular. However, the absence of a thorough charging infrastructure prevents the wide-scale adoption of EVs. Therefore, creating effective and efficient EV chargers is essential for the market's success. A MATLAB-based simulation of an EV charger is described in this paper. A power electronic converter, a control algorithm, and a battery management system are all included in the simulation model. In order to charge the EV battery, the converter transforms AC electricity from the grid into DC power. The charging of the battery is made safe and effective by the control algorithm. The battery management system keeps an eye on how fully charged the battery is and guards against either overcharging or undercharging. The simulation model is tested under different scenarios, including varying input voltages and loads, and the performance is evaluated in terms of charging time and efficiency. The simulation results show that the designed charger is capable of charging an EV battery within a reasonable time while maintaining a high level of efficiency. Overall, this paper demonstrates the feasibility and effectiveness of using MATLAB to design and simulate an EV charger. The simulation results can be used to optimize the charger's design and improve its performance, leading to the development of more efficient and reliable EV charging systems.

Keywords: Electric vehicles (EVs), Electrical vehicle supply equipment (EVSE), DC-DC converter, Lithium battery, Transformer, EV charger

1. Introduction

The infrastructure required to accommodate the increasing number of electric vehicles on the road today must include an EV charger. Electric cars (EVs) are a more and more common choice for drivers as the need for environmentally friendly transportation options rises. However, a strong charging infrastructure is necessary to guarantee that drivers have access to dependable and practical charging options in order to fully realise the potential of EVs.

A device called an electric car charger provides electric energy to an electric vehicle so that it can recharge its batteries. A charging cable is normally used to connect the car to the charging station during the charging process to transfer electrical energy from the power grid to the batteries of the electric vehicle.

An electric vehicle (EV) charger is an apparatus that provides electric energy to an EV to recharge its batteries. A charging cable is normally used to connect the car to the charging station during the charging process to transfer electrical

[1]sagar.boorgula@gmail.com, [2]drnrsvarma@gmail.com, [3]g.dhasharatha@gmail.com, [4]chenchireddy.kalagotla@gmail.com

energy from the power grid to the batteries of the electric vehicle.

There are several different types of EV chargers, including Level 1, Level 2, and Level 3 (sometimes referred to as DC fast charging) chargers, each of which has a different charging rate and power required. An electric car can be fully charged in as little as 24 hours using level 1 chargers, which are generally used for home charging. In public spaces like parking lots, workplaces, and retail malls, level 2 chargers can offer quicker charging times. The quickest and most potent chargers are Level 3 chargers, which can recharge an electric car in as little as 30 minutes.

The demand for EV charging infrastructure has expanded as a result of the expansion of the EV market. Investments in EV charging infrastructure are being made by governments, companies, and private citizens in order to facilitate the switch to electric mobility and lower emissions from fossil fuel-powered vehicles.

Range anxiety, or the worry that an EV will run out of power before reaching its destination, is one of the major obstacles to the general adoption of EVs. It is essential to have a robust charging infrastructure if you want to allay this worry and give drivers the assurance they need to move to EVs.

There are several places where charging infrastructure can be installed, including homes, offices, public parking lots, and roads. The most typical way to charge an EV is at home, but public charging stations are becoming more and more prevalent.

The need for EV chargers will rise as the number of electric vehicles on the road rises. Investments in EV charging infrastructure are being made by governments, companies, and private citizens in order to facilitate the switch to electric mobility and lower emissions from fossil fuel-powered vehicles.

Many EV charging stations have smart technology installed that enables them to interact with the grid, control charging loads, and give drivers real-time updates on charging availability and status. Energy suppliers may be able to better control the grid and incorporate renewable energy sources into the system with the help of this technology[1].

Infrastructure for EV charging plays a significant role in the overall energy transition, which aims to lower greenhouse gas emissions and move towards renewable energy sources. The installation of EV chargers can aid in lowering emissions from the transportation industry, a major source of greenhouse gas emissions worldwide. The demand for EV charging infrastructure rises along with the EV market[2]. Investing in charging infrastructure can boost economic growth and the EV market's expansion while also generating new jobs.

Governments in some areas are offering incentives for the installation of EV charging infrastructure. For instance, the federal government in the United States offers tax credits for the building of EV charging stations, and some states provide additional incentives to promote the deployment of charging infrastructure. Infrastructure for EV charging, nevertheless, is not without problems. The expense of installing charging infrastructure, which can be high, especially for fast-charging stations, is one of the key obstacles. The requirement for interoperability between several charging networks presents another difficulty. Some of the problems with EV charging infrastructure may be solved by the development of new technologies, like wireless charging[3]. A physical cable is not required for wireless charging, which might make charging more convenient and lower the cost of installing charging infrastructure. The advancement of EV technology will have a significant impact on the future of EV charging infrastructure[4].

2. Topology

Single phase full wave bridge converter

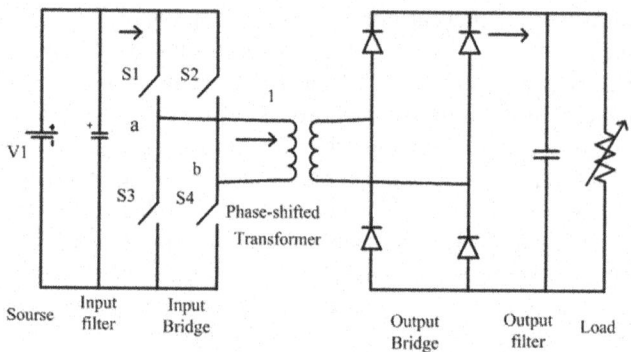

Fig. 62.1 Single phase full wave converter

The output dc voltage across the resistive load is given by

$$V_{0(av)} = \frac{1}{T} \int_0^T v_o(\omega t) d\omega t$$

$$V_{0(av)} = \frac{1}{\pi} \int_\alpha^{\pi+\alpha} V_m \sin \omega t$$

$$d\omega t = \frac{V_m}{\pi} [-\cos \omega t]_\alpha^{\pi+\alpha}$$

$$V_0 = \frac{2V_m}{\pi} \cos \alpha$$

$$V_{0(rms)} = \left[\frac{1}{T} \int_0^T v_0^2(\omega t) d\omega t \right]^{\frac{1}{2}}$$

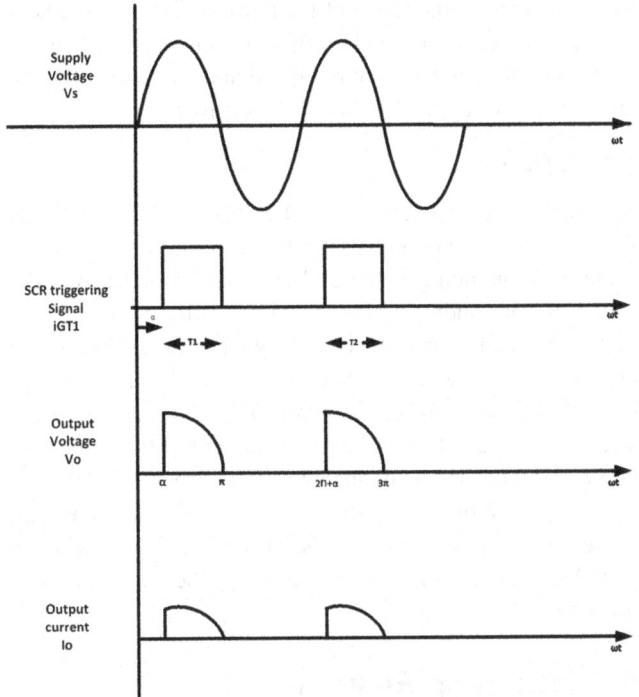

Fig. 62.2 Output waveform for single phase full wave converter

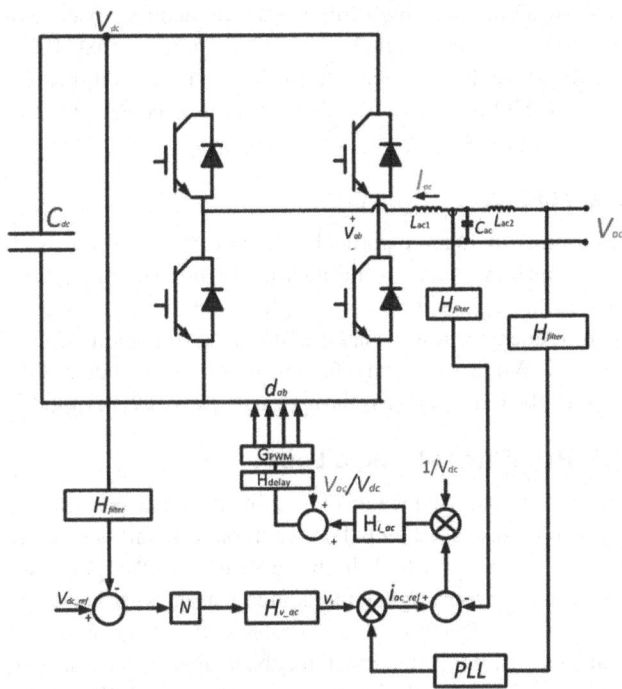

Fig. 62.3 Proposed bidirectional digital control system

$$V_{0(rms)} = \left\{ \frac{1}{T} \left[\int_{0}^{\frac{T}{2}} V_s^2 \, dt + \int_{\frac{T}{2}}^{T} \left(-V_S \right)^2 \, dt \right] \right\}^{\frac{1}{2}}$$

$$= \left\{ \frac{V_s^2}{T} \left[\int_{0}^{\frac{T}{2}} dt + \int_{\frac{T}{2}}^{T} dt \right] \right\}^{\frac{1}{2}} = V_s$$

3. Control Scheme

The circuit diagram (Fig. 62.3) shows the proposed bidirectional digital control system, to obtain a high power factor and loosely adjust the dc-link voltage at a high voltage (500 V), the controller for the ac/dc stage is employed. The voltage loop and current loop both control the Iac and Vdc for the ac/dc stage, respectively. The measured Vdc and reference Vdc_ref are compared in the voltage loop. After that, a notch filter is used to remove the ripple at 120 Hz from the error signal. The ac voltage loop compensator Hvac processes the filtered error signal. The sinusoidal reference for the inner current loop will be produced by multiplying the compensator output Vc by the PLL output. It is simpler

to construct a compensator since the transfer function won't be affected by variations in the dc link voltage to the term 1/Vdc's use in the current loop to cancel the coupled term in the power stage transfer function. The disturbance from the input voltage Vac is compensated for by the feed forward term Vac/Vdc and the current loop compensator Hiac[5].

3.1 IGBT

Insulated Gate Bipolar Transistor is referred to as IGBT. It is an electric 3-terminal semiconductor device with quick switching capabilities and great efficiency. The two primary purposes of a transistor, a tiny electronic component, are as follows. It can boost signals and serves as a switch for lighting circuits. Depending on their additional utility or particular use, transistors can be classified into various categories. IGBT, MOSFETs, and BJT (Bipolar Junction Transistor) are the most widely utilised transistors. Each has its preferences and advantages over the other, including BJTs and MOSFETs.

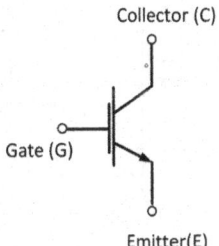

Fig. 62.4 Symbol of IGBT

Low switching loss, high I/P impedance, and the absence of secondary breakdown make MOSFETs superior than BJTs, which favour low on-state drops. The IGBT combines the BJT and MOSFET in a way that maximises the performance of each transistor.

3.2 H Filter

High-pass filters are a particular kind of filter that rejects all low-frequency signals while allowing high-frequency signals to pass through without any amplitude attenuation. Signals with frequencies lower than the filter's cut-off frequency are blocked. While all signals that are higher in frequency than the cut-off frequency pass through with their full amplitude.

3.3 PLL (Phase-Locked Loop)

The phase and frequency of the input signal are tracked using a phase-locked loop. For synchronous communication, it is a really helpful tool. In the receiver, the PLL generates a coherent carrier signal for demodulation after acquiring the carrier frequency in the suppressed carrier mode of transmission. Like a normal feedback mechanism, a PLL functions. It adjusts the output frequency of the VCO until it syncs with the input signal or until it matches the frequency of the input signal.

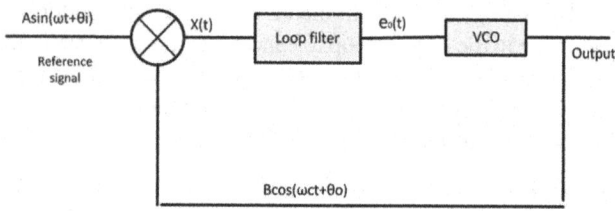

Fig. 62.5   Circuit diagram of LLC filter

3.4 GPWM (Giga Pulse Width Modulation)

Gigabit Pulse Width Modulation is known as GPWM. It is a technology that enables the adjustment of a fan's or a pump's speed by altering the width of the power pulses that are sent to it. This is accomplished by employing an integrated circuit to regulate the fan's or pump's speed and, consequently, the amount of cooling the CPU or GPU receives.

3.5 H Delay

The best all-around option for all edits and effects is the H Delay analogue echo processor. You can utilise a variety of its effects, including delay and modulation, to change the sound of your audio. It also includes a noise-reduction tool that successfully mutes the noise in your vocals. You can add effects, such delay, to any specific section of your song with the H Delay analogue echo processor. Additionally, you can select the degree of the effect. You may adjust the amount of delay using the H Delay analogue echo processor. You can tune the delay time in milliseconds, just like the echo boy processor, to get flawless vocals. There is a display window included with it where you may enter the timing of any impact.

4. Simulation Results

Figure 62.6 shows the simulink model of Electrical hybrid vechile charger in this circuit diagram, the DC input is conected to a bridge rectifier circuit, which is composed of four IGBT or an MOSFET. The diodes are arranged in a bridge configuration, which allows the DC input to be converted to the pulslating DC voltage.

The bridge rectifier is connected to the transformer,the transformer allows the DC-DC converter to provide isolation between the input and output circuits, which is useful for applications where there is a need to protect sensitive electronic components from voltage transients or electrical noise[6]. Additionally, transformers can improve the efficiency of the DC-DC converter by reducing losses due to resistive elements in the circuit.

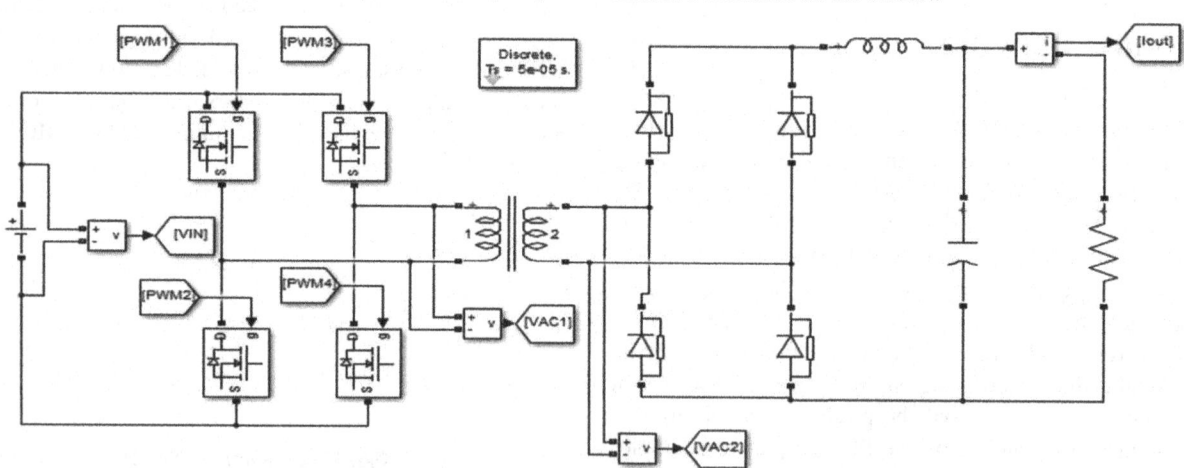

Fig. 62.6   Simulink model of electrical hybrid vehicle charger

The output of the bridge rectifier circuit is then connected to a DC-DC converter circuit, which is composed of an inductor, a capacitor, and a switch[7]. The switch is typically a Diodes, which allows the DC voltage to be converted into a high-frequency AC voltage.

The inductor and capacitor are used to filter the AC voltage and produce a smooth DC voltage at the output.

Overall, the EHV charger model is a simple circuit that uses a bridge rectifier circuit and a DC-DC converter circuit to convert AC power into DC power and then boost the DC voltage to recharge the battery of the EHV[8].

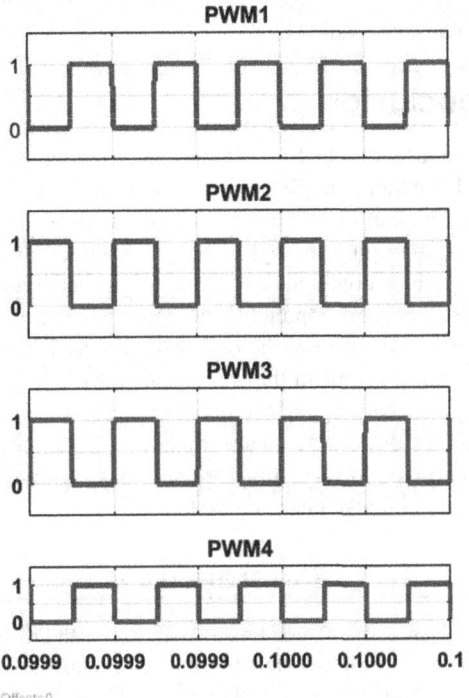

Fig. 62.7 Output of DC converter

Figure 62.7 shows the output of DC converter when switch 1 and 3 in ON position its starts from 0(zero) position, and other switches in off mode. And when switch 2 and 4 in ON position its starts from 1(one) position, and other switches in off mode.

Figure 62.8 shows the output of the battery, the battery output of an EHV charger refers to the voltage and current that the charger supplies to the battery of an electric vehicle during the charging process.

SOC (State of Charge) refers to the amount of charge that is currently stored in a battery, SOC can be determined by measuring the voltage, current and temperature of the battery.

It is not accurate to say that the battery current in an EHV (Extreme High Voltage) charger is negative. In fact, the

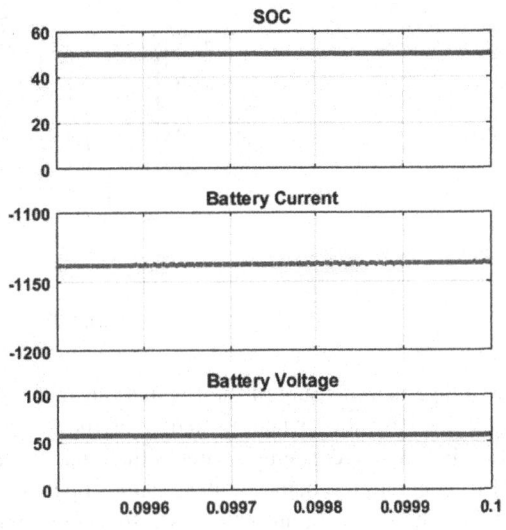

Fig. 62.8 Battery output

battery current in an EHV charger can be either positive or negative, depending on whether the battery is being charged or discharged.

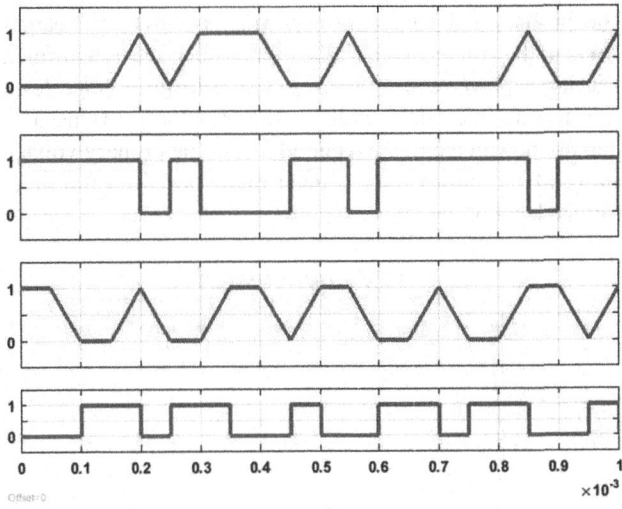

Fig. 62.9 Input to a different DC output voltage

The input voltage to a different DC output voltage. The output voltage can be either higher or lower than the input voltage, depending on the type of DC to DC converter.

The output of a DC to DC converter can be explained in terms of its voltage regulation and ripple voltage.

Voltage regulation refers to the ability of the DC to DC converter to maintain a stable output voltage despite changes in the input voltage or load[9]. A well-designed DC to DC converter will have a high level of voltage regulation, ensuring that the output voltage remains within a tight range of values.

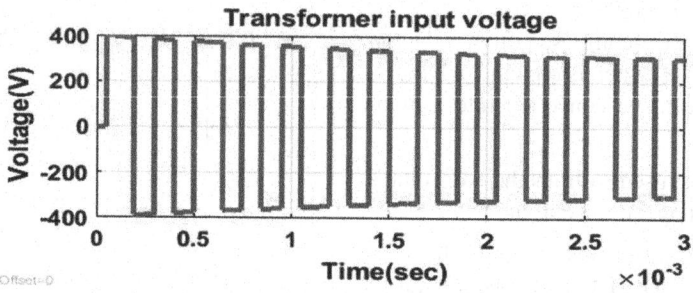

Fig. 62.10 Transformer primary voltage

Ripple voltage is a measure of the fluctuations in the output voltage caused by the switching action of the DC to DC converter[10]. As the converter switches the input voltage on and off to generate the output voltage, small fluctuations in the output voltage can occur, which is called ripple voltage. The amount of ripple voltage present in the output of a DC to DC converter depends on the type of converter, the switching frequency, and the load.

Figure 62.10 shows the transformer primary voltage, the transformer is designed to step up the voltage of the input power supply to the required EHV level.

The primary voltage of a transformer in an EHV charger refers to the voltage level on the input side of the transformer. The input power supply is usually at a lower voltage level than the desired EHV level, so the transformer is used to step up the voltage to the required level. The primary voltage rating of the transformer is determined based on the input voltage level of the power supply.

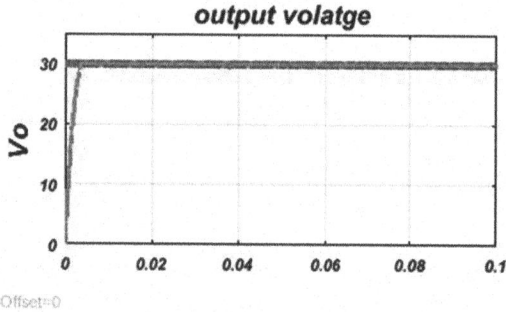

Fig. 62.11 The output voltage of a DC to DC converter

The output voltage of a DC to DC converter depends on the specific design of the converter and can be either higher or lower than the input voltage.

The output voltage of a DC to DC converter can also be regulated by using a feedback mechanism. A feedback mechanism compares the output voltage to a reference voltage and adjusts the duty cycle of the switch to maintain a constant output voltage even if there are variations in the input voltage or load[11].

5. Conclusion

Since the use of electric vehicles is growing daily, it is essential that their charging systems be strong and dependable so that users can drive electric vehicles without any issues. For charging plug-in hybrid electric automobiles, the article provides conductive AC charging procedures. In this study, the theory of conductive charging is examined, and a thorough examination of the on-board charger is performed, allowing one to confirm the legal states for the beginning of charging and the range of the maximum current limit, which is dictated by duty cycle. In accordance with standards set by the automotive industry, this document demonstrates how to construct conductive charging systems.

References

1. M. Salem et al., "Three-Phase Series Resonant DC-DC Boost Converter With Double LLC Resonant Tanks and Variable Frequency Control," in IEEE Access, vol. 8, pp. 22386–22399, 2020, doi: 10.1109/ACCESS.2020.2969546.
2. H. -J. Choi, K. -W. Heo and J. -H. Jung, "A Hybrid Switching Modulation of Isolated Bidirectional DC-DC Converter for Energy Storage System in DC Microgrid," in IEEE Access, vol. 10, pp. 6555–6568, 2022, doi: 10.1109/ACCESS.2021.3138988.
3. S. A. Gorji, H. G. Sahebi, M. Ektesabi and A. B. Rad, "Topologies and Control Schemes of Bidirectional DC–DC Power Converters: An Overview," in IEEE Access, vol. 7, pp. 117997–118019, 2019, doi: 10.1109/ACCESS.2019.2937239
4. M. V. Soares and Y. R. de Novaes, "MMC Based Hybrid Switched Capacitor DC-DC Converter," in IEEE Open Journal of Power Electronics, vol. 3, pp. 142–152, 2022, doi: 10.1109/OJPEL.2022.3154726.
5. A. K. Singh, A. K. Mishra, K. K. Gupta and Y. P. Siwakoti, "High Voltage Gain Bidirectional DC-DC Converters for Supercapacitor Assisted Electric Vehicles: A Review," in CPSS Transactions on Power Electronics and Applications,

vol. 7, no. 4, pp. 386–398, December 2022, doi:10.24295/CPSSTPEA.2022.00035.

6. Z. Sun and S. Bae, "Multiple-Input Soft-Switching DC–DC Converter to Connect Renewable Energy Sources in a DC Microgrid," in IEEE Access, vol. 10, pp. 128380128391, 2022, doi:10.1109/ACCESS.2022.3227439.

7. A. Elserougi, I. Abdelsalam, A. Massoud and S. Ahmed, "A Non-Isolated Hybrid-Modular DC-DC Converter for DC Grids: Small-Signal Modeling and Control," in IEEE Access, vol. 7, pp. 132459–132471, 2019, doi: 10.1109/ACCESS.2019.2941249.

8. M. R. Haque, K. M. A. Salam and M. A. Razzak, "A Modified PI-Controller Based High Current Density DC–DC Converter for EV Charging Applications," in IEEE Access, vol. 11, pp. 27246–27266, 2023, doi: 10.1109/ACCESS.2023.3258181.

9. H. -J. Choi, K. -W. Heo and J. -H. Jung, "A Hybrid Switching Modulation of Isolated Bidirectional DC-DC Converter for Energy Storage System in DC Microgrid," in IEEE Access, vol. 10, pp. 6555–6568, 2022, doi: 10.1109/ACCESS.2021.3138988.

10. J. Yang, R. Li, K. Ma, Y. Wang and P. Xu, "Analysis and Design of Cascaded DC-DC Converter Based Battery Energy Storage System With Distributed Multimode Control in Data Center Application," in CPSS Transactions on Power Electronics and Applications, vol. 7, no. 3, pp. 308–318, September 2022, doi: 10.24295/CPSSTPEA.2022.00028.

11. B. Vidyasagar, Dr. S S. Tulasiram "DWT with Enhanced ANN Technique for detecting and classifying the faults of synchronous generator" The Intelligent Networks and System Society, ISSN: 3185-2118, volume 9, Issue 4 Dec 2016, JAPAN

12. Dhasharatha, G., et al. "Design and Implementation of Three-phase Three Level NPC Inverter." *2023* 7th International Conference on Trends in Electronics and Informatics (ICOEI). IEEE, 2023

13. Sreejyothi, Khammampati R., et al. "Bidirectional Battery Charger Circuit using Buck/Boost Converter." 2022 6th International Conference on Electronics, Communication and Aerospace Technology. IEEE, 2022.

14. Boorgula, Vidyasagar, and S. S. Tulasi Ram. "Incipient Fault Diagnosis in Stator Winding of Synchronous Generator: A CMFFLC Technique." IETE Journal of Research 65.5 (2019): 667–678.

15. Vidyasagar, B., et al. "Different Types of Faults Detection and Identification in Synchronous Generator Using MFO-based FL Techniques."

Note: All the figures in this chapter were test data and experimental analysis as part of authors' research work.

Design of Highly Isolated Wearable Textile MIMO Antenna for Portable Wireless Applications in Sub-6GHz Band

63

**Shaik Mahaboob Peer[1], Madduri Vinay[2],
Kaniganti Pavan Kumar[3]**
Electronics and Communication, Engineering,
Sreenidhi Institute of Science and Technology, Hyderabad, India.

P. Pradeep[4]
Assistant Professor, Electronics and Communication Engineering,
Sreenidhi Institute of Science and Technology, Hyderabad, India

C. N. Sujatha[5]
Professor, Electronics and Communication Engineering,
Sreenidhi Institute of Science and Technology, Hyderabad, India

Abstract: For use in portable wireless applications, this study provides the design and modeling of an atypical Octagon layered MIMO patch antenna. The proposed antenna consists of an irregular octagon-shaped patch with two edges cut by circles, one edge is fixed with a 50-ohm feedline, two patches are connected with a neutralization line in between them placed a parasitic strip line, and partial ground with a small hole connected with etched stub in the shape of plus sign stub in the bottom of the antenna. The substrate utilized is constructed using denim fabric, having a dielectric constant of 1.6 and a thickness of 1mm. The proposed antenna is analyzed and optimized using the ANSYS HFSS tool. Moreover, the effect of the slot on the ground plane and parasitic strip line is used in the design to enhance the performance, high isolation, and impedance matching between antenna elements. The dimensions of the jean's substrate are 40mm x 50 mm. The impedance bandwidth of this proposed antenna is greater than 4.5 GHz over the operating frequency from 3.5 GHz – 5.5 GHz. The antenna under consideration exhibits a peak gain of 4.78 dB, ensuring strong signal reception. Additionally, it offers high isolation, measuring approximately 57 dB across the operating band. The average Specific Absorption Rate (SAR) value, a measure of absorbed electromagnetic energy, is recorded at 0.46 W/kg. These performance characteristics make the proposed antenna highly suitable for portable and wearable wireless applications within the sub-6 GHz frequency range. To support the deployment of 5G in various scenarios, including portable wireless applications, innovative antenna designs are being explored.

Keywords: MIMO, Portable, Wearable, Wireless, Jeans, Highly isolated, ANSYS HFSS, Irregular octagon, Patch antenna, SAR

1. Introduction

The most recent wireless technology known as 5G communication offers improved data speeds, reduced latency, and greater capacity when compared to earlier generations. It operates in a frequency range called sub-6GHz, which includes frequencies below 6 gigahertz. To support the deployment of 5G in various scenarios, including portable wireless applications, innovative antenna designs are being explored.

Wearable antennas have been developed as a result of the quick expansion of portable wireless devices and the rising

[1]mahaboobpeermunna786@gmail.com, [2]maddurivinay123@gmail.com, [3]kanigantipavankumar@gmail.com, [4]pradeeppendli@gmail.com, [5]cnsujatha@gmail.com

need for high-speed wireless communication. Compact and lightweight wearable antennas are ideal for use in portable and wearable gadgets like smart watches, activity trackers, and medical monitoring equipment. One of the key challenges in designing wearable antennas is to achieve **high isolation between multiple antennas operating in close proximity**. However, the proximity of the antennas can lead to mutual coupling and interference, which can degrade the antenna performance and reduce the system's capacity. In response to this obstacle, the development of wearable MIMO(Multiple Input Multiple Output) antennas with excellent isolation have emerged as a vibrant field of study. MIMO antennas refer to antenna systems that employ multiple antennas for signal transmission and reception. Through the utilization of MIMO techniques, these systems enhance wireless communication by boosting data rates, capacity, and reliability. The objective of this design is to develop compact, low-profile, and highly isolated antennas that can operate near each other while maintaining good radiation performance and system capacity.

This paper proposes a compact MIMO antenna design for wearable applications in portable wireless systems, aiming to achieve high isolation between the antennas. The design incorporates techniques such as frequency reconfiguration, polarization diversity, and structures for the electromagnetic band gap that decrease mutual coupling. Simulation results demonstrate the effectiveness of the design, showcasing its performance in the between 3.5 and 5.5 GHz in frequency. The proposed antenna system utilizes microstrip patch elements with an irregular octagon shape to reduce size while maintaining good performance. A stub in the shape of a plus sign stub on the ground plane enhances isolation. The design employs Jean's material substrate compatible with portable and wearable applications. Results from simulation and testing in ANSYS HFSS confirm that the antenna performs well over the specified frequency range and has exceptional isolation of over 57 dB. This design offers an efficient solution for highly isolated MIMO antenna systems suitable for portable and wearable applications. The 5G mid-band spectrum refers to a range of frequencies spanning from 3.5 GHz to 5.5 GHz. This specific frequency range is highly desirable for implementing 5G networks because it strikes a balance between coverage area and data speeds.

2. Literature Survey

In [1], a highly isolated dual-polarized MIMO antenna which operates at 2.4 GHz and features two resonant modes providing different polarizations. In [2], a design combining a defected ground structure and a parasitic element is investigated to achieve high isolation, suitable for compact and high-performance antennas in WiMAX systems.

Reference [3] and [4] introduces a design with two identical antenna elements on a single substrate, achieving over 30 dB isolation at 2.4 GHz and 5.2 GHz using a defective ground structure and offering excellent isolation exceeding 35 dB across the entire frequency range respectively.

In [7], a highly isolated MIMO antenna system for 5G bands with over 20 dB isolation is presented, while [8] focuses on a compact planar MIMO antenna design for LTE and WLAN applications, emphasizing high isolation between the antenna elements.

Regarding specific antenna designs, [13] showcases a circular board operating at 2.4 to 2.49 GHz with a gain of 4.2 dBi and isolation over 15 dB. In [14], a square board operating at 2.37 to 2.52 GHz demonstrates port isolation exceeding 20 dB and is designed for GPS and distress signal frequency applications.

A square board in [15] operates at 2.3 to 2.8 GHz, focusing on circularly polarized MIMO ground radiation antennas for wearables, featuring a gain of 2.79 dBi and 12 dB isolation. Similarly, [16] discusses a rectangular board for wearables, operating at 2.367 to 2.53 GHz, with a gain of 5.8 dB and port isolation of at least 20 dB.

In [17], a square board operating at 3.3 to 3.6 GHz utilizes substrate-integrated waveguide (SIW) technology for a dual-band MIMO antenna design, aimed at wireless communication applications. [18] presents a rectangular board operating at 2.4 to 8.0 GHz, offering wide band coverage and low mutual coupling between antenna elements, with a gain of 4.4 dB and maximum port isolation of 53 dB.

Furthermore, [19] investigates the design and specific absorption rate (SAR) analysis of wearable antennas on different parts of the human body, while [20] focuses on SAR values when wearable antenna textiles are in use, contributing to understanding electromagnetic exposure and safety considerations for wearable antenna systems.

3. Structure and Design of the Antenna

Figure 63.1 shows the suggested antenna arrangement and gives the antenna geometry from the top. Two atypically copper patch components in the form of octagons make up the MIMO antenna design, which is mounted on a cheap and easily accessible material known as Jean's substrate. The permittivity and loss tangent of the Jeans substrate is 1.6 and 0.02, respectively. The Jeans substrate is configured to have a height (h) of 1 mm. The substrate's dielectric characteristics are ascertained using the methods described in Paper [18]. These idealized substrate characteristics were discovered by simulations using two simple octagon patch antenna

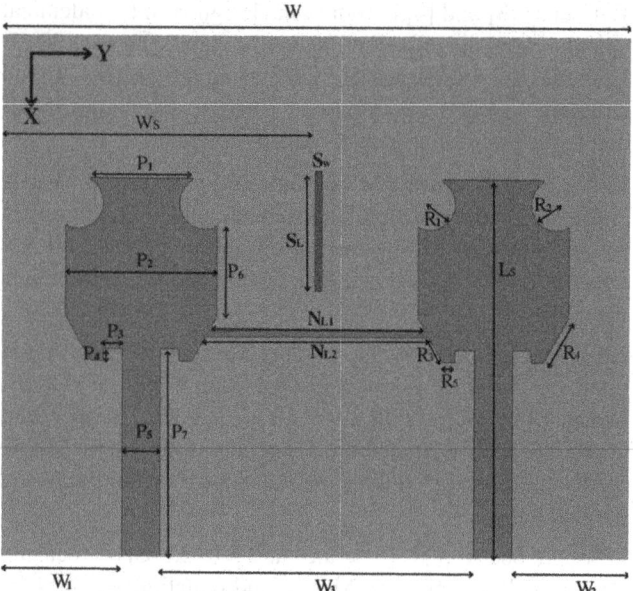

Fig. 63.1 Top view of the arrangement of the MIMO antenna

components mounted on a single Jeans substrate. The antenna is $0.75\lambda \times 0.61\lambda \times 0.015\lambda$ in size overall. As shown in Fig. 63.2, the suggested MIMO construction incorporates stubs in the shape of plus sign stubs placed on the ground to improve port isolation. Approximately 0.0625 inches between the two antenna components and Fig. 63.2's stub in the shape of plus sign stubs on the ground are what is responsible for the strong isolation between the ports. The separation between the two antenna elements is approximately 0.0625λ, and the significant isolation achieved between the ports is attributed to the presence of stubs in the shape of plus sign stubs on the ground surface, as depicted in Fig. 63.2.

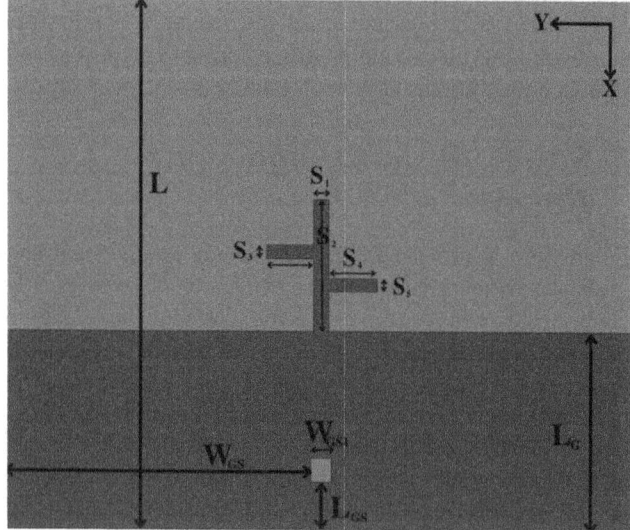

Fig. 63.2 Bottom view of the arrangement of the MIMO antenna

The MIMO antenna described in the paper operates over a frequency range from 3.5 GHz to 5.5 GHz, with a central frequency of 4.5 GHz, as shown in Fig. 63.3. The antenna's frequency range covers both the S-band, which spans from 2 GHz to 4 GHz and the C-band, which ranges from 4 GHz to 8 GHz. This wide frequency range ensures the antenna is suitable for a broad spectrum of communication systems, enabling compatibility with diverse applications. Table 63.1 presents the essential characteristics of the MIMO antenna, outlining its specific measurements and dimensions.

3.1 Enhancing Inter-Element Isolation in Two Irregular Octagon Antennas through the Use of Parasitic Strip and Neutralization Line

A parametric analysis technique involves the addition of a parasitic strip, an extra conductive element positioned near the primary radiating element of an antenna with the shape in Fig. 63.1. By strategically incorporating a parasitic strip into each irregular octagon antenna, the current distribution and electromagnetic fields surrounding the antenna are modified. This alteration influences important antenna characteristics such as radiation pattern, impedance, and bandwidth. The main objective of the parasitic strip is to boost the effectiveness of the antenna by enhancing elements like gain, bandwidth, and radiation pattern.

In addition to the parasitic strip, another conductive element called the neutralization line, also known as a slot line or floating strip, is integrated into the antenna design. The neutralization line is carefully positioned to mitigate or neutralize the mutual coupling between the two irregular octagon antennas. When antennas are placed in close proximity, mutual coupling can occur, leading to interference and performance degradation, particularly in terms of isolation between the antenna elements. By incorporating a neutralization line, the undesired coupling effect is reduced, resulting in improved isolation between the two antennas and enhancing inter-element performance.

3.2 Series-connected Ground-Linked MIMO Antenna featuring a Stub in the Shape of a Plus Sign

We conducted a comprehensive examination of various adjustments to the parameters in order to reach the ultimate design. Our strategy entails using a MIMO antenna that includes two patch components made of amorphous material, specifically Jeans fabric, which are placed on the upper surface of the substrate. By simulating the antenna's performance using Ansys HFSS, we determined that it operates within the frequency range of 3.5 GHz to 5.5 GHz. As shown in Fig. 63.2, a stub shaped like a plus sign is included on the ground plane to ensure very low mutual coupling between the

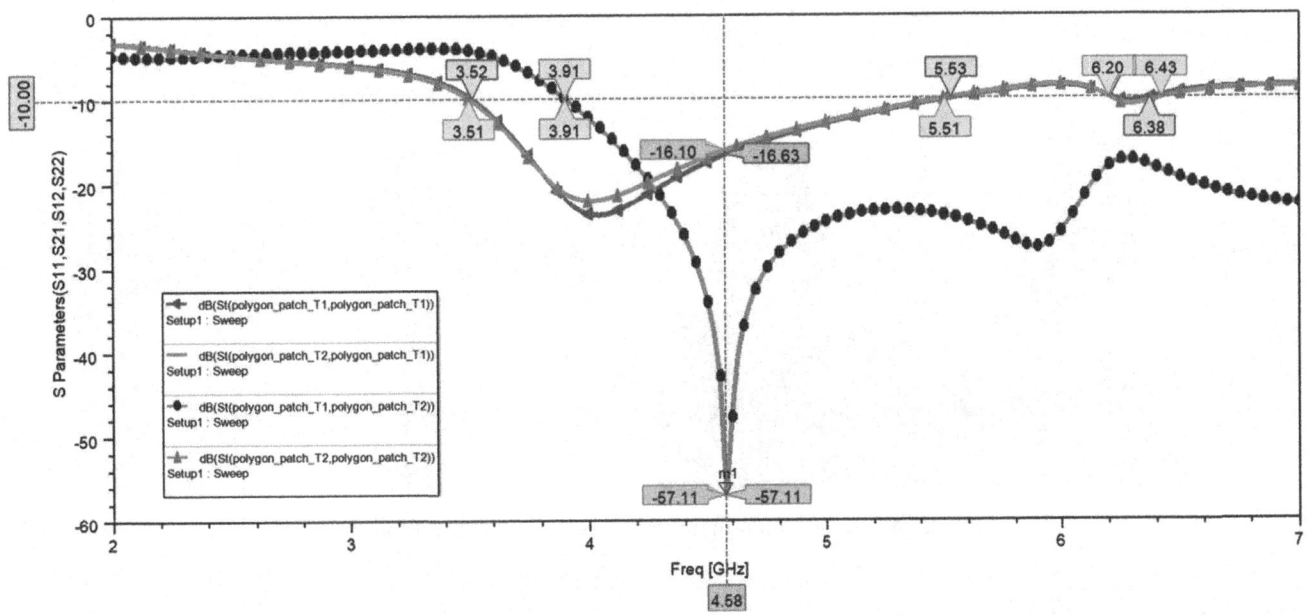

Fig. 63.3   S-parameters of antenna with plus sign stub on ground plane

Table 63.1   Characteristics of the suggested MIMO antenna dimensions

Name of the *Variable*	Length(mm)	Name of the *Variable*	Length(mm)
P_1	8	R_1	2
P_2	12	R_2	2
P_3	1.5	R_3	2.2
P_4	0.9	R_4	4
P_5	3	R_5	1
P_6	6.93	S_w	0.6
P_7	16	S_L	9
L_G	15	L_s	29.1
W	50	W_s	24.6
L	40	N_{L1}	17.2
W_{GS}	24	N_{L2}	17.7
W_{GS1}	2	S_1	1.2
L_{GS}	3.6	S_2	10
W_1	9.5	S_3	1
W_2	9.5	S_4	3.7
W_3	25	S_5	1

ports. Figure 63.2 and Table 63.1 both provide the optimized parameter values for the two stubs that resemble plus signs. This design strategy relies on the microstrip filtering principle, wherein the combination of stub and gap creates a capacitive and inductive setting. This generates a comparable stop band filter within the structure. The electromagnetic coupling between the two radiators is significantly reduced and strong port isolation is ensured by carefully positioning the stubs in the form of plus signs on the ground plane.

4. Results with Explanations

The subsequent sections provide a discussion of the simulated results and the performance evaluations.

4.1 S-Parameters Analysis

Figure 3 displays the simulation outcomes of the prototype's scattering parameters. The simulation was conducted within a frequency range of 3.5 GHz to 5.5 GHz to ensure that the return loss (S11) remained below -10 dB. To enhance the -10 dB bandwidth, the initial design included additional plus sign elements positioned on the lower surface of the MIMO antenna. By modifying the space separating the visible tip of the stub and the surface it is connected to, the effective inductance and capacitance were modified, resulting in a decrease in the antenna's quality factor and an improvement in bandwidth. The addition of the stubs also led to a measured port isolation of up to 57 dB at 4.58 GHz, thus providing further confirmation of the accuracy of the simulation results. When evaluating an antenna, return loss plays a significant role and should be considered. It is closely tied to impedance matching and the theory of maximum power transfer. Return loss indicates how effectively an antenna can transmit power from one source to another antenna. Within the spectrum of frequencies of 3.5 GHz to 5.5 GHz, as depicted in Fig. 63.3, the return loss, represented by S11, achieves values below -10 dB.

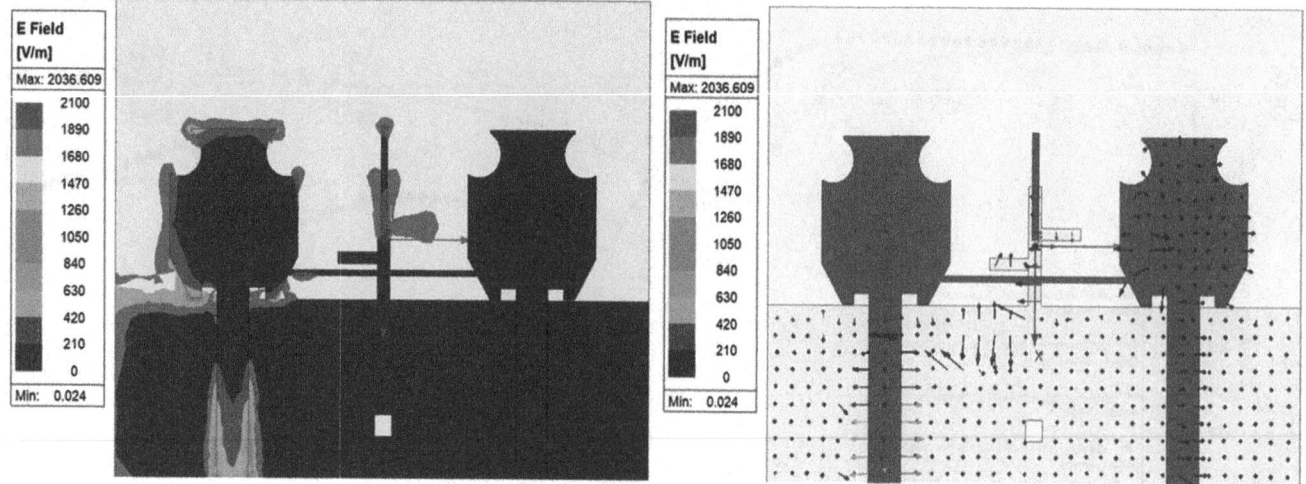

Fig. 63.4 Surface current distributions (A) Magnitude (B) Vector

4.2 Distribution of Surface Current

The MIMO antenna system's surface current distributions, operating at 4.5 GHz, are contained within the frequency range suitable for its intended use. The objective was to analyze and comprehend the decrease in mutual interference between the ports of the antenna. Furthermore, the study aimed to identify the most favourable position for the stub placed on the ground surface. The researchers discovered that incorporating plus sign-shaped stubs on the ground surface had a notable impact on reducing the coupling current between the antenna elements. As a result, this arrangement achieved outstanding port isolation. These findings are illustrated in Figure 4, which showcases the surface current distributions using visualization techniques such as magnitude current density plots (Fig. 63.4A) and vector field plots (Fig. 63.4B).

4.3 Behaviour of the Antenna on the Flat-Body Phantom

The presence of a human body in close proximity to the antenna negatively impacts its performance because the body's varied composition and ability to absorb signals play a role in this interference. Therefore, to evaluate the antenna's effectiveness, it is first tested in an open space without any obstructions before conducting on-body tests on flat sections. In order to simulate these on-body tests, a four-layer body phantom consisting of bone, muscle, fat, and skin is created using Ansys HFSS, as shown in Fig. 63.5. The physical dimensions of the human body layers, as described in reference [18], are 12 mm for bone, 23 mm for muscle, 8 mm for fat, and 2 mm for skin.

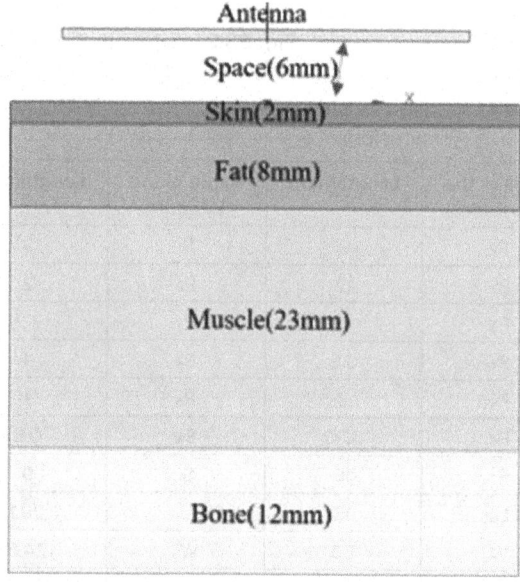

Fig. 63.5 The Phantom's body is equipped with an antenna

4.4 SAR Distribution

The Specific Absorption Rate (SAR) quantifies the speed at which biological tissue, including the human body, absorbs electromagnetic energy. In the context of antennas, the SAR distribution refers to the distribution of the absorbed energy in the body that is exposed to electromagnetic radiation from the antenna. Several factors, such as the frequency of operation, the separation between the antenna and the human body, and the geometry of the antenna can influence the SAR distribution in an antenna. Generally, the SAR is highest in

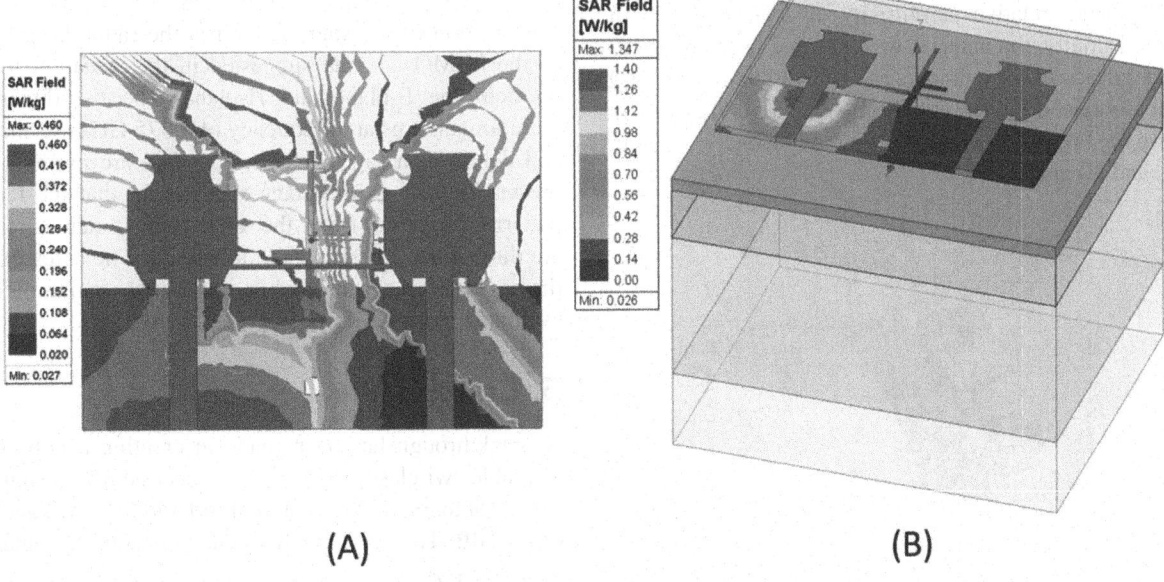

Fig. 63.6   SAR distribution (A) Off-body (B) On-body with space 6mm from phantom

areas of the body that are closest to the antenna and in regions where the electromagnetic field is strongest. On average, the SAR value must not surpass 1.6 W/Kg. As shown in Fig. 63.6. In situations where the antenna is not in direct contact with the body, the suggested antenna demonstrates a specific absorption rate (SAR) of 0.46 W/Kg.

In the realm of software, the physical form is replicated through a virtual entity known as a phantom, which comprises four distinct tiers, with each level representing a specific material. The placement of the antenna on the body determines which of these four levels is assigned to the corresponding region in the simulation. Table 63.2 shows how the simulation process's SAR value is impacted by the Defected Ground Structure. At a specific location, the fantastic outcome gradually becomes visible.

Table 63.2   Simulated result antenna with the Phantom

S. No.	Space(mm)	SAR (W/kg)
1	0	3.656
2	2	2.483
3	4	1.831
4	5	1.57
5	6	1.347
6	8	1.079
7	10	0.914
8	20	0.569

4.5 Gain

Gain is a single figure that combines directivity and efficiency. In the case of transmission, the gain must account for an antenna's ability to effectively convert its input power into radio waves travelling in a particular direction. In contrast, gain indicates the antenna's efficiency in converting received waves into electrical power during the reception. In Fig. 63.7, it can be observed that the MIMO antenna attains a peak realized gain of around 4.78 dB when operating at a frequency of 3.8 GHz.

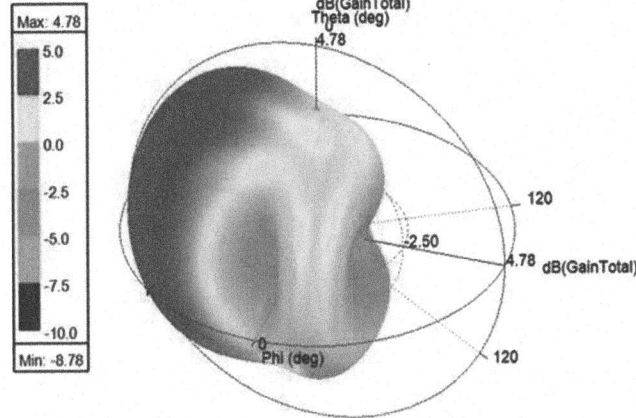

Fig. 63.7   Antenna gain

4.6 Bending Analysis

To evaluate how the proposed antenna performs when subjected to bending, an experiment was conducted where the antenna was placed around a virtual cylinder made of PVC plastic (relative permittivity of 2.7) as shown in Fig. 63.8. Initially, the antenna was attached horizontally to the outer surface of the cylinder. The S11 results in Fig. 63.9 were then obtained by bending the antenna around cylinders as shown in Fig. 63.8 with varying radii, namely 40 mm, 60 mm, and 80 mm.

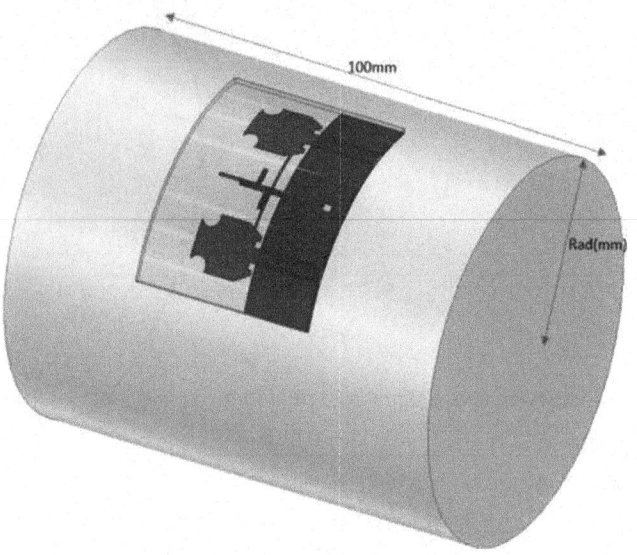

Fig. 63.8 Bending analysis

4.7 Analysis of Radiating Properties

This section of the study examines the radiation properties of the MIMO antenna suggested in the research, focusing on both the E-plane and H-plane. The two-dimensional radiation patterns at a frequency of 4.575 GHz are simulated and depicted in Fig. 63.10, where they are normalized for presentation. It is important to highlight that the radiation patterns observed from the underlying antenna structure are directional. The E-plane radiation can be assessed when the angle phi is set to 0, while the H-plane radiation can be evaluated when the angle phi is set to 90.

5. Conclusion

A breakthrough has been made in creating a portable and wearable wireless solution by successfully designing a MIMO antenna with exceptional isolation of 57 dB and a gain of 4.78 dB. The operating frequency range of this antenna is 3.5 to 5.5 GHz, which is ideal for wireless applications. The antenna has better S-parameter performance and other important parameters, making it an ideal choice for portable wireless devices. The design of highly isolated textile MIMO antennas for portable wireless applications in the sub-6GHz band enables improved connectivity and performance in 5G communication. These antennas leverage MIMO technology while considering the unique characteristics of textiles

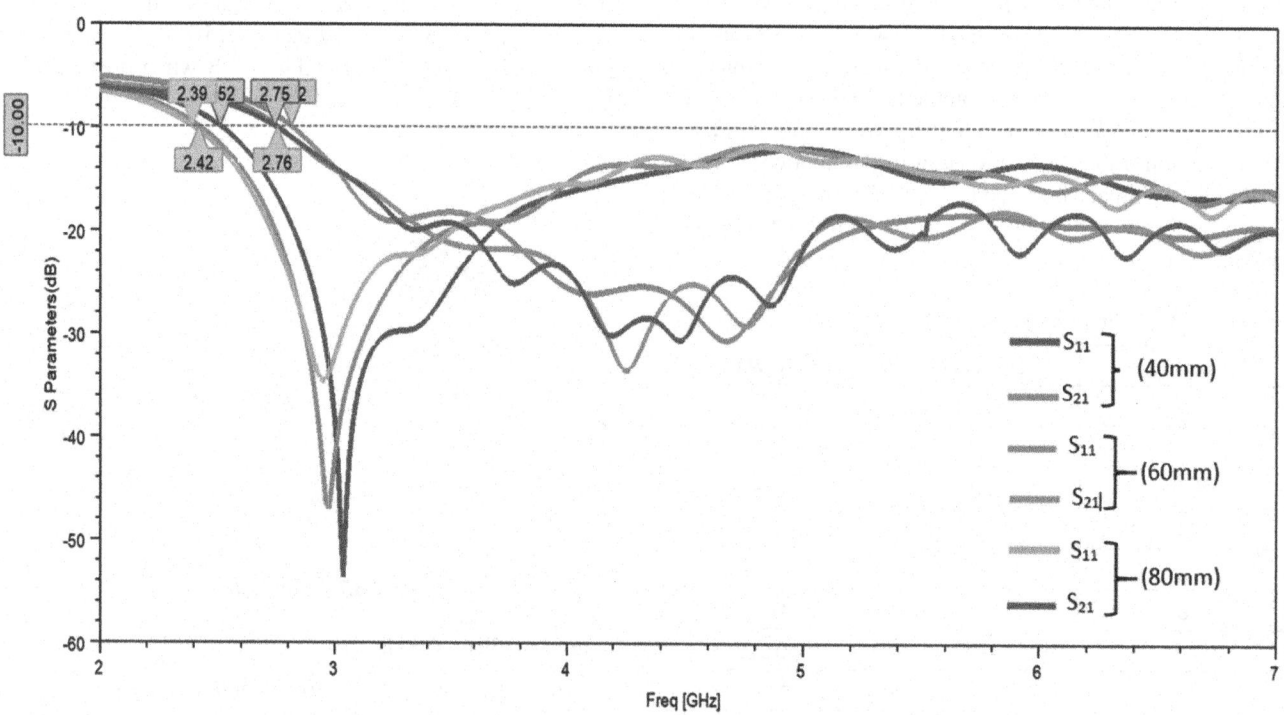

Fig. 63.9 Bending analysis with different radii of curvature: (i) 40mm, (ii) 60mm, and (iii) 80mm

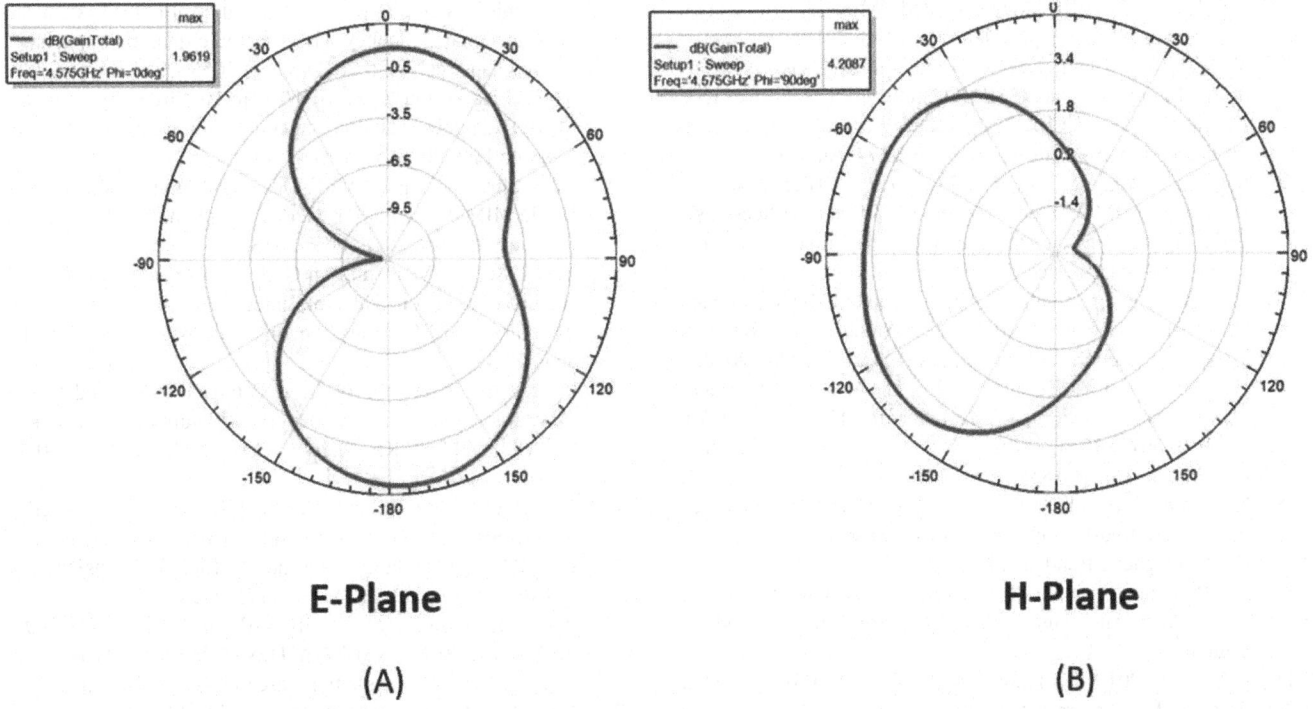

E-Plane

(A)

H-Plane

(B)

Fig. 63.10 Normalized radiation patterns of the suggested antenna at 4.575 GHz, illustrating (A) the electric plane (E-Plane) and (B) the magnetic plane (H- Plane)

Table 63.3 Evaluation alongside some of the comparable recent studies

Design	Board (mm²)	Op. Freq (GHz)	Max gain	Port Iso. (dB)
[13]	$\pi \times (21.1^2)$	2.4–2.49	4.2 dBi	≤ 15
[14]	30 × 30	2.37–2.52	NA	≥ 20 12
[15]	38.1 × 38.1	2.3–2.8	2.79 dBi	LB 20
[16]	92.3 × 101.6	2.367–2.53, 5.147–5.863	5.8dB	UB 35
[17]	40 × 40	3.3–3.6, 4.5–5.0	NA	≥17
[18]	40 × 70	2.4–8.0	4.4 dB	Min. 22 Max. 53
Prop.	40 × 50	3.5–5.5	4.78 dB	Min. 4 Max. 57

to provide flexible, lightweight, and effective wireless communication solutions.

The MIMO antenna design has undergone optimization, considering parameters such as isolation, bandwidth, return loss, and radiation pattern. The antenna configuration consists of two identical radiators positioned on a ground plane, separated by a distance of 0.06. An excitation is applied to the antenna through a 50-ohm microstrip feedline. To enhance isolation, a decoupling structure is incorporated between the radiators. Simulation results demonstrate that the designed antenna exhibits exceptional performance, characterized by high isolation, minimal return loss, and wide bandwidth. The antenna also demonstrates favorable radiation characteristics, including Minimal cross-polarization rates and optimal radiation effectiveness. Moreover, the antenna is compact and lightweight, making it well-suited for portable wireless applications. This antenna design presents a promising solution for achieving enhanced data rates, reliable communication, and improved signal quality in portable wireless devices.

References

1. H. Chen, X. L. Liu, and X. J. Li. "A Highly Isolated Dual-Polarized MIMO Antenna with Two Resonant Modes for WLAN Applications". IEEE Transactions on Antennas and Propagation, vol. 66, no. 10, pp. 5213–5217, Oct. 2018.
2. M. S. Haque, M. U. Ali. "A Compact Highly Isolated MIMO Antenna for WiMAX Applications". Progress in Electromagnetics Research C, vol. 89, pp. 37–47, 2019.
3. W. S. Chen and K. L. Wong. "Compact and Highly Isolated Dual-Band MIMO Antenna for Portable Wireless Applications". IEEE Antennas and Wireless Propagation Letters, vol. 16, pp. 2809–2812, 2017.
4. H. Chen, X. L. Liu, and X. J. Li. "A Compact Dual-Polarized Highly Isolated MIMO Antenna for Portable Wireless

Applications". IEEE Antennas and Wireless Propagation Letters, vol. 17, no. 3, pp. 448–451, March 2018.

5. Y. Wang, L. Chen, and K. Yang. "Design and Analysis of a Highly Iso- lated MIMO Antenna System for LTE Applications". IEEE Transactions on Antennas and Propagation, vol. 65, no. 8, pp. 4369–4374, Aug. 2017.

6. S. Chen, L. Yang, Y. Huang, and L. Chen. "A Highly Isolated Dual-Band MIMO Antenna System for WLAN Applications". IEEE Transactions on Antennas and Propagation, vol. 65, no. 12, pp. 6586–6591, Dec. 2017.

7. J. Li, X. Li, and Y. Zhang. "A Highly Isolated MIMO Antenna System for 5G Applications". IEEE Antennas and Wireless Propagation Letters, vol. 17, no. 5, pp. 829–832, May 2018.

8. H. Chen, X. L. Liu, and X. J. Li. "A Compact Planar MIMO Antenna with High Isolation for LTE and WLAN Applications". IEEE Antennas and Wireless Propagation Letters, vol. 15, pp. 1835–1838, 2016.

9. N. Hassan et al. (2017). "Isolation Enhancement of Planar Inverted-F Antenna MIMO System by Utilizing Electromagnetic Bandgap Structures"

10. A. Rahman et al. (2015)."Mutual Coupling Reduction for MIMO An- tennas Using Electromagnetic Bandgap Structures"

11. L. Li et al. (2019)."Isolation Enhancement of MIMO Antenna Array by Using Corrugated Ground Plane"

12. C. Huang et al. (2017). "High Isolation MIMO Antenna with Low Cross- Polarization Using Frequency Selective Surface"

13. Wen D, Hao Y, Munoz MO, Wang H, Zhou H. "A compact and low- profile MIMO antenna using a miniature circular high impedance surface for wearable applications." IEEE Trans Antennas Prop ag. 2018; 66(1): 96–104.

14. Alkhamis R, Wigle J, Song H. "Global positioning system and distress signal frequency wrist wearable dual-band antenna." Microw Opt Tech Lett. 2017; 59(8): 2055–2064.

15. Qu L, Piao H, Qu Y, Kim HH, Kim H. Circularly polarised MIMO ground radiation antennas for wearable devices. Electron Lett. 2018; 54(4): 189–190.

16. Li H, Sun S, Wang B, Wu F. Design of compact single-layer textile MIMO antenna for wearable applications. IEEE Trans Antennas Propag. 2018; 66(6): 3136–3141.

17. Yan S, Soh PJ, Vandenbosch GAE. Dual-band textile MIMO antenna based on substrate-integrated waveguide (SIW) technology. IEEE Trans Antennas Propag. 2015; 63(11): 4640–4647.

18. Biswas AK, Chakraborty U. A compact wide band textile MIMO antenna with very low mutual coupling for wearable applications. IntJ RF Microw Comput Aided Eng. 2019; e21769.

19. Ali, U., et al., Design and SAR analysis of wearable antenna on various parts of human body, using conventional and artificial ground planes," Journal of Electrical Engineering and Technology, Vol. 12, No. 1, 317328, Jan. 2017.

20. Harris, Hasri Ainun et al. "DESIGN AND IMPLEMENTATION OF WEARABLE ANTENNA TEXTILE FOR ISM BAND." Progress In Electromagnetics Research C (2022): n. pag.

21. Tighezza M, Rahim SKA, Islam MT. Flexible wideband antenna for 5G applications. Microw Opt Technol Lett. 2017; 60: 38–44.

Note: All the figures and tables in this chapter were authors' test data and experimental analysis as part of their research work.

Brain Tumor Based on Berkley Wavelet Transformation by Using Image Processing

64

C. Anna Palagan[1], G. Sirisha[2]

Professor in ECE, Teegala Krishna Reddy Engineering College,
Meerpet, Hyderabad, Telangana, India

D. Ramadevi[3], Muthyala V. V. S. Chowdary[4]

Assistant Professor in ECE, Teegala Krishna Reddy Engineering College,
Meerpet, Hyderabad, Telangana, India

Abstract: Brain tumours are much more dangerous than the other type of disease. These diseases are identified by using many technologies. Engineers have been actively developing tools to detect tumors and to process medical images. Medical image segmentation is a powerful tool that is often used to detect tumors. Many scientists and researchers are working to develop and add more features to this tool. This project is about detecting Brain tumors from various images using MATLAB the embedding algorithm proposed by least significant bits (LSB) technique while the Experimental result show that the feature for the watermark image is high and its PSNR value in proposed method is incredibly higher compare than another method.

Keywords: Disease identification, Segmentation, LSB, Watermark image, PSNR

1. Introduction

The rapeutic incorporates the space of medicinal services is even more progressively utilizing the open systems to trade electronic patient record among the emergency clinics and well-being focuses. The significant territory utilizing the telemedicine it is offer the critical electronic patient's record going through information trade. The Information and correspondence innovation have just given increment the spaces of the tele-diagnosis and tele-counsel for the specialists. While the correspondence of electronic patient records (EPR) through open systems is not verify.

This paper, we described the brain tumours detection using morphological operation and watermarking process using LSB and distinctive information hiding method must become heavy data hide rate using IWT technique will develop an unique image either distortion since the distinct form of image behind the secrets of data can be extracted These

method using practical to a broad range of dissimilar image successfully [1]. The MRI Images playing the vital role in the detection of brain tumor identification and treat a patient for healthy lifeand supportive for a doctor to finding earlier step of brain tumor disease[4].

2. Related Works

S. Majina Baby, V. Meena have described it can include a huge quantity of data compared with the obtainable noisy less information hiding method and convince the imperceptibility condition[1].The basic image has two-dimensional signal through an F(x, y) mathematical function x and y gives the value of horizontal and vertical coordinates [2]. It has proposed morphology image pre-processing for thinning algorithms to briefly describe the binary morphological operation. While the proposed work using of binary morphology since versatility and comparatively rapid execution [3]. The process

---
[1]annapalagan@tkrec.ac.in, [2]gshirisha4@gmail.com.com, [3]dubalarama@gmail.com, [4]mvvsvenkatchowdary716@gmail.com

of "Pre-processing" is to get better the quality of image by remove the unnecessary part of the images[4].

The Brain tumor detection are used on MRI brain images is a difficult function as the multifarious structure of the brain images. It is an irregular cell development of brain. [5].Detect the intensity of discontinuities of a digital image and process of classify as well as placed on sharping discontinuities in an image [6]. The criminal copy, change, tamper, copyright protection consists of major problem in the speed of computer usage [7]. It can describe the morphological operations such as erosion, dilation, opening, closing, boundary extraction and region filling [8]. While the Edge detection proposed with pre-processing approach involve compute the histogram, find out the total number of peaks and suppress irrelevant peaks [9]and also [10,11]research a pattern continue the works.

3. Proposed System

While the suggestion of system can provide a more accurate result relate with identification and classification of disease, Here the captured image is first pre-processed next pixel intensity values detection and then moved to tumours detection. The features are extracted from the image by binary morphing operations.

From the three or four decades, research in the field of tumours recognition and classification contributed greatly identified diseased data will be watermark by the wrapped images used for LSB method it is possible to more secure and unrisky the rapeutic data while to get the data and image will be processed to extract it.

3.1 Gray Scale Conversion

Gray scale images consist of the rate of each pixel is distinct sample which carries only on the intensity of information and the value of each pixel range from 0 to 255, its pixel value represent as black and white.

3.2 Gaussian Smoothing

While the 2-D distribution has a function of `point-spread' and also accomplish by convolution because the images are store as a group of discrete pixel we should to generate a distinct estimate to the Gaussian function

3.3 Binarization

It converts the gray image to twofold image by replace all pixel values in the input images with luminance greater than levels with the Value represent 1 (white) and replace all other values 0 (black).

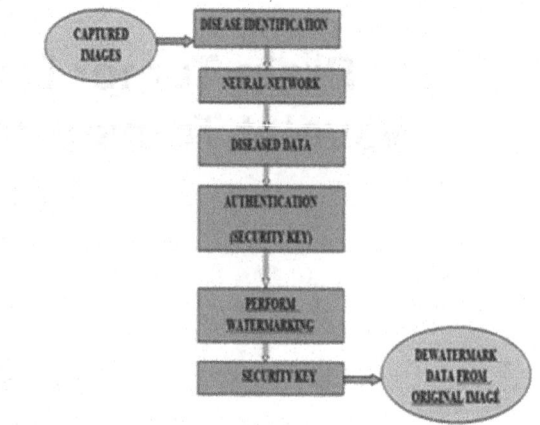

Fig. 64.1 Block diagram of proposed work

Source: Made by author

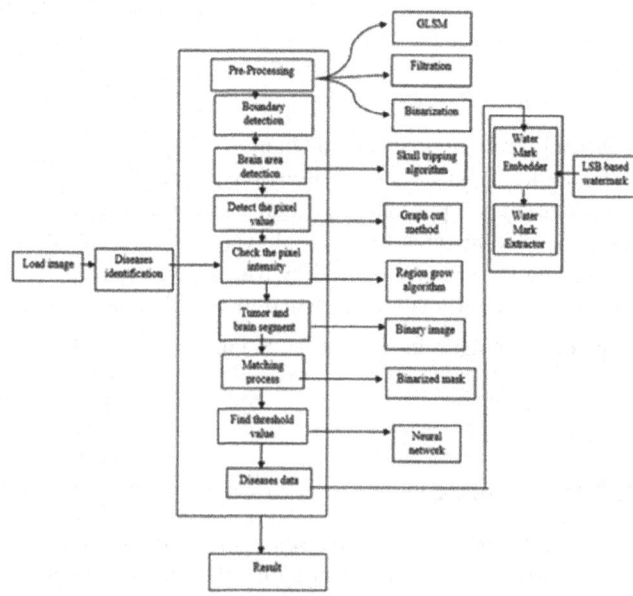

Fig. 64.2 System architecture [2]

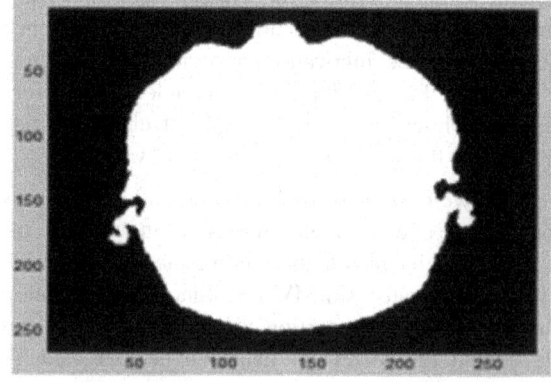

Fig. 64.3 Binarized mask [4]

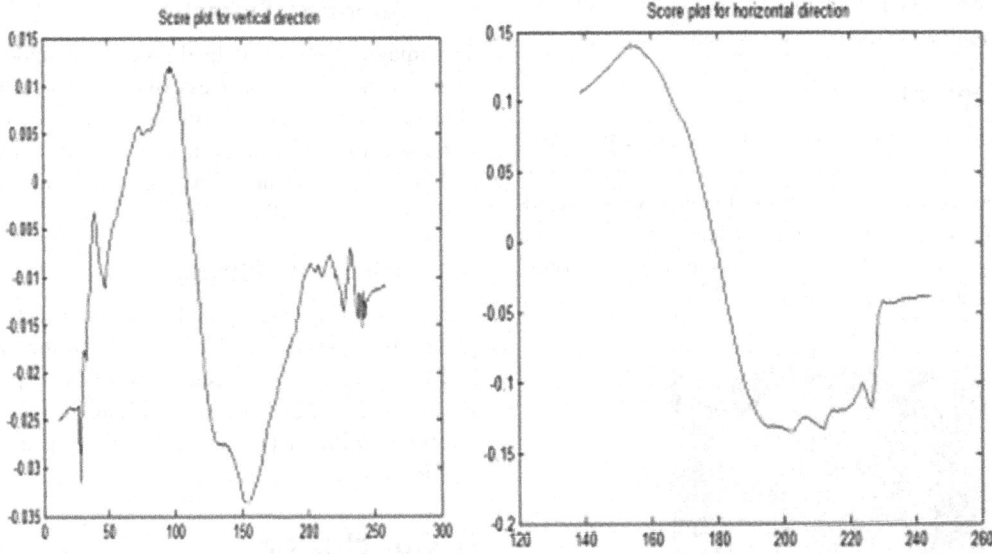

Fig. 64.4 Image segmentation on captured images on 'x' axis represent as pixel values(intensity) and 'y' axis represent as saturated threshold values by using graph cut segment method

Source: Made by author

3.4 Skull Stripping Algorithm

This algorithm describes an beginning step was designed to reduce non-brain tissues from brain images for various medical application and analysis with its correctness and swiftness are consider as the key factor in the brain image segmentation and analysis.

3.5 Image Segmentation Methods

In this section we are discussed about the image segmentation method, they are number of methods using in this method.

Graph cut method

The principle items out of a picture utilizing a division technique dependent on chart cuts. The scheme of hubs is equal to the arrangement of pixels in the image. To portion of the article, we set edges with high weight from the source to the client seeds in the item and from the foundation client seeds to the sink.

Region growing method

In this method mainly used to be on image segment and classify as a pixel-found image segmentation process given that it includes the choice of primary kernel points and determines whether a pixel value of image on neighbors must be further to the region.

3.6 Least Significant Bit Algorithm-LSB, RGB, BMP

This algorithm is mainly using development of adjusting the minimum significant bit pixel of the captured images

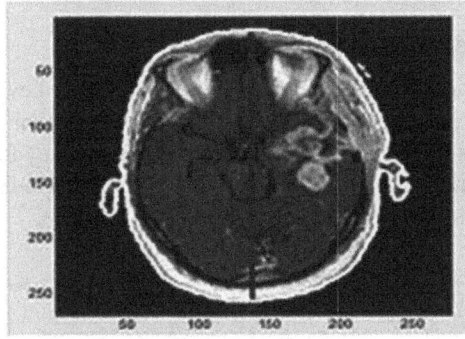

Fig. 64.5 Image segmented based on pixel values [4]

while 24 bit of image represented as the color of each one of the component are red,blue and green are changed while hide the secret message within an image of BMP file using LSB algorithms it requires a large image which is used as a cover. So, it uses LSB substitution for embedding the data into images.

4. Brain Tumor for Image Marking

The brain image up loader module has used to store the image in the database. It will help full to find the image in the future. This image has only uploaded the admin only. He has only the full authority to use the system so he can only upload the image for the certain persons in the data base. Once if we enter the image, we can predict the image with the higher performance in the future. Those images are store in the database server so it will be in very secure format and as well the binary encrypted format so that the intruder should not change or modify the image in our system.

4.1 Disease Detection

It has two stages. In the first stage, the irrelevant parts are removed from the image by using graph cut and region grow algorithm. In the second stage, threshold values are detected by using neural network technique, Disease identification using MRI images for neutral networks and skull stripping algorithms to identify brain tumours.

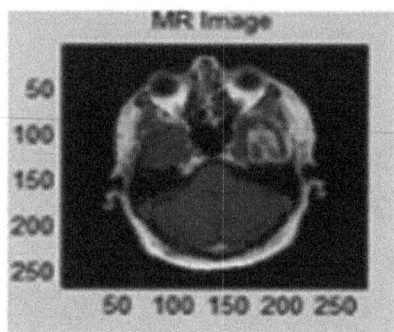

Fig. 64.6   Disease identification for brain MRI images [4]

4.2 Watermark Embedding

Watermark is embedded with the use of secret key for the security purpose, in this module the LSB algorithm is used, once this LSB algorithm is used to create secret key for avoid unauthorized access and easily to diagnosis and telediagnosis for exchanging information easily.

4.3 Watermark Extraction

The image is retrieved in the dewater mark process and it involves the reverse process of watermark. In case of LSB algorithm, the image is extracted using receiver's private key. After identify brain tumour disease then enter security key and perform watermark images and re-enter key to dewater mark images with disease name.

4.4 Histogram of Images

It describes as frequency and pixels force esteems. While the picture histogram the 'x' unit speaks to the dim level power and the 'y' unit as the recurrence of these powers. The test results on Graphical portrayal of caught pictures to implant information into separated sick picture to look at on higher PSNR values.

5. Conclusion

This paper has implemented the neural network operations and Finally, images are segmented by using various techniques (Graph cut and Region grow method) using a innovative LSB watermark techniques for images and proved that different operation provide enhanced PSNR value and also using data hiding techniques to determine for a extensive range of application on the area (i.e)safe medicinal images information scheme, rule enforcement, e-government and digital image verification, secret communication.

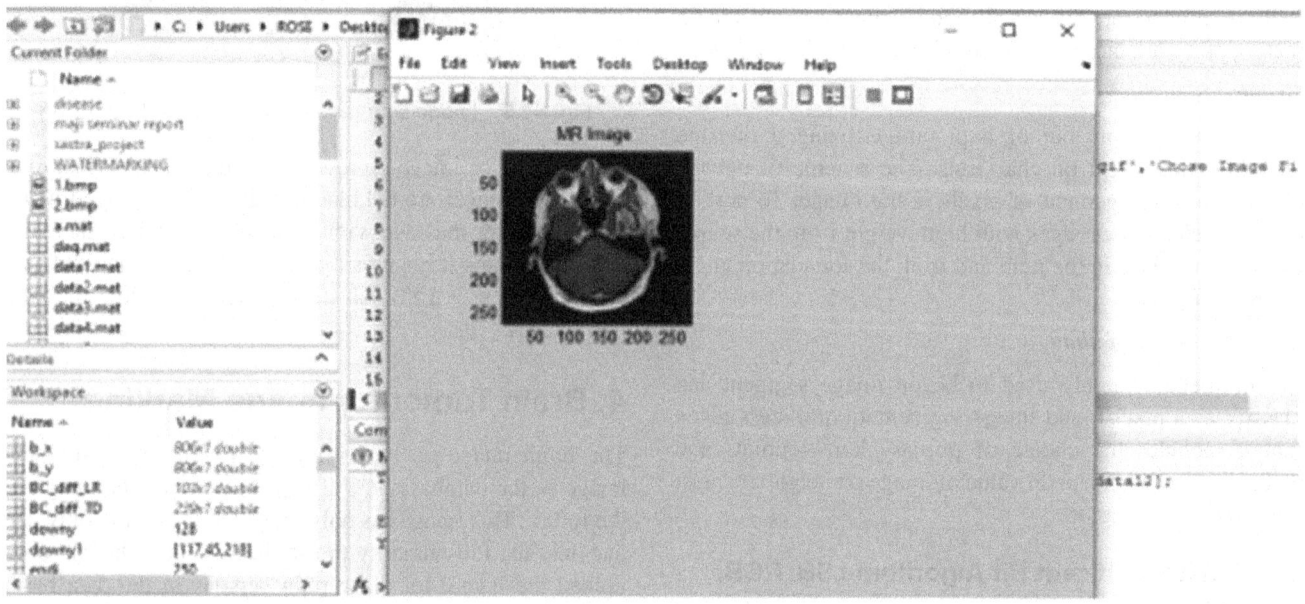

Fig. 64.7   Prediction of brain tumors images

Source: Made by author

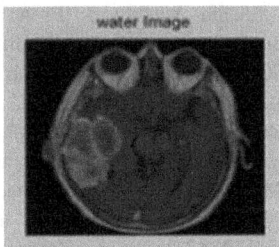

Fig. 64.8   Information hiding with captured image [4]

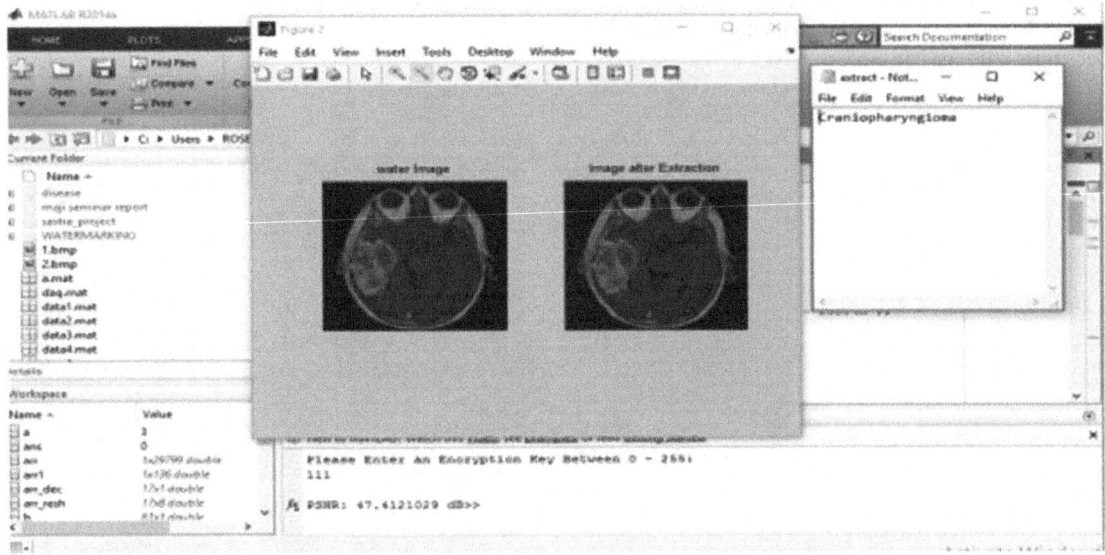

Fig. 64.9   Recover diseased image and related information

Source: Made by author

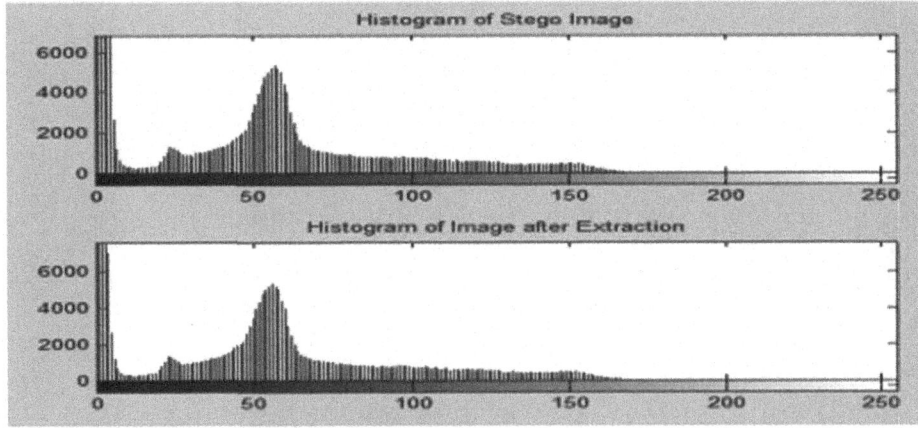

Fig. 64.10   Performance comparision

Source: Made by author

References

1. S. Majina Babya,*, V.Meenab "secured medical information using an intelligent and reversible watermarking technique" International Journal of Pure and Applied Mathematics Volume 119 No. 7 2018, 613–622.

2. R.C. Gonzalez and R.E. Woods, "Digital Image Processing", third edition, PHI publication, 2008.

3. Peter., "Morphology Image Pre – Processing for thinning Algorithms", International Journal of Innovative Research in Electrical, Electronics, Instrumentation and Control Engineering, Volume 5, No 1, 2007, pp 131–138.

4. Vipin Y. Borole1, Sunil S. Nimbhore2, Dr. Seema S. Kawthekar2 "Image Processing Techniques for Brain Tumor Detection": A Review Volume 4, Issue 5(2), September - October 2015.

5. Jayachandra, N. and Kamal. A. R. N. B., "Detection of the Effect Of Sharpening of Color Image Sharpened by Frequency Domain Filters and Identification of Rio using Morphology Mathematical Property Image", International Journal of Advanced Research in Computer Science, Volume 4, No 11, December 2013, pp 17–25.

6. Muthu krishnan, M. and Radha. M., "Edge Detection Techniques for Image Segementation", International Journal of Computer Science &Information Technology, Volume 3, No 6, December 2011, pp 259–267.

7. F. A. P. Petitcolas, R. J. Anderson, and M. G. Kuhn, "Information hiding-a survey," Proceedings of the IEEE, vol. 87, pp 1062–1078, 1999.

8. Raid, A.M, Khefr, W.M., "Image Restoration Based on Morphological Operations", International Journal of Computer Applications, Volume 4, June 2014, pp 9–21.

9. Zahhad, M., Gharieb, R. R., Ahamed, M. S., "Edge Detection with a Preprocessing Apporach", Journal of Signal and Information Processing, Volume 5, May 2014, pp: 123–134.

10. Bartunek, S. J. and Nilsson, M., "Adaptive Finger Prinst Image Enhancement with Emphasis on Pre-processing of data" IEEE, 2011, pp: 1–13.

Enhancing Visual Features Investigating the Synergy of Gamma Correction and Wavelet Transform in Image Enhancement

65

M. Renu Babu[1]

Asst. Professor of ECE, Teegala Krishna Reddy Engineering College,
Meerpett, Hyderabad, TS, India

S. China Venkateswarlu[2]

Professor of ECE, Institute of Aeronautical Engineering,
Dundigal, Hyderabad, TS, India

Deshoju Vemana chary[3]

Professor of ECE, Teegala, Krishna Reddy Engineering College,
Meerpett, Hyderabad, TS, India

Radhamma Erigela[4]

Associate Professor of ECE, Teegala, Krishna Reddy Engineering College,
Meerpett, Hyderabad, TS, India

Abstract: Image Enhancement using both gamma and wavelet transforms and implemented the programs using Python and OpenCV in Google Colab. OpenCV is indeed a popular and powerful library for image processing, and it provides a wide range of functions and tools to work with images effectively. Using gamma and wavelet transforms together for image enhancement can lead to more effective and visually appealing results. Gamma correction helps adjust the brightness levels of an image, while wavelet transforms allow for multi-scale analysis and selective amplification or suppression of different frequency components.

Keywords: Image enhancement, Gamma, Wavelet, Histogram, Contrast, Blur, OpenCV

1. Introduction

Now a days Digital Image Processing and elevated determined communication move toward had a huge region. To addition up, openCV library acts as a very important role. It's a versatile, open source policy that can be use in nearly every figure related enterprise. It will be examined the five most significant use objects where OpenCV papers provide a solution. As fact that, image processing always involves enhancing or changing a given image. Sometimes to expedite the process, other times to integrate it with the computer vision system. As it turns out, processor apparition frequently uses image processing techniques. That's why both these

discipline are strongly inter-linked. In this case, these are being enhanced images using the gamma and wavelet transforms, and are also demonstrating with graphs. An essential phase in digital image processing is image improvement, which aims to enhance a picture's aesthetic appeal by emphasizing key details and minimizing noise or artifacts.

One common approach to image enhancement involves combining multiple techniques to achieve better results. In this case, the paper titled "Image Enhancement via Combined Gamma and Wavelet Transform: An Experimental Analysis" proposes a method that utilizes both gamma correction and wavelet transform for image enhancement. Image enhancement acts as a very important role in industrial

---
[1]renubabum@tkrec.ac.in, [2]c.venkateswarlu@iare.ac.in, [3]vemanad@tkrec.ac.in, [4]radha1977@gmail.com

and commercial applications to improve the quality of an attributes.

2. Literature Survey

At present, in the image processing and in the programming area Open CV is the most important Computer Vision library. In OpenCV, there are many algorithms which it can be used in the image processing and have the inbuilt functions. And also done the histogram, thresholding, Face detection, edge detection using OpenCV using the inbuilt functions. OpenCV is same like as C programming.[1]

Image processing and computer vision are widely used today to increase one's understanding of picture processing. Many initiatives in the field of image processing and related issues are served by the OpenCV library. It can be achieved more image clarity than the original image by utilizing OpenCV for image improvement.[2]

The basic goal of the fields of image processing and computer vision is to give machines the ability to see like humans do. This viewpoint makes it possible to define image processing as the conversion of digital images from the human visual system (HVS). It should be processed the digital photos in order to get the most realistic results. Image processing is utilized in a variety of application areas. Image enhancements, or referred to as image processing, improving the quality of the image to get more information about it from 380 to 700 nanometers, are one of the most prevalent application cases. This spectrum is known as visible light.[1].

2.1 Existing Method

The existing method is related to the work based on image enhancement using gamma transform.

This process starts with an input image, which is then transformed into many forms, such as edge detection, gray scale, histogram, dilation, and other filters using the gamma transform, to give us an output image with good visual clarity

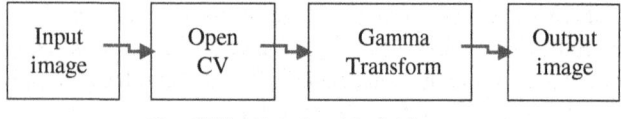

Fig. 65.1 Existing block diagram

2.2 Problem Identification

In the existing method there is a problem in quality of the image using gamma transform.

Additionally, there will be a lot of noise present in images taken with cameras; in order to obtain noise-free images; The transforms of OpenCV are used picture noise, often known as silent but inescapable fluctuations, is a side effect of image

capturing that cannot be avoided. If the light entering the lens is not aligned with the sensors, picture noise may result in a digital camera.

There will always be some sort of visual noise, even if it is not immediately apparent. Any type of electronic system that receives noise distributes it to the things it produces. As the images are broadcast via channels, they become deformed with impulse noise as a result of the noisy channels. Filters are therefore necessary before processing in order to remove sounds. Image noise can be minimized by using a variety of filters[5]

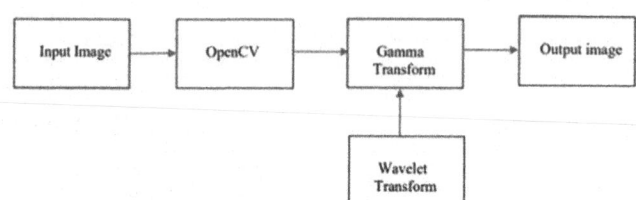

Fig. 65.2 Proposed block diagram

2.3 Proposed Method

It has been provided an approach that makes use of the gamma and wavelet transforms. These have been done using OpenCV which have libraries and inbuilt functions. First input image is given to the OpenCV and using both the gamma and wavelet transforms, the final image is obtained.. Using gamma and wavelet transforms together for image enhancement can lead to more effective and visually appealing results. Gamma correction helps adjust the brightness levels of an image, while wavelet transforms allow for multi-scale analysis and selective amplification or suppression of different frequency components.[6]

Wavelet Transform

A wavelet is a wave -like oscillation with amplitude that begins at zero, increases, and then decreases back to zero. It can typically be visualized as a "brief oscillation" like one recorded by a seismograph or heart monitor.

In terms of time and frequency, wavelets are functions that are concentrated around a specific location. The shortcomings of the Fourier method are overcome with this transformation technique.

Fourier transformation, although it deals with frequencies, does not provide temporal details.

Heisenberg's Uncertainty Principle states that either high frequency resolution and poor temporal resolution exist, or the opposite is true.

Wavelet based transition extraction for image segmentation.

Gamma Correction: Gamma correction is a nonlinear process that modifies an image's brightness levels. It aids in resolving

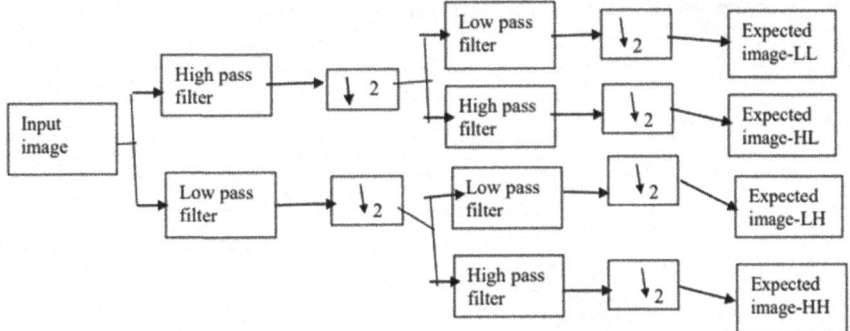

Fig. 65.3 Decomposition of an image 2-D discrete wavelet transform

the asymmetry in the relationship between pixel values and the perceived intensity by the human eye. The dark regions of the image can be made lighter while the details in the bright areas are preserved by using gamma correction.

Wavelet Transform: A mathematical technique for analyzing and processing signals or images at various scales is the wavelet transform. The image is divided up into many frequency components, enabling a multi-resolution examination. By deliberately amplifying or suppressing particular frequency components, wavelet transform can be used to improve photographs and highlight particular image details.

3. Software Used

3.1 Google Colab

To be more explicit, Colab is a completely cloud-based, free Jupiter notebook environment. The notebooks you create can be simultaneously modified by your team members, exactly like you edit documents in Google Docs, and most significantly, it doesn't require any setup. Many well-known machine learning libraries are supported by Colab and are simple to load in your notebook.

Google Colab used as a programmer to carry out the following tasks. Create and run Python code

- Document your code that supports mathematical equations
- Create/Upload/Share notebooks
- Import/Save notebooks from/to Google Drive
- Import/Publish notebooks from GitHub
- Import external datasets e.g. from Kaggle
- Integrate PyTorch, TensorFlow, Keras, OpenCV

3.2 OpenCV Python

A large open-source library for image processing, machine learning, and computer vision is called OpenCV.

Processing in Python, C++, Java, and many other programming languages are supported by OpenCV. It can analyze pictures and movies to find faces, objects, and even human handwriting.

The number of weapons in your arsenal rises when it is merged with other libraries, such as the highly efficient library for numerical operations known as Numpy, so that any operations that Numpy can perform can be combined with OpenCV.

With the aid of a vast array of OpenCV-programs and projects, this OpenCV tutorial will teach you how to process images in many ways, from basic to advanced, including operations on images and videos.

Practical Setup

This diagram illustrates the actual configuration used to carry out the programs for picture improvement within Google Colab. Here, Putting in the image's code location and input an image from mounting the Google Drive. As a result, The code is executed, and the result is produced photographs using several methods. The solution that is suggested in the research combines these two approaches to improve the aesthetic appeal of photographs. The combined approach's precise steps might be as follows:

Preprocessing remove the intensity or luminance component from the color information, convert the input image from RGB to the proper color space (such as YUV or Lab). By doing this, the color integrity is maintained and the enhancement only impacts the intensity information. Gamma correction should be applied to the image's intensity component. The amount of amplification can be managed by adjusting the gamma value. Gamma values that are greater than one (> 1) will emphasize the dark portions, whereas values less than one (1) will emphasize the bright areas.[8]

Perform wavelet decomposition on the image's gamma-corrected intensity component using the wavelet transform. The image is broken down into various frequency sub-bands

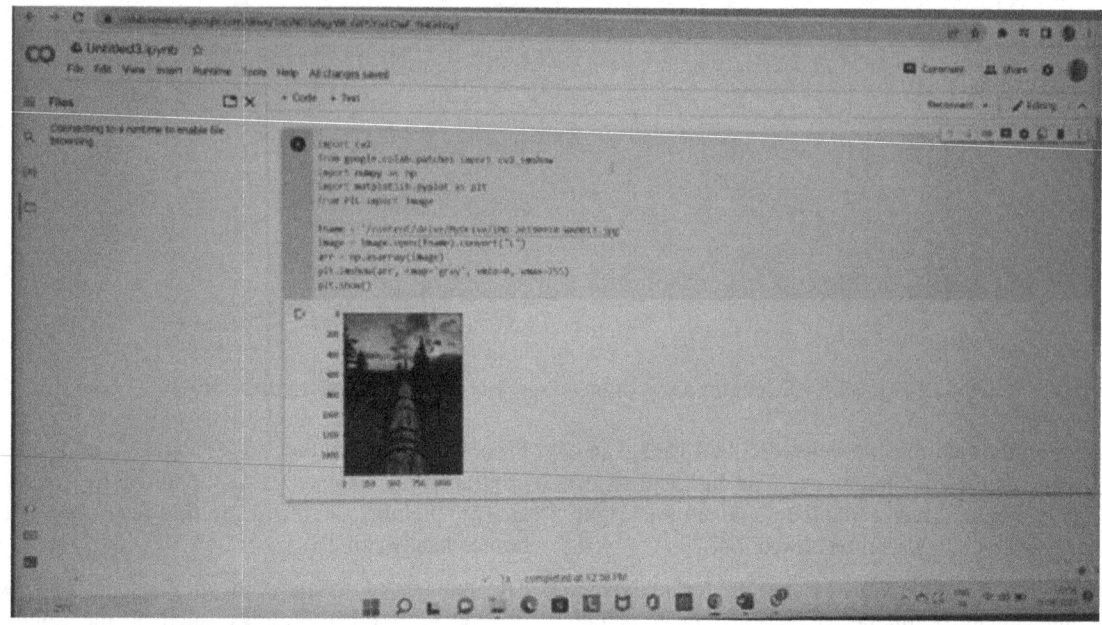

Fig. 65.4 Realistic setup used in running the image-enhancement programs

using the wavelet transform. High-frequency sub-bands often carry local features and edges, while low-frequency sub-bands typically contain global image information. Enhancement in the Wavelet Domain: To highlight particular details in the image, amplify or suppress certain wavelet sub-bands. This could entail modifying the low-frequency sub-bands to emphasize global features and boosting the amplitudes in the high-frequency sub-bands to emphasize edges or textures. Inverse Wavelet Transform: Reconstruct the enhanced intensity component of the image by performing the inverse wavelet transform, combining the modified wavelet sub-bands. Post-processing: Merge the enhanced intensity component with the original color information to obtain the final enhanced image.[9]

4. Results and Discussion

Input Image   Output Image

Fig. 65.5 It was noticed that a color image was changed to a grayscale image

From the Fig. 65.5, it has been observed that an input image is converted into gray scale image and also observed that way of plot the graph from that gray scale image.

Table 65.1 Observed that an input image with typical values the image is converted into gray scale image

Original Image	Typical values	Performance	Remarks
	[250000, 200]		The original image is converted into **gray scale** image.

From Table 65.1, it has been observed that an input image with typical values the image is converted into gray scale image using some libraries in the code, the image is converted and it can be seen in the above Table 65.1 and the status is gray scale image.

image = cv2.imread('/content/drive/MyDrive/S　2023-06-19 233443.jpg').

grayImage = cv2.cvtColor(image, cv2.COLOR_ BGR2GRAY) cv2_imshow(grayImage)

From the above Fig. 65.2, It was noticed that transformation of an input image into an edge detection image.

And also, observed that in an output image there is an edge like, outer line.

Table 65.2 An input image with some typical values was seen to be transformed into an edge detection image

Original Image	Typical values	Performance	Remarks
	[100, 200]		The original is converted into **edge detection** image.

From the above Table 65.2, we observed that an input image with some typical values, it is converted into edge detection image. Using Canny filter, edge detection can be done in the code.

To implement the code and to place in an image the google drive mounted and insert an image.

img = cv2.imread('/content/drive/MyDrive/Screenshot 2022-06-19 233443.jpg',0)

edge_det = cv2.Canny(img,100,200)

cv2_imshow(edge_det)

Input Image   Output Image   Graph

Fig. 65.6 Observed that an input image is converted into blur image

From the above Fig. 65.6, image that was the input became blurry.

Table 65.3 Observed that an image is converted into blur image

Original Image	Typical Values	Performance	Remarks
	[1.0, 1.0], [250000, 200]		The original Image is converted into **blur** image.

From the above Table 65.3, image is converted into blur image.

With some typical values and plot the graph as shown above.

In this, technique Gaussian filter used to blur an image and also some filters involved to blur an image.

img = cv2.imread('/content/drive/MyDrive/Screenshot 2022-06-19 233443.jpg')

blur = cv2.blur(img,(5,5))

cv2_imshow(blur)

cv2_imshow(img)

Input Image   Output Image

Fig. 65.7 Observed that an image is converted into erosion image

From the above Fig. 65.7, an image is converted into erosion image. It is a technique of morphological method.

Table 65.4 Observed that an image is converted into erosion image

Original Image	Typical Values	Performance	Remarks
	[5, 5]		The original image is converted into **erosion** as output image.

From the above Table 65.4, an image is converted into erosion image.

In which it shrinks an pixel of an image. And also used kernel functions for the erosion technique.

With some typical values it can be executed the code in the google colab and an image inserted with the help of mounting the google drive.

img = cv2.imread('/content/drive/MyDrive/Screenshot 2022-06-19 233443.jpg')

kernel = np.ones((5,5),np.uint8)

img_erosion = cv2.erode(img, kernel, iterations=1)

cv2_imshow(img)

cv2_imshow(img_erosion)

Input Image   Output Image

Fig. 65.8 Pbserved that an image is converted into dilation image

From the above Fig. 65.8, image is converted into dilation image. It is also a technique of morphological technique.

Table 65.5 Observed that an image is converted into dilation image

Original Image	Typical Values	Performance	Remarks
	[5, 5]		The original image is converted into **dilation** as output image.

From Table 65.5, an image is converted into dilation image.

Using some values in the code dilation image generated. In this kernel function used for the dilation image.

In the code, inserted image from google drive.

img = cv2.imread('/content/drive/MyDrive/Screenshot 2022-06-19 233443.jpg')

kernel = np.ones((5,5),np.uint8)

img_dilation = cv2.dilate(img, kernel, iterations=1)

cv2_imshow(img)

cv2_imshow(img_dilation)

Input Image   Output Image   Graph

Fig. 65.9 Observed that an image is converted into histogram equalization

From the above Fig. 65.9, an image is converted into histogram equalization.

Table 65.6 Observed that an image is converted into histogram equalization

Original Image	Typical Values	Performance	Remarks
	[250, 2500], [250, 100000]		Original image is converted into **histogram equalization** as output image.

From the above Table 65.6, an image is converted into histogram equalization with some typical values.

Using some inbuilt functions and libraries execute the code.

In this, inserted image from google drive.

img = cv2.imread('/content/drive/MyDrive/Screenshot 2022-06-19 233443.jpg', 0)

equ = cv2.equalizeHist(img)

res = np.hstack((img, equ))

cv2_imshow(equ)

cv2.waitKey(0)

cv2.destroyAllWindows()

Input Image   Output Image

Fig. 65.10 Observed that an image is converted into contrast stretching

From the above Fig. 65.10, image is converted into contrast stretching. Contrast Stretching which means it converts an image with high quality.

From above Table 65.7, an image is converted into contrast. With some values it can be done the execution of program. In this output is the image with high quality.

Table 65.7 An image is converted into contrast

Original Image	Typical Values	Performance	Remarks
	[0,255]		Original image is converted into **contrast stretching** as output image.

```
img = cv2.imread('/content/drive/MyDrive/Screenshot 2022-
    06-19 233443.jpg')

original = img.copy()

xp = [0, 64, 128, 192, 255]

fp = [0, 16, 128, 240, 255]

x = np.arange(256)

table = np.interp(x, xp, fp).astype('uint8')

img = cv2.LUT(img, table)

cv2_imshow(original)

cv2_imshow(img)

cv2.waitKey(0)

cv2.destroyAllWindows()
```

Here's a general outline of how you can perform image enhancement using both gamma and wavelet transforms with OpenCV: Read the Image: Load the image you want to enhance using the cv2.imread() function from OpenCV. Make sure to convert the image to the appropriate color space (e.g., RGB, BGR, grayscale) depending on the specific enhancement you want to perform. Gamma Correction: Apply gamma correction to the image using the cv2.LUT() function or other appropriate methods in OpenCV. You can experiment with different gamma values (> 1 for darkening, < 1 for brightening) to achieve the desired enhancement. Wavelet Transform: Perform wavelet decomposition on the gamma-corrected image using OpenCV's functions like cv2.dwt() or other wavelet transformation methods provided by OpenCV. Enhancement in Wavelet Domain: Amplify or suppress certain wavelet sub-bands to enhance specific details in the image. You can achieve this by directly modifying the wavelet coefficients. Inverse Wavelet Transform: Reconstruct the enhanced image by performing the inverse wavelet transform using OpenCV's functions like cv2.idwt(). Post-processing: If necessary, convert the enhanced image back to the appropriate color space (e.g., RGB or grayscale) and save or display the final output. It's important to note that the specific implementation details may vary depending on the Python libraries and versions you are using, as well as the actual image enhancement techniques employed. The

process might also differ based on whether you are enhancing grayscale or color images.

5. Conclusion and Future Scope

It can be concluded that by running some image processing programs using OpenCV. Additionally, got enhanced images by using certain techniques. The visual quality in this case is good. Both gamma and wavelet transforms used for the image enhancement and performed this with few lines of code.

The image processing is essential in day-to-day life. Human lives have been changed or impacted by improvement and advancements in the field of science and technology.

Additionally, these image processing techniques are essential in a variety of domains, including computer vision, face detection, image enhancement.

6. Acknowledgment

The proposed research work is implemented in the Simulation and Design Laboratory, Teegala Krishna Reddy Engineering College, Hyderabad. The authors of this paper express their gratitude to the management of this institute for the support and encouragement given.

References

1. https://www.ijrreset.com/fileserve.php?FID=8952.
2. Face Detection and Recognition using OpenCV, Article.
3. Facial Recognition using OpenCV, Image Enhancement on OpenCV based on the Tools: Python 2.7 February 2, 2019.
4. Sweety Deswal, Shailender Gupta and Bharat International Journal of Signal Processing, Image Processing, and Pattern Recognition, "Bhushan: A Survey of Different Bilateral Filtering Techniques" Vol.8, No. 3(2015).
5. Image Filtering Methods: A Review Ramanjeet Kaur 1, Er. Ram Singh 21, and Assistant Professor, Department of Computer Engineering, Punjabi University Patiala, Punjab, India.
6. MelikaMostaghim, Elnaz Ghodousi and Farshad "Image Smoothing Using Non- Linear Filters A Comparative Study" IEEE 2014 .
7. D. Z. P. F. L. I. Abdalla Mohamed Hambal, "Image Noise Reduction and Filtering Techniques, "International Journal of Science and Research, 2015.
8. Python 2.7 is used for image enhancement with OpenCV. V. Ravi, Ch. Rajendra Prasad, S. Sanjay Kumar, P. Ram Chandra Rao, and others from the Department of ECE at S. R. Engineering College in Warangal,Telangana, India.

Note: All figures and tables in this chapter were authors' test data and experimental analysis as part of their research work.

Numerical study of MHD Stagnation-Point Flow of Nanofluid Flow Past a 3-D Sinusoidal Cylinder with Thermophoresis and Brownian Motion Effects

66

G. Kathyayani[1]

Department of Applied Mathematics,
Yogi Vemana University, Kadapa, India

P. Venkata Subrahmanyam[2]

Department of Applied Mathematics,
Yogi Vemana University, Kadapa, India

Abstract: The concept of nanofluids aggregating with the help of nanoparticles is gaining industrial interest, such as crossflow heat exchangers and geothermal panels. Process quality needs to be improved continuously in polymer and conversion processes due to the transport phenomena that occur during the stagnation zone. The numerical study is devoted to investigating an MHD stagnation-point flow of a nanofluid flow over a circular sinusoidal cylinder that has a steady three-dimensional incompressible flow. With the appropriate similarity transformations, the partial differential equations can be changed to boundary value ODE's by including the Runge-Kutta fourth order technique and the shooting strategy in the numerical solution of the nonlinear system. The thermophoresis, magnetic and Brownian motion effects occur in the transport equations. Based on variables of interest, such as the magnetic parameter, thermophoresis parameter, Lewis number, Brown's motion parameter, and Prandtl number, influence on the temperature, velocity, and nanoparticle concentration profiles are analyzed. Numerical values of physical quantities such as drag force in x and y directions, diffusion mass flux, and heat flux parameters are employed

Keywords: 3D stagnation-point flow, Two-phase nanofluid, Magnetic field, Brownian motion, Sinusoidal cylinder

1. Introduction

Nanofluid convective transport models have been tested by several researchers [1-5]. Generally, nanofluids are two-phase mixtures containing nanoparticles as part of the solid phase. A two-phase flow theory based on conventional particles may not apply to nanofluid flow characteristics, given their nanoscale size. However, it should be noted that several factors that may affect a nanofluid may include friction between fluid-solid particles, gravity; the phenomenon of sedimentation, Brownian diffusion, and dispersion. The slip effect of velocity between fluid and particles must be considered when computing nanofluid flows. Nanofluid studies may be more accurate when using a two-phase approach because it takes into account fluid-solid motions.

By using a model that accounts for Brownian motion as well as thermophoresis and convective nanofluid transport, Buongiorno [6] conducted a comprehensive investigation into nanofluid kinematics. It was Buongiorno [6] who proposed an equilibrium model for the momentum, heat transfer and mass in nanofluids that includes two components, four equations, and is non-homogeneous. Convective transport in nanofluids was examined by Buongiorno [6] using a Brownian motion and thermophoresis model. A discussion carried out by Rasool et al. [7] which discussed electro-magnetohydrodynamic nano liquid flow across a vertical Riga plate with zero mass flux. By heat flux model named as Cattaneo–Christov, they analyzed the flow and heat characteristics. Mishra et al. [8] studied the mixed convective nanofluid flow of Walters liquid B past a stretching surface with momentum slip effects. In

[1]kathyagk@yogivemanauniversity.ac.in, [2]pvsmsubbu@gmail.com

their study, Buongiorno's nanofluid model, nanoparticles' boundary conditions are passively controlled by the particles. Later, Puneeth et al. [9] examined mixed convective 3D flow towards a bidirectional stretchable sheet under the consideration of both the Tiwari and Das models for the flow model and Buongiorno's model. An integrated numerical technique with a shooting scheme for flow over a porous surface was proposed by Rasheed et al. [10], which includes a suction or injection and Buongiorno nanofluid model. Mebarek-Oudina et al. [11] utilized the stretchable rotating disk to scrutinize the nanofluid heat transfer of magnetized fluid flow using the Buongiorno model. They noticed that an increasing magnetic field parameter value degrades mass transport rate while increasing the Schmidt number increases it.

Hydromagnetic nanofluid flow past a circular sinusoidal cylinder in combination with an adapted Buongiorno nanofluid model had not been investigated yet, based on a literature review. By addressing this gap, this work aims to fill it. This research work further advances the real-world application of this research by deploying experimentally derived functions of effective dynamic viscosity and effective thermal conductivity. Major works has been studied on different nanofluids past a stretching cylinder with various external thermal forces. The reasons mentioned above motivated us to accomplish the present study. In the present manuscript, the numerical investigating an MHD flow at its stagnation point for a nanofluid flow over a circular sinusoidal cylinder that has a steady three-dimensional incompressible flow has been addressed. The graphical representation have been exhibited with the influence of Nb, Nt, M, c and Sc of flow of the stagnation point's boundary layer nanofluid flow past a 3-D Sinusoidal Cylinder.

2. Mathematical Formulation

A steady stagnation-point flow heat and mass transfer in three-dimensional incompressible nanofluid is considered to flow over a cylinder with a sinusoidal variation of its radius (Fig. 1). Throughout the radius of the cylinder, there are points of stagnation (points A, B, and C) at each extreme of the radius. The fluid flow particles velocity profiles in the direction of x, y, z is considered with u, v, and w, respectively. It is important to note that the stagnation point in the Cartesian coordinate system Oxyz lies at the origin of the system and velocity profiles are taken at the form:

$$u_e = ax, \quad v_e = aby, \quad w_e = -(a+b)z \qquad (1)$$

Streamlines can be described by the equation $x = \delta y^{\frac{1}{c}}$, where c is the fraction of the slope of the stream velocities and expressed as $c = \dfrac{b}{a}$ is nodal point with $0 < c < 1$, saddle point

with $-1 < c < 0$ and finally $c = 0$ refers for plane flow, δ is constant, which gives a particular streamline. The following equations can be used to represent the boundary layer over the sinusoidal cylinder in accordance with the above-mentioned assumptions in order to model the boundary layer over the cylinder. This is done after removing the pressure with the help of Bernoulli equation. [12, 13, 14, 15].

$$\frac{\partial u}{\partial x} + \frac{\partial v}{\partial y} + \frac{\partial w}{\partial z} = 0 \qquad (2)$$

$$u\frac{\partial u}{\partial x} + v\frac{\partial u}{\partial y} + w\frac{\partial u}{\partial z} = a^2 x + \upsilon\frac{\partial^2 u}{\partial z^2} - \frac{\sigma B_0^2}{\rho}(u - ax) \qquad (3)$$

$$u\frac{\partial v}{\partial x} + v\frac{\partial v}{\partial y} + w\frac{\partial v}{\partial z} = b^2 y + \upsilon\frac{\partial^2 v}{\partial z^2} - \frac{\sigma B_0^2}{\rho}(v - by) \qquad (4)$$

$$u\frac{\partial T}{\partial x} + v\frac{\partial T}{\partial y} + w\frac{\partial T}{\partial z} = \alpha\frac{\partial^2 T}{\partial z^2} + \tau\left[D_B\frac{\partial T}{\partial z}\frac{\partial C}{\partial z} + \frac{D_T}{T_\infty}\left(\frac{\partial T}{\partial z}\right)^2\right]$$
$$(5)$$

$$u\frac{\partial C}{\partial x} + v\frac{\partial C}{\partial y} + w\frac{\partial C}{\partial z} = D_B\frac{\partial^2 C}{\partial z^2} + \frac{D_T}{T_\infty}\frac{\partial^2 T}{\partial z^2} \qquad (6)$$

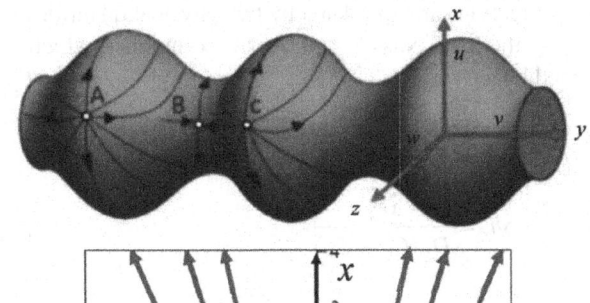

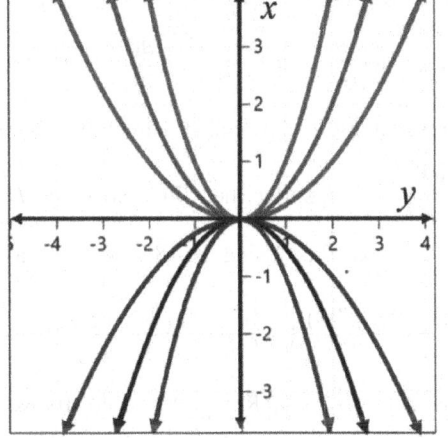

Fig. 66.1 Schematic model and the streamlines

As a result of these boundary conditions, the formula is as follows:

$$u = v = w = 0, T = T_w, C = C_w \quad at \quad z = 0$$
$$u \to u_e, v \to v_e, T \to T_\infty, C \to C \quad as \quad z \to \infty \qquad (7)$$

There is a similarity solution to each equation of the form (see. Ref. [15]).

$$u = axf'(\eta), v = byg'(\eta), w = -\sqrt{av}\left(f(\eta) + cg(\eta)\right),$$

$$\frac{T - T_\infty}{T_w - T_\infty} = \theta(\eta) \; \phi(n) = \frac{C - C_\infty}{C_w - C_\infty}, \eta = z\sqrt{\frac{a}{v}}, \tag{8}$$

It is worth noting that derivative with respect to η denotes the primes. Utilizing similarity transformations (8), the differential equations (2)–(6) are reduced to ordinary differential equations, which are nonlinear in character.

$$f''' + (f + cg)f'' - (f')^2 + 1 - M(f' - 1) = 0 \tag{9}$$

$$g''' + (f + cg)g'' - c(g')^2 + c - M(g' - 1) = 0 \tag{10}$$

$$\frac{1}{Pr}\theta'' + (f + cg)\theta' + Nb\,\theta'\phi' + Nt(\theta')^2 = 0 \tag{11}$$

$$\phi'' + Sc(f + g)\phi' + \frac{Nt}{Nb}\theta'' = 0 \tag{12}$$

The transformed boundary conditions are

$$f = 0, f' = 0, g = 0, g' = 0, \theta = 1, \phi = 1 \quad \text{at} \quad \eta = 0$$

$$f' = 0, g' = 0, \theta = 0, \phi = 0 \quad\quad \text{at} \quad \eta = \infty \tag{13}$$

The interested quantities are friction coefficients in the x and y directions are correlated by two physical quantities C_{fx} and C_{fy}, the local Nusselt and Sherwood number, which are defined by:

$$C_{fx} = \frac{\tau_{wx}}{\rho u_e^2}, C_{fy} = \frac{\tau_{wy}}{\rho v_e^2}, Nu_x = \frac{xq_w}{k(T_w - T_\infty)},$$

$$Sh_x = \frac{xs_w}{D_B(C_w - C_\infty)} \tag{14}$$

Where $\tau_{wx} = \mu \dfrac{\partial u}{\partial z}\bigg|_{z=0}$ and $\tau_{wy} = \mu \dfrac{\partial v}{\partial z}\bigg|_{z=0}$ are the surface shear stress along the x and y directions, respectively,

$q_w = -k\dfrac{\partial T}{\partial z}\bigg|_{z=0}$ is the surface heat flux, and $s_w = -D_B\dfrac{\partial C}{\partial z}\bigg|_{z=0}$ is the surface mass flux. Using Eq. (8), we obtain as:

$$C_{fx}\left[\mathrm{Re}_x\right]^{\frac{1}{2}} = f''(0), \left(\frac{x}{y}\right)C_{fy}\left[\mathrm{Re}_x\right]^{\frac{1}{2}} = cg''(0),$$

$$\left[\mathrm{Re}_x\right]^{\frac{-1}{2}} Nu_x = -\theta'(0), \left[\mathrm{Re}_x\right]^{\frac{-1}{2}} Sh_x = -\phi'(0), \mathrm{Re}_x = \frac{u_e x}{v} \tag{15}$$

Kuznetsov and Nield [17] expressed $\left[\mathrm{Re}_x\right]^{\frac{-1}{2}} Nu_x$ and $\left[\mathrm{Re}_x\right]^{\frac{-1}{2}} Sh_x$ as the Nusselt number and Sherwood number respectively.

3. Method of The Solution

In this computational section, we combine the Runge-Kutta method with the shooting scheme to solve the coupled nonlinear ODEs (9) to (12) of Magnetohydrodynamic (MHD) Boundary Layer flow of stagnation-point nanofluid flow past a 3-D Sinusoidal Cylinder with boundary conditions (13). we are considered for [0,7] as the domain of the boundary value problem in terms of [0, ∞). Interestingly, the results are unaffected after $\eta = 7$ in computational Runge-Kutta method with the shooting scheme. In order for the computation method to work, the governing nonlinear ODEs (9) to (12) must be transformed into system of first-order ODEs.

The first-order ODE system is defined as follows:

$$f = y_1, f' = y_2, f'' = y_3, f''' = y_3'$$

$$g = y_4, g' = y_5, g'' = y_6, g''' = y_6'$$

$$\theta = y_7, \theta' = y_8, \theta'' = y_8'$$

$$\phi = y_9, \phi' = y_{10}, \phi'' = y_{10}'$$

$$y_1' = y_2,$$

$$y_2' = y_3,$$

$$y_3' = -(cy_3 y_4 + y_1 y_3 - y_2^2 + 1 - M(y_2 - 1))$$

$$y_4' = y_5,$$

$$y_5' = y_6,$$

$$y_6' = -(cy_6 y_4 + y_1 y_6 - cy_5^2 + c - M(y_5 - 1)),$$

$$y_7' = y_8,$$

$$y_8' = -Pr(y_8 y_1 + cy_8 y_4 + Nty_8^2 + Nby_8 y_{10})$$

$$y_9' = y_{10},$$

$$y_{10}' = -\left(Sc(y_{10}y_1 + cy_{10}y_4) + \frac{Nt}{Nb}y_8'\right)$$

$$y_1(0) = 0, y_2(0) = 0, y_3(0) = p_0, y_4(0) = 0\,y_5(0) = 0,$$

$$y_6(0) = p_1, y_7(0) = 1, y_8(0) = p_2, y_9(0) = 1, y_{10}(0) = p_3.$$

The unknown initial estimates P_0, P_1, P_2 and P_3 are computed by Newton-Raphson method with convergence tolerance to $\varepsilon = 10^{-9}$.

Table 66.1 shows our numerical scheme validation results for dimensionless secondary velocity profile, and skin friction.

Table 66.1 Comparison of the values of $f''(0)$, $g'(0)$ for base fluid water

ω	Wang [16]		Present Study	
	$f''(0)$	$g'(0)$	$f''(0)$	$g'(0)$
0.00	-1.0000	0.00000	-1.0000	0.0000000
0.50	-1.1384	-0.5128	-1.1384	0.5128213
1.00	-1.3250	-0.8371	-1.3250	0.8371426
2.00	-1.6523	-1.2873	-1.6524	1.2873230

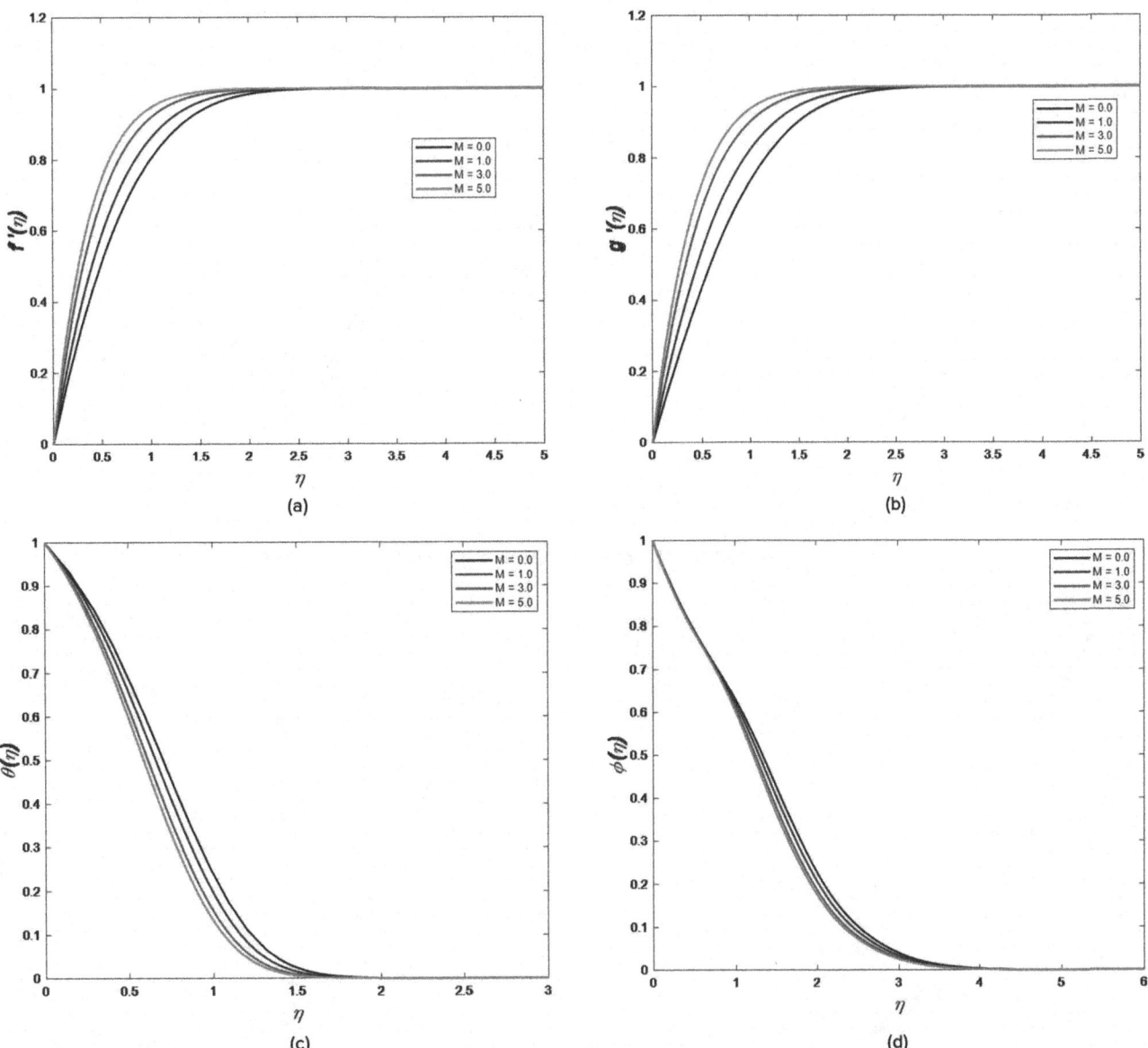

Fig. 66.2 The effect of M on f'(η), g'(η), θ(η) and ϕ(η)

The results agree significantly with Wang [16]. The present computational method is therefore suitable for studying the heat and fluid flow of the present problem with substantial confidence.

4. Results and Discussion

This section illustrates the numerically computed graphical outcomes and thermo-physical description of relevant terms over the nanofluid velocity profiles in x and y directions, temperature profile, nanofluid concentration profile and physical quantities. We perform a computational investigation of the flow of a nanofluid at its stagnation point

across a 3-D sinusoidal cylinder in a magnetohydrodynamic (MHD) boundary layer, taking into account the effects of thermophoresis and Brownian motion. Figure 66.2(a)-(d) highlights the impact of magnetic parameter M. As M rises, improving the velocity field in both the x and y axes and has the effect of decreasing the temperature and concentration profile is noted. Physically, the intensity of a magnetic field grows as charge is accelerated in its path. Consequently, the momentum boundary improves and the thermal and concentration boundary layers thin out. The thickness of the thermal boundary layer beyond the cylinder is strongly affected by the magnetic parameter. Figure 66.3(a-b) displays the effects of Nb on the energy and species fields. Particle

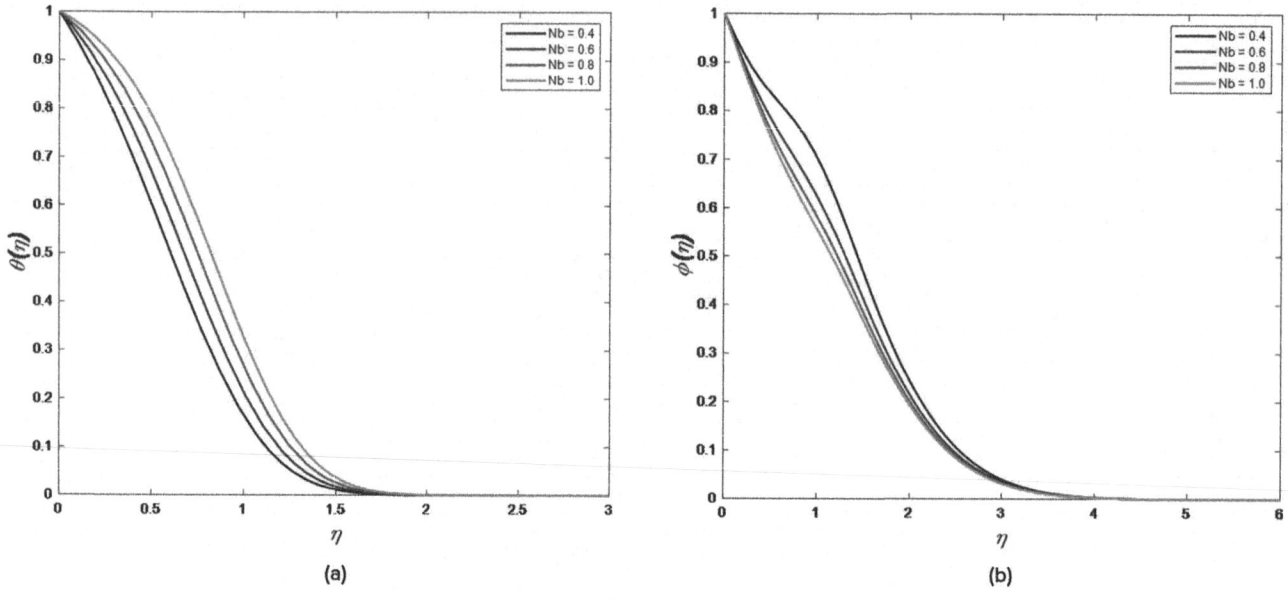

Fig. 66.3 The effect of Nb on $\theta(\eta)$ and $\phi(\eta)$

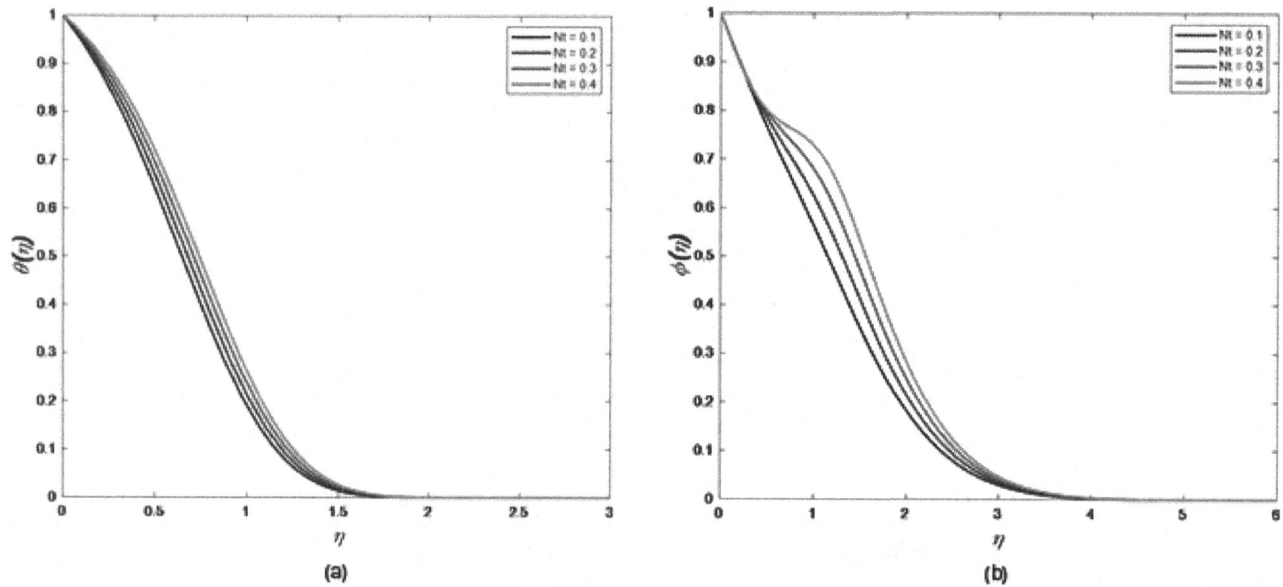

Fig. 66.4 The effect of Nt on $\theta(\eta)$ and $\phi(\eta)$

collisions quickly alter the Brownian motion (also known as pedesis). The particles in a fluid are in continual motion because of Brownian movement. As a result, colloidal solutions are more stable since their particles are kept in suspension. When the Brownian diffusion parameter is large, particles move quickly into the less dense area because of the high mass-transfer rate. Here, as Nb increases, the energy curve flattens out because less energy is required to generate vibrations in the fluid, leading to a more equivalent distribution of concentration. The temperature profile rises

as the Brownian motion parameter rises. Figure 66.4(a) and (b) illustrate how a thermophoretic parameter, Nt, affects temperature and concentration. An important relationship between Nt, a parameter of thermophoresis, and the resulting temperature and concentration fields was discovered. Therefore, as the value of the Thermophoresis parameter rises, so does the energy distribution. Final temperature reports indicate that the thermal boundary layer thickens with increasing Nt and Nb. Physically, when nanoparticles spin and wiggle erratically, they increase the system's kinetic

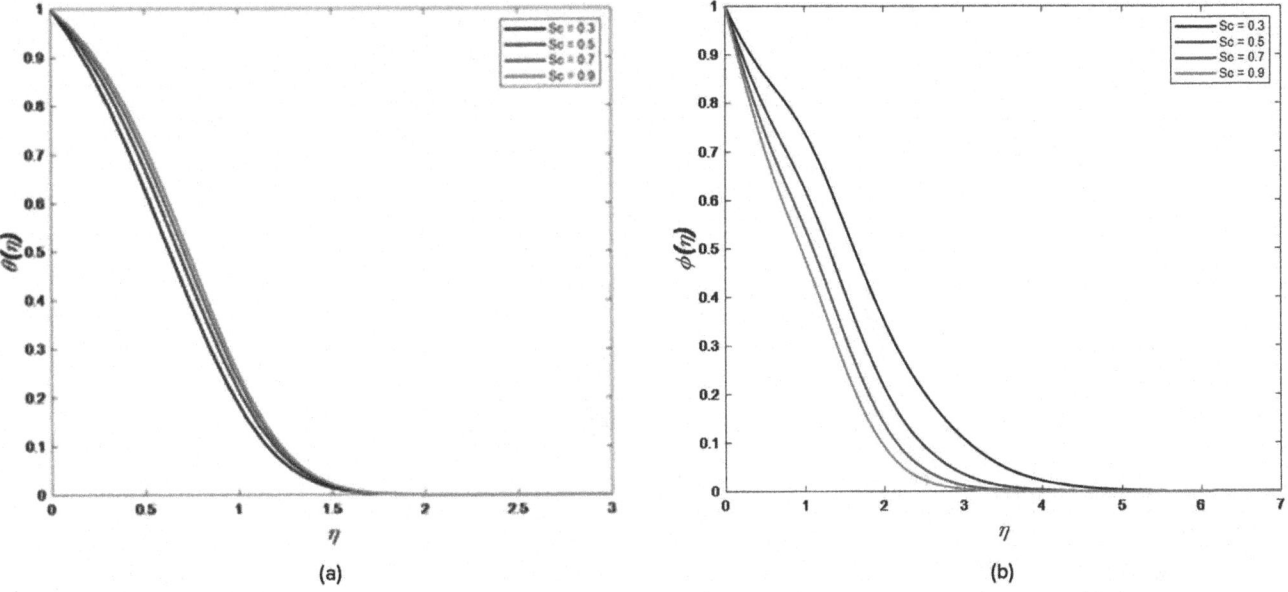

Fig. 66.5 The effect of Sc on $\theta(\eta)$ and $\phi(\eta)$

energy, which is related to the system's internal energy. Figure 66.5(a) and 66.5(b) depicted the impact of Sc on both temperature and concentration, respectively. An increase in Sc increases thermal conductivity and consequently the rate at which thermal energy transferred. When Sc is increased, density and mass diffusivity decrease but dynamic viscosity increases, as a result, the concentration profile is much reduced. The effect of saddle/nodal point parameter 'c' on the velocities, temperatures, and concentrations is shown in Figure 66.6(a-d). Two velocity components grow with stream line parameter 'c' and the temperature and the nanoparticle volume declines.

Figure 66.7 shows the dimensionless quantities   verses saddle/nodal point parameter c under the influence of the magnetic field and absence of magnetic field. When varying the parameter (c) as $-0.5 \le c \le 0.5$, drag force along the x-direction and y-direction increases the linearly. From the figure 7a,   the increases with presence of magnetic field and minimum value are notice in absence of magnetic field while this tendency is reverse in the case of Fig. 66.8 has been presented heat and mass flux verses the saddle/nodal indicative point parameter c both when a magnetic field is present and when one is not. We find an upward trend in nodal point area c (0, 0.5], and a parabolic trend in c [–0.5, 0). Obviously, the maximum local mass transfer and heat transfer is observed for the case of hydromagnetic fluid. The Brownian motion Nb impact on local Nusselt and Sherwood quantities verses stronger magnetic field has been presented in Fig. 66.9 graphs against Nb, are displayed against M in Fig. 66.9. As Nb grows, drops and rises, is noted. The

thermophoresis Nt impact on local Nusselt and Sherwood quantities verses stronger magnetic field has been presented in Fig. 66.10 graphs against Nb, are displayed against M in Fig. 66.10 against Nt, are displayed against M in Fig. 66.10. As Nt rises, drops and grows, is noted.

5. Final Remarks

In this research, we perform a computational investigation of the flow of a nanofluid at its stagnation point across a 3-D sinusoidal cylinder in a magnetohydrodynamic (MHD) boundary layer, considering the impact of the magnetic field (M), thermophoresis (Nt), saddle/nodal indicative point parameter (c) and Brownian motion (Nb). Nonlinear partial differential equations that regulate a system may be simplified by using a similarity transformation.

Here are the most important findings:

(a) The temperature profiles decrease with M and 'c,' and the thermal boundary layer thins down; nevertheless, when Nb, Nt, and Sc values increase, the thermal boundary layer thickens.

(b) As M, Nb, Nt, Sc, and 'c' are increased, the concentration profiles decrease, and the concentration boundary layer thickens.

(c) By raising the velocity in both the $f'(\eta)$ and $g'(\eta)$ components, it has been demonstrated that increasing the magnetic parameter M's value enhances the momentum boundary layer.

(d) The saddle/nodal indicative point parameter c increases from – 0.5 to 0.5. the ski friction coefficients and heat

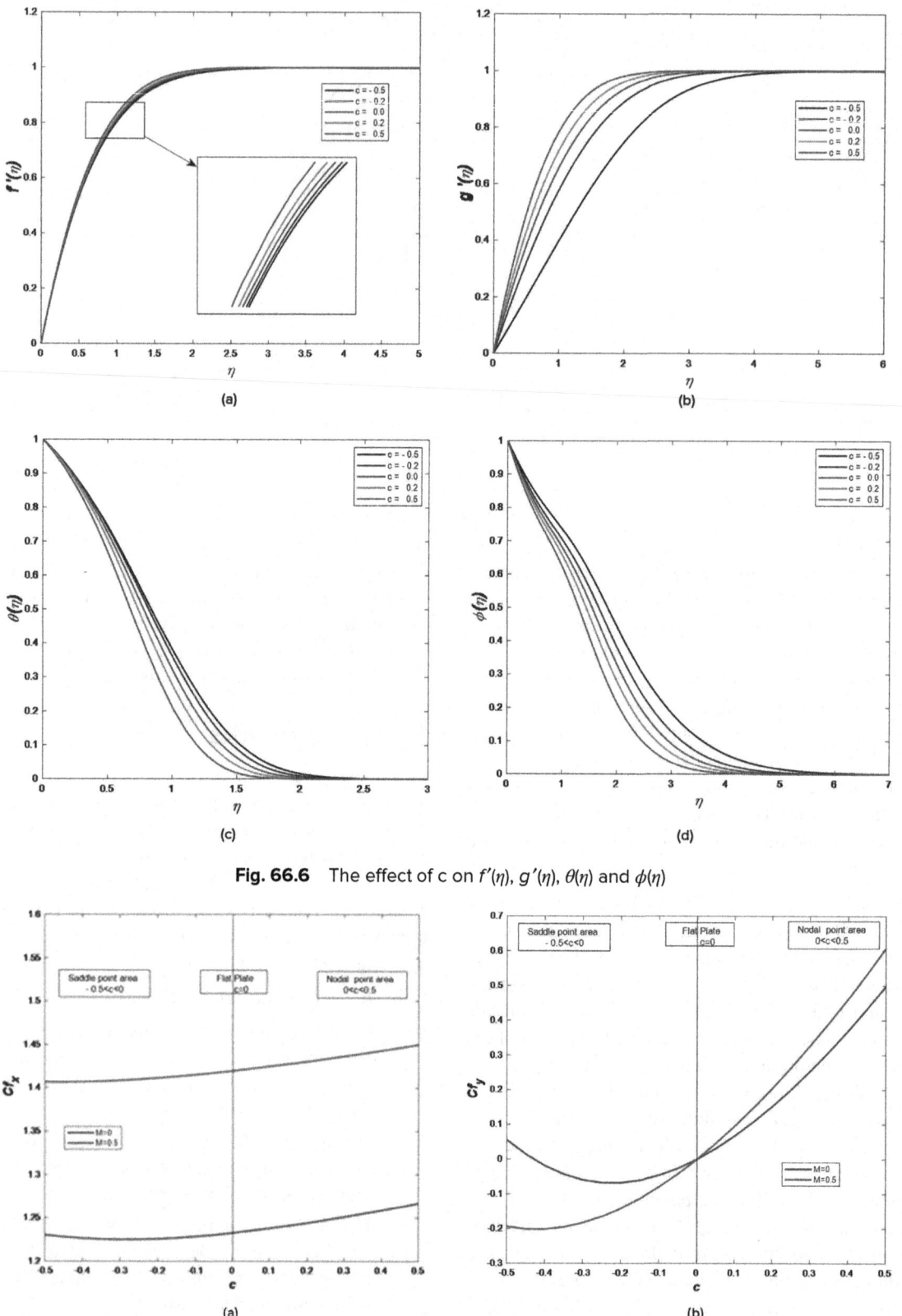

Fig. 66.6 The effect of c on $f'(\eta)$, $g'(\eta)$, $\theta(\eta)$ and $\phi(\eta)$

Fig. 66.7 The skin friction coefficients versus the nodal/saddle indicative parameter c for magnetic field effect

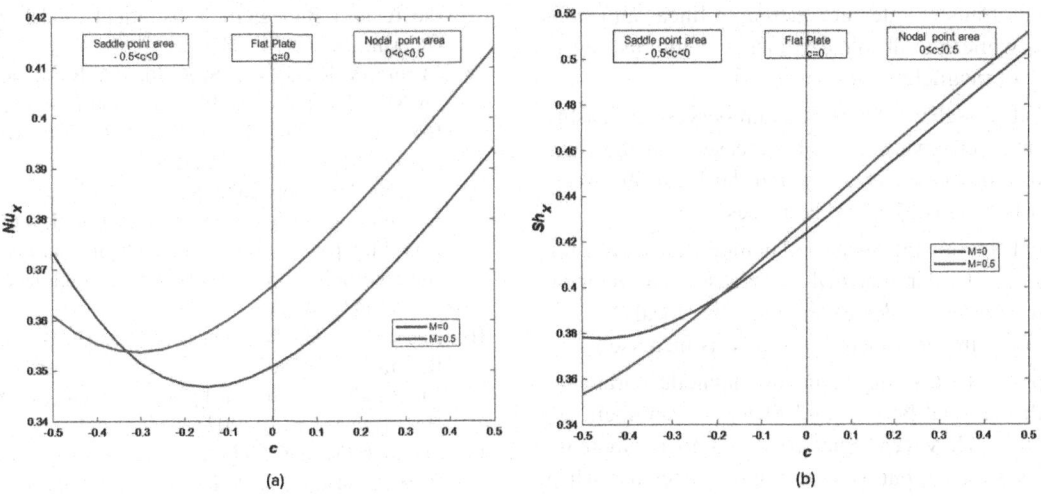

Fig. 66.8   The Nusselt and Sherwood versus the nodal/saddle indicative parameter c for magnetic field effect

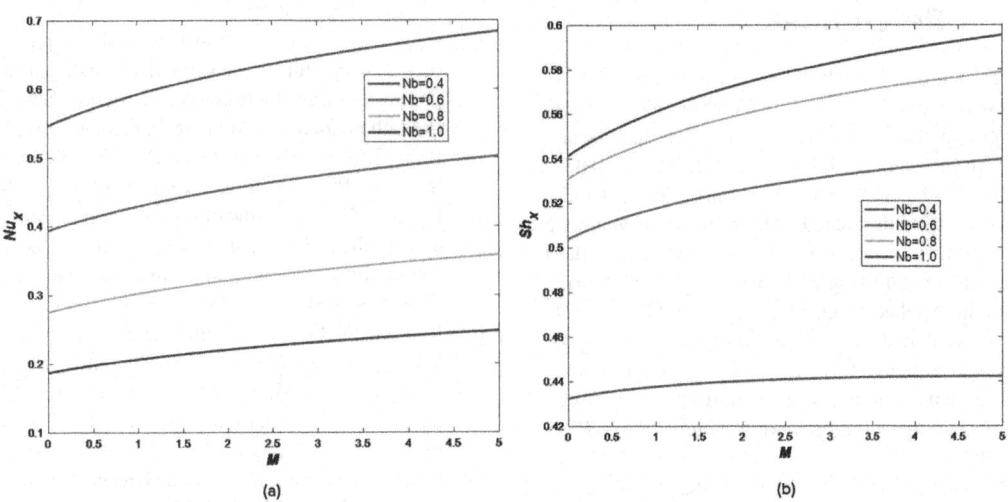

Fig. 66.9   The Nusselt and Sherwood versus versus M for Nb impact

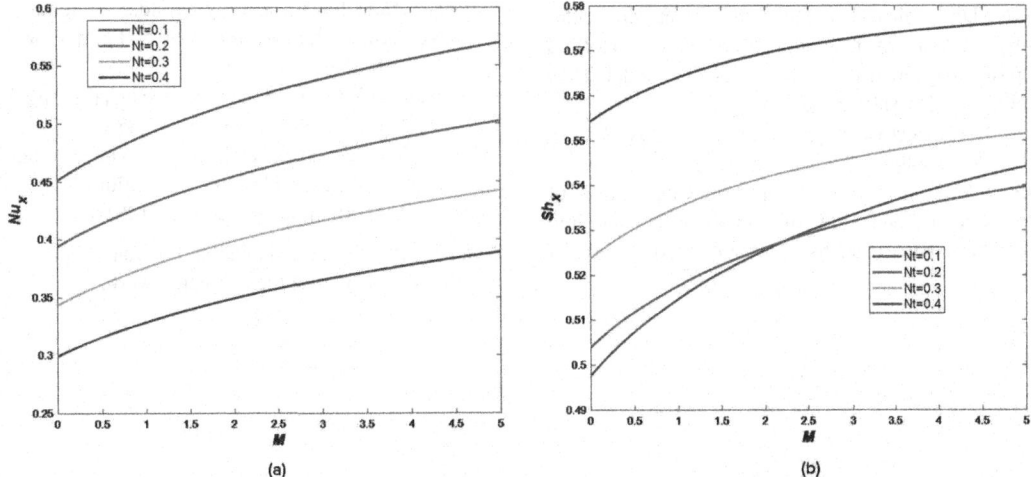

Fig. 66.10   The Nusselt and Sherwood versus versus M for Nt impact

and mass transfer rates are increases linear and non-linearly while, the absolute values are diminishing in absence of magnetic field.

(e) The local Nusselt and Sherwood numbers are increasing as the magnetic parameter M increases, but the local Nusselt number is decreasing and the local Sherwood number is increasing as Nb increases.

(f) The local Nusselt and Sherwood numbers are increasing as the magnetic parameter M is increased, but the local Nusselt number is decreasing and the local Nusselt Sherwood number is increasing as Nt is increased.

(g) In response to a rising value of magnetic parameter M, both the local Nusselt and Sherwood numbers are improving. However, Sherwood numbers increase when Nt is raised, but Nusselt numbers decrease when Nt is raised.

References

1. Venkatadri, K., Gaffar, S. A., Babu, C. S., & Fazuruddin, S., MHD radiative heat transfer analysis of Carreau nanofluid flow past over a vertical plate: a numerical study., Nanoscience and Technology: An International Journal, 12(4) (2021) 81–103.

2. Venkatadri, K., shobha, A., Venkata Lakshmi, C., Ramachandra Prasad, V., & Hidayathulla Khan, B. M., Influence of Magnetic Wire Positions on free convection of Fe3O4-Water nanofluid in a Square Enclosure Utilizing with MAC Algorithm, Journal of Computational Applied Mechanics, 51(2) (2020) 323–331.

3. Vedavathi, N., Venkatadri, K., Fazuruddin, S., & Raju, G. S. S., Natural convection flow in semi-trapezoidal porous enclosure filled with alumina-water nanofluid using Tiwari and das' nanofluid model.,Engineering Transactions, 70(4) (2022) 303–318.

4. Vedavathi, N., Dharmaiah, G., Venkatadri, K. & Gaffar, S., Numerical study of radiative non-Darcy nanofluid flow over a stretching sheet with a convective Nield conditions and energy activation. Nonlinear Engineering, 10(1) (2021) 159–176.

5. Vedavathi, N., Dharmaiah, G., Abdul Gaffar, S., & Venkatadri, K., Entropy analysis of magnetohydrodynamic nanofluid transport past an inverted cone: Buongiorno's model. Heat Transfer, 50(4) (2021) 3119–3153.

6. Buongiorno J. Convective transport of nanofluids. J Heat Transf. 128: (2006) 240–250

7. Rasool, G., Wakif, A. Numerical spectral examination of EMHD mixed convective flow of second-grade nanofluid towards a vertical Riga plate using an advanced version of the revised Buongiorno's nanofluid model. J Therm Anal Calorim, 143 (2021) 2379–2393.

8. Mishra, M.K., Seth, G.S. & Sharma, R. Navier's Slip Effect on Mixed Convection Flow of Non-Newtonian Nanofluid: Buongiorno's Model with Passive Control Approach. Int. J. Appl. Comput. Math 5 (2019) 107.

9. Puneeth, V., Manjunatha, S., Madhukesh, J. K., & Ramesh, G. K., Three dimensional mixed convection flow of hybrid casson nanofluid past a non-linear stretching surface: A modified Buongiorno's model aspects. Chaos, Solitons & Fractals, 152 (2021) 111428.

10. Rasheed, H. U., Zeeshan, Islam, S., Ali, B., Shah, Q., & Ali, R., Implementation of shooting technique for Buongiorno nanofluid model driven by a continuous permeable surface. Heat Transfer, 52(4) (2023) 3119–3134.

11. Mebarek-Oudina, F., Preeti, Sabu, A. S., Vaidya, H., Lewis, R. W., Areekara, S., & Ismail, A. I., Hydromagnetic flow of magnetite–water nanofluid utilizing adapted Buongiorno model. International Journal of Modern Physics B, (Accepted 30 November 2022).

12. Kathyayani G, R.Lakshmi Devi. MHD Free convective flow of a Jeffrey fluid in a vertical channel partially filled with porous medium. International Journal of Scientific Research in Mathematical and Statistical Sciences 5(6) (2018) 336–342.

13. Raju. S. S. R., Renuka Devi, R. L. V., Asogwa, K. K., Raju, S. S. K., Raju, C. S. K., Kathyayani. G., & Siva Kumar N, Falkner–Skan slip flow of non-Newtonian fluid over a moving and nonlinearly radiated wedge with variable heat source/sink and viscous dissipation. International Journal of Modern Physics B, (2023) 2450004.

14. Kathyayani G, R. Lakshmi Devi, Numerical Study on The Effect of Linear/Non-Linearly Stretching Sheet with Suction or Injection of MHD Mixed Convection Jeffrey Fluid Flow in A Vertical Stagnation-Point of A Porous Medium in The Presence of Thermal Radiation and Chemical Reaction. Bulletin of Pure and Applied Sciences Section E – Maths& Stat. 38E(2), pp. 550–562(2019).

15. Yousefi, M., Dinarvand, S., Eftekhari Yazdi, M. and Pop, I., "Stagnation-point flow of an aqueous titania-copper hybrid nanofluid toward a wavy cylinder", International Journal of Numerical Methods for Heat & Fluid Flow, 28 (7) (2018) 1716–1735.

16. Wang., C. Y. Stretching a surface in a rotating fluid. Z. Angew. Math. Phys., 39(2) (1988) 177–185.

17. A. V. Kuznetsov and D. A. Nield, "Natral Convective Boundary Layer Flow of a Nanofluid Past a Vertical plate", Int. J. Thermal Sci. 49, 243–247(2010).

Note: All the figures and table in this chapter were authors self data and the work is part of their research work.

Network Slicing in 5G for the Emerging Commercial Needs and Networking Challenges

67

SwarnaKamalam Vaddi[1]

Research Scholar, Department of CSE,
Koneru Lakshmaiah Education Foundation, Vaddeswaram,
Andhra Pradesh, India

Venkata Vara Prasad Padyala[2]

Assoc. Professor, Department of CSE,
Koneru Lakshmaiah Education Foundation, Vaddeswaram,
Andhra Pradesh, India

Abstract: Because of the widespread use of devices, applications, and services, client requests and expectations for network service providers', quality of service (QoS) have significantly increased. Experts in network architecture and optimization are doing a remarkable investigation. But despite this, the dynamic network environment keeps presenting fresh problems that need to be successfully resolved by today's networks. Joining existing networks results in increased capacity and coverage. According to the experts, mobility management is currently being studied in an effort to construct the current model more adaptable, user-centered, convenient and service-centric. Along with offering faster speeds and lower latency, 5G networks also offer enhanced availability, extraordinarily high capacity, enhanced connectivity, and increased stability. The infrastructure of the network needs to be more dynamic and agile than compared to the past in order to support demanding application requirements. If done correctly, Network Slicing may even be able to satisfy the current, strict requirements of the application for network model. The ultimate objective of this study is to optimize the use of existing network resources while also enhancing the level of service offered with the help of current mobility management technologies. The advancement of Network Function Virtualization (NFV) and Software-Defined Networking technologies (SDN) is crucial. Aparadigmof5GNetworkscalled"network slicing" allows for the integration of different networks. Network slicing is crucial for effectively meeting the needs of the numerous use cases for networks due to rising need for high data speeds, bandwidth utilization and minimal latency.

Keywords: 5Gsystems, Network slicing, NFV, SDN,QoS

1. Introduction

5G has widely been hailed as a significant catalyst for the emergence of network slicing. One that puts out a novel idea: the increased flexibility of running several logical end-to-end networks on a single shared infrastructure. In essence, this is dividing up the network into distinct sections for various users and/or use cases. Slice-ability is significantly easier to do with 4G and 5G, which increases accessibility. With 5G infrastructures in place, service providers could eventually be able to dedicate specific areas of their network to fulfil the unique demands of their clients, scaling up or down services. In general, examples could include linking and managing autonomous vehicles in a transportation fleet, allowing the Internet of Things (IoT) in a manufacturing context, and isolating AI-driven video analytics from point-of-sale data in a retail setting. In a factory, a 5G network slice might assist autonomous forklifts to help maintain connectivity and keep

[1]swarna.vaddi@gmail.com, [2]varaprasad_cse@kluniversity.in

adjacent industrial workers safe even when there is a spike in communications traffic from other regions of the factory.

As you may understand, dynamic network slicing necessities an underpinning infrastructure that is not only highly flexible and automated but also provides comprehensive visibility of the entire network. This is essential due to the significantly diverse resource requirements, be it compute, storage, or networking demands, and the additional challenge of simultaneously managing distinct network life cycles.

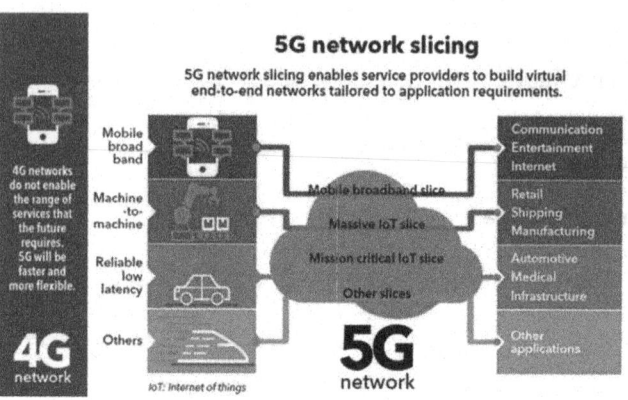

Fig. 67.1   Architecture [2]

A sophisticated application concept that is gaining popularity for the Internet of Things is network slicing. Network slicing is impacted by a number of variables, such as resource availability, physical infrastructure, crucial enablers, and security[1].Among the technology advancements in recent decades are cloud computing, edge computing, virtualization, the Internet of Things, Software-Defined Networking, and smart services, to name just a few.

The presence of various logical and physical network infrastructure needs allows for the segmentation of networks to facilitate smart services. When intelligent transportation technologies are implemented, VANET may record auto accidents involving autonomous driving[2].

It is essential that you alert it right away following a collision to reduce the amount of damage. It's a good thing that smart agricultural services have fewer latency needs than smart transportation services do. Slices of the intelligent transport systems (ITS) must be postponed even further if agricultural intelligent transport system(AIS) slices are delayed (L. U. Khan, 2019). Network slicing can currently be handled by a number of smart services[3].

Software-Defined Networking, network service virtualization, and cloud computing technologies enable 5Gnetwork slicing. It's doable to utilizing a network slicing architecture to offer adaptable solutions to a wide range of business circumstances and network traffic groups operating on the same networks [4]. Network slicing technology offers a wide range of

applications, including voice communication, intelligent transportation systems, and health care. Various services for many different uses. Figure 67.1's network slicing diagram demonstrates the procedure. Network slicing was a method for combining numerous virtual networks onto a single physical networking infrastructure[5] when it came to virtual networks.

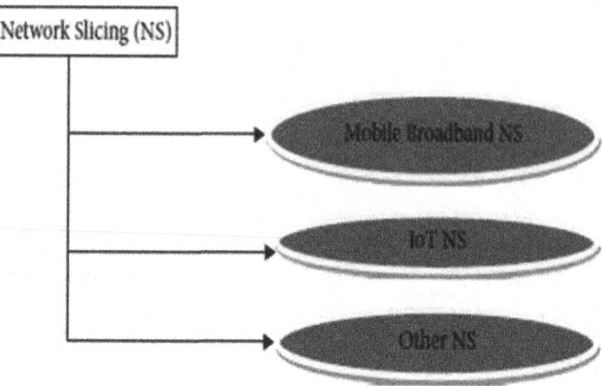

Fig. 67.2   Network slicing in 5G network functionality [5]

The software architecture acts as a model for facilitating communication among the members of a development team. Since the software industry is growing quickly, examination and maintaining work may start nearly immediately after an application has been built and released. Through my interactions with many team leaders, project managers, and solution architects over the course of several years[6], I came to understand that change management in software is a challenging task. As a result, it's vital to emphasize the value of quality assurance at every stage of the software development process to ensure that dependability as well as performance are not compromised in any manner. An increased knowledge of the network slicing architecture, the 5G network architecture, and the efficiency of network slicing in terms of performance, efficient use of resources, and adaptability will be made possible by this effort.

The three essential elements of this work are:

1. Presenting the network slicing concept;
2. Understanding various use cases that must be served by the network slices in order to understand various types of network slice requirements; and
3. Show casing enabler techniques, which include understanding a variety of techniques that will enable the implementation of network slicing in the 5G network.

Based on the research done, several historical and modern systems can be compared and studied in order to envision the design of network slices architecture. The 5G network nodes, network slicing use cases, and enabling techniques that form the 5G Network Slicing are briefly addressed in this article. Each key element needed for network slicing was defined, studied, and illustrated with concrete instances. Future researchers will be able to quickly identify the parts of the system that can be reused, need improvement, or require alternative methods and methodologies and validation of the main entities of the current system.

2. Literaturereview

2.1 5G Network and Technology

The objective and requirements for 5G networks are outlined in X. Li et al. (2016). These standards consist of the demand for better latency, greater data rates, and support for extensive inter connection. It also provides insight into the challenges associated with building a 5G network, such as the deployment of infrastructure and the demand for fresh spectrum allocation. A. Osseiran et al. (2014) present a summary of the technology prerequisites and drawbacks associated with 5G networks.

It emphasizes the relevance of new radio access technologies like massive MIMO and millimeter-wave for permitting greater data speeds. It also illustrates the need for inventive network architectures that facilitate an extensive selection of services and applications. K. Yang et al. (2019) offer an in-depth review of the potential and challenges associated with 5G networks. It investigates the potential impacts that 5G might have on an array of other industries, including those of healthcare, transportation, and entertainment. It also highlights some of the difficulties that accompany creating 5G networks, including those pertaining to spectrum availability, equipment deployment, and privacy and security concerns.

The authors of Taleb et al. (2018) address challenges associated with 5G networks' privacy and security. Such challenges include, for instance, the need for new security measures to defend against cyber-attacks and the necessity of privacy-preserving data management methods. This also illustrates an opportunity for the creation of novel technologies, like block chain and holomorphic encryption that enhance both security and privacy. The study by Z. Ahmed et al. (2020) includes an explanation of the technological features of 5G networks as well as the anticipated uses of these networks. It explores how 5G might affect a variety of sectors, including as healthcare, transportation, and entertainment, and it highlights issues related to the deployment of infrastructure and the distribution of spectrum that arise with the development of 5G networks. In addition, it highlights the potential for innovative and new services and applications, such driverless automobiles and smart homes.

An important element of 5G networks and a key enabler technology, 5G network slicing is presented in-depth analysis by H. Chen et al. (2020). It examines the concept of network slicing, which enables many virtual networks to share space on a single physical network infrastructure, and it focuses on the advantages of network slicing, including improved service agility, resource utilization, and service quality. Additionally, a description of the network slicing research issues, including resource management, security, and orchestration, is provided in this paper.

M. Alam et al. (2020) present an extensive examination of the research issues and new developments connected to 5G networks. It looks at how new network architectures, radio access technologies, and supporting technologies are required to serve a wide range of services and applications. The paper highlights the potential effects that 5G could have on a variety of industries, including healthcare, transportation, and the entertainment industry, and discusses the necessity for industry and academic institutions to collaborate for the purpose to deal with the research challenges associated with 5G networks. The study by Kheirkhah et al. (2019) examines the opportunities and constraints of 5G wireless networks for the use of augmented and virtual reality applications. It highlights the necessity for higher rates of data, shorter latency, as well as greater dependability in order to provide immersive and interesting experiences in augmented and virtual reality. The study also analyses the possible effects of 5G on a number of sectors, including education, entertainment, and healthcare. It also emphasizes the necessity of ongoing research and development in order to support cutting-edge virtual and augmented reality services and applications on 5G networks. The study additionally looks at how 5G might affect other sectors, including healthcare, entertainment, and education. These scholarly works provide a thorough review of the potential and issues related to 5G networks. They also highlight the need for ongoing study and development to enable innovative and new applications and services. They also examine the effects that 5G networks might have on various industries and emphasize the need for cooperation between business, government, and academic institutions to enable the successful execution of 5G networks. The investigation in to challenges associated with 5Gnetworks, such as preserving user privacy and enhancing energy efficiency, are also thoroughly covered.

2.2 5G Network Slicing and Techniques

Network slicing is described as "an approach to create and proactively operate virtual functionally-discrete end-to-end networks over common physical infrastructure". Theoretically, operators can now provide a variety of networks, each adapted to the specific requirements of a different customer, and to be launched, managed, and decommissioned as needed on the same common infrastructure. As a result, the requirement for specialized physical appliances for particular networks

Table 67.1 Key features and opportunities associated with 5g networks

Author	Feature and Description	Methodology
M. M. Alam et al. (2020)	Collaboration: Theeffectivedeploymentof5Gnetworks and the facilitation of new and cutting-edge applications and services depend on collaboration between business, government, and academia.	Surveys, case studies, and reviews of the literature.
N. Kheirkhah et al. (2019)	Augmented and virtual reality: Virtual and augmented reality apps now have new options thanks to 5G networks, providing immersive and engaging experiences.	Simulation,modeling, and experimentation
J. Liu et al. (2019)	Energy efficiency: As 5G networks require large energy consumption to support increased data rates and wide spread connectivity,energy efficiency is one of their main challenges.	Simulation,modeling, and experimentation.
M. M. Alam et al. (2020)	Security: The numerous apps and services that5G networks. Enable raise Security and privacy issues, which pose a substantial barrier to their mainstream adoption management	Risk evaluation, threat analysis, and the implementation of security measures.
J. Liu et al. (2019)	Edge computing: Edge computing eases pressure on the central network by relocating processing to the edge of the network, nearer to the end users. This opens the door for low-latency applications.	Simulation, modeling, and experimentation
H. Chen et al. (2020)	Network slicing: by allowing multiple virtual networks to share the infrastructure of a single physical network, network slicing improves resource efficiency, service agility, and quality of service.	Modeling and simulation, experimentation, and Prototyping.

Source: Made by author

is no longer necessary, giving operators better control over their services and the ability to supply them more affordably. Network slicing offers considerable benefits to both operators and business clients.

Regarding the first aspect mentioned, it primarily revolves around the ability to provide service with agility,flexibility, and customization. Ultimately, enterprise customers also benefit from these qualities. The design and implementation of network slicing in fifth generation networks have been the subject of several research papers. For instance, Zhang et al.'s research from 2021 suggested are volutionary paradigm for network slicing that allows for fine-grained control over network resources including bandwidth, latency, and dependability. SDN and NFV are combined in the proposed framework to offer a scalable and adaptable platform for network slicing. Chiang et al. (2021) suggested a reinforcement learning-based method for dynamic network slicing in5G networks in another paper. The suggested method optimizes network slicing choices based on current network conditions and application needs using a deep reinforcement learning algorithm. Results revealed that in terms of network performance and resource utilization, the suggested approach performed better than currently used static network slicing techniques. Numerous research publications have concentrated on network slicing difficulties in 5G networks in addition to network slicing techniques. In a report published in2020, Ma et al., for instance, cited resource allocation, slice isolation, and slice orchestration as three major difficulties in network slicing. In order to overcome these difficulties and enable efficient and effective network slicing in 5G networks, a thorough framework for network

slicing was provided in the article. Li et al.'s other research from 2021 focused on the security issues with network slicing in 5G networks. The study proposed a network slicing architecture that unifies different security techniques, such as access control, authentication,and encryption. According to the findings, the suggested architecture offers solid security for network slicing in 5G networks. These studies underline the significance of network slicing in 5G networks and the demand for cutting-edge methods and frameworks to make network slicing efficient and effective. The suggested methods and frame works can assist in resolving network slicing-related difficulties and enable the development of specialized virtual networks that are tailored to the needs of various applications and services. In addition, a number of research have been done on how network slicing is really implemented in 5G networks. In research published in 2021, Chen et al., for instance, suggested a software-defined network (SDN)-based method for dynamic network slicing in 5G networks. The suggested method makes use of a central controller to control network slices and dynamically distribute resources according to application needs. The findings demonstrated that, in comparison to current static network slicing strategies, the suggested approach can increase network performance and decrease resource wastage. There have been several research on the economic benefits of network slicing in 5G networks in addition to these studies. For instance, Liu et al. (2021) suggested an auction-based resource allocation approach for network slicing. According to the needs of their applications, service providers can bid for network resources using the suggested process, and there sources are then assigned to the highest bidder. The outcomes demonstrated that the

Table 67.2 5G network slicing and techniques

Technique	Methodology	Pros	Cons	Results	Use Case
Static slicing	Resources are allotted for each slice in advance.	Simple, foreseeable, and little overhead	Limited flexibility, inefficient resource utilization	Suitable for applications with low variability	IoT networks, industrial automation
Dynamic slicing	Real-time resource allocation depending on demand	Highly flexible, efficient resource utilization	Complex, expensive, and challenging to handle	Suitable for applications with considerable variability	Video streaming, online gaming
SDN based slicing	Centralized management, high flexibility	Limited scalability, complex implementation	Appropriate for small-scale networks	Appropriate for small-scale networks	Enterprise networks, campus networks
NFV based slicing	Uses NFV to construct and manage network slicing	High scalability, cost-effective	Limited flexibility, complex implementation	Appropriate for large-scale networks	Cloud data Centers and mobile operator networks
Hybrid slicing	Combination of static, dynamic, SDN, and NFV techniques	Offers a mix between flexibility and effectiveness	Complex, requires sophisticated management	Suitable for a variety of usage situation	Smart cities, healthcare

Source: Made by author

suggested mechanism can increase resource utilization while lowering service providers' expenses.

3. Functionalities of 5G Slices

The 5G network has the capacity to satisfy a variety of purposes. Three major categories can be used to group the services:

1. Enhanced Mobile Broad band (eMBB)
2. Ultra-reliable and Low-latency Communications (uRLLC)
3. Massive Machine Type Communications (mMTC).

Enhanced Mobile Broadband, or eMBB, provides mobile data connectivity to customers in one of three configurations: dense clusters of users, highly mobile users, or users distributed over significant areas. Huge arrays of multiple input, multiple output (MIMO) antennas and a variety of spectrums starting at conventional 4G wavelengths and continuing into the millimeter band are its key defining features.

Massive Machine-Type Communications, ormMTC, services are designed to support a large number of devices in a constrained space with the expectation that they will produce little data (a few tens of bytes per second) and be able to withstand significant latency (upto10 seconds round trip). Furthermore, the specifications require that data transmission and reception utilize little power in order for devices to have long battery lives. In connection with that, the 5G New Radio specification aimsfor1milliondevicestobe supported per square kilometer.

3.1 Existing Methodology

Ultra Reliable Little-Latency Communications, or URLLC, uses 5G to offer secure communications with latencies of 1 millisecond (ms), high dependability with little, or even zero, packet loss. To accomplish this, avariety of physical device advancements on MIMO antenna assemblies, simultaneous manipulation of several frequency bands, packet coding and processing techniques, and enhanced signal handling are all used.

4. Major Findings

Some important conclusions from the assessment of the literature and analysis of current approaches for 5G network slicing are as follows:

1. A possible solution for providing optimal resource allocation in 5G networks is network slicing.
2. The use of network slicing can accommodate a large variety of applications with different requirements.
3. To solve the difficulties involved with network slicing, a variety of approaches have been proposed, including static, dynamic, SDN-based, NFV-based, and hybrid.
4. Each methodology has benefits and drawbacks, and the optimal one to utilize may vary depending on the specific use case and application requirements.
5. Network performance can be enhanced by applying advanced techniques like reinforcement learning and machine learning to network slicing decisions.
6. In network slicing, security is a crucial concern, and strong security procedures must be put in place to

Table 67.3 Comparing the existing techniques for 5G networkslicing

Reference	Pros	Cons	Methodology
Iu et al. (2021)	-Provides service providers the opportunity to place bids for network resources based on the needs of their applications. -Increase resource efficiency and lowers expenses for service providers	Only includes auction-based mechanisms	Proposed an auction-based resource allocation system for network slicing
Yang et al. (2021)	Improves network speed and user happiness by using machine learning techniques to optimize network slicing choices depending on current network conditions and user preferences.	Only supports federated learning-based approaches	Presented a federated learning-based strategy for network slicing
Chen et al. (2021)	Creates use of SDN to dynamically assign network resources according to application needs. Enhances network efficiency and decreases resource waste	Confined to SDN-based methods	Suggested an SDN- based strategy for dynamic network slicing.
Li et al. (2021)	-Addresses the security issues posed by network slicing in 5G networks -Offers strong network slicing security	-No specific drawbacks are noted	Suggested a reliable and effective network slicing architecture
Zhang et al. (2021)	Allows for precise control of network resources	No major drawbacks are stated	Presented a brand-new network slicing framework.

Source: Made by author

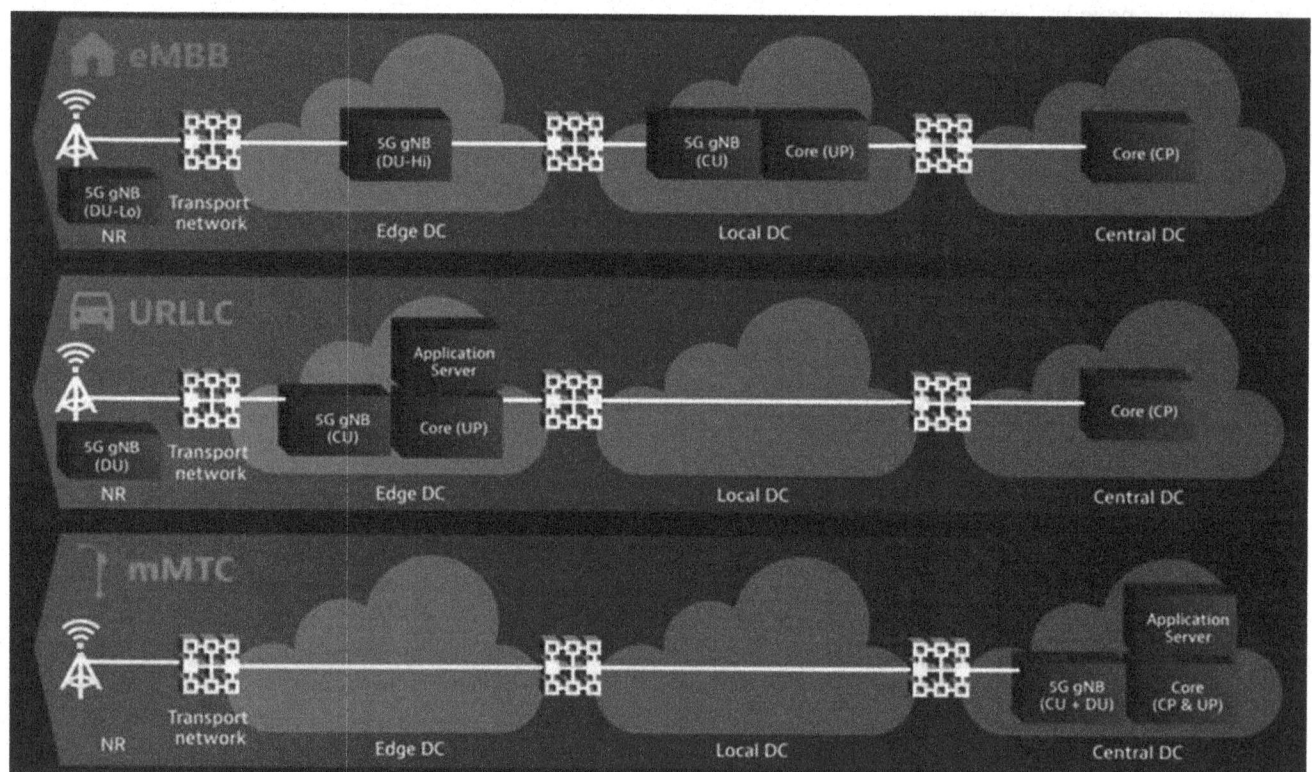

Fig. 67.3 Slicing functionality [15]

guarantee the confidentiality and integrity of network slices.

7. Although network slicing has many advantages, it also makes networks more complex and calls for sophisticated resource allocation algorithms.

8. In order to enable efficient and effective network slicing in 5G networks, further research is required to overcome the difficulties associated with network slicing and to develop new approaches and frameworks.

5. Challenges

According to numerous study publications, there are a number of difficulties with the deployment and execution of 5G networks. Some of the main challenges are:

Spectrum Availability: One of the primary impediments facing 5G networks is the spectrum's availability. Higher frequency bands, which are needed for 5G networks but have a limited supply, would need new spectrum allocation. Political and regulatory difficulties could come from this, and operators might pay more as a result.

Infrastructure Deployment: Setting up 5G infrastructure is difficult since it necessitates considerable investments in new infrastructure, like tiny cells and fiber optic connections. Numerous constraints, including a lack of suitable locations for antennas and towers and high construction costs, might make it difficult to deploy infrastructure.

Security and privacy: The 5G networks' greater connectivity and data interchange raise fresh security and privacy issues. 5G networks are more susceptible to cyber-attacks like Distributed Denial of Service(DDoS) assaults because of their high speed and low latency. This necessitates the installation of new security mechanisms including end-to-end encryption, secure booting, and secure device provisioning.

Interoperability: Interoperability is hampered by the combination of many network technologies, including 5G and Wi-Fi. This is crucial for managing various device and application kinds as well as for smooth hand overs between various networks.

Energy Efficiency: The installation of 5G networks uses a lot of energy, which poses problems for sustainability and economic viability. Small cells and the growing number of gadgets necessitate effective methods for managing power and gathering energy.

To ensure the successful deployment and execution of 5G networks, these issues underscore the necessity for a coordinated effort across stakeholders, including governments, regulators, operators, and vendors.

6. Features

The essential characteristics of 5G networks that set them apart from earlier generations of wireless networks have been outlined in research publications. The following are some of the key characteristics:

High Data Rates: Compared to earlier generations, 5G networks offer much faster data transfer rates. Research indicates that peak data rates on 5G networks can reach 20 Gbps, which is around 20 times faster than peak data rates on 4G networks. Faster downloads, streaming, and other data-intensive applications are made possible as a result.

Low Latency: The latency of 5G networks—the amount of time it takes for data to move between devices—is extremely low. In comparison to 4G networks, which typically have latency of 30 to 50 milliseconds, research has revealed that 5G networks can reach latencies as low as 1 milli second. The low latency is crucial for tasks such as remote surgery and autonomous vehicles, where real-time performance is of utmost importance.

Massive Connectivity: Compared to earlier generations, 5G networks are capable of supporting afar higher density of devices. In comparison to 4G networks, which can only support upto 100,000 devices per square kilometer, 5G networks can support up to 1 million devices per square kilometer. To handle the rising number of IoT devices and smart city applications, this capacity must be enhanced.

Network Slicing: Network slicing, which allows for the development of numerous virtual networks on a common physical infrastructure, is supported by 5G networks. Service level agreements (SLAs) and distinctive characteristics can vary from virtual network to virtual network. This enables improved service customization for various apps and users as well as more effective use of network resources.

Edge computing: Edge computing is supported by 5G networks and enables data to be processed and stored near to the sources of the data. Due to the decrease in latency, real-time applications like augmented and virtual reality work better.

Overall, these 5G network characteristics have the potential to support cutting-edge services and novel applications like autonomous driving, smart cities, and virtual reality.

7. Applications

There are many uses for 5G network slicing across numerous industries, including:

Health Care: Medical devices and systems can be connected virtually thanks to 5G network slicing. For instance, a dedicated virtual network with low latency and great dependability can be used for remote surgery, medical imaging, and patient monitoring.

Smart Cities: 5G network slicing may make it possible to design apps for smart cities. In emergencies, virtual network scan be built to give priority to important services and programs.

Autonomous Vehicles: A virtual network specifically for autonomous vehicles can be built using 5Gnetwork slicing. Real-time communication and coordination between infrastructure, linked devices, and moving objects can be supported by this technology.

Gaming and entertainment: 5G network slicing can support the creation of interactive and immersive gaming and entertainment experiences. To prioritize bandwidth and low latency for high-quality streaming and real-time gaming, virtual networks can be built.

Industrial Automation: Applications for industrial automation, such robotics and remote monitoring, can be supported by 5G network slicing.

8. Conclusion

A promising method called 5G network slicing enables network operators to build many virtual networks with various performance characteristics on a common physical infrastructure. Numerous advantages,including increased flexibility, higher network performance, and enhanced user experience, are providedbythistechnology. Tosolvethedifficultiesconnectedwith5Gnetworkslicing, a number of methodologies and strategies have been put out in the literature, including deep learning, game theory, reinforcement learning, and heuristic algorithms. When it comes to maximizing network resources and enhancing network performance, these strategies have demonstrated encouraging outcomes. To fully realize the potential of5G network slicing, there are still a few issues that need to be resolved. These include challenges with complexity, scalability, and security and privacy. To create efficient remedies to these problems, more research is required. In general, 5G network slicing has the power to completely change how we access and utilize network services. We may anticipate more creative applications and use cases in the future as this technology is developed further.

References

1. Aijaz, A., & Memon, Q. (2017). 5G networks: Challenges and research trends. Journal of Network and Computer Applications, 97, 1–21.
2. Zhang, Y., Yu, F. R., & Leung, V. C. M. (2019).A survey on 5G networks for the internet of things: Communication technologies and challenges. IEEE Internet of Things Journal, 6(3), 4203–4219.
3. Giannoulakis, I., Anastasopoulos, M., & Tassiulas, L. (2019). Machine learning for network slicing: A review. IEEE Communications Surveys &Tutorials, 21(4), 3604–3634.
4. Li, X., &Yu, H. (2018). Software defined network function virtualization: A survey. IEEE Access, 6, 3189–3206.
5. Orsini, R., & Chowdhury, K. R. (2018). 5G network slicing: A survey. IEEE Communications Magazine, 56(4), 112–119.
6. Zhao, J., & Zhang, H.(2020).A survey of reinforcement learning for 5G networks: Foundations, applications, and future directions. IEEE Communications Surveys & Tutorials, 22(1), 625–658.
7. Zhang, Q., Wang, S., Chen, Y., Mao, S., & Leung, V. C. M. (2019). Security and privacy in 5G networks: Challenges and solutions. IEEE Wireless Communications, 26(1), 87–93.
8. Rost, P., Banchs, A., & Gavras, A. (2016). Network slicing to enable scalability and flexibility in 5G mobile networks. IEEE Communications Magazine, 54(2), 18–24.
9. Salsano, S., Liu, H., & Contreras, L. M. (2018). Network slicing for 5G: Challenges and opportunities.
10. Lin, Y., Li, X., & Li, Q. (2020). A comprehensive survey on network slicing in 5G networks. IEEE Wireless Communications, 27(1), 16–23.
11. Afolabi, I., Li, Z., & Chen, H. H. (2020). Intelligent network slicing for 5G and beyond: Opportunities, challenges, and future directions. IEEE Communications Magazine, 58(4), 28–34.
12. Gharbaoui, M., & Chowdhury, K. R. (2021). A survey of AI-based techniques for 5G network slicing. IEEE Transactions on Network and Service Management, 18(1), 87–100.
13. Zhang, Y., Guo, W., & Li, B. (2019). Network slicing based on deep reinforcement learning in 5Gnetworks. IEEE Communications Magazine, 57(5), 120–126.
14. Kim, Y., Choi, S., & Lee, S. (2020). Network slicing in 5G: A comprehensive survey of algorithms and solutions. IEEE Communications Surveys &Tutorials,22(3), 1980-2015.
15. Sun, Q., Cheng, J., Chen, J., Huang, L., & Zhang, Y. (2021). A survey on machine learning-based network slicing techniques for 5G and beyond networks. IEEE Transactions on Network and Service Management, 18(1), 58–71.
16. Li, Q., Li, X., & Zhang, Y. (2019). AI-enabled network slicing for 5G and beyond: challenges and opportunities. IEEE Communications Magazine, 57(3), 13–19.
17. Jiang, X., Cheng, L., & Zhou, L. (2021). Machine learning-based 5G network slicing: State-of-the-art and future challenges. Journal of Network and Computer Applications, 177, 102923.
18. Yin, X., Xu, C., Li, Y., & Li, W. (2019). Network slicing in 5G: a survey based on the service requirements. IEEE Access, 7, 151515–151532.
19. Hu, Y., Xiao, Y., & Yang, Y. (2021). A survey on network slicing for 5G networks. Journal of Ambient Intelligence and Humanized Computing, 12(8), 8075–8090.
20. Wei, J., Zhang, Q., & Zhang, Q. (2019). Research on network slicing technology in 5G mobile communication era. Journal of Physics: Conference Series, 1333(1), 012035.
21. Li, Q., Li, X., Wang, Y., Zhang, Y., & Jia, W. (2019). Multi-objective optimization for network slicing in 5G and beyond: A comprehensive survey. IEEE Communications Magazine, 57(8), 22–28.
22. Bhushan, N., Li, J., Malladi, D., Gilmore, R., Brenner, D., & Damnjanovic, A. (2014). Network densification: The dominant theme for wireless evolution into 5G. IEEE Communications Magazine, 52(2), 82–89.
23. Li, M., Li, X., Li, X., Li, M., Zhang, Y., & Li, B. (2019). Multi-dimensional network slicing optimization in 5G: A survey. IEEE Communications Magazine, 57(4), 70–75.
24. Wang, C., Wang, Q., Jin, D., & Li, Y. (2020). An efficient network slicing strategy for 5G and beyond: A survey. IEEE Access, 8, 152416–15243

Digital Electronic Voting Machine Using Verilog

68

Tangelapalli Swapna[1]

Assistant Professor, Dept. of ECE,
Sreenidhi Institute of Science and Technology,
Hyderabad, Telangana, India

Dharmavaram Asha Devi[2]

Professor, Dept. of ECE,
Sreenidhi Institute of Science and Technology,
Hyderabad, Telangana, India

Manu Gupta[3]

Assistant Professsor, Dept. of ECM,
Sreenidhi Institute of Science and Technology,
Hyderabad, Telangana, India

Abstract: In place of the ballot papers and boxes that were formerly used in traditional voting methods, votes are now recorded using a straightforward electronic device called an electronic voting machine (EVM). The capacity to vote, or simply the right to cast a ballot, is the cornerstone of democracy. In the past, whether it was a state or center-level election, a voter would stamp the name of their favorite candidate, fold the ballot paper following the instructions, and then deposit it in the ballot box. This is a time-consuming, difficult process that is prone to mistakes. Until automated voting machines dramatically changed the election landscape, this setup was in place. Nowadays, ballot paper, ballot boxes, stamps, etc. are replaced by a simple box known as the ballot unit of voting machines. Since they are more difficult to lose, falsify, or trade than traditional tokens or knowledge-based IDs, biometric identifiers are regarded as more reliable for recognizing individuals. So, the most modern technology, especially the biometric system, has to be used to improve the electronic voting system. The proposed work on digital voting equipment makes it possible to conduct elections in a timely, secure, and effective manner. An algorithm, a flowchart, and the code required to put the logic into action and activate it were all created as part of the design process for this device. The suggested digital DEVM was created on Xilinx Vivado 2018.3v using Verilog HDL, and it may be used in real-time applications on an FPGA board. This article provides a comprehensive examination of voting methods, issues, and contrasts between biometric DEVMs and other voting procedures.

Keywords: Ballot box, EVM, FPGA, Verilog HDL

1. Introduction

The sole factor used by voters to select their representatives in any vote-based system is the casting of a ballot. To guarantee that only a just and deserving candidate is chosen based entirely on public opinion, this entire process must be carried out with the utmost care. Voters would simply place a stamp next to the name of the challenger they believed should win under the poll paper technique that was once used to hold elections. However, the outcomes of this method were

[1]swapnat@sreenidhi.edu.in, [2]ashadevi@sreenidhi.edu.in, [3]manugupta@sreenidhi.edu.in

often unjust, and the vote counting was usually problematic. Manual voting carries a significant risk of manipulation, which can take the form of things like booth capture, phony voting, and more. Electronic voting technology is exceedingly dependable, secure, and precise. Each of these disparities was addressed by the development of an electronic voting system. Electronic voting machines (EVMs) were first put on the market with the assurance that they would be safe and dependable, but over time, they too were compromised, highlighting the need for changes to EVMs. The concept of a basic electronic voting system with a removable memory card wasn't well received, though, because quick access to the memory card may convert all of the votes to a different harmful code and date could be destroyed. A simple, secure method that requires little of your time is polling using an electronic mechanical mechanism (EVM).

The Lok Sabha and Assembly elections now use Electronic Mechanical Devices (EVMs), which only allow each person to cast one vote. However, given that voting devices won't function in elections like GRAM PANCHAYATH and COOPERATIVE SOCIETIES, where each voter casts a single ballot for one candidate, The article also describes a PROGRAMMABLE ELECTRONIC mechanical gadget that can accommodate one or several votes, depending on demand.

EVMs provide a mode management feature that allows them to be set up to accept more than one vote from each elector, depending on the kind of election. The main benefit of using this type of EVM is to prevent illegitimate votes, especially in elections for cooperative societies, because each voter must cast a ballot for nine candidates. To operate an electronic voting system more effectively, we consequently require a framework. With the awareness of the challenges associated with regulating control signals, a more efficient way of developing an electronic voting machine in Verilog HDL using Xilinx Vivado can be implemented on FPGA (Field Programmable Gate Array) hardware. A secret key that is computerized in nature and challenging to hack is also included in this execution.

2. Literature Survey

Smith, T.F., and Waterman, M.S. [1] claim that, To mechanically register votes without handling actual ballots, a simple gadget called an electronic mechanical device could be employed. An essential element of every democracy is the right to vote. Before, voters would stamp the names of the candidates on their ballots to indicate the ones they thought should win. This method can occasionally be unfair, is time-consuming, prone to errors, and is frequency one.

Abeesh A, Amal Prakash [2] developed EVM Using PIC microcontroller along with GSM module, Fingure print module,LCD etc, making the voting process efficient , fast and secure The human element of this strategy, which is most vulnerable to error and disaster, is replaced by technology. Flexibility will increase, and wrong decisions won't be made. To demonstrate their vote, an elector must submit a fingerprint. So voting is exclusive.

Farzaliyev, V., Krips, K., [3] worked on internet voting machine using a microcontroller for single purpose user controlled EVM. It acts as open source voting client for electronic voting purpose.

D. A. Kumar and T. U. S. Begum [4] explains the advantages of biometrics sensors which are considered to be more relaiable which can enhance the security of EVM.

Exploiting secured physical and mechanical devices Recently, biometrics have become required. This machine is made with contemporary engineering. The human element of this strategy, which is most vulnerable to error and disaster, is replaced by technology. Flexibility will increase, and wrong decisions won't be made. To demonstrate their vote, an elector must submit a fingerprint. So voting is exclusive. A voting machine may be portable as well, much like the micro vote box, which is convenient to take wherever safely. This machine will have a significant social influence during elections in tiny SOCIETIES. We can design and operate voting machines according to our needs thanks to the abundance of nano, medium, and big FPGAs that are readily available on the market. With the help of these technologies, we can design voting machines.

Electronic voting machines (EVMs) are a solid, safe, and absolutely lovely method of polling that requires less of your time. Currently, the electromechanical voting machines (EVMs) used in Lok Sabha, Vidhan Sabha, and Assembly elections only allow one vote per voter. However, in elections like Gram panchayat, when each voter casts a vote for a single candidate, their voting machines don't function well. Therefore, important elections taking place in all countries using EVM are completely secure and more accurate than any other voting equipment on the market. EVMs can be built on FPGA that will address the majority of market issues.

3. Necessity of the Proposed EVM Technology

Technological developments have significantly altered democracies during the past twenty years. The use and impact of digital technology have been heavily debated concerning the usage of Electronic Voting Machines (EVMs) in the election of political leaders. Compared to a pen-and-paper system, technology offers a lot greater potential to empower people, amplify their voices, and allow them to hold governments accountable. Elections that are free and fair result in a political mandate that forms the basis for governance in electoral democracies. Better and more efficient voting procedures support democratic institutions.

The literature on democracy and development claims that greater representation, which offers a voice to the underprivileged and vulnerable segments of society, fosters growth. The efficiency and quick turnaround times of EVMs are crucial when they are used for larger populations. The importance of the most recent general election in India serves as proof of how EVM technology lowers the incidence of electoral fraud and simplifies the voting process. The election saw participation from a record-breaking 67 percent of the approximately 900 million registered voters in the 542 parliamentary seats. In a large-scale democracy with a complex multi-party system, electoral fraud is undoubtedly a serious concern. However, over time, the use of electronic voting machines (EVMs) in India's electoral process has boosted voters' perceptions that their votes matter for democratic governance and election outcomes. Since polling sites were regularly looted and ballot boxes overfilled, voter turnout was notably high while using the paper ballot system.

EVMs assisted in reducing this risk by incorporating a vital feature—registering just five votes per minute. Election fraud requires the longer-term takeover of polling stations. Additionally, the data reveals a significant decline in electoral fraud in politically sensitive states where election tampering required repeated voting. This voting information is ubiquitous[6] and needs to be maintained secure[5] before the announcement of the final declaration.

4. Proposed System

Voting on paper was a labor-intensive, prone-to-mistakes method that took a very long time. The least time-consuming method of polling is with an electronic mechanical mechanism (EVM). The suggested digital EVM was created using Verilog HDL and then put into action using an Artix 7 FPGA.

The design is implemented using Verilog HDL and the functionality is verified on FPGA.

4.1 Verilog Programming

Hardware description languages like Verilog are analogous to software programming languages because of how they may be used to specify signal sensitivity and propagation durations. There are two kinds of assignment operators: blocking assignments (=) and non-blocking assignments

(<=). Because the assignment is non-blocking, designers can specify a state-machine update without having to define and employ temporary storage variables. Since these concepts are a component of the semantics of the Verilog language, designers may easily produce relatively simple and quick descriptions of large circuits. When Verilog was initially released in 1984, circuit designers were already using graphical schematic capture tools and custom software programs to describe and simulate electrical circuits. For these designers, Verilog provided a sizable productivity boost.

The Verilog language's syntax was influenced by the C programming language, which was already extensively used to create technical software. Verilog's preprocessor is straightforward (albeit less so than that of ANSI C or C++), case-sensitive, and similar to that of C. The control flow keywords if/else, for, while, case, and others are comparable, and the order of their operators is compatible with C. The minimum bit widths for variable declarations, the way procedural blocks are separated (Verilog uses begin/end rather than curly brackets), and many other small distinctions are examples of syntactic variances.

Verilog requirements state that variables must have a defined size. In C, these sizes can be inferred from the variable's 'type' (for instance, an integer type might be 32 bits).

In a Verilog design, the modules are arranged hierarchically. The design hierarchy is included inside the collection of input, output, and bidirectional ports that are used by modules to connect with one another. Concurrent and sequential statement blocks, instances of other modules (sub-hierarchies), and net/variable declarations (wire, reg, integer, etc.) are all examples of internal components that may be found in a module. A begin/end block inserts and executes successive statements one after the other. Because the blocks themselves are executed concurrently, Verilog is a dataflow language.

A portion of the Verilog language's assertions can be synthesized. Verilog modules that follow the RTL (register-transfer level) coding standard can be physically actualized using synthesis software. The (abstract) Verilog source is algorithmically transformed into a netlist, which is a logically equivalent description that only contains the fundamental logic primitives (AND, OR, NOT, flip-flops, etc.) that are supported by a specific FPGA or VLSI technology. Additional changes to the netlist lead to a blueprint for the manufacture of a circuit (such as a photo mask set for an ASIC or a bitstream file for an FPGA).

4.2 FPGA

FPGAs (field-programmable gate arrays) are very adaptable integrated circuits that provide significant benefits to embedded system development. Unlike traditional ASICs, FPGAs may be customized using hardware description languages (HDL) after production, offering flexibility and lowering development costs. They are made up of an array of programmable logic blocks connected by reconfigurable interconnects, allowing for seamless wiring of blocks to accomplish complex functions or operate as simple logic gates like AND and XOR. Furthermore, these logic blocks frequently include memory elements, which enhance their capabilities.

FPGAs' importance in embedded system development stems from their ability to ease hardware and software co-creation. Because FPGAs and HDLs are programmable, designers can begin software development alongside hardware development. This parallel approach speeds up the development process and offers more flexibility.

The advantages of using FPGAs continue with their adaptability, enabling multiple design iterations without requiring new chips to be fabricated. Therefore, these are reconfigurable[7] devices. This reduces design time and supports efficient prototyping and testing of various system architectures. Moreover, FPGAs permit hardware testing and validation, enabling designers to thoroughly assess their designs before committing to expensive ASIC production.

FPGAs also offer cost advantages as they have lower non-recurring engineering (NRE) costs compared to ASICs, making them suitable for low to medium-volume applications. Their rapid prototyping capabilities allow for quick implementation and evaluation of new ideas, ideal for research and development projects. By embracing FPGA technology, developers can significantly reduce time-to-market for embedded systems, gaining a competitive edge in dynamic industries.

5. Methodology

The election process is depicted in the flowcharts. It describes the voting procedure. It also marks the announcement of the electoral process's winner.

The voting process starts with power enable. When power is in ON condition the message will be displayed on the system stating the system is ready to scan for polling the votes. Before the scanning of votes, the voter has to register with Adhar identity or fingerprint identity. If the voter is registered then the message is to be displayed as "voter is authenticated". The authentication process is done the system will allow the voter to vote.

The vote will be taken for a count then the process will terminate. If the voter try to cast the vote again with the same identity then the message has to be be displayed as "your vote

already cast". If the voter is not registered then the message is to be displayed as "fingerprint not matched" and the voter has to go back to the registering step. The loop goes on to iterate until the voter got registered as illustrated in Fig. 68.1.

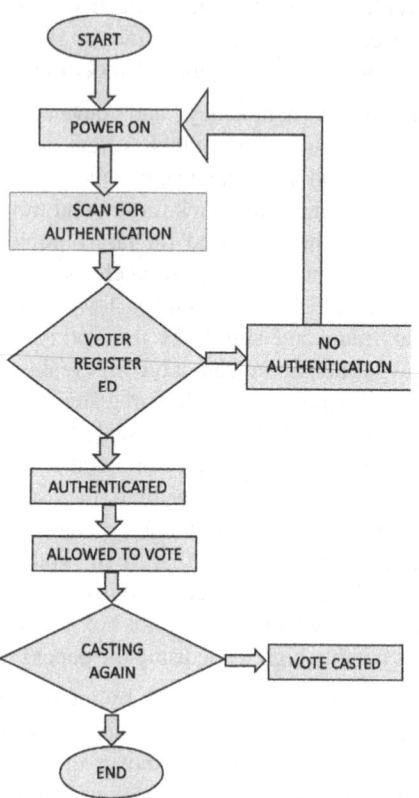

Fig. 68.1   Flow chart for the process of elections using digital electronic voting machine (DEVM)

Source: Made by author

The winner of elections will be considered with the number of parties which are participating in the voting process. Here the number of parties participating the election is considered as three. It is going to compare with the first two parties votes, if the first party got votes more than the second party then it compares with the third party. If the third party got less than the first party then it is stated as the first party won the elections. If the first party got fewer votes than the second party then it compares with the third party and if the second party got more votes than the third then it is confirmed as the second party has won the elections. If both the conditions are false then it is going to be stated as a third party has won the elections. This process is expressed using simple if-else conditions denoted in the flowchart shown in Fig. 68.2. After the counting process is done the process gets stopped and the winner is declared.

In the proposed work, the digital electronic voting machine is implemented in Verilog HDL in behavioural model. The various input and output ports of the proposed system are

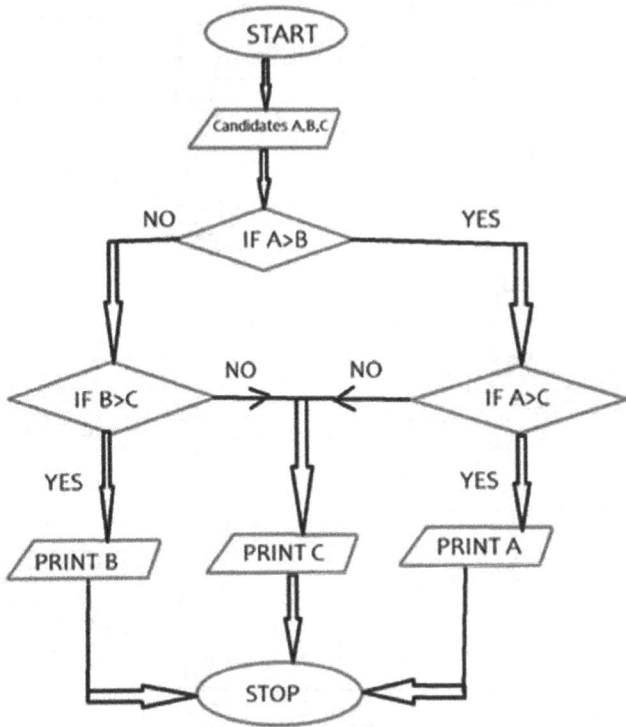

Fig. 68.2   Flow chart for declaration of winner

Source: Made by author

clk, voter_switch, voting_en, opled, invalid, TotalVotes, TRS,BJP,CONG, and winner. Here three parties have been considered, TRS, BJP and CONG taken as an example of Telangana State in the year 2023. As it is a real time embedded application, clock 100MHz is used. Voter_switch is meant for the selection of the parties: 1 represents TRS, 2 represent BJP and 3 represent CONG. If the voter is a valid voter after the verification of adhar and voter ID card voting_en is enabled. opled is an output indication to view the voting of an individual voter to which he/she has casted the vote. If any mismatch of adhar and voter_ID or try to give second time vote, then invalid will be enabled and the voter cannot vote. TotalVotes is an output to indicate the results of total votes casted by the voters. The RTL schematic is illustrated in Fig.6. The proposed work is presented mainly the functional verification of DVEM.

6. Results and Discussion

In the proposed work, contestants A, B, and C are considered for Telangana State elections with names TRS, BJP, and Congress respectively. Fig. 68.3 shows that the simulation results at 42.2ns. Voter_en, clk nad voter_switch are taken to be as input. When the voter_en and clk are high, the voter can do poling.

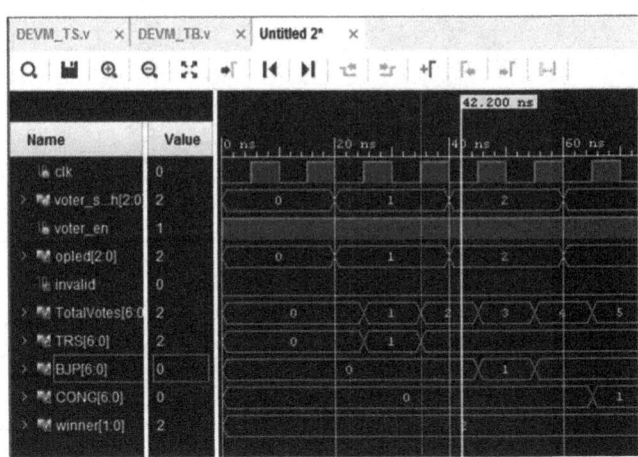

Fig. 68.3   Voting results in the middle of the polling time

Source: Made by author

As per the cursor marking, 42.2 ns, it shows, TRS secured 2 votes, and BJP and CONG secured zero votes. At this time the TRS is in the leading position as per the votes cast.

Fig. 68.4 shows the simulation results up to 1000 ns time. Indicating that TRS secured 6, BJP scored 4 and CONG score is 88 as per the test inputs given in the test bench. Therefore, the total number of votes polled is 98 and declaring CONG as the winner with maximum votes.

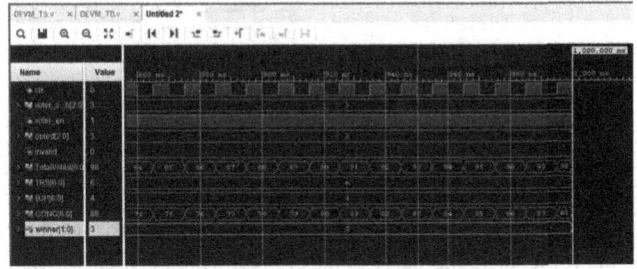

Fig. 68.4   Simulation results up to 1000 ns

Source: Made by author

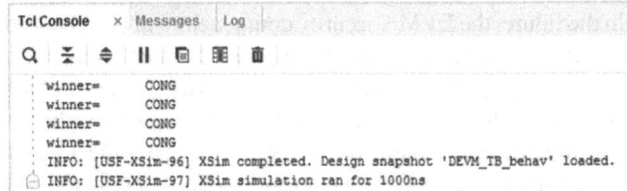

Fig. 68.5   Results displayed at Tcl console window

Source: Made by author

7. Conclusion

In the proposed work, the RTL design and verification of the Digital Electronic Voting Machine is demonstrated. The

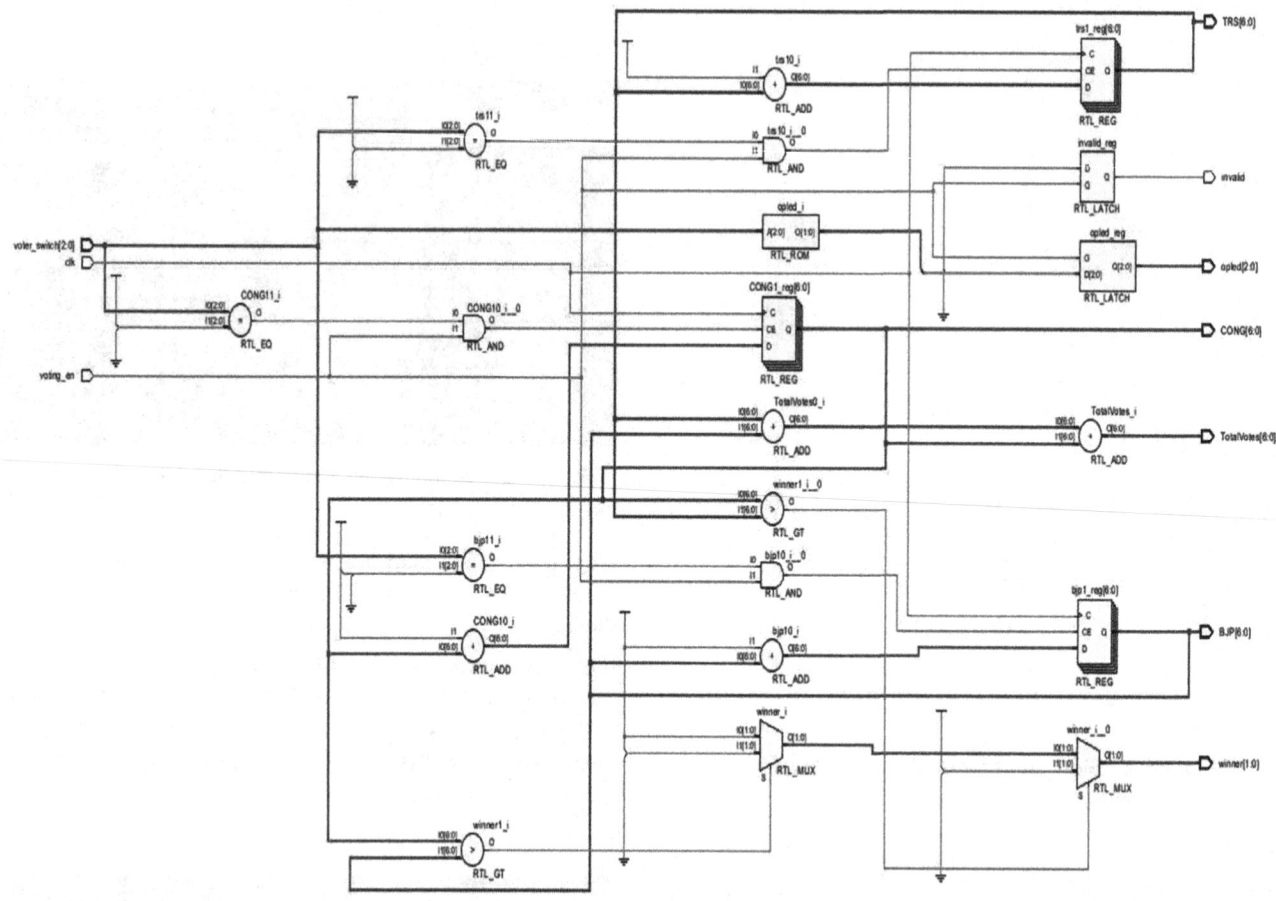

Fig. 68.6 RTL design of DEVM [9]

Xilinx-based electronic voting system conformed to election criteria such as registering all eligible voters and candidates in the first round of voting and allowing the voter to cast their vote for a particular party of his choice, which is then verified by the electorate in the second voting round. The last phase evaluates all of the lawful votes gathered from various parties to choose the winner of the election.

In the future, the EVM's security component will be enhanced by adding bio-metric. One of the project's main elements is theor finger print scanner. EVM's security feature, which requires each voter to use a special voting card and his or her fingerprint as a password and as part of the project. After the computer has confirmed the voter's identification, the poll worker will let them cast their ballot if they are considered eligible. The PRSTV Election Procedure, which is frequently utilised in elections for the offices of the president, vice president, and Vidhan parishads where voters would cast multiple ballots, can also be efficiently implemented using the proposed method on FPGAs.

Concerns have been raised concerning the future usage of the election programme in light of the arrival of digital electronic voting machines (DEVMs). We have already seen evidence of increasing fraud and inaccurate results caused by the manipulation of both traditional voting methods and EVMs. The Digital Electronic Voting Machine, on the other hand, will deliver exact results without any tampering. Voters will need dependable voting technology in the future so that the best candidate can lead the public; hence, these voting machines will be employed more frequently.

Adding bio-metric, fingerprint, or eye iris detection sensors, security can be enhanced, and also multi vote system with a hierarchical voting system using data logger [8] and the limited time for the voting process by keeping inbuilt real time clock [9] can be implemented which is considered to be a future scope.

References

1. T.F. Smith and M.S. Waterman, Identification of Common Molecular Subsequences, 1981, J. Mol. Biol. 147195–197.

2. Abeesh A I, Amal Prakash P, Arun R Pillai, Ashams H S, Dhanya M, Seena R, 2017, Electronic Voting Machine Authentication

using Biometric Information, INTERNATIONAL JOURNAL OF ENGINEERING RESEARCH & TECHNOLOGY (IJERT) NCETET – 2017 (Volume 5 – Issue 16)

3. Farzaliyev, V., Krips, K., & Willemson, J. (2021, March). Developing a personal voting machine for the Estonian internet voting system. In Proceedings of the 36th Annual ACM Symposium on Applied Computing (pp. 1607–1616).

4. D. A. Kumar and T. U. S. Begum, "Electronic voting machine A review," International Conference on Pattern Recognition, Informatics and Medical Engineering (PRIME-2012), Salem, India, 2012, pp. 41-48, doi: 10.1109/ICPRIME.2012.6208285.

5. Babu, M.S., Raj, K.B., Devi, D.A. (2021). Data Security and Sensitive Data Protection using Privacy by Design Technique. In: Haldorai, A., Ramu, A., Mohanram, S., Chen, MY. (eds) 2nd EAI International Conference on Big Data Innovation for Sustainable Cognitive Computing. EAI/Springer Innovations in Communication and Computing. Springer, Cham. https://doi.org/10.1007/978-3-030-47560-4_14

6. Suresh Babu, M., Bhavana Raj, K., Asha Devi, D., "Future trends of business intelligence and big data analytics in ubiquitous environment", International Journal of Engineering and Advanced Technology, 8(3 Special Issue), pp. 773–778, 2019.

7. Dharmavaram Asha Devi, Tirumala Satya Savithri and Sai Sugun.L, "Design and Implementation of Real Time Data Acquisition System using Reconfigurable SoC" International Journal of Advanced Computer Science and Applications(IJACSA), 11(9), pp. 325–331, 2020.

8. M. Bhavani and D. A. Devi, "Design of smart Monitor for automobiles using FPGA based Data Logger," *2019 International Conference on Communication and Electronics Systems (ICCES)*, Coimbatore, India, 2019, pp. 1940–1945, doi: 10.1109/ICCES45898.2019.9002034.

9. Devi, D.A., Rani, N.S., "Design and Implementation of custom IP for Real Time Clock on Reconfigurable Device", Proceedings of the 3rd International Conference on Inventive Systems and Control, ICISC 2019, pp. 414–418, 9036428, 2019.

Live Sketch for Computer Vision 69

Manu Gupta[1]

Department of Electronics and Computer Engineering,
Sreenidhi Institute of Science & Technology, Hyderabad, India

Tangelapalli Swapna[2], Dharmavaram Asha Devi[3]

Department of Electronics and, Communication Engineering,
Sreenidhi Institute of Science & Technology, Hyderabad, India

Abstract: This research presents a novel computer vision-based system for real-time live sketching using a webcam. The aim of this study is to enable users, regardless of their artistic abilities, to engage in spontaneous and creative sketching, thereby fostering artistic expression and interactive experiences in a digital environment. The proposed system employs image processing to detect and analyze human gestures and movements captured through the webcam. In particular, the use of pose estimation and key point detection algorithms enables the system to accurately track the user's hand movements and interpret them as digital sketch strokes on a virtual canvas. The system is designed to be user-friendly, allowing seamless interaction between the user and the digital sketching interface. The evaluation of the system involved a series of user studies to assess its usability and overall performance. Results indicated that the proposed approach achieved impressive real-time sketching capabilities, demonstrating its potential as an engaging tool for digital artists, and designers alike. This study introduces a useful and approachable technique for live sketching using common cameras, which advances the field of interactive digital art.

Keywords: Image processing, Image sharpening, Edge detection, Python, NumPy, OpenCV

I. Introduction

This Live sketching with a webcam is an exciting and innovative approach that blends traditional artistry with modern technology. It offers a dynamic and interactive way to create visual masterpieces, bringing art to life in real time. This intriguing fusion allows artists, enthusiasts, and even novices to explore their creativity, experiment with different styles, and connect with their audience like never before[1].

The creative possibilities of live sketching are boundless. Artists can experiment with different mediums, styles, and techniques, transcending the boundaries of physical artistry. Whether it's creating vibrant portraits, capturing landscapes, or crafting abstract interpretations, live sketching allows artists to embrace spontaneity and fluidity in their work[2].

Advancements in webcam technology have transformed ordinary webcams into powerful tools for artistic expression. Artists can now harness the capabilities of high-resolution cameras, real-time image processing, and specialized software to instantly transform live video feeds into captivating sketches. The process involves analyzing the video stream frame-by-frame, extracting key elements, and rendering them into expressive and artistic forms. One of the key attractions of live sketching with a webcam is its interactivity. Artists can engage with their viewers in real-time, showcasing their creative process as it unfolds[3]. This live connection builds a unique bond with the audience, making them active participants in the artistic journey. Viewers can offer suggestions, witness the artist's decisions, and witness the evolution of a blank canvas into a mesmerizing sketch.

---
[1]manugupta5416@gmail.com, [2]swapnat@sreenidhi.edu.in, [3]ashadevi.d@rediff.com

Additionally, live sketching with a webcam has found its place in various domains beyond traditional art. It has become a popular form of entertainment in live streaming platforms, where artists from around the world host drawing sessions, providing entertainment, education, and inspiration to a global audience[4]. Moreover, this technology has practical applications in fields like education, where instructors can demonstrate complex concepts through visual representation, making learning engaging and memorable. Finally, live sketching with a webcam opens up a world of artistic exploration and interaction. It merges the timeless charm of traditional art with the dynamic nature of technology, creating an enriching and immersive experience for both artists and their audiences. As this captivating form of expression continues to evolve, it promises to redefine how art is perceived, produced, and shared in the digital age.

2. Literature Review

Many methods are proposed in the literature for sketch generation. In the work proposed by Jiatao et al. [5], a method for generating pen and ink drawings from photos is developed. They used wavelet transform for edge detection in the photos. Zhou and Li [6] proposed a method for sketch drawings from personal images using Sobel and Laplacian operations. Khayan and Khoenkaw[7] present a method for the generation of a sketch of the landscape from the photo using vertical and horizontal edge detection based on Sobel operator. An approach for sketch generation based on pixels is presented by Ahmad et al.[8]. In this method, texture maps are applied to produce output sketches. Tong et al. [9] proposed an approach for pencil sketch drawing using texture rendering. A method using a Canny edge detector for pencil sketch generation is presented by Ryota et al.[10].A technique for line drawing based on likelihoodfunction estimation is designed by Son et al.[11]. Lu et al.[12] proposed method for drawing production using sketch and tone. A histogram of gradients-based technique is developed by Hu and Collomosse [13] for image sketch.

3. Proposed Method

An approach for generating live sketches of people is developed in the proposed work. In this method firstly, the video frames are captured from the webcam. Next, the faces are detected in every image frame, and finally, the designed sketch filters are implemented to detect the faces.

The Histogram of Oriented Gradients (HOG) approach is utilized for face detection in the proposed method. The HOG approach counts gradient orientation occurrences in a particular area of a picture. It is a popular feature description model. To explain the form and look of an object, it examines the distribution of edge orientations within the object. In order to divide an image into smaller cells, the HOG method first determines the gradient magnitude and orientation for each pixel in the image. A distinct Histogram would be produced by the HOG for each region in the input image. The phrase "Histogram of Oriented Gradients" describes the histograms created using the gradients and orientations of the pixel values.

The noise reduction in input images is performed using Gaussian blurring. It is a low-pass filter that reduces the high frequency. In order to identify the edges in the input image, the Canny edge detector is employed. The noise in a picture is smoothed using canny edge detection, which determines the strength and direction of the edge for each pixel, using a Gaussian kernel and linear filtering.

The following is a list of the steps that are involved:

1. Install necessary libraries: Ensure you have Python installed with all the following libraries: NumPy (for numerical operations). dlib (for face detection and facial landmark estimation). OpenCV (for capturing video frames from the webcam and image processing). Matplotlib (for visualizing the sketch).

2. Capture video using the webcam: Webcam will the use the OpenCV library on each feed and capture frames continuously.

3. Face detection: Identifying faces in each frame using the Histogram of Oriented Gradients(HOG) method.

4. Facial landmark estimation: To enhance the face sketch, we can use facial landmark estimation to get the facial key elements positions.

5. Sketch filter: Next, edge detection is carried out using the canny edge detection method and the image is converted to a binary image.

4. Software Requirements

To implement the proposed work NumPy (Numerical Python) and OpenCV libraries in Python are used [14,15]. These libraries are explained as follows:

NumPy in Python is the most useful library and it is used for numerical computing. It facilitates the computation of massive, multidimensional arrays and matrices, as well as a variety of complex mathematical operations on these arrays.. NumPy is an essential package in Python for scientific computing and is widely used in various areas such as data analysis, machine learning and scientific research. The key features of NumPy include N-dimensional arrays, Array slicing and indexing, Integration with C/C++ and Fortran. NumPy is a crucial library for Python's scientific computing since it offers a wide variety of functions and techniques for working well with arrays.

Real-time computer vision applications use a Python package for open-source computer vision and machine learning called OpenCV (Open-Source Computer Vision). It is mostly used in various fields such as video and image processing, object detection, facial recognition, augmented reality and more. OpenCV is penned in C++ and has an intersection for various programming languages, including Python. In Python, OpenCV is available as the cv2 module. In the proposed work OpenCV is used to perform image filtering and enhancement, capture frames using camera and camera calibration.

5. Implementation

The proposed work is implemented in Python and the operating systems can be either Windows or Linux. The implementation steps are as follows.

In this proposed work to convert an image into a live sketch, firstly the image feed is captured using the webcam, and next, different operations are performed using OpenCV and NumPy[16,17]. This includes the following three steps.

Step 1: Importing Modules

We need to pre-install all the libraries to start the process.

Step 2: To convert the frame to Sketch we define the function

To sketch from the frame, we will go through the steps:

(i) Change the Image into a gray image because the gray image has no intensity information and simplifies the sketching process.

(ii) Assign Gaussian Blur to the gray image captured as it reduces the noise and smoothens the image.

(iii) Assign Canny Edge Detection to detect the edges of the blurry image.

(iv) Create a binary picture from the edge-detected image, with the background represented by white pixels and the edges by black pixels.

Step 3: Live sketching process

(i) The code runs with the infinite loop to continuously capture frames from the webcam and reads a frame from the webcam captured. We use ret as a variable and it is a Boolean indicating the frame was captured successfully. Frame refers to the actual image data.

(ii) (cv2imshow) this combines the original frame and sketched frame horizontally using np.h stack and displays the result .cv2 wait key waits for key press 1m second if the pressed key is Q the program stops.

6. Results and Discussion

The live sketch using a webcam can be a fascinating and versatile application, offering several benefits and use cases.

The execution process is illustrated in the flowchart shown in Fig. 69.1. By utilizing advanced image processing techniques and computer vision, it allows us to create real-time sketches or artistic representations of the scenes captured by the webcam. Some possible applications include:

Artistic Expression: The live sketch system can serve as a powerful tool for artists, allowing them to instantly create digital sketches or paintings from live scenes. It enhances artistic expression and encourages creative exploration.

Interactive Entertainment: Incorporating the live sketch into various interactive applications, games, or virtual reality experiences can add an engaging and immersive element to entertainment platforms.

Educational Tool: It can be employed as an educational tool. This system can provide visual feedback and real-time demonstrations, making learning more engaging and effective.

Facial Expression Analysis: In the context of facial recognition and emotion analysis, the live sketch system can be used to detect and represent facial expressions, enabling applications in psychology, human-computer interaction, and entertainment.

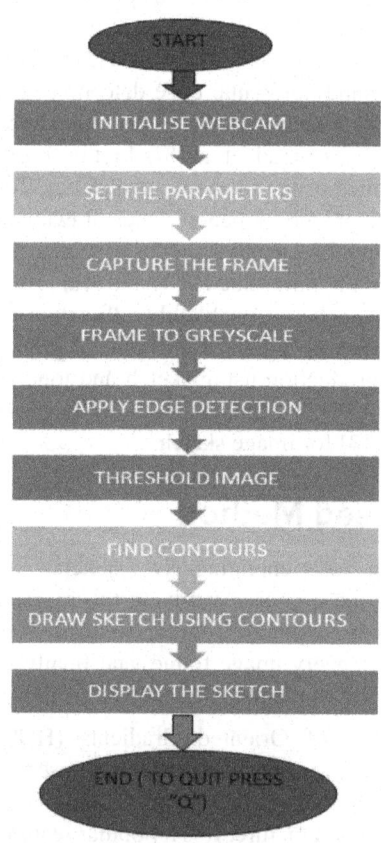

Fig. 69.1 Execution process

Augmented Reality: Combining live sketches with augmented reality technology can create unique and innovative experiences, blurring the lines between real and virtual worlds.

User Interface Design: The system can potentially find applications in user interface design and prototyping, allowing designers to quickly sketch out ideas and visualize potential user interactions.

Virtual Filters and Effects: Leveraging live sketching in video conferencing or social media applications can enable users to apply virtual filters and effects in real time, enhancing the visual communication experience.

The examples of input images captured from the webcam are demonstrated in Fig. 69.2(a) and Fig. 69.2(b). Next, a grayscale image was generated for edge detection as illustrated in Fig. 69.3(a) and Fig. 69.3(b). Finally, the

(a)

(b)

Fig. 69.2 Examples of raw image

(a)

(b)

Fig. 69.3 Examples of the gray scale image

(a)

(b)

Fig. 69.4 Examples of sketched image

output sketch image obtained after the HOG method and edge detection using the proposed model is illustrated in Fig. 69.4(a) and Fig. 69.4(b). As observed the proposed method produces sketches with clear contours and the background is suppressed.

7. Conclusion and Future Scope

In this study, the method to generate the live sketch from a given input picture or scene captured by the webcam instantaneously is presented. In this method, the histogram of gradient method is implemented for face detection in input image frames. The canny edge detection technique is used for detecting edges in the image. And finally converted into binary images generating live sketch.

The future scope includes providing features such as texture, shadow and shading. The proposed method can be extended to video sequences.

References

1. T. Bui, L. Ribeiro, M. Ponti and J. Collomosse, "Generalisation and sharing in triplet convnets for sketch based visual search", CoRR Abs, 2016.
2. J. Collomosse, T. Bui, M. Wilber, C. Fang and H. Jin, "Sketching with style: Visual search with sketches and aesthetic context", Proc. ICCV, 2017.
3. Ren, J., & Xu, K "Real-time and Continuous User Gesture Tracking for Sketch Interaction." IEEE International Conference on Robotics and Automation (ICRA), 2018.
4. R. Munir, M. Rehman, S. G. Murtaza, and M. Arsalan, "A Real Time Image Sketching Algorithm for Color Images," in 2020 International
5. Conference on Computing, Electronics & Communications Engineering (ICCECE).
6. Song, Jiatao, Zheru Chi, Jilin Liu, and Hong Fu. "Automatic generation of pen-and-ink drawings from photos." In 2004 International Conference on Image Processing, 2004. ICIP'04., vol. 2, pp. 1185–1188. IEEE, 2004.
7. Zhou, Jin, and Baoxin Li. "Automatic generation of pencil-sketch like drawings from personal photos." In 2005 IEEE International Conference on Multimedia and Expo, pp. 1026–1029. IEEE, 2005.
8. Khayan, Atiporn, and Paween Khoenkaw. "Automatic Pencil Sketch Landscape Image Generation From Photograph." In 2021 Joint International Conference on Digital Arts, Media and Technology with ECTI Northern Section Conference on Electrical, Electronics, Computer and Telecommunication Engineering, pp. 27–30. IEEE, 2021.
9. Ahmad, Azhan, Somnuk Phon-Amnuaisuk, and Peter D. Shannon. "Emulating Pencil Sketches from 2D Images." In Recent Advances on Soft Computing and Data Mining: Proceedings of The First International Conference on Soft Computing and Data Mining (SCDM-2014) Universiti Tun Hussein Onn Malaysia, Johor, Malaysia June 16th-18th, 2014, pp. 571-580. Springer International Publishing, 2014.
10. Tong, Zhengyan, Xuanhong Chen, Bingbing Ni, and Xiaohang Wang. "Sketch generation with drawing process guided by vector flow and grayscale." In Proceedings of the AAAI Conference on Artificial Intelligence, vol. 35, no. 1, pp. 609–616. 2021.
11. Okawa, Ryota, Hiromi Yoshida, and Youji Iiguni. "Automatic pencil sketch generation by using canny edges." In 2017 fifteenth IAPR international conference on machine vision applications (MVA), pp. 282285. IEEE, 2017.
12. M. Son, H. Kang, Y. Lee and S. Lee, "Abstract line drawings from 2d images," Pacific Conference on Computer Graphics and Applications, pp. 333–342, 2007.
13. Lu, Cewu, Li Xu, and Jiaya Jia. "Combining sketch and tone for pencil drawing production." In Proceedings of the symposium on nonphotorealistic animation and rendering, pp. 65–73. 2012.
14. Rui Hu and John Collomosse, "A performance evaluation of gradient field HOG descriptor for sketch based image retrieval", Computer Vision and Image Understanding (CVIU), vol. 117, no. 7, pp. 790–806, 2013.
15. Guan, Yurong, Fei Zhou, and Jing Zhou. "Research and practice of image processing based on python." In Journal of Physics: Conference Series, vol. 1345, no. 2, p. 022018. IOP Publishing, 2019.
16. Naveenkumar, Mahamkali, and Ayyasamy Vadivel. "OpenCV for computer vision applications." In Proceedings of National Conference on Big Data and Cloud Computing (NCBDC'15), pp. 52–56. 2015.
17. Sharma, Ayushi, Jyotsna Pathak, Muskan Prakash, and J. N. Singh. "Object detection using opencv and python." In 2021 3rd International Conference on Advances in Computing, Communication Control and Networking (ICAC3N), pp. 501–505. IEEE, 2021.
18. Dharmavaram Asha Devi, Niharika Reddy Kathula, Gopinath Kalluri, and Leela Sai Bondalapati, "Design and Implementation of Image Processing Application with Zynq SoC, ISSN (2210-142X) Int. J. Com. Dig. Sys.14, No. 1 (Jul-23).

Note: All the figures in this chapter were made by the author.

Printed in the United States
by Baker & Taylor Publisher Services